Principles of Mineral Processing

Edited by
Maurice C. Fuerstenau and Kenneth N. Han

Published by
Society for Mining, Metallurgy, and Exploration, Inc.

Society for Mining, Metallurgy, and Exploration, Inc. (SME)
12999 East Adam Aircraft Circle
Englewood, Colorado 80112
(303) 948-4200 / (800) 763-3132
www.smenet.org

SME advances the worldwide mining and minerals community through information exchange and professional development. With members in more than 70 countries, SME is the world's largest association of mining and minerals professionals.

ISBN 10: 0-87335-167-3
ISBN 13: 978-0-87335-167-6
Ebook: 978-0-87335-291-8

Library of Congress Cataloging-in-Publication Data

Principles of mineral processing / [edited by] Maurice C. Fuerstenau
p. cm.
Includes bibilographical references and index.
ISBN 0-87335-167-3
1. Ore dressing. 2. Hydrometallurgy. I. Fuerstenau, Maurice C.

TN500.P66 2003
622'.7--dc21 2002042938

Contents

Authors

Frank F. Aplan
Distinguished Professor Emeritus
Mineral Processing Section
The Pennsylvania State University
University Park, PA

Hendrick Colijn
Consulting Engineer
Pisgah Forest, NC

Donald A. Dahlstrom
Professor Emeritus
Department of Chemical Engineering
The University of Utah
Salt Lake City, UT

Brian C. Flintoff
Senior Vice President
Metso Minerals
Kelowna, BC

Maurice C. Fuerstenau
Professor of Metallurgy
Metallurgical and Materials Engineering
The University of Nevada
Reno, NV

Donald W. Gentry
President & CEO
PolyMet Mining Corporation
Golden, CO

Stoyan N. Groudev
Professor
Department of Engineering Geoecology
University of Mining and Geology
Studenski grad-Durvenitza
Sofia, Bulgaria

Kenneth N. Han
Regents Distinguished Professor
Metallurgical and Materials Engineering
South Dakota School of Mines and Technology
Rapid City, SD

John A. Herbst
General Manager
Svedala Minerals Industry Inc.
Kealakekua, HI

Richard Hogg
Professor Emeritus
Mineral Processing Section
The Pennsylvania State University
University Park, PA

Matthew J. Hreber
Professor Emeritus
Department of Mining Engineering
Colorado School of Mines
Golden, CO

Frank S. Knoll
Consultant
Carpco Division
Outokumpu Technology Inc.
Jacksonville, FL

James E. Lawver
Deceased

Yi Chang Lo
Retired
GS Industries
Kansas City, MO

Robert D. Morrison
Technical Director
Julius Kruttschnitt Mineral Research Center
Indoorroopilly, Queensland, Australia

Andrew L. Mular
Professor Emeritus
Mining and Mineral Processing Engineering
The University of British Columbia
Vancouver, BC

J. Mark Richardson
President
J. K. Technology/Contract Support Services
Red Bluff, CA

Ross W. Smith
Professor Emeritus
Metallurgical and Materials Engineering
The University of Nevada
Reno, NV

Ponisseril Somasundaran
Von Duddleson Krumb Professor
Mineral Engineering Department
Columbia University
New York, NY

Partha Venkatraman
Research & Development Manager
Carpco Division
Outokumpu Technology Inc.
Jacksonville, FL

Preface

The world is faced with opportunities and challenges that require ever-increasing amounts of raw materials to fuel various industrial sectors, and, at the same time, meet environmental constraints associated with excavating and processing these raw materials. In addition, gradual depletion of mineral resources and the necessity of handling more complex forms of resources, primary and secondary, have led to challenges in the development of state-of-the-art technologies that are metallurgically efficient and environmentally friendly. Unquestionably, technology advances are the key to sustaining a sufficient supply of necessary raw materials.

To advance the technology in the production of material resources, nations look to practicing and future engineers. Current and future mineral processing engineers must obtain sound and rigorous training in the sciences and technologies that are essential for effective resource development. Many industrial and academic leaders have recognized the need for more textbooks and references in this important area. This was the driving force for writing a comprehensive reference book that covers mineral processing and hydrometallurgical extraction.

This book was written first to serve students who are studying mineral processing and hydrometallurgy under various titles. We also hope that the book will serve as a valuable reference to many industrial practitioners in the mineral processing field. In the chapters that follow, you will find first principles that govern various unit operations in mineral processing and hydrometallurgy, along with examples to illustrate how fundamental principles can be used in real-world applications. In general, the volume covers topics in the order of the usual processing sequence. Comminution, the breakage of rocks and other materials, is covered in such a way that the fundamental principles can be used not only in mineral processing but also in other relevant areas such as chemical engineering and pharmaceutical fields.

Understanding the characteristics of particles and the separation of particulate materials from one another is of ultimate importance. Separation technologies based on properties such as magnetism, electrical properties, and surface properties of various minerals are present along with industrial examples.

Because most mineral processing unit operations take place in water as a medium, understanding how solids can best be separated from water is of industrial importance. Efficiently using water during effective solid–liquid separation is often vital to the success of the overall mineral beneficiation operation.

With computer application technologies continuing to emerge rapidly, the mineral industry has made tremendous advances in its industrial production. Plant automation and control often play a vital role in the overall success of the plant operation. The chapter on comminution covers some of these innovations in automation.

Once desired minerals are recovered from the undesired portion of an ore deposit, chemical treatment to unlock the desired metal elements from various minerals is necessary. Hydrometallurgical treatment for the chemical release of metal elements from various minerals is presented along with fundamental water chemistry and kinetic principles.

We are fortunate that many world-class authorities in various areas of mineral processing have joined this endeavor, and we thank them for their participation. We would also like to take this opportunity to thank the staff of the Society for Mining, Metallurgy, and Exploration, Inc., for their support in producing this book.

CHAPTER 1

Introduction

Maurice C. Fuerstenau and Kenneth N. Han

The term *mineral processing* is used in a broad sense throughout this book to analyze and describe the unit operations involved in upgrading and recovering minerals or metals from ores. The field of mineral processing is based on many fields of science and engineering. Humanities and social science have also become an integral part of this technology because mineral processing, like many other technologies, is carried out to improve human welfare. In addition, environmental science and engineering have become inseparable components; the steps involved in mineral processing have to be founded not only on sound scientific and technological bases but on environmentally acceptable grounds as well.

GOALS AND BASICS OF MINERAL PROCESSING

In the traditional sense, mineral processing is regarded as the processing of ores or other materials to yield concentrated products. Most of the processes involve physical concentration procedures during which the chemical nature of the mineral(s) in question does not change. In hydrometallurgical processing, however, chemical reactions invariably occur; these systems are operated at ambient or elevated temperatures depending on the kinetics of the processes.

The ultimate goal in the production of metals is to yield metals in their purest form. Mineral processing plays an integral part in achieving this objective. Figure 1.1 shows a generalized flow diagram for metals extraction from mining (step 1) through chemical processing. Steps 2 and 3 involve physical processing and steps 5 and 7 involve low-temperature chemical processing (hydrometallurgy). All four steps are considered part of mineral processing. High-temperature smelting and refining (pyrometallurgy), steps 4 and 6, are not included under the heading of mineral processing.

Table 1.1 specifies processing routes from ore to pure metal for a number of metals. Note that processing routes can be quite different and that more than one route may be possible for many of these metals. For example, in the extraction of copper or gold from low-grade ores, dump or heap leaching is commonly practiced. The choice of this leaching practice is frequently driven by the overall economics of the operation. Because crushing and grinding of ores are quite expensive, leaching of ores in large sizes is attractive compared to the leaching of finely ground ores, even though the overall recovery of metals from the leaching of fine particles is, in general, much greater than that obtained with large particles. The introduction of this innovative leaching process has made feasible the mining of many mineral deposits that could not be processed economically through conventional technologies.

METALLURGICAL EFFICIENCY

One of the most important and basic concepts in mineral processing is metallurgical efficiency. Two terms are commonly used to describe the efficiency of metallurgical processes: *recovery* and *grade*. These phenomena are illustrated in the generalized process presented in Figure 1.2. In this example, 100 tph of ore are being fed into a concentration operation that produces 4.5 tph of concentrate and

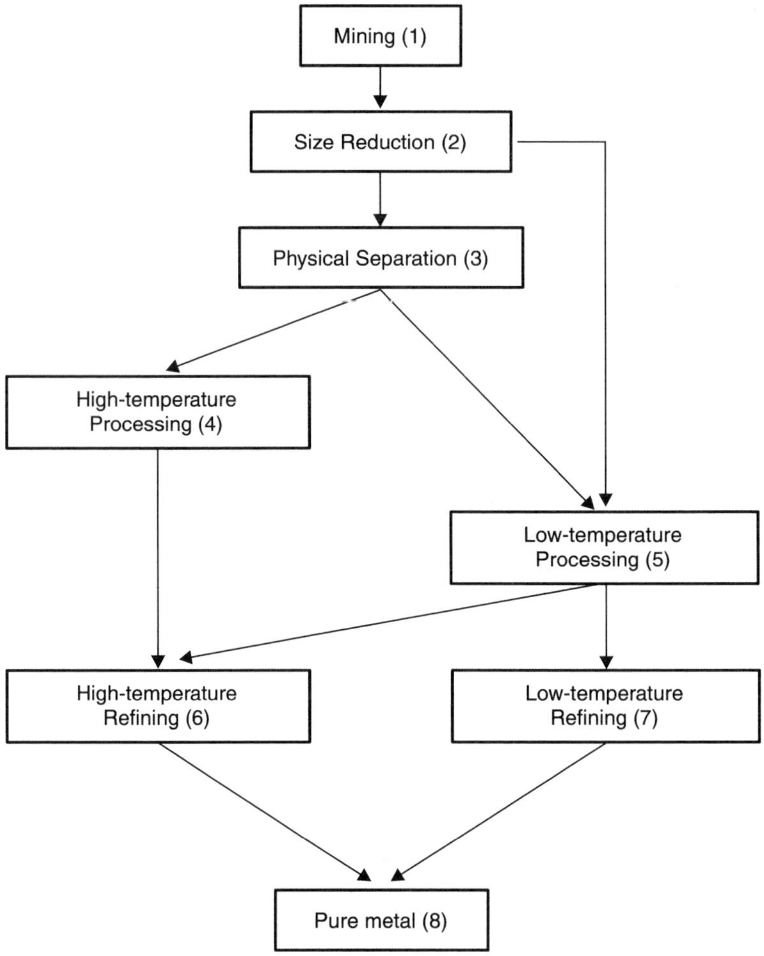

FIGURE 1.1 Generalized flowchart of extraction of metals

TABLE 1.1 Processing sequence(s) for a number of selected metals

		Steps Involved in the Processing Route (see Figure 1.1)							
Metal	**Associated Major Minerals**	**1**	**2**	**3**	**4**	**5**	**6**	**7**	**8**
Iron	Hematite, Fe_2O_3; magnetite, Fe_3O_4	x	x	x	x		x		x
Aluminum	Gibbsite, Al_2O_3-$3H_2O$; diaspore, $Al2_2O_3xH_2O$	x	x			x	x		x
Copper	Chalcopyrite, CuFeS2; chalcocite, Cu2S	x	x	x	x	x		x	x
Zinc	Sphalerite, ZnS	x	x	x	x		x		x
		x	x	x	x	x		x	x
Lead	Galena, PbS	x	x	x	x		x		x
Gold	Native gold, Au	x	x	x		x	x		x
		x	x*			x	x		x
Platinum	Native platinum, Pt; platinum sulfides	x	x	x		x		x	x
Silver	Native silver, Ag	x	x	x		x	x		x

*Only crushing is practices; grinding is usually omitted.

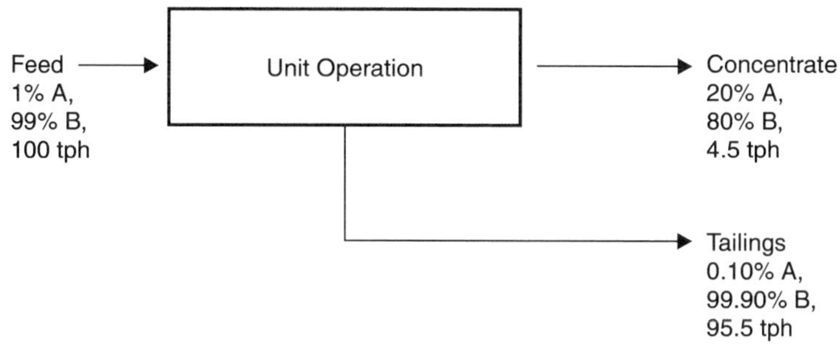

FIGURE 1.2 A simple material balance for a unit operation

TABLE 1.2 U.S. total and recycled supply of selected metals in 1996

Metal	Total Supply, million t metal content	Recycled Supply, million t metal content	% Recycled
Iron and steel	183	72	39
Aluminum	8.34	3.29	39
Copper	3.70	1.30	35.1
Lead	1.63	1.09	66.8
Zinc	1.45	0.379	26.1
Chromium	0.48	0.098	20.5
Magnesium	0.205	0.0709	35
Gold	516 t*	150 t*	29

Source: U.S. Bureau of Mines (1997).
*Value for 1995.

95.5 tph of tailings. In upgrading this process, then, 1.0 tph of the desired material, A, is introduced into the unit operation and 0.9 tph (4.5 × 0.2) of this material reports to the concentrate, resulting in 90% recovery (0.9/1.0 × 100). The grade of the mineral, A, has been improved from 1% to 20%. The term *percent recovery* refers to the percentage of the valuable material reporting to the concentrate with reference to the amount of this material in the feed. Note that obtaining the highest possible recovery is not necessarily the best approach in a concentration process. High recovery without acceptable grade will lead to an unsalable product and is therefore unsatisfactory.

Mineral processing engineers are responsible for optimizing processes to yield the highest possible recovery with acceptable purity (grade) for the buyers or engineers who will treat this concentrate further to extract the metal values. To achieve this goal, economic assessments of all possible technological alternatives must be conducted.

ECONOMIC CONCERNS

Table 1.2 summarizes the total U.S. supply and recycled supply of selected metals in 1996. The total supply of iron and steel includes supply from primary and secondary sources as well as imports; these two metals represent by far the largest of commodities produced and consumed, followed by aluminum, copper, and lead. Note that the recycled supply of these metals from processing scrap is strikingly high. In addition, the tonnage of precious metals consumed is rather small. However, because of the high prices of precious metals, their monetary value is substantial. For example, the monetary value of 516 t of gold was $12.8 billion in 1996, compared to $10.7 billion for 5.3 million t of copper and lead.

TABLE 1.3 Abundance of various elements in the Earth's crust compared to annual U.S. consumption

Element	Relative Abundance, %	U.S. Consumption, st/year
Fe	5	1.28×10^8
Al	8.13	5.4×10^6
Cu	7×10^{-3}	2.3×10^6
Zn	8×10^{-3}	1.0×10^5
Pb	1.5×10^{-3}	1.2×10^6
Au	1.0×10^{-7}	113
Ag	2.0×10^{-6}	4.52×10^3

Source: U.S. Bureau of Mines (1990).

Table 1.3 lists the relative abundance of various metals in the Earth's crust. Most metals are present in extremely small concentrations in nature, and none of these metals can be recovered economically at these concentrations. Rock that contains metals at these concentrations is not ore; ore is rock that can be processed at a profit. An average copper ore, for example, may contain 0.3% to 0.5% copper. Even this material cannot be treated economically at high temperature without prior concentration. There is no way that rock containing 10 lb of copper and 1,990 lb of valueless material can be heated to 1,300°C and treated to recover this quantity of metal economically. Concentrating the ore by froth flotation to approximately 25% or more copper results in a product that can be smelted and refined profitably.

UNIT OPERATIONS

Numerous steps, called unit operations, are involved in achieving the goal of extracting minerals and metals from ores in their purest possible form. These steps include

- *Size reduction.* The process of crushing and grinding ores is known as comminution. The purpose of the comminution process is threefold: (1) to liberate valuable minerals from the ore matrix, (2) to increase surface area for high reactivity, and (3) to facilitate the transport of ore particles between unit operations.
- *Size separation.* Crushed and ground products generally require classification by particle size. Sizing can be accomplished by using classifiers, screens, or water elutriators. Screens are used for coarse particulate sizing; cyclones are used with fine particulates.
- *Concentration.* Physicochemical properties of minerals and other solids are used in concentration operations. Froth flotation, gravity concentration, and magnetic and electrostatic concentration are used extensively in the industry.
  - *Froth flotation.* The surface properties of minerals (composition and electrical charge) are used in combination with collectors, which are heterogeneous compounds containing a polar component and a nonpolar component for selective separations of minerals. The nonpolar hydrocarbon chain provides hydrophobicity to the mineral after adsorption of the polar portion of the collector on the surface.
  - *Gravity concentration.* Differences in the density of minerals are used to effect separations of one mineral from another. Equipment available includes jigs, shaking tables, and spirals. Heavy medium is also used to facilitate separation of heavy minerals from light minerals.
  - *Magnetic and electrostatic concentration.* Differences in magnetic susceptibility and electrical conductivity of minerals are utilized in processing operations when applicable.

- *Dewatering.* Most mineral processing operations are conducted in the presence of water. Solids must be separated from water for metal production. This is accomplished with thickeners and filters.
- *Aqueous dissolution.* Many metals are recovered from ores by dissolving the desired metal(s)—in a process termed *leaching*—with various lixiviants in the presence of oxygen. Following leaching, the dissolved metals can be concentrated by carbon adsorption, ion exchange, or solvent extraction. Purified and concentrated metals may be recovered from solution with a number of reduction techniques, including cementation and electrowinning.

EXAMPLES OF MINERAL PROCESSING OPERATIONS

Figure 1.3 shows a typical flowsheet for crushing and sizing rock in a quarrying operation. Run-of-mine ore can be present as lumps as large as 1.5 m (5 ft) in diameter. In this figure's example, 91.4-cm (3-ft) lumps of rock are fed to a crusher that reduces the material to 20.3 cm (8 in.) or less in diameter. After screening to remove rock that is less than 57.2 mm (2¼ in.) in size, rock between the sizes of 57.2 mm (2¼ in.) and 20.3 cm (8 in.) is further reduced in size by a gyratory crusher. The product from this step is then classified by screening to the desired product for sale.

Figure 1.4 shows an integrated circuit demonstrating crushing, grinding, size separation, and gravity concentration of a tin ore. Initial size separation is effected with a grizzly set at 1½-in. Oversize material is fed to a jaw crusher set at 1½-in., and the crushed product is, then, further reduced in size to 20 mesh by ball milling. The −20-mesh material is classified by hydrocyclones set at 150 mesh, and the −150-mesh material is sent to shaking tables to concentrate the heavy tin mineral, cassiterite. The middlings in this process receive additional treatment. The concentrate from this operation is reground and sized at 200 mesh. Two-stage vanning is used to produce a fine tin concentrate.

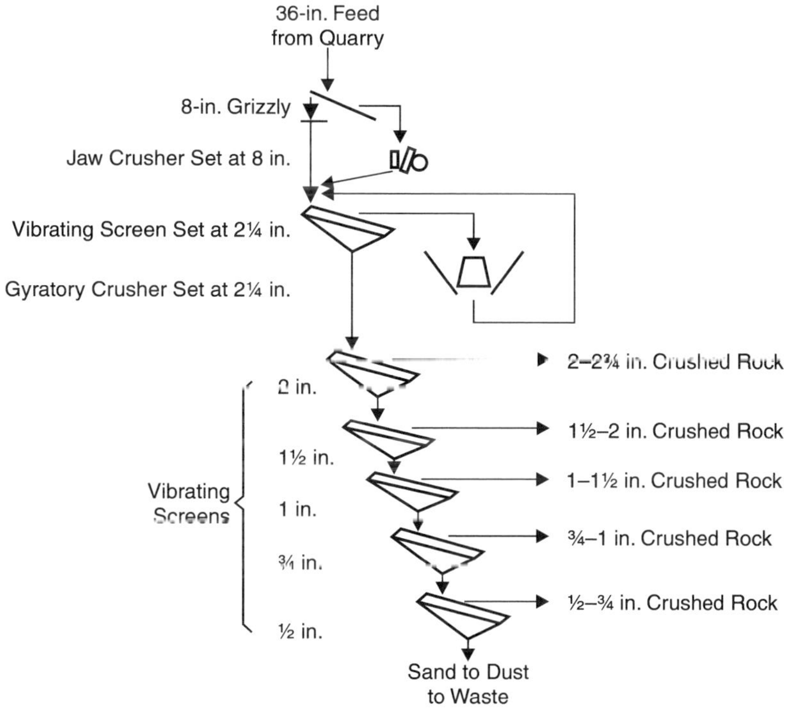

FIGURE 1.3 Flowsheet for crushing and grading rock

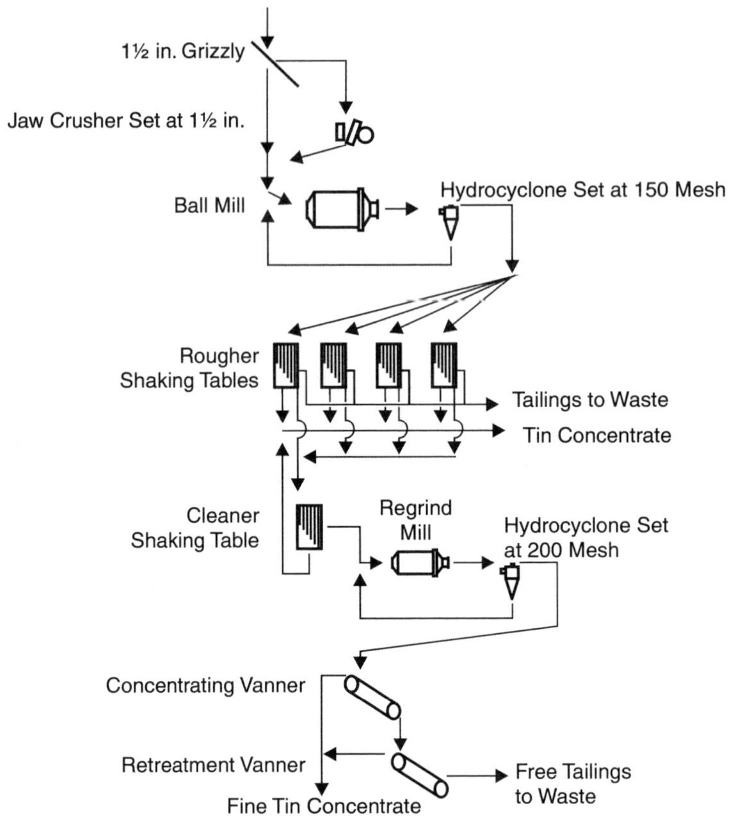

FIGURE 1.4 Flowsheet for the gravity concentration of a tin ore

The flowsheet describing the flotation processing of a copper ore containing chalcopyrite and molybdenite is shown in Figure 1.5. After grinding and classification, pulp is fed to rougher flotation. The rougher tailings are thickened and sent to a tailings dam. The rougher concentrate is classified, and the oversize is reground. Cyclone overflow is fed to cleaner flotation, and the cleaner concentrate is recleaned. Cleaner tailings are recycled back to rougher flotation, and the recleaner concentrate is thickened and sent to the molybdenum recovery plant for further processing. In this operation, the feed contains 0.32% Cu and 0.03% Mo. Rougher concentrate, cleaner concentrate, and recleaner concentrate contain 7%–9% Cu, 18% Cu, and 25% Cu, respectively. Recleaner concentrate also contains 2%–3% Mo.

Figure 1.6 depicts a flowsheet for processing free-milling oxidized gold ore. The kinetics of gold leaching is slow, and gold ores are frequently ground to less than about 75 µm before leaching. Even then, one day is usually required in the leaching step. In this process, run-of-mine ore is crushed and ground. The ball mill discharge in subjected to gravity concentration to recover the larger particles of free gold. The tailings from this operation are thickened, and the underflow from the thickeners is then subjected to cyanide leaching. In some instances, ores may contain oxygen-consuming minerals, such as pyrrhotite and marcasite, and a preaeration step may be conducted ahead of cyanide leaching.

Heap leaching has revolutionized the gold mining industry. Low-grade oxidized ores containing approximately 0.03 oz gold per short ton of ore can be processed with this technology, whereas they could not be processed by the higher cost grinding/agitation leaching (milling) process. Figure 1.7 presents a simplified flowsheet of heap leaching. As the figure shows, run-of-mine ore may or may not be crushed. If crushing is done, the ore is generally crushed to <2 in. in diameter.

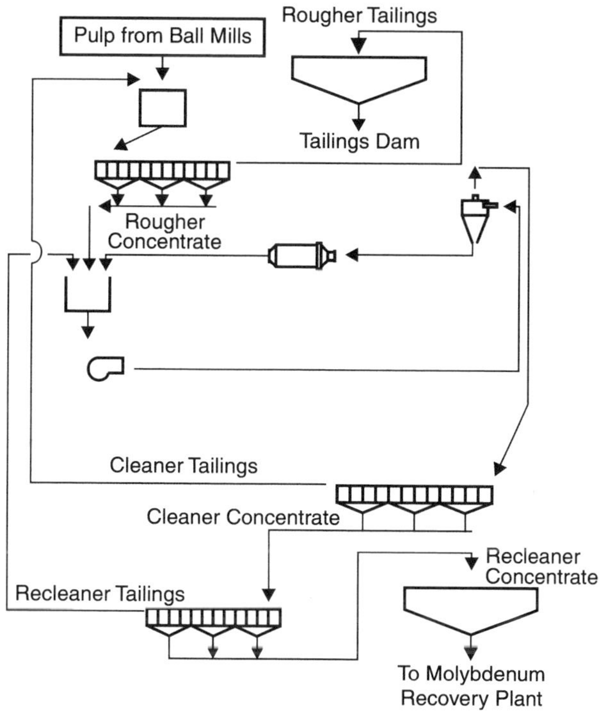

FIGURE 1.5 Flowsheet for the flotation of copper sulfide ore

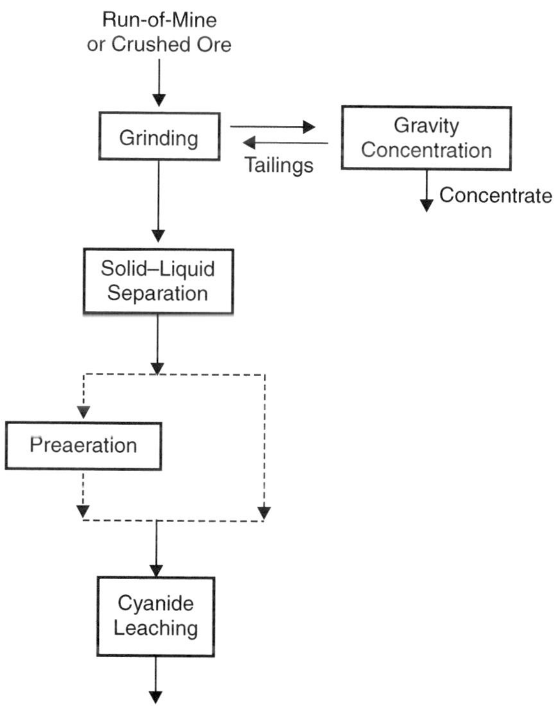

FIGURE 1.6 Flowsheet options for grinding and agitated leaching of free-milling oxidized gold ores

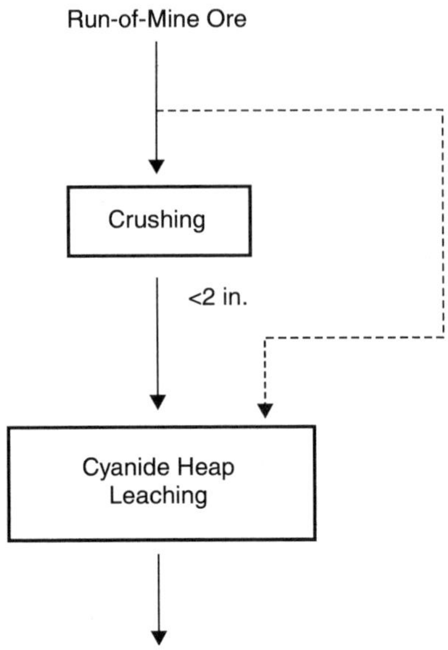

FIGURE 1.7 Flowsheet for heap leaching of oxidized gold ores

ENVIRONMENTAL CONSEQUENCES OF MINERAL PROCESSING

During the course of mining and metal extraction, an unavoidable consequence is that the environment will be disturbed. The hills and valleys will be excavated during mining, and because strong reagents are used in solubilization processes, the rocks and water involved will be contaminated. The mineral industry is very conscious of these phenomena and spends large amounts of capital on remediating the environment and neutralizing toxic wastes.

BIBLIOGRAPHY

Gaudin, A.M. 1939. *Principles of Mineral Dressing*. New York: McGraw-Hill.

Janes, C.J., and L.M. Johnson. 1976. The Duval Sierrita Concentrator. In *Flotation*. Edited by M.C. Fuerstenau. New York: AIME.

Marsden, J., and I. House. 1992. *The Chemistry of Gold Extraction*. New York: Ellis Horwood.

U.S. Bureau of Mines. 1990. *Mineral Commodity Summaries*. Washington, D.C.: U.S. Bureau of Mines.

——. 1997. *Mineral Commodity Summaries*. Washington, D.C.: U.S. Bureau of Mines.

CHAPTER 2

Particle Characterization

Richard Hogg

Particulate materials—dry powders as well as liquid or gas suspensions—play an increasingly important role in modern society. Most industrial processes involve particulates in some stage of the operation, perhaps as raw materials, as products, as unwanted by-products of wear, or simply as atmospheric dust. Particle systems are especially important in mineral processing—a field that deals almost exclusively with particulates, from run-of-mine ore to final concentrate. The objective of a mineral processing operation is to take an input stream of particles with a given set of characteristics, modify those characteristics, and separate the material into product streams, each with its own set of specified characteristics.

Obviously, characterization is critical to the operation, assessment, and control of mineral processing unit operations and systems. The primary aims of this chapter are to address the goals of particle characterization for mineral processing applications in light of practical constraints, to discuss general schemes for representing particle characteristics, and to describe and evaluate the various techniques available for measuring particle characteristics.

Fine particle systems are a distinct class of materials whose behavior is often determined more by their particulate characteristics than by the bulk properties of the actual solids. Of these characteristics, the distributions of size, shape, and structure are especially important, and their evaluation is a vital step in process control and product specification. The characteristics are not usually single valued. Each particle has its own set of characteristics; the system of particles is described by the distributions of the different characteristics. The use of average values may be appropriate in some cases; in others, it may be quite inadequate. In addition to the individual particle characteristics, there are bulk properties that belong to the particle *system*. To some extent, these bulk properties are determined by the complete set of individual characteristics, but they may also depend on the relative arrangement of the particles in space and on interactions among particles and with any intervening medium (air, water, etc.).

PARTICLE CHARACTERISTICS

Two subsets of individual particle characteristics can be considered: basic and derived. Basic characteristics represent a minimum set that, when taken together, completely define the particle. By definition, the basic characteristics include

- Size
- Shape
- Composition (chemical and mineralogical)
- Structure (single component or composite; arrangement of constituent phases including pores)

Examples of derived characteristics include

- Density
- Optical characteristics: color, refractive index, reflectance

- Electromagnetic characteristics: conductivity, magnetic susceptibility
- Thermal characteristics: conductivity, heat capacity
- Chemical characteristics: solubility, reactivity
- Mechanical characteristics: strength, Young's modulus, Poisson's ratio

Derived characteristics are—in principle, at least—fixed by and dependent on the set of basic characteristics. In other words, all the characteristics just listed are determined by the size, shape, composition, and structure of an individual particle.

The bulk properties of a particle system include

- Surface area
- Reactivity
- Toxicity

These are essentially determined by the set of basic, individual characteristics and by (1) bulk density and porosity, (2) homogeneity, and (3) rheology. The latter features depend additionally on the spatial arrangement of the particles and on interactions among them. Bulk properties are, by definition, single valued, but they may depend on the state of the system as well as on its content.

Distributions of Particle Characteristics

Individual characteristics generally vary from particle to particle and can be represented by distributions. In general, the distributions can be expressed as discrete values or continuous functions in either incremental or cumulative form. For some characteristic p (e.g., size, shape, composition), the incremental distribution can be defined as a set of discrete values:

$$q_i = \text{the fraction of particles for which } p \text{ has the specific value } p_i \qquad \textbf{(Eq. 2.1)}$$

or as a continuous variable:

$$q(p)dp = \text{the fraction for which } p \text{ lies in the range } p \text{ to } p + dp \qquad \textbf{(Eq. 2.2)}$$

The cumulative distribution is defined as the fraction for which p is less than some specific value. Thus, for the discrete case,

$$Q_i = \sum_{j=1}^{i} q_j \qquad \textbf{(Eq. 2.3)}$$

and the continuous equivalent is

$$Q(p) = \int_0^p q(p)dp \qquad \textbf{(Eq. 2.4)}$$

Distributions of particle characteristics are, for the most part, inherently continuous; that is, not restricted to specific values. It is often convenient, however, to consider discrete classes of particles, in which case

$$q_i = \int_{p_i}^{p_{i+1}} q(p)dp \qquad \textbf{(Eq. 2.5)}$$

In practice, it is often necessary to consider variations in more than one characteristic; for example, size *and* composition. Denoting these characteristics by p, r, s, for example, the variations can be described by using what is called the joint distribution:

$$q_{ijk...} = \text{the fraction of particles for which } p = p_i,\ r = r_j,\ s = s_k,\ ..., \text{ etc.} \qquad \textbf{(Eq. 2.6)}$$

Although it becomes apparent that all four of the basic characteristics can vary in typical ore samples, it is normally practical to consider only two at the most (e.g., size and composition). In this case, a useful alternative is to introduce the conditional distribution:

$$f_p(r_j) = f_j = \text{fraction of particles with a given value of } p \text{ for which } r = r_j \qquad \textbf{(Eq. 2.7)}$$

and the marginal distribution:

$$q(p_i) = q_i = \text{the fraction of all particles for which } p = p_i \text{ regardless of the value of } r \qquad \textbf{(Eq. 2.8)}$$

The conditional and marginal distributions are related to the joint distribution, q_{ij}, through

$$q_{ij} = f_p(r_j) \cdot q(p_i)$$

For particle systems, various characteristics are commonly expressed relative to particle size. Thus, the particle size distribution is used as the marginal distribution, with the distributions of other characteristics (shape, composition, etc.) as conditionals. Ores and coal can often be regarded as binary mixtures of values and gangue. Because these components typically vary significantly in density, particle density is widely used as an indicator of particle composition. This practice is especially appropriate when gravity separations are to be used for beneficiation. An example of a size/density distribution for coal is given in Table 2.1 and in Figures 2.1 and 2.2. Figure 2.1 shows the overall size distribution (marginal) and the size distribution for 1.25 specific-gravity material (conditional). Table 2.1 and Figure 2.2 give the joint distribution.

Description of Particle Characteristics

To describe the characteristics of a particle, it is generally desirable to assign them numerical values. These values should be clearly defined, unique, and measurable in practice. Satisfying these criteria is not simple, and problems arise for each of these various characteristics.

Particle Size and Shape. It is well known that the behavior of systems of fine particles is strongly dependent on the sizes of the individual particles and that the size effects become increasingly important as the particles become progressively smaller. Despite the obvious importance of particle size, however, the evaluation and even precise definition of particle size are far from simple tasks. In general, we want to express the size of a particle as a single, linear dimension and refer to, for example, a 6-ft boulder, a 1-in. pebble, and a 10-micron particle. The problem is, which linear dimension do we use?

Only in the case of simple shapes, such as spheres or cubes, can we identify a single dimension that adequately characterizes particle size. Note, however, that even in these cases we must specify the

TABLE 2.1 Example of a size/specific-gravity distribution for coal: Weight percent (q_{ij}) values

Specific Gravity (ρ_j)	Size (x_i)							
	6 in. × 3 in.	3 in. × 1⅝ in.	1⅝ in. × ½ in.	½ in. × ¼ in.	¼ in. × 8 Mesh	8 × 14 Mesh	14 × 48 Mesh	48 Mesh × 0
1.3 (float)	4.44	5.73	16.30	8.75	9.75	4.91	5.58	2.32
1.3 × 1.4	1.62	2.09	5.63	3.26	3.05	0.94	0.77	0.33
1.4 × 1.5	0.64	0.68	1.40	0.84	0.83	0.32	0.25	0.12
1.5 × 1.6	0.43	0.46	0.86	0.34	0.43	0.17	0.16	0.08
1.6 × 1.7	0.21	0.26	0.49	0.17	0.22	0.09	0.08	0.04
1.7 × 1.8	0.13	0.18	0.37	0.11	0.16	0.06	0.06	0.04
1.8 (sink)	4.13	2.79	3.55	0.82	1.26	0.51	0.40	0.28
Total	11.6	12.2	28.6	14.3	15.8	7.0	7.3	3.2

Source: Data from Sokaski, Jacobson, and Geer 1963.

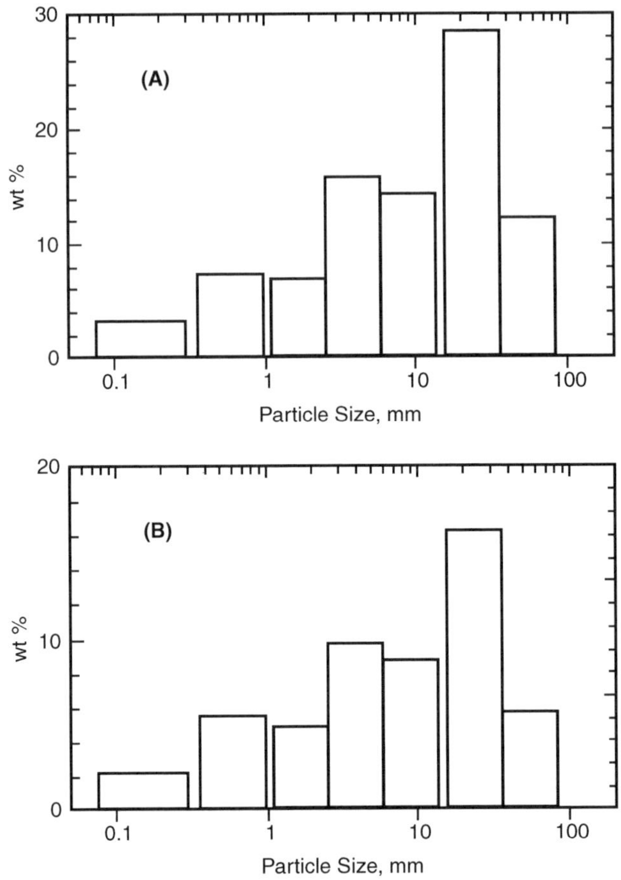

FIGURE 2.1 Example of size and specific-gravity distributions for coal: (A) overall size distribution (marginal); (B) size distribution for 1.25 specific gravity (conditional)

shape of the particle and which particular dimension is being used (diameter of sphere; side, face diagonal, etc., of a cube; and so on). In general, we cannot define a particle's size without first describing its shape. A more unique description of size could be obtained by considering the mass or volume of the particle. However, because mass and volume can rarely be measured directly, at least for very fine particles, and because the behavior of the particles depends on their shape, little advantage is gained from this approach.

Describing particle shape is difficult; taking quantitative measurements is even more so. Regular shapes such as spheres, cubes, or tetrahedra can be described and quantified, but real particles very rarely fall into such categories and are most commonly described as "irregular." In principle, any shape can be described by fitting a mathematical function to it. For example, a two-dimensional image can be fitted to a Fourier's series. For example,

$$r(\theta) = a_0 + a_1 \sin \theta + a_2 \sin 2\theta + \ldots \quad \textbf{(Eq. 2.9)}$$

where r is the radial vector, at some angle θ, from the center of the image to some point on the periphery. The complete set of coefficients ($a_0, a_1, a_2, \ldots$), in effect, defines the shape of the image. Because each coefficient is likely to be different for each particle, applying this approach to real systems is rarely practical.

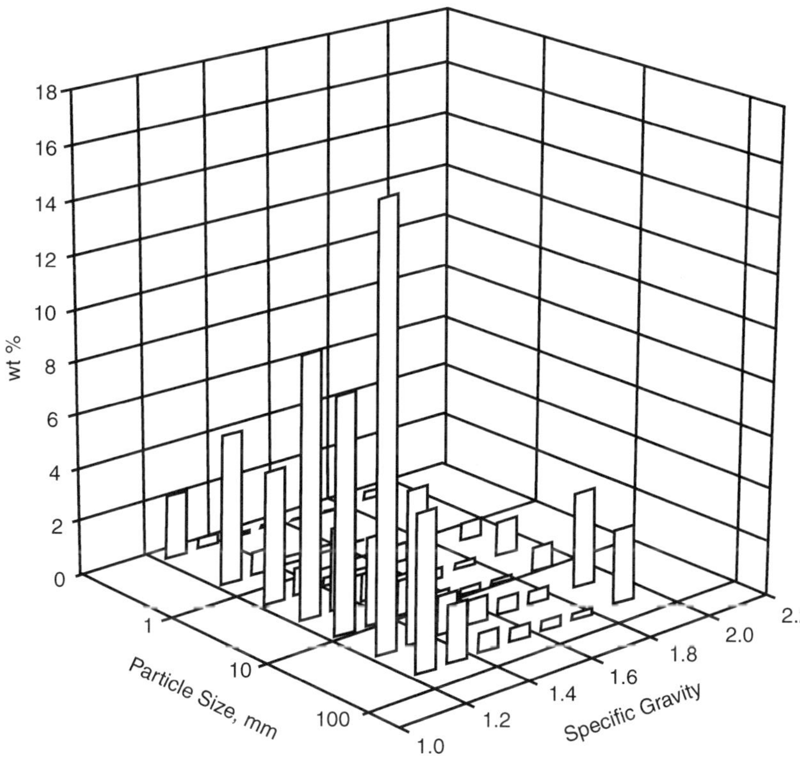

FIGURE 2.2 Example of a joint size specific-gravity distribution for coal

In some cases, associating particles with general shape classes, such as ellipsoids, is a useful method. The lengths of the major axes define the size and shape of the "particle." This approach is appropriate for minerals that tend to occur as plates (e.g., clays) or needles (e.g., asbestos). For other shapes, the value of this method will often be outweighed by the difficulty in obtaining the necessary measurements.

One simple, practical solution to these problems is to combine the effects of size and shape and to characterize particles in terms of an equivalent, simple shape (usually a sphere) with a given dimension. Thus, we say that a particle behaves as though it were a sphere of diameter *d*. There are several important consequences of the use of this simplified approach. In the first place, we must recognize that the definition of size will itself depend on the method by which that size is determined. If an irregular particle is sized using sieves, we can say that the particle acts as though it were a sphere whose diameter lies between two sieve opening sizes. However, in a sedimentation device, the same particle may behave as a sphere of a quite different diameter. Thus, we must specify not only the "size" of the particle but also the method by which the size was obtained. Differences between these sizes can be ascribed to particle shape; ratios of the sizes obtained by different methods are often called shape factors.

The use of an equivalent spherical diameter involves the implied assumption that all particles in a given system have essentially the same shape. Although this is often a reasonable assumption, there can obviously be cases where it is not valid. In these instances, variations in particle shape would manifest themselves as apparent variations in size.

From a practical standpoint, the uncertainty in the definition of particle size places some important restrictions on the choice of sizing methods:

1. Direct comparisons can be made only between sizes determined by the same kind of technique (sieving, microscopy, sedimentation, etc.).
2. For systems containing a broad range of sizes, we will want to use, as much as possible, the same technique for all sizes. When this is not possible, we must pay particular attention to the evaluation of the appropriate conversion factors (e.g., sieve "size" to sedimentation "size"). When more than one method is used, there should be as much overlap as possible in the ranges covered by each.
3. In choosing a sizing method, consideration should be given to matching the method to the particular application for which the size information is desired. Thus, if we wish to characterize the particles in a liquid suspension, we would try to use a method that evaluates the behavior of the particles in a liquid; for example, a sedimentation method. In this way, we can automatically compensate for some of the uncertainties in the meaning of size and shape.

Particle Composition and Structure. The chemical composition of a particle can be uniquely defined and can sometimes be represented by the value of a derived characteristic, such as density, color, or magnetic susceptibility. The distribution of composition or its surrogates (e.g., density) can be determined by particle-by-particle analysis or by appropriate separation methods (gravity, magnetic, color sorting, etc.). Particle structure presents a more difficult problem. The same composition can arise in an infinite number of ways: homogeneous or composite (binary, ternary, etc.). A binary composite, for instance, can consist of two attached grains or one component dispersed within a matrix of the other. A dispersed component can have any number of possible grain size distributions within individual particles. Because of this complexity, there is little value in attempting to establish a general scheme for describing the distributions of particle structure. One approach is to define a set of discrete particle types that can be distinguished in practice and are useful indicators. The distributions of the other characteristics–size, shape, and overall composition–can then be evaluated for each type.

MATHEMATICAL TREATMENT OF PARTICLE DISTRIBUTIONS

The distributions of particle characteristics are similar to and subject to the same constraints as probability distributions. Many of the concepts and terminology used in probability and statistics can be directly applied to particle systems. The treatment presented here makes use of definitions and terminology established at the University of Karlsruhe (Rumpf and Ebert 1964).

To describe distributions of particle characteristics, we must represent

- The value of the characteristic itself
- The relative amount of material that has that value

Representation of Particle Characteristics

Particle size can be represented as a linear dimension, an area (surface area or projected area), a volume, or a mass. The relationships among these different representations depend on particle shape and, in the case of mass, on density. Thus, for a sphere of diameter x and density ρ, the surface area, A_s, is given by

$$A_s = \pi x^2 \quad \textbf{(Eq. 2.10)}$$

The projected area, A_p, is given by

$$A_p = \frac{\pi x^2}{4} \quad \textbf{(Eq. 2.11)}$$

The volume, V, is given by

$$V = \frac{\pi x^3}{6} \quad \textbf{(Eq. 2.12)}$$

The mass, m, is given by

$$m = \frac{\pi \rho x^3}{6} \quad \textbf{(Eq. 2.13)}$$

For particles of arbitrary shape, the relationships for area (A) and volume (V) can be written as follows:

$$A = k_2 x^2 \quad \textbf{(Eq. 2.14)}$$

$$V = k_3 x^3 \quad \textbf{(Eq. 2.15)}$$

where k_2 and k_3 are shape factors defined by Eqs. 2.14 and 2.15, respectively.

Particle shape distributions require that shape be represented by a numerical factor. The factors k_2 and k_3 defined above are obvious choices. Other factors, such as aspect ratios ("length" to "width" of elongated particles), can also be used. When more than one factor is used to describe shape, the joint distributions of all the factors must be considered.

Particle composition can be described by using a set of composition variables. For n_c chemical constituents, a minimum of $n_c - 1$ variables must be specified. A similar approach can be applied to particle structure by using appropriately defined structure "types." Again, the joint distributions of composition and structure must be evaluated.

Representation of Particle Quantity

The quantity of particulate material that possesses specific values of the characteristics (size, shape, etc.) can be represented in various ways. For a system of particles, there will generally be some number n_i that are essentially identical—i.e., have the same size, x_i; the same shape factors,$(k_2)_i$ and $(k_3)_i$; the same density, ρ_i, for example. The quantity of these particles can be represented by

- Total number: n_i
- Total length: $n_i x_i$
- Total area: $n_i (k_2)_i x_i^2$
- Total volume: $n_i (k_3)_i x_i^3$
- Total mass: $n_i \rho_i (k_3)_i x_i^3$

In general, the fractional quantities can be expressed as

$$(q_r)_i = \frac{n_i (k_r)_i x_i^r}{\sum n_i (k_r)_i x_i^r} \quad \textbf{(Eq. 2.16)}$$

with $r = 0, 1, 2, 3$ corresponding to the number, length, area, and volume fractions, respectively, and k_0 and $k_1 = 1$ by definition. The number, length, area, and volume fractions are all clearly different when there are variations in particle size. The area and volume fractions depend on particle shape. The volume and mass fractions differ only if there are variations in particle density.

Particle Size Distributions

The distribution of particle size is of major importance in mineral processing. The behavior of particles in crushing and grinding circuits, concentration operations, and solid–liquid separations is strongly dependent on size. Furthermore, the range of sizes in a single process stream is typically very large and can include particles that vary in diameter from 1 m (3.3 ft) to less than 1 μm (10^{-6} m).

It was noted in a previous section that particle size can be expressed in a variety of ways: diameter, area, volume, or mass. However, regardless of how the size is measured, the almost universal practice is to present size as a linear (very often, equivalent sphere) diameter x. Size is inherently a continuous variable and data are commonly classified into appropriate size intervals. In this chapter, we will define x_i as the upper boundary of size interval i and select x_1 as the maximum size present.

This approach is convenient because a maximum size is generally easier to establish than a minimum size (which may be too small to detect or to measure with any accuracy). On the basis of this definition, the interval width is given by

$$\Delta x_i = x_i - x_{i+1} \qquad \textbf{(Eq. 2.17)}$$

We will also select the interval boundaries so as to fix the width of the interval relative to the size it represents. This is accomplished by choosing interval boundaries in a geometric progression:

$$x_{i+1} = \frac{x_i}{r_s} \qquad \textbf{(Eq. 2.18)}$$

where r_s is a constant. This approach is consistent with, for example, standard sieve series such as the U.S. or Tyler standards, where each successively coarser sieve differs from the previous one by a constant ratio of $2^{1/4}$ (1.189). In other words, each sieve opening is about 19% larger than that of the next finer sieve in the series. The principal advantage of this use of geometric intervals is that the same amount of detail is provided at each point on the scale.

It is obvious from Eqs. 2.18 that

$$x_2 = \frac{x_i}{r_s} \qquad \textbf{(Eq. 2.19)}$$

and

$$x_3 = \frac{x_2}{r_s} = \frac{x_1}{r_s^{\,2}} \qquad \textbf{(Eq. 2.20)}$$

so that, in general,

$$x_i = \frac{x_1}{r_s^{\,i-1}} \qquad \textbf{(Eq. 2.21)}$$

Also, from Eqs. 2.17 and 2.21,

$$\Delta x_i = \frac{\Delta x_i}{r_s^{\,i-1}} \qquad \textbf{(Eq. 2.22)}$$

Equations 2.21 and 2.22 can be useful in manipulating particle size data.

Using the interval boundaries as established above, we can define the incremental size distribution, $(q_r)_i$, as follows:

$$(q_r)_i = \text{“}x^r\text{” fraction whose size lies between } x_i \text{ and } x_{i+1} \qquad \textbf{(Eq. 2.23)}$$

where $r = 0, 1, 2, 3$ again corresponds to the number, length, area, and volume fractions, respectively (as described earlier), so that q_0 represents the number fraction, q_3 represents volume fraction, and so on. The cumulative form is called the particle size distribution function, $Q_r(x_i)$, defined by

$$Q_r(x_i) = \text{“}x^r\text{” fraction for which } x < x_i \qquad \textbf{(Eq. 2.24)}$$

It follows from the definition of the interval boundaries that

$$Q_r(x_i) = \sum_{j=1}^{n} (q_r)_j \qquad \textbf{(Eq. 2.25)}$$

where n represents the "sink" interval that includes all particles smaller than x_n. For this particular interval,

$$Q_r(x_i) = q_r(n) \qquad \textbf{(Eq. 2.26)}$$

Because $i = 1$ represents the largest particle present,

$$Q_r(x_i) = 1 \qquad \textbf{(Eq. 2.27)}$$

and $(Q_r)_i$ can also be represented by

$$Q_r(x_i) = 1 - \sum_{j=1}^{i-1} (q_r)_j \qquad \textbf{(Eq. 2.28)}$$

The values of $Q_r(x_i)$ also represent points on the continuous form of the distribution function $Q_r(x)$. The latter can be used to define a particle size density function, $q_r(x)$, such that

$$q_r(x) = \frac{dQ_r(x)}{dx} \qquad \textbf{(Eq. 2.29)}$$

and

$$Q_r(x) = \int_0^x q_r(x)dx \qquad \textbf{(Eq. 2.30)}$$

The physical meaning of the density function is that

$$q_r(x)dx = \text{“}x^r\text{” fraction whose size falls between } x \text{ and } x + dx \qquad \textbf{(Eq. 2.31)}$$

It should be noted that the incremental distribution $(q_r)_i$ is not directly equivalent to the density function $q_r(x)$; instead, we have

$$(q_r)_i = \int_{x_{i+1}}^{x_i} q_r(x)dx \qquad \textbf{(Eq. 2.32)}$$

and

$$q_r(x_i^*) \approx \frac{(q_r)_i}{\Delta x_i} \qquad \textbf{(Eq. 2.33)}$$

where x_i* refers to some size in the interval x_i to x_{i+1}. The discrete, incremental distributions and the continuous density function are proportional to each other only when the interval widths are constant (linear intervals); they are not proportional for the more commonly used geometric intervals. Examples of a distribution function and the corresponding density function and incremental distribution are given in Figure 2.3.

Transformations. Later in this chapter we will show that different methods for measuring particle size distribution involve different quantity representations. Counting methods usually give number distributions ($q_0(x)$), whereas gravimetric methods give mass or volume distributions ($q_3(x)$), and some optical methods give area distributions ($q_2(x)$). For comparison purposes, as well as in many applications, transforming from one type of distribution to another (e.g., from number to volume) is often necessary. The general formula for transforming a distribution based on x^r fraction to one based on x^t is as follows (Leschonski 1984):

$$q_t(x) = \frac{k_r x^{t-r} q_r(x)}{\int_0^\infty k_r x^{t-r} q_r(x)dx} \qquad \textbf{(Eq. 2.34)}$$

If the shape factors, k_r, are independent of size, they can be eliminated from Eq. 2.34, leading to

$$q_t(x) = \frac{x^{t-r} q_r(x)}{\int_0^\infty x^{t-r} q_r(x)dx} \qquad \textbf{(Eq. 2.35)}$$

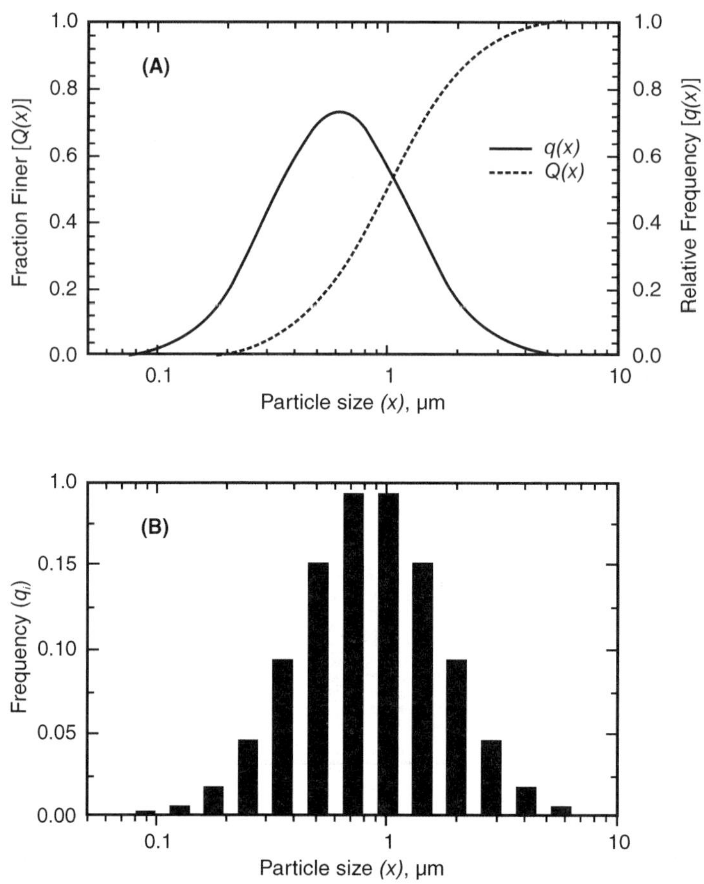

FIGURE 2.3 Example of particle size distributions: (A) continuous density and distribution functions; (B) discrete incremental distribution (histogram)

or, in discrete form,

$$(q_t)_i = \frac{x_i^{t-r}(q_r)_i}{\sum_{i=1}^{n} x_i^{t-r}(q_r)_i} \quad \textbf{(Eq. 2.36)}$$

Equations 2.35 and 2.36 can readily be applied to real data. We should note that using Eqs. 2.35 and 2.36 does not require that the shape factors be the same for all particles; instead, it requires merely that there are no systematic variations with size. In other words, the *average* shape factors should be the same for all sizes. Although there are obviously cases where this requirement is not satisfied—delamination of clays, for example—most systems of particles do not show significant variations in shape with size.

A more serious problem in the use of these transformations is the need, in effect, to extrapolate to "zero" in order to integrate (Eq. 2.35) or sum (Eq. 2.36) over all sizes (0 to ∞ or $i = 1$ to n). The problem is usually not serious when t is greater than r (e.g., in transforming from number to volume), but it can be critical in the reverse transformation ($t < r$; e.g., volume to number). Specifically, the problem lies in determining the exact form of $q_r(x)$ for integration or in assigning an appropriate mean value of size in the sink interval ($i = n$). An example illustrating this problem is given in Appendix 2.1. The only realistic solution to the problem is to extend the range of reliable measurement to finer sizes.

Transformation of a volume distribution ($q_3(x)$) to a mass distribution ($q_3'(x)$) can be accomplished by using the formula

$$q_3'(x) = \frac{\rho(x)q_3(x)}{\int_0^\infty \rho(x)q_3(x)dx} \quad \textbf{(Eq. 2.37)}$$

where $\rho(x)$ is the average density of a particle of size x. If the density is independent of size, the mass and volume distributions are identical. Variations in density with size can become significant when the degree of liberation of different minerals changes with size. This is a common occurrence in mineral processing systems where grinding is widely used for enhancing liberation.

Average Sizes. The use of average sizes can be convenient, but caution should be exercised and the average should be carefully specified. Any average is an indicator of the location of the size distribution within the size spectrum. However, its value is also influenced by the width or spread of the distribution. The nature and extent of this effect depend on the particular average being used.

A great many different average or characteristic sizes can be defined, such as median, mode, or mean. The specific surface area, which will be discussed later in this chapter, also represents a kind of average (but inverse) size. The values of these averages may vary widely depending on the particular definition and the form of the distribution they represent.

The *median size* in a distribution is that size that splits the distribution into two equal parts; that is, half of the material is finer and half is coarser. In general, the median size can be defined by $x_{50,r}$ such that

$$Q_r(x_{50,r}) = \frac{1}{2} \quad \textbf{(Eq. 2.38)}$$

The value of the median depends on which distribution it refers to ($r = 0$, number; $r = 1$, length; etc.). In general,

$$x_{50,s} > x_{50,t} \text{ for } s > t \quad \textbf{(Eq. 2.39)}$$

The difference between values increases with increasing spread of the distribution.

The *mode* of a distribution, x_{mr}—sometimes referred to as the most frequent size—corresponds to the peak in the density function $q_r(x)$. Again, the mode's value depends on whether $r = 0$, 1, 2, or 3. In addition,

$$x_{ms} > x_{mt} \text{ for } s > t \quad \textbf{(Eq. 2.40)}$$

Distributions with more than one maximum are said to be multimodal. Bimodal distributions (two maxima) are quite common. They occur in mixtures of particle systems (e.g., sand and gravel) and, under certain circumstances, can be generated in size reduction and agglomeration processes (Hogg in press; Rattanakawin and Hogg 1998).

Mean sizes represent a group of averages defined by the *moments* of a size distribution. The kth moment of the size distribution $q_r(x)$ is defined by

$$M_{k,r} = \int_0^\infty x^k q_r(x)dx \quad \textbf{(Eq. 2.41)}$$

and represents the quantity x^k averaged using the "r" distribution. For $k = 1$, 2, or 3 and using appropriate shape factors, the moments correspond to mean diameter, area, or volume, respectively. Thus, for example,

$M_{1,3}$ = volume mean diameter; i.e., particle diameter averaged with respect to the volume distribution

$k_3M_{3,0}$ = number mean volume; i.e., particle volume averaged with respect to the number distribution. In this case, the moment represents the mean value of x^3; the shape factor is necessary to convert to an actual volume.

Values of k are not, however, restricted to 1, 2, or 3. Other values, including negative numbers, are equally valid and are often encountered in practice. The zeroth moment ($k = 0$) is identically equal to unity, regardless of r, because Eq. 2.41 then expresses the fraction of particles that have any size between zero and infinity; that is, all of them. Negative values of k simply represent averages of $1/x$, $1/x^2$, and so on.

The integral in the denominator of Eq. 2.35 can be written as the moment $M_{t-r,r}$. Useful relationships among the various moments are discussed in more detail by Leschonski (1984). The moments can be expressed as mean sizes (which are indicated with an overbar above the x term) via the equation

$$\bar{x}_{k,r} = (M_{k,r})^{1/k} \quad \textbf{(Eq. 2.42)}$$

so that, for example,

$\bar{x}_{3,0} = (M_{3,0})^{1/3}$ = number mean volume diameter; i.e., the diameter corresponding to the number mean volume defined above. The shape factors appear implicitly on both sides of Eq. 2.42 and cancel out.

Heywood (1963) defined several such mean sizes, all of which can be expressed as moments of the size distribution (Leschonski 1984). The actual values of $\bar{x}_{k,r}$ depend on k and r and on the form of the distribution. In general, the values increase with increasing k or r. Specifically,

$$\bar{x}_{k,r} \leq \bar{x}_{k+i,r+j} \text{ for } i \text{ and } j \text{ both} \geq 0 \quad \textbf{(Eq. 2.43)}$$

Specific surface area, defined as the surface area per unit volume (S_v) or per unit mass (S_m), also represents an average (but inverse) size. The volume and mass specific surface areas are related through the equation

$$S_v = \rho S_m \quad \textbf{(Eq. 2.44)}$$

S_m has units of area/mass (usually square meters per gram), whereas S_v is an inverse size (e.g., per micrometer [μm^{-1}]). The hybrid unit of square meters per cubic centimeter (m^2/cm^3) is numerically equal to units of per micrometer.

For particles of uniform size,

$$S_v = \frac{k_2 x^2}{k_3 x^3} = \frac{k_{23}}{x} \quad \textbf{(Eq. 2.45)}$$

where k_{23} is called the specific surface shape factor, defined as the ratio k_2/k_3. For spheres, $k_2 = \pi$ and $k_3 = \pi/6$, so that $k_{23} = 6$.

More generally, for systems with a distribution of sizes,

$$S_v = \frac{\int_0^\infty k_2 q_0(x)dx}{\int_0^\infty k_3 x^3 q_0(x)dx} \quad \textbf{(Eq. 2.46)}$$

If the shape factors are independent of size,

$$S_v = k_{23}\frac{M_{2,0}}{M_{3,0}} \quad \textbf{(Eq. 2.47)}$$

Applying the transformation formula, Eq. 2.35, to the moments in Eq. 2.47 leads to

$$M_{2,0} = \frac{M_{-1,3}}{M_{-3,3}} \quad \textbf{(Eq. 2.48)}$$

and

$$M_{3,0} = \frac{1}{M_{-3,3}} \qquad \textbf{(Eq. 2.49)}$$

so that Eq. 2.47 can be replaced by the more convenient

$$S_v = k_{23} M_{-1,3} \qquad \textbf{(Eq. 2.50)}$$

The specific surface mean diameter, $\bar{x}_{-1,3}$ is defined in the usual way (i.e., by using Eq. 2.42):

$$\bar{x}_{-1,3} = (M_{-1,3})^{-1} \qquad \textbf{(Eq. 2.51)}$$

That is,

$$\bar{x}_{-1,3} = \frac{k_{23}}{S_v} \qquad \textbf{(Eq. 2.52)}$$

If the shape factors and density are independent of size, $\bar{x}_{-1,3}$ can be expressed in terms of the mass specific surface area; i.e.,

$$\bar{x}_{-1,3} = \frac{k_{23}}{\rho S_m} \qquad \textbf{(Eq. 2.53)}$$

Because $k_{23} = 6$ for spheres, an equivalent-sphere specific surface diameter can be defined by

$$(\bar{x}_{-1,3})_{ES} = \frac{6}{\rho S_m} = \frac{6}{S_v} \qquad \textbf{(Eq. 2.54)}$$

This form is useful because specific surface area can be measured directly.

Figure 2.4 shows an example of a fairly typical size distribution. Some of its associated averages are given in Table 2.2. The more-than-tenfold range in the values for different averages for the same distribution clearly illustrates the potential ambiguity involved in the unqualified use of average sizes.

Algebraic Forms. It is often useful (e.g., for application to process models) to fit specific algebraic functions to particle size distribution data. Typically, these functions have two parameters that can be adjusted to provide the best fit to a set of experimental data. The values of the parameters provide an improved means of summarizing the actual distribution as compared to using a single, average size.

It should be emphasized that, in general, there is no particular form that is expected, theoretically, to describe size distribution data. For example, there are no equivalents to the binomial, Poisson, and normal distributions of probability and statistics. However, some functional forms have been found to give a reasonable fit to some sets of data. These are simply equations that

- Increase monotonically from 0 to 1
- Can fit data reasonably well, usually with only two adjustable parameters
- Are reasonably simple to apply

The distribution types discussed in the following paragraphs are (1) the Gaudin–Schuhmann distribution, (2) the Rosin–Rammler distribution, and (3) the logarithmic-normal (or log-normal) distribution.

The *Gaudin–Schuhmann distribution* expresses the mass (volume) distribution function by a simple power law

$$Q_3(x) = \begin{cases} \left(\dfrac{x}{k_s}\right)^{\alpha} & \text{for } x \le k_s \\ 1 & \text{for } x \ge k_s \end{cases} \qquad \textbf{(Eq. 2.55)}$$

where

k_s = the size modulus, which locates the distribution in the overall size spectrum

α = the distribution modulus, which is an inverse measure of the spread of the distribution

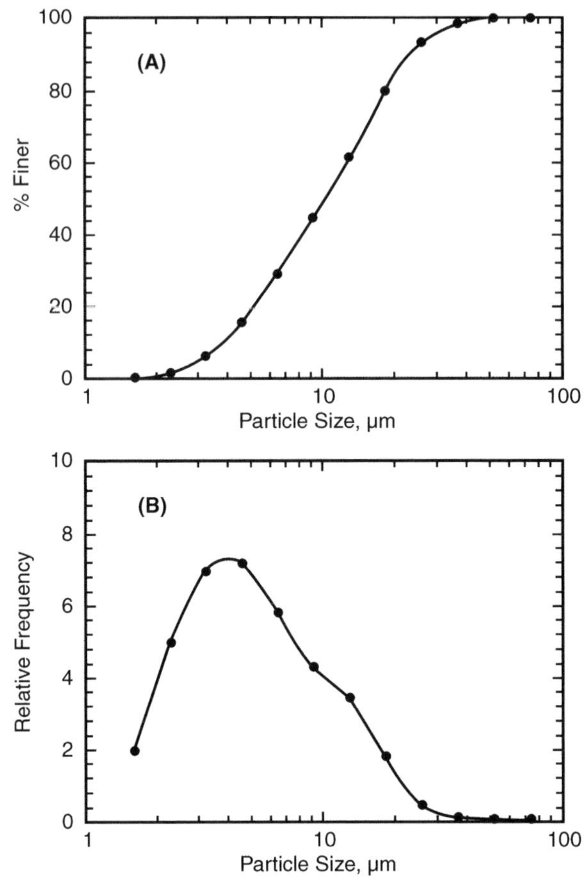

FIGURE 2.4 Example of experimental particle size distribution: (A) distribution function $Q(x)$; (B) density function $q(x)$

TABLE 2.2 Average particle sizes corresponding to Figure 2.4

Average	Value, μm
Volume median diameter, $x_{50,3}$	10.35
Volume mode, x_{m3}	3
Volume mean diameter, $\bar{x}_{1,3}$	12.55
Specific surface mean diameter, $\bar{x}_{-1,3}$	7.76
Number mean volume diameter, $\bar{x}_{3,0}$	1.10

For materials that conform to the Gaudin–Schuhmann equation, a straight line is obtained by plotting the cumulative fraction (or cumulative percentage) finer than the stated particle size versus that particle size on log-log paper. These plots are often called Schuhmann plots; an example is given in Figure 2.5. The slope of the straight line is equal to the distribution modulus, α, and the size at which the (extrapolated) straight line crosses $Q_3 = 1$ (or cumulative percent finer = 100%) is the size modulus, k_s.

It should be emphasized that both the Gaudin and Schuhmann plots are based on the Gaudin–Schuhmann equation (Eq. 2.55). However, while the cumulative Schuhmann plot can be used for

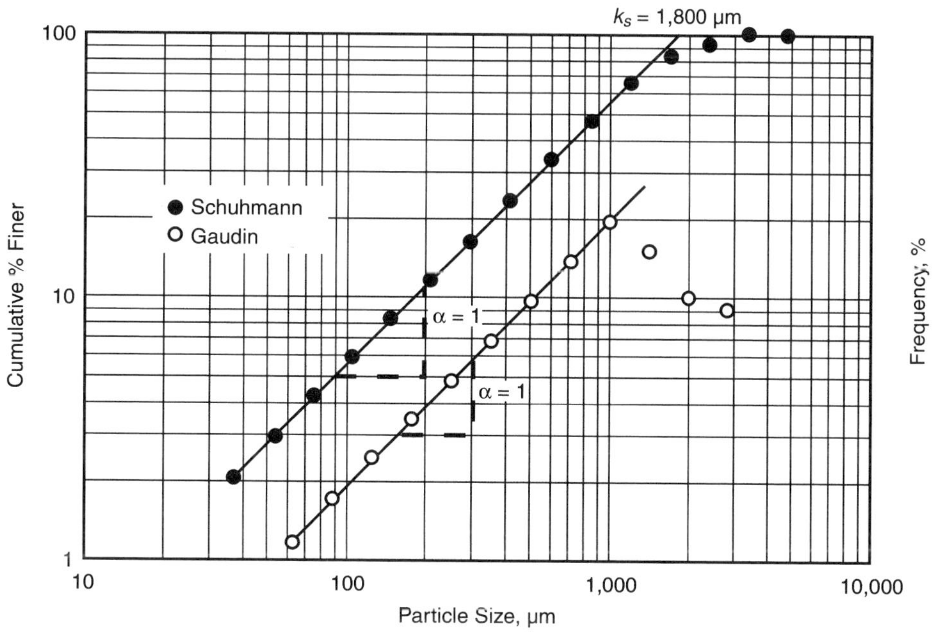

FIGURE 2.5 Gaudin–Schuhmann size distribution

any set of data, the Gaudin plot is appropriate only for data arranged in a geometric series of size intervals.

It follows from Eq. 2.29 and 2.55 that the corresponding density function is given by

$$q_3(x) = \begin{cases} \left(\frac{\alpha}{k_s}\right)\left(\frac{x}{k_s}\right)^{\alpha-1} & \text{for } x \le k_s \\ 0 & \text{for } x > k_s \end{cases} \quad \text{(Eq. 2.56)}$$

A log-log plot of frequency versus size would therefore yield a straight line of slope $\alpha - 1$.

However, for materials that follow this distribution, the following special and more useful kind of frequency distribution can be used. If the experimental data are given in the form of weight or volume fraction in discrete size intervals, and if the size intervals are arranged in a geometric progression (sieving data are normally generated in this form, for example), the weight fraction in some interval x_i to x_{i+1} will be given by

$$(q_3)_i - Q_3(x_i) \quad Q_3(x_{i+1}) \quad \text{(Eq. 2.57)}$$

From Eq. 2.55,

$$(q_3)_i = \left(\frac{x_i}{k_s}\right)^{\alpha} - \left(\frac{x_{i+1}}{k_s}\right)^{\alpha} \quad \text{(Eq. 2.58)}$$

or

$$(q_3)_i = \left(\frac{x_i}{k_s}\right)^{\alpha}(1 - r_s^{-\alpha}) \quad \text{(Eq. 2.59)}$$

where $r_s = x_i/x_{i+1}$.

For size intervals arranged in a geometric progression, r_s is constant and a log-log plot of the weight fraction in the size interval versus some characteristic size in the interval should give a straight line of slope α. These plots are often known as Gaudin plots; their major utility lies in their high sensitivity to discrepancies in the individual weights. The cumulative Schuhmann plot tends to smooth out variations; the Gaudin plot tends to emphasize them, which makes it extremely useful for detecting sources of error. Typical Gaudin and Schuhmann plots are illustrated in Figure 2.5.

In practice, the Gaudin–Schuhmann distribution appears to give remarkable agreement with the size distributions of a wide variety of crushed minerals. Typically, good agreement is found for the finer sizes, with some deviation at the coarser end of the distribution. For most systems, values of the distribution modulus, α, seem to lie between 0.5 and 1.5; the size modulus, k_s, of course, depends on the extent of grinding. In many cases, α appears to be constant for a given material in a given grinding machine.

The moments of the Gaudin–Schuhmann distribution can be obtained by integrating Eq. 2.41 after substituting from Eq. 2.56. The result is

$$M_{k,3} = \left(\frac{\alpha}{\alpha + k}\right) k_s^{\,k} \qquad \textbf{(Eq. 2.60)}$$

The values obtained using Eq. 2.60 are reasonable only if $\alpha + k > 0$. This has important consequences in estimating average sizes or specific surface area or in transforming from the volume distribution to the area, length, or number distributions. For example, to estimate specific surface area, the moment $M_{-1,3}$ is required (see Eq. 2.50). Thus, $k = -1$ and reasonable values are obtained only if $\alpha > 1$. Yet, as noted above, values of $\alpha < 1$ are frequently observed. The problem is even more serious in attempting to transform to the number distribution, $Q_0(x)$. In this case, the required moment is $M_{-3,3}$ (see Eqs. 2.35 and 2.41) and the transformation can be carried out only if $\alpha > 3$. These values are, in fact, quite rare. As discussed previously (see "Transformations"), these problems arise through the implicit extrapolation of the Gaudin–Schuhmann distribution to zero size; that is, in the integration of Eq. 2.41 from $x = 0$. Clearly, it is mathematically impossible for the Gaudin–Schuhmann distribution to be valid as size approaches zero. At very fine sizes, the slope of the distribution must increase. Indeed, this behavior has been observed and has been attributed to the approach to a "grind limit" (Schönert 1986; Cho, Waters, and Hogg 1996).

In principle, the problem can be solved by introducing a minimum size, x_o, and replacing Eq. 2.55 with

$$Q_3(x) = \begin{cases} \left(\dfrac{x - x_o}{k_s - x_o}\right)^{\alpha} & \text{for } x \le k_s \\ 1 & \text{for } x \ge k_s \end{cases} \qquad \textbf{(Eq. 2.61)}$$

However, this introduces a third parameter, x_o, which can be estimated only by trial and error. Transformations and calculations of specific surface area, for example, are extremely sensitive to the value selected for x_o.

The *Rosin–Rammler distribution* describes the mass (volume) distribution function in exponential form as

$$Q_3(x) = 1 - \exp\left[-\left(\frac{x}{k_r}\right)^{m}\right] \qquad \textbf{(Eq. 2.62)}$$

where m and k_r are the distribution and size moduli, respectively. Eq. 2.62 can be inverted to give

$$\log\log\left(\frac{1}{1 - Q_3}\right) = m \log x - m \log k_r - \log 2.303 \qquad \textbf{(Eq. 2.63)}$$

A plot of log log $[1/(1 - Q_3)]$ versus log x should therefore yield a straight line of slope m (see Figure 2.6). The size parameter, k_r, can be obtained directly from the size at which the straight line crosses $Q_3 = 63.21\%$. By rearranging Eq. 2.62, it can be shown that

$$m \approx \frac{2}{\log\,(k_r/x_1)} \qquad \textbf{(Eq. 2.64)}$$

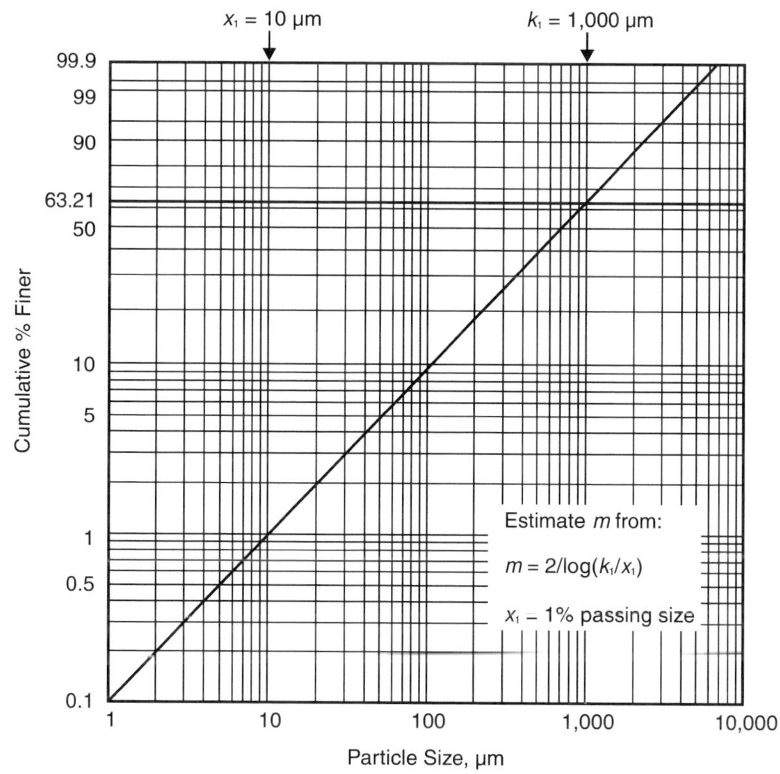

FIGURE 2.6 Rosin–Rammler size distribution

where x_1 is the 1% passing size (i.e., the size for which $Q_3 = 0.01$, or 1%). Eq. 2.64 provides a simplified means of estimating m. The alternative–direct measurement of the slope of the line–often leads to confusion and incorrect calculation. Special Rosin–Rammler graph paper is available commercially.

From Eqs. 2.29 and 2.62, the density function is given by

$$q_3(x) = \left(\frac{m}{k_r}\right)\left(\frac{x}{k_r}\right)^{m-1} \exp\left[-\left(\frac{x}{k_r}\right)^m\right] \qquad \textbf{(Eq. 2.65)}$$

The density function passes through a maximum only if $m > 1$.

By expanding the exponential term as a power series, it can be shown that, for $x << k_r$

$$Q_3(x) \cong \left(\frac{x}{k_r}\right)^m \quad \text{for } x << k_r \qquad \textbf{(Eq. 2.66)}$$

Thus, for the very fine sizes, the Rosin–Rammler distribution reduces to the same form as the Gaudin–Schuhmann distribution with $\alpha = m$ and $k_s = k_r$.

The moments can be obtained by using Eqs. 2.41 and 2.65. The general result is

$$M_{k,3} = k_r^{\,k}\,\Gamma\left(\frac{k}{m} + 1\right) \qquad \textbf{(Eq. 2.67)}$$

where the gamma (Γ) function is defined by

$$\Gamma(a+1) = \int_0^\infty t^a e^{-t}\,dt \qquad \textbf{(Eq. 2.68)}$$

Tables of gamma functions are readily available (e.g., Lide [1998/9]). The relationship $\Gamma(a + 1) = a\Gamma(a)$ is often useful. As for the Gaudin–Schuhmann distribution, the moments are finite only for $k > -m$, and the same restrictions apply to transformations and the calculation of averages.

In practice, the Rosin–Rammler distribution frequently gives a better fit to the coarse end of the distribution than does the Gaudin–Schuhmann equation. Its main disadvantage lies in the more complex form of the distribution function and the subsequent need for special plotting paper.

The *logarithmic-normal distribution* is obtained by applying the normal (Gaussian) distribution to the particle size on a logarithmic rather than a linear scale. The normal distribution is very commonly encountered in mathematical statistics and can be derived theoretically for a number of random processes, but it is rarely applicable to particle size distributions. The normal curve is symmetric on a linear scale, whereas size distributions are almost invariably skewed. For materials with a reasonably broad size distribution (extending over two or more orders of magnitude), direct application of the normal distribution would imply the existence of negative size.

The log-normal density function can be expressed as

$$q_r(x) = \frac{1}{x(\sqrt{2\pi})\ln\sigma_g}\exp\left[-\frac{1}{2}\left(\frac{\ln x - \ln x_{50,r}}{\ln \sigma_g}\right)\right] \qquad \textbf{(Eq. 2.69)}$$

where

$\sigma_g =$ the geometric standard deviation

$x_{50,r} =$ the median size (sometimes defined as the geometric mean size)

Applying Eq. 2.4 leads to the following expression for the distribution function:

$$Q_r(x) = \frac{1}{\sqrt{2\pi}}\int_{-\infty}^{X} e^{-\beta/2}d\beta \qquad \textbf{(Eq. 2.70)}$$

where β is a dummy variable and

$$X \equiv \frac{\ln x - \ln x_{50,r}}{\ln \sigma_g} \qquad \textbf{(Eq. 2.71)}$$

Extensive tabulations of the integral in Eq. 2.70 are available in statistical tables or handbooks. Special probability or log probability graph papers are also available. Similarly, probability scales are available on many plotting software packages. The median size, $x_{50,r}$, can be determined directly from the intercept at $Q_r = 0.5$ (50%). The log-normal standard deviation, σ_g, can be obtained from the intercepts at $Q_r = 0.16$ (16%), 0.5 (50%), and 0.84 (84%) via the equation

$$\sigma_g = \frac{x_{50}}{x_{16}} = \frac{x_{84}}{x_{50}} = \sqrt{\frac{x_{84}}{x_{16}}} \qquad \textbf{(Eq. 2.72)}$$

The values obtained in this way are, of course, valid only if the log probability plot yields a straight line. An example of a log probability plot is given in Figure 2.7.

One property of the log-normal distribution is that if $Q_r(x)$ is log normal, so is $Q_s(x)$ with the same value of σ. Thus, log probability plots of the number, length, area, and volume distributions give a set of *parallel* straight lines. The median sizes are related through

$$x_{50,r} = x_{50,s}\exp\left[(r - s)(\ln\sigma)^2\right] \qquad \textbf{(Eq. 2.73)}$$

For example, the relationship between the number median ($r = 0$) and the volume median ($r = 3$) is

$$x_{50,3} = x_{50,0}\exp\left[3(\ln\sigma)^2\right] \qquad \textbf{(Eq. 2.74)}$$

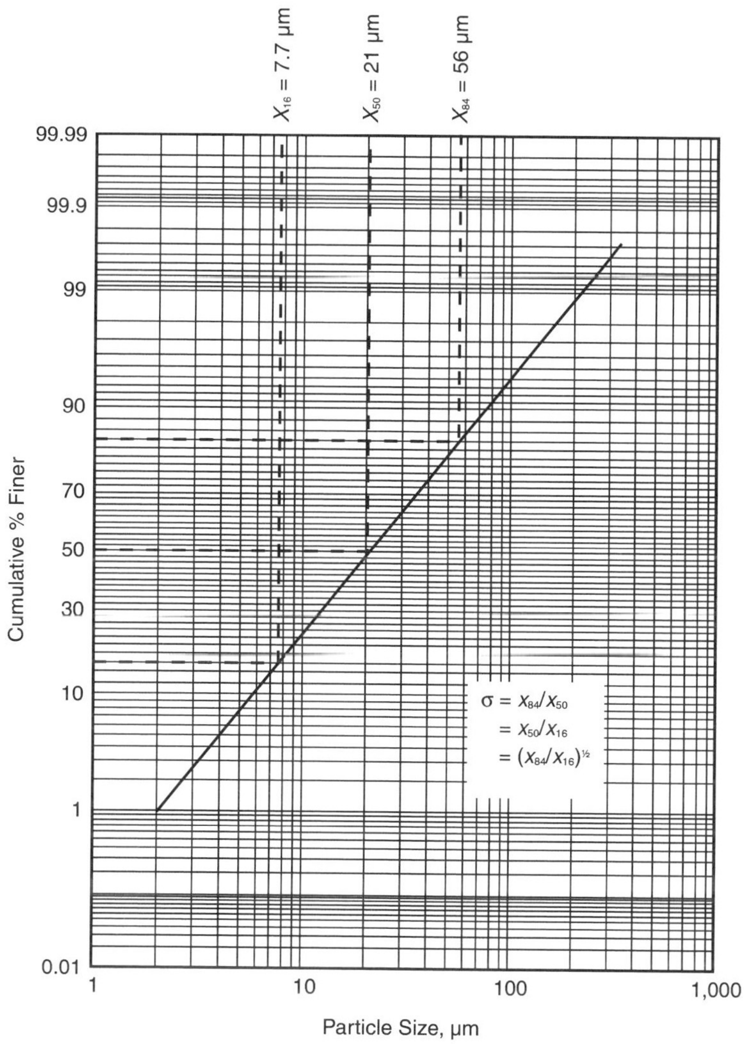

FIGURE 2.7 Log-normal size distribution

It should be noted that transformations from number to volume and area to volume, for example, involve a translation of the curve in the Q_r direction rather than the x direction. Thus, a small deviation from log normal at, say, the fine end of the volume distribution can appear as a significant deviation in the center of the number distribution (see Figure 2.8).

In contrast to the Gaudin–Schuhmann and Rosin–Rammler distributions, the moments $M_{k,r}$ of the log-normal distribution are all finite, regardless of the values of k and r. The moments can all be determined from

$$M_{k,r} = x_{50,r}^{k} \exp\left[\frac{1}{2}(k \ln \sigma)^{2}\right] \quad \textbf{(Eq. 2.75)}$$

In practice, the log-normal distribution is found to apply reasonably well to a variety of particulate materials, including fine clays and finely ground (<50 µm), unclassified powders. Furthermore, the mathematically well-behaved nature of this distribution gives it an advantage over the Gaudin–Schuhmann

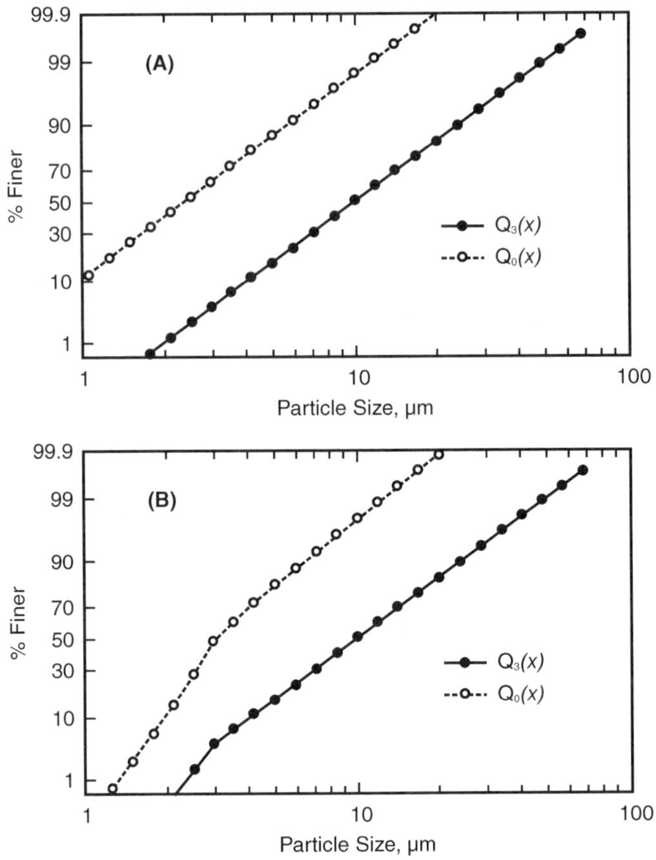

FIGURE 2.8 Transformation of volume distribution to number distribution: (A) log-normal distribution; (B) log normal with deviation at the fine end of the volume distribution

and Rosin–Rammler equations for some applications, even though the latter may give a "better" representation of the particle size data.

Particle Shape Distributions

Description and measurement of particle shape, especially quantitative evaluation of shape distributions, are seldom carried out in mineral processing applications. Nevertheless, the procedures discussed previously for presenting, transforming, and manipulating particle size distributions can be equally applied to shape. Because measurements are usually made on individual particle images, data are generated as number distributions, generally at (approximately) constant size. If a measured shape parameter s can be related to the volume shape factor (k_3) as defined by Eq. 2.15, transformation from the number distribution to the volume distribution can be accomplished via

$$q_3(s) = \frac{k_3(s)q_o(s)}{\int_s k_3(s)q_o(s)ds} \quad \textbf{(Eq. 2.76)}$$

where the integration is carried out over all possible values of the parameter s. Averages and moments, then, can be defined and evaluated as for size distributions. Because of the scarcity of information, standard forms have not been established for particle shape distributions. However, some data for ground minerals appear to conform quite closely to the log-normal distribution (Kaya, Kumar, and Hogg 1996).

Distribution of Particle Composition and Structure

Formal presentations of the distribution of composition and structure are rarely encountered. However, density (specific gravity) distributions are widely used, especially in coal processing. Again, the procedures used for size distributions are generally applicable. The principal difference is that, because particle volume does not depend on composition, the number and volume distributions (at fixed size) are identical. The mass distribution q'_3 (which is typically measured) does differ. Transformation can be accomplished by using

$$q_0(\rho) = q_3(\rho) = \frac{q'_3(\rho)/\rho}{\int_\rho q'_3(\rho)d\rho/\rho} \quad \text{(Eq. 2.77)}$$

Although attempts have been made to use functional forms to fit density distribution data, standard expressions analogous to those used for size distributions have not yet been established.

MEASUREMENT OF PARTICLE CHARACTERISTICS

Before beginning to characterize a particle system, we must (1) obtain a representative sample of the material, (2) prepare the sample for analysis, and (3) select the most appropriate analytical procedure. Each of these steps can be critical in terms of the reliability and utility of the results obtained from the analysis.

Sampling

The first step in any analytical procedure is to obtain an appropriate sample of the material to be analyzed. Two important questions should be asked before selecting the sample:

1. How large should the sample be?
2. How can we be sure that the sample is truly representative of the material to be analyzed?

As quite distinct problems, these two should not be confused. Just because a sample is large enough does not mean it is necessarily representative; a representative sample may still be too small. Sample requirements for particulate materials have been discussed in detail by Gy (1982). Some simple guidelines will be presented here.

Sample Size. For particulate materials, the primary criterion for establishing the required sample size is that all kinds (sizes, shapes, etc.) of particles should be adequately represented in the sample. Because particles are discrete entities, sample size is thus dictated by the *number* size distribution of the material being sampled. Consider, for example, a system of quartz particles (specific gravity 2.65) that contains 10% by weight of 1-cm particles. Each 1-cm particle would weigh about 1.4 g, so that a sample weight of 14 g would be necessary to ensure a reasonable likelihood of containing even one of these particles. Obviously, this weight would be quite inadequate because there would be a high probability of taking such a sample and finding none of those particles (0%) or two of them (20% by weight). If the sample size were increased tenfold, to 140 g, we would expect to find 10 of the 1-cm particles in any sample, and the errors introduced by the chance inclusion of one extra or one fewer would be correspondingly reduced; that is, a sample containing only 9 of them would analyze at 8.9% rather than 10%. Sample size, then, should be based on statistical criteria such that the errors introduced by random variations in the numbers of different particles included are acceptably small.

Based on analysis of the statistics of random mixtures, the required sample weight for particle size analysis can be estimated by computing a set of values of the quantity M_i via

$$M_i = \frac{4}{\varepsilon^2}\left[\left(\frac{1}{q_i} - 2\right)w_i + \bar{w}\right] \quad \textbf{(Eq. 2.78)}$$

where

q_i = the weight fraction in size class i

w_i = the mean weight of a single particle in that class

ε = a specified tolerance on the estimated value of q_i

The term $\bar{w}$ in Eq. 2.78 is the overall mean particle weight and is given by

$$\bar{w} = \sum_i q_i w_i \quad \textbf{(Eq. 2.79)}$$

Equation 2.78 gives the sample weight M_i required to determine q_i to an accuracy ε at the 95% confidence level.

In general, there will be a minimum sample weight, M_i, for each class in the distribution, and the required sample weight, M, will be the largest of these M_i values. In practice, however, the values of q_i are not known until the analysis is complete. It is necessary, therefore, to use an initial estimate to determine the sample weight. If we recognize (1) that the largest particles will normally give the maximum value of M_i and (2) that for spherical particles,

$$w_i = \frac{\pi}{6} x_i^3 \rho \quad \textbf{(Eq. 2.80)}$$

where x_i and ρ_i are the mean particle size and density in class i, we can obtain an initial estimate from

$$M \approx \frac{40\rho x_m^3}{\varepsilon^2} \quad \textbf{(Eq. 2.81)}$$

where x_m is the maximum size (in centimeters) present. Eq. 2.81 is based on the arbitrary designation of the maximum size as the average of the class that contains the coarsest 5% (by weight) of the distribution. Some fairly typical examples of minimum sample sizes for given maximum particle sizes are given in Table 2.3.

After the analysis, when the actual values of q_i are known, Eq. 2.78 can be used to check the adequacy of the sample size used. Alternatively, Eq. 2.78 can be inverted to evaluate the relative errors, ε_i, associated with each measured q_i value. Thus,

$$\varepsilon_i = 2\sqrt{\frac{\left(\frac{1}{q_i} - 2\right)w_i + \bar{w}}{M}} \quad \textbf{(Eq. 2.82)}$$

Equation 2.82 then defines the 95% confidence interval on the measurements.

TABLE 2.3 Minimum sample size requirements

Maximum Particle Size	Corresponding Minimum Sample Size
100 µm	40 mg
1 mm	4 g
1 cm	40 kg
10 cm	40 t

Note: Calculations based on quartz (ρ = 2.65 g/cm^3) with Gaudin–Schuhmann distribution (α = 1), 10 weight percent in the top size interval, and allowable error of ±5%.

TABLE 2.4 Sample size requirements for screen analysis on coal (specific gravity = 1.40)

Size Range (Tyler Mesh)	Mean Size (x_i), mm	Weight Percent (q_i)	Particle Weight (w_i)	q_iw_i, mg	Required Sample Weight*	Expected Error for 1-kg Sample %†
+3/8 in.	11.2	0.034	1.03 g	0.35	1.21 t‡	350
3/8 in. × 3	8.0	0.315	375 mg	1.18	47 kg	69
3 × 4	5.66	1.48	133 mg	1.97	3.5 kg	19
4 × 6	4.00	4.07	47 mg	1.91	427 g	6.5
6 × 8	2.83	7.63	16.6 mg	1.27	77 g	2.8
8 × 10	2.00	10.88	5.9 mg	0.64	20 g	1.4
10 × 14	1.41	12.37	2.1 mg	0.26	8.2 g	<1
14 × 20	1.00	12.38	733 μg	0.09	4.9 g	<1
20 × 28	0.71	11.34	259 μg	0.03	3.8 g	<1
28 × 35	0.50	9.53	92 μg	0.01	3.4 g	<1
35 × 48	0.35	7.41	31 μg	0.00	3.2 g	<1
–48	—	22.56	—	—	3.1 g	<1
		Total = 100.00		w = 7.71 mg		

*Calculated using Eq. 2.78 with ε = 0.1 (10%).
†Calculated using Eq. 2.82.
‡If the 3/8-in. screen is eliminated, the sample size needed for +3 mesh increases to 49.9 kg. If both 3/8-in. and 3-mesh screens are omitted, for example, the sample size for +4 mesh increases to 4.0 kg.

Table 2.4 presents an example of the use of Eqs. 2.78 and 2.82 to determine sample size requirements and evaluate sampling errors. The table clearly shows that a very large sample would be needed to determine the size distribution (column 3) with a maximum error of ±10%. This requirement arises from the need to determine a very small amount (0.034%) of the coarsest fraction (+3/8 in.) with reasonable accuracy. For many applications, the fact that 0.034% is coarser than 3/8 in. would be immaterial; the 3/8-in. screen could be eliminated from the analysis, giving a total of 0.349% (0.034 + 0.315) +3 mesh. In this case, the required sample size would be reduced from 1.2 t to about 50 kg. If the 3-mesh screen were also eliminated, the sample size requirement would fall to about 4 kg. In other words, accurate determination of small quantities, especially at the coarse end of the size distribution, can require excessively large samples to be used. The need for such accuracy should be considered carefully. Conversely, the determination of such small quantities using more normal sample sizes can be subject to significant error.

Tables 2.3 and 2.4 clearly demonstrate that sample size constraints can be significant for coarse particles and can present major difficulties in characterizing run-of-mine ores and primary crusher products, among others. On the other hand, the constraints become insignificant for very fine particles. Thus, for subsieve material (<40 μm), samples of less than 1 mg are usually statistically sufficient. In such cases, the sample size can usually be determined entirely from the requirements of the specific technique being used for the analysis.

The presence of coarse particles mandates large samples. However, complete analysis of the entire sample is not always necessary. If the purpose of the sample is simply a bulk assay, the sample can be immediately crushed to a finer size and a much smaller subsample taken for analysis. For example, if the coal described in Table 2.4 were being sampled for proximate analysis only, a bulk sample of several kilograms would still be required, but this sample could be crushed to <100 mesh, for example, whereupon a subsample of a few grams would be adequate for the actual analysis.

When information on individual sizes is needed, successive subsampling and analysis can be used. For example, the coarsest fraction can be removed by screening and being analyzed. The sample size needed for the undersize material is less, and an appropriate subsample can be used. Two or three such steps will usually be sufficient to minimize the need to analyze large samples.

Sampling Procedures. Once the required sample size is determined, the next step is to choose a sampling procedure in which the different kinds of particles are selected entirely without bias. In principle, this simply requires complete mixing of the bulk material before the sample is taken. Unfortunately, complete mixing is often impractical, especially when very large quantities are involved. Furthermore, particulate materials are notoriously difficult to mix because of the tendency for different kinds of particles to segregate, particularly for relatively free-flowing materials with wide variations in particle size. Specific procedures for sampling from both batch and continuous-flow systems have been described in detail by Gy (1982) and Allen (1997).

In *grab sampling,* the simplest method of all, a scoop or shovel is used to take the appropriate quantity of material, essentially at random, from the bulk. Grab sampling is satisfactory only if the bulk material can be thoroughly mixed, which may in fact be the case for reasonably small quantities of fairly cohesive (which usually means fine) powders. A series of grab samples taken from different locations, or with intermediate mixing of the bulk, can offer significant improvement over the single sample and is often the only practical alternative for very large populations. Care should be taken to avoid biasing the sample toward the surfaces of, for example, large piles. If each sample from a series is analyzed separately rather than combined into a single analysis sample, information can be acquired on relative homogeneity, and segregation, among other factors, in the material and on the extent to which a sampling problem actually exists.

Cone-and-quarter sampling, which is widely practiced, presents certain advantages. The procedure involves mixing and turning the material over with a scoop or shovel and piling it into a conical heap. The heap is divided into roughly equal quarters; two opposite quarters are removed, and the remaining two are remixed. The procedure is repeated until the material is reduced in quantity to the desired sample size. Advantages of this approach are (1) the entire batch is subject to the sampling procedure with minimal operator bias, (2) no special equipment is needed, and (3) the method can be applied to very large quantities (by using front-end loaders, for example). However, accumulation of fines as a result of segregation during heap formation can lead to biasing of the sample. The procedure is rather tedious and time-consuming, and operators are often tempted to take shortcuts, which can increase bias.

Sample splitters (riffles) are mechanical devices used to divide a material into two or more parts in a random fashion. The simple chute splitter uses a series of alternately directed chutes to separate the material into two parts. Repeated applications can be used for further subdivision as required. The procedure is simple and effective but is usually limited to fairly small quantities. Loss of fines can present problems for "dusty" materials. Spinning rifflers in which the material is fed slowly, usually from a vibrating feeder, into a series of collection vessels on a rotating table are attractive for fine powders because the generation of a dust cloud can often be minimized. Before any kind of mechanical splitter or sample reducer is used, its design should be evaluated carefully to ensure that segregation does not lead to sample bias. An improperly designed sample splitter can easily become a classifier!

In sampling from very large populations (e.g., stockpiles), grab sampling is essentially the only option. In these cases, several (as many as possible) individual samples should be taken. To minimize bias, the entire volume of material should be conceptually divided into a regular, three-dimensional grid. Sampling locations should then be selected at random from the grid points. Although it is not necessary for each individual sample to satisfy the size requirement, the combination of all samples must. Obviously, each individual sample must be substantially larger than the coarsest particles present.

Sampling from slurries, particularly settling slurries, is especially difficult and frequently leads to biased results. For small batches, it is sometimes best to filter and dry the entire batch and then use a dry sampling procedure. Care must be taken to avoid loss of fines that may pass through a filter or be trapped in the filter medium. Nonsettling slurries can simply be mixed thoroughly and sampled as for homogeneous liquids. Settling slurries can be subjected to vigorous agitation, then sampled in the same way.

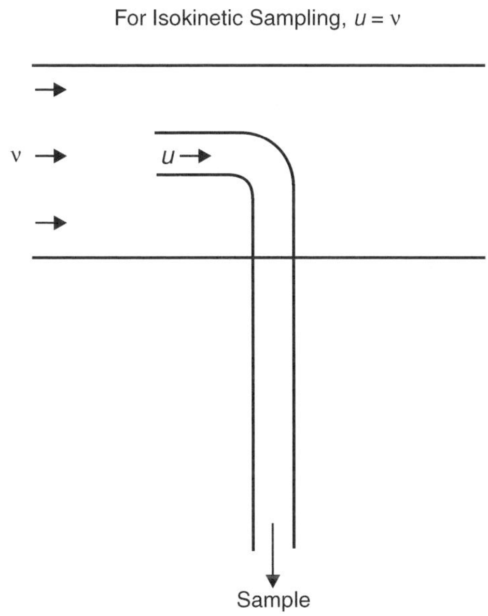

FIGURE 2.9 Isokinetic sampling

Another approach is to circulate the slurry at high rate through a pump and sample from the flow stream. In sampling from flow streams–such as conveyors and slurry pipelines–the preferred method is to divert the entire stream for a short time rather than splitting off part of the stream. If repeated samples are taken, time variations and cycling, among other factors, can be detected. For high-volume flows of suspended particles where diverting the entire stream is impractical, isokinetic sampling should be used. The procedure for isokinetic sampling is illustrated schematically in Figure 2.9. Samples are withdrawn from the stream through a probe located as shown. The rate of withdrawal is adjusted, by using a pump, so as to ensure that the inlet velocity is the same as the flow velocity in the main stream.

Measurement of Particle Size

The size distribution of a particulate material is a description of the relative abundance of the different sizes present. Previously in this chapter we pointed out that, for the kinds of irregular particles typically encountered in mineral processing systems, particle "size" cannot be uniquely defined. In most procedures for evaluating size distribution, size is arbitrarily defined on the basis of response to some particular process, such as passage through an aperture, settling in a fluid, or scattering of light. The actual measurements involved in the analysis are of the relative quantities of material that give a specific response. The relationship between response and actual size is generally known only for spheres, so that measurements give an estimate of the distribution of *equivalent spherical diameter*. Because deviations from the spherical shape will have different effects on the response to different processes, size distribution estimates obtained for the same material but by different techniques cannot be expected to agree exactly, even in the absence of measurement error. Such discrepancies become especially important when more than one technique must be employed to span a broad range of sizes. The problems of data interpretation in these cases will be discussed at some length in the final part of this chapter.

It is also important to recall that different analytical procedures use different measures of particle quantity, such as mass, volume, and number. Direct comparison is possible only on a common basis, and appropriate transformation of the distributions must be carried out where necessary. (See the previous section entitled "Transformations.") The importance of specifying the basis of the distribution when reporting data cannot be overemphasized.

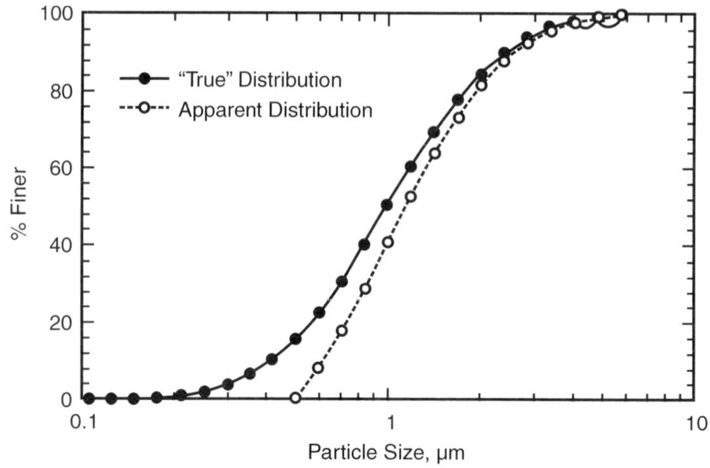

FIGURE 2.10 Graph demonstrating the effect of a lower detection limit on measurement of particle size distribution

Limitations of Sizing Techniques. Essentially all methods for particle size measurement are limited with respect to range of applicability. The nature of this limitation, however, is not the same for all techniques. In general, we can identify two principal kinds of size limitations:

- A limit of measurement
- A limit of detection

In the first case, measurement is limited to a specific range of sizes, but the existence of particles outside that range is recognized and quantified. For example, conventional sieving gives data on material coarser than, for example, 400 mesh (37 μm), but also specifies the amount of undersize present. In the second case (limit of detection), on the other hand, material outside of the range is not detected; the method yields an apparent distribution that is based on the assumption that all particles present fall in the measurement range. Thus, the measured (apparent) distribution refers only to an unspecified fraction of the material and is not an estimate of the true distribution over a limited part of the size spectrum. Figure 2.10 demonstrates the effects of detection limits. In this figure, the apparent distribution does not account for particles below a certain size, although such particles are indeed present.

In general, sizing methods that are based on the collective response of an assemblage of particles are subject only to measurement limits, although there may be effective detection limits in specific applications. On the other hand, those techniques that involve the response of individual particles (e.g., the use of particle counters) are invariably subject to detection limits because there can always be some particle that is too small to see. Clearly, measurement limits cause less serious problems than detection limits. Data obtained from methods that involve detection limits should always be questioned when significant quantities are reported at sizes close to the limit.

Resolution. Different procedures for size analysis vary in their ability to discriminate between sizes. For example, standard sieves can distinguish between particles that vary in size by more than $(2^{1/4} - 1)$; that is, 19%. Direct measurement methods such as microscopy can, in principle, detect even smaller differences. Other techniques, however, have substantially lower resolution. As a general rule, methods based on individual particle measurement (e.g., the use of particle counters) have high resolution (but are prone to detection limits), whereas those that involve overall system response (e.g., sedimentation and light scattering) have lower resolution (but are often subject only to measurement limits).

Resolution is especially important for very narrow size distributions–low-resolution techniques will tend to overstate the width of the distribution. For broad distributions, errors at different sizes tend to cancel each other out.

Dispersion of Fine Powders in Fluids. Virtually all methods for subsieve analysis require that particles be completely dispersed in a fluid medium. Inadequate dispersion can lead to very serious errors in measured size distributions. Dispersion in air is usually quite difficult, and liquid dispersion media are generally preferred.

Dispersion of solid particles in a liquid can generally be considered as a three-step process (Parfitt 1973): wetting, deaggregation, and stabilization.

Wetting of the solid surfaces by the liquid is a necessary prerequisite to dispersion. For hydrophilic solids, such as quartz and most other oxide minerals, wetting is usually spontaneous and no special precautions are needed. Other solids such as coal, sulfide minerals, and many organics are less easily wetted by water, and it may be necessary to add a wetting agent to reduce the surface tension of the water and promote wetting. Alternatively, a nonaqueous liquid (e.g., hydrocarbon) can be used instead of water.

Deaggregation is necessary to break up the small agglomerates that remain when a dry powder is incorporated into a liquid. This objective can usually be accomplished by mechanical agitation, although, of course, too much agitation could lead to breakage of the individual particles themselves. A brief period of ultrasonic treatment is often found to be particularly helpful in breaking up very small agglomerates.

Stabilization is almost always necessary to prevent reagglomeration of the dispersed particles. Stabilization is usually accomplished by ensuring that there are adequate repulsive forces caused by surface charges on the particles and solvation forces resulting from the presence of adsorbed films on the particle surfaces. Dispersing agents generally function by controlling surface charges or by increasing the solvation forces through adsorption. Electrolytes, especially those containing polyvalent ions, have a serious effect on the electrical forces between particles. Other impurities, such as organics, can also tend to promote aggregation of the particles. Consequently, it is important to ensure that glassware, mixer impellers, and other lab equipment are strictly clean and to use only distilled water and high-purity reagents.

The simplest procedure for evaluating the completeness of particle dispersion is to examine a drop of the suspension under a microscope. For submicron particles, where the individual particles or even small agglomerates may be difficult to resolve, the long-term stability of the suspension can be used as a criterion. Measuring supernatant turbidity after a fixed period of settling is one useful approach. The optimum combination of dispersion procedure and reagent addition is the one that gives the highest turbidity. Visual observation of relative clarity will often suffice if equipment for turbidity measurement is not available. An alternative approach is to use the actual sizing method results as the dispersion criterion–the "best" dispersion will generally give the finest distribution.

Sieving. The sieving methods are the most widely used means for sizing particles coarser than about 37 μm (400 mesh). Conventional (woven wire) sieves are available with aperture sizes down to about 25 μm (see Table 2.5), and so-called micromesh sieves can be obtained with apertures as small as 5 μm.

The Tyler and U.S. Standard sieve series are most commonly used for size analysis. Both of these use a geometric progression of sieve apertures with a constant ratio of $2^{1/4}$ between adjacent members. It is common practice to omit the intermediate sieves, leaving a $2^{1/2}$ ratio. Testing sieves manufactured according to the two standards are essentially interchangeable, but it is important to note that the "mesh" designations are not always the same; for instance, a Tyler 12 mesh is equivalent to 14 mesh on the U.S. scale; a Tyler 14 corresponds to a U.S. 16 (see Table 2.5).

Size analysis by sieving is normally carried out by using a stack of standard sieves with opening sizes that decrease progressively from top to bottom. The sample should be weighed accurately before it is placed on the top (coarsest) sieve. The use of a mechanical shaking device is generally

TABLE 2.5 Size of standard test sieves

Tyler Series				U.S. Series		
	1.189 Ratio		1.414 Ratio		1.189 Ratio	
mm	Opening, in.	Mesh	Opening, in.	mm	Opening, in.	Mesh
26.67	1.050	—	1.050	26.9	1.06	1.06 in.
22.43	0.883	—	—	22.6	0.875	7/8 in.
18.85	0.742	—	0.742	19.0	0.750	3/4 in.
15.85	0.624	—	—	16.0	0.625	5/8 in.
13.33	0.525	—	0.525	13.4	0.530	0.530 in.
11.20	0.441	—	—	11.2	0.438	7/16 in.
9.423	0.371	—	0.371	9.51	0.375	3/8 in.
7.925	0.312	2½	—	8.00	0.312	5/16 in.
6.680	0.263	3	0.263	6.73	0.265	0.265 in.
5.613	0.221	3½	—	5.66	0.223	No. 3½
4.699	0.185	4	0.185	4.76	0.187	4
3.962	0.156	5	—	4.00	0.157	5
3.327	0.131	6	0.131	3.36	0.132	6
2.794	0.110	7	—	2.83	0.111	7
2.362	0.093	8	0.093	2.38	0.0937	8
1.981	0.078	9	—	2.00	0.0787	10
1.651	0.065	10	0.065	1.68	0.0661	12
1.397	0.055	12	—	1.41	0.0555	14
1.168	0.046	14	0.046	1.19	0.0469	16
0.991	0.390	16	—	1.00	0.0394	18
0.833	0.0328	20	0.0328	0.841	0.0331	20
0.701	0.0276	24	—	0.707	0.0280	25
0.589	0.0232	28	0.0232	0.595	0.0232	30
0.495	0.0195	32	—	0.500	0.0197	35
0.417	0.0164	35	0.0164	0.420	0.0165	40
0.351	0.0138	42	—	0.354	0.0138	45
0.295	0.0116	48	0.0116	0.297	0.0117	50
0.246	0.0097	60	—	0.250	0.0098	60
0.208	0.0082	65	0.0082	0.210	0.0083	70
0.175	0.0069	80	—	0.177	0.0070	80
0.147	0.0058	100	0.0058	0.149	0.0059	100
0.124	0.0049	115	—	0.125	0.0049	120
0.104	0.0041	150	0.0041	0.105	0.0041	140
0.088	0.0035	170	—	0.088	0.0035	170
0.074	0.0029	200	0.0029	0.074	0.0029	200
0.063	0.0024	250	—	0.063	0.0024	230
0.053	0.0021	270	0.0021	0.053	0.0021	270
0.044	0.0017	325	—	0.044	0.0017	325
0.037	0.0015	400	0.0015	0.037	0.0015	400
—	—	—	—	0.031	0.0012	450
—	—	—	—	0.026	0.0010	500

recommended. Various sieve-shaking systems are available that can accommodate up to about 12 individual sieves. After the appropriate period of shaking (see the discussion of sieving kinetics that follows), the particles retained on each sieve are removed and weighed. Gentle brushing of the *underside* of each sieve can aid in releasing particles trapped in the apertures. For analytical purposes, it is often convenient to weigh the particles from the different sieves cumulatively; that is, by adding particles from the second sieve to those previously weighed from the first and so on. The advantages of this approach are that errors do not accumulate and a mistaken reading for one sieve is automatically corrected at the next.

Comparing the total weight collected (including that on the bottom pan) with the original sample weight is a good way to check the overall procedure. Some weight loss is inevitable as a result of adhesion of fines to sieve surfaces and sticking of particles in openings. These losses should not exceed 1%. Significant weight gain is an indication of weighing errors or the inclusion of material remaining from previous tests. In any case, such data should be discarded and the test repeated. Observed small weight losses can be handled by expressing the individual weights as fractions of either (1) the actual final weight or (2) the initial (sample) weight. The first approach, in effect, distributes the errors proportionately among all sizes. On the other hand, the use of the initial weight, with the discrepancy assigned to the "pan," meaning to sizes finer than the finest sieve used, involves the implicit assumption that all losses are caused by fines adhering to surfaces or becoming airborne. The first approach is most commonly adopted; the second may be appropriate for very dusty materials.

The particle size determined for sieving experiments is defined as the minimum square aperture through which the particle will pass. For irregular particles, size refers to the particle's smallest cross-sectional area. It is important to recognize that although a particle that has passed through a sieve is definitely smaller than that size, one that has not passed is not necessarily larger. Irregular, "near-size" particles may require several attempts before their orientation is such that they can pass through the aperture. Thus, it is necessary to allow sufficient time for sieving to reach completion while recognizing that excessive shaking can lead to abrasion of the particles; that is, to size reduction.

Sieving of an assembly of particles to determine its size distribution is inherently a kinetic process. As the sieves are shaken, layers of particles are presented to the surface of each screen. If these particles are small enough and in the correct orientation, and if no other particles are obstructing the opening, they will pass to the next finer sieve.

The kinetics of sieving are illustrated in Figure 2.11. Ideally, the curve consists of three regions: (1) an initial steep portion, where particles, much finer than the sieve aperture, pass through rapidly; (2) a region of decreasing rate corresponding to the slow passage of near-size particles; and (3) a final horizontal line signifying the endpoint of the process. Abrasion of the particles leads to a continuous decrease in weight retained, as shown in the figure. Blinding of the sieve caused by agglomeration of the particles, sticking of fine particles to the mesh, or bridging of particles across the aperture can lead to curves that approach an apparent, but erroneous, endpoint. Blinding is an especially important problem in sieving and should be watched for carefully. In many cases, blinded sieves can be cleared by gentle brushing of the *underside* of the sieve. In severe cases, resorting to wet sieving methods may be necessary. A common procedure is to wet sieve at the finest size, dry the oversize, and then carry out a normal (dry) sieving operation on this oversize material.

The required sieving time is determined by the near-size particles. A reasonable supposition is that the rate of passage of near-size material is proportional to the number of openings in the sieve and inversely proportional to the number of oversize particles (which tend to block the openings). The number of openings is proportional to the overall open area (which generally decreases with decreasing aperture size) divided by the area of each aperture (proportional to aperture size squared). The number of oversize particles is proportional to the volume of oversize divided by size cubed. It follows that the rate

- Decreases with decreasing sieve aperture
- Decreases with increased loading
- Decreases with increased oversize fraction

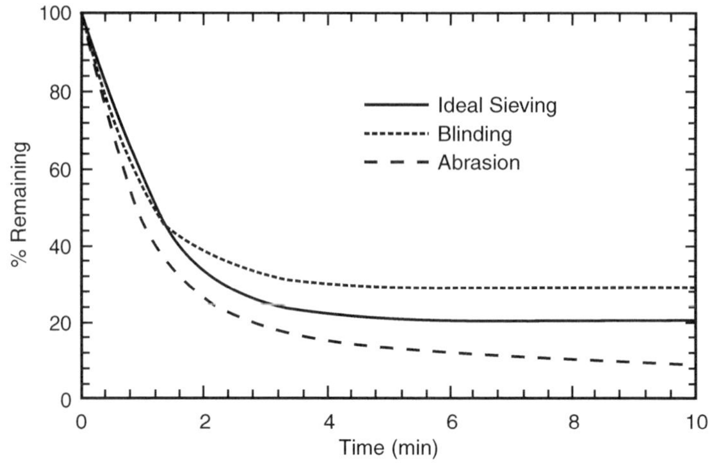

FIGURE 2.11 Sieving kinetics

As a general rule, sample size should be decreased in proportion to sieve opening. Stacking or nesting of sieves, as noted previously, can offer a significant benefit by decreasing the amount of material (especially oversize) presented to the finer sieves.

For accurate sieving, the kinetics of the process should always be investigated before the actual analysis is carried out. Failure to do so can lead to erroneous results caused by incomplete sieving or excessive abrasion of the particles. For routine work, preliminary investigations to establish an appropriate sieving schedule are recommended.

Standard testing sieves and various types of mechanical sieve shakers are readily available from most laboratory supply houses. Obviously the optimum sieving schedule may vary widely from one type of shaker to another. Wet ultrasonic sieving methods have been used successfully down to about 5 µm. However, the use of micromesh sieving is generally limited by the high cost and relatively short life of the sieves.

Microscopy. Microscope methods are highly attractive in that they involve direct observation of the particles and, through the combination of optical and electron microscopes, are virtually unlimited with respect to size. Consequently, they are extremely useful for qualitative or semiquantitative assessment of the average size and approximate range of sizes present in a distribution. However, these methods are *not* generally recommended for the quantitative evaluation of size distributions, especially for materials in which a broad range of sizes is present. The major problems with microscope methods for quantitative size analysis are

1. Errors can result because of the use of very small samples—a 1-mg sample of 10-µm particles would typically contain almost 1 million particles.
2. Slide preparation is difficult yet critical.
3. Particle counting is tedious, time-consuming, and subject to bias.
4. For broad size distributions, it may be necessary to use several different magnifications. This introduces problems in matching the different parts of the distribution. At any given magnification, there will be a size limit below which particles cannot be resolved.

Although these problems are by no means insurmountable, other methods are usually considerably simpler and often more reliable for determining size distributions.

Various different size definitions are used for characterizing microscope images. These include

- *Feret's diameter*, defined as the distance between two tangents on opposite sides of the particle, parallel to some arbitrarily fixed direction

- *Martin's diameter*, the length of the line that bisects the image of the particle, parallel to some fixed direction
- The *projected area diameter*, defined as the diameter of a circle that has the same area as the image

The most widely used of these is the projected area diameter. Because particles on a slide generally lie in some "stable rest plane," the image generally corresponds to the largest cross section of the particle. Consequently, microscope methods usually report a larger size than, for example, sieving.

Particle sizing and counting can be carried out directly on the image or on photomicrographs. Direct calibration through the use of a stage micrometer is recommended. Various kinds of standard graticules and micrometer eyepieces are available for manual counting, but their use has largely been superseded by computerized image-analysis systems. These systems eliminate the problems of, for example, operator fatigue associated with manual counting. They also result in more accurate sizing of individual particles. However, the problems of resolution (see item 4 in the list on the previous page) remain, especially for the broad size distributions that typically prevail in mineral processing systems. Furthermore, because of some loss of the ability to distinguish particles from agglomerates or to recognize touching particles as separate entities, slide preparation becomes even more critical.

A frequently asked question is, "How many particles do I need to count?" The answer, of course, depends on how much error can be tolerated, as well as on the form of the size distribution. For a set of discrete size classes, the errors associated with each depend on how many particles were counted in that particular class. As a rough guide, the 95% confidence interval on the number of particles in a class can be estimated as $n_i \pm 2\sqrt{n_i}$, where n_i is the number counted.

Slide preparation for microscope size analysis is subject to two important requirements: (1) the sample (which is extremely small) must be representative of the material, and (2) the individuals must be uniformly distributed, at an appropriate density, over the slide. For relatively coarse particles (>50 μm), spreading a thin layer of the dry powder directly onto the slide is often sufficient. This approach is not usually practical for much finer material because of the tendency for agglomerates to form, and so initial dispersion in a liquid is usually preferred.

A common procedure is to disperse the powder in a mixture of a low-viscosity, volatile liquid and a plastic or resin. After a drop of the suspension has been placed on the slide, the liquid evaporates, leaving the particles mounted in a plastic film. Suitable mixtures include collodion and amyl acetate, ethyl cellulose and toluene, Canada balsam and xylol, rubber and xylol, and polystyrene and xylene.

Freeze-drying techniques have been successfully applied to the preparation of slides for evaluation by scanning electron microscopy (SEM). The powder is dispersed in water, and a drop of the suspension is placed on the slide. A glass cover is placed over the drop, which is then frozen, very rapidly, in liquid nitrogen. The cover is removed, and the ice is sublimed away under vacuum.

Nuclepore membrane filters (Nuclepore Corp., Pleasanton, Calif.) have also been used to prepare powder specimens for SEM evaluation (Dumm 1986). These filters consist of thin, flat sheets of plastic (polycarbonate or polyester) with circular holes of controlled size (made by an irradiation/chemical etching process). By filtering an appropriate amount of very dilute suspension through the membrane, a layer of particles can be deposited on the surface.

Data analysis involves more than simply counting and sizing particles. For reasonably narrow size distributions, counting particles (at a suitable magnification) and constructing the number distribution are both relatively simple. In practice, making accurate measurements for size ranges greater than about 10:1 is usually not feasible. The lower end of the range is limited by resolution; counting statistics provide the upper limit. Because the particle systems encountered in mineral processing typically cover size ranges of 100:1 or more, counting particles at more than one magnification is generally necessary—at low magnifications to ensure that a statistically sufficient number of large particles is counted, and at higher magnifications so as to resolve the finer particles. Unfortunately, the results of counting at different magnification levels are not directly comparable because the area of the field of view necessarily changes with magnification. Matching procedures are, therefore, required for construction of the complete size distribution.

The simplest matching procedure is to express the counts at each magnification as particles per unit area of the original slide. In this form, the results should be directly comparable and should give essentially identical counts in overlapping size classes. However, this is often found not to be the case, probably because of nonuniform particle density on the slide. An alternative approach is to use the actual counts in overlapping classes to determine normalization factors. The procedure is to:

1. Obtain particle counts at sufficient different magnifications to cover the complete size range and ensure significant overlap between each.
2. Select a common set of size classes to cover the range and classify the data from each magnification accordingly.
3. Select those classes (a) that contain a reasonable number of particles, based on the criteria outlined above, at *two* adjacent magnifications and (b) in which particles are large enough to be *clearly* resolved.
4. Using the classes selected in step 3, determine the ratios of the counts in the same classes at two magnifications; that is,

$$R = \frac{\text{count at lower magnification}}{\text{count at higher magnification}}$$

5. Determine the average value of R for all of these overlapping classes and multiply each of the high-magnification counts by this value.
6. Construct the complete distribution from the normalized data. Use the value corresponding to the largest number of *actual* counts as the *best* value for each size class.

An example of the normalization procedure, for 13 size classes at two magnifications, is given in Table 2.6. Note that the number of counts, at high magnification, in the overlapping classes (9 and 10) is still much too low. The 95% confidence interval for the 10 particles counted, at high magnification, in class 9 would be approximately 10 ± 6. For class 8 it would be 4 ± 4, making any estimate of the normalization factor R meaningless. For statistically reliable results, we would need to count more particles (more fields of view) at the high magnification or introduce an intermediate magnification to result in greater overlap.

Sedimentation. Sedimentation methods are widely used for particle size analysis in the range from 0.5 to 50 µm. This range can be extended to smaller particles by using centrifuges. Most of the methods are based on the use of Stokes' law to describe the settling velocity, v, of a particle of diameter x and density ρ_p, settling, under laminar flow conditions, in a fluid of viscosity μ and density ρ_f. Thus,

$$v = \frac{(\rho_p - \rho_f)gx^2}{18\mu} \quad \textbf{(Eq. 2.83)}$$

where g is the gravitational acceleration constant.

Equation 2.83 is valid for spheres. For irregular particles, Stokes' law can be used to define the so-called Stokes' diameter as the diameter of a sphere of the same density as the particle that would have the same settling velocity in the same fluid. By considering the forces acting on a settling particle, it can be shown that

$$x_{st}^2 = \frac{x_v^3}{x_d} \quad \textbf{(Eq. 2.84)}$$

where

x_{st} = Stokes' diameter

x_v = the volume diameter (the diameter of a sphere that has the same volume as the particle)

x_d = the drag diameter (the diameter of a sphere that experiences the same viscous drag as the particle)

The drag diameter can be roughly correlated with the projected area diameter.

TABLE 2.6 Example of microscope data normalization

Size Class	Particles Actually Counted: Low Magnification	High Magnification	Normalization Factor (R)	Normalized (High-magnification) Counts	Best Values
1 (coarse)	0	0	—	0	0
2	0	0	—	0	0
3	1	0	—	0	1*
4	29	2	—	59	29
5	40	1	—	30	40
6	59	2	—	59	59
7	91	2	—	59	91
8	149	4	—	119	149
9	272	10	27.2	297	272
10	577	18	32.1	535	577
11	TSTR†	53	—	1,574	1,574
12	TSTR	242	—	7,187	7,187
13 (fine)	TSTR	625	—	18,563	18,563
Total	1,218	961	$\bar{R} = 29.7$		28,542

Note: Actual number counted = 2,179; effective number counted = 28,542.
*Statistically unreliable value.
†TSTR = too small to resolve.

The use of Stokes' law to relate settling velocity to size is subject to several limitations. Stokes' law applies provided that the Reynolds number, Re, is small. Re is defined as follows:

$$\mathrm{Re} = \frac{\rho_f v x}{\mu} \quad \textbf{(Eq. 2.85)}$$

In practice, a value of Re ≤ 0.2 is usually considered acceptable. For quartz particles (ρ = 2.65 g/cm^3) settling in water (density 1.0 g/cm^3 and viscosity 1 centipoise), this Reynolds number criterion leads to an upper particle size limit of about 60 µm. Progressively larger errors will be obtained if Stokes' law is applied for coarser particles, although alternative expressions are available that can be used in such cases (Concha and Almendra 1979). In reality, however, the upper limit is determined more by settling velocities than by the applicability of Stokes' law. A 60-µm quartz particle settles in water at about 0.3 cm/s, giving a time of about 1 min to settle through 20 cm. Disturbances caused by mixing of the suspension generally make the use of shorter settling times impractical.

Very small particles suspended in a fluid are also subject to random, thermal motion caused by the bombardment by fluid molecules. This leads to diffusive transport of the particles, which tends to oppose sedimentation. The effective diffusion coefficient, D, is related to particle size through the Stokes–Einstein equation:

$$D = \frac{kT}{3\pi\mu x} \quad \textbf{(Eq. 2.86)}$$

where

k = Boltzmann's constant

T = absolute temperature

For fine particles in water, diffusion becomes significant at sizes less than about 0.1 µm (Chung and Hogg 1985). This size, then, represents a lower limit to the applicability of sizing methods based on gravity settling. Again, however, the actual lower limit for gravity sedimentation is established more by the very low settling velocity (<1 cm/week for 0.1-μm quartz particles in water). Convection currents, caused by even the tiniest of temperature gradients, can easily lead to much larger velocities than this value.

Errors can also arise because of the effects of container walls and interference between adjacent particles (hindered settling). Wall effects can be minimized by making the container very large relative to the size of the particles. Hindered settling, which typically leads to reduced settling velocity (reduced apparent size), can be avoided by minimizing solids concentration. A maximum of about 1% by volume is usually recommended; even lower concentrations are preferred.

Sedimentation methods for particle size analysis involve determining the amount of material settling as a function of time. Different techniques can be classified according to the initial conditions and the kind of quantity measurement used. The initial condition can be a homogeneous suspension or a "line-start" in which the dispersed particles are initially floated as a thin layer on top of clear suspension fluid. Quantity measurement can be incremental (the concentration of particles at a given level below the surface of the suspension) or cumulative (accumulation of sediment at the bottom of the container). Generally, the incremental and line-start methods give the simplest data analysis, whereas the cumulative and homogeneous-suspension methods are often simpler experimentally.

According to Stokes' law, particles of size x settle at velocity v given by Eq. 2.83. It follows that, at any time t, a particle of size x must have settled through height h, where

$$v = \frac{h}{t} \quad \text{(Eq. 2.87)}$$

Thus, from Eq. 2.83 and 2.87,

$$x = k\sqrt{\frac{h}{t}} \quad \text{(Eq. 2.88)}$$

where

$$k \equiv \left[\frac{18\mu}{(\rho_p - \rho_f)g}\right] \quad \text{(Eq. 2.89)}$$

For line-start methods, any particle found at depth h below the surface after settling time t must be of size x. Particles at the bottom of the container must be of size x or larger.

Incremental/line-start methods are rarely used for gravity sedimentation but have been used in centrifugal analyses. Because all particles start from the same level, the measured concentration at depth h and time t (i.e., $C(h, t)$), relative to the total concentration (i.e., C_0), is a direct measure of the volume density function, $q_3(x)$. Thus,

$$\frac{C(h,t)}{C_0} = q_3(x) \quad \text{(Eq. 2.90)}$$

where x is given by Eq. 2.88 and 2.89.

Incremental/homogeneous methods are widely used for sedimentation size analysis. In this case, particles larger than x will have settled below the level h. The relative concentration at h and t is therefore the fraction finer than size x; that is,

$$\frac{C(h,t)}{C_0} = Q_3(x) \quad \text{(Eq. 2.91)}$$

Cumulative/line-start methods also give a direct measure of the distribution function $Q_3(x)$. At time t, particles larger than x will have reached the bottom of the container (depth h), and smaller particles will still be in suspension. Thus, if the accumulated mass of sediment, w, is measured as a function of time,

$$\frac{w(t)}{w_\infty} = 1 - Q_3(x) \quad \text{(Eq. 2.92)}$$

where

$w(t)$ = weight fraction of particles that have accumulated at depth h in time t

w_∞ = accumulated mass at at depth h and infinite time

Cumulative/homogeneous methods involve determining the accumulation of sediment at the bottom of the container (depth h). It can be shown (Allen 1997) that

$$Q_3(x) = 1 - w(t) + t\frac{dw(t)}{dt} \quad \textbf{(Eq. 2.93)}$$

Sedimentation size analyzers that make use of these principles are available commercially. Concentration measurements are usually by weight, so normally the mass distribution is what is obtained. Because of the limitations discussed above, the methods are generally applicable to particles larger than about 0.5 to 1.0 μm. However, this is usually a measurement limit only. The final reading gives the fraction finer than that particular size.

The *Andreasen pipette* consists of a thin, capillary pipette with its tip fixed at a known depth in a vertical cylinder containing an initially homogeneous suspension. Small samples of known volume are drawn off—for example, after 1, 2, 4, 8 min, etc.—and the solids content is determined, usually by drying and weighing. Typically, the cylinder holds about 500 mL of suspension, and samples are of 10 mL each. Principal sources of error arise from the disturbance of the suspension during sampling and the occurrence of convection currents caused by small temperature variations. The system should be carefully thermostated for accurate analysis at fine sizes. Because of the relatively high solids concentrations (about 5% by weight) and long settling times (up to 24 h) typically used, dispersion and suspension stabilization are especially critical in these analyses. The main advantages of the Andreasen pipette are that it is cheap and quite readily available. An experienced operator can obtain very reliable and accurate results. Indeed, the device is often used as a standard for comparison with other techniques.

Hydrometers can also be used for incremental analysis, because suspension density is a direct measure of solids concentration. However, the need to use quite high concentrations to provide measurable density changes means that hindered-settling conditions usually prevail. Problems also arise in specifying the exact location (h) at which the measured density applies. For these reasons, the use of hydrometers is not recommended.

Photosedimentometers use a light beam to estimate the solids concentration at a known depth. Unfortunately, however, the attenuation of light in the presence of particles depends on size as well as on concentration. Corrections can be made by using light-scattering theory, or the method can be simplified by neglecting the size dependence of light attenuation. In the latter case, the results should be regarded as comparative rather than absolute, especially for sizes smaller than about 2 μm. Advantages of photosedimentation are that no disturbance of the suspension caused by sampling occurs and that very dilute suspensions can be used. Because scattering is based on the cross-sectional area of the particles, photosedimentometers provide a measure of the *area* distribution, $Q_2(x)$. Transformation to the volume distribution, $Q_3(x)$, can be accomplished as described previously.

X-ray sedimentometers operate on the same principle as the photosedimentometers but use a collimated x-ray beam to determine concentration. Because the attenuation of the x-ray beam depends on mass and is independent of particle size, a direct measure of the mass distribution is obtained. The commercial instruments generally incorporate an additional modification that allows the time for an analysis to be substantially reduced. Fifty-micrometer particles settle at a rate of the order of 10 cm/min. To allow steady sedimentation conditions to be established, therefore, the sampling depth should be several centimeters. However, particles of 1 μm or less take many hours to settle this distance. To reduce the test time for such small particles, small values of h must be used. This is accomplished by scanning the x-ray beam over the cell during the course of a run, usually by raising the cell. In this way, both h and t are varied simultaneously. By reducing the distance over which the finer particles must settle, the effective range of the instrument is increased. The lower limit, however, is still governed by thermal diffusion (Brownian motion). The major disadvantage of x-ray sedimentation is that, for materials with low x-ray absorption coefficients (e.g., "light" elements), high concentrations must be used, which lead to hindered-settling phenomena.

The *sedimentation balance* uses a recording balance to determine the accumulation of sediment as a function of time. The arrangement proposed by Leschonski (Pretorius and Mandersloot 1967) is the recommended design. Because all of the particles are collected on a pan suspended across the entire section of the sedimentation column, the clear liquid in which the pan is suspended stays at constant density. This avoids the buoyancy and convection effects that occur when the pan is suspended in the sedimentation column.

The particle size distribution is obtained from Eq. 2.93. The value of $w(t)$ is determined directly from the recorder trace, and its time derivative, dw/dt, is obtained by drawing tangents to the curve. The need to differentiate the curve is often regarded as a major disadvantage of the method. A more serious problem arises from the need to evaluate the final weight settled. This can lead, in effect, to a lower *detection* limit. Advantages are that the technique can be made semiautomatic, that there is little disturbance of the suspension after the run has commenced, and that suspensions of low solids content (typically about 0.5 g of solids in about 500 cm^3 of fluid) can be used.

Centrifugal methods can be used to extend the useful range of sedimentation methods by increasing the effective "gravitational" force acting on a particle. By analogy to Eq. 2.83, the settling velocity of a particle at a distance r from the center of rotation of a centrifuge is given by

$$v = \frac{(\rho_p - \rho_f)\omega^2 r x^2}{18\mu} \qquad \textbf{(Eq. 2.94)}$$

where ω is the angular velocity of the centrifuge.

It should be noted that, for this case, the terminal velocity is not constant; instead, it depends on r, the position of the particle in the tube. Because we know that

$$v = \frac{dr}{dt} \qquad \textbf{(Eq. 2.95)}$$

Equation 2.94 can be rewritten as

$$\frac{dr}{r} = \frac{(\rho_p - \rho_f)\omega^2 x^2 dt}{18\mu} \qquad \textbf{(Eq. 2.96)}$$

If, at time zero, all particles are in a thin layer at the surface of the fluid (i.e., at $r = r_0$), Eq. 2.96 can be integrated to give

$$ln\ \frac{r}{r_0} = \frac{(\rho_p - \rho_f)\omega^2 x^2 t}{18\mu} \qquad \textbf{(Eq. 2.97)}$$

Thus, after time t, all particles at r will be of size x such that

$$x = \left[\frac{18\mu\ ln\ (r/r_0)}{(\rho_p - \rho_f)\omega^2 t}\right]^{1/2} \qquad \textbf{(Eq. 2.98)}$$

If the line-start approach is used, r_0 is the same for all particles and Eq. 2.98 can be used instead of Eq. 2.88 to calculate particle size. More complicated analyses are required for initially homogeneous suspensions (Allen 1997). Additional corrections are required for small-radius centrifuges to account for the fact that, in a centrifugal field, particles travel radially rather than in the straight lines that occur in gravity sedimentation. Both tube-type and special disk-shaped centrifuge bowls have been used for centrifugal size analysis. Centrifugal sedimentation size analyzers are available commercially, but their use has generally been limited to research applications.

Light Scattering. The scattering of light by suspensions of fine particles is strongly size dependent. Light-scattering measurements are, therefore, extremely attractive for particle size analysis. Some of the advantages of light-scattering methods are

- The method is not limited to any particular range of sizes.
- The method can be applied in situ, minimizing sampling problems and disturbance of the material.
- Measurements can be made on particles suspended in either liquid or gas.
- Measurements can be made essentially instantaneously.

Because of the complex nature of the scattering phenomenon, its application to particle size analysis was severely limited until the availability of low-cost microprocessors made it feasible to perform the necessary calculations on-line.

The theory of light scattering is extremely complicated, and only a very brief outline is given here. Detailed descriptions are given by van der Hulst (1957), Kerker (1969), and Bohren and Huffman (1983). A beam of light can be regarded as an oscillating electric field. When the beam strikes a solid particle, its positively and negatively charged parts (the nuclei and electrons, for example) are subject to opposite forces that lead to the formation of an oscillating dipole. This dipole is itself a source of electromagnetic radiation; that is, of "scattered" light.

If the particle is very small compared to the wavelength of the incident radiation, it acts as a point source of scattering. In this case, relatively simple relationships can be derived to account for the angular variation in scattered intensity. This is the basis of the well-known Rayleigh theory (Strutt 1871) of light scattering. For larger particles, each point on the surface of the particle will act as a source of scattering, and the theory must account for the combined effects of all such points. The general theory was developed by Mie (1908) and can be applied to all particles regardless of size. The effects of particle shape are much less well understood; the Mie theory applies to spheres, although extensions to other simple shapes have been developed.

From the Mie theory, it is possible to predict the intensity of scattered light when the particle is viewed from any angle. The angular scattering patterns depend, in general, on the relative refractive index, m, and relative particle size, α (which equals $\pi x/\lambda$, where λ is the wavelength of the incident light). In principle, then, it is possible to determine the size of a particle by comparing the observed scattering pattern with the theoretical forms. Other properties, such as the relative polarization of the incident and scattered beams, can also be described theoretically and used to estimate particle size.

The light-scattering theories describe the effects of a single particle. In practice, of course, we are generally concerned with many particles, which normally vary in size and shape. Light scattering by an assembly of particles is naturally more complex than that caused by a single particle because of differences among the particles, as well as the possibility of multiple scattering–light scattered by one particle is scattered again by another. The latter effect can be minimized by taking measurements at the lowest possible concentration or by extrapolating to zero concentration. Under these conditions, the effects of each particle can be assumed to be additive. The major problem in the use of light scattering for particle size analysis lies in describing the behavior of heterogeneous mixtures of particles.

Several approaches can be taken to the problem of heterogeneous particulate systems:

1. The particles can be examined one at a time–this approach is used widely in optical particle counters (see the "Automatic Particle Counters" section later in this chapter).
2. Light scattering can be used in conjunction with a separation process, such as sedimentation, so that essentially uniform size fractions are examined optically. This is the basis of the photosedimentation techniques (as discussed earlier), where light scattering is used to determine concentration rather than size.

3. Observed scattering behavior can be compared with theoretical models based on assumed forms for the size distribution. Light-scattering data are then used to estimate the parameters of the distribution.

Approaches 1 and 2 unfortunately lose some of the potential advantages of the light-scattering technique—in particular, the possibility of instantaneous measurement and the essentially unlimited range of application. Approach 3 was traditionally limited chiefly by the complex form of the scattering functions, but it has become relatively simple because of the use of modern computer systems.

Fraunhofer diffraction is the basis for most of the modern light-scattering size analysis systems. When a suspension of fine particles is illuminated by a beam of light, a Fraunhofer diffraction pattern is produced. This pattern consists of a series of concentric rings whose radii are each uniquely related to a particular particle size. Typically, these instruments use laser illumination of particle suspensions; the diffracted light is focused onto a suitable detector, and the relative intensities at different angles, corresponding to the diffraction rings, are measured. The intensity at each angle includes contributions from each of the particle sizes that may be present. By comparing the measured intensity distribution to that predicted theoretically, an estimate of the particle size distribution (normally the volume distribution, $Q_3(x)$) can be obtained. In effect, each angular measurement yields one point on the size distribution. Most of the instruments can be used either on liquid suspensions (normally water) or on dry (airborne) particles. The diffraction technique can be applied to particles greater than about 2 µm; extension to finer sizes (~0.1 µm) can be accomplished using a combination of Fraunhofer diffraction and some other measurement, such as polarization ratio or right-angle scattering, at a series of different wavelengths. The extent to which these limits are of measurement or detection is not entirely clear. Smaller particles certainly scatter light, although the intensity tends to be low. How their contribution is accounted for depends on the particular algorithm used in the system. This information is, of course, proprietary.

Because the size measurement is based on cross-sectional area, light-scattering instruments tend to report a somewhat larger size than other devices for the same particles—typically about 20% larger. Because the distribution is estimated from the integrated response of the entire particle system, the resolution of these systems is lower than that obtained by methods that look at individuals. Very narrow size distributions typically appear broader, for example. Compensation effects generally reduce the problem for broader distributions.

Dynamic Light Scattering. Also known as quasi-elastic light scattering (QELS) or photon correlation spectroscopy (PCS), dynamic light scattering is based on the time variations of the scattered light rather than on the average intensity. The intensity of scattered light from suspended particles is subject to fluctuations caused by interference effects resulting from the random, Brownian motion of the particles. Autocorrelation procedures, in which the intensity at time t is correlated with the value at some previous time t' (which equals $t - \Delta t$), are used to obtain a characteristic "decay" time for the fluctuations. The decay time is related to the Brownian diffusion coefficient of the particles, which in turn is related to particle size through the Stokes–Einstein equation (given earlier as Eq. 2.86):

$$D = \frac{kT}{3\pi\mu x}$$

This approach is especially attractive for submicron particles because of the inverse relationship between the diffusion coefficient and particle size. Several instruments are available commercially that offer measurement ranges of about 5 nm to 5 µm. The simple measurement gives a mean diffusion coefficient, corresponding to a mean particle size. Typically, the distribution of sizes is estimated by assuming some general form for the distribution of decay times and using optimization procedures to estimate the distribution's parameters. These values are then used to determine the approximate size distribution. As a consequence, the resolution associated with this technique is rather poor, especially for broad size distributions.

Automatic Particle Counters. Automatic counting systems represent an important class of sizing devices. For the most part, these instruments count and size individual particles, in liquid or gas suspension, as the particles pass through a "sensing zone." Sizing is usually accomplished by measuring the effect of the particles on the optical or electrical properties of the carrier fluid.

Optical counters, which detect and size particles as they pass through a light beam, are available for both gas and liquid suspensions. Three types of optical systems are in common use:

- Near-forward scattering systems use mirror and lens combinations to collect light scattered within a cone of fixed solid angle in the direction of travel of the incident beam. A light trap is employed to eliminate directly transmitted, incident radiation.
- Right-angle scattering systems use a detector placed at right angles to the path of the incident beam.
- Light obscuration systems are designed to observe the shadow cast by the particle. Size is estimated from the reduction in intensity of the transmitted beam.

Equipment manufacturers say that right-angle scattering gives the best size resolution for particles of fixed composition and refractive index. When the particles may be of variable composition, the near-forward scattering system is generally preferred. Light obscuration methods are normally reserved for opaque particles suspended in a liquid; under these conditions, the technique is said to have several advantages over the scattering systems. These advantages arise chiefly from elimination of problems caused by variations in shape, color, and refractive index, among others.

It should be emphasized that these are not absolute methods of size analysis. The optical signals are converted to particle size by calibration of the instruments with standards of known size. Because light scattering is strongly dependent on the refractive index of the particle–as well as on the particle's size–considerable error can arise from variations in refractive index from one particle to another. Similarly, differences in refractive index between the particles being measured and the calibration standard can lead to uncertainty in the definition of the measured size.

Several optical particle-counting systems are available commercially. Instruments using the near-forward scattering principle are used primarily for airborne particles, although systems have also been designed for liquid suspensions. Typically, the instruments have a lower limit of detection of around 0.3 μm, below which background noise tends to dominate. Gas flow rates of up to 5 L/s and particle concentrations up to 3,000 particles/cm^3 can be handled. Instruments that apply right-angle scattering are also available and are generally used for airborne particles in the range of 0.3 to 10 μm; flow rates up to 3 L/min and concentrations up to 100 particles/cm^3 can be accommodated. Instruments using light obscuration on liquid suspensions are used primarily for sizing particles larger than about 2 μm at concentrations up to about 1,000 particles/cm^3. Special detectors are available to extend the range into the submicron region. Other approaches include the use of a scanning laser beam focused in the interior of a sample jar to count and size particles, by using near-forward scattering, in a "sensitive zone." Scanning lasers focused inside a sample container (or process vessel, pipe, etc.) have also been developed in which particle size is inferred from the time taken for a particle to cross the laser beam rather than from a scattered intensity measurement. The manufacturers claim that these systems are applicable for particle sizes ranging from 0.5 to 250 μm, with extension to 1,000 μm, as well as for solids concentrations up to 30%. The latter type of usage makes the system especially attractive for on-line analysis.

Electrical sensing methods, such as the Coulter Counter (Coulter Electronics, Hialeah, Fla.), determine particle size from measurements of the electrical conductivity of an electrolyte solution that contains suspended particles flowing through a small orifice. Each time a particle passes through the orifice, it displaces fluid and hence increases the effective electrical resistance of the orifice. If electrodes are placed on either side of the orifice, a voltage pulse is generated as each particle passes. It can be shown that the change in resistance–and consequently the size of the voltage pulse–are roughly in proportion to the volume of the particle, so the reported size is a volume diameter. Size limitations are determined by the size of the orifice. An upper limit in particle diameter of about 40% of the orifice diameter is generally considered sufficient to minimize blockage of the opening; a size range (maximum to minimum) of about 32:1 is possible for any given orifice. Thus, for a 100-μm orifice, the upper limit is about 40 μm, while the 32:1 size range sets the lower limit at just over 1 μm. Orifices ranging in diameter from 10 to 1,000 μm are available. In practice, an absolute lower limit of about 0.8 μm is found; below this size, heat generation in the orifice leads to a rapid increase in background noise. Although the observed voltage pulse is a measure of particle volume, electrical resistance counters are not absolute instruments. Calibration against standards of known size is necessary. However, the voltage pulse is independent of other properties of the particle (unlike light scattering, which also depends on refractive index), so that problems of heterogeneity of the material, for example, are probably insignificant. Calibration at more than one size is recommended, however.

Electrical resistance counters are highly reproducible when operated carefully. Analysis is extremely rapid provided only one orifice is needed, and the sensitivity is excellent. These instruments have found application in a wide range of industries. For materials with a broad range of sizes, problems can arise because the upper size limit may require the sample to be separated at some intermediate size to prevent blockage of the orifice. The separate fractions must then be analyzed independently by using an appropriate orifice for each. This can be a difficult proposition, especially if the separation must be made in the subsieve size range (i.e., less than 20 μm), because the separation must be quantitative to permit recombination of the parts of the distribution.

Because they measure individual particles, one at a time, particle counters generally have excellent resolution. All counters, however, suffer from detection limits (see the previous section entitled "Limitations of Sizing Techniques"). Because of the problem of distinguishing the particle signal from the background noise, there is always some size that is too small to be detected. Thus, all particles are assumed to be coarser than the limit, whereas, in fact, a large proportion smaller than this size may be present. Size distributions obtained with counters therefore tend to be biased toward the larger sizes. The effect can be corrected for, in principle, by comparing the apparent volume counted with the known solids concentration, determined independently. Because of the difficulty in estimating particle volume (for example, because of unknown shape factors), this approach is generally useful only as a qualitative measure of the relative amount of undetected, undersize material.

Most particle counters are also subject to a maximum concentration limit that is caused by the so-called coincidence problem–multiple particles passing through the sensing zone simultaneously are counted as one (larger) particle. The maximum concentration is set so as to maintain the probability of coincidence at an acceptable level. The combination of coincidence and detection limits can be especially troublesome because the presence of undersize particles affects the background noise level to an unknown extent.

Automatic particle counters are ideal for evaluating relatively narrow size distributions because of the very high resolution that can be obtained. However, their applicability to broad distributions, such as those typically found in ground mineral systems, is reduced by the detection limit problem.

Surface Area Measurement. The total surface area of a collection of particles can be arbitrarily divided into the external surface and the internal surface caused by pores and cracks in the individual particles. A third area class is the geometric area, which consists of a smoothed envelope around the particles, neglecting fine-scale surface roughness. These different types of surface area are reflected in variations among area estimates based on different measurement techniques. The geometric area is determined almost entirely by the size of the particles and can be calculated directly from the particle size distribution. The external area includes surface roughness but excludes the contribution from the surfaces of internal pores. Again, the distinction between external features (roughness) and internal cracks and fissures is somewhat arbitrary. The resistance to fluid flow around individual particles or through packed beds is determined by the external area. Adsorption from gases or liquids is possible on all accessible surfaces, including open pores, and is determined by the total surface area (external and internal).

In general,

geometric area < external area < total area

For microporous materials such as coal, the internal area is quite commonly very much larger than the external area. As a result, wide variations exist among different area measurements and the total area is essentially independent of particle size.

Geometric Surface Area. An estimate of the geometric surface area of a powder can be obtained directly from size distribution measurements. According to Eqs. 2.41, 2.50, and 2.54, the specific (geometric) surface area is given by

$$S_v = k_{23} \int_0^\infty x^{-1} q_3(x) dx = \frac{6}{(\bar{x}_{-1,3})_{ES}} \qquad \textbf{(Eq. 2.99)}$$

Permeametry. The external surface area of a bed of particles can be estimated from the bed's permeability to fluid flow. The flow of a fluid through a packed bed of particles can be described by regarding the pores in the bed as a bundle of capillaries and assuming that flow through each individual capillary takes place according to Poiseuille's equation:

$$u_m = \frac{d^2}{32\mu} \frac{\Delta p}{L} \qquad \textbf{(Eq. 2.100)}$$

where

u_m = mean fluid velocity in the pore
d = diameter of the pore
μ = viscosity of the fluid
Δp = pressure drop across a length L of pore

Geometric arguments can then be used to show that the effective pore diameter, d, is given by

$$d = 4\left(\frac{\varepsilon}{1-\varepsilon}\right) \frac{1}{\rho S_m} \qquad \textbf{(Eq. 2.101)}$$

where

ε = the fractional porosity of the bed
S_m = the mass specific surface area, as before (here the assumption is that the surface area of the pores is identical to that of the particles)

The mean fluid velocity in a pore, μ_m, is, of course, not amenable to direct measurement. The approach velocity, u, on the other hand, can be measured readily and is simply related to μ_m through

$$u = \varepsilon u_m \qquad \textbf{(Eq. 2.102)}$$

Combining Eqs. 2.100, 2.101, and 2.102 leads to the well-known Carman–Kozeny equation for flow through porous media (Carman 1956; Kozeny 1927):

$$u = \frac{1}{k\mu\rho S_m^2} \cdot \frac{\varepsilon^3}{(1-\varepsilon)^2} \cdot \frac{\Delta p}{L} \qquad \textbf{(Eq. 2.103)}$$

where k is a correction factor that reflects the tortuosity of the pores in the packed bed. Carman has shown that, for most particulate systems, $k \cong 5$. Eq. 2.103 applies to the case of laminar flow through the bed and is valid when the mean free path of the fluid molecules is small compared to the pore diameter; that is, for liquids or gases at normal pressures flowing through beds of coarse particles (>0.1 μm).

If the flow rate and pressure drop through a packed bed of known porosity and dimensions are measured, Eq. 2.103 can be used to calculate the specific surface area of the material. Because the technique involves fluid flow around the particles, the measurement is of external surface area. Several devices for the measurement of specific surface area by gas (air) permeability are available commercially.

The Fisher Subsieve Sizer (Fisher Scientific, Pittsburgh, Pa.; see Gooden and Smith [1940]) uses a standardized procedure for preparation of the sample bed. Pressure drop and gas flow rate are determined by means of a combination of calibrated jets and a manometer. The instrument is supplied with a calculator chart, which is devised to read the specific surface mean diameter, $\bar{x}_{sv}$, directly. The method is rapid (<20 minutes), is quite reproducible, and can be applied to materials with mean sizes ranging from 5 to 50 µm.

The Blaine Permeameter (ASTM 1992) uses displaced fluid to force gas through the particle bed. The flow velocity is determined by the time taken to pass a known volume of the gas. Because the pressure drop across the bed varies during the course of the test, a modified version of Eq. 2.103 is used, in combination with calibration based on a standard cement sample, to calculate the specific surface area of the sample. The device is widely used for routine testing and quality control, particularly in the cement industry. As with the Fisher Subsieve Sizer, bed preparation is critical and standardized procedures must be followed.

Gas Adsorption. The amount of gas that can be adsorbed on the surfaces of a powder gives a measure of the total (accessible) surface area. Only for nonporous particles with relatively smooth surfaces can this measurement realistically be related to a mean particle size (by using, for example, Eq. 2.53). Gas adsorption areas are most commonly used as direct characteristics of the powder.

If the results of adsorption measurements are plotted as volume adsorbed versus equilibrium pressure (at constant temperature), the resulting curves are known as adsorption isotherms. Several theoretical models have been developed to describe the shape of typical adsorption isotherms mathematically.

The BET Isotherm (Brunnauer, Emmett, and Teller 1938) is the most widely used of these models. It is based on the equilibrium between the gas phase and the adsorbed film on the surface. The BET equation can conveniently be written as

$$\frac{x}{v(1-x)} = \frac{1}{cv_m} + \frac{(c-1)x}{cv_m} \qquad \textbf{(Eq. 2.104)}$$

where

x = p/p_o (by definition)
p = gas pressure
p_o = saturation vapor pressure of the gas at the particular temperature of the experiment
v = volume of gas adsorbed
c = a constant
v_m = the monolayer volume, i.e., the volume of adsorbed gas needed to form a single layer (one molecule thick) on the solid surface

The BET equation predicts that a plot of the quantity $x/[v(1-x)]$ versus x should give a straight line. The slope of the line is equal to $(c-1)/(cv_m)$, and the intercept at $x = 0$ is equal to $1/(cv_m)$. From the values of the slope and intercept, the monolayer volume (v_m) can be calculated. The specific surface area of the solid can then be determined from

$$S_m = \frac{v_m N \sigma_o}{m v_o} \qquad \textbf{(Eq. 2.105)}$$

where

S_m = the specific surface area, in cm^2/g
N = Avogadro's number = 6.034×10^{23}
σ_o = the area occupied by one molecule of the gas (= 16.2 Å^2 for N_2)
m = the mass of the sample, in grams

v_o = the molar volume of the gas (= 22,400 cm^3 at standard temperature and pressure [STP])

The characteristic isotherm for any particular adsorbate gas is obtained by plotting the fractional surface coverage, v/v_m, versus relative pressure, p/p_o, for a series of solids whose surface areas are known. Typically, the values fall on a single smooth curve for a given gas on a wide variety of solids. The implication of this result is that, at any relative pressure, the amount of gas adsorbed is directly proportional to the surface area of the solid and is independent of the solid's nature. Consequently, any expression, theoretical or empirical, that describes adsorption data must contain some parameter that is proportional to surface area.

The t-plot technique (Lippens, Linsen, and de Boer 1964) is an extension of the concept of a characteristic isotherm. The quantity v/v_m represents the number of monolayers adsorbed on the surface and is, in turn, proportional to the effective thickness, t, of the adsorbed film. Thus,

$$v = \frac{v_o(\text{gas})}{v_o(\text{film})} St \qquad \textbf{(Eq. 2.106)}$$

where

$v_o(\text{gas})$ = molar volume of the gas
$v_o(\text{film})$ = molar volume of the film
S = total surface area of the adsorbent solid

Furthermore, thickness t is a unique function of relative pressure for any given gas. Values of t at different relative pressures have been tabulated. (See, for example, Adamson 1997.)

The t-plot is obtained by plotting the measured volume of gas adsorbed at different relative pressures against the t values corresponding to those pressures. The resulting plot should be a straight line whose slope is $S[v_o(\text{gas})/v_o(\text{film})]$. Thus, for nitrogen at 78 K, the slope would be $S \times 0.0646$ (for S in square meters, v in cubic centimeters, and t in angstroms).

The t-plot can also give information on the structure of the solid surface. An ideal solid will give a linear t-plot, whereas a microporous solid will show a flattening of the t-plot, for high surface coverages, because of the reduction in available area as the pores become filled. Alternatively, capillary condensation in somewhat larger pores can lead to an increase in the slope of the plot.

Allen (1997) has extensively reviewed the experimental procedures available for the evaluation of gas adsorption isotherms for determining surface area. These methods can be classified into three groups: volumetric, gravimetric, and continuous flow.

Volumetric methods represent the classical approach. Many variations of volumetric apparatuses are described in the literature. However, they all employ the same principles. A known weight of solid is "cleaned" by being heated and exposed to a high vacuum. The "dead space"–the volume surrounding the solid–is determined by expanding a known volume of nonadsorbing gas (usually helium) into the experimental system. After evacuation of this gas, a known amount of adsorbate gas is allowed to come into contact with the solid until constant (equilibrium) pressure is achieved. From a knowledge of the dead space and the initial and final (equilibrium) pressures, it is possible to calculate the volume of gas adsorbed. To determine further values for the isotherm, more gas is allowed to come into contact with the solid. In this way several experimental points can be obtained to define the isotherm over a range of relative pressures.

For solids of relatively large surface area (>1 m^2/g), the most commonly used adsorbate is nitrogen at liquid nitrogen temperature (77 K). Full details of an apparatus and methods of calculation are given in a British standard (British Standards Institution 1969). Traditionally, many laboratories have designed and built their own volumetric adsorption equipment. However, several commercial units are available, including computerized, fully automatic systems.

Gravimetric methods offer two main advantages: (1) the volume of the apparatus is of no importance and (2) the amount of gas adsorbed is measured by direct weighing. Generally, the sample is suspended in an inert container from a delicate quartz spring. The extension of the spring is measured by a cathetometer, and the weight adsorbed is determined from calibrations. If available, a sensitive electrobalance can be substituted for the quartz spring.

Continuous flow techniques (Nelsen and Eggersten 1958) use gas chromatography to determine the composition of a mixture of the adsorbate gas and a nonadsorbing carrier gas before and after passing over a sample of powder. The major advantage of the continuous flow systems is that the amount of gas adsorbed is measured directly rather than by difference. Other advantages are speed, simplicity of operation, ambient pressure measurements (no vacuum system), and elimination of dead-space corrections. Total surface areas down to about 1,000 cm^2 can be measured readily, but background signals caused by thermal diffusion effects become significant at smaller areas. Desorption isotherms can be evaluated by this technique (Karp, Lowell, and Mustacciuolo 1972), but the volumetric methods are generally preferred for such measurements.

Measurement of Particle Shape

Particle shape evaluation is generally carried out by computerized image analysis. Typically, a set of coordinate points corresponding to the perimeter of each particle image is obtained. These can be used, for example, to determine Fourier coefficients. An alternative approach (Dumm 1989; Kumar 1998; Kaya, Kumar, and Hogg 1996) is to fit the perimeter points to a general polygon that consists of intersecting, straight-line sequences of adjacent points. In this way, the amount of information required to describe each particle can be reduced very substantially with little or no loss of resolution. The intersection points on the polygon can be used, in turn, to define specific shape descriptors, such as radial variability (the mean square deviation from a circle) and angular variability (a measure of the variation in the angle between adjacent edges) (Dumm 1989; Kumar 1998).

Differences in the apparent size distribution as determined by different techniques—such as light scattering and sedimentation—can generally be ascribed to particle shape. Because the measurements normally provide an equivalent-sphere diameter, the results generally agree closely for spherical particles. Observed differences for irregular particles presumably give a measure of the departure from the ideal spherical shape. However, such differences have yet to be related to specific, physical shape characteristics, such as aspect ratios.

Measurement of Particle Composition. The chemical composition of individual particles can be determined by scanning electron microscopy with energy dispersive x-ray spectroscopy (SEM-EDS) methods (Dumm, Hogg, and Austin 1991). In practice, however, this method is seriously limited by calibration problems and the need to apply corrections to account for "light" elements that are not detected by most EDS systems. These corrections become especially unreliable for particles smaller than about 10 µm. Mineralogical composition can be evaluated by optical microscopy for relatively large particles and by transmission electron microscopy for very fine material.

Density Measurement

Buoyancy measurements can be taken to determine the density of individual particles. If a particle is weighed in air and in a wetting liquid of known density, its density can be obtained from

$$\rho = \frac{w_a \rho_l}{w_a - w_l} \qquad \textbf{(Eq. 2.107)}$$

where

w_a = the weight in air

ρ_l = the density of the liquid

w_l = the weight in liquid

The method can be used down to quite small sizes if a sensitive microbalance is used.

Density distributions are generally determined by sink-float analysis using a series of heavy liquids. The method is widely used in so-called washability analysis of coal (ASTM 1984). Its application to fine sizes (<100 μm) generally requires that centrifuges be used to ensure proper separation of fine particles of density close to that of the suspending liquid that settle (or float) very slowly (Dumm and Hogg 1988). Suitable heavy liquids include halogenated hydrocarbons (e.g., methylene iodide; ρ = 3.3 g/cm^3) and aqueous solutions of heavy metal salts.

COMPARISON AND INTERCONVERSION OF PARTICLE SIZE DATA

Different size measurement methods generally give different results because of the different definitions of size employed. Correlation of the data is necessary so as to compare results from different kinds of measurements. More important, using more than one method is often necessary to cover the range of sizes present in a given system. For example, a common practice is to use sieving down to 37 μm (400 mesh) and to carry out light scattering or sedimentation measurements on the subsieve material to extend the distribution to 1 μm or finer.

Conversion factors (sometimes referred to as shape factors) can be obtained by direct comparison of different measurements on the same material. An example of such a comparison, for fine coal particles, is given in Figure 2.12. Table 2.7 lists the corresponding conversion factors, f_{AB}, defined by

$$f_{AB} = \frac{x_{50,A}}{x_{50,B}} \quad \textbf{(Eq. 2.108)}$$

where $x_{50,A}$ and $x_{50,B}$ are the median sizes obtained by methods A and B (arbitrary designations for the different measurement methods).

When one method is used to extend the range of another, the conversion factor can be obtained from a region of overlap between the two. In the particular case of extending sieving, the recommended approach is to analyze a narrow sieve fraction (e.g., 270 × 400 mesh) by the other (subsieve) method so as to obtain the appropriate conversion factor. An example of the use of this approach is given in Appendix 2.2.

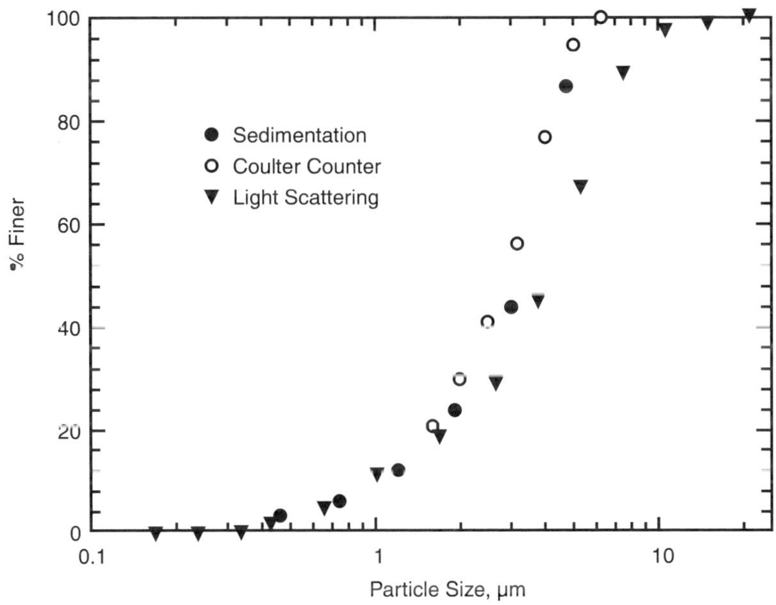

FIGURE 2.12 Comparison of size distribution for fine coal measured by different methods

TABLE 2.7 Conversion factors for different sizing methods on fine coal: Data corresponding to Figure 2.12

Method	Median Size, μm	Conversion Factor (to Volume Diameter)
Coulter Counter	2.90	1.00
Sedimentation	3.23	1.11
Light scattering	4.04	1.39

Source: Data from Dumm (1986).

APPENDIX 2.1: MOMENT DETERMINATION AND QUANTITY TRANSFORMATION FROM EXPERIMENTAL DATA

A fairly typical size distribution, as might be obtained by sieving, is illustrated in Figure 2.13. The volume distribution, $Q_3(x)$, can be transformed to the number distribution, $Q_0(x)$, by using the specific form of Eq. 2.36:

$$(q_0)_i = \frac{\bar{x}_i^{-3}(q_3)_i}{\sum_{i=1}^{n} \bar{x}i^{-3}(q_3)_i} \qquad \textbf{(Eq. 2.109)}$$

The arithmetic mean of each size interval, including the sink interval, is used in the calculations. The steps in the procedure are outlined in Table 2.8. The data extend only to 38 μm (400 mesh); $Q_3(x)$ values for finer sizes are obtained by extrapolating the linear portion of the curve shown in Figure 2.13. The values shown in italics in the table columns are based on this extrapolation. The summations given at the bottom of column VI are the corresponding values of the moment $M_{-3,3}$.

The estimated number distributions based on the direct and extrapolated volume distribution are shown in Figure 2.14. The figure clearly shows that the calculated moments and the transformation depend heavily on the assignment of a value for the mean size in the sink interval. (The extrapolation procedure, in effect, assigns a specific form to the distribution within the original sink interval [<38 μm].) In the absence of actual data extending the distribution, it is not feasible to obtain reliable estimates of the moments $M_{k,3}$ for $k < 0$ or of the distributions $Q_r(x)$ for $r < 3$.

APPENDIX 2.2: COMBINATION OF SIEVE AND SUBSIEVE SIZE DATA

A previous section of this chapter noted that different analytical procedures yield different measures of particle size. This fact is especially important when subsieve sizing methods, such as sedimentation or light scattering, are used to extend sieving data to finer sizes. The recommended procedure is to calibrate the subsieve method against sieving for the material of interest by conducting a subsieve size analysis on a specially prepared, narrow sieve fraction of the material. For example, if sieving is normally carried out down to 37 μm (400 mesh) and is to be extended by subsieve sizing of the −400 mesh fraction, calibration can be performed on a sample of 325 × 400 mesh material. The sample should be prepared by wet sieving at 400 mesh (to ensure removal of any adhering fines) followed by dry sieving at both sizes (to ensure the correct placement of near-size particles).

An example of a calibration test by light scattering on a 325 × 400 mesh (44 × 37 μm) sample of quartz is given in Figure 2.15. The median (sieve) size for this material can be estimated from the geometric mean of the interval; that is, $(44 \times 37)^{1/2} = 40.4$ μm. Figure 2.15 shows that the light-scattering data give a coarser median size (46.8 μm versus 40.4 μm) and also increase the spread of the distribution from a single $2^{1/4}$ (1.19) ratio to about 4 or 5. The increased spread is a reflection of the relatively low resolution of the light-scattering methods, as well as of the greater variations in area diameter (which depends on all three dimensions of a particle) than in sieve diameter (which is dominated by

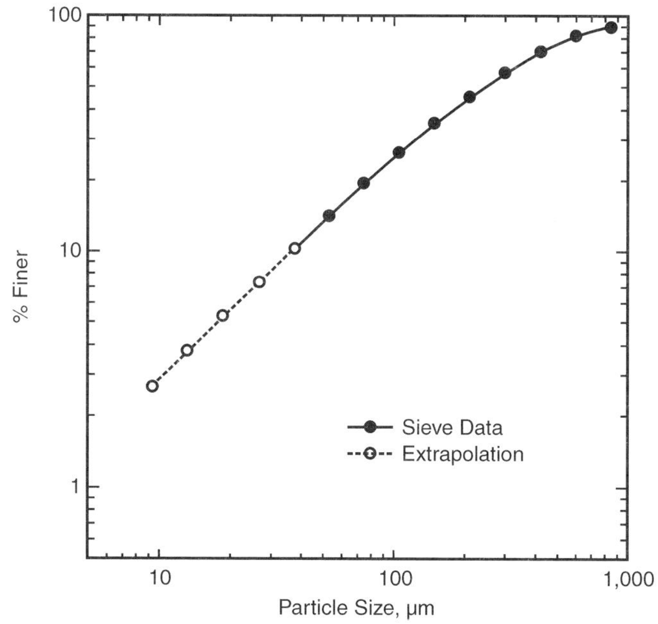

FIGURE 2.13 Typical sieve size distribution (20 to 400 U.S. mesh showing extrapolation to below 10 μm)

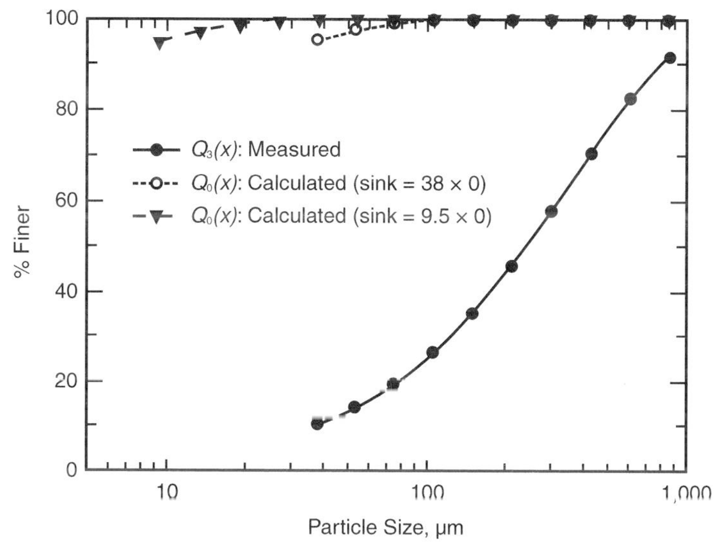

Note: The percent finer is plotted on a linear scale.

FIGURE 2.14 Transformation of volume distribution to number distribution based on (1) assumption of arithmetic mean size (19 μm) in the sink interval (<38 μm) and (2) extrapolation of the distribution function $Q_3(x)$ to include additional size intervals down to 9.5 μm

TABLE 2.8 Example of transformation from volume distribution to number distribution

I. Size Interval (i)	II. Upper Boundary (x_i), µm	III. Mean Class Size ($\bar{x}_i$), µm	IV. $Q_3(x_i)$ (data)	V. $(q_3)_i$*	VI. $(q_3)_i/\bar{x}_i^3$	VII. $(q_0)_i$ Sink = 38 µm × 0†	VIII. $Q_0(x_i)$ Sink = 38 µm × 0‡	IX. $(q_0)_i$ Sink = 9.5 µm × 0†	X. $Q_0(x_i)$ Sink = 9.5 µm × 0‡
1	1,000	925	1.000	0.088	1.11×10^{-10}	7.05×10^{-6}	1.000	4.18×10^{-7}	1.000
2	850	725	0.912	0.092	2.41×10^{-10}	1.53×10^{-5}	1.000	9.07×10^{-7}	1.000
3	600	512.5	0.82	0.117	8.69×10^{-10}	5.51×10^{-5}	1.000	3.27×10^{-6}	1.000
4	425	362.5	0.703	0.127	2.67×10^{-9}	1.69×10^{-4}	1.000	1.00×10^{-5}	1.000
5	300	256	0.576	0.122	7.27×10^{-9}	4.61×10^{-4}	1.000	2.73×10^{-5}	1.000
6	212	181	0.454	0.105	1.77×10^{-8}	1.12×10^{-3}	1.000	6.65×10^{-5}	1.000
7	150	128	0.349	0.088	4.2×10^{-8}	2.65×10^{-3}	1.000	1.58×10^{-4}	1.000
8	106	90.5	0.261	0.068	9.17×10^{-8}	5.81×10^{-3}	1.000	3.45×10^{-4}	1.000
9	75	64	0.193	0.052	1.98×10^{-7}	1.25×10^{-2}	0.997	7.45×10^{-4}	*1.000*
10	53	45.5	0.141	0.038	4.03×10^{-7}	2.55×10^{-2}	0.977	1.52×10^{-3}	*0.999*
11	38	19	0.103	0.103	1.50×10^{-5};	9.52×10^{-1}	0.952	*3.19×10^{-3}*	*0.997*
		32.45		*0.029*	*8.49×10^{-7}*				
12	26.9	*22.95*	*0.074*	*0.021*	*1.74×10^{-6}*			*6.53×10^{-3}*	*0.994*
13	19	*16.2*	*0.053*	*0.015*	*3.53×10^{-6}*			*1.33×10^{-2}*	*0.988*
14	13.4	*11.45*	*0.038*	*0.011*	*7.33×10^{-6}*			*2.75×10^{-2}*	*0.974*
15	9.5	*4.75*	*0.027*	*0.027*	*2.52×10^{-4}*			*9.47×10^{-1}*	*0.947*
				$\Sigma = 1.00$	$\Sigma = M_{-3,3}$ (1.58×10^{-5}; *2.66×10^{-4}*)	$\Sigma = 1.00$		$\Sigma = 1.00$	

Note: Values in italics are based on extrapolation from 38 to 9.5 µm.

* Column V calculated from data values given in column IV by using this formula: $(q_3)_i = Q_3(x_i) - Q_3(x_{i+1})$; e.g., 0.088 = 1.000 – 0.912.

† Columns VII and IX given by column VI divided by the appropriate value of the moment $M_{-3,3}$; i.e., 1.58×10^{-5} for column VII; *2.66×10^{-4}* for column IX.

‡ Columns VIII and X obtained by accumulating values from columns VII and IX; e.g., (column VIII) $0.977 = 9.52 \times 10^{-1} + 2.55 \times 10^{-2}$.

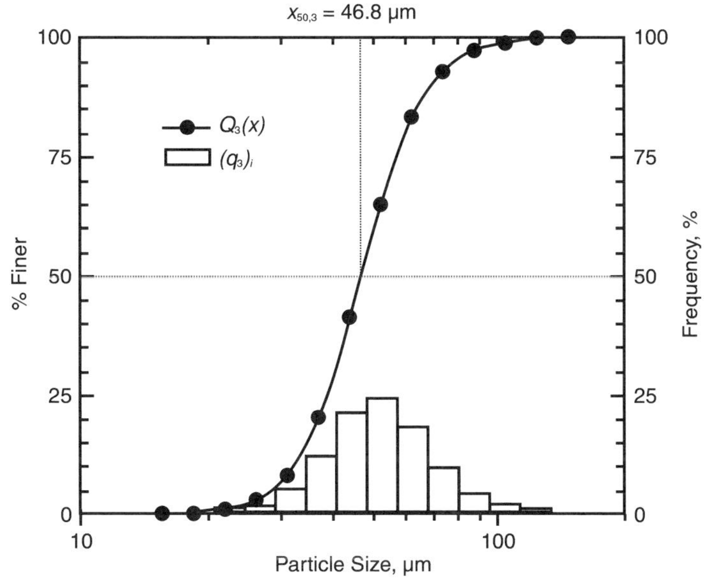

FIGURE 2.15 Subsieve (light scattering) size data for 325 × 400 U.S. mesh quartz particles

the two smaller dimensions of an irregular particle). These differences have been discussed at length by Austin and Shah (1983) and Austin et al. (1988), who have shown that the use of a simple factor—$x_{50,\text{subsieve}}/\bar{x}$, where $\bar{x}$ is the geometric mean size of the sieve interval—is sufficient to connect the sieve and subsieve parts of the overall size distribution. An example of the application of this factor is shown in Figure 2.16 and Table 2.9.

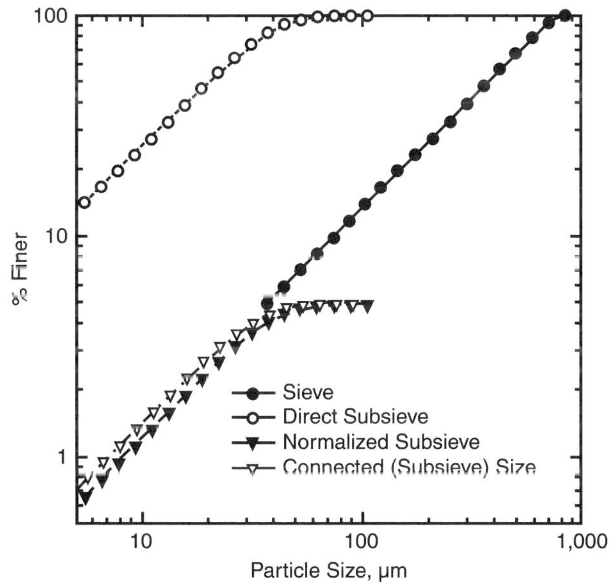

FIGURE 2.16 Example of combination of sieve and subsieve data to obtain an overall size distribution

TABLE 2.9 Example of procedure for combining sieve and subsieve data

I. Size μm	II. Percent Finer (Sieve)	III. Percent Finer (Subsieve)	IV. Normalized Subsieve	V. Corrected (Subsieve) Size
837	100			
704	93.87			
592	78.93			
497.8	66.37			
418.6	55.81			
352	46.93			
296	39.47			
248.9	33.19			
209.3	27.91			
176	23.47			
148	19.73			
124.5	16.60			
104.7	13.96	100.00	4.93	90.21
88	11.73	99.80	4.92	75.82
74	9.87	99.29	4.90	63.76
62.23	8.30	97.93	4.83	53.62
52.33	6.98	95.01	4.69	45.09
44	5.87	89.88	4.43	37.91
37	4.93	82.43	4.07	31.88
31.11		73.29	3.62	26.80
26.16		63.02	3.11	22.54
22		54.42	2.68	18.96
18.5		46.04	2.27	15.94
15.56		38.81	1.91	13.41
13.08		32.69	1.61	11.27
11		27.53	1.36	9.48
9.25		23.19	1.14	7.97
7.778		19.54	0.96	6.70
6.541		16.46	0.81	5.64
5.5		13.86	0.68	4.74
4.625		11.63	0.57	3.98
3.889		9.71	0.48	3.35
3.27		8.04	0.40	2.82
2.75		6.60	0.33	2.37
1.945		5.39	0.27	1.68
1.635		4.40	0.22	1.41
1.375		3.56	0.18	1.18
1.156		2.79	0.14	1.00

The general procedure can be outlined as follows:

1. Conduct a standard sieve size analysis on the sample to be evaluated. The wet-dry procedure indicated earlier (see the section entitled "Sieving") is recommended, especially if there is a substantial quantity of subsieve material.
2. Analyze the subsieve material by the selected method (sedimentation, light scattering, etc.).
3. Prepare a narrow sieve fraction (e.g., 325 × 400 mesh) by using the procedure outlined above, and analyze by the same subsieve sizing method.
4. Determine the subsieve correction factor, $x_{50}/\bar{x}$ (= 1.16 for the example given in Figure 2.16 and Table 2.9).
5. Normalize the cumulative subsieve data from step 2 by using the fraction passing the finest sieve used in step 1; i.e., multiply $(Q_3(x))_{\text{subsieve}}$ by $(Q_3(x_{\min}))_{\text{sieve}}$.
6. Adjust the subsieve size scale according to the factor obtained in step 4; i.e., $x_{\text{corrected subsieve}} = x_{\text{subsieve}}/(x_{50}/\bar{x})$.

The steps involved in combining the sieve and subsieve data are given in Table 2.9 and are illustrated in Figure 2.16. The plateau regions in the normalized and corrected subsieve data simply reflect the fact that these values represent the subsieve material only. The overall distribution is estimated from a smooth curve connecting both sets of the data.

For distributions for which both parts (sieve and subsieve) can be plotted as parallel straight lines, a simple alternative procedure is to adjust the subsieve size scale so as to align the normalized subsieve data to the sieving distribution. The relative adjustment required provides an estimate of the correction factor. Generally, however, the direct calibration procedure outlined in steps 1 through 6 is preferred.

REFERENCES

Adamson, A.W. 1997. *Physical Chemistry of Surfaces*. 6th ed. New York: John Wiley & Sons.

Allen, T. 1997. *Particle Size Measurement*. 5th ed. London: Chapman and Hall.

ASTM (American Society for Testing and Materials). 1984. *Standard Test Method for Determining the Washability Characteristics of Coal*. D4371-84. Philadelphia, Pa.: ASTM.

——. 1992. *Standard Method of Test for Fineness of Portland Cement by Air Permeability Apparatus*. C204-92. Philadelphia, Pa.: ASTM.

Austin, L.G., and I. Shah. 1983. A Method for Inter-conversion of Microtrac and Sieve Size Distributions. *Powder Technol.,* 35:271–278.

Austin, L.G., O. Trass, T.F. Dumm, and V.R. Koka. 1988. A Rapid Method for Determination of Changes in Shape of Comminuted Particles Using a Laser Diffractometer. *Particle Charact.,* 5:13–15.

Bohren, C.F., and D.R. Huffman. 1983. *Absorption and Scattering of Light by Small Particles.* New York: Wiley Interscience.

British Standards Institution. 1969. *British Standard BS 4359: Part 1.* London: British Standards Institution.

Brunnauer, S., P.H. Emmett, and E. Teller. 1938. Adsorption of Gases in Multi-molecular Layers. *J. Am. Chem. Soc.,* 60:309–319.

Carman, P.C. 1956. *Flow of Gases Through Porous Media.* London: Butterworths.

Cho, H., M.A. Waters, and R. Hogg. 1996. Investigation of the Grind Limit in Stirred-media Milling. *Int. J. Miner. Process.,* 44/45:607–615.

Chung, H.S., and R. Hogg. 1985. The Effect of Brownian Motion on Particle Size Analysis by Sedimentation. *Powder Technol.,* 41:211–216.

Concha, F., and E.R. Almendra. 1979. Settling Velocities of Particulate Systems, I: Settling Velocities of Individual Spherical Particles. *Int. J. Miner. Process.,* 5:349–367.

Dumm, T.F. 1986. *An Evaluation of Techniques for Characterizing Respirable Coal Dust*. Master's thesis, Pennsylvania State University, University Park, Pa.

——. 1989. *Characterization of Size/Composition and Shape of Fine Coal and Mineral Particles*. Ph.D. diss., Pennsylvania State University, University Park, Pa.

Dumm, T.F., and R. Hogg. 1988. Washability of Ultrafine Coal. *Min. and Metall. Proc.,* 5:25–32.

Dumm, T.F., R. Hogg, and L.G. Austin. 1991. Limitations of SEM/EDS Techniques for Particle-by-Particle Analysis in Respirable Coal Dust. In *Proceedings, 3rd Symposium on Respirable Dust in the Mineral Industries.* Edited by R.L. Frantz and R.V. Ramani. Littleton, Colo.: SME.

Gooden, E.L., and C.M. Smith. 1940. Measuring the Average Particle Diameter of Powders. *Ind. Eng. Chem. (Anal. Ed.),* 12:479–482.

Gy, P. 1982. *Sampling of Particulate Materials, Theory and Practice*. 2nd revised ed. New York: Elsevier.

Heywood, H. 1963. Evaluation of Powders. *Pharm. J.,* 191:291–293.

Hogg, R. In press. Breakage Mechanisms and Mill Performance in Ultrafine Grinding. *Powder Technol.*

Karp, S., S. Lowell, and A. Mustacciuolo. 1972. Continuous Flow Measurement of Desorption Isotherms. *Analytical Chemistry,* 44:2395–2397.

Kaya, E., S. Kumar, and R. Hogg. 1996. Particle Shape Characterization Using an Image Analysis Technique. In *Changing Scopes in Mineral Processing.* Edited by M. Kemal, V. Arslan, A. Akar, and M. Canbazoglu. Rotterdam: A.A. Balkema.

Kerker, M. 1969. *The Scattering of Light and Other Electromagnetic Radiation.* New York: Academic Press.

Kozeny, J. 1927. Über Kapillare Leitung des Wasser im Boden. *Sitzungsber. Akad. Wiss. Wien,* 136:271–306.

Kumar, S. 1998. *Characterization of Particle Shape*. Master's thesis, Pennsylvania State University, University Park, Pa.

Leschonski, K. 1984. Representation and Evaluation of Particle Size Analysis Data. *Particle Charact.,* 1:89–95.

Lide, D.R., ed. 1998/9. *CRC Handbook of Chemistry and Physics.* 79th ed. Boca Raton, Fla.: CRC Press.

Lippens, B.C., B.G. Linsen, and J.H. de Boer. 1964. Studies on Pore Systems in Catalysts. I. The Adsorption of Nitrogen, Apparatus and Calculation. *J. Catalysis,* 3:32–37.

Mie, G. 1908. Contributions to the Optics of Turbid Media, Especially Colloidal Metal Solutions. *Ann. Physik,* 25:377–445.

Nelsen, F.M., and F.T. Eggersten. 1958. Determination of Surface Area: Adsorption Measurement by a Continuous Flow Method. *Analytical Chemistry,* 30:1387–1390.

Parfitt, G.D. 1973. *Dispersion of Powders in Liquids.* 2nd ed. New York: John Wiley & Sons.

Pretorius, S.T., and W.G.B. Mandersloot. 1967. The Leschonski Modification of the Sartorius Sedimentation Balance for Particle-size Analysis. *Powder Technol.,* 1:23–27.

Rattanakawin, C., and R. Hogg. 1998. Aggregate Size Distributions in Flocculation. Paper presented at the 72nd American Chemical Society Colloid and Surface Science Symposium, The Pennsylvania State University, June 21–24.

Rumpf, H., and K.F. Ebert. 1964. Darstellung von Korngroβenverteilungen und Berechnung der Spezifischen Oberflache. *Chem. Ing. Tech.,* 36:523–537.

Schönert, K. 1986. Advances in the Physical Fundamentals of Comminution. In *Advances in Mineral Processing.* Edited by P. Somasundaran. Littleton, Colo.: SME.

Sokaski, M., P.S. Jacobson, and M.R. Geer. 1963. *Performance of Baum Jigs in Treating Rocky Mountain Coals*. USBM RI 6306. Seattle, Wash.: U.S. Bureau of Mines.

Strutt, J.W. (Lord Rayleigh). 1871. On the Light from the Sky, Its Polarization and Colour. *Phil. Mag.,* 41:107–120.

van der Hulst, H.C. 1957. *Scattering of Light by Small Particles.* New York: John Wiley & Sons.

CHAPTER 3

Size Reduction and Liberation

John A. Herbst, Yi Chang Lo, and Brian Flintoff

INTRODUCTION

What Is Comminution and Why Is It Important?

Comminution is a process whereby particulate materials are reduced by blasting, crushing, and grinding to the product sizes required for downstream processing or end use. In mineral processing, comminution operations are used to ensure that valuable constituents are physically liberated from waste constituents before physical or chemical separations are attempted.

The United States currently uses about 15 billion kWh per year for blasting, crushing, and grinding minerals of all types. This energy constitutes about 1% of the total electric power produced in the United States (the corresponding percentage on a worldwide basis is estimated to be about 2%). Energy used for comminution in the year 2000 was distributed according to mineral commodity group approximately as shown in Table 3.1. These commodities account for about 60% of the total energy consumed, with the top two alone, copper ore and iron ore, consuming almost half of the total.

In addition to the energy that is directly used by devices, an additional 1.8 billion kWh contained in size-reduction consumables. This is the energy required to produce the approximately 500,000 tons of steel that are consumed in media, liners, and other wear parts in current comminution devices. The importance of this contribution to the total energy requirement for a given commodity is determined by the abrasiveness of the material being comminuted, the corrosiveness of the environment inside the devices, and the required fineness of product.

One of the greatest challenges faced by mineral processing professionals today is the efficient design and operation of industrial comminution circuits. This is the case because the energy intensive comminution operations use on the order of 50% of a mineral processing plant's operating costs and often carry an even larger percentage of the capital cost price tag for a plant. These expenses, combined with the fact that the average energy efficiency of current comminution devices is something less than 5%, clearly point to the desirability of using these devices wisely and creatively to increase the profitability of minerals operations.

TABLE 3.1 Highest ranking comminution energy consumers by mineral type

Rank	Commodity Energy	10^9 kWh
1	Copper ore	3.6
2	Iron ore	3.3
3	Phosphate ores	1.3
4	Clay	0.5
5	Titanium ores	0.3

FIGURE 3.1 Photo of state-of-the-art semiautogenous grinding mill (left) and high-fidelity simulation of charge motion and breaking action inside the mill (right)

A state-of-the-art comminution device, such as the semiautogenous grinding mill shown in Figure 3.1, uses impact energy derived from tumbling balls and large rocks to reduce a copper ore from a feed size of a few hundred millimeters down to a product size of a few tens of millimeters at a rate of 5,000 metric tons/hour, while consuming 16 MW of electric power and using on the order of 5 tons per hour (tph) of balls and liners. Enough is known about such a device to do realistic simulations of performance (Figure 3.1, lower right), which can be used for optimal equipment design and operation.

In this chapter, comminution technology is described with respect to both theory and practice. The basic principles for size reduction processes begin with fundamentals of breakage. Detailed consideration of the phenomena involved in the fracture of individual particles forms a basis for evaluating energy consumption in size reduction. In addition to the fundamental physics of particle fracture, the theory of comminution includes the quantitative description of the rate of breakage of an assembly of particles. Mathematical modeling at both the macroscale and at the microscale is shown to be an important tool for understanding and improving comminution systems.

The most significant aspects to be gleaned from reviewing comminution theory are (1) breakage of ore particles requires forces and energy that strongly depend on size and composition; (2) breakage of small particles requires extremely large forces, (3) accurate prediction of the size distribution of broken fragments for a given energy input has required the development of detailed models, and (4) increased classification efficiency contributes significantly to improved energy economy in comminution technology.

This chapter describes traditional and emerging size reduction processes for mineral processing. Size reduction practice is discussed for different types of comminution equipment. For each type of comminution device, design criteria and operating conditions are specified, together with the suitability of the device for a particular application.

Almost without exception, the efficiency of size reduction processes depends strongly on grinding circuit design and embedded automatic controls to handle disturbances. For this reason, modeling, design, and control methods as tools for optimization are also discussed in this chapter.

FUNDAMENTALS OF PARTICLE BREAKAGE

How Do Particles Break and What Do Their Progeny Look Like?

Size reduction devices in use today break particles by applying various types of force to assemblages of particles. The results of each breakage event cannot be predicted as, for example, a chemical reaction in which reactants go to products in exact stoichiometric amounts. Because many such events occur, it is not surprising that it is difficult to forecast the behavior of commercial devices. Currently, to make such predictions we must use either strictly empirical relationships, such as Bond's energy-size equation (Bond 1952), or phenomenological ones, such as population balance or multiphysics models (Herbst 2000). The most accurate of these get their form from theory and account for particle size distributions in a bookkeeping fashion.

However, even this approach fails to give all the "hows" of particle breakage, which are necessary to make extrapolations for a process involving so many events. Nonetheless, a great stride forward occurs when we understand the fundamentals of breakage as they pertain to crushers and grinding mills. Important work is taking place in the development of first principles, multiphysics models that use discrete element and finite element modeling techniques. In this section, we will examine single and multiple breakage events so that the reader can develop a semiquantitative understanding of this complex process. In the same way that computer simulation is widely used in comminution practice, we will use simulations extensively here to illustrate principles.

Single Particle Breakage

Even the results of a single particle breakage event are not totally predictable because of the extremely large number of variables that affect the outcome. Therefore, those attempting to understand this process have directed their attention to the breakage of single particles under specific sets of conditions. These studies have focused on the strength of a particle, the energy consumed in a breakage event, and the size distribution of progeny (or "daughter fragments") produced in such an event. The behavior observed in the laboratory often gives important insights into the complex behavior of commercial size reduction devices.

Particle Strength and Breakage Energy Requirement. The strength of a particle is the applied stress at the first breaking point. Careful photographic measurements plus an ultrafast load cell measurement such as that shown in Figure 3.2 allow strength and breaking energy to be determined. In this case breaking strength is the force per unit area of a particle cross section at the point of first fracture, whereas breaking energy is the work that must be done on the particle to get it to fracture. It is important to realize that the actual strengths of materials are much lower than their theoretical strengths. The theoretical strength of a material can be estimated from its modulus of elasticity, Y, by

$$\sigma(\text{Actual}) \approx \left(\frac{1}{20}\right)Y \text{ to } \left(\frac{1}{10}\right)Y \qquad \text{(Eq. 3.1)}$$

The actual and the theoretical strengths for selected materials are compared in Table 3.2.

The underlying assumption of the theoretical strength is that the material is homogeneous. However, flaws are always present in normal bulk materials as lattice faults, grain boundaries, and microcracks. The latter are particularly important in material fragmented by blasting before arriving at a mineral processing operation. Stress concentrations at these flaws are much greater than in other portions of the body. Owing to the higher stress levels, fracture can initiate at these points. Thus, the actual strength is lower because of the presence of these flaws. Fracture of materials occurs with the initiation and propagation of cracks. This process is illustrated in Figure 3.3.

The energy consumed in the fracture or breakage of materials goes to the extension of these cracks. Some of the energy consumed by cracks is caused by the creation of a new surface, γ (specific surface free energy), and to the plastic deformation of material near the crack tip. Both of these terms

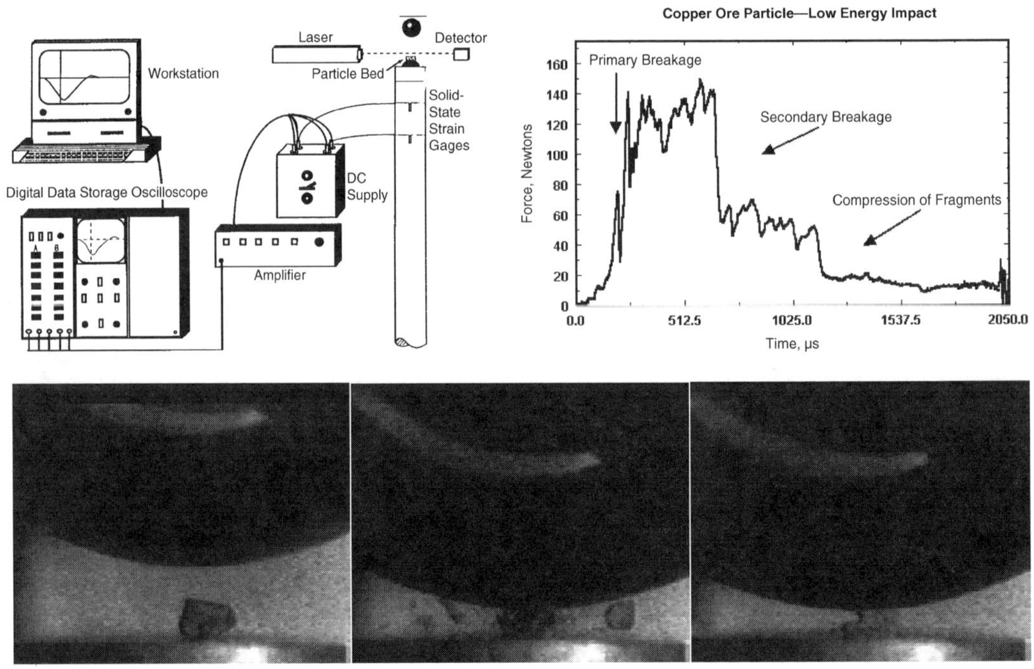

Source: Höfler, A. 1990.

FIGURE 3.2 Measuring the strength and breaking energy of a copper ore particle using an ultrafast load cell

TABLE 3.2 Actual and theoretical strengths of some materials

Material	σ (Actual) kg_F/cm^2	σ (Theoretical) kg_F/cm^2
NaCl	50–200	$2 \cdot 10^4 - 4 \cdot 10^4$
Glass	500–2,000	$3.5 \cdot 10^4 - 7 \cdot 10^4$
Steel	3,000–8,000	$10^5 - 2 \cdot 10^5$

contribute to the specific crack surface energy, β, which is the energy required per unit of crack surface produced. The specific crack surface energy, β, and the specific surface free energy, γ, for various types of materials are given in Table 3.3.

Typically β is more than 1,000 times the value of γ. This results from the large amount of energy that goes into plastic deformation of material at the tip of a crack as it propagates through a solid. If a crack is to propagate, two conditions must be satisfied: the force condition and the energy condition. The force condition requires that the tensile stress exceed the molecular strength at the tip of a crack. The stress at the tip is a maximum, with σ_{max} given by

$$\sigma_{max} = \sigma_{\infty}(1 + 2\sqrt{a/\rho}) \quad \textbf{(Eq. 3.2)}$$

where a is the length of a crack and ρ is the crack radius at the tip.

The maximum stress at the tip of the crack greatly exceeds the stresses placed on the body. Note, for example, for a crack of length where a = 10 μm and crack radius is $\rho \cong 10$Å, $\sigma_{max} \cong 200\ \sigma_{\infty}$. The leverage on the breaking force at the crack tip is an impressive 200 times!

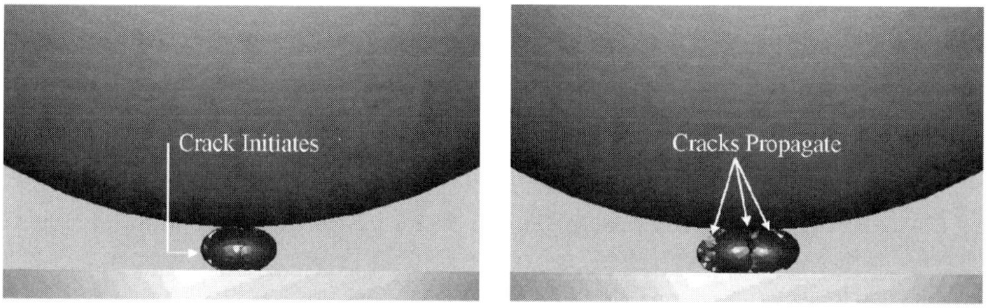

FIGURE 3.3 Three-dimensional particle breakage simulation of a ball impacting a particle showing crack initiation and propagation

TABLE 3.3 Specific crack surface energy, β, and specific surface free energy, γ

Material	β ergs/cm^2	γ ergs/cm^2
NaCl (or other ionic solids)	10^4	300
Glass	10^4	1,000
Plastics	10^5	20–200
Metals	10^6–10^8	500–3,000

Once a crack forms, an energy balance requires that the energy to propagate the crack be available from the stress field surrounding the crack. An analysis for the case in which the elastic stress field is the only energy source was first provided by Irwin (1961).

The energy loss of the stress field owing to a differential crack extension of amount $2a$ is the crack extension energy, G, and is given by

$$G = \frac{-1}{2}\frac{\delta u}{\delta a} \quad \text{(Eq. 3.3)}$$

The differential energy consumed by the crack, δu, is $\beta 4\delta a$, where β is the specific fracture surface energy (energy consumption of the crack per unit area).

The energy condition in this case requires that the crack extension energy must exceed the specific fracture energy, which requires that

$$\beta < \frac{1}{4}\frac{\delta u}{\delta a} = \frac{1}{2}G \quad \text{(Eq. 3.4)}$$

The importance of this result is that only half or less of the energy released from the stress field during the propagation of a crack is available for doing work, such as the creation of a new surface or plastic deformation of material at the crack tip.

From elastic theory, the geometry of specimens can be related to G by

$$G = \frac{\pi\sigma_\infty^2 a}{E}(1 - \nu^2) \quad \text{(Eq. 3.5)}$$

where σ_∞ is the external stress and ν is Poisson's ratio. If the applied stress σ_∞ is constant during the propagation of the crack, G increases as a increases. Thus, the energy released from the stress field is always increasing, and if β is constant or does not increase as fast as G, the energy condition $\beta < G/2$ is always satisfied, and a crack will propagate once it is initiated.

The initiation of crack motion is the critical process in fracture physics; Griffith (1920) analyzed the crack initiation process. As an acknowledgement of his contributions, the microcracks that act as sites for crack motion initiation are often called Griffith cracks. In this analysis, he considered only the specific surface free energy, γ, and he assumed only elastic deformation behavior. The energy from the stress field required to increase the crack length, $2a$, is

$$\delta u = 4\gamma\delta a \tag{Eq. 3.6}$$

Rumpf (1961) provided a more complete energy balance for a crack than did Griffith. The sources of energy are

1. External forces
2. The stress field caused by the external forces
3. Residual internal stresses caused by structural flaws, thermal treatment, etc.
4. Thermal energy of constituents
5. Chemical reactions or adsorption at the crack tip or fracture surfaces

The consumption of energy is caused by

1. Creation of a new surface
2. Plastic deformation around the crack tip
3. Alteration of material structure in the vicinity of the crack
4. Electrical phenomena resulting from charge separation or discharge (emission)
5. Endothermic chemical reactions or adsorption at the crack tip or on fracture surfaces
6. Kinetic energy of elastic waves

For all these reasons, the total energy required is often more than 100 times as great as the energy required to produce a new surface under ideal conditions.

These considerations apply to the breakage of any material under any loading condition. Some of the variables of loading events that have the largest effects are manner of loading, particle size, particle composition, and environment.

Although loading can be accomplished in several ways, the most basic are two-surface loading and one-surface loading. Two- and one-surface loading events are simulated for spheres in Figure 3.4. The breaking strength of a particle is less when it is subjected to two or more forces than when it is stressed by a single force. In general, the probability of fracture increases (corresponding to a reduction in breaking strength) as the number of contacting forces increases, as shown by Schönert (1980; Figure 3.5).

The term "compression loading" is generally used for the two-surface loading of particles. In compression loading, the stresses nearest the contact area are the most important in causing cracks (refer again to Figure 3.4). This type of crack pattern occurs when the contact time of the force is greater than the transit time of an elastic wave through the particle. For most commercial comminution devices, the contact time is much longer than the transit time, and this situation is termed "slow compression loading." The term "impact loading" is used when particles are impacted against a surface. At very high impact velocities, the dynamic effect becomes important, as the impact time is smaller than the transit time of elastic waves sweeping across the particle. In comminution devices, these high velocities are rarely encountered so that the stresses near the contact surfaces are most important, and, as such, impact loading corresponds to one-surface fast compression loading. The velocity of impact in impact crushers is between 20 and 200 m/s, and in tumbling mills, it is up to 20 m/s. With an impact velocity of 200 m/s, the impact time is 10 times that of the transit time.

The probability of fracture when a particle is loaded in a particular manner is very sensitive to particle size. The probability of fracture for several particle sizes is plotted against the energy input per unit mass in Figure 3.6. The probabilities shown are approximately normally distributed with respect to

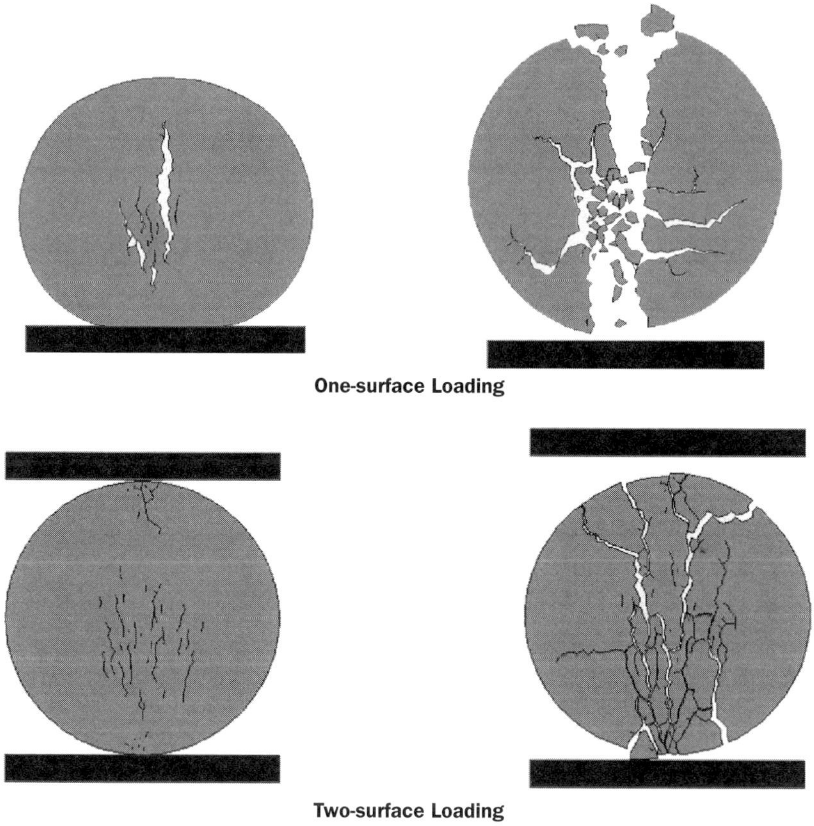

Source: Potapov and Campbell 1996.

FIGURE 3.4 Snapshots of simulations of one-surface and two-surface loading events upon crack initiation (left) and final fragmentation state (right)

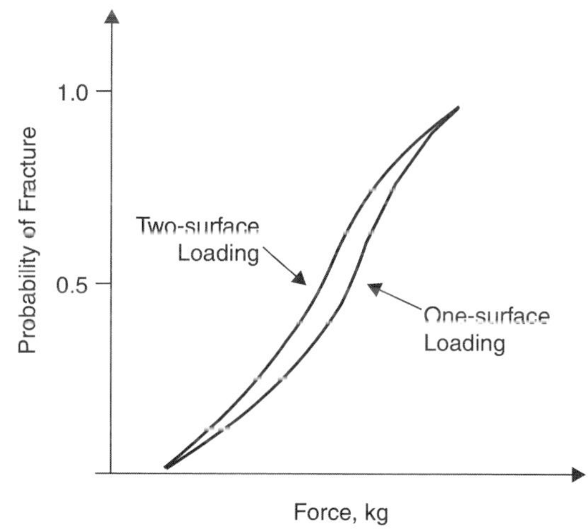

Source: Schönert 1980.

FIGURE 3.5 The breaking strength for two-surface and one-surface loading

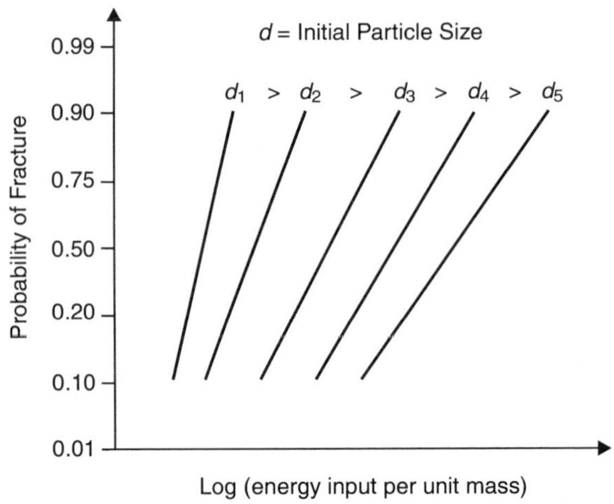

FIGURE 3.6 The breaking strength for different particle sizes

the log of the energy input for 0.1 < probability < 0.9. Figure 3.6 also shows that the range of breaking strengths for particles of the same initial size increases as the initial particle size decreases. This is because of the depletion of flaws with decreasing particle size. At larger particle sizes, most particles will have at least one, and perhaps many, major flaws of the same magnitude, so the breaking strengths of almost all particles are equal. However, in the smaller particles the major flaws will not be so evenly distributed so that the spread in stress levels needed to cause cracks is wider, which results in a wide distribution of strengths.

For irregularly shaped particles, two other modes of breakage, chipping and abrasion, can play an important role in determining their comminution behavior. Figure 3.7 illustrates that in a chipping event, sharp edges and corners break off, leaving the remainder of the parent particle intact. Abrasion, on the other hand, results from the simultaneous application of shear and normal forces to produce very fine debris. This leaves behind a parent whose shape is basically unaltered.

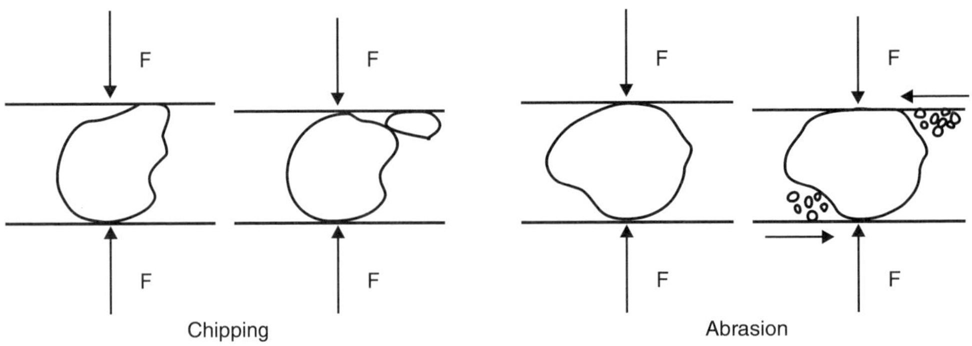

FIGURE 3.7 Chipping and abrasion modes of breakage for irregular particle shapes

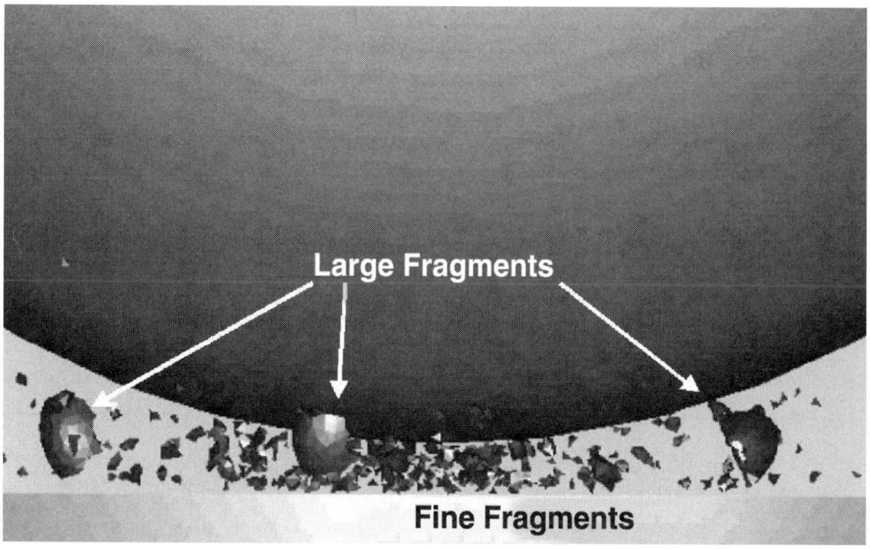

FIGURE 3.8 Progeny created in single particle breakage event simulation started in Figure 3.3

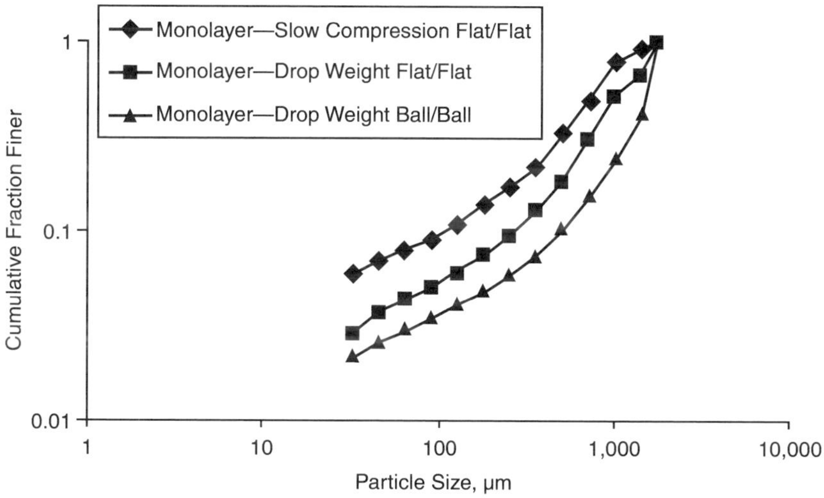

Source: Cho 1987.

FIGURE 3.9 Progeny created in a monolayer of particles exposed to 1.7 kWh/t by slow compression and drop weight (flat/flat and ball/ball)

Size Distributions of Product (Progeny) Fragments. When a particle breaks, a few large fragments are created along with a suite of fine fragments that are much smaller than the major fragments. This point is illustrated in the next stage of the simulation (begun in Figure 3.3) shown in Figure 3.8.

The size distributions of the progeny particles from the same size and composition of parent exhibit a strong dependence on the manner and intensity of loading. This is illustrated in Figure 3.9 where we can see that even though energy input is constant, changes in the manner of loading (rate and geometry) have a significant effect on the efficiency of energy utilization for fragmentation. Slow compression flat/flat is most efficient, while drop weight cases are less efficient because of a higher

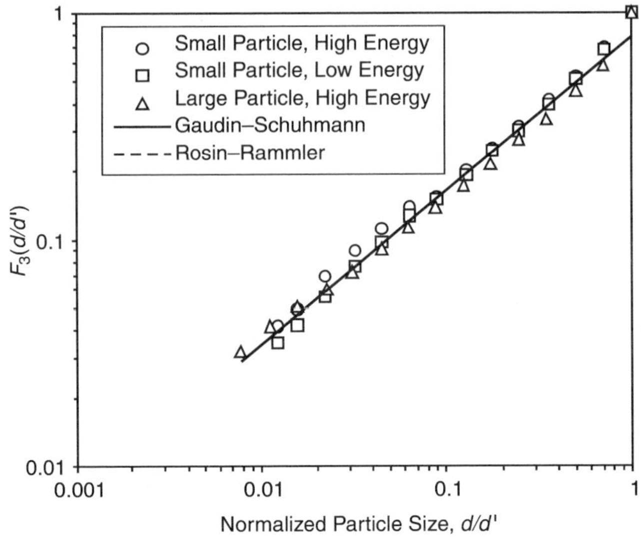

FIGURE 3.10 Self-similar distributions of progeny fragments for quartz testing in ball/ball loading

rate of loading; flat/flat, which confines particle motion is more efficient than ball/ball, which allows significant particle motion.

For compressive loading the size distributions that result from different energy inputs exhibit a high degree of similarity. Colinear plots of F_3 (d/d') (cumulative weight fraction of progeny smaller than relative size d/d') plotted against d/d' are shown in Figure 3.10.

The remarkable aspect of this phenomenon is that this co-linearity holds even for different initial particle sizes. Thus, with compressive loading that occurs in tumbling mills (ball mills, pebble mills, semiautogenous and autogenous mills), the resulting size distributions for a multitude of breakage events also normalize; that is, they become self-similar.

Because of this property, empirical size distributions have been developed (Herbst and Sepulveda 1985). These fragment size distributions include the Gaudin-Schuhmann distribution

$$F_3(d/d') = \left(\frac{d}{d_{\mathrm{max}}}\right)^{\omega} \qquad \textbf{(Eq. 3.7)}$$

where d_{max} = constant × d', and the Rosin–Rammler distribution

$$F_3(d/d') = 1 - \exp\left(-k\left(\frac{d}{d'}\right)^{\omega}\right) \qquad \textbf{(Eq. 3.8)}$$

A fit of these empirical models to breakage data is shown in Figure 3.6. They describe some cases quite well but, in general, they have deficiencies that make them unacceptable to describe size distribution for all cases.

For impact loading, the progeny size distributions typically cannot be normalized. For this reason the above-size distribution relations are not capable of describing the results of most impact-loading fracture situations. Likewise, chipping and abrasion modes of breakage do not produce self-similar distributions. Examples of the progeny size distributions for these cases are shown in Figure 3.11.

Breakage of Multiphase Particles. Virtually all the ores treated in mineral processing operations are composed of multiphase particles. In fact, the release of one phase from another or "liberation" is a principal goal of virtually all related comminution operations. Figure 3.12 shows a photomicrograph of fragments from a two-component locked particle. Note that in some instances a single-phase particle–

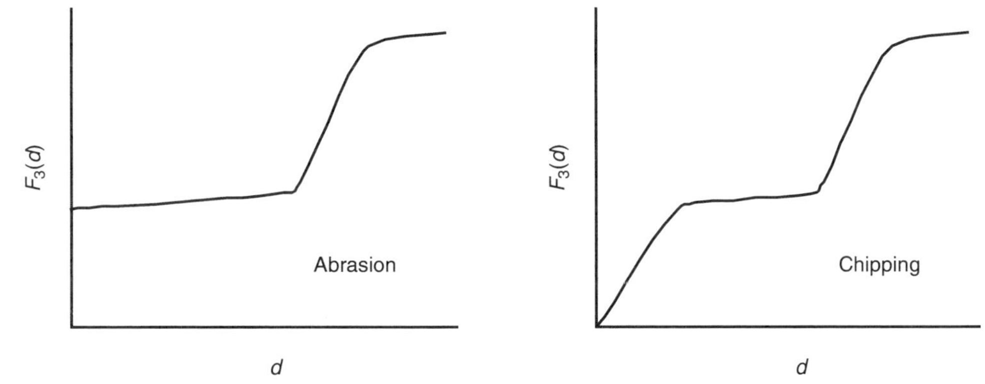

FIGURE 3.11 Progeny size distribution for abrasion and chipping events

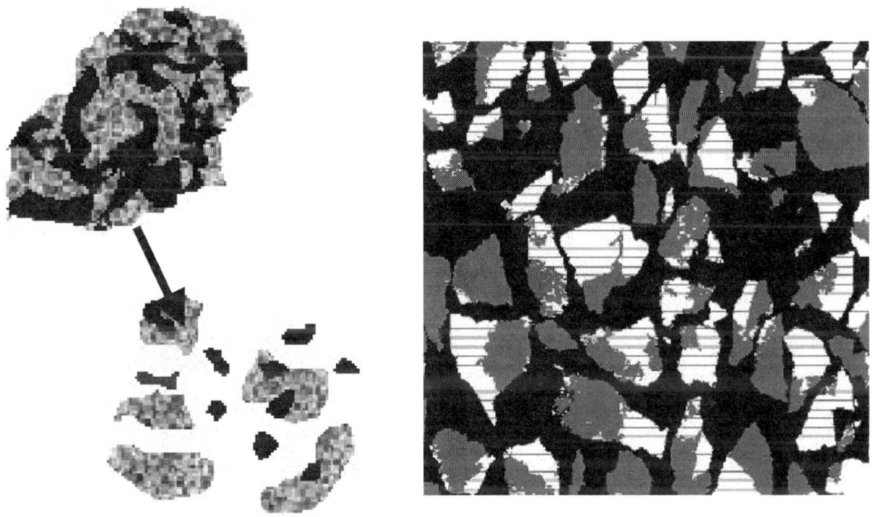

FIGURE 3.12 Liberation event (left) and photomicrograph of fragments (right)

which represents completely liberated fragments of valuable or gangue components—occurs. In most cases, however, the fragments remain locked (with portions of each component existing in the same particle).

Multiphase particles have various degrees of complexity depending on the type and extent of mineral intergrowth (ore texture) that is displayed. Their breakage behavior (strength, fracture energy, progeny size, and type) depends on the mechanical properties of the individual phases and the texture. From a liberation point of view, the most important issue is what happens to a propagating crack when it reaches a phase boundary. If the crack continues to propagate unabated across the boundary into the adjacent phase, a condition called "random liberation" occurs. If each phase has similar mechanical properties, the crack patterns (and therefore progeny size distributions) will be the same, independent of the texture. By its random nature, this condition requires very fine sizes to achieve a high degree of liberation; that is, release of valuable and gangue phases. If, on the other hand, a crack upon reaching a phase boundary (with different mechanical properties) changes direction and propagates along the

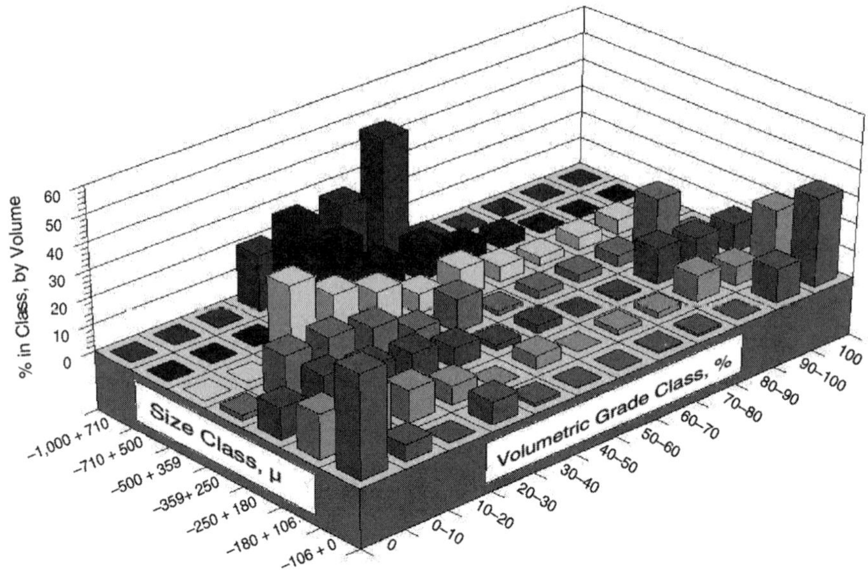

Source: Schneider 1995.

FIGURE 3.13 Grade distribution (frequency by size and composition) for broken fragments

grain boundary, the condition is called "selective liberation." In this case less size reduction is required to achieve a certain level of liberation. At the extreme of selective liberation, individual mineral grains are carved out of a parent particle as a result of crack propagation along the interphase boundaries. The liberation process can be represented by grade distribution plots on a size-by-size basis, as shown in Figure 3.13.

Note that each progeny size has a distribution of grades. The degree of liberation (amount of free component 1 and free component 2 represented by the corresponding peaks at or near 0% and 100%) increases as the progeny size decreases. The overall "degree of liberation" resulting from particle breakage can be determined experimentally in a variety of ways. The most appropriate method depends on the physical properties of the components. For example, if the optical properties of the components are significantly different (as in Figure 3.8), automated optical image analysis with stereological transformation is a good method. If density differences are large, gravimetric methods applied to individual size fractions can be very successful.

Multiparticle Breakage

Commercial comminution devices obviously do not break particles one at a time. To process ores at hundreds or even thousands of tons per hour, these devices must work on large assemblies of particles. The semiautogenous mill shown in Figure 3.1 contains about 180 tons of ore. Unfortunately, the resulting multiparticle breakage events occurring in these cases are both more complex and less efficient than single-particle events. Compare Figure 3.14 with Figures 3.3 and 3.8.

Types of Particle Interaction. The complexity and additional inefficiency of multiparticle events arise from the fact that particles interact during breakage. It is generally believed that the amount of breakage in an assembly of particles depends on how energy is dissipated (useful and wasted) in the assembly and its distribution into the various particle types. Figure 3.15 shows that the efficiency of breakage (relative to single-particle slow-compression-loading events) decreases as the number of (equal-size) particle layers in a bed of particles undergoing compression breakage increases.

FIGURE 3.14 Simulation of a multiparticle breakage event

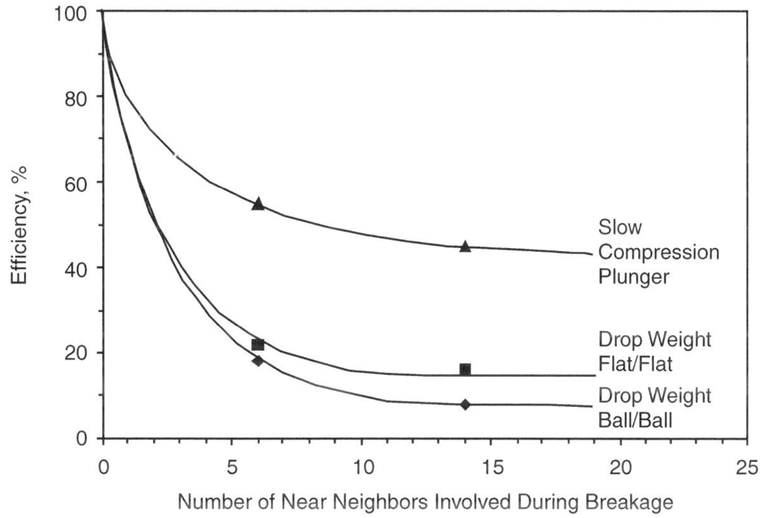

Source: Cho 1987.

FIGURE 3.15 Efficiency for various loading conditions

Note that the efficiency drops well below 100% to 50% or less as the number of particles in an event increases. Note also that different geometries of loading produce different curves. The principal reason for the loss in efficiency in all cases is frictional losses between particles. The case of drop weight ball/ball with near neighbors greater than 15 (~5% efficiency) corresponds closest to conditions in a commercial tumbling mill, whereas the case of slow compression plunger with near neighbors greater than 10 (~50% efficiency) corresponds to conditions in a commercial fine-crushing device.

Other important forms of interaction are small particles with large particles and hard particles with soft particles. When hard particles surround a soft particle, the hard particle contact points increase the probability of breakage of the soft particle. On the other hand, the effectiveness of hard particle breakage is reduced because there are fewer near neighbors with sufficient strength to load the hard particle to fracture. From an energy standpoint, applying breaking forces to a mixture of hard and soft particles results in an energy split that favors the soft particles and produces more breakage than

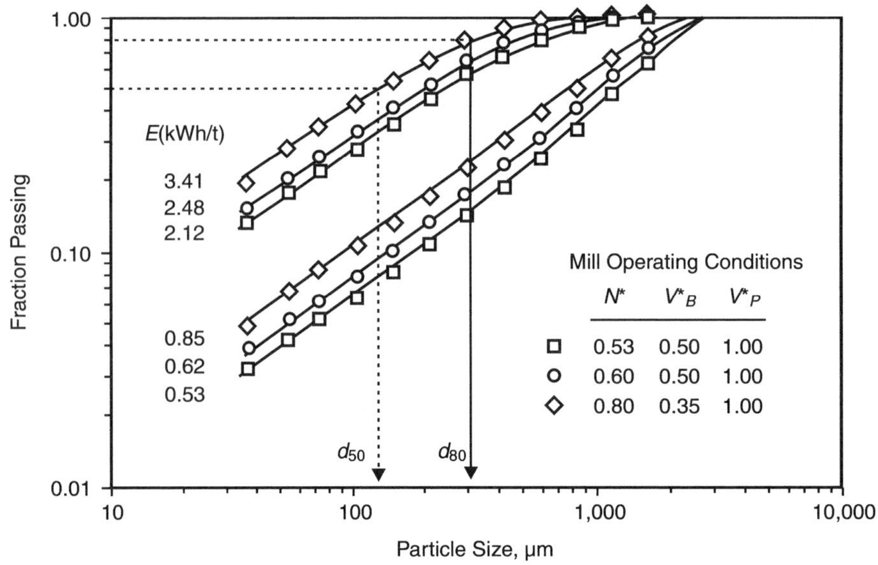

FIGURE 3.16 Product size distributions for ball milling with various speeds, ball loads, and particle fillings

would occur if these particles had the same energy per mass applied to them without the presence of hard particles. When small particles surround large particles, the number of contact points for loading initially increases and that increase lowers the breaking strength of the large particle. However, when the small particles are extremely fine, interparticle frictional losses cushion the coarse particle and hinder breakage.

Quantitative Descriptions of Multiparticle Breakage. Because the application of energy is the driving force for all breakage events, it is natural that mineral processors would try to write relationships between product size and energy input. Prior to 1960, relatively simple relationships were used to relate a change in a point on a product size distribution (e.g., the 80% passing size or the 50% passing size) to the increase in the energy input to a mass of particles (joules/gram or kWh/ton). Typical plots of product size distribution changes with energy input for ball mill grinding under different conditions are shown in Figure 3.16.

In Figure 3.16, $N^* = N/N_C$ is the fraction of the critical speed, N_C, at which the tumbling mill is rotated. N_C is the speed (in rpm) at which a ball of diameter $d_B(m)$ will begin to centrifuge in a mill of diameter $D(m)$. This can be calculated from

$$N_C = \frac{42.2}{\sqrt{D - d_B}} \quad \textbf{(Eq. 3.9)}$$

where $V_B{}^* = \dfrac{V_B}{V_M}$, $V_P{}^* = \dfrac{V_P}{V_I}$, and $V_I = \varepsilon V_B$, and ε is the porosity of the ball charge ($\varepsilon \cong 0.4$).

The single-point energy-size relationships developed during this time are generally of the form

$$\frac{d\overline{E}}{d(d^*)} = -C(d^*)^{-\alpha} \quad \textbf{(Eq. 3.10)}$$

The most widely celebrated of these are those of Rittinger ($\alpha = 2$), Kick ($\alpha = 1$), Bond ($\alpha = 1.5$), and Charles ($\alpha \geq 1$).

TABLE 3.4 Bond work indices for selected ores and minerals

Solid	W_I [kWh/t] [μm]$^{0.5}$	Solid	W_I [kWh/t] [μm]$^{0.5}$
Barite	4.73	Magnetite	9.97
Bauxite	8.78	Manganese ore	12.20
Coke	15.13	Nickel ore	13.65
Copper ore	12.72	Phosphate rock	9.92
Diorite	20.90	Pyrite ore	8.93
Dolomite	11.27	Pyrrhotite ore	9.57
Feldspar	10.80	Quartzite	9.58
Fluorspar	8.91	Rutile ore	12.68
Gold ore	14.93	Taconite	14.61
Hematite	12.84	Tin ore	10.90
Lead ore	11.90	Titanium ore	12.33
Lead-zinc ore	10.93	Zinc ore	11.56
Limestone	12.74		

Source: Chemical Engineering, Vol. 69, No. 2, 103–108 (1962).

Among these, the Bond relationship has the integrated form

$$\bar{E} = \frac{k}{(0.5)}\left(\frac{1}{\sqrt{d_F}} - \frac{1}{\sqrt{d_P}}\right) \tag{Eq. 3.11}$$

or equivalently, in the form originally presented by Bond (1952),

$$W = 10W_I\left(\frac{1}{\sqrt{F_{80}}} - \frac{1}{\sqrt{P_{80}}}\right) \tag{Eq. 3.12}$$

where F_{80} and P_{80} are the 80% passing sizes for the feed and product and W_I is the Bond Work Index. Equation 3.10 has been widely used for equipment design. Values of the Bond Work Index for several ores and minerals are presented in Table 3.4.

In the middle to late 1960s, mineral processors realized that it was frequently important to be able to predict the size distribution of an entire product rather than a single point, such as the d_{80} or d_{50}. One solution was to combine, in a de facto fashion, the energy-size relationships with the empirical size distributions such as the Gaudin–Schuhmann and Rosin–Rammler in Eq. 3.10. This goal was accomplished by requiring that the values of $\alpha - 1$ and ω are the same and calculating the value of the index k or k^1, which makes the energy consumption and d_{80} values match for one or more experimental points. An alternative approach to computing size distributions uses a framework that is phenomenologically correct and therefore more appealing in a fundamental sense. This approach initially invoked a probability of breakage for each size i (or size class i) in the population, p_i, and a distribution of daughter fragments, b_{ij}, from the breakage of size j particles into size i. This process of breakage and redistribution is shown conceptually in Figure 3.17 (Broadbent and Callcott 1956).

This conceptual process can be written mathematically for each size fraction i (d_i to d_{i+1} for $i = 1$ to n) as

$$m_{P,i} = m_{F,i} - p_i m_{F,i} + \sum_{j=1}^{i} b_{ij} p_j m_{F,j} \tag{Eq. 3.13}$$

or in compact matrix notation the product size distribution vector (a column vector of length n) m_P can be computed from the feed size vector by a linear transformation involving the set of breakage probabilities and distributions of fragments as follows:

$$m_P = [I - (I - B)P]m_F \tag{Eq. 3.14}$$

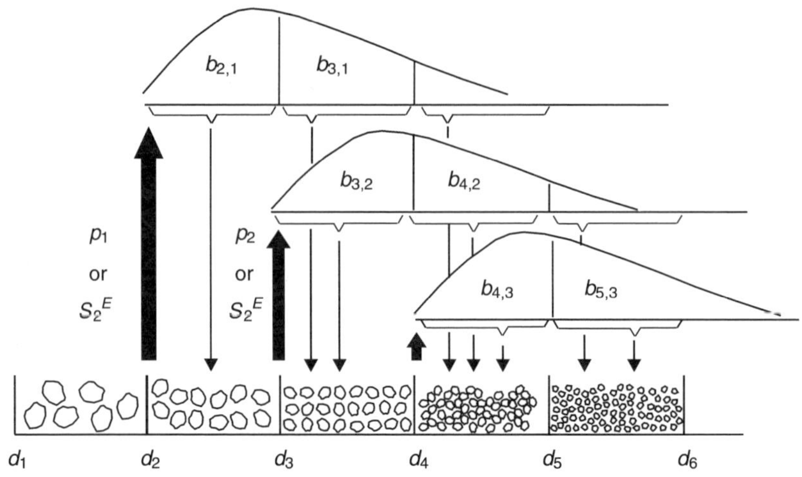

FIGURE 3.17 Conceptual view of population balance accounting

where I is the unitary matrix, P is the breakage probability matrix (diagonal consisting of elements p_i to p_n), and B is the redistribution matrix (a lower triangular matrix consisting of a column b_{ij} *(i = j to n)* for each parent size *j*).

The practical problem associated with applying Eq. 3.14 is that P and B are not known a priori. On the basis of the preceding fundamental considerations, these values depend in a complex way on particle size, composition, and loading environment, and each is somehow related to energy input, which does not even appear in the equation. In addition, energy is applied to particles in a basically continuous manner as they move through a comminution device. Thus, dwell time or rate of energy application must be important. By the middle 1960s or early 1970s, modelers were focusing on time continuous forms of Eq. 3.11 referred to as population balance models (Gaudin and Meloy 1962; Austin 1973; Herbst, Grandy, and Mika 1971). This type of model in size-discrete form with no flow in and out of the comminution device is

$$\frac{dHm(t)}{dt} = -[\mathrm{I} - \mathrm{B}]\mathrm{S}(t)Hm(t) \qquad \textbf{(Eq. 3.15)}$$

where $m(t)$ is the product vector at any time t, H is the hold-up mass of particles in the device, B is the breakage function matrix, and S(t) is the selection function matrix that contains the fractional rates of breakage for each size. The values of S and B can be estimated from experimental data (Herbst, Rajamani, and Kinneberg 1977) and the set of batch-grinding equations represented by Eq. 3.14 can be solved numerically.

In this case, if the probability of breakage is constant over time, the mass fractions of each size can be found by solving each of the differential equations to yield

$$\begin{aligned} m_1(t) &= e^{-S_1 t}\, m_1(0) \\ m_2(t) &= e^{-S_2 t}\, m_2(0) + b_{21}(1 - e^{-S_1 t})m_1(0) \\ m(t) &= \exp(-[\mathrm{I} - \mathrm{B}]St)v(0) \end{aligned} \qquad \textbf{(Eq. 3.16)}$$

where m_1 (0), m_2 (0) ... m_n (0) is the set of mass fractions in each of the size intervals in the feed.

Figure 3.18 shows a test of the constant probability assumption for dry ball milling for a variety of operating conditions (speed and load). Note that the disappearance for top-size material for each condition is linear on a semilog plot as required by Eq. 3.13. Note also that the time-based selection function (the slope of the plot) varies strongly with operating conditions. Clearly a unified method of calculating changes in the selection function with mill speed, load, and particle filling is highly desirable.

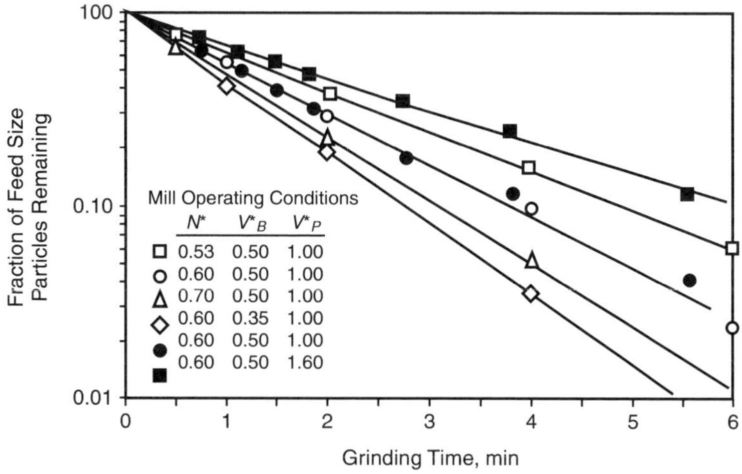

FIGURE 3.18 Efficiency for various loading conditions

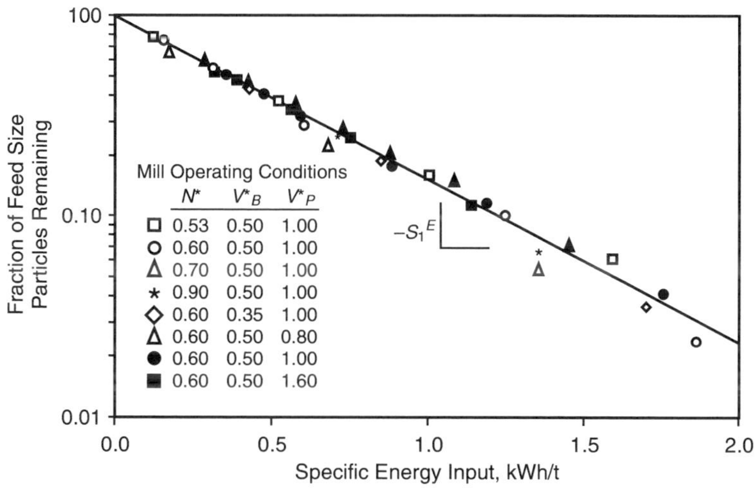

Source: Herbst and Fuerstenau 1973.

FIGURE 3.19 Energy-normalized feed disappearance kinetics

In 1973, Herbst and Fuerstenau reported that the time-based population balance equations could be energy normalized by the transformation

$$S_i(t) = S_I^E \frac{P}{H} \quad \textbf{(Eq. 3.17)}$$

This is illustrated in Figure 3.19 by plotting the time-based data from Figure 3.18 against $\bar{E} = Pt/H$, which collapses all the data to a single line of slope S_1^E. In this case, S_1^E, which is called the energy-specific selection function, is virtually independent of mill operating conditions and therefore acts as a material constant.

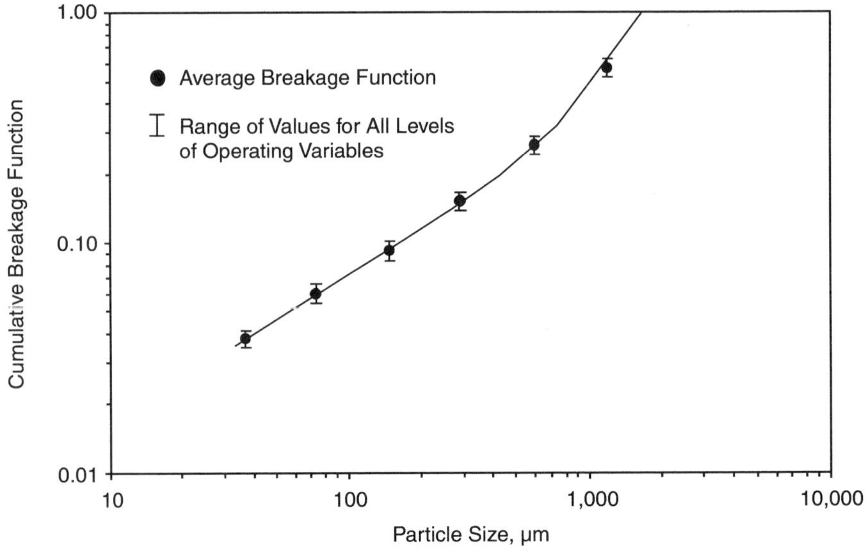

FIGURE 3.20 Average breakage functions for all operating variables

If in addition we recognize that because $\bar{E} = E/H = \int_o^t (P^{dt}/H)$

$$\overline{dE} = \left(\frac{P}{H}\right)dt \qquad \textbf{(Eq. 3.18)}$$

and that the breakage function is basically independent of operating conditions (as seen in Figure 3.20), Eq. 3.14 can be transformed to obtain

$$\frac{dm\bar{E}}{d\bar{E}} = -[\mathrm{I} - \mathrm{B}]\mathrm{S}^{E}m(\bar{E}) \qquad \textbf{(Eq. 3.19)}$$

which is usually referred to as the energy-normalized batch-grinding model. Figure 3.19 shows a test of the energy-normalized model for a wide range of batch ball milling conditions. This energy-size distribution equation predicts the grinding behavior using only one set of selection functions (the specific selection functions; Figure 3.21). It, like the Bond equation, has been found to be very useful for scale-up design. In fact, it has been shown that the Bond model arises as a special case of the energy-size distribution model when $b_{ij}s_j^E \alpha d_i^{0.5}$. For this special case, the specific selection function can be calculated from the work index or vice versa

$$S_I^E = \frac{(\ln 5)d_1^{\ 0.5}}{(100)^{0.5}W_I} \text{ or } W_I = \frac{(\ln 5)d_1^{\ 0.5}}{(100)^{0.5}S_I^E} \qquad \textbf{(Eq. 3.20)}$$

As we will see later, Eq. 3.18 is very useful for scale-up predictions as well as circuit simulations.

Extensions of the above concepts have been successfully applied to describing breakage in virtually all types of comminution systems ranging from blasting through to fine grinding (Pate and Herbst 1999). The challenge that must be overcome in applying these population balance models to new situations centers on the need to know the distribution of forces applied to particles and the associated energy utilization that, when linked to mechanical properties, allow the prediction of energy-based selection and breakage functions. Recent multiphysics modeling efforts and fundamental breakage tests are providing many of the necessary links (Nordell, Potapov, and Herbst 2001).

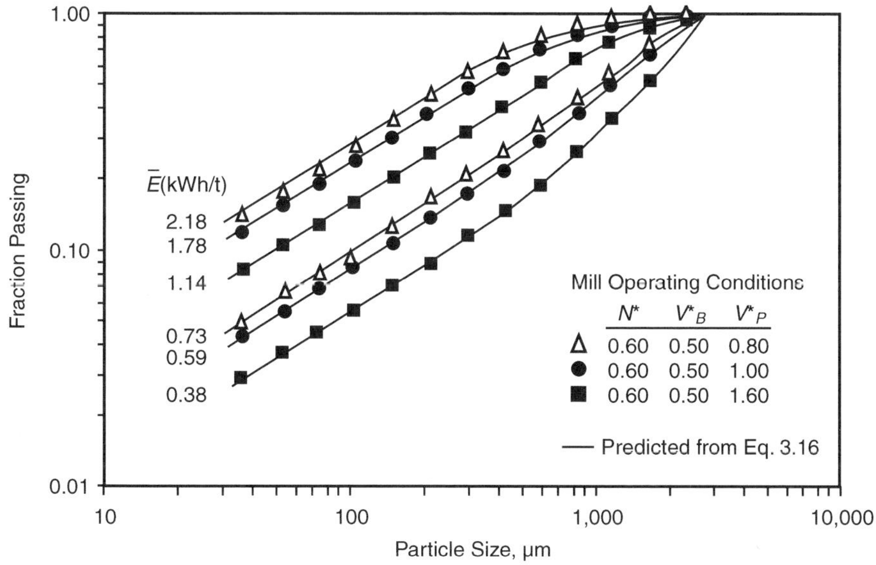

FIGURE 3.21 Comparison of experimentally observed size distributions and energy-normalized predictions for various combinations of operating variables

COMMINUTION EQUIPMENT

Which Devices Are Used for Which Tasks?

In practice, size reduction of mineral particle assemblies is accomplished stepwise in commercial devices. The complicating features in the breakage of large assemblies of particles are the different particle sizes and strengths and the sometimes random way in which stress is applied in the various steps. In mining operations, blasting with explosives is usually the first step in the comminution sequence. During blasting, high-velocity gas pressure pulses created by the explosives load blocks of ores to provide initial fragmentation. The next step in mineral processing operations is typically crushing, which is accomplished by slow compression of large particles (greater than 1 cm) against rigid surfaces. The mechanism of breakage in the crushing step is contrasted with subsequent size reduction by tumbling or stirred mills in which breakage of smaller particles (less than 1 cm) is accomplished by a combination of impact, chipping, and abrasion events caused by energy transferred from grinding media, such as balls, rods, or large particles. The characteristics of the product required in a given application determines what device or series of devices is required.

Crushing is performed in one or more stages with small reduction ratios (d_{80}^F / d_{80}^P is between 3 and 6 per stage). Practically speaking, the reduction ratio represents the ratio of feed size opening (the gape) to the discharge size opening (the set). The first stage, primary crushing, usually produces a product that contains particles finer than 10 cm with an attendant energy consumption of less than 0.5 kWh/t. The energy efficiency is on the order of 80%. Secondary crushing can generally achieve size reduction to less than 1 cm with an energy consumption of less than 1.0 kWh/t. Here the efficiency is closer to 50%.

The next stage of size reduction for most mineral processing operations is accomplished by wet grinding in rotating cylindrical vessels termed "tumbling mills." In these mills, particle breakage occurs by compression, chipping, and abrasion caused by the tumbling action of the grinding media. Preliminary grinding can be done with a rod mill, in which case the grinding media consists of an assortment of rods, a ball mill using balls, an autogenous mill that uses no grinding media, or a semiautogenous mill that uses a light load of balls. The product from primary grinding can be as fine as 300 μm. The energy required can be between 5.0 and 25.0 kWh/t depending on the ore and the product size. The energy efficiency of these devices ranges from 15% down to 3%. Sometimes a special crusher called a high-compression grinding roll is used at this stage; it has an efficiency of up to 30%.

Final stages of grinding are normally accomplished in tumbling ball mills or stirred mills. Here it is possible to reach product sizes of a few microns but at costs of as much as 50 kWh/t. In this case, the energy efficiency may drop to as low as 1% when based on the energy for single-particle slow-compression loading.

In this section descriptions of different types of comminution devices are given together with specific design criteria and operating characteristics. For the past quarter century, the size of comminution equipment and associated drives required for commercial-scale comminution tasks has been selected on the basis of the specific energy (energy per unit mass of product) necessary to reduce a feed material to the desired product size. The choice of specific energy as a scale-up criterion is based on two important premises: (1) equipment of different sizes delivering the same specific energy will yield identical products when fed the same feed material, and (2) existing equipment size/power draft relationships are accurate enough to allow an equipment size that will deliver the necessary energy at the design throughput to be selected.

Whenever possible, the specific energy requirement for a given feed to product transformation is determined in such a way as to minimize the design risk. In other words, the value is determined from an existing full-scale operation or from a pilot-plant circuit that is operated in a fashion similar to that anticipated for the commercial installation. When commercial- or pilot-scale data are not available, design engineers often use the Bond energy size reduction equation, or its equivalent, to estimate specific energy requirements. Recently other techniques involving population balance model parameters have been shown to reduce design risk.

Crushing Devices

The types, sizes, and number of crushers employed in a complete reduction system will vary with such factors as the volume of ore to be processed, ore hardness, and the size of the feed and product.

Gyratory Crushers. Primary crushers are heavy-duty machines run in open circuit (sometimes in conjunction with scalping screens or grizzlies). They handle dry run-of-mine feed material as large as 1 m. There are two main types of primary crushers—gyratory crushers and jaw crushers. Gyratory crushers are the most common for new operations.

Secondary crushers are lighter-duty and include cone crushers, roll crushers, and impact crushers. Generally, the feed to these machines will be less than 15 cm, and secondary crushing is usually done on dry feed. Cone crushers are similar to gyratory crushers, but differ in that the shorter spindle of the cone is not suspended but is supported from below by a universal bearing. Also, the bowl does not flare as in a gyratory crusher. Cone crushers are generally the preferred type of secondary crusher because of their high reduction ratios and low wear rates. However, impact crushers are used successfully for relatively nonabrasive materials such as coal and limestone. Frequently, size reduction with secondary crushers is accomplished in closed circuit with vibrating screens for size separation.

The gyratory crusher is used as a primary and secondary stage crusher. The cone crusher is used as a secondary, tertiary, and quaternary crusher. The action of a typical gyratory-type crusher is illustrated in Figure 3.22. In gyratory crushers the crushing process comprises reduction by compression between two confining faces and a subsequent freeing movement during which the material settles by gravity until it is caught and subjected to further compression and again released. The particles are subjected to maximum breaking forces when they are on the side with the minimum gape. Table 3.5 shows nominal tonnages for gyratory crushers that range from 1,600–7,600 tph depending on feed opening and open-side setting.

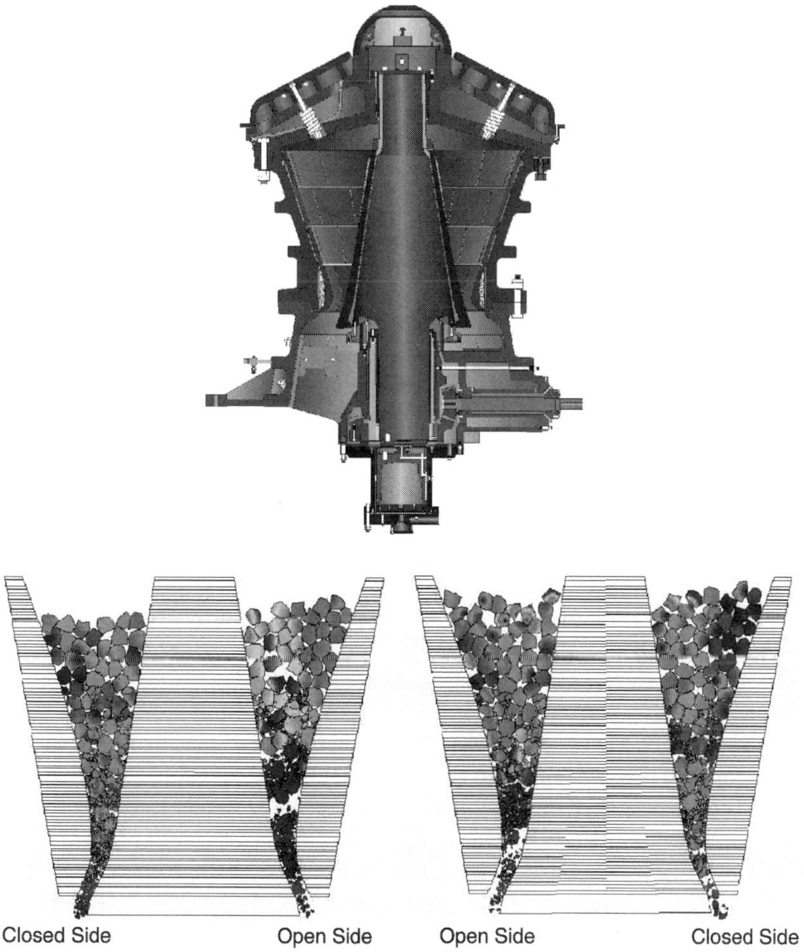

Source: Conveyor Dynamics, Inc.

FIGURE 3.22 Cutaway of Superior MKII gyratory crusher and snapshots of simulation of primary crusher breakage at two points in the travel of the mantle

TABLE 3.5 Nominal gyratory crusher capacities (tph) for various crusher size and open-side settings (Superior MKII)

Size	Feed Opening, mm (in.)	Pinion, rpm	Max. kW (hp)	Crusher Capacity, tph										
				Open-side Settings of Discharge Opening, mm										
				125	140	150	165	175	190	200	215	230	240	250
42–65	1,065 (42)	600	375 (500)		1,635	1,880	2,100	2,320						
50–65	1,270 (50)	600	375 (500)			2,445	2,625	2,760						
54–75	1,370 (54)	600	450 (600)			2,555	2,855	3,025	3,215	3,385				
62–75	1,575 (62)	600	450 (600)			2,575	3,080	3,280	3,660	3,720				
60–89	1,525 (60)	600	600 (800)				4,100	4,360	4,805	5,005	5,280	5,550		
60–110	1,525 (60)	514	1,000 (1400)					5,575	5,845	6,080	6,550	6,910	7,235	7,605

Source: Metso Minerals.

FIGURE 3.23 Cutaway of cone crusher

In the normal gyratory, the crushing stroke or travel of the head usually has an important bearing on the size of the finished product, although this factor is subject to modification when a parallel sizing zone is adopted. The movement of the head in the cone crusher is similar to that in the ordinary gyratory with an exception—toward the bottom of the cone, the head travels through a much greater distance and gyrates much faster (see cutaway in Figure 3.23). The long movement changes the crushing stroke from slow compression to fast compression, and the increased clearance on the freeing stroke allows the material to fall away vertically after each impact. Table 3.6 shows nominal product size distributions (top) and capacities (bottom) for different cone crusher sizes. Nominal tonnages in this case range from 60–1,850 tph depending on crusher size and closed-side setting.

Impact Crushers. Impact devices are often used for fine crushing. One type, a hammer mill, uses hammers rotating at high speed to break particles. Another type, an autogenous impact crusher, accelerates particles with a rotor causing them to impact a curtain of falling particles (Figure 3.24). The latter has a significant advantage from a wear point of view.

Tumbling Mill Grinding Devices

The various grinding devices used in the industry are distinguished in terms of the manner by which energy is introduced into the system and in terms of capacity and particle transport into and out of the mill. Each device is characterized with respect to particle size range, design relationships, wear, and efficiency of energy utilization. For some grinding devices, well-defined design relationships have not been established, and in these cases detailed data on typical installations are given when available. Similarly, wear data for some grinding devices are not always available, especially for the more recently developed mills that have not been used extensively on an industrial scale. In addition, the wear characterization is difficult to generalize because it is highly dependent both on the nature of the feed and the materials of construction. The efficiency characterization is the free-crushing efficiency and represents that portion of the total energy consumption that would be required for single-particle fracture under slow compression.

TABLE 3.6 Nominal cone crusher product size distributions (% passing) and capacities (mtph) for various closed-side settings (CSS) and crusher sizes

	Product Size Distributions (% Passing for Various Closed-side Settings)				
Sieve Size, mm	**CSS = 50 mm**	**CSS = 38 mm**	**CSS = 25 mm**	**CSS = 19 mm**	**CSS = 13 mm**
90	97–100	100			
75	92–98	99–100	100		
50	67–81	86–94	99–100		
38	54–64	68–78	92–98	100	100
25	38–45	48–54	65–80	94–98	99–100
19	30–35	37–42	51–62	82–90	96–99
16	25–29	31–35	43–53	73–82	92–97
13	22–25	26–29	35–44	63–73	83–93
10	18–21	22–24	28–34	52–61	70–91
6	13–14	15–16	19–34	36–44	50–57
Crusher Size	**Capacity, tph**				
MP1000	1,830–2,420	1,375–1,750	915–1,210	720–900	615–730
MP800	1,460–1,935	1,100–1,285	735–980	580–690	495–585
HP800	785–1,200	600–950	495–730	435–545	325–425
HP500	580–725	445–555	365–455	320–400	230–290
HP400	465–580	360–450	295–370	255–320	185–230
HP300		300–380	230–280	200–240	150–185
HP200		210–250	170–220	150–190	120–160
HP100			85–115	75–100	60–90

Source: Metso Minerals.

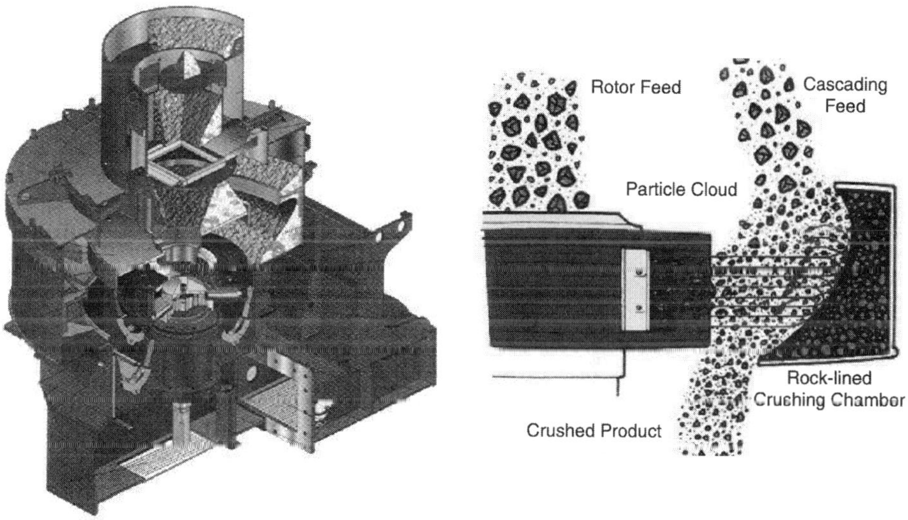

Source: Metso Minerals

FIGURE 3.24 Autogenous impact crushing

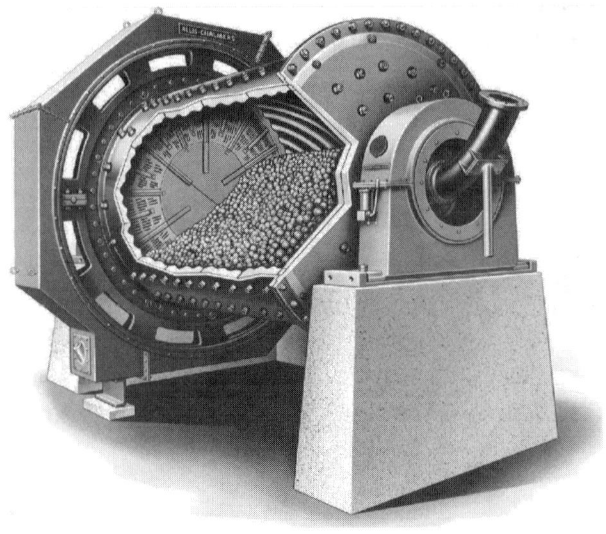

FIGURE 3.25 Cutaway of ball mill with chute feeder and grate discharge

Ball Mill. Intermediate and fine size reduction by grinding is frequently achieved in a ball mill in which the length of the cylindrical shell is usually 1 to 1.5 times the shell diameter. Ball mills of greater length are termed "tube mills," and when hard pebbles rather than steel balls are used for the grinding media, the mills are known as "pebble mills." In general, ball mills can be operated either wet or dry and are capable of producing products on the order of 100 μm. This duty represents reduction ratios as great as 100.

The ball mill, an intermediate and fine-grinding device, is a tumbling drum with a 40% to 50% filling of balls (usually steel or steel alloys; Figure 3.25). The material that is to be ground fills the voids between the balls. The tumbling balls capture the particles in ball/ball or ball/liner events and load them to the point of fracture. Very large tonnages can be ground with these devices because they are very effective material handling devices. The feed can be dry, with less than 3% moisture to minimize ball coating, or a slurry can be used containing 20% to 40% water by weight. Ball mills are employed in either primary or secondary grinding applications. In primary applications, they receive their feed from crushers, and in secondary applications, they receive their feed from rod mills, autogenous mills, or semiautogenous mills. Regrind mills in mineral processing operations are usually ball mills, because the feed for these applications is typically quite fine. Ball mills are sometimes used in single-stage grinding, receiving crusher product. The circuits of these mills are often closed with classifiers at high-circulating loads.

These loads maximize throughput at a desired product size. The characteristics of ball mills are summarized in Table 3.7, which lists typical feed and product sizes. The size of the mill required to achieve a given task—that is, the diameter (D) inside the liners—can be calculated from the design relationships given by Rowland and Kjos (1978). The design parameters that must be specified are

- The size in micrometers at which 80% of the material is passing for the feed, F_{80}, and the product, P_{80}
- The Bond Work Index, W_i(kWh/t), of the material as determined by ball mill grindability tests (Bond 1961) or S_i^E values determined from energy-monitored tests that yield full-size distribution predictions (Herbst and Fuerstenau 1980)
- The length-to-diameter (L/D) ratio
- The fraction of critical speed, *Fcs*
- The feed rate (tph)

TABLE 3.7 Summary of ball mill characteristics

Particle Size Range

Feed size:	<1 cm (10,000 µm)
Product size:	>0.002 cm (20 µm)

Design Relationships*

Power draft: $P(kW) = 2.347\ \rho_{balls}\left(\frac{L}{D}\right)D^{3.3}V_p{*}(3.2 - 3V_p{*})N{*}\left(1 - \frac{.1}{2^{9-10N*}}\right)$

Energy size reduction relationship: $W(\text{kWh/t}) = W_i\left(\frac{10}{\sqrt{P_{80}}} - \frac{10}{\sqrt{F_{80}}}\right)$

W_i values in table

Population balance relationship: $S_i = S_i^E\left(\frac{P}{H}\right)$, b_{ij} constant

Wear Characteristics—Wet Ball Mills

Balls	kg/kWh = 0.175 $(A_i - 0.015)^{1/3}$
Liners	kg/kWh = 0.013 $(A_i - 0.015)^{0.3}$

Wear Characteristics—Dry Ball Mills

Balls	kg/kWh = 0.023 $\sqrt{A_i}$
Liners	kg/kWh = 0.0023 $\sqrt{A_i}$

Efficiency

5%

Material Characteristics

Material	Abrasion Index A_i
Copper ore	0.0950
Taconite	0.6837
Limestone	0.0256
Clinker	0.0409
Coal	11.37

*The bulk density of balls, ρ_{balls}, is expressed in t/ft^3.

The liner- and ball-wear equations are typically written in terms of an abrasion index (Bond 1963). The calculated liner and ball wear is expressed in kilograms per kilowatt-hour (kg/kWh), and when multiplied by the specific power (kWh/t), the wear rates are given in kilograms per ton of feed. The wear in dry ball mills is approximately one-tenth of that in wet ball mills because of the inhibition of corrosion. The efficiency of ball mills as measured relative to single particle slow compression loading is about 5%. Abrasion indices for five materials are also listed in Table 3.7.

The L/D ratios of ball mills range from slightly less than 1:1 to something greater than 2:1. The tube and compartment ball mills commonly used in the cement industry have L/D ratios 2.75:1 or more. The fraction of critical speed that the mill turns depends on the application, and most mills operate at around 75% of critical speed. Increased speed generally means increased power, but as the simulations presented in Figure 3.26 show, it can also produce more wasted ball impacts on the liners above the toe, causing more wear and less breakage.

Currently ball mills are built up to a diameter of 26 ft. These mills are installed at many locations around the world and frequently require more than 10 MW to operate.

There are three principal forms of discharge mechanism. In the overflow ball mill, the ground product overflows through the discharge end trunnion. A diaphragm ball mill has a grate at the

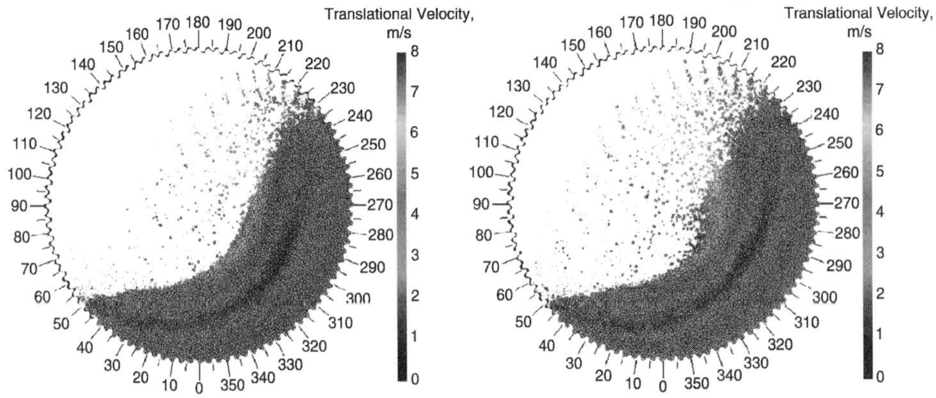

FIGURE 3.26 Ball motion simulations for two mill speeds

discharge end (Figure 3.25). The product flows through the slots in the grate. Pulp lifters may be used to discharge the product through the trunnion, or peripheral ports may be used to discharge the product.

The majority of grinding balls are forged carbon or alloy steels. Generally, they are spherical, but other shapes have been used. The choice of the top (or recharge) ball size can be made using empirical equations developed by Bond (1958) or Azzaroni (1984) or by using special batch-grinding tests interpreted in the content of population balance models (Lo and Herbst 1986). The effect of changes in ball size on specific selection functions has been found to be different for different materials. A ball size-correction method can be used along with the specific selection function scale-up method to determine the best ball size. To do this, a set of "ball size tests" are performed in a batch mill from which the specific selection function dependence on ball size can be determined. Then, the mill capacities used to produce desired product size can be predicted by simulation using the kinetic parameter corresponding to the different ball sizes.

The mill liners used are constructed from cast alloy steels, wear-resistant cast irons, or polymer (rubber) and polymer metal combinations. The mill liner shapes often recommended in new mills are double-wave liners when balls less than 2.5 in. are used and single-wave liners when larger balls are used. Replaceable metal lifter bars are sometimes used. End liners are usually ribbed or employ replaceable lifters.

The typical mill-motor coupling is a pinion and gear. On larger mills two motors may be used, and in that arrangement two pinions drive one gear on the mill. Synchronous motors are well suited to the ball mill, because the power draw is almost constant. Induction, squirrel cage, and slip ring motors are also used. A high-speed motor running 600 to 1,000 rpm requires a speed reducer between the motor and pinion shaft. The "gearless" drive has been installed at a number of locations around the world.

Autogenous/Semiautogenous Mills

Autogenous and semiautogenous mills represent a relatively new type of tumbling mill that, under certain conditions, can replace size reduction equipment used for secondary crushing as well as primary and final grinding. Basically, the breakage mechanism is similar to that found in other tumbling mills. The unique feature of this device is that the coarse ore particles themselves are used as the grinding media, not unlike a pebble mill in which the pebbles are generated naturally from the ore body. In this regard, autogenous grinding is to be applied to ores with necessary characteristics.

The autogenous mill itself is a coarse-grinding device, consisting of tumbling drum with a 25% to 40% volume filling of ore (Figure 3.27). Metallic or manufactured grinding media is not used. Autogenous mills are fed run-of-mine ore or primary crusher product that is −10 in. Inside the mill, large

FIGURE 3.27 **Large autogenous mills installed on the Minnesota Iron Range**

FIGURE 3.28 **Semiautogenous mill with snapshot of simulation showing rocks (tetrahedral) and balls (spheres) and how they move**

pieces break into smaller pieces a few inches in size. These natural pebbles act as the grinding media in the autogenous mill. The main modes of breakage are thought to be impact breakage and abrasion.

Semiautogenous milling results when a small amount of steel balls, 3% to 20% of mill volume, is added to the mill charge. The addition of a small ball charge to an autogenous mill changes the nature of the mill performance considerably. In this case, major design modifications may be required to carry the additional charge. Generally, the addition of a ball charge increases the mill capacity significantly but increases operating costs for balls and power (Figure 3.28).

TABLE 3.8 Summary of autogenous mill characteristics

Particle Size Range

Feed size: <25.4 cm (250,400 μm)

Product size: >0.02 cm (200 μm)

Design Relationships

Power draft: 1) Modified Rowland and Kjos Model

2) Morrell Model

Energy size reduction relationship: $OW_i = \dfrac{W}{\left(\dfrac{10}{\sqrt{P_{80}}} - \dfrac{10}{\sqrt{F_{80}}}\right)}$

Population balance: 1) JK drop weight tests

2) Svedala pilot tests

Efficiency

3%

Material Characteristics

Material	Work Index, W_i (kWh/t) $(\mu m)^{0.5}$	Operating Work Index, OW_i (kWh/t) $(\mu m)^{0.5}$
Copper ore	13.13	18.3–21.0
Taconite	14.87	22.4–29.8

Many circuit configurations are possible, but essentially the autogenous mill is operated as a single-stage primary mill, or it can be followed by secondary pebble or ball milling. The autogenous mill is often operated in closed circuit with a trommel screen or external vibrating screen classifying the discharge. Circulating loads are low compared with those in ball mill circuits, because autogenous mills do not benefit from high-circulating loads in the same way ball mills do. Intermediate crushers are sometimes used to crush the largest pieces in the recycle stream.

Table 3.8 summarizes the characteristics of autogenous/semiautogenous mills. Typical feed and product sizes are presented. The design relationships are those necessary to calculate size of the mill from the test data and input parameters. Unlike the ball mill, the product size cannot be determined beforehand. The product size of an autogenous mill depends strongly on the nature of the ore being ground so that tests must determine the competency of the ore and its "natural" particle size. The power information from pilot-plant tests is then used to calculate the diameter (in feet) inside the liners from the power draft relationship using these input parameters: the charge density, ρc(ton/cubic foot) and the L/D ratio. Mill dimensions are inside the liners and expressed in feet. The operating work index, OW_i (kWh/t), is calculated from the autogenous circuit specific energy consumption, W(kWh/t), and the 80% passing size in the feed, F_{80} (μm).

As an alternative, population balance design methods, which use single-particle impact plus abrasion tests or 6-ft batch tests, can be used to estimate breakage and selection functions; existing power equations can be used for sizing. A means of predicting liner wear rates (in pounds per kilowatt-hour [lb/kWh]) or, for semiautogenous mills, ball wear rates (lb/kWh), is to use test mill wear rate (lb/kWh) data. The steel wear rates for selected autogenous and semiautogenous mills are given in Table 3.9. The efficiency of the autogenous mill is lower than the efficiency of a ball mill because of the inherent low efficiency of rock breaking rock versus balls breaking rock. A factor of 1.5 is a common value of the ratio of the autogenous operating work index to the laboratory Bond Work Index. The operating work index range for copper ore is 1.3 to 1.6 times as great as the Bond Work Index. The range of the operating work index for taconite is 1.5 to 2.0.

TABLE 3.9 Ball and liner consumption for selected autogenous and semiautogenous mills

	Ball Charge % Mill Vol.	Balls, lb/t Feed	Liners, lb/t Feed	Energy Consumption, kWh/t	Balls, lb/kWh	Liners, lb/kWh
Lornex	3–8	0.62	0.0967	5.8	0.107	0.017
Pima	8	0.8	0.157	7.9	0.101	0.020
Island copper						
(single-stage)	—	2.25	0.34	21.45	0.105	0.016
(two-stage)	—	1.6	0.19	17.2	0.093	0.011
Similkameen	8.5	0.85	0.26	19.5	0.040	0.13
Dofasco	—	—	0.294	12.2	—	0.024
Savage River	—	—	0.09	12.6	—	0.007

Source: MacPherson and Turner 1978.

The typical length-to-diameter ratio in mills in North America is 0.30:1 to 0.35:1. In Europe and South Africa, these ratios range from 1:1 to 2:1. Manufacturers have different ideas about the relative importance of impact breakage and abrasion, and this is reflected in the L/D ratio. North American manufacturers believe that impact breakage is important, so large diameters are used to provide a long downward path and consequent high velocities.

The mill volume fraction of charge is 0.30 to 0.35 in short mills and 0.45 to 0.50 in long mills (Digre 1979). The mill speed is usually between 73% and 78% of critical speed and is typically controlled by a variable speed drive. Autogenous/semiautogenous mills are larger than ball mills on a volume basis. This results from the low density charge in the mill. Thus, on a volume basis, the autogenous mill is not able to pull as much power as a ball mill. The addition of a ball charge to an autogenous mill will increase its power draw and its capacity. The world's largest semiautogenous mills are 40-ft diameter × 19-ft long and draw up to 20 MW.

The experience of operators has been that autogenous circuits require more energy (consistent with the lower efficiency of particle self-breakage) than conventional circuits to obtain the same product size and throughput. The additional energy may be as great as 100% of the energy used by a conventional circuit.

The autogenous/semiautogenous mill is fed by a feed chute on rollers. The opening of autogenous mills is much larger to accommodate the larger feed size. Large lifters are used to lift the charge high up in the mill. They are incorporated in the design as rail and double rail liners. Lifter bars may be bolted to the shell through the liner. The height of lifters is 5 to 12 in. The best height and profile can be predicted with 3-dimensional, discrete element methods (DEM) simulations (Herbst and Nordell 2001). Figure 3.29 shows snapshots of simulations of charge motion for newly installed lifters in a 34-ft-diameter mill with two different release angles. The motion simulations show that the lower release angle causes balls and particles to impact high above the toe of the charge (at the 8:30 or 9:00 position); the higher release angle moves more of the impacts back into the charge where they can be effective for breakage. These high-fidelity simulations provided predictions of wear profile, power, and throughput as they evolved over the lifetime of the liner. In a plant trial, the higher release angle yielded the expected power and wear as well as a throughput increase of 4.4% relative to the lower release angle.

Because of high-impact forces, Cr-Mo liners and lifters are used. NiHard or similar alloys break up in high-impact areas after a ball charge is added. The wear of liners and lifters in autogenous/semiautogenous mills is higher than in ball mills.

Barratt (1979) compared semiautogenous and conventional circuits at plants that have these two types of circuits side by side. The ball costs in the semiautogenous circuits were 19% less than in the conventional circuits and, similarly, the liner costs were 4% less. The reduction of liner costs for a semiautogenous circuit comes from the elimination of the liner costs for secondary and tertiary crushers.

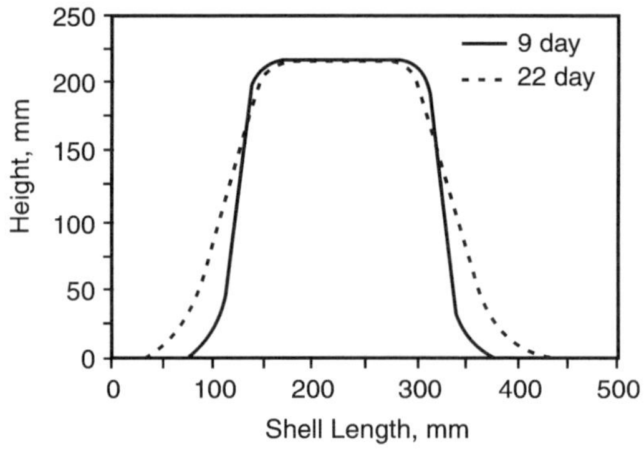

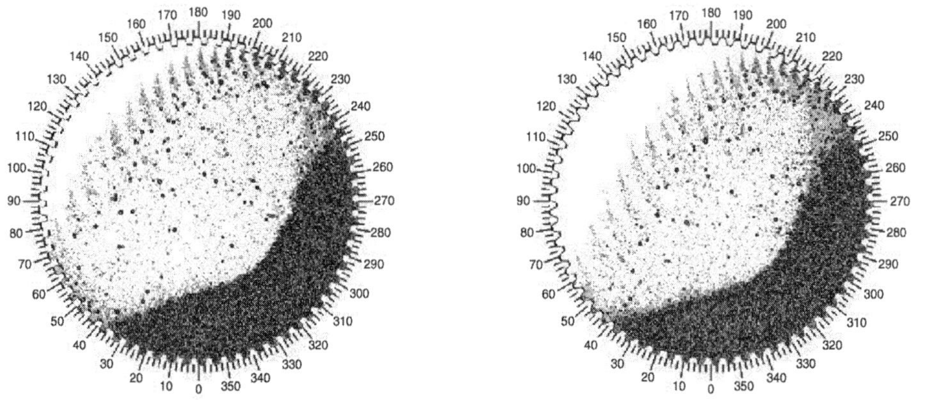

FIGURE 3.29 Three-dimensional DEM simulations of charge motion with lower release angle lifter (left) and higher release angle lifter (right) profiles

The discharge mechanism is more elaborate for an autogenous or a semiautogenous mill than in a ball or rod mill. A grate discharge is used by most or all operators. Radial slots ranging in size from 10 mm to 100 mm are used in the grate. A snapshot of a high-fidelity simulation of the action of such a discharge mechanism is shown in Figure 3.30. After flowing through the slots, the product is raised to the discharge end trunnion by pulp lifters. These pulp lifters have an action very much like that of a water wheel. Often, a trommel screen is attached to the trunnion. The screen, 6- to 8-ft long, has openings ½ to ¾ in. in diameter.

High-pressure Grinding Mills

As mentioned previously, single-particle compressive size reduction is very effective in minimizing energy requirements and uses only a fraction of the energy of conventional ball mill grinding. Schönert (1979) compared the specific energy demands of single-particle compressive comminution at 2.5 to 5.8 kWh, and single-particle impact comminution at 23 to 56 kWh. As noted previously, achieving large throughputs with single-particle compressive comminuting is not technically feasible for small particle sizes, but compressive comminuting can be achieved in a bed. Under these conditions particle-particle interactions do increase the energy required although the total is far less than that required for ball mill grinding. For high-pressure grinding rolls (Figure 3.31), the feed particles are accelerated into the opening by dropping them from a height of up to 15 ft. Thus, friction losses are reduced because the rollers have less work to do accelerating the particles up to the peripheral speed. The degree of comminution is influenced by the stress. The product particle size decreases as the stress, and consequently, the specific energy, are raised.

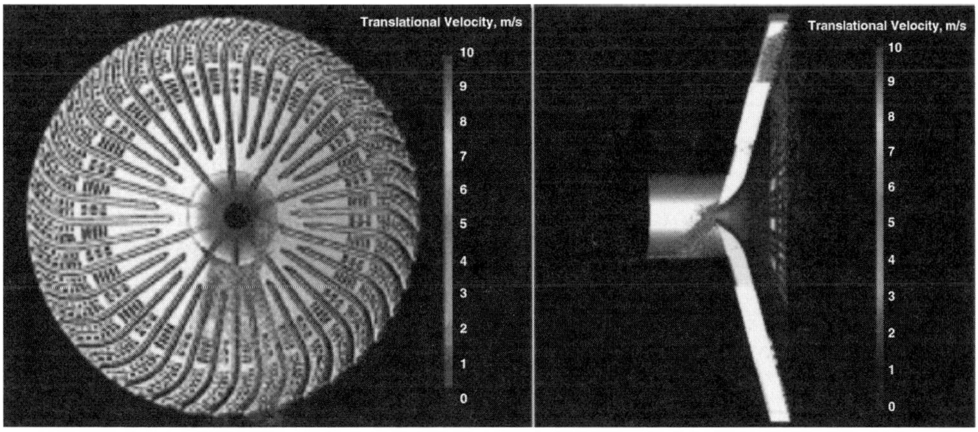

FIGURE 3.30 Snapshots of simulation of semiautogenous mill discharge mechanism

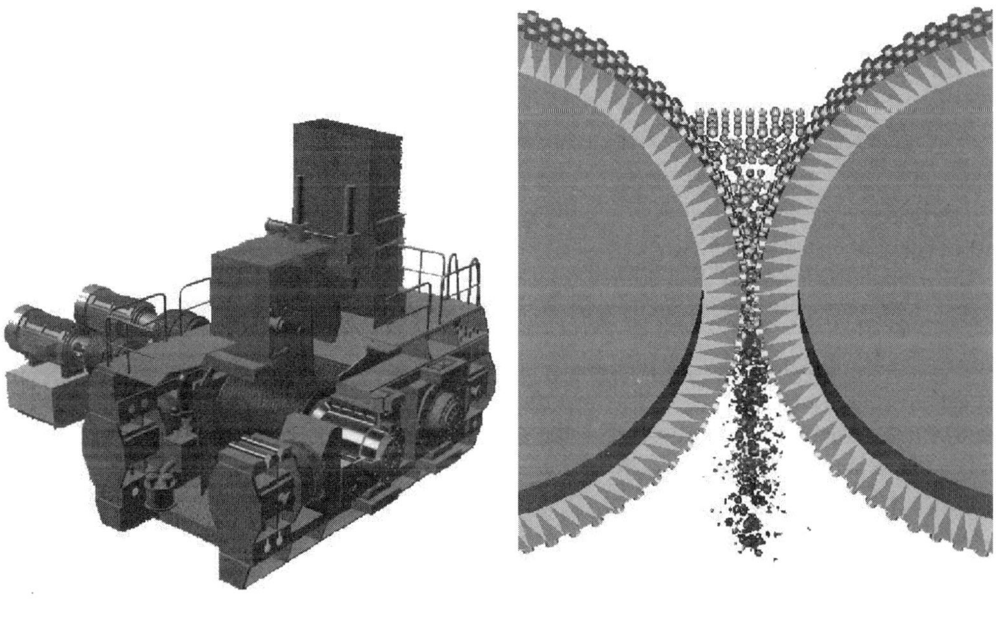

Source: Conveyor Dynamics, Inc.

FIGURE 3.31 High-pressure grinding rolls apparatus and snapshot of simulaton showing particle breakage in the device

At high stresses, the product is packed into briquettes (agglomerates). The agglomerates are easily dispersed when the roll pressure is 30–50 MPa; when higher pressures are used, deagglomeration in a ball mill is needed to disperse the strong agglomerates. Typical roll diameters are 200 to 1,000 mm, and the roll pressures are 125 to 375 MPa.

Interparticle friction in the bed causes some of the particles to remain unbroken (Schönert 1979), so the deagglomerated product is classified in such a way that oversize particles are returned to make another pass through the rolls. A low degree of comminution is more efficient due to reduced interparticle friction, but the recycle ratio has to be increased because less finished product results from lowering the roll pressure.

TABLE 3.10 Nominal operating conditions and capacities for various sizes of high-pressure grinding rolls

Type	Units	10.0–120/120	12.5–140/140	15.0–140/160	20.0–170/180
Roll diameter, D	mm	1,200	1,400	1,400	1,700
Roll width, W	mm	1,200	1,400	1,600	1,800
Total force, F	KN	10,000	12,500	15,000	20,000
Max. specific force, F_{sp}	N/mm^2	6.9	6.4	6.7	6.5
Max. speed, v	m/s	1.55	1.80	1.80	2.2
Max. throughput, M	t/h	480	770	880	1,470
Max. power, P	kW	2 × 1,000	2 × 1,500	2 × 2,000	2 × 3,000

Source: KHD Humboldt Wedag AG.

The main design quantities for the rolls are specific force, $F_{sp} = F/DW$ (where F is the force, D is the roll diameter, and W is the roll length) and mass flow of particles; $\dot{M} = uWd_g\rho$ (where u is the linear velocity of the roll at the surface, d_g is the gap width, and ρ is the bulk density of the material being ground). Table 3.10 shows nominal performance parameters for different roll sizes.

Stirred Grinding Mills

As the product size required from tumbling ball mills becomes finer, the optimum ball size decreases. At a certain size, however, the small balls (moving under the influence of gravitational forces in a tumbling mill) can no longer provide sufficient impact force for particle fracture. Stirred mills increase the impact forces by inducing larger momentum changes than the tumbling mill.

The stirred ball mill is a fine-grinding device capable of producing product particle sizes less than 1 µm. A cylindrical vessel is charged with grinding balls made of steel or ceramic and the material to be ground. The charge is stirred by means of a rotating central shaft with either flights or arms extending into the charge as shown in Figure 3.32. Different devices are stirred with different rotational speeds ranging from, for example, 20 rpm up to 2,000 rpm. The most intense agitation occurs just in front and just to the rear of the moving flights or impellers. The velocity of the charge is highest in this area and decreases to zero at the central shaft and walls.

In vertical operation, slurry can be introduced into the bottom or top of the vessel. For horizontal orientations, the mill is fed under pressure at one end and discharged through a retaining ring at the other end. The stirred ball mill can be close circuited with a classification device appropriate for the small product particle size.

The power intensity (energy input per unit volume) of these mills is larger than that of conventional tumbling ball mills. Values of up to 100 times greater volume-specific energy have been reported. The high power intensity is attributed to the relatively large impeller speeds. Extremely long residence times in tumbling ball mills for product particle sizes smaller than 1 µm are significantly reduced in the stirred ball mill. This is attractive in terms of the volume of installed mills and comparatively short retention times. The specific energy (kWh/t) requirements are approximately equal for the stirred ball mill and the tumbling ball mill for a product particle size greater than 20 µm. The characteristics of stirred ball mills are summarized in Table 3.11.

The power characteristics of the stirred mills are similar to those of the turbine mixer. The design parameters for the equation given in the summary are the mill volume, V; the diameter of the grinding balls, d_B; the density of the grinding media balls, ρ_B, and the shaft angular velocity, N. The empirical energy-size relationship that has been shown to be most appropriate (Sepulveda 1980) is the Charles equation. The parameters of this equation are the constant, $A(\text{kWh}/t\,(\mu\text{m})^{1.8})$; the median size of the product (µm); the median size of the feed (µm); and the exponent (ω). The constants, A and α, are given in the summary for various minerals and coal in water.

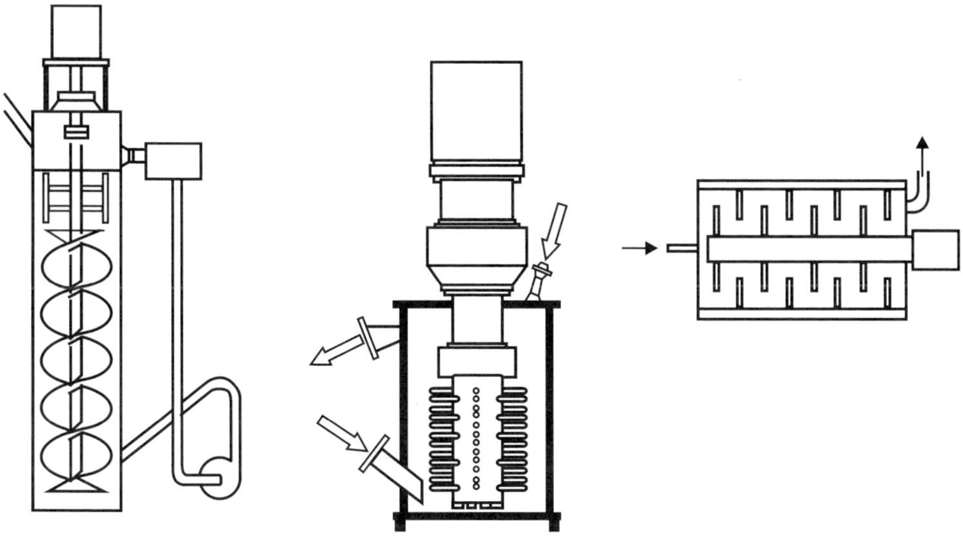

FIGURE 3.32 Various types of stirred mills for different operating speed ranges

TABLE 3.11 Summary of stirred ball mill characteristics

Particle Size Range

Feed size: <150 μm

Product size: >0.2 μm

Design Relationships

Power draft: $P(\text{kw}) = C\,V^{\alpha}N^{\beta}d_B^{\gamma}\rho_B^{\delta}$

Energy size reduction relationship: $\bar{E} = A(d^{-\omega}_{\text{Median},P} - d^{-\omega}_{\text{Median},F})$

Efficiency

1%–5%

Material Characteristics	**Charles' Constant (A)** $\left(\frac{\text{kWh}}{t}(\mu m)^{1.8}\right)$	**Exponent (ω)**
Chalcopyrite	460	1.8
Limestone	500	1.8
Quartz	920	1.8
Coal (Montana Rosebud)	3,000	1.8

The Vertimill, shown in Figure 3.33, is a narrow cylindrical shell that stands vertically. At the center of this stationary cylinder is a drive shaft on which are mounted a series of flights. The mill is partially filled with steel balls up to 1.5 in. in diameter. During operation, the flights are turned by the shaft at speed of 25–30 rpm drag, mixing the balls. Simulations such as those shown give insights into ball and slurry velocities and impact energies. Any feed particle present is crushed by shear and impact forces when it is caught between moving balls that impart sufficient energy to induce fracture.

The feed slurry is generally pumped through a valve at the bottom of the mill. As the feed particles work their way up the mill and mix with the balls, they are ground by intense attrition and

FIGURE 3.33 Model 1250 Vertimill and DEM simulation of ball motion

compression. The ground product overflows at the top and is collected. In general, finer product can be obtained by using smaller balls. For maximum reduction ratio, however, the balls should be no smaller than seven times the feed particle diameter. A vertimill is capable of grinding 100-µm feed particles down to 0.3–1.5 µm.

The Draiswerke mill is a closed design, high-speed, stirred ball mill. The grinding chamber (Figure 3.31) of the Draiswerke mill is 80%–90% filled with small balls, normally with a diameter of 1–10 mm. In the center of the chamber, a high-speed agitator with a number of pinned discs rotates and accelerates the balls up to 100 to 150 times gravity. To further activate the grinding balls, rows of stationary pins are fastened radially to the inner wall of the chamber. Actual grinding occurs within the immediate vicinity of the pin tips where the greatest difference in velocity and high shear forces exist. As the material is discharged from the mill after processing, a frictional gap separator or screen cartridge retains the grinding media. The process variables, such as solids feed rate, liquid feed rate, agitator speed and separator gap, may be controlled proportionally or independently of each other to meet the optimum product requirement. The largest Draiswerke mills are 2,500-L mills with 3,000-hp motors.

COMMINUTION CIRCUITS

How Comminution Equipment Is Arranged Optimally

In practice individual pieces of comminution equipment are almost never used alone, but rather appropriate "stages" of size reduction are used in a plant to transform the material from the run-of-mine size

TABLE 3.12 Normal size range and approximate energy efficiencies for various devices

Device	Normal Size Range, mm	Approximate Efficiency, %
Explosive	∞–1,000	70
Gyratory crusher	1,000–200	80
Cone crusher	200–20	60
Autogenous/semiautogenous	200–2	3
Rod mill	20–5	7
Ball mill	5–0.2	5
Stirred mills	0.2–0.001	1.5
High-pressure grinding rolls	20–1	20–30

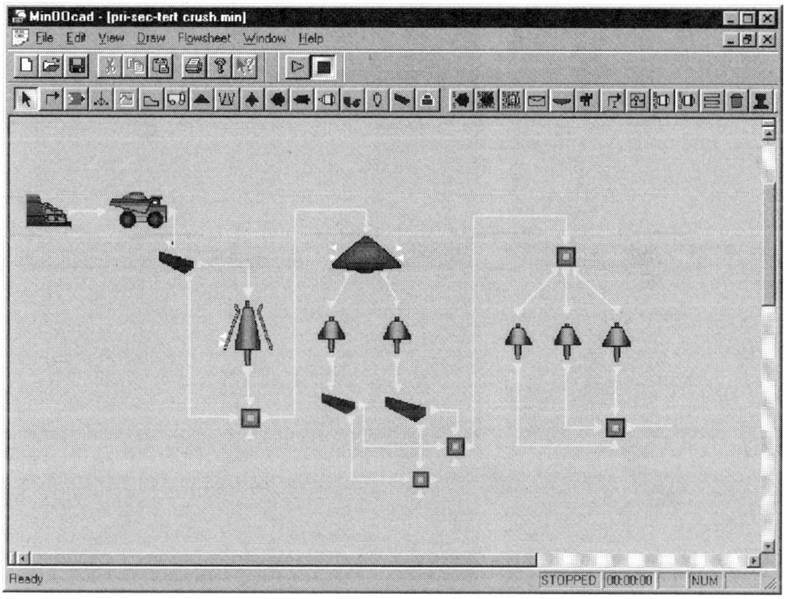

FIGURE 3.34 Primary, secondary, and tertiary crushing circuit

to the final product size. The appropriate selection of equipment on a stage-by-stage basis is determined by feed size, ore type, tonnage, and final product size. Table 3.12 lists the size ranges within which various comminution methods operate most efficiently. The fact that the inherent efficiency of some devices is higher than others causes circuit designers to select equipment that produces a favorable overall efficiency.

In addition to taking into account inherent device efficiency in circuit design, the use of interstage size separation equipment will be shown in this section. A key aspect of achieving high overall efficiency is to remove product size material as soon as possible after it is created. Material that is already "finished" takes up energy and interferes with the breakage of coarse particles. In addition, the finished material becomes overground. Thus efficient size separation using screens, hydrocyclones, and other classifiers is a critical part of circuit design. Some typical crushing and grinding circuits are shown in Figures 3.34 to 3.38.

Figure 3.34 shows a circuit that takes run-of-mine size material (up to 1,000-mm pieces) through three stages of crushing to result in a product of 10 mm. Grizzlies with large spacings are sometimes

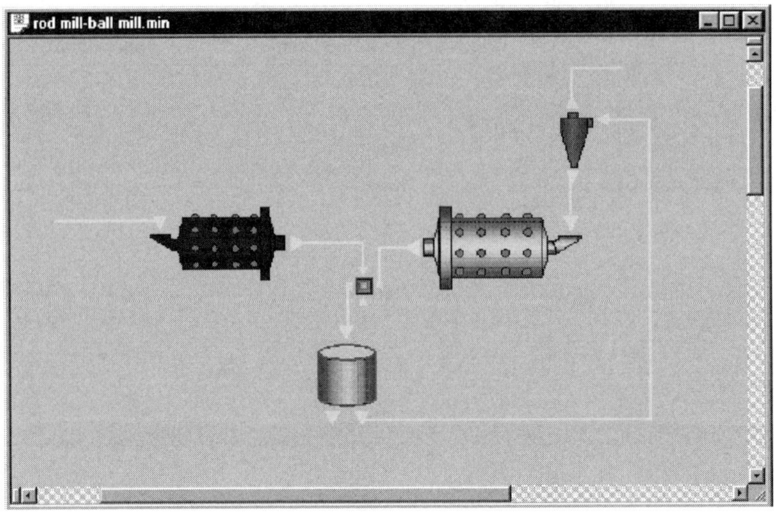

FIGURE 3.35 **Rod mill and ball mill circuit**

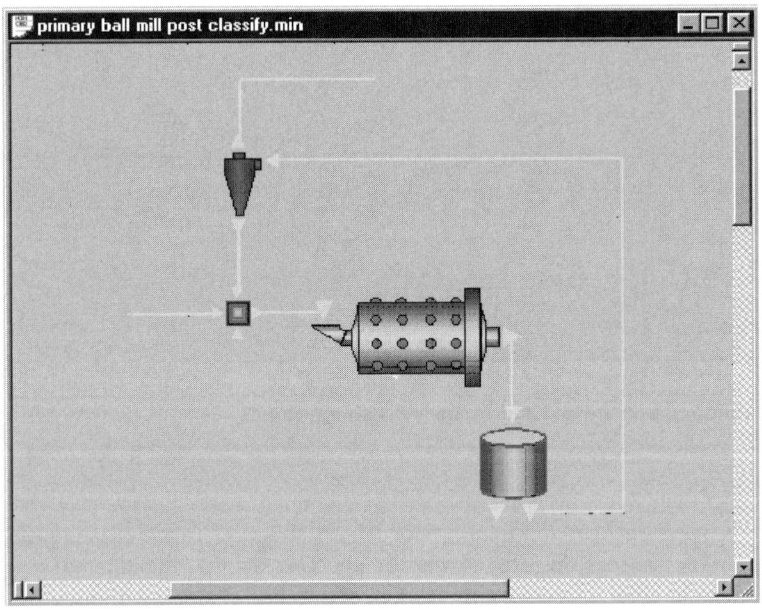

FIGURE 3.36 **Single-stage ball mill circuit**

used to protect the primary (gyratory) crusher from extremely large rock pieces and other foreign matter that might disrupt its operation. Primary crushers are almost always operated in open circuit (without a size separation step to return oversize material). The primary crusher product usually goes into coarse ore storage to buffer the other size reduction steps against big variations in mine production. The two secondary (cone) crushers are operated in closed circuit with screens to improve their efficiency and to control the top size into the three tertiary (also cone) crushers.

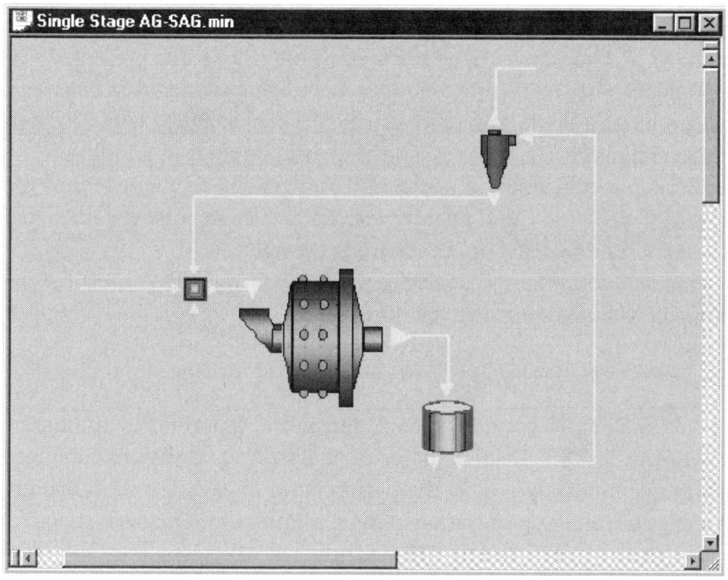

FIGURE 3.37 Single-stage autogenous/semiautogenous circuit

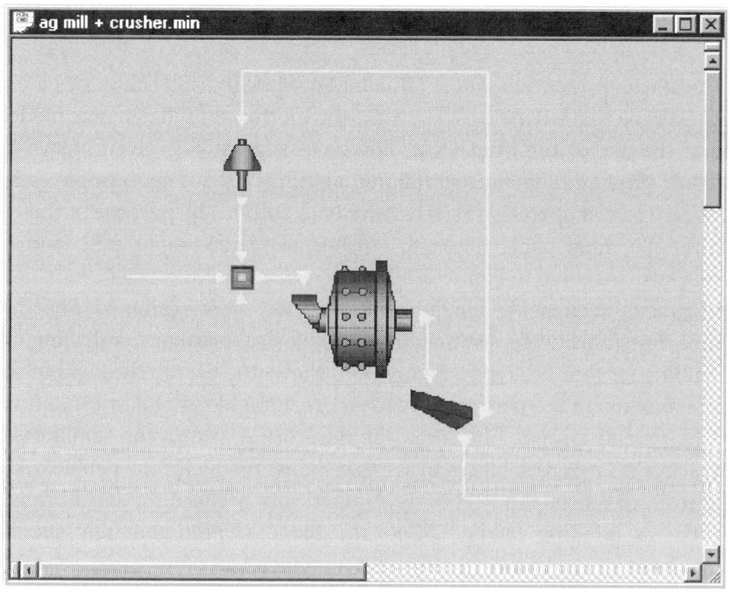

FIGURE 3.38 Autogenous/semiautogenous circuit with pebble crushing

Figure 3.35 shows a typical (circa 1960) grinding circuit, which may follow the circuit in Figure 3.34. Here, the tertiary crusher product (10–15 mm) is fed to a rod mill (operated in open circuit) to yield a discharge of 3–6 mm followed by a ball mill operated in (reverse or pre-) closed circuit with a cyclone cluster. The cyclone overflow contains a product in the range of 0.2 to 0.05 mm.

Figure 3.36 shows a primary ball mill circuit. In this case, a tertiary crusher product is fed to the ball mill. The ball mill discharge flows into a sump that is in turn pumped to the cyclone cluster for normal or postclassification. The coarse stream is returned to the sump and the fine stream forms the product.

Figure 3.37 shows an alternative to the more conventional circuits presented above. In this case, primary crusher product comes to an autogenous/semiautogenous mill via a coarse-ore stockpile. The crusher product is ground in a single stage and the product postclassified with a cyclone. Advantages of this circuit are that it eliminates two stages of crushing and one stage of grinding.

Figure 3.38 shows a modification of a standard autogenous/semiautogenous grind for ores that produce “hard-to-grind” pebbles. These pebbles are fed to a cone crusher rather than just returning them to the mill where they may build up, overloading the mill.

Many other circuits options exist, especially those involving stirred mills for finer grinding and pregrinding treatments with high-pressure grinding rolls.

Circuit Simulation

Model-based decision making is becoming more and more important to the mining industry. The ability to model the behavior of individual pieces of equipment and then to combine these models in such a way that they quantitatively predict the performance of circuits and ultimately entire plants is becoming increasingly critical for optimization, design, and control. The most effective of these simulators is based on population balance models. The performance of each piece of comminution equipment can be modeled with the general macroscopic conservation equation

$$\text{IN} - \text{OUT} + \text{GENERATION} = \text{ACCUMULATION} \quad \textbf{(Eq. 3.21)}$$

Continuing with the notion of Eq. 3.21 in matrix form

$$M_F m_F - M_o m_o - [\mathrm{I} - \mathrm{B}]\mathrm{S}Hm = \frac{d[Hm]}{dt} \quad \textbf{(Eq. 3.22)}$$

Three of the most widely used simulators (JK SimMet, ModSim, and USIMPAC) are based on steady-state models (ACCUMULATION = 0 in Eq. 3.21) while the fourth uses full dynamic models (MinOOcad).

An example of the use of the MinOOcad flowsheet simulator is given below. In this example, aspects of both steady-state and dynamic simulation are illustrated. The example concerns a mine-to-mill simulation for an iron ore operation (Herbst and Pate 2001). The purpose of the work was to optimize throughput for the mine bank, haulage, crusher, semiautogenous mill, and ball mill system configured in MinOOcad as shown in Figure 3.39.

Blending and control strategies to maximize productivity were evaluated. Specific selection functions and breakage functions were estimated for explosive breakage, crushing, semiautogenous milling, and ball milling for four ore types. Figure 3.40 shows the overall throughput for the comminution operations as a function of the percent of hardest ore, for a blend of hardest and softest ores.

The soft ore blasts and crushes finer than the hard ore A, but in the semiautogenous mill, it is harder because it lacks the large rock pieces to act as grinding media for the pebbles. As more and more hard ore is added to B, the throughput goes up dramatically because the hard ore is providing large rock pieces to serve as grinding media. When the blend is predominately the hard ore A, the throughput declines because the breakage rates of A are much less than B.

Figure 3.41 shows a simple blending strategy for two ore types over 16 hours that produces an average blend of 30% A and 70% B during each 4-hour period.

The variations in the instantaneous blend are large and the effect on grinding significant. The performances of each of the three alternative control strategies explored for this example were predicted with MinOOcad as shown in Figure 3.42.

The first shown is a constant feed-rate strategy (290 mtph, reduced to 145 mtph for 2 hours when overload occurs). The second is the throughput resulting from a constant filling strategy (maintaining constant filling by manipulating feed rate with a volume set point of 32%). The third is the production resulting from a feed-forward model-based strategy based on the specified truck haulage schedule

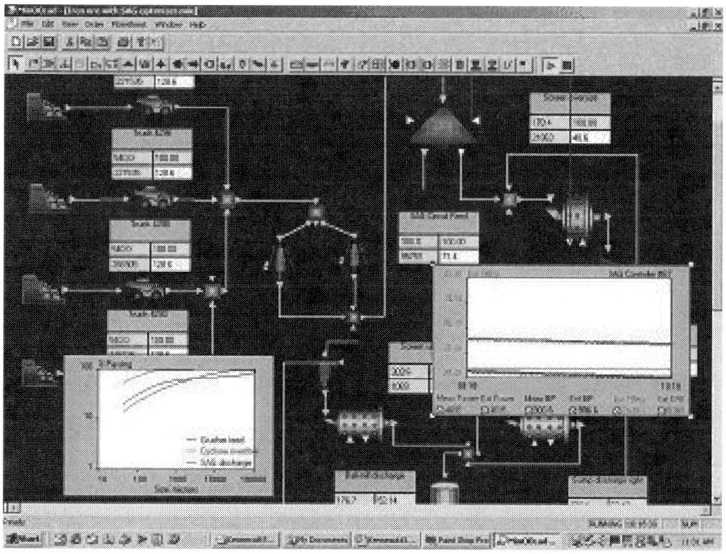

FIGURE 3.39 Autogenous/semiautogenous object inserted into full mine-to-mill flowsheet for dynamic simulation

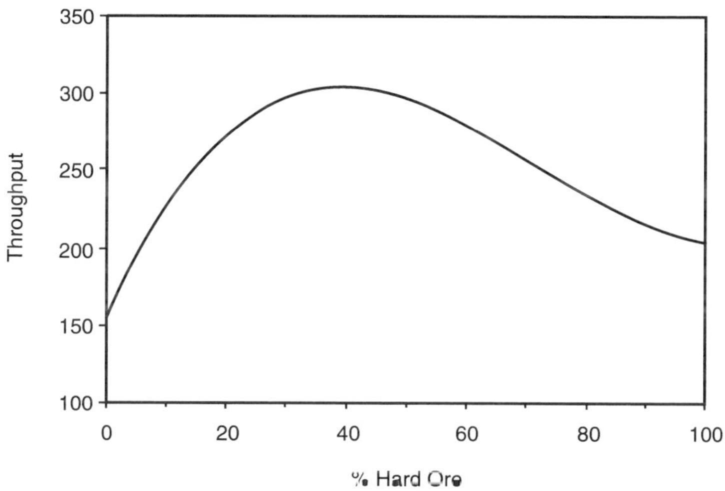

FIGURE 3.40 Semiautogenous mill throughput as a function of feed composition

required to achieve the blend as shown in Figure 3.40. The figure of merit used to compare strategies is total tons through the semiautogenous mill during the two-shift period. The results are presented in Table 3.13.

In this specific case, the constant filling strategy is 6.1% more effective than the constant feed-rate strategy, and the schedule feed-forward strategy is 16.3% more effective than the constant feed-rate strategy. The significance here is that the dynamic simulator has permitted multicomponent behavior to be realistically predicted in an off-line setting. The trends for different levels of control sophistication are as expected (see next section), but in this case, quantitative predictions of performance differences are made available for economic evaluation.

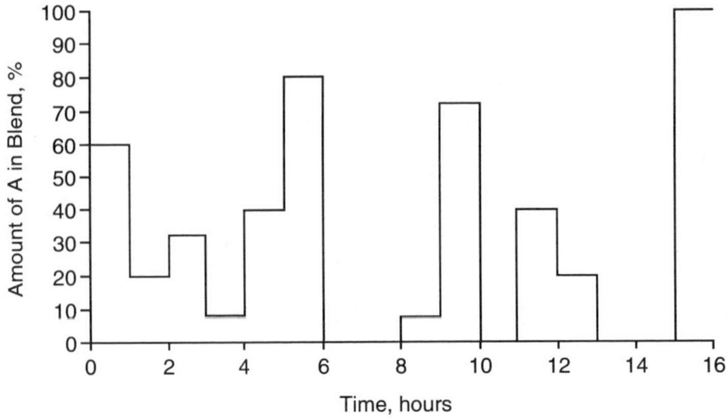

FIGURE 3.41 **Blend schedule for a 16-hour period**

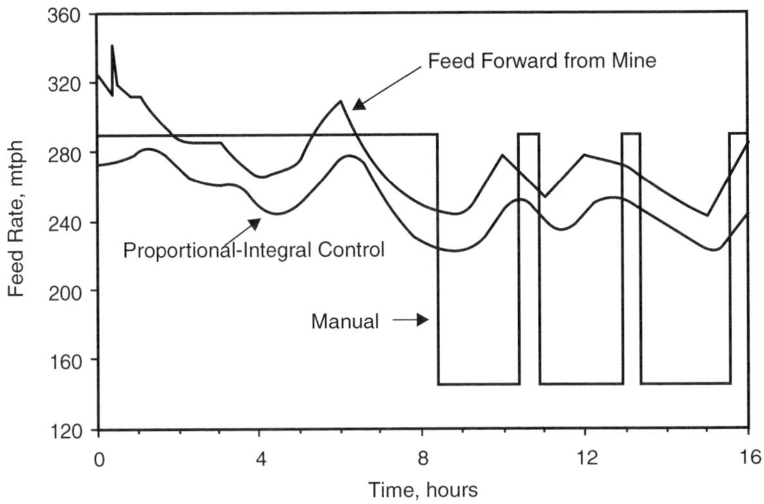

FIGURE 3.42 **Semiautogenous mill production rate for various control strategies**

TABLE 3.13 **Total tons milled during a 16-hour period**

Strategy	Number of Overloads	Total Tons Milled
Constant feed rate	3	3,750
Constant filling	0	3,980
Truck schedule feed forward	0	4,360

PROCESS CONTROL IN COMMINUTION

Controlling Size Reduction and Liberation Against Disturbances

Process control is an essential component of any comminution system. The widespread adoption of automation in mineral processing plants began more than four decades ago, when rudimentary regulatory strategies for regulation of ore, water, and slurry were successfully deployed on single-loop analog

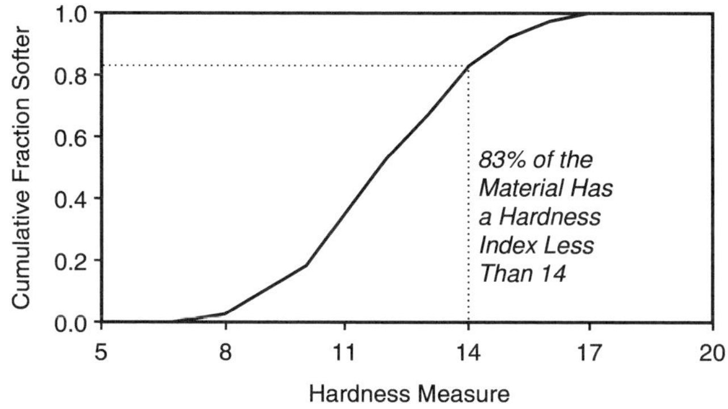

FIGURE 3.43 Cumulative probability density function for ore hardness from testing

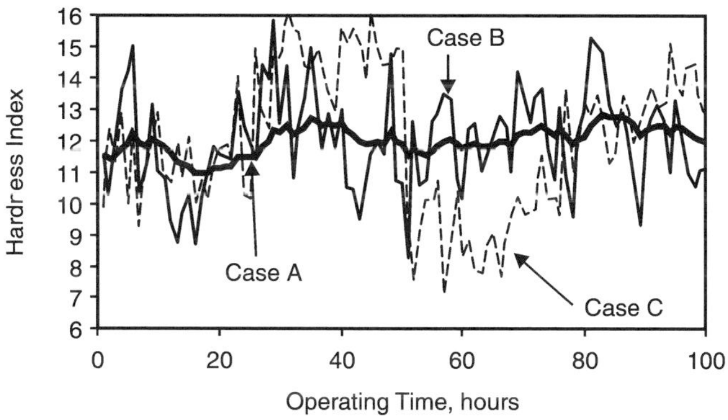

FIGURE 3.44 Temporal variation of ore hardness

controllers. Virtually all plants built today have a sophisticated digital control system that enables all basic control functions, providing the human–machine interface (HMI), and acting as the gateway to plant management information systems, which couple process and business controls. In addition, most new plants adopt advanced process control applications to deal with the multivariable nature of process optimization in real time. Although the journey has not always been smooth, the industry has increasingly embraced process control as one of the most capital-effective investments available in the pursuit of lower costs and increased revenues.

The need for process control is made evident in early stages of the comminution circuit design process. To illustrate, Figure 3.43 shows a hypothetical distribution of a hardness index in an ore body. The need to adapt to changing feed conditions is readily apparent. If the plant is designed to produce the desired product size for a hardness of 14 at design tonnage, the ore will be harder 17% of the time; unless the tonnage is reduced, the product size will be too coarse and possibly some loss in liberation will occur. On the other hand, 83% of the time, the ore will be a softer, finer product, and more liberation will possibly result.

The key implicit assumption here is that a process control system will allow the circuit to achieve steady-state targets and overall operational stability in the face of feed-ore variations.

These temporal changes in ore characteristics in the feed to a comminution process are called "disturbances" because nothing can be done by the operator or control system to modulate them. The nature (frequency and amplitude) of these disturbances will dictate the severity of the control problem and the complexity of the solution. For example, Figure 3.44 illustrates three possible variants of

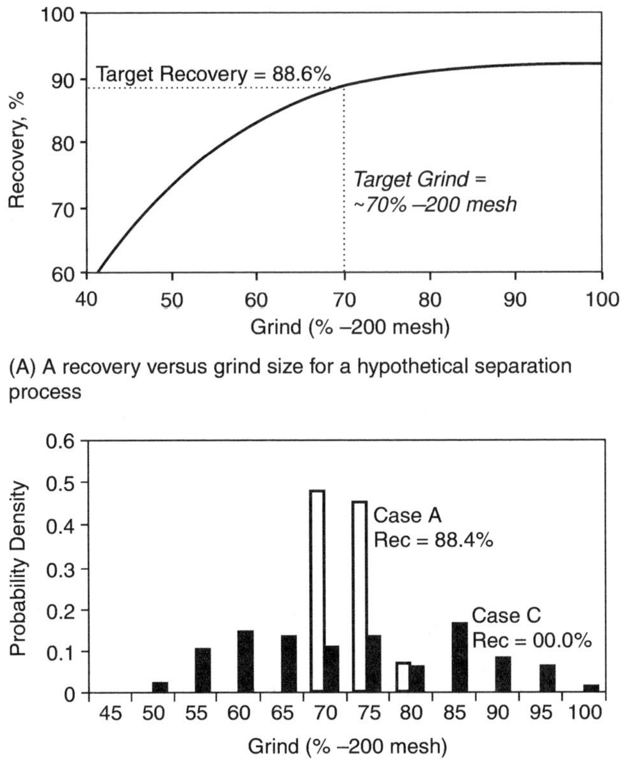

(A) A recovery versus grind size for a hypothetical separation process

(B) Grind-size variations for the disturbance scenarios A and C in Figure 3.44

FIGURE 3.45 Illustration of the impact of disturbances in comminution on downstream separation processes

circuit feed for the ore characterized in Figure 3.43. Case A illustrates a well-blended feed; case B shows greater short-term variability; and case C shows longer-term variability. As Murphy's law would correctly predict, disturbances, such as those in cases B and C, are more common in practice. In fact, all of these disturbances coexist.

Disturbances will arise in ore hardness (e.g., different ore types and genesis modes), feed size (e.g., blasting practices and stockpile segregation), and liberation requirements (e.g., requiring a change in grind). It is also quite common to see disturbances arising from internal sources; for example, a sump pump that intermittently delivers a dilute slurry to a cyclone feed pump box, or a mechanical feeder prone to flow interruptions. Although the comminution process may well be stable to these fluctuations without intervention, control systems are normally required to ensure stability and to enhance the overall economic performance of the process.

To continue with the illustration above, Figure 3.45A shows a hypothetical grind-size recovery curve. If we were able to maintain a target product size of 70% at 200 mesh, we would expect to see a valuable metal recovery of 88.6%. However, if the feed tonnage is constant, the grind will vary based on the nature of the disturbances. Combining the information of Figures 3.44 and 3.45A produces Figure 3.45B. As we would expect, the greater variability in hardness for case C produces a wider distribution in grind size to the separation circuit. Figure 3.45B also shows the overall recovery expected under each open-loop operating scenario. (Open loop implies no intervention by the operator or the control system.) Clearly, the more stable feed of case A provides a recovery much closer to the optimal value shown in Figure 3.45A, while case C incurs an ~2% recovery loss.

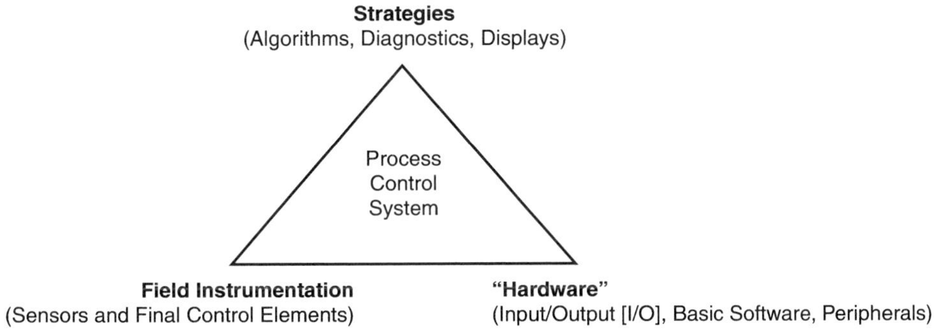

Source: Adapted from Herbst and Bascur 1984.

FIGURE 3.46 The control triad

In this hypothetical case, if a control system were to be applied to maintain the grind at the average or target value, the total tonnage treated would be essentially the same, but a 2% recovery gain would be seen for case C. The latter number is more or less typical of the recovery gains associated with supervisory control applications. Throughput increases frequently lie in the range of 3% to 15%. The magnitude of these numbers underscores the attractiveness of such investments.

Industrial process control systems are powerful tools for maintaining process stability and ensuring optimum economic performance in the face of disturbances. The complexity of the control strategy depends in part on the complexity of the process (the so-called "resiliency factor," articulated by Morari 1983), and in part on the nature of the disturbances. Estimates of the nature of disturbances are increasingly available at the design stage, opening new avenues for the a priori design of control strategies. Operationally, efforts to mitigate disturbances upstream (at the mine or crusher) will simplify the control requirements, although the spatial variability of ore characteristics often precludes effective blending.

The combination of the magnitude and frequency of the disturbances will also have an impact on control requirements. Simply put, minor amplitude variations (as in case A above) are easier to handle. Similarly, low-frequency disturbances can often be very effectively rejected by control, while the process effectively filters very high-frequency disturbances. Those lying in the intermediate range can usually be rejected to a greater degree (the shorter the frequency). This frequency range is related to the time constant of the process. For example, the dynamics of a response to a feed-hardness change in a grinding circuit are much slower than the change in water flow in a pipe to a response in supply pressure.

The Control Triad

The control triad illustrated in Figure 3.46 provides a useful framework for this overview. This schematic conveys the message that an effective process control system will include the proper blend of field instrumentation, hardware, and control strategies.

Figure 3.47 provides a practical illustration of the control triad in the context of a simple water flow regulation loop. In this instance, the field instrumentation consists of the orifice meter and ball valve. The hardware comprises the input/output (I/O) subassembly, the computer with the basic software, and the HMI. The strategy employs a simple well-tuned Proportional Integral and Differential (PID) Actions control law, where the operator determines the set point or target for the water flow rate.

Of course, there are many important and related subjects that are beyond the scope of this discussion. These include signal filtering, sampling intervals, loop tuning, and dead-time compensation. The interested reader should consult one of the many good textbooks in this area (e.g., Seborg, Edgar, and Mellichamp 1983).

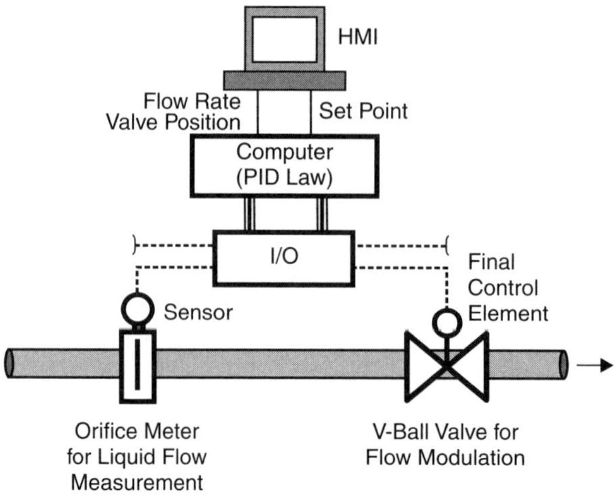

FIGURE 3.47 A simple flow control loop

Instrumentation. Because final control elements are largely restricted to devices that control position (e.g., valves, knife, or flop gates) or electrical motor speed (e.g., feeder, pump, and mill), this section will focus on sensors, offering a much broader range of devices. The first law of process control–"All control starts with measurement, and the quality of control can be no better than the quality of measurement," –or in the vernacular–"Garbage in, garbage out"–helps validate this choice.

Table 3.14 lists some of the more common sensors used to monitor comminution circuit equipment. (Because there are many manufacturers of competing instruments, we have elected to distinguish instruments on the basis of the technology employed to make the measurement.) Although the list is not exhaustive, it does show that there is a good capacity for measurement in such process systems.

It is evident from this table that the process control system designer often faces a problem related to choices. In other words, what kind of technology is best suited for the measurement problem at hand, and, which vendors manufacture proven products employing this technology? In the lower level stabilizing loops typically associated with the regulation of ore, slurry, reagent, and water flows, the preferred sensors are generally well established. For example, electronic belt scales are the sensor of choice for measuring solids mass flow on a conveyor belt.

Instrumentation provides the interface between the process and the control strategies. The proper selection, installation, and maintenance of these field devices is essential to ensure that the benefits associated with process control applications are sustained for the life of the project. Ongoing sensor development efforts also means that the process control engineer needs to stay abreast of measurement technology, looking for opportunities to further develop or enhance the performance of a process control system.

Hardware. Control hardware most frequently encountered in mineral processing plants are the Distributed Control Systems (DCS; e.g., Bailey, Foxboro, Fisher-Rosemount) or the programmable logic controllers (PLCs; e.g., Modicon, Allen Bradley, GE). In many plants hybrid architectures involving a combination of DCS and PLC technologies are common. Figure 3.48 is an illustration of such a hybrid structure and shows the hardware layout for a typical process control system in the mid-1990s. This picture is expected to change in the coming years as smart instruments and equipment displace the more traditional I/O interfaces. Moreover, as bandwidth increases, the likelihood of delivering control applications over the Internet increases, and remote hardware and application maintenance and development support will be simplified.

TABLE 3.14 Sensor technology in comminution systems

Factor Measured	Technology Employed
Levels	
Bin (solids)	Ultrasonic devices, laser devices, load cells, mechanical devices
Tank (slurry/water)	Ultrasonic devices, capacitance probes, differential pressure devices, conductivity probes, mechanical devices
Motor power	Current transducer (+ conversion), power transducer, torque meter
Flow	
Solids flow	Electronic belt scale, nuclear belt scale, impact meter
Slurry flow	Magnetic units, ultrasonic units
Water flow	Vortex-shedding devices, turbine meters, differential pressure devices
Moisture	
Dry solids	Microwave units
Slurries	Radiation gages, U tubes, differential pressure devices
Pressure	Diaphragm devices
Vibration	Accelerometers
Temperature	Thermocouples, resistance thermal devices, infrared imaging
Particle size	
Dry solids	Image analysis techniques
Slurries	Ultrasonic devices, mechanical (caliper) devices, soft sensors
pH	Specialized electrodes, conductivity probes
Tramp metal	Magnetic field devices
Mill load	Power-based devices, acoustics, load cells, strain gages, soft sensors, conductivity
Speed	Tachometer

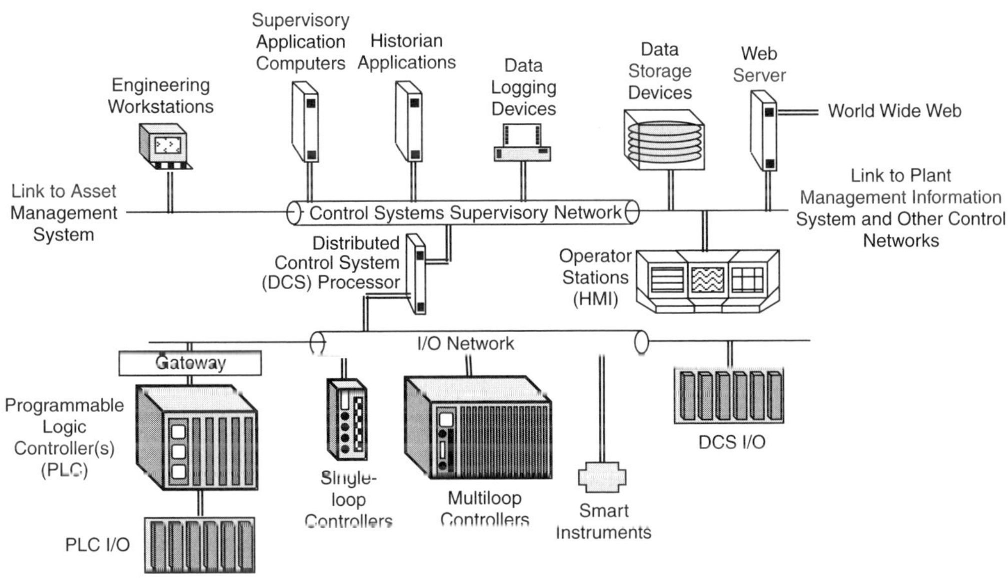

Source: Flintoff and Mular 1992.

FIGURE 3.48 Components of a distributed control system

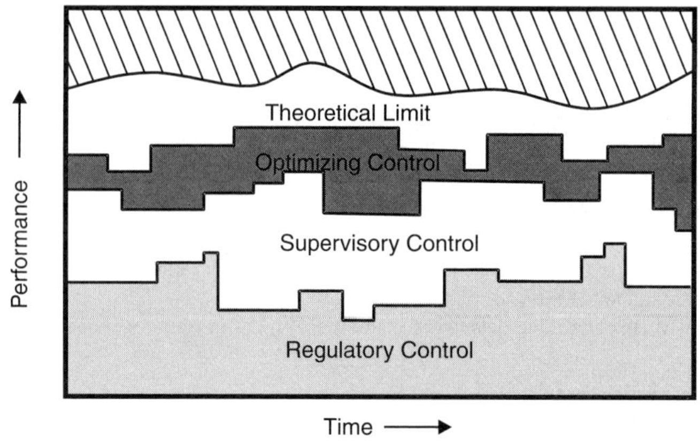

FIGURE 3.49 The levels of continuous control: Area under the curve represents improvements possible from better control

To provide some notion of scale, the I/O count (i.e., the total number of discrete, analog, and digital inputs and outputs) will range from about 2,000 for simpler, smaller plants to 6,000 for larger, more complex operations.

It is interesting that over the years the functionality and pricing of the PLC and the DCS have converged, yet two camps of mine operators have steadfastly maintained their loyalty to one approach or the other. More and more, new operations are inclined to choose one system or the other to host their process control functions. The implication is that one can develop a successful process control system using either approach, and probably at about the same overall cost. In the case of the PLC, the hardware itself is relatively inexpensive, but the system integration and configuration engineering are relatively expensive. The opposite is true for the DCS. The benefits of the latter approach may emerge in the simplification of ongoing system development.

In addition to these two major classes of control hardware, other options are sometimes encountered, such as multiloop stand-alone controllers, PC-based systems, and Supervisory Control and Data Acquisition (SCADA) systems. These tend to be associated with small process control problems, and in the case of SCADA, remote monitoring and control.

Control Strategies. Although there may be elements of discrete I/O in control strategies (e.g., opening or closing cyclones to maintain header pressure), they are normally designed to continuously modulate manipulable variables (e.g., feeder speed, valve position, pump or mill speed) to ensure that the controlled variables (e.g., ore flow, water flow, tank level, particle size) are at or near the set-point value. For that reason the emphasis in this section is on continuous control, and Figure 3.49 conveniently summarizes the levels of continuous control, while offering insight into the structure of control strategies.

Figure 3.49 illustrates that there is some performance benefit associated with each level of control strategy. It also carries the important implicit message that control strategies are hierarchical. That is, one cannot build an effective supervisory strategy if the regulatory strategies underpinning it are ineffective. This is not merely a point of academic interest, because numerous studies in all process industries have highlighted unexpectedly poor performance of the low-level controls. A similar argument can be advanced for optimizing controls, which must rely on the supervisory level.

The definitions of what fits where in this hierarchy are debatable, but some general attributes are associated with each level.

Regulatory Control Strategies

- Strategies are mostly implemented with feedback loops aimed at stabilizing process inputs, such as ore and water flows, or bin and tank levels.
- These loops almost always involve the PID controller, generally using only the Proportional-Integral-Differential (PID) functions.
- Typical control intervals range from one to a few seconds.
- They are always implemented on the DCS or PLC operating software.
- Occasionally dead-time compensation and gain scheduling are required (e.g., long conveyor belts and multiple feeders).
- For highly nonlinear systems, it may be necessary to resort to other control options, such as fuzzy logic or self-tuning controllers.

Supervisory Control Strategies

- These strategies calculate set points for the regulatory strategies in the pursuit of some operational objective, such as maximum throughput subject to a maximum particle constraint.
- The simplest form may be the cascade PID loop that delivers a set point to the associated regulatory loop.
- Typical control intervals range from a few seconds to a couple of minutes.
- They are almost always implemented on a dedicated PC on the DCS/PLC network.
- Strategies are frequently multivariable in nature; for example, attempting to control grinding circuit product particle size and recirculating load (see example below).
- They are prone to interaction problems; i.e., the multivariable nature of the problem leads to competition among the supervisory loops.
- They often require sophisticated approaches, such as heuristic, model-based, or blended approaches.

Optimizing Control Strategies

- These strategies calculate operating objectives for the supervisory strategies, based on some economic objective function.
- Typical control intervals range from a few minutes to an hour.
- They are almost always implemented on a dedicated PC on the DCS/PLC network.
- They tend to employ optimization techniques based on plant experimentation (SSDEVOP) or analytical techniques (e.g., multivariable search) that employ adapted process models.
- They tend to look at the coordination of several circuits to ensure that local optimization of each does not lead to the suboptimization of the whole plant.

Using a primary ball mill circuit, we can create a simple illustration of the structure outlined in Figure 3.47. Figure 3.50A shows the process flow, instrumentation layout, and typical regulatory controls for such a circuit. It should be noted that control strategies are usually documented through a combination of loop narratives and Process/Piping and Instrumentation Diagrams. In both cases, there are standards for preparation one should use, but in this example a quasi-PID representation is employed. There are four regulatory loops to stabilize inputs and internal variables. R1 is a PID loop that measures tonnage, W, and regulates feeder speed, VS, to maintain the set point, entered by the operator. R2 and R3 are PID water flow stabilization loops that ensure flow set points are maintained in the presence of variation in supply pressure. R4 is a sump-level control PID loop that would ensure the tank does not run dry or overflow.

There are a number of possible operating objectives for such a circuit, but here we assume that the goal is to maintain the product particle size at some set point, and to manage the circulating load so as to ensure maximum ore feed rate while avoiding a ball mill overload. The approach adopted for this

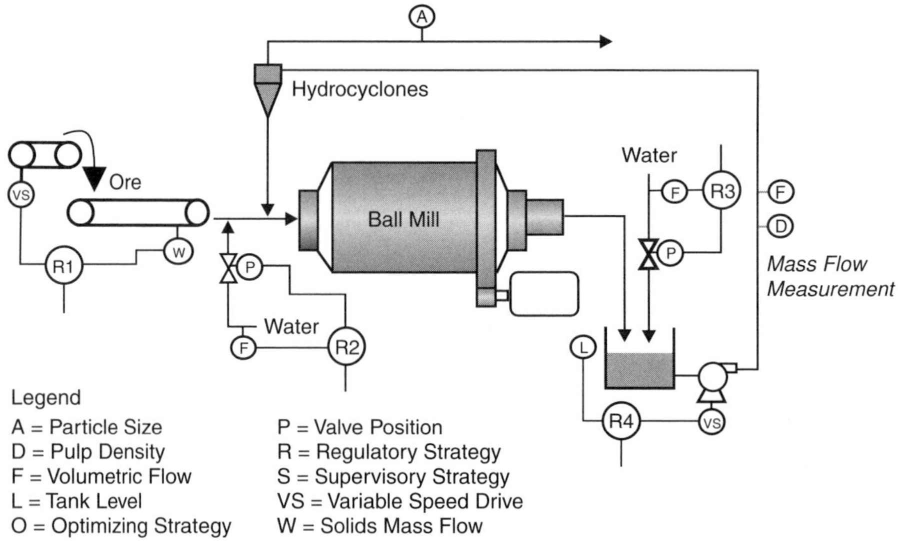

FIGURE 3.50A An illustration of regulatory control in a primary ball mill circuit

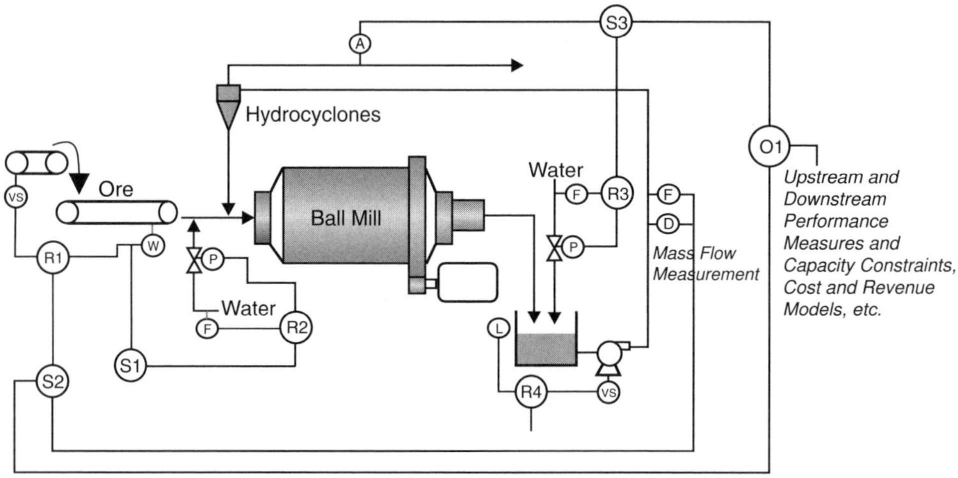

FIGURE 3.50B An illustration of supervisory and optimizing control in a primary ball mill circuit

example is to employ cascade PID loops, one delivering a set point to the sump water addition to maintain particle size, and the other sending a set point to the tonnage loop to maintain circulating load. This is shown in Figure 3.50B. Here S1 is the standard ratio controller aimed at maintaining a constant slurry density in the fresh feed to the mill, which could just as easily have been shown at the regulatory level. S2 is the supervisory cascade loop to regulate circulating load, and S3 is the particle size cascade loop. For completeness, the optimizing strategy is depicted as O1, and it endeavors to ensure that maximum revenue is generated across the plant by avoiding capacity imbalances that would lead to downtime, and by continually reevaluating the optimum grinding circuit throughput and particle size for maximum plant net revenue.

Efforts to implement this kind of supervisory and regulatory control strategy were quite common in the late 1970s and early 1980s, because the control hardware essentially limited the control engineer to PID tools, and some higher level coding (such as BASIC and FORTRAN). One can immediately imagine the problems arising from cascade loop competition because changes in sump water to control particle size affects recirculating load, and changes in ore feed rate to regulate recirculating load affects particle size. Clearly, this is a multivariable problem involving interactions, one which does not lend itself well to a classic multiloop PID approach. The solution was often to detune one of the loops, compromising control performance and paying the consequences to achieve reasonable circuit stability.

During the past 20 years or so, multivariable approaches have been used to solve this problem. Decoupling, model predictive control, model-based optimal control, fuzzy expert systems, and model-based expert control have all been successfully demonstrated in plant environments. Perhaps because of the apparent simplicity with respect to control strategy maintenance and development, fuzzy expert systems are probably the most common platform for developing such supervisory control strategies.

The development of a successful control strategy requires a very good understanding of process dynamics and operating characteristics, a broad knowledge of control tools, and the ability to clearly articulate the control problem; that is, what needs to be done and where the priorities lie. At a minimum, it will comprise a blend of regulatory and supervisory techniques that must work well, both independently and together. Although it has not been discussed previously, good operator training and strategy documentation are also required to achieve the benefits attached to the investment. In the past, the mineral processing control community has been rather poor at the last two steps, and as a consequence, has frequently been condemned to repeat history when new process control people are brought on-board.

Case Studies

To conclude the discussion on process control, two case studies are presented. One is taken from crushing and the other from grinding, reflecting the general focus of this chapter.

Crushing Case Study. In the new millennium crusher control applications in mineral processing will be largely restricted to primary and autogenous/semiautogenous pebble crusher applications. Nevertheless, there are still some crushing plants in operation, and from a pedagogical perspective, some interesting lessons can be learned from the control work that was performed in these types of operations.

There are a number of articles on crushing-plant control in the technical literature (Norby and Hales 1986; Flintoff and Edwards 1992; Manlapig, Thornton, and Gonzalez 1987). The example presented here is for a secondary crusher at the Brenda Mines concentrator and is described in more detail in Flintoff and Edwards (1992).

Figure 3.51 provides the process and instrumentation layout required for the example. (The reader should understand that, as in the case of Figure 3.47, much of the instrumentation used for monitoring and control of this circuit has been omitted to enhance clarity.) Because the secondary crushing circuit was proving to be a bottleneck for overall plant production, the general objective was to increase throughput.

A number of control problems in this circuit rendered more traditional methods ineffective, including the following:

- Given the scale of the equipment, there was a significant dead time between the weigh scale and the feeders, and between the weigh scale and the crusher.
- Stockpile segregation meant varying particle size distributions from each feeder; i.e., the tonnage corresponding to maximum throughput or maximum power varied depending on the feeder configuration in use.
- Vibrating feeders were prone to both hang-ups and sloughs of material onto the belt.

To tackle these problems, the control engineers began by employing a regulatory loop to control solids mass flow by manipulating the vibration frequency of the feeders. Because PID was ineffective, a

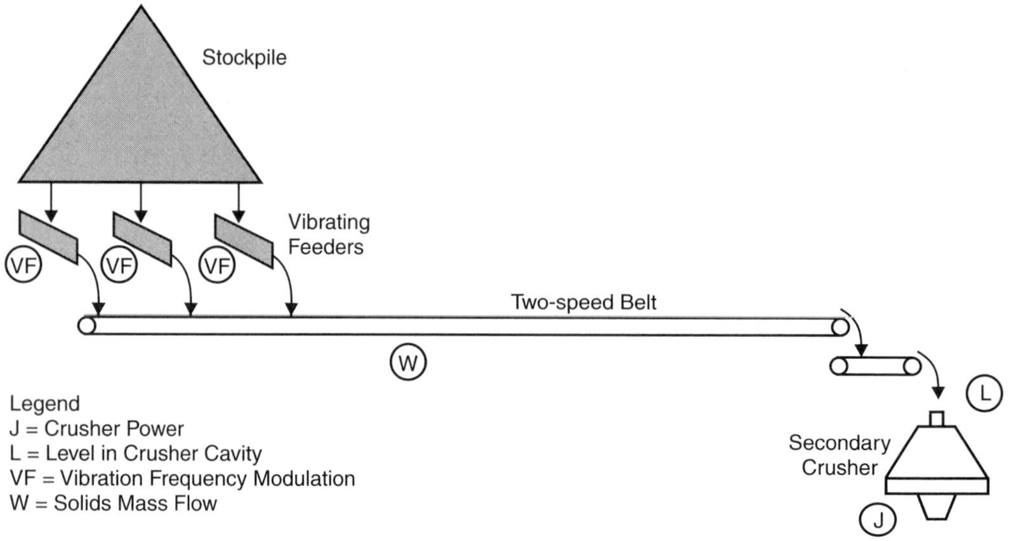

FIGURE 3.51 Secondary crusher control process and instrumentation layout

dead-time compensation scheme (Dahlin algorithm) was employed to improve the control performance. In addition, and because the number and specific configuration of the feeders could be changed, the regulatory control algorithm also included scheduling for process gain (i.e., based on the number of feeders running) and the estimated dead time (i.e., based on the specific configuration of feeders running). This proved to be a very effective approach to achieve good regulatory behavior, and it is of some interest to see that similar techniques are now used on large semiautogenous mill feed systems, which suffer the same control problems (Vien et al. 2000)

Having solved the regulatory problem, the supervisory strategy was to ensure maximum throughput. Because this crusher treated a scalped primary crusher discharge (~200 mm × 19 mm), level control in the cavity is not generally a suitable means of control, because coarse hard feed will lead to a plugged crusher. (However, level is monitored to prevent spillage and to aid in control when the ore is fine, soft, or both.) In this particular instance, maximum throughput generally relates to maximum power draw.

Given the variability in size and ore hardness, it is intuitive that the relationship between power and tonnage is nonlinear. Approaches ranging from fuzzy expert control to self-tuning controllers have been applied to this nonlinear problem, but the control engineers in this instance elected to use a clever implementation of Model-Reference Adaptive Control (MRAC). Concisely, to accommodate the dead time between the weigh scale and the crusher, as well as the variable power versus the tonnage relationship, an empirical model was employed to predict power from tonnage. Using historical data from the crushing circuit, the appropriate model form was deduced to be

$$P_{t+k} = A_t(b_0 + b_1 \dot{M}_t^{\,b_2}) \qquad \textbf{(Eq. 3.23)}$$

where

P_{t+k} = the predicted value of power k time steps ahead of the current time t (where k quantifies the dead time between the weigh scale and the crusher)

A_t = the adaptive parameter

b_0, b_1, b_2 = model coefficients deduced in off-line studies

$\dot{M}_t$ = tonnage measured at time t

MRAC usually requires a least-squares on-line adaptation of the variable parameters, but the central processing unit and random access memory limitations in the control hardware led this group to use a similar approach based on a first order digital filter:

$$A_t = (1 - \alpha_F)\left[\frac{P_{t-1}}{P_{t-1}}\right] + \alpha_F A_{t-1} \quad \textbf{(Eq. 3.24)}$$

where

α_F = digital filtering constant

P_{t-1} = measured power at time $t - 1$

To complete the MRAC installation, the engineers chose to remove tonnage from the regulatory loop and substitute power. In other words, tonnage was used only for power prediction purposes, and the dead-time compensation controller was effectively regulating power.

An analysis of production data before and after the installation of this control strategy revealed a 15% production increase with the MRAC approach. The financial benefit is difficult to resolve, although a testimonial by the lead control engineer follows (see Chapter 2 and Chris Larsen, Brenda Mines Ltd., as quoted in Flintoff and Mular 1992): "The incremental value of production due to computer control in the most successful areas of crushing and grinding ranges from \$3 million to \$5 million per year, depending on metal prices."

Before concluding this example, there are some additional observations that need to be made. One of the leaders of modern advanced control research has observed that some extremely important heuristics are required for good overall controller performance, although this area has not attracted much attention from researchers (see Åström 1986). Because this comment remains true today, the practicing control engineer needs to be cognizant of such a requirement. These heuristics are "wrapped" around algorithms, such as the one outlined above; they have been variously called safety jackets, supervision safety nets, or watchdog software. In the context of the case study described here, one example should suffice for illustrative purposes.

Previously, it was indicated that vibrating feeders are prone to sloughing. A large pile of rock on the belt would cause the controller to make a quick reduction in feeder speed, but it would soon return to near normal, once the pile passes the weigh scale. However, this large pile of rock may well be sufficient to plug the crusher. The operator, who would be responsible for cleanup, would soon switch the power controller into manual mode, citing excessive downtime, for example. To circumvent this eventuality, a watchdog function monitors the weight profile on the belt, and when a large pile is discovered, it will suspend the supervisory controls, slow the feeders, decrease the belt speed, and hold this condition until the pile of material is known to have passed through the crusher, whereupon normal control is restarted. There are numerous other examples of watchdog control in this specific case, and their existence is one of the principal drivers behind the embrace of fuzzy expert systems as the platform of choice for supervisory level automation in mineral processing.

Grinding Case Study. With their relatively high capital and operating cost, grinding circuits have been the focus of much of the attention in industrial process control for the past four decades. There are other contributing factors, such as the fact that these circuits are fairly well understood from a phenomenological point of view, and that in many plants grinding turns out to be the bottleneck in economic optimization. This latter point often leads to maximum-throughput strategies, which have the additional complication of accommodating the physical capacity constraints of the equipment.

Figure 3.52 is a simplified representation of the flowsheet for the case study. (Details of this application are provided by Samskog et al. 1996.) The control objective is to maximize throughput while maintaining a product particle size dictated by downstream production processes.

This particular circuit was well instrumented and had very good regulatory controls and well-trained operators using modern control hardware with an excellent HMI. Nevertheless, frequent changes in ore hardness tended to lead to conservative operation to avoid overloads. More specifically,

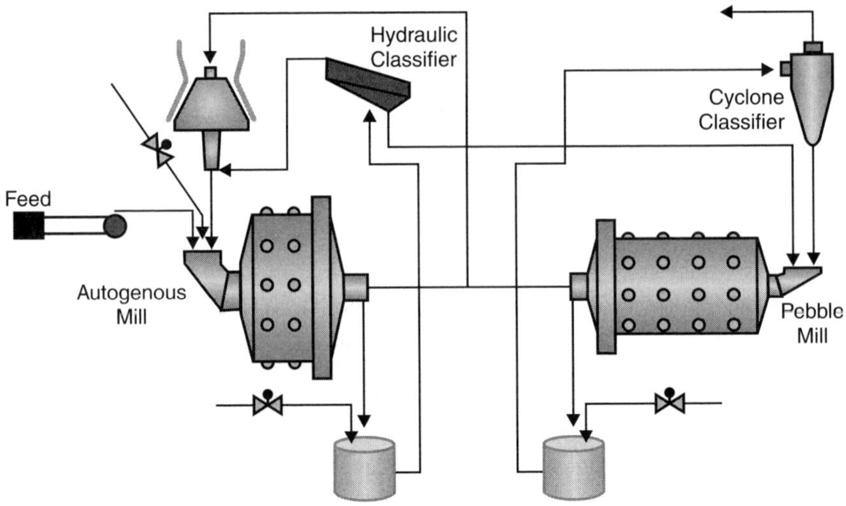

FIGURE 3.52 Flowsheet for autogenous and pebble mill circuit

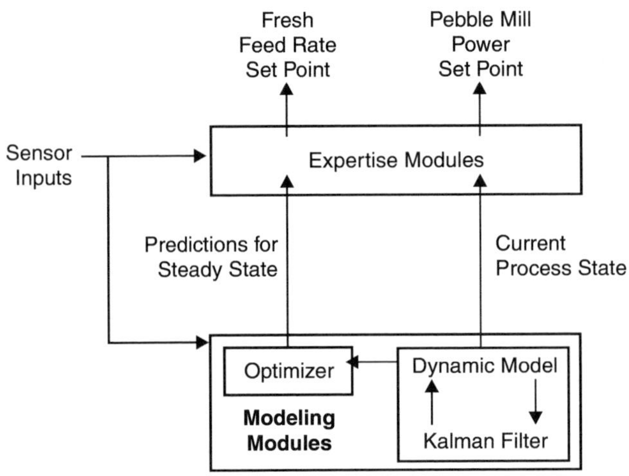

Source: Samskog et al. 1996.

FIGURE 3.53 Software structure of the MBEC system

in soft ore that is lean in coarse-grinding media, the autogenous mill operates with low power draw and a high circulating load of pebbles. The manual supervisory strategy was to set the fresh feed rate and to change it only when the pebble recycle stream reached high or low values. In cases where the ore was hard, the autogenous mill tended to run at high loads and high power draws, with low pebble recycle. Once again, the manual supervisory strategy was to run at a conservative feed rate to avoid overloading the autogenous mill (i.e., high charge levels). In both cases, it was difficult to maintain product particle size, and opportunities to increase the feed rate were often missed.

In this case a Model-Based Expert Control (MBEC) supervisory strategy was implemented. The basic structure for such an algorithm is shown in Figure 3.53. It consists of the heuristics encoded in the expertise modules, as well as deep process knowledge, encoded in the phenomenological mathematical models.

This structure utilizes the models to estimate the state of the process, including the prediction of many variables that would otherwise not be measurable (i.e., a soft sensor). Because there are temporal changes in the feed and equipment characteristics, the model is adapted to ensure a minimum of plant-model mismatch. Both of these functions are accomplished by embedding the model in an extended Kalman Filter (Pate and Herbst 1999). One of the outputs of this modeling module is the current process state, which can be used directly by the expertise modules. The other is a well-tuned dynamic model, which can be used with the process state information in the optimization modeling module. In this case, the optimizer simply integrates the model to predict the steady-state results if no disturbances were to enter the system, under any particular combination of regulatory loop set points. The results of these steady-state predictions can then be used to recommend regulatory loop set points that will achieve optimal grinding performance.

Because the grinding circuit cannot be perfectly modeled, and because there is a need for watchdog functionality, the expert modules play an important role in the supervisory control strategy. A particularly important aspect is to filter the set points coming from the optimizer, because these are based exclusively on model calculations, which in turn can be sensitive to sensor problems in the field signals.

Extensive on-off testing of the MBEC supervisory strategy against the manual model of operation demonstrated a 6% improvement in grinding circuit throughput and a much lower variance on product particle size distribution. The economic impact was not disclosed, although the payback period was said to be a few months.

To conclude, it is worth noting that the approach taken in this latter case study is rapidly becoming the industry standard. The use of process models is highly recommended, and despite any intuitive belief to the contrary, the model-based component does not necessarily imply a need for many sensors (see Broussard and Guyot 2001, for example.) Modifying an old adage, the governing phrase here may be that "An equation is worth a thousand rules." For supervisory control in comminution, best practices mean a mix of expert systems and mathematical models. The exclusive use of one or the other will likely lead to suboptimal results.

FINANCIAL ASPECTS OF COMMINUTION

Costs Associated with and Profit Derived from Size Reduction and Liberation

As mentioned in the introduction to this chapter, the cost of comminution operations is typically a very significant proportion of the total cost for mineral processing. Comminution costs are conveniently divided into two parts: Capital costs (the original cost of equipment and its installation) and operating costs (the day-to-day costs associated with power, wear parts, maintenance, and labor provided by operators).

Typical capital costs are shown in Tables 3.15 and 3.16 for a copper ore and an iron ore operation, respectively. The total investment cost for the copper crushing and grinding plant is about $48.8 million, expressed in 2001 dollars. The total investment cost for the iron ore crushing and grinding plant is about $54.2 million, expressed in 2001 dollars.

It should be kept in mind that the hardness of these two ore types differ by 113% and that the final product size of each is different. Copper ore product size required for flotation is about 80% passing 100 mesh, whereas the iron ore size required for making iron ore pellets is approximately 80% passing 270 mesh. Note that in each case the installed cost is roughly 1.5 times the equipment cost. The installed cost includes foundations, buildings, wiring, and basic regulatory controls. In addition, the cost of equipment varies roughly as the size reduction effort; that is, it must deliver the cost of the maximum power it is capable of drawing during breaking operations.

Finally, operating costs vary with comminution device type, ore type, feed size and product size, local energy and labor costs, media and wear protection materials used, and equipment operating

TABLE 3.15 Approximate investment costs for an 85,000-tpd copper crushing and grinding plant

	Installed Cost, \$M/unit	Equipment Cost, \$M/unit	Power Draw, kW/unit
Crushers			
Gyratory	4.8	3.5	450
Shorthead	3.1	2.2	300
Ball mills	7.0	4.8	2,500
Autogenous mills	12.7	8.7	6,000

TABLE 3.16 Approximate investment costs for a 25,000-tpd iron ore crushing and grinding plant

	Installed Cost, \$M/unit	Equipment Cost, \$M/unit	Power Draw, kW/unit
Crushers			
Gyratory	2.7	1.9	450
Cone	1.5	1.1	250
Rod mills	3.9	2.7	2,000
Ball mills	6.5	4.5	4,500

TABLE 3.17 Approximate operating costs for an 85,000-tpd copper ore crushing and grinding plant

	Operating Cost (\$/day/unit)			
Equipment	**Liners**	**Media**	**Energy**	**Maintenance**
Crushers	1,890	—	1,260	3,494
Ball mills	305	2,883	3,055	611
Autogenous mills	535	—	7,484	3,322

TABLE 3.18 Approximate operating costs for a 25,000-tpd iron ore crushing and grinding plant

	Operating Cost (\$/day/unit)			
Equipment	**Liners**	**Media**	**Energy**	**Maintenance**
Crushers	430	—	290	800
Rod mills	350	2,000	1,120	500
Ball mills	130	1,550	5,600	1,120

modes and maintenance programs. Typical copper and iron ore costs are presented in Tables 3.17 and 3.18.

A more general division of operating costs is shown by equipment type in Figure 3.54. Note that the relative costs for energy and wear parts, liners and grinding media are different for different mill types.

When taken in total, these costs (capital plus operating) seem to be huge, but an analysis of the net present value (which takes into account the comminution plant revenue minus operating costs discounted to their current value) of a comminution operation for a 10-year period for an 85,000-tpd copper ore operation and a 25,000-tpd iron ore operation in 2001 in North America, for example, predicts profitable performance. As long as the metal values of the ores are high enough and the overall finenesses of the product required for separation are not too small, the overall economics of comminution will always be quite favorable.

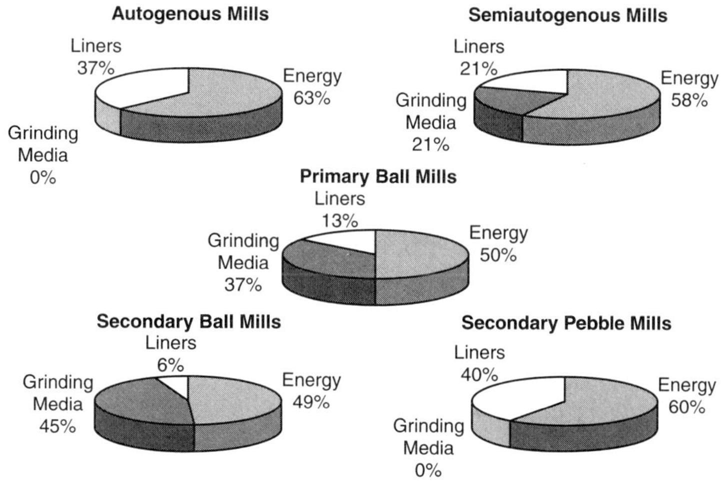

FIGURE 3.54 Pie graph representation of the distribution of operating costs for different devices and applications

SYMBOL GLOSSARY

Latin Symbols

A_i	abrasion index
A_t	adaptive parameter in crusher at time t
a	crack length (l)
B	nxn matrix of breakage functions
b_{ij}	fraction of daughter fragments from the breakage of parents of size I into progeny of size j or breakage function
b_0, b_1, b_2	empirical constants
D	mill size (l)
d, d'	particle size (l)
d_{50}	50% passing size (l)
d_{80}	80% passing size (l)
d_B	ball size (l)
d_g	gape size (l)
E	specific energy input ($e/m = l^2/t^2$)
e	energy (ml^2/t^2)
F	force (ml/t^2)
F_3	cumulative mass fraction finer
G	crack extension energy ($e/l = l/t^2$)
H	hold-up mass (m)
k	dimensionless constant
L	mill length (l)
M	mass flow rate (m/t)
M	nx1 vector of mass fractions
m_i	mass fraction in ith size interval
N	mill rotation speed (rev/t)
N^*	fraction of critical speed
P	power ($e/t = ml^2/t^3$)
p_i	probability of breakage of particle in ith size class

Latin Symbols

S	nxu diagonal matrix of selection functions
S_i	time-based selection function (t^{-1})
S_i^E	energy-based selection function (m/e)
t	time (t)
V	velocity (l/t)
V_B	volume of balls (l^3)
V_B*	fraction of mill volume occupied by balls
V_M	volume of mill (l^3)
V_p	volume of particles (l^3)
V_P*	fraction of interstitial ball volume occupied by particles
W	work ($e/m = l^2/t^2$)
w	roller width (l)
W_i	Bond Work Index (e/m $(l)^{0.5}$)
Y	module of elasticity (stress/strain = m/lt^2)

Greek Symbols

α	Exponent in energy-size relationship
β	Specific fracture surface energy (e/area = m/t^2)
γ	Specific surface free energy (e/area = m/t^2)
ε	Porosity
ρ	Crack radius (l)
σ	Stress (f/area = m/lt^2)
ν	Poisson's ratio

ACKNOWLEDGMENTS

The authors of this chapter wish to acknowledge the support of A. Potapov, X. Qiu, and L. Nordell for DEM simulations, W.T. Pate for flowsheet simulations, and J. Lichter for input on stirred milling.

BIBLIOGRAPHY

Åström, K.J. 1986. Expert Control. *Automatica*, 22(3):277–286.

Austin, L.G. 1973. Understanding Ball Mill Sizing. *Ind. Eng. Chem. Process Design and Development*, 12(2):121.

Azzaroni, E. 1984. Calculo de la Tasa de Molienda en Molinos de Bolas. In *IV Simposium Sobre Molienda*, Vina del Mar, Chile: 287.

Barratt, D.J. 1979. Semi-autogenous Grinding–A Comparison with the Conventional Route. *CIM Bulletin*, 11:74.

Barrientos, R., and M. Telias. 1997. Nuevos Sensores en el Circuito SAG. In *Proc. SAG Workshop*, Vina del Mar, Chile.

Bond, F.C. 1952. The Third Theory of Comminution. *Trans. AIME*, 193:484.

——. 1958. Grinding Ball Size Selection. *Mining Engineering*.

——. 1961. Crushing and Grinding Calculation. Reprints from *Chemical Engineering*.

Bond, F.C. 1963. Metals Wear in Crushing and Grinding. In *American Institute of Chemical Engineers, 54th meeting*; also A-C Bulletin 07P1701. Milwaukee, Wisc.: Allis-Chalmers.

Broadbent, S.R., and T.G. Callcott. 1956. A Matrix Analysis of Processes Involving Particle Assemblies. *Phil. Trans*.

Broussaud, A., and O. Guyot. 2001. Experience of Advanced Control of AG and SAG Mills with Comprehensive or Limited Instrumentation. In *Proc. Int'l AG and SAG Grinding Technology 2001*. Edited by D.J. Barratt, Allan and A. Mular. Vancouver, B.C.

Cho, K. 1987. *Breakage Mechanisms in Size Reduction*. Ph.D. diss. University of Utah, Salt Lake City.

Connell, J. 1988. The Laws of Process Control. Process Industries in Canada: 4.

Digre, M. 1979. Autogenous Mill Design Factor. In *Proceedings Autogenous Grinding Seminar*. Trondheim, Norway.

Flintoff, B., and R. Edwards. 1992. *Process Control in Crushing in Comminution–Theory and Practice*, Littleton, Colo.: SME.

Flintoff, B., and A. Mular, eds. 1992. *A Practical Guide to Process Controls in the Minerals Industry*. Vancouver, B.C.: Gastown Printers.

Gaudin, A.M., and T.P. Meloy. 1962. Model and a Comminution Distribution Equation for Repeated Fracture. *Trans. AIME*, 223:43.

Griffith, A. 1920. *Phil. Trans. R. Soc.*, 221A:163.

Herbst, J.A. 2000. Model Based Decision Making for Mineral Processing–A Maturing Technology, In *Control 2000*. Littleton, Colo.: SME.

Herbst J., and O. Bascur. 1984. Mineral Processing Control in the 1980s: Realities and Dreams. In *Control '84*. Edited by J.A. Herbst, D.B. George and K.U.S. Sastry. Littleton, Colo.: SME.

Herbst, J.A., and D.W. Fuerstenau. 1972. The Influence of Mill Speed and Ball Loading on the Parameters of the Batch Grinding Equation. *Trans. AIME*, 252:169.

——. 1973. Mathematical Simulation of Dry Ball Milling Using Specific Power Information. *Trans. AIME*, 254:373.

——. 1980. Scale-up Procedure for Continuous Grinding Mill Design Using Population Balance Models. *Int. J. Miner. Process.*, 7(1):1.

Herbst, J.A., G.A. Grandy, and T.S. Mika. 1971. On the Development and Use of Lumped Parameter Models for Continuous Open- and Closed-circuit Grinding Systems. *Transactions of the Institution of Mining and Metallurgy*, 80:C193.

Herbst, J.A., and L. Nordell. 2001. Optimization of the Design of SAG Mill Internals Using High Fidelity Simulation. In *SAG 2001*.

Herbst, J.A., and W.T. Pate. 2001. Dynamic Modeling and Simulation of SAG/AG Circuits With MinOOcad: Off-line and On-line Applications. In *SAG 2001*.

Herbst J., W. Pate, and E. Oblad. 1989. Experiences in the Use of Model Based Expert Control Systems in Autogenous and Semi Autogenous Grinding Circuits. In *Proc. Advances in AG and SAG Grinding Technology*. Vol. 2. Edited by A.L. Mular and G.E. Agar. Vancouver, B.C.: .

Herbst, J.A., K. Rajamani, and D.J. Kinneberg. 1977. ESTIMILL A Program for Grinding Simulation and Parameter Estimation with Linear Models. University of Utah, Salt Lake City, Utah.

Herbst, J.A., and J.L. Sepulveda. 1985. Particle Size Analysis. In *SME Mineral Processing Handbook*. Edited by N.L. Weiss. New York: AIME.

Höfler, A. 1990. *Fundamental Breakage Studies of Mineral Particles with an Ultrafast Load Cell Device*. Ph.D. diss. University of Utah, Salt Lake City.

Irwin, G.R. 1961. Plastic Zone Near a Crack and Fracture Toughness. In *Sagamore Research Conference Proceedings*, Vol. 4.

Lo, Y.C., and J.A. Herbst. 1986. Consideration of Ball Size Effects in the Population Balance Approach to Mill Scale-Up. In *Advances in Mineral Processing*. Edited by P. Somasundaran. New York: AIME.

MacPherson, A.R., and R.R. Turner. 1978. Autogenous Grinding from Test Work to Purchase of a Commercial Unit. In *Mineral Processing in Plant Design*. Edited by A.L. Mular and R.B. Bhappu. New York: AIME.

Manlapig E., A. Thornton, and G. Gonzalez. 1987. Application of Adaptive Control in the Copper Concentrator, Mt. Isa Mines, Copper 1987, Vol. 2. 389–407.

Morari, M. 1983. Design of Resilient Processing Plants III, *Chem. Engng. Sci.*, 38(11):1881–1891.

Norby, C., and L. Hales. 1986. *Crushing Control at Kennecott's Ray Mines Division*. Preprint. New York: AIME.

Nordell, L., A. Potapov, and J.A. Herbst. 2001. Comminution Simulation Using Discrete Element Method (DEM) Approach—From Single Particle Breakage to Full-scale SAG Mill Operation. In *SAG 2001*.

Pate, W.T., and J.A. Herbst. 1999. *MinOOcad Manual*. Kelowna, B.C.: Svedala Process Technology.

Potapov, A., and C. Campbell. 1996. A Three-dimensional Simulation of Brittle Solid Fracture. *Int. J. Mod. Phys. C*, 7:(5):717–729.

Rowland, C.A., and D.M. Kjos. 1978. Rod and Ball Mills. In *Mineral Processing Plant Design*. Edited by A.L. Mular and R.B. Bhappu. New York: AIME.

Rumpf, H. 1961. Material Prufung, 3:253.

Samskog, P.O., P. Soderman, J. Bjorkman, O. Guyot, and A. Broussaud. 1996. Model-Based Control of Autogenous and Pebble Mills at LKAB Kiruna KA2 Concentrator. In *Proc. Intl. AG and SAG Grinding Technology*. Vol. 2. Edited by A.L. Mular, D.J. Barratt, and D.A. Knight. Vancouver, B.C.

Schneider, C. 1995. Measurement and Calculation of Liberation in Continuous Grinding Circuits. Ph.D. diss. University of Utah, Salt Lake City.

Schönert, K. 1979. Aspects of the Physics of Breakage Relevant to Comminution, Fourth Tewksburg Symposium, Melbourne, Australia.

——. 1980. *Zerkleinern*. Institut fur Mechanische Verfahren Stechnik der Universitat Karlsruhe, Germany.

——. 1995. Comminution from Theory to Practice. In *Proc. of the XIX Int. Mineral Proc. Congress*, San Franciso, CA. Littleton, Colo.: SME, 1:7.

Seborg, D., T. Edgar, and D. Mellichamp. 1983. *Process Dynamics and Control*. New York: John Wiley & Sons.

Sepulveda, J.L. 1981. A Detailed Study on Stirred Ball Milling. Ph.D. diss., University of Utah, Salt Lake City.

Vien A., J. Palomino, P. Gonzalez, and R. Perry. 2000. Multiple Feeder Control. In *Proc. 32rd AGM Can. Min. Proc.* Montreal, Quebec: Canadian Institute of Mining, Metallurgy and Petroleum.

CHAPTER 4

Size Separation

Andrew L. Mular

INTRODUCTION

Size separation is the parceling of particulate material on the basis of size (Luckie 1984). In mineral processing plants, such parceling means that the transfer of material unsuited for a specific processing step (such as the transfer of fines to a primary jaw crusher or the transfer of oversize to flotation) is avoided to improve the performance or efficiency of equipment or metallurgical processes. Devices employed for size separation may be screens (grizzlies, fixed screens, revolving screens, shaking screens, and vibrating screens) or classifiers (nonmechanical classifiers, mechanical classifiers, cyclone classifiers, and pneumatic classifiers). Screens allow certain particles to pass through screen apertures, whereas classifiers act on particles suspended in a medium to separate them based on differences in characteristics such as particle size and specific gravity. In general, classifiers behave like imperfect screens.

Mineral processing circuits employ sizing devices for various reasons (Figure 4.1). Thus, a primary jaw crusher is fed with oversize from a grizzly to minimize packing by fines in the crushing chamber (Figure 4.1A). Secondary crusher and tertiary crusher discharge is conveyed to double-deck vibrating screens, which produce undersize for fine ore bins and oversize for tertiary crusher feed. This sequence eliminates packing by fines in tertiary crushers, reduces production of finely sized material, and (often) maximizes circuit throughput (Figure 4.1B). A hydrocyclone treats rod and ball mill discharge slurry to produce an overflow (fines stream) for flotation and an underflow (coarse stream) for ball mill feed. This sequence provides finished flotation feed immediately, minimizes slimes production, and permits higher circuit throughput (Figure 4.1C). An air cyclone treats dust from a screening plant collection system to eliminate fine particles from air and recover fine-valuable minerals otherwise lost to the atmosphere. Here, the air cyclone acts as a predust collector to reduce the dust load to the bag house (Figure 4.1D).

Size separators may serve a variety of other purposes in mineral processing, such as improving grinding circuit efficiency by continuously removing the product of final size from the circuit, improving metallurgical performance by desliming before flotation, or classifying the feed to a tabling operation. Applications include dewatering, trash removal, conveying, and media recovery. In other fields, size-separation devices produce narrow-size fractions of material for purposes such as road building and dam construction. Sizing devices may be placed in series or in parallel (i.e., multiple staging) for greater sizing efficiency or capacity. Sizing efficiency and capacity are related, as is discussed in subsequent sections of this chapter.

Size-separation devices commonly are used with certain size ranges (Figure 4.2). Generally, screening devices are used to make coarser separations and classifiers are employed for finer ones. However, size ranges can overlap substantially.

Size distributions and mass balances associated with corresponding feed and product streams of typical sizing devices used in large-tonnage mineral processing plants are shown in Figure 4.3. The coarse product of screens is called the "oversize," whereas the fine product is referred to as the

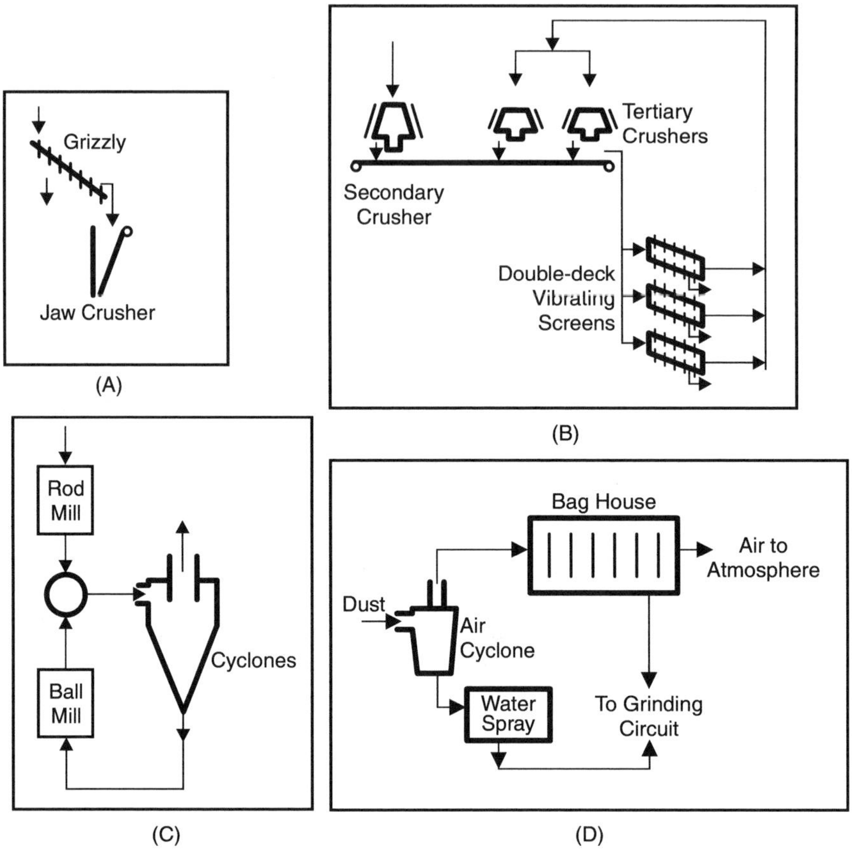

FIGURE 4.1 Applications of sizing devices

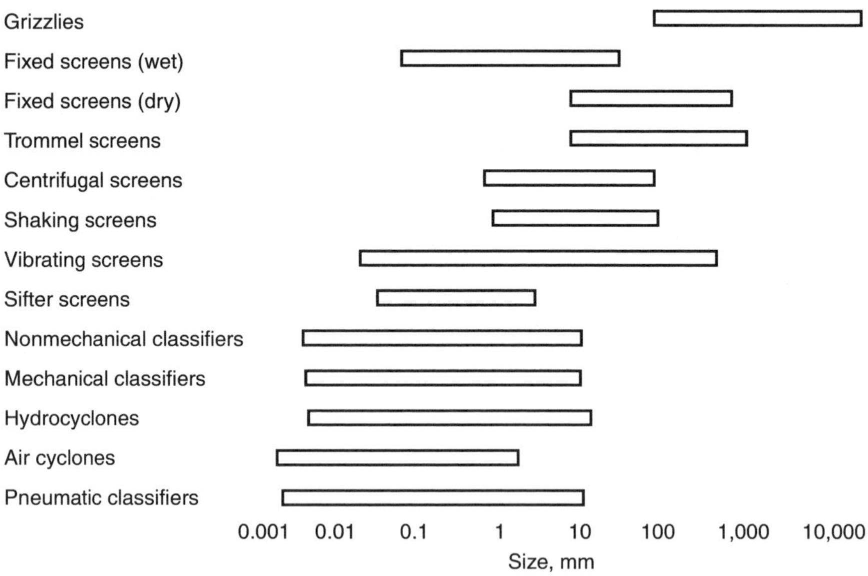

FIGURE 4.2 Typical size ranges treated by common size-separating devices

(A)

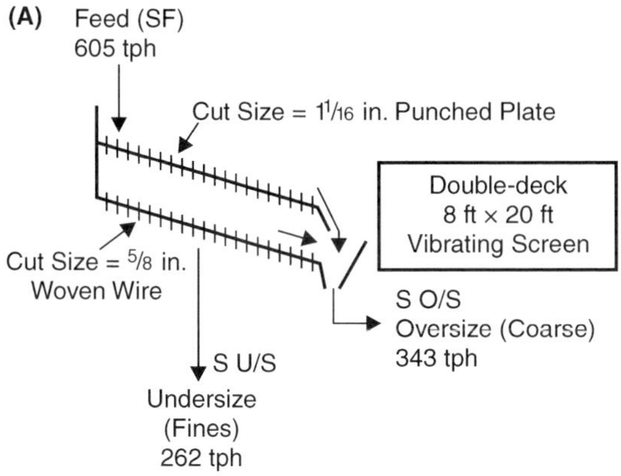

% Retained:

Size, cm	SF	S U/S	S O/S
21.76	0	0	0
10.88	0.063	0	0.118
5.44	13.856	0	26.085
2.72	19.322	0	36.374
1.36	28.692	22.874	33.826
0.68	15.478	30.944	1.83
0.34	8.774	17.914	0.708
0.17	4.717	9.65	0.363
0.085	3.088	6.319	0.236
0.0425	2.074	4.245	0.159
0.0212	1.389	2.843	0.106
0.0106	0.918	1.879	0.07
0.0053	0.601	1.23	0.046
–0.0053	1.027	2.102	0.078

% Passing:

Size, cm	SF	S U/S	S O/S
21.76	100	100	100
10.88	99.937	100	99.882
5.44	86.081	100	73.797
2.72	66.758	100	37.423
1.36	38.066	77.126	3.567
0.68	22.588	46.182	1.767
0.34	13.814	26.268	1.058
0.17	9.098	18.618	0.695
0.085	6.01	12.299	0.459
0.0425	3.935	8.054	0.301
0.0212	2.546	5.211	0.195
0.0106	1.628	1.628	0.124
0.0053	1.027	2.102	0.078

(B)

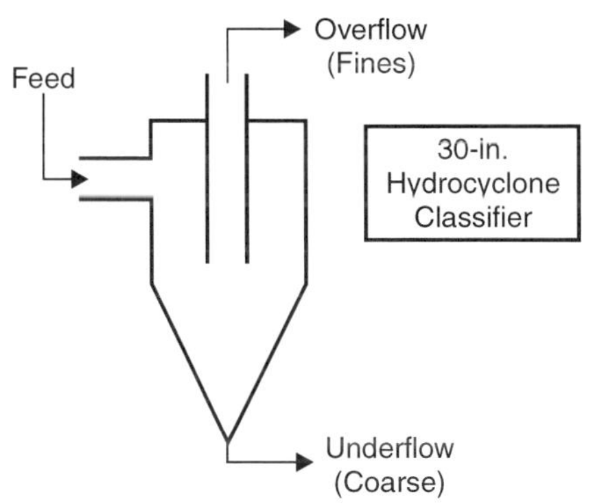

	Percent Retained on Size		
Size, mesh	Cyclone Feed	Overflow	Underflow
3	0.01	0	0.01
6	0.29	0	0.40
12	3.02	0.16	4.02
20	9.77	0.36	13.10
40	21.29	1.82	28.17
70	26.36	18.34	29.19
140	14.03	22.40	11.00
270	7.09	14.38	4.51
–270	17.54	42.48	8.74

	Main Balance, tph		
Constituent	Cyclone Feed	Overflow	Underflow
Solids, tph:	495.79	129.42	366.37
Water, tph:	297.9	181.40	116.50
Slurry, tph:	793.69	310.82	482.87

Cyclone diameter = 30 in., Vortex = 9 in., Spigot = 5.5 in.

FIGURE 4.3 Size distributions and balances around size separators

"undersize." The coarse product of classifiers that treat slurry is called the "underflow," and the fine product is called the "overflow."

LABORATORY SIZE SEPARATION

Laboratory sizing devices are employed to obtain size fractions or size splits for a variety of studies, such as to determine the degree of liberation of an ore, to assess the effect of size on efficiency or performance of processing equipment, to measure particle size distributions, and to ascertain whether size specifications at various points in mineral processing circuits are maintained. In any case, the definition of particle size becomes important (Malghan and Mular 1982) because the "size" may be dependent on the sizing device employed.

Size of Single Particles and Size Distributions of Particle Assemblies

Traditionally, the size of a single particle has referred to a single dimension that is used to determine particle spatial extent (Herbst and Sepulveda 1985). Thus, the size of a sphere is its diameter, d, where its area is proportional to d^2 and its volume is proportional to d^3. Area and volume are important size variables related to d, where the constants of proportionality (π for area and $\pi/6$ for volume) are called shape factors. For a particle of irregular shape, particle diameter depends on the sizing method. Shape

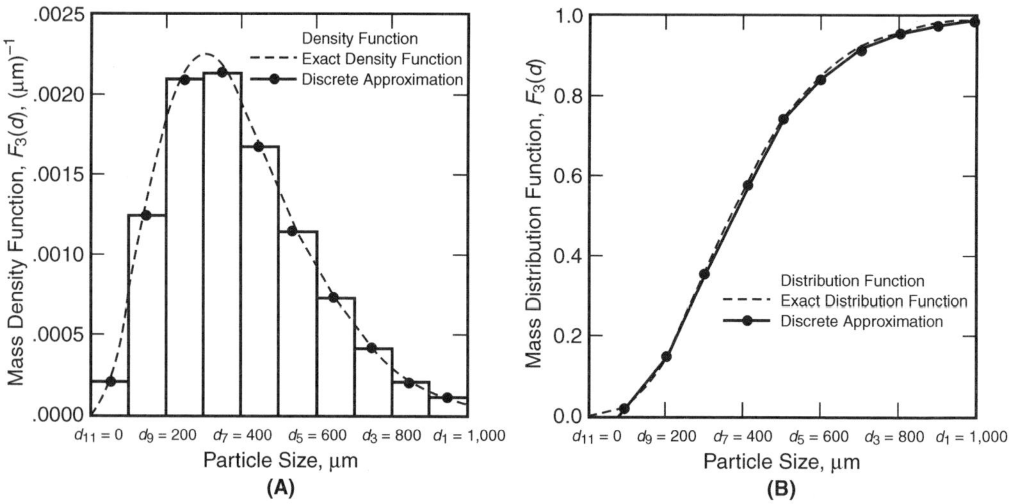

Source: Herbst and Sepulveda 1985.

FIGURE 4.4 Plots illustrating (A) discrete approximations to density function and (B) distribution function

factor ratios vary with diameter (Herbst and Sepulveda 1985) and additional shape factors, such as sphericity and circularity (where sphericity is the ratio of the surface area of a sphere to actual surface area of a particle of equal volume and circularity is the ratio of the perimeter of a circle to the actual perimeter of a particle of equal projected area) may be useful. Any one of several possible diameters of a real particle, as defined in Chapter 1, can be used in mineral processing. For practical purposes, sieve diameter and Stokesian settling diameter are the measures of particle size employed in subsequent sections of this chapter. Sieve diameter is the aperture of a square mesh sieve that just retains the particle; Stokesian settling diameter is the diameter of a sphere of the same specific gravity and settling velocity as that of the real particle settling in the same liquid (water) under laminar flow.

When an assembly of particles is produced such as by comminution, the assembly has a distribution of sizes. Hence, it becomes important to specify the quantity (e.g., mass) of particles of a given size in the assembly. A suitable plot of quantity versus size, often with quantity expressed as a mass percentage, will represent the relative frequency of occurrence of mass percentage of particles of a given size (a density function). These curves are readily transformed to a cumulative form, which is widely used in mineral processing and is analogous to a cumulative distribution function encountered in statistical texts. Figures 4.4(A) and 4.4(B) illustrate the essential idea, where the dotted curves represent exact functions superimposed on discrete approximations (Herbst and Sepulveda 1985).

Empirical equations (Herbst and Sepulveda 1985; Harris 1968; Gaudin and Meloy 1962; Schuhmann 1948; Bergstrom 1966) can be fitted to cumulative distribution plots of discrete data. Table 4.1 shows the more common size-distribution equations that have been used. The one that best fits a set of data is the one to select. If points are plotted on log-log paper:

1. Use Eq. 4.1, Table 4.1, if Y (cumulative weight fraction finer than size X) versus X looks straight. X is the X-intercept at $Y = 1$ and m is the slope $(lnY_i - lnY_j)/(lnX_i - lnX_j)$.
2. Use Eq. 4.2, Table 4.1, if $ln(1/(1 - Y)$ versus X looks straight. The value of n is the slope $(ln(ln[1/(1 - Y_i)] - ln(ln[1/(1 - Y_j)])/ln(X_i/X_j)$ and $p = \exp[(lnR)/n]$ with R being the value of $ln[1/(1 - Y)]$ at $X = 1$.
3. Use Eq. 4.3, Table 4.1, if $1 - Y$ versus $1 - X/X_o$ looks straight. r is the slope $[ln(1 - Y_i)/(1 - Y_j)]/[ln(1 - X_i/X_o)/(1 - X_j/X_o)]$.

TABLE 4.1 Common size-distribution equations, where *Y* is the cumulative weight fraction finer than size *X*

Name	Equation	
Gates (1915), Gaudin (1926), Schuhmann (1948)	$Y = \left(\frac{X}{K}\right)^m$	**(Eq. 4.1)**
Rosin–Rammler (see Harris 1968)	$Y = 1 - \exp - \left(\frac{X}{P}\right)^n$	**(Eq. 4.2)**
Gaudin and Meloy (1962)	$Y = 1 - \left(1 - \frac{X}{X_o}\right)^r$	**(Eq. 4.3)**
Bergstrom (1966)	$Y = \left[1 - \left(1 - \frac{X}{X_o}\right)^r\right]^q$	**(Eq. 4.4)**
Log Normal (Herbst and Sepulveda 1985)	$Y = \dfrac{\sum^{x} \exp[-\log(X \times u) \cdot 2(\log t)^2]}{X\sqrt{2\pi}\,\log t}$	**(Eq. 4.5)**
Parameters		
Gates, Gaudin, Schuhmann	*K* related to coarsest size *m* measures spread in distribution	
Rosin–Rammler	*p* is related to coarsest size *n* measures spread in distribution	
Gaudin and Meloy	X_o is coarsest size *r* is degree of fragmentation	
Bergstrom	X_o is coarsest size *r* and *q* is degree of fragmentation	
Harris	X_o is coarsest size *s* and *r* are constants	
Log Normal	*u* is geometric mean size *t* is standard deviation	

4. Use nonlinear least-squares fitting techniques (Bergstrom 1966) to find values of constants in Eq .4.4 and Eq. 4.5, Table 4.1. Graphical methods are available, however (Harris 1968; Bergstrom 1966).
5. Use log-probability paper with size on the log scale instead of log-log paper.

Laboratory Screening

Screening is one of the oldest sizing methods known. In ancient times, for example, woven baskets were employed for hand screening (Taggart 1951). Several devices are available for modern laboratory screening, depending on the purpose, the size range of interest, and the amount of sized material desired. To determine size distributions or to obtain small amounts of narrow-size fractions, circular full- or half-size sieves are available for use with Ro-tap shakers. When large amounts (100 kg and more of material coarser than about 0.1 mm) are desired, Gilson screens or their equivalent may be employed. In size ranges below about 0.037 mm, small amounts of sized fractions are obtainable by using microsieves. Wet screening is necessary to obtain larger amounts of material of fine size.

Ro-tap Shaker with Nest of Sieves. Figure 4.5 shows a typical Ro-tap shaker complete with a nest of sieves, and Figure 4.6 shows the principle of laboratory screening (W.S. Tyler Company 1973; Kelly and Spottiswood 1982). Each sieve is 8 inches in diameter and has square openings (apertures). Screen size is specified by either a linear dimension of a square opening or by mesh (Note: the mesh of the screen is the number of openings per linear inch). Various mesh designations are in use, a common

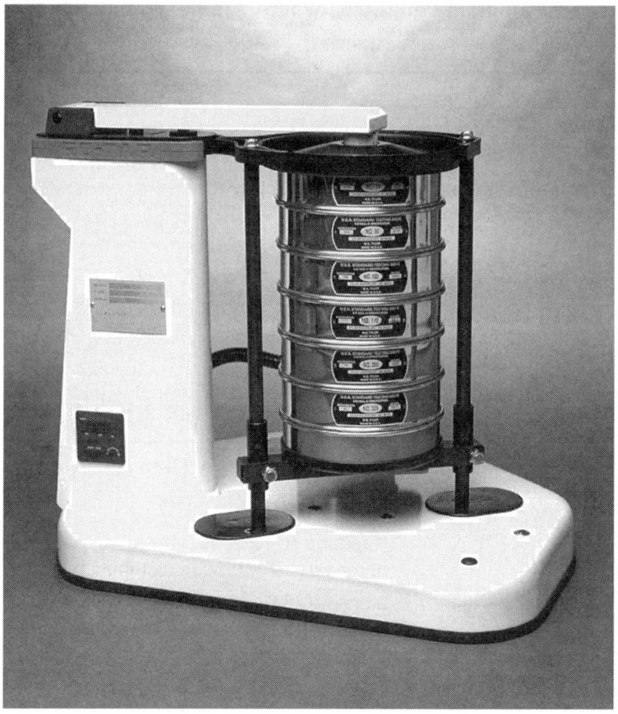

Source: W.S. Tyler Company.

FIGURE 4.5 Ro-tap testing sieve shaker

Source: W.S. Tyler Company.

FIGURE 4.6 Laboratory screening

TABLE 4.2 Comparison of Tyler and U.S. Standard sieve series

Tyler Mesh	Size, mm	U.S. Mesh	Size, mm
4	4.699	4	4.75
6	3.327	6	3.35
8	2.362	8	2.36
10	1.651	12	1.70
14	1.168	16	1.18
20	0.833	20	0.850
28	0.589	30	0.600
35	0.417	40	0.425
48	0.298	50	0.300
65	0.212	70	0.212
100	0.147	100	0.150
150	0.104	140	0.106
200	0.074	200	0.075
270	0.052	270	0.053
400	0.037	400	0.038

feature being that the size of an opening can vary by a definite ratio when sieves are placed in a nest. If the top screen in the nest is of size X_1, each sieve below decreases in a ratio of R such that the ith screen is of size $X_i = X_1(R)^{i-1}$. Typical values for R include $1/\sqrt{2}$ and 1/2. Obviously, a definite ratio is not a requirement for many purposes. When size distributions are desired, a definite R is an asset. Table 4.2 compares the various sieve series that are available.

The Ro-tap supplies a circular motion to sieves with a periodic tap on the lid. Motion is uniform and leads to reproducible results. A sample is placed on the top screen and sieved for up to 40 minutes. Fines will collect in the pan. The amount of sample to sieve must be chosen to avoid blinding of the finest screen in the nest, so that coarser screens will be unaffected. On an 8-in. sieve, the weight, W_{i+1} (in grams), of a single layer of average particles passing size X_i that covers the i+1 screen of size X_{i+1} at 35% voidage is approximately $W_{i+1} = 25.6\, X_{i+1}\, \rho_s$, where ρ_s is the specific gravity of the particles (Pryor 1965). For example, −150 mesh particles (Tyler mesh) of specific gravity 2.7 cover the 200 mesh sieve to a depth of one particle when $W_{.075} = 25.6 \times .075 \times 2.7 = 5.2$ g. Note that the sieve ratio in the formula is $1/\sqrt{2}$. Referring to Table 4.2, U.S. mesh size and mesh number are related approximately by $X = 21.7/(M)^{1.07}$, where M is mesh and X is size in mm. This approximation is useful for some purposes. Table 4.3 shows the results of a screen analysis of a 300-g sample of −1-mm material of specific gravity 2.8. The finest screen employed was 270 mesh. Time of screening was 30 minutes. Both weight percent retained on a given sieve and the cumulative weight percent passing a given sieve have been calculated. The size ratio was $1/\sqrt{2}$. To calculate cumulative percent coarser than a given size, subtract each Y_i from 100. Each Y_i has been calculated from the expression

$$Y_i = 100 - \sum_{j=1}^{i} y_j \qquad \textbf{(Eq. 4.6)}$$

where y_j represents the weight percent retained on the jth screen. Each y_j has been calculated from $y_j = 100[W_j/(W_1 + W_2 + W_3 + ... + W_{10} + W_p)]$, where the denominator term represents the sum of all weights retained, including that on the pan, W_p. The denominator should add to 300 g, but some losses can be expected. The denominator used was 299.00 g, the actual sum of the weights. The 1-g loss could have been assigned to the pan (on the assumption that dust losses occurred) or in any reasonable manner as desired.

TABLE 4.3 Example of screen analysis calculation

i or *j*	U.S. Sieves, mesh	Screen Opening, mm	Weight W_j Retained on Size, g	Weight % Retained y_j, on Size	Cumulative % Finer Y_i, than Size	Cumulative % Coarser than Size
1	16	1.18	0	0	100.00	0
2	20	0.850	0.45	0.15	99.85	0.15
3	30	0.600	2.72	0.91	98.94	1.06
4	40	0.425	25.30	8.46	90.48	9.52
5	50	0.300	58.60	19.60	70.88	29.12
6	70	0.212	60.40	20.20	50.68	39.32
7	100	0.150	45.45	15.20	35.48	64.52
8	140	0.106	31.99	10.70	24.78	75.22
9	200	0.075	20.21	6.76	18.02	81.98
10	270	0.053	12.95	4.33	13.69	86.31
pan	pan	pan	40.93	13.69	0	100.00
			299.00			

Source: Gilson Company, Inc.

FIGURE 4.7 Gilson Screen

Below about 200 mesh, wet screening is employed for greater accuracy. It is not uncommon to wet screen a sample on 200, 270, and 400 mesh sieves. Note that coarser protective sieves may be necessary to avoid damage to fine sieves. The −400 mesh fraction is saved for further sizing, while the +200 mesh fraction is dried and subsequently dry screened in the Ro-tap. Screening error arises when a sample "clumps" or "cakes" after preparation for screening by drying. Clumping is often associated with reagents or soluble salts, or both, that do not evaporate with water. In these cases, wet screening must be considered. In extreme situations, solids have been washed in organic solvents.

Gilson Screen or Equivalent. When large amounts of sample must be sized, higher capacity screening equipment is required. Figure 4.7 is a photograph of a Gilson Screen that is typical of such devices (e.g., the Tylab Tester and others). Up to seven square sieves, 18 in. on a side, are nested inside. The nest is vibrated mechanically to give high efficiency where speed can be adjusted.

The Gilson Screen has been employed for coarse screen analyses of large samples. Screen edges must be sealed first to prevent dust losses.

For large-capacity wet screening, rotary screens have been employed. Such units are usually center fed, and a circular action causes oversize to be discharged through a discharge spout at the periphery. As many as four screen decks can be mounted. Both frequency and amplitude are adjustable.

Microsieves. For screening in finer size ranges (0.037 mm and finer), microsieves can be nested together and mounted in an apparatus that is enclosed in a special housing to minimize dust losses. Great care must be exercised to avoid blinding and destruction of the sieves. Both wet and dry techniques are possible, although extremely small amounts of material are necessarily involved.

SEDIMENTATION SIZING METHODS

Sedimentation sizing techniques rely on the settling behavior in fluids of particles acted on by force fields (e.g., gravitational and/or centrifugal). Size distributions are measurable, and both large and small amounts of sized material are obtainable.

Sedimentation in Gravitational Field. This technique normally assumes that particles settle under laminar flow conditions (Stokesian settling). The terminal settling velocity, v_m(cm/s), of a spherical particle of density, ρ_s(gm/cc), settling in a fluid of viscosity, μ(poise), and density, ρ(gm/cc), is

$$v_m = \frac{d^2(\rho_s - \rho)g}{18\ \mu} \qquad \textbf{(Eq. 4.7)}$$

where g is the acceleration caused by gravity (980 cm/s²) and d(cm) is the Stokesian settling diameter. A real particle of similar specific gravity settling in the same fluid at the same velocity is judged to be of diameter d. With spheres, laminar flow exists when the particle Reynolds number is about 1 or less (i.e., drag coefficient of 24 or more) for practical purposes. For more precise calculations, use a Reynolds number for nonspheres of 0.2 or less (Allen 1975). The size above which a spherical particle will not settle in laminar flow is estimated from

$$d_m = [18N_{\mathrm{Re}}\mu^2/(\rho_s - \rho)\rho g]^{1/3} \qquad \textbf{(Eq. 4.8)}$$

where N_{Re}, the Reynolds number, is taken as 1 for spheres and 0.2 for nonspheres. For spheres of specific gravity 1.7 settling in water of specific gravity 1 and viscosity 0.01 poise, d_m is about 103 µm.

Size distributions in the range of 5 to 60 µm are often measured by means of an Andreasen Pipette (Schuhmann 1948; Maghan and Mular 1982; Herbst and Sepulveda 1985). Figure 4.8 shows the essential characteristics. A 1% or 2% suspension of sample by weight (5 to 15 g) is prepared, where small amounts of dispersant may be added if required. Note that a correction for dispersant weight may be necessary. A standard mixing procedure is followed after dilution to the 20-cm mark. Samples are withdrawn in 10-cc increments at 1 minute and subsequently in a 2 to 1 progression to ensure a root 2 progression in particle size. After each aliquot, the distance S is measured. At settling time, t, the concentration, C_t, of solids at depth S is the same as the initial concentration of all particles of settling velocities less than S/t. Hence, the ratio, C_t/C_o, is the cumulative weight fraction finer than the size d that is calculated from the Stokes' equation

$$d = [18\ \mu\ S/(\rho_s - \rho)gt]^{0.5} \qquad \textbf{(Eq. 4.9)}$$

where S is the distance from the top of the suspension at time, t, to the tip of the pipette (zero mark) with S in centimeters and t in seconds. Remember, S will change after each 10-cc aliquot is taken; C_t is the weight in grams of dry sample per 10 cc of aliquot taken at time t; C_o is the weight in grams per 10 cc of aliquot in the initial homogeneous suspension.

Large amounts of sized material can be obtained by fractionation techniques that employ a settler/decanter unit. Figure 4.9 shows a beaker/siphon arrangement, where a uniform suspension is allowed to settle for a time, t, and then decanted at a distance, S, below the suspension surface. Thus, the residue

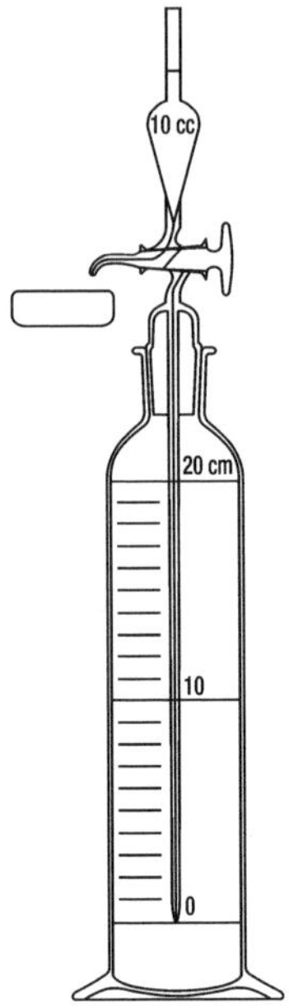

FIGURE 4.8 Andreasen Pipette

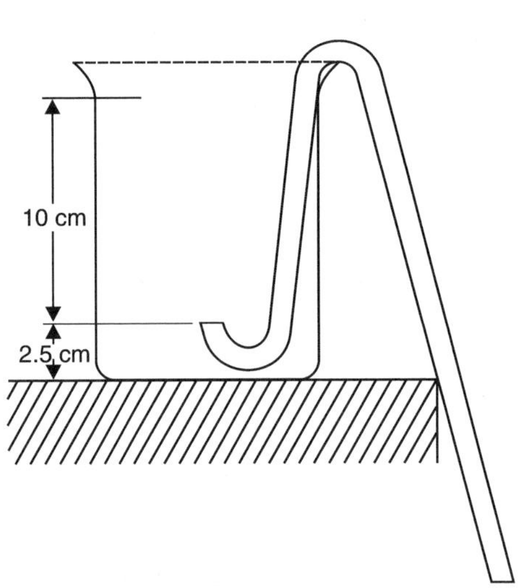

Source: Pryor 1965, with permission of Kluwer Academic Publishers.

FIGURE 4.9 Simple elutriator

will contain all particles whose velocities were greater than S/t plus a small amount of particles with velocities less than S/t. To eliminate the fines, the residue is resuspended, and the procedure is repeated as often as may be required to obtain a supernatant clear of residue. Values of S/t are calculated from the Stokes' equation for the range of diameters involved. Often, settling velocities are doubled in progression from the smallest to largest d, thus providing a root 2 progression in size. The finest size fraction is removed first, the settling time is halved, and the next fraction removed, and so on.

A variety of elutriation devices are available (Schuhmann 1948; Pryor 1965), including devices where air substitutes as the liquid. The Haultain Infrasizer is an example of the latter (Schuhmann

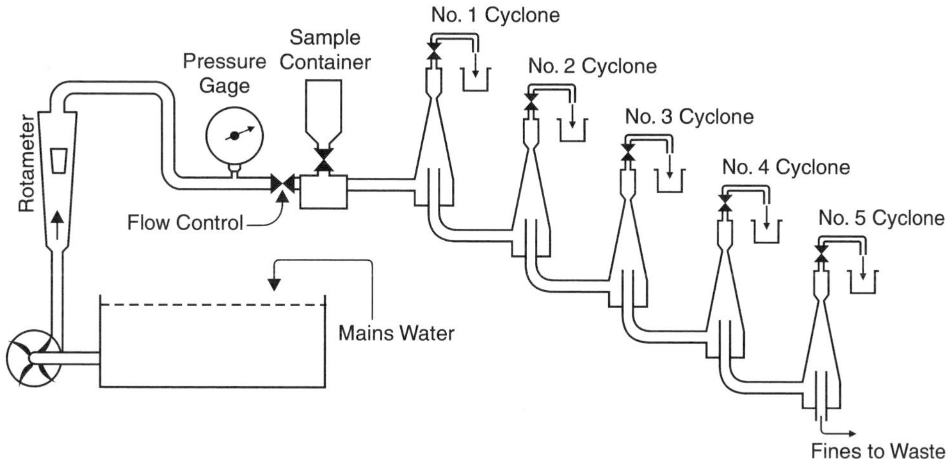

Source: Cyclosizer Instruction Manual, Warman Equipment Ltd., Weir Warman LTD.

FIGURE 4.10 Cyclosizer

1948) which utilizes seven steel, cone-shaped tubes connected in series. The tube diameters are proportioned by root 2 and strung side by side with rubber tubing connections.

Sedimentation in Centrifugal Field. The Bahco Microparticle Classifier (Allen 1975) and the Warman Cyclosizer (Warman 1965) are examples of devices that employ centrifugal forces for sizing. The former treats dry samples in a spiral vortex of air created by a spinning disk. Particles are drawn inward against centrifugal force acting outward.

The cyclosizer is a connected series of hydrocyclones with varying inlet and vortex diameters chosen to obtain size fractions in the 9- to 45-µm range. Each cyclone is inverted and has been fitted with closed apex chambers to permit repetitive sorting/washing. A sharp split is obtained as a result. Figure 4.10 shows the major components of the cyclosizer. The principles of hydrocyclones are discussed in subsequent sections of this chapter.

Diameter Reconciliation. It must be obvious that the screen diameter of a particle is not the same as its Stokesian settling diameter. Because the screen diameter is widely used, it is not unusual to determine a conversion factor, f_c, so that $d_s = f_c d_m$, where d_m is Stokesian diameter and d_s is screen diameter. The technique is described in Chapter 1.

INDUSTRIAL SCREENING

Screening is one of the oldest of unit operations and is used in many industries worldwide. Many screening devices and a variety of screening surfaces are available in the marketplace. Choice depends on the size range involved, the nature of the application, the desired capacity, and the corresponding efficiency of the screen. Screen performance, which is measured in several ways, becomes important and is amenable to mathematical description. Most large-scale screening operations are continuous. Screen deck replacement and normal maintenance influence the operating expense, which is relevant to performance criteria.

Classes of Screens

Industrial screens are categorized in Table 4.4 by mode of operation or motion; typical uses are also listed (Matthews 1985a). Photographs of typical vibrating screens encountered in the mineral industry are provided in Figure 4.11.

TABLE 4.4 Classes of industrial screens after Matthews (1985a)

Screen Class	Mode of Operation and Motion	Typical Uses
Grizzly, stationary	Level or inclined parallel rails, bars, rods with definite spacing; may be tapered.	Scalping of coarse rock preceding crushers, bins, belts.
Grizzly, moving	Vibrating/moving discs, rollers, spaced bars.	Same as above.
Vibrating screen, horizontal	Mechanical/electromagnetic drives; horizontal screens; deck motion pulsed up/forward, then backward/down.	Limited headroom, dewatering, close sizing.
Vibrating screen, inclined	Mechanical/electromagnetic drives; inclined screens, deck motion circular to stratify bed.	Crushing circuits, scalping, high capacity.
Shaking screen, oscillating	Large stroke, slow speed linear oscillation.	0.5 in. + 60 mesh, light, free flowing.
Shaking screen, reciprocating	Horizontal, linear motion, 1- to 4-in. stroke, 30–200 rpm, deck slightly inclined.	Conveying, size separation, sizing large lumps.
High-speed screen	Mechanical and electromagnetic drives; speeds of 3,000 rpm or cpm; inclined deck.	For fine and ultrafine screening.
Revolving screen	Inclined trommel, scrubber or barrel; cylindrical, rotating wire cloth and perforated plate; 15–20 rpm; open at ends.	Scrub, wash, rough size; placer mining; capacity and efficiency low.
Sifter screen	Motion is circular, gyratory, or spiral at screen plane.	4–200 mesh and finer.
Centrifugal screen	Rotating, gyratory motion to vertical; cylindrical screen; fines pass through wall, coarse moves to bottom.	Wet/dry, –0.5 in. to 35 mesh.
Sieve bend Dutch State Mines (DSM)	Parallel bars/wires at right angle to flow. Slope 50° to horizontal.	Wet scalping and dewatering from 10 mesh and finer.

Major components of vibrating screen systems are the screening surface, the vibrating assembly, the base frame, the support frame, the vibrating frame, the motor or drive assembly, and the feed box or distributor. Auxiliaries include feed chutes, dust enclosures, conveyor belts, and dust collection systems.

Vibrating screens are widely used in crushing circuits that have either a mechanical or an electromagnetic drive arrangement. Figure 4.12 shows an electromagnetic drive section, and Table 4.5 (Matthews 1985a) summarizes ways in which mechanical vibration can be generated for various screening applications.

Screening Media

Matthews (1985a) stresses that the most important element of any screen is the screening medium (the surface) where stratification and separation take place. Screening surfaces can be placed into one of three general categories (Taggart 1945, Matthews 1985b): woven wire screen (cloth), perforated screen plate, and profile wire or bar. Figure 4.13 shows woven wire screen types and weaves in general use (Matthews 1974), and Figure 4.14 shows perforated plate and profile wire (or bar) shapes (Matthews 1985b).

Woven wire screen accounts for 75% of sales. For very intensive use and coarse sizes, perforated plate is often employed, but when finer sizing is desired, profile wire is selected (Figure 4.14). Many materials have been used to make screen surfaces, including brass, copper, bronze, aluminum, monel, nickel, stainless steel, abrasion-resistant alloy steels, high-carbon steels, rubber, and synthetics such as reinforced polyurethane (tyrethane). A screen surface must withstand the stresses and loads applied to it and maintain a high degree of resistance to abrasion and corrosion. Once the aperture size and

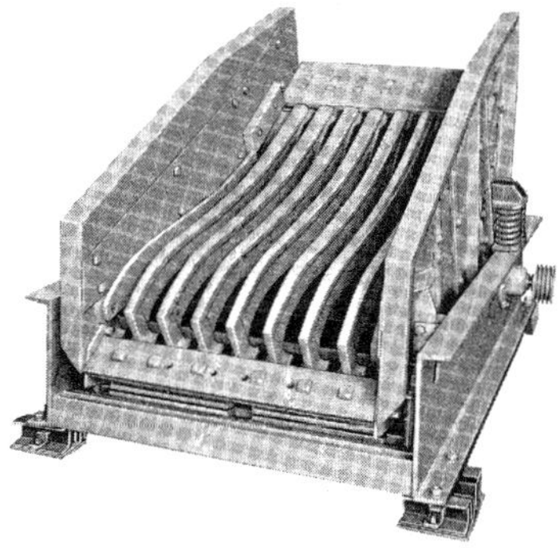

Source: Metso Minerals.

FIGURE 4.11A Vibrating grizzly

Source: McNally Pittsburgh.

FIGURE 4.11B Horizontal vibrating screen

Source: Derrick Corporation.

FIGURE 4.11C High-frequency vibrating screen

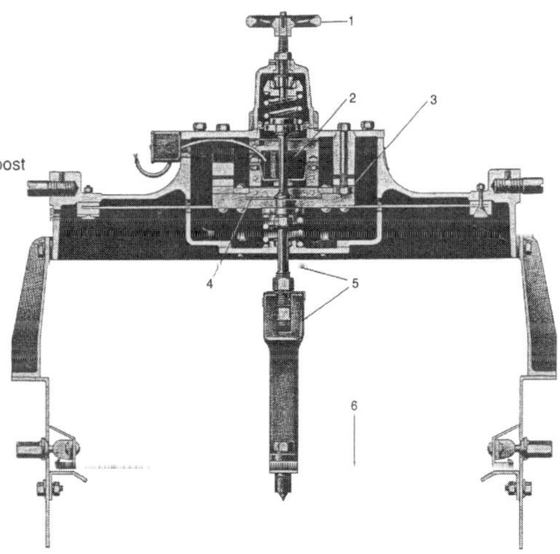

Source: CE Tyler.

FIGURE 4.12 Cross section of an electromagnetic drive for a screen

TABLE 4.5 Vibrating screen motions

Motion	Characteristics	Applications
Vibrating: 1 shaft, 2 bearings	Unbalanced pulley type—one concentric shaft with adjustable counterweights and two bearings. Circle-throw motion produces an oscillating vibration. Stroke may be varied by adjusting the counterweights.	Generally used on light-duty screens.
Vibrating: 1 shaft, 2 bearings	Commonly designated as "2-bearing." One eccentric shaft with adjustable counterweights and 2 or 4 bearings. Circle-throw motion produces vibration. Stroke may be varied by adjusting the counterweight.	Used on light- and heavy-duty inclined vibrating screens.
Vibrating: 1 shaft, 4 bearings	Commonly designated as "positive-stroke" or "4-bearing." One double-eccentric shaft with 2 sets of bearings. One set supports screen frame; other shaft. Stroke cannot be varied except by changing shaft. Double set of bearings and double eccentric of shaft produces a positive motion that is not dampened by load on screen deck. In most designs, shaft is on center of gravity of screen box.	Used on heavy-duty inclined vibrating screens.
Reciprocating: 2 shafts, 4 bearings	Commonly designated a "4-bearing." Two shafts, eccentric or weighted, counter-rotating in phase produce a positive straight-line motion. By operating slightly out of phase, the stroke is inclined.	Used on horizontal vibrating screens and some conveyors.
Vibrating (flow): top mounting	Vibrator mounted on top of frame produces elliptical motion. Flow rotation of vibrator produces stroke card indicated.	Used for rough screening where high rate of feed is needed. Flow rotation moves material faster, increases capacity, lowers efficiency.
Vibrating (counterflow): top mounting	Vibrator mounted on top of frame produces elliptical motion. Counterflow rotation of vibrator produces stroke card indicated.	Used for more efficient size separation. Counterflow holds material on screen longer, given deeper bed, but reduces capacity.
Vibrating: center mounting	Vibrator mounted centrally between size frames produces circular motion. Rotation of vibrator may be "flow" or "counterflow." Flow gives higher capacity, lower efficiency, and vice versa with counterflow.	Generally, used for heavy-duty screen of inclined type.
Reciprocating: inclined vibrator	Vibrator mounted above (or below) frame with slight inclination of axis to use positive straight-line motion to move material along the screen surface. Used for horizontal screens.	Used for close separation of medium-sized material. Dewatering or media recovery. Use for limited headroom installations.
Reciprocating: unphased vibrator	Vibrator mounted above (or below) frame. Straight-line motion is obtained by setting one eccentric to lead the other. Phase adjustment determines the angle of inclination of straight-line force.	Used for same applications as inclined vibrator.

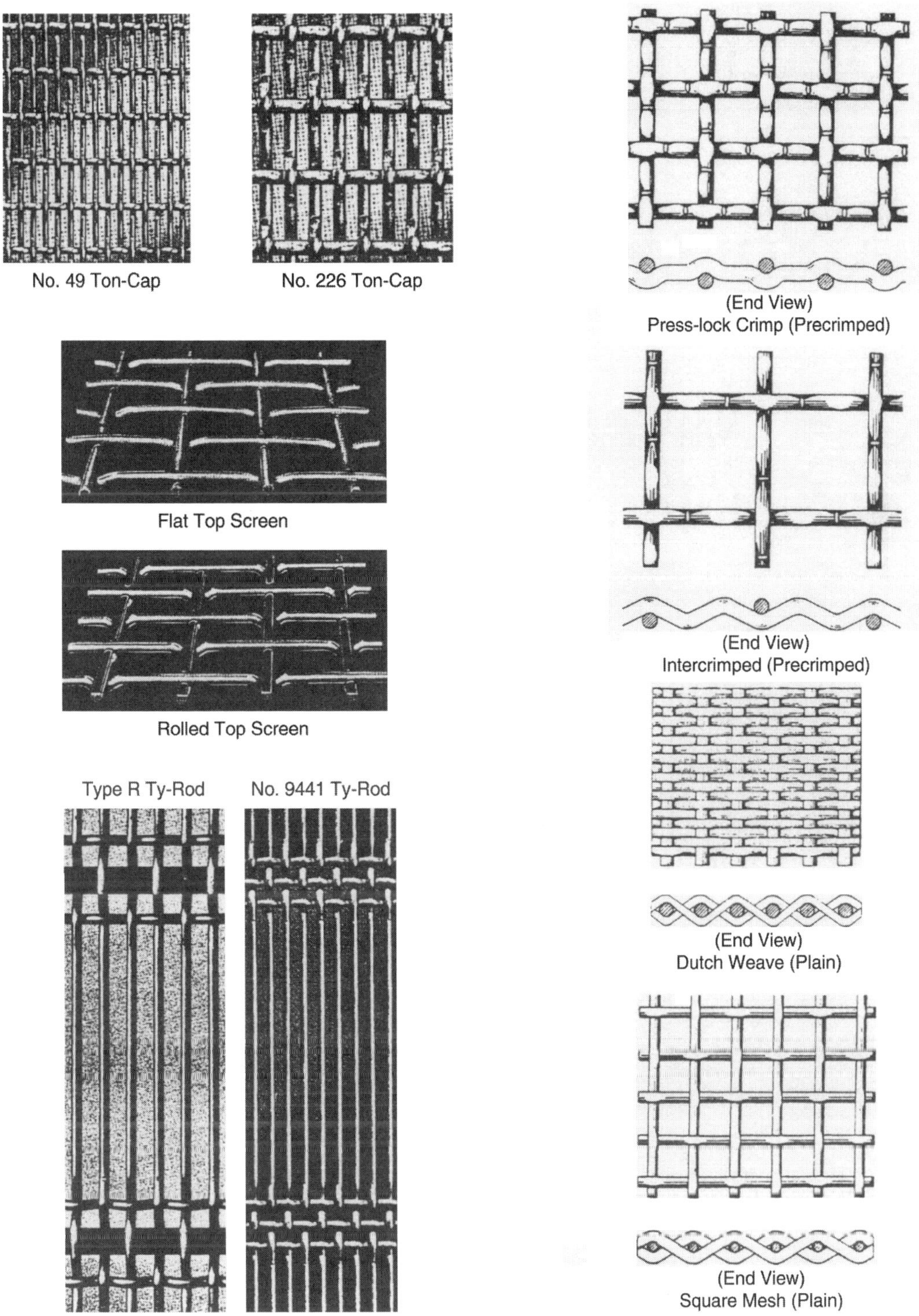

Source: CE Tyler.

FIGURE 4.13 Woven wire screen types and weaves

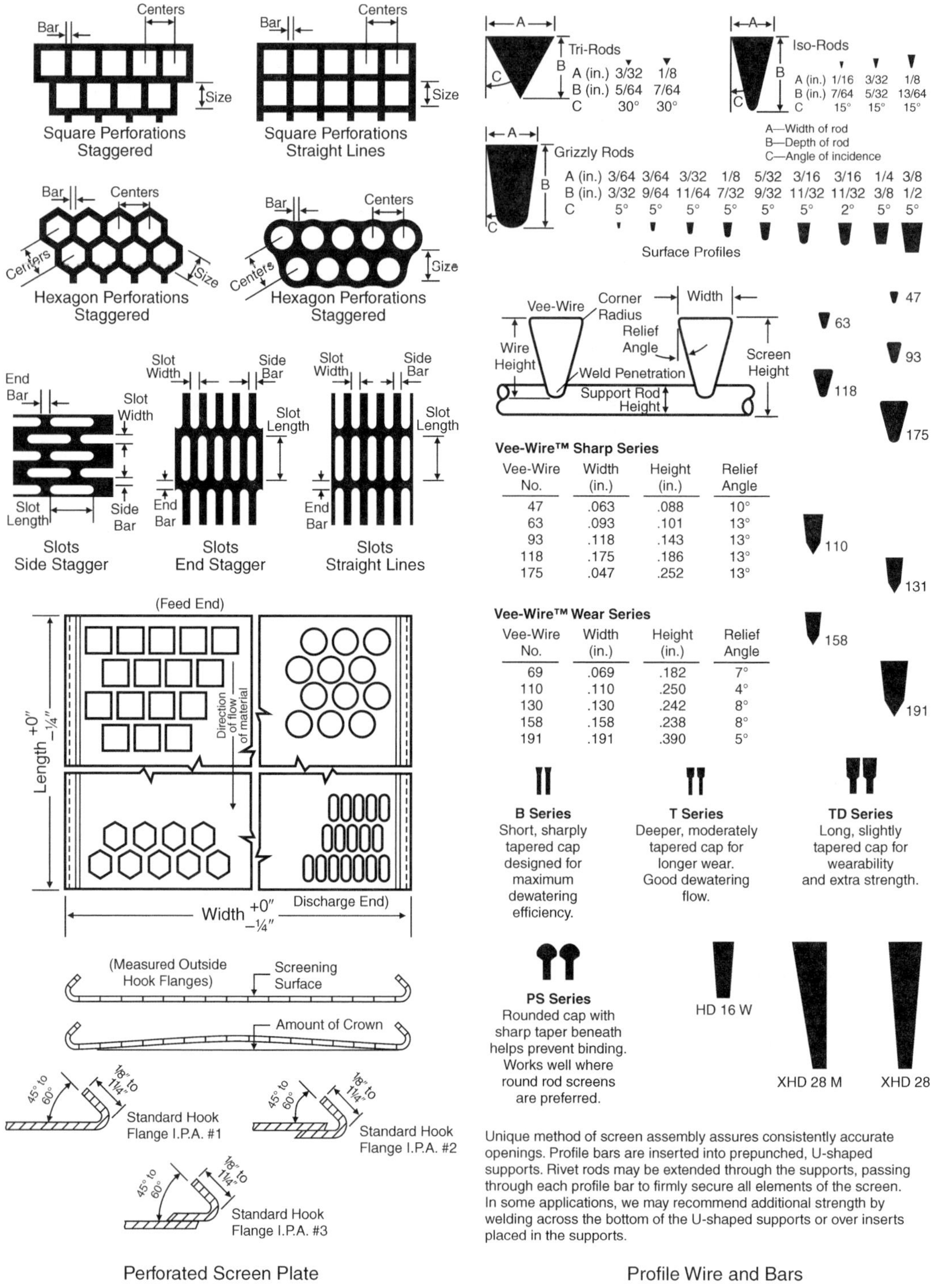

Vee-Wire™ Sharp Series

Vee-Wire No.	Width (in.)	Height (in.)	Relief Angle
47	.063	.088	10°
63	.093	.101	13°
93	.118	.143	13°
118	.175	.186	13°
175	.047	.252	13°

Vee-Wire™ Wear Series

Vee-Wire No.	Width (in.)	Height (in.)	Relief Angle
69	.069	.182	7°
110	.110	.250	4°
130	.130	.242	8°
158	.158	.238	8°
191	.191	.390	5°

Source: Matthews 1985b.

FIGURE 4.14 Perforated screen plate and shapes of profile wire and bar

capacity characteristics are determined and a screen is fully operational, the "best" screen surface is one that never needs replacing. In practice, the goal is a minimal replacement cost per unit of throughput. For example, carbon steel screen is consumed at a rate as measured by a replacement cost of *C* dollars/ton/year. A synthetic material does not wear out as fast (perhaps lasting three or more times as long) but is more expensive to purchase. If *C* for the synthetic is greater, carbon steel will continue to be the material of choice.

Square mesh surfaces are often selected for coarse applications if accurate sizing is necessary or if particles are slabby. However, on an incline the effective square mesh aperture and capacity may be reduced. On the other hand, rectangular mesh surfaces of comparable sizing will exhibit a higher capacity, because the proportion of open area is greater. Moreover, rectangular surfaces are not as susceptible to blinding (i.e., the plugging of openings with near-mesh particles or wet, sticky ones; the latter also cling to decrease the effective aperture) and are suited for needle-like particles, for high-moisture ores, and for ores with a high clay content. The flow of feed can be either parallel or perpendicular to the longer dimension of the mesh. Parallel flow of high-moisture or clayey ores allows a higher capacity and reduces blinding. Perpendicular flow of dry ore is less apt to blind screens, which then have a longer life and a higher efficiency. When blinding is severe, special screening decks should be considered. A heated deck is useful for fine, high-moisture ore. Ball decks rely on rubber balls bouncing against a screen bottom to loosen material. As a last resort, water sprays are recommended.

Perforated screen plates make coarse separations and are useful as an upper deck screen to reduce wear and damage to a lower deck screen of smaller aperture. Plates are more expensive, but they resist wear and have a long life, less blinding, higher efficiency, and a high degree of accuracy in sizing. Screen openings of less than about 3/4 in. have an even smaller percentage of open area. An incline further reduces effective aperture.

Profile wire (rods or bars) has been used for coarse screening, for dewatering applications, and for special screen assemblies (such as cone shapes). Wire in parallel with the flow is most common, but transversely placed wire is effective for wet screening (e.g., the sieve bend) in the fine size ranges. Profile wire surfaces are not widely used in crushing circuits but may be used in grinding circuits to avoid producing slimes from friable ores.

Efficiency of Screens

The performance of a mill's vibrating screen is often neglected until productivity must be increased. In this situation, changes in screen aperture size, feed rate (analogous to circulating load in most cases), feed size, frequency and amplitude of the vibrator, screen efficiency, and other factors will be studied to determine whether throughput can be increased or operating costs decreased, or both. Screen capacity and efficiency become extremely important.

Fractional Efficiency. Fractional screen efficiency can be defined with the aid of Figure 4.15, which shows how steady-state screening produces a variation in the mass of particles falling through the screen along its length.

The recovery of a narrow size fraction of feed that reports to the coarse stream is the fractional recovery, R_i , to oversize. Hence,

$$R_i = 100\left(\frac{Oo_i}{Ff_i}\right) \quad \textbf{(Eq. 4.10)}$$

In Figure 4.15, o_i and f_i are weight fractions retained between any two sieves that constitute interval *i* in the oversize and feed streams, respectively (sieves are assumed to follow a definite size ratio). The selection of screen opening size, x_i , to associate with *i* is largely arbitrary. Thus x_i can refer to the opening size of the lower sieve (on which the weight fraction is retained) or to the opening size of the upper sieve (just above the lower one) or to the geometric mean (or arithmetic mean) opening size of the two sieves.

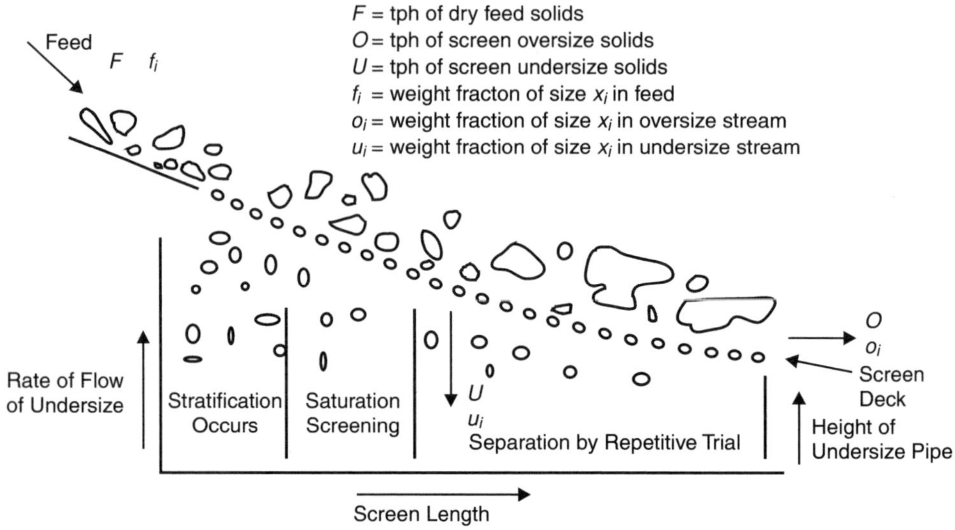

FIGURE 4.15 Variation in the mass falling through a screen along its length

From a mass balance around a steady-state screen at each interval i,

$$\frac{O}{F} = \frac{f_i - u_i}{o_i - u_i} \quad \textbf{(Eq. 4.11)}$$

where u_i is the undersize weight fraction retained in interval i. Remember that when i is less than the interval corresponding to the cut size, u_i will be zero (i starts with the value of 1 at the coarsest interval).

Referring to screen data shown in Figure 4.3, R_i values may be calculated for each i and plotted against a suitable x_i. Figure 4.16 is typical of such plots for relatively dry feed (less than about 3% moisture) treated on vibrating screens, where the x_i selected is the geometric mean size plotted on a log scale. The curve does not go to zero because fine material short-circuits to the oversize stream. Fine, moist particles cling to oversize, whereas fine, dry particles transfer to a dust collector. Such particles do not participate in the sizing process and are considered to have bypassed to the coarse stream. (If coarse solids bypass to the fines stream, the device has malfunctioned—a screen has torn or a hydrocyclone has developed rope condition.) The amount bypassed is approximately the distance 0 to B on Figure 4.16. Hence, $R_i = B + (1 - B)\, f(x_i)$ and rearranging,

$$\frac{R_i - B}{1 - B} = f(x_i) = R_{ic} \quad \textbf{(Eq. 4.12)}$$

where

B = an adjustment that is related to short-circuiting of fines to the oversize

R_{ic} = the reduced or corrected fractional efficiency at size x_i

$f(x_i)$ = the functional form involving x_i and other variables that influence the curve

The fractional efficiency curve has been referred to as either a fractional recovery or a fractional performance curve. It has a steep slope near the screen aperture size. For moist ores, the tail curves upward quite dramatically. Thus, only when feed is relatively dry and the dust collection system is stable is B approximately a constant for a given set of conditions.

Gross Efficiency. Fractional efficiency curves are difficult to determine experimentally. Hence, either the gross efficiency of undersize removal from the oversize stream (Nichols 1982) or the gross

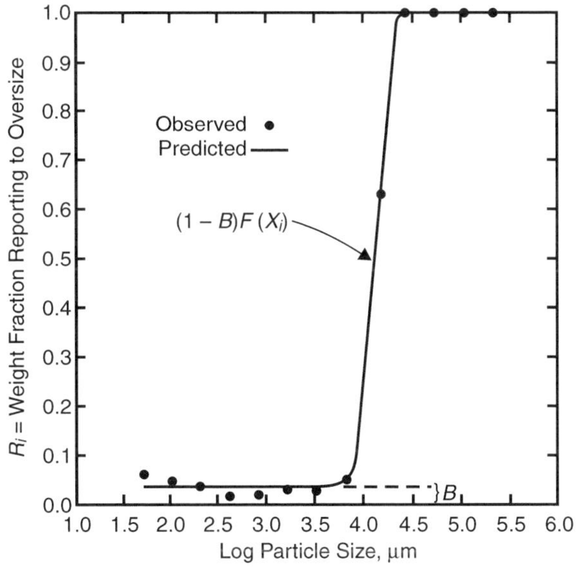

FIGURE 4.16 Uncorrected fractional recovery curve for a vibrating screen

efficiency of undersize recovery is measured. The efficiency of undersize removal from the oversize is found at steady state from

$$E_o = [100]\frac{\text{mass flow rate of solids coarser than screen size in feed stream}}{\text{mass flow rate of solids in the oversize stream}} \quad \text{(Eq. 4.13)}$$

If x is the effective screen aperture size (cut size or screen size)

$$E_o = 100\left[\frac{F(1-f_x)}{O}\right] = 100(1-o_x) \quad \text{(Eq. 4.14)}$$

where

F = stph of feed ore

O = stph of oversize solids discharging as screen oversize

f_x = cumulative weight fraction of feed finer than x

o_x = cumulative weight fraction of oversize finer than x

The efficiency of undersize recovery is determined from

$$E_u = \left[100\frac{\text{mass flow rate of solids in the undersize stream}}{\text{mass flow rate of solids finer than screen size in feed stream}}\right] \quad \text{(Eq. 4.15)}$$

Thus

$$E_u = 100\left(\frac{U}{Ff_x}\right) = 100\left[\frac{f_x \quad o_x}{f_x(1-o_x)}\right] \quad \text{(Eq. 4.16)}$$

where U is the mass flow rate of solids (stph) in the undersize stream.

Screen Performance and Efficiency

Screen performance or efficiency is influenced by design variables such as screen area and open area, aperture size and shape, slope of screen deck, and deck motion; and by operating variables such as particle size, shape, and distribution, feed rate and bed depth of solids, and moisture content of the feed.

Screen Area and Open Area. Other things being equal, the capacity of a vibrating screen varies directly with screen area. For a given area, capacity is proportional to screen width, whereas gross efficiency is proportional to screen length. Length is about two or three times the width, because at some point an increase in length will not influence efficiency. A given screen area develops its best capacity and efficiency when the material at the discharge end is all oversize and one layer deep (high efficiency). Screens may operate at lower efficiency to obtain higher capacity.

The area of a screen is limited by the strength of the screen deck, which must be able to handle heavy loads in motion. In turn, deck strength depends on the proportion of the screen area that is open to particle passage. The open area can range between 12% and 90%, depending on the characteristics of the screen and its projected usage. Capacity is directly proportional to open area, and efficiency is expected to increase likewise. Open area is chosen to minimize the possibility of screen rupture or other damage. Thus, when wire cloth is employed, wire diameter should provide the maximum open area consistent with the strength required for the application. The possibility of blinding should be considered because large-diameter wire may induce moist particles to plug openings.

Effective screen area is less than actual to allow for tension rails, center clamps, and apertures blocked by support bars. If effective area is not stated by the manufacturer, actual area is commonly reduced by 10%.

Aperture Size and Shape. The capacity of a screen will decrease with a decrease in aperture size. At a fixed capacity, screen efficiency likewise will decrease as aperture size decreases. As aperture size decreases, wire diameter tends to decrease to maintain the same percentage of open area. When strength must be retained as aperture size decreases, wire diameter may remain about the same, so that the open area must decrease and capacity and efficiency are reduced. Blinding increases as aperture size decreases, particularly with dry feeds containing abundant near-mesh-size particles or with feeds high in moisture. Dry screening below about 6 mesh is not a commercial success, because capacities are too low for mineral processing plants. However, wet screening at these meshes (e.g., DSM screens) is used, particularly in processing iron and coal ores.

Aperture shape (square, rectangular, round, or slotted) strongly influences screen performance. Rectangular or slotted openings offer more open area and are less apt to be blinded by most ores, thereby increasing capacity and efficiency. However, square and round openings permit a more accurate split at the cut size of interest. In general, apertures are staggered to prevent particles from riding on screen material too long before encountering an aperture.

In general, industrial screens do not split particles precisely at their aperture size but rather somewhat below it. For instance, a screen with 2-in. square openings will split at a cut size slightly below 2 in. This smaller actual cut size reflects factors such as aperture shape, deck slope, bed fluidization, rate of travel, particle shape, size feed distribution, and screen motion.

For square openings, if X_S is the desired cut size in inches, $X = 0.0238 + 1.155\ X_S$, where X is the aperture size (in inches) of a commercial screen as calculated from published data (Matthews 1985b). Thus, if the cut size must be 0.75 in., the plant-scale screen should have an aperture size of 0.89 in. X_S is equivalent to the effective screen aperture, which, for slotted and rectangular screens, varies with deck slope. Relationships between X and X_S for other than square openings are available (Matthews 1985b).

Slope of Screen Deck. When the discharge end of a screen deck is inclined down from the horizontal, material cascades more rapidly down the slope and either passes through an opening or over the screen surface in accord with some probability. Hence, the capacity of a screen must increase as deck slope increases. Efficiency will remain constant or increase up to a critical slope and then decrease as slope increases. Coincidentally, an increase in deck slope will decrease the effective aperture size at a given feed rate. In crushing plants, screen decks are commonly installed at angles of 20°–25° below the horizontal; 20° is most common. Material travels on circle throw screens with counterflow rotation at rates approximately proportional to inclination. Thus, at 20° the flow rate is 80 ft/min, whereas at

22° it is 100 ft/min. However, other factors also influence flow rates: speed, throw, type of screen surface, and aperture size.

Deck Motion (Speed and Throw). Vibration of screen decks is produced by either circular or elliptical motion, in which the vibrator rotates in a flow or counterflow direction at amplitudes of 3–15 mm and shaft speeds of 900–1,200 rpm. Frequencies of 700–1,000 cycles/min are normal, and high-speed devices attain frequencies of 3,600 cycles/min. Trommels do not vibrate—they rotate, whereas sifter screens can combine vibratory and rotary motions. In mineral processing, inclined vibrating screens are by far the most popular; the vibration lifts and stratifies the particles and conveys them on the incline. Speed, slope, and direction of rotation affect blinding of screens by near-size particles.

The amplitude (throw) of vibration strongly influences blinding. Too small a throw will permit near-size particles to plug openings; too large a throw will keep them away from the screen surface and reduce efficiency. Large throws reduce bearing life, which varies inversely with the 10/3 power of the bearing load (Gluck 1965a).

Rotation in the flow direction increases capacity by increasing the rate of flow of material, but efficiency may be reduced. Counterflow rotation (reversing the motor) tends to retard the rate of flow on the screen (which increases the efficiency but decreases capacity) and may create blinding. Blinding may be compensated for by increasing deck slope.

Speed (frequency of vibration) produces a lifting component for stratification and a conveying component in which the screen pulls back at the end of a cycle. High speeds go with small throws, and low speeds go with large throws. This compromise reduces bearing wear. Generally, large throws are required for screening coarse material or when bed load is large. For screening fine material, small throws with higher speeds are best. Crushed material calls for throws larger than those required for rounded material.

Particle Size, Shape, and Distribution. For a given screen aperture, both screening rate and passage probability (hence capacity) will increase as particle size decreases, although particle shape will modify the effect. For a given design, screen capacity will be less for acicular particles than for more rounded ones. For a given material, the size distribution defines the proportion of fines, near-size particles, and oversize particles present on the screen in relation to aperture size.

Near-size particles are those whose sizes are ½ to 1½ times the size of the aperture, so that a small particle for one size of screen will be a near-size particle for a smaller size of screen. Oversize particles never pass through a screen; small particles (fines) pass through rapidly. Thus, near-size particle passage is the rate-determining step in screening, because if they are poorly aligned, near-size particles will not pass through. To obtain higher screening rates (higher capacities), exposure of fines and near-size particles must be maximized, and the quantity of oversize must be reduced by, for instance, an upper screen deck of coarser size.

Large, rapid variations in the proportion of oversize and near-size particles in the feed stream will influence the load on the screen. Efficiency may suffer accordingly, so that the feed size distribution should be stabilized if possible.

Solids Feed Rate and Bed Depth. For a screen of fixed throw, speed, and aperture, bed depth depends on factors such as feed rate, deck slope, feed size distribution, and direction of rotation (either with or counter to flow). At steady state, large particles on top of the bed prevent finer ones from bouncing around, thereby keeping them close to the screen surface. Likewise, oversize helps to push near-size particles along or through the screen to reduce blinding. There is an optimum bed thickness (which increases with feed rate), because efficiency increases with feed rate and then falls off. For a given feed rate, F (stph), screen width is selected to maintain bed depth at the discharge end, so that screen width determines capacity. Bed depth at the discharge end should not exceed n times the screen opening (in.), where $n = 2 + 0.02 \times$ bulk density. Here, the bulk density is given in pounds per cubic feet of ore. Bed depth, D (in.), may be estimated from this formula:

$$D = \frac{400F}{TW(\text{bulk density})} \qquad \text{(Eq. 4.17)}$$

where T (ft/min) is the rate of travel of the bed material and W (ft) is the effective width of the screen. At the feed end, D may be estimated by letting F represent the feed rate to the screen.

Another way to calculate bed depth follows:

Screen Length, ft	Depth of Dry Rock Bed
6 to 10	1.5 to 2.0 times average particle size
12 to 16	2.0 to 2.5 times average particle size
20 to 24	2.5 to 3.0 times average particle size

Bed retention time is estimated as $t = L_S/T$ = retention time in minutes, and L_S (ft) is the effective screen length. Rate of travel, T, varies with deck slope and motion characteristics.

On inclined circle throw screens in counterflow rotation, between angles, A, of 18° and 25°, T is approximated by

$$T = -120 + 10\,A$$

Thus, if $A = 20°$, $T = 80$ ft/min. In flow rotation, T will be somewhat higher. In production environments, bed depth is adjusted by manipulation of the feed rate where possible. If capacity cannot be sacrificed, deck slope may be modified to obtain the desired depth.

To stabilize bed depth as much as possible, the feed size distribution should be as constant as possible.

Moisture Content of Feed. Screen feed that has a moisture content in excess of a few percent or a high clay content may blind screens or reduce efficiency or capacity. Moisture causes fine particles to stick to oversize. In addition, fines may agglomerate in the presence of clay, which acts as a binder. Agglomerates that reach a size equivalent to half that of apertures may block the apertures and reduce capacity.

Moist fines may adhere to screen wire and decrease effective screen aperture. In exceptional cases, apertures may be totally closed by adhesive fines. Screen cloth may be heated and rubber ball trays may be used to help remedy severe problems. Wet screening may be considered.

Vibrating Screens

Most methods for selecting vibrating screens have evolved from basic capacity data acquired from full-scale screening equipment, where modifying factors (multipliers) were necessary to force a match with actual operation. At least four methods are available. One employs the flow rate of oversize; another is based on direct experimentation with scaleable screening equipment. Of more interest are the other two methods, the feed rate method (Nichols 1982; Colman 1963; Gluck 1965a) and the throughput method (Matthews 1985a; Colman 1972, 1980, and 1985). The former was employed by Allis Minerals (Colman 1963; currently Svedala Industries) and is effectively similar to the throughput method, except that E_o is used for screen efficiency and "screen feed rate" is the term that is modified by factors. A detailed description of the feed-rate method is available (Gluck 1965a and 1965b).

The throughput method, based on the mass flow rate of screen undersize, uses E_u for screen efficiency and is employed by Nordberg (Colman 1972). It is recommended by the Vibrating Screen Manufacturers Association, although any manufacturer may develop modifiers that are most appropriate for its own screens. In this method, an effective screen area is calculated from input data that permit the selection of a basic capacity and corresponding modifying factors. In general, once effective screen area is determined, consideration is given to width, length, severity of duty, support structures, feeding arrangements, and screen enclosures that collect dust.

Gross screen efficiency, E_o, can be related to the capacity of a screen (Figure 4.17; Nichols 1982, Colman 1963). In turn, E_u, which is the recovery of fines in the feed to the undersize stream, is related to E_o. Referring to efficiency definitions previously defined,

$$E_u = \frac{E_o - (1 - f_x)}{E_o f_x} \qquad \text{(Eq. 4.18)}$$

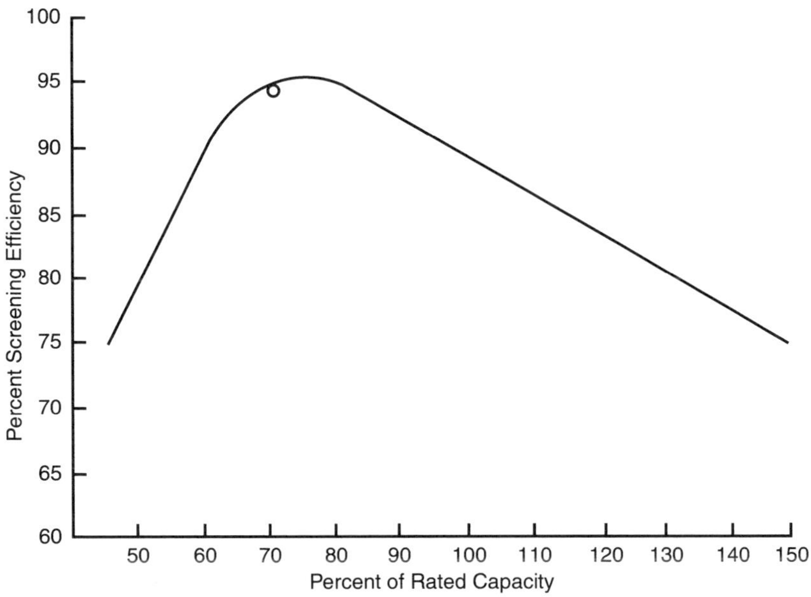

Source: Colman 1963; Nichols 1982.

FIGURE 4.17 Screen efficiency as affected by rated capacity

The efficiencies E_o and E_u differ through the term f_x, which is the cumulative weight fraction in the screen feed that is finer than the effective aperture size, x. E_u must not be confused with E_o. Depending on the value of f_x, large differences are possible.

Estimating Area of Vibrating Screens by the Throughput Method. This discussion of the throughput method is based on Matthews 1985b, page 3E-11. The screen area, A (ft^2) is estimated from

$$A = \frac{U}{CF_t F_o F_e F_d F_{oa} F_s F_w} \text{ ft}^2 \qquad \textbf{(Eq. 4.19)}$$

where

U = undersize in the dry feed (stph)

C = base capacity (stph) through the screen per square foot , at 95% efficiency E_s

F_f = fines factor, to account for the difficulty of screening the percentage of material passing-openings equal to half the aperture size

F_o = oversize factor, to account for difficulty of stratification in the presence of material coarser than the aperture size but estimated in terms of the percentage of material finer than the aperture size (see Chart B of Figure 4.18).

F_e = efficiency factor, to account for desired efficiency (Chart B, Figure 4.18)

F_d = deck factor, to allow for area lost on lower deck (Chart C, Figure 4.18)

F_{oa} = open area factor, equal to the ratio of the percentage of open area used to a standard percentage of open area. This ratio is used to develop the C versus screen opening curve (Chart A, Figure 4.18). The standard varies with screen opening (Chart E, Figure 4.18).

F_s = slot factor, to account for the influence of shape with the long dimension parallel to the flow direction (Chart F, Figure 4.18)

F_w = wet screening factor (Chart D, Figure 4.18)

Chart A—Screen capacity per square foot

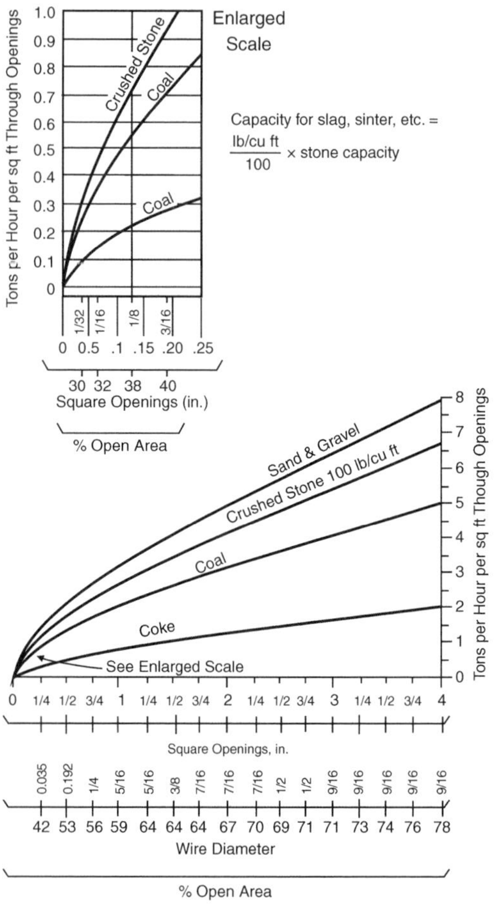

Chart B

%	Factor		
	Fines, F_f	Oversize, F_o	Efficiency, F_e
0	0.44		
10	0.55	1.05	
20	0.70	1.01	
30	0.80	0.98	
40	1.00	0.95	
50	1.20	0.90	
60	1.40	0.86	
70	1.80	0.80	
80	2.20	0.70	1.75
85	2.50	0.64	1.50
90	3.00	0.55	1.25
95	3.75	0.40	1.00

Chart C

Decks	Deck Factor, F_d
Top	1.00
2nd	0.90
3rd	0.75

Source: Matthews 1985a.

Chart D

Opening Size, sq in.	Wet Screening Factors		
	Limiting Moisture*	F_w	F_{wa}†
1/32 or less	0%	1.25	1.10
1/16	1%	3.00	2.00
1/8	1%	3.50	2.50
3/16	2%	3.50	2.50
5/16	4%	3.00	2.00
3/8	4%	2.50	1.50
1/2	6%	1.75	1.30
3/4	6%	1.35	1.20
1–2	6%	1.25	1.10
1–1/2	No limit	1.25	1.10
+2	No limit	1.0	1.0

*When screening dry: If moisture exceeds this limit, must consider special aperture constructions.

†Use F_{wa} when uncertain about maximum spray water being available, or being used efficiently.

Chart E

Open Area Factor—% Open Area = P
For the More Common Apertures

Type of Aperture	Formula for % Open Area, P
Rectangular Openings	P = % open area; d = diam of wire, or horizontal width of bar (for plate); a = clear opening dimension. $P = \frac{\text{Open area} \times 100}{\text{Total area}} = \frac{100(a_1)(a_2)}{(a_1 + d_1)(a_2 + d_2)}$
Square Openings Specified by Opening Size	$P = \frac{100\,a^2}{(a + d)^2} = 100\left(\frac{a}{a + d}\right)^2$ $a_1 = a_2 = a$, $d_1 = d_2 = d$
Square Openings Specified in Mesh (m)	$P = 100a^2m^2$ $m = \frac{1}{a + d}$
Parallel Rod Decks	$P = \frac{100a}{(a + d)}$
Special Weaves (Ty-Rod, Nonblind, etc.)	Assuming $a_2 = a_1$: $P = 100\left[\frac{a_1(a_3 + 2a_1)}{(a_2 + 2a_1 + 3d_2)(a_1 + d_1)}\right]$

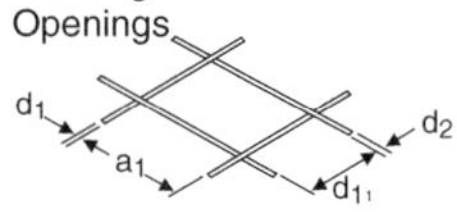

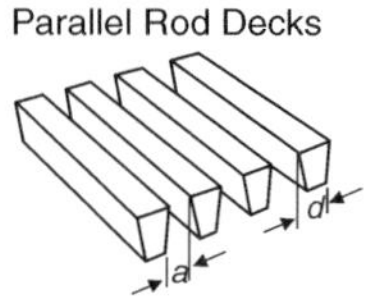

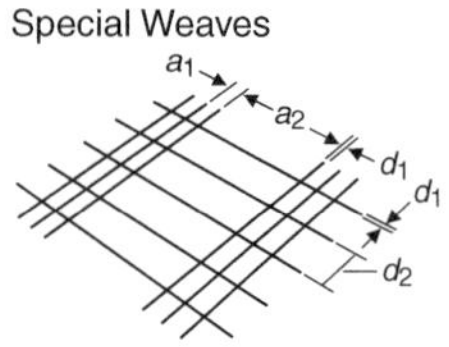

Chart F—Slotted opening factor

Typical Screen Media	Length/Width Ratio [(a_2/a_1), Chart E]	Slotted Opening Factor, F_s
Square and slightly rectangular openings	Less than 2	1.0
Rectangular openings, Ton-Cap	Equal to or greater than 2, but less than 4	1.1
Slotted openings Ty-Rod nonblind	Equal to or greater than 4, but less than 25	1.2
Parallel rod decks	Equal to or great than 25	1.4

FIGURE 4.18 Base capacities and modifying factors for throughput method

Note that an efficiency factor has been incorporated, such that screen area is reduced to maintain the same capacity at any desired lower efficiency. An efficiency greater than 95% is not considered practical.

Figure 4.18 shows the base capacity curves for various materials relative to crushed stone of bulk density 100 lb/ft^3 and modifying factors. The capacity of other material, C, (stph per ft^2) is

$$C = \left[\frac{\text{bulk density}}{100}\right] (\text{crushed stone capacity}) \tag{Eq. 4.20}$$

Bulk densities of various materials (Colman 1963) are shown in Table 4.6.

TABLE 4.6 Bulk densities of various materials

Material Loosely Piled	Weight/ft^2
Alum	33
Ashes, cinders	40–45
Basalt	96
Bauxite	85
Cement, clinker	95
Cement, portland	90
Charcoal	10–14
Chips, wood	18
Clay and gravel, dry	100
Coal, anthracite	47–58
Coal, bituminous	40–54
Coke	23–32
Dolomite	109
Feldspar	100
Fuller's earth	42
Gneiss	96
Granite	96
Greenstone, hornblende	107
Gypsum	75
Ilmenite	120
Iron ore, hematite	130–160
Iron pyrite ore	165
Lime, gypsum	66–75
Limestone	95–100
Marble	95
Mica	100
Phosphate rock	75
Porphyry	103
Quartz	95
Rock salt	94
Sand and gravel, dry	90–105
Sandstone	82
Shale	92
Slag	98–117
Sulfur ore	87
Taconite	150–200
Talc	109
Trap rock	109

Source: CFS, Inc.

Estimating Screen Width, Length, and Deck Angle. Once screen area, A, has been estimated, then screen width, W (ft), and screen length, L (ft), must be determined. A length-to-width ratio of 2 or 3 to 1 must be maintained. W is chosen to maximize capacity; length is chosen to maximize efficiency.

The width, W, can be estimated from

$$W = \frac{400F}{DT(\text{bulk density})} \tag{Eq. 4.21}$$

Bed depth, D, at the oversize discharge end must be less than or equal to

$$D^* = [2 + 0.02(\text{bulk density})]\ X_S \tag{Eq. 4.22}$$

where X_S is the desired cut size. Values for T may be approximated for inclined circle throw machines in counterflow rotation from

$$T = -120 + 10A \tag{Eq. 4.23}$$

where A (degrees) is the angle of the deck inclination. In flow rotation, T will be about 10% higher.

Length, L (ft), is found from

$$L = \frac{A}{W} \tag{Eq. 4.24}$$

where A is the estimated deck area. Values for L and W should be matched as well as possible with off-the-shelf machines to keep costs down.

Starting deck angles, A, at various ideal oversize flow rates, F (stph), and various screen widths, W (in.), are estimated from

$$A = 15.5\sqrt{\frac{F}{W}} \tag{Eq. 4.25}$$

for angles of about 16°–28° and for circle throw machines. Standard widths are 24, 36, 48, 60, 72, 84, and 96 in., and A should be rounded to the nearest whole number.

Screen Type, Installation, and Dust Collection

Most vibrating screens employed in mineral processing are horizontal or inclined; the latter are widely used in crushing plants. Horizontal screens are ideal for medium-range sizing and for liquid–solid separations because they require less headroom.

Manufacturer's installation procedures are supplied with each machine. Designs should allow for small adjustments in deck inclination, and the supporting structure must withstand resonant frequencies caused by vibration. Proper screen alignment is requisite to proper operation.

Screen feed (usually provided by devices such as conveyors, feeders, or fine ore bins) must be uniformly distributed to each screen with respect to mass flow and feed size. The design of the distribution system, then, is critical.

As part of a dust collection system, a screen commonly is enclosed in a housing, which must permit easy access for maintenance and replacement of deck panels. Such enclosures must be designed to avoid trapping dust within them.

Sizing a Vibrating Screen Classifier: Example

An example problem in sizing a vibrating screen classifier can be solved using the throughput method, as follows. The solution should be examined carefully to ensure that the reader understands the source of factors employed. Figure 4.18 presents the factors used.

1. Estimate the area, width, length, and deck slope of a double-deck, inclined circle throw, counterflow vibrating screen.

   An ore with the following feed size distribution (Table 4.7) is to be screened by means of inclined vibrating screens.

TABLE 4.7 Feed size distribution

Size		Feed Cumulative % Passing Size by Weight
cm	in.	
15.39	6.06	100
10.88	4.28	99
7.69	3.03	94
5.44	2.14	86.1
3.85	1.52	78.2
2.72	1.07	66.8
1.92	0.76	50.4
1.36	0.535	38.1
0.96	0.379	29.5
0.68	0.267	22.6
0.48	0.189	17.8
0.34	0.134	10.1
0.24	0.095	10.1
0.17	0.067	9.1

The following information is available about the feed and the screening operation:

- Final cut size = 0.535 in.
- Bulk density of feed = 102 lb/ft^3
- Feed rate to screen = 600 stph
- Dry screening
- Efficiency of undersize recovery, E_u = 90%

An upper deck with an aperture of 1.07 in. is to be used to protect a lower deck and help to reduce the overall dimensions of the machine.

2. Calculate the feed size distribution to deck 2 from the feed size distribution to deck 1:

   Table 4.8 shows the calculation from the feed size distribution to deck 1.

3. Calculate basic capacity and modifying factors for deck 1 using Figure 4.18, after first determining the undersize mass flow rate.

   (a) Undersize flow rate, U:

   U = feed rate times (percentage passing 1.07 in.)/100

   $U = 600 \times 0.668 = 401$ stph

   Note: oversize flow rate = 600 − 401 = 199 stph at 100% efficiency

TABLE 4.8 Feed size distribution to deck 1

Size		Feed Cumulative % Passing Size by Weight
cm	in.	
2.72	1.07	100 × 66.8/66.8 = 100
1.92	0.76	100 × 50.4/66.8 = 75.4
1.36	0.535	100 × 38.1/66.8 = 57.0
0.96	0.379	100 × 29.5/66.8 = 44.2
0.68	0.267	100 × 22.6/66.8 = 33.8
0.48	0.189	100 × 17.8/66.8 = 26.6
0.34	0.134	100 × 10.1/66.8 = 20.7
0.24	0.095	100 × 10.1/66.8 = 15.1
0.17	0.067	100 × 9.1/66.8 = 13.6

(b) Basic capacity, C:

From Figure 4.18, $C_{stone} = 2.9$ at 59% open area

$C = (102/100) \times C_{stone}$

$C = 2.96$

where 102 is the bulk density and 100 is the reference density of stone.

(c) Modifying factors:

Fines factor

Half-size of aperture = 1.07/2 = 0.535 in.

Percentage passing half-size = 38.1% from size analysis of feed to deck 1

From Chart B of Figure 4.18, interpolate to find factor as follows:

$$(1 - 0.8)/(40 - 30) = 0.02$$
$$0.02(38.1 - 30) = 0.16$$
$$0.8 + 0.16 = 0.96 = F_f$$

Oversize factor

Size of aperture = 1.07 in.

Percentage passing 1.07 in. = 66.8% from size analysis of feed to deck 1

From Chart B of Figure 4.18, interpolate to find factor as follows:

$$(0.86 - 0.8)/(70 - 60) = 0.006$$
$$0.006(66.8 - 60) = 0.04$$
$$0.86 - 0.04 = 0.82 = F_o$$

Efficiency factor

From Chart B, Figure 4.18, $F_e = 1.25$ at 90% efficiency

Deck factor

From Chart C of Figure 4.18, $F_d = 1$ for deck 1

Open area factor

From Chart E of Figure 4.18, F_{oa} = (percentage of open area)/59, where 59 is taken from capacity figure at 1.07-in. opening.

Use punched plate with square, staggered openings that have a percentage of open area = 52. (Refer to screen deck tables, Chart E of Figure 4.18.)

Hence, $F_{oa} = 52/59 = 0.88$

Slot factor

From Chart F of Figure 4.18, $F_s = 1.0$

Wet screening factor

For dry operation, from Chart D of Figure 4.18, $F_w = 1.0$

4. Calculate the area of deck 1 from base capacity, undersize flow rate and factors found in step 3.

$$A = \frac{U}{CF_f F_o F_e F_d F_{oa} F_s F_w} = \frac{401}{2.96 \cdot 0.96 \cdot 0.82 \cdot 1.25 \cdot 1 \cdot 0.88 \cdot 1 \cdot 1} = 156 \qquad \textbf{(Eq. 4.26)}$$

$A = 156$ ft^2

5. Calculate base capacity and modifying factors for deck 2 using Figure 4.18, after first determining the undersize flow rate.

(a) Undersize flow rate, U:

U = (feed rate to deck 2) × (percentage passing 0.535 in.)/100

$U = 401 \times 0.57$; i.e., 57% passes 0.535 in. (size analysis of deck 2 feed)

$U = 229$ stph

Note: Oversize flow rate from deck 2 = 172 stph at 100% efficiency

(b) Basic capacity:

From Chart A of Figure 4.18, C_{stone} = 1.8 at 53% open area (deck 2 reference area)

$C = (102/100) \times 1.8$

$C = 1.84$

(c) Modifying factors:

Fines factor

Half-size of aperture = 0.535/2 = 0.267 in.

Percentage passing 0.267 in. = 33.8% from size analysis of deck 2 feed

From Chart B of Figure 4.18, interpolate to find factor as follows:

$$\begin{aligned} (1 - 0.8)/(40 - 30) &= 0.02 \\ 0.02(33.8 - 30) &= 0.08 \\ 0.8 + 0.08 &= 0.88 = F_f \end{aligned}$$

Oversize factor

Size of aperture = 0.535 in.

Percentage passing 0.535 in. = 57.0% from size analysis of feed to deck 2

From Chart B of Figure 4.18, interpolate to find factor as follows:

$$\begin{aligned} (0.9 - 0.86)/(60 - 50) &= 0.004 \\ 0.004(57 - 50) &= 0.03 \\ 0.9 - 0.03 &= 0.87 = F_o \end{aligned}$$

Efficiency factor

From Chart B of Figure 4.18, F_e = 1.25 at 90% efficiency

Deck factor

From Chart C of Figure 4.18, F_d = 0.90 for deck 2

Open area factor

From Chart E of Figure 4.18, F_{oa} = (percentage of open area)/53 where 53 is taken from capacity figure at 0.535-in. opening

Referring to screen deck tables (chart E of Figure 4.18) use heavy-duty ton-cap rectangular screen with percentage of open area = 56.

Hence, F_{oa} = 56/53 = 1.06

Slot factor

From Chart F of Figure 4.18, F_s = 1.1

Wet screening factor

For dry operation, from Chart D of Figure 4.18, F_w = 1.0

6. Now calculate the area of deck 2 from base capacity, undersize flow rate and factors found in step 5.

$$A = \frac{U}{CF_f F_o F_e F_d F_{oa} F_s F_w} = \frac{229}{1.84 \cdot 0.88 \cdot 0.87 \cdot 1.25 \cdot 0.90 \cdot 1.06 \cdot 1.1 \cdot 1} = 124 \text{ ft}^2 \qquad \textbf{(Eq. 4.27)}$$

7. Estimate the width, length, and deck angle for a double-deck vibrating screen.

(a) Width:

The deck that determines the screen area is deck 1. The feed end takes 600 stph, and the screen discharges 199 stph of oversize. Hence, the deck area required is approximately 160 ft^2. An 8-ft by 20-ft screen should do the job, but bed depth at the oversize discharge end must be less than or equal to (2 + 0.02(bulk density)) × (screen aperture). Thus bed depth must not exceed

$$D^* = (2 + 0.02 \times 102)1.07 = 4.32 \text{ in.}$$

Bed depth is estimated from $D = 400\ F/(WT$ (bulk density)). $F = 199$ stph, W is in feet, bulk density = 102 lb/ft^3, and $T = -120 + 10\ A$; A is deck slope. Assume this slope is 22°, so that $T = 100$ ft/min. Thus,

$$D = 400 \times 199/(8 \text{ x } 100 \times 102) = 0.98 \text{ in.}$$

The estimate of bed depth is well below critical. An 8-ft width is thus acceptable.

(b) Length:

Because deck area is about 160 ft^2, length must be 160/8 = 20 ft. The ratio of length to width is 20/8 = 2.5, which is within the acceptable range of this ratio.

(c) Deck angle:

At this stage, deck angle has been estimated at 22°. This angle can be checked using the following equation:

$$A = 15.5\sqrt{\frac{F}{W}} = 15.5\sqrt{\frac{199}{96}} = 22.3° \qquad \textbf{(Eq. 4.28)}$$

where $W = 96$ in. and $F = 199$ stph.

The estimate of A is close enough to that used in step 7a.

(d) The upper deck is punched plate that contains staggered square openings with 52% open area. The lower deck has rectangular openings of 0.535 in. by 3.0 in. and 56% open area.

SIZE CLASSIFICATION

Size classifiers separate particles of various sizes, shapes, and specific gravities in fluids (e.g., water or air) under the influence of gravitational or centrifugal forces. In principle, such devices should make a size split based on particle size rather than other properties. Unfortunately, the split is always imperfect. Measures of the performance of size classifiers are similar to those employed for screens, except that the definition of cut size is not as simple. Separations are normally made between about 20 and 325 mesh, although some pneumatic devices size readily to below 95% passing 0.010 mm.

Classification devices attempt to take advantage of the following aspects of particle behavior.

1. Smaller particles fall more slowly in fluids than do larger ones.
2. In free vortex motion (i.e., cyclones), centrifugal forces have greater influence on large particles and lesser influence on small particles.
3. Small particles, having less inertia, tend to behave like the suspending medium or fluid.
4. Larger particles require higher conveying velocity for coarse separation.
5. Collision frequency increases with particle size.

To take advantage of these phenomena various mechanical components (such as rakes, spiral arms, vanes, spindles, and baffles) are used, as are means to regulate the direction of fluid flow. Because hydrocyclones are extensively used in the mineral processing industry, they are discussed in detail in the sections that follow.

Size classifiers are distinguished from each other, initially, on whether the fluid employed is air or water. Those that rely on water include nonmechanical classifiers (surface sorters such as cones, and hydraulic classifiers such as the Richards hindered settler), mechanical classifiers (spiral classifiers), and hydrocyclones. This last type of device uses centrifugal forces, another distinguishing feature. Pneumatic (air) classifiers rely on a suitable interplay between the force of gravity and drag forces, and in many devices collision forces and centrifugal forces, to effect size separation in air. Figure 4.19 categorizes size classifiers.

Nonmechanical Classifiers

Nonmechanical classifiers include surface sorters (horizontal flow devices without mechanical components, such as spiral arms) like cones, and hydraulic classifiers (machines with water deliberately added to create a vertical flow or rising current) that function under free or hindered-settling conditions.

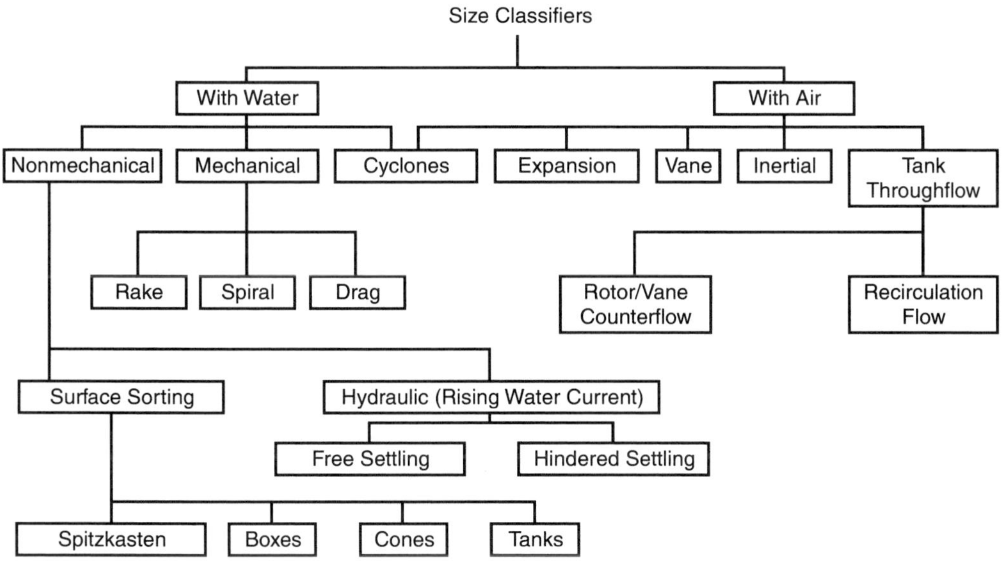

FIGURE 4.19 Categories of size classifiers

Surface Sorters. Cones are often used for fine sizing (desliming), although at very high feed rates, they can be effective for scalping out coarse particles (1/8 in. and larger) that may damage downstream processing units. Hydraulic classifiers are used for coarse separations. Both classes of machines rely on principles involved in the settling in fluids of particles acted on by gravitational force. Surface sorters are not discussed further in this chapter; selected hydraulic devices are considered in the following sections.

Hydraulic Classifiers. Hydraulic classifiers use additional water (called "hydraulic water") introduced to oppose the settling direction of particles in the separation zone. Hydraulic water is the major variable manipulated to control the split. If a particle in the separation zone of the classifier settles downward with velocity V_p, hydraulic water with velocity V_w will eventually act as a rising current to oppose the direction of particle fall. The net velocity of the particle will become $V_n = V_w - V_p$. In the first instance, hydraulic classifiers sort out and group particles on the basis of differences in specific gravity. Consequently, they are mineral separation devices. However, for feeds of essentially uniform specific gravity, they classify according to differences in particle size, and they do so with high efficiency and low maintenance cost. They can be designed to operate as either free-settling or hindered settling units.

Free-settling Hydraulic Classifiers. Free-settling hydraulic classifiers have sorting columns that are uniform in cross section throughout their column length. They are either of the tank or the launder type as typified by the Evans unit (Gaudin 1939; Figure 4.20). Water is introduced through pipes F (see Figure 4.20) and controlled by valves. The flow is either over weirs (point E) or through spigots G. Openings at B and C are adjustable to manipulate upward velocities. Faster-settling particles discharge through G, while slower ones are carried to the next box in line.

Free-settling hydraulic classifiers are still in use in the form of tanks or columns (e.g., elutriators). However, their capacity-to-size ratios are not large, and they take up too much space for the higher-capacity plant of today.

Hindered-settling Hydraulic Classifiers. Hindered-settling classifiers differ from free-settling units in that the sorting column is constricted, either gradually or abruptly, near the bottom end. The constriction increases the upward velocity of hydraulic water relative to fluid velocity above the

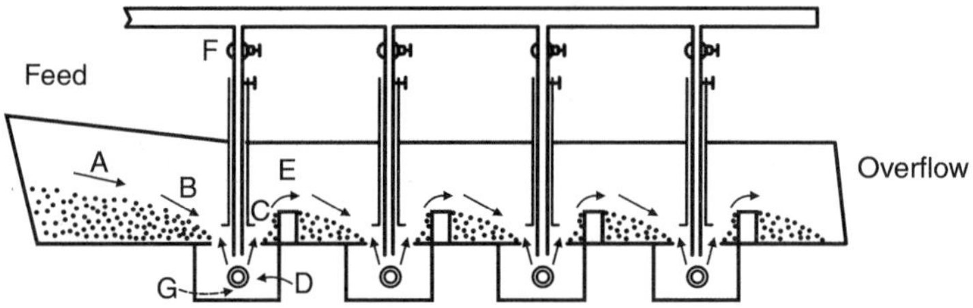

Source: Gaudin 1939.

FIGURE 4.20 Evans free-settling classifier

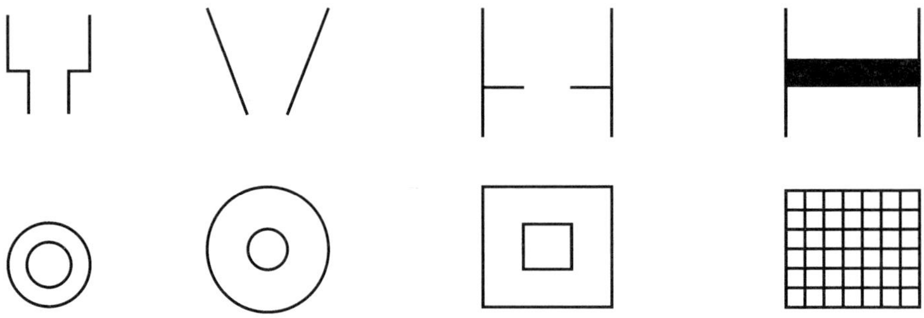

Source: Taggart 1945.

FIGURE 4.21 Constrictions for hindered settlers

constriction. Particles of certain size and density combinations (i.e., heaviness combinations) will begin to accumulate just above the constriction to form a quicksand-like column—a fluidized bed.

The pressure at the top of the bed is less than it is at the bottom, so particles within the column teeter (particles at the center rise repeatedly from the bottom to the top of the column, and they fall down again at the sides). The teeter column behaves almost as a heavy liquid. Light particles cannot pass, the heaviest pass through, and those in between are retained in the column. In consequence, hindered-settling devices can separate minerals of different specific gravities far better than free-settling units. Moreover, for particles of essentially uniform density, hindered-settling units can be effective sizing devices. Constriction shapes that have been used (Taggart 1945) are shown in Figure 4.21.

Across a stable teeter bed, the superficial velocity, $V_w = Q/A$, of water necessary to maintain the fluid bed of bed voidage, ε, is found from

$$V_w = \frac{Q}{A} = V_p \varepsilon^n \qquad \textbf{(Eq. 4.29)}$$

For spheres at N_{Re} larger than 500, $n = 2.4$; for N_{Re} between 1 and 500, $n = 4.4N_{Re}^{-0.1}$; between N_{Re} values of 0.2 and 1, $n = 4.4N_{Re}^{-0.03}$. Thus, the velocity of fluidization that is needed to form a teeter bed of particles of a given size d can be estimated (remember, V_p is a function of d). The ratio of teeter chamber area to constriction area influences the separation size.

Hindered-settling devices are either of the launder or tank type. A version of the latter is called a siphon sizer (Gaudin 1963; Figure 4.22). Feed enters through a feed line (not shown in the figure) and is distributed throughout a free-settling zone above the constriction near the wall. To maintain the teeter column, hydraulic water is added through a network of perforated pipes (not shown) at the bottom. Heavy particles pass through the bed and collect at the bottom, where they form a dense suspension that is drawn out of the tank by a siphon line. The removal rate is detected by means of a "superelevation" tube that has a float inside to detect level (a measure of bed depth at the bottom). If

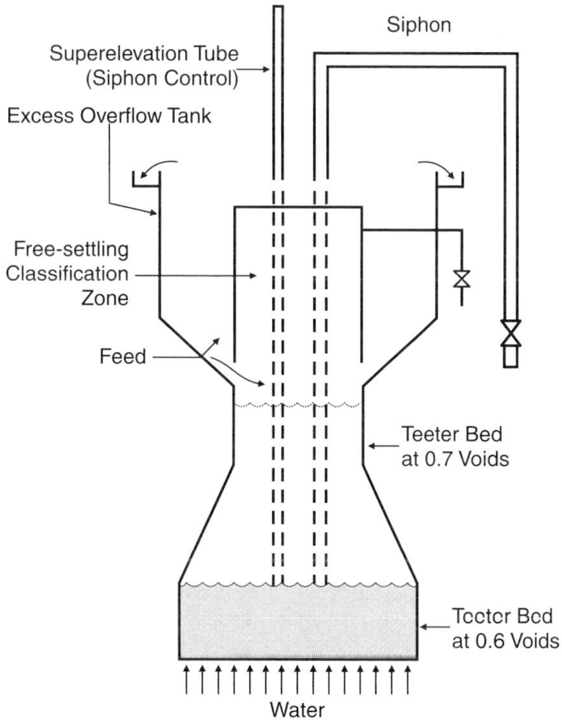

Source: Gaudin 1963.

FIGURE 4.22 A siphon sizer

the level becomes too high, an automatic valve on the siphon line opens to increase flow; if the level decreases, the valve will reduce the flow to build up the bed again. Light particles enter the overflow launder at the top. Inside the top section, an inverted cylinder with its bottom end open has been inserted with about 0.5 ft of clearance between the bottom edge and the constriction wall. Water is metered into this section to keep a constant head and to serve as hydraulic water that maintains free-settling conditions between the tank wall and the outer wall of the cylinder.

Hindered-settling hydraulic classifiers have a small capacity-to-size ratio that makes them unattractive for large-tonnage operations. In certain situations, however, such as in circuits where tonnages are low, these devices have potential for sizing material in preparation for gravity separation.

Mechanical Classifiers

Mechanical classifiers have moving parts that agitate the pulp and help move the underflow out of the separation zone. Current flow in these classifiers may be either horizontal (as in rake or spiral classifiers) or vertical (as in bowl or tank classifiers). Hydraulic water may or may not be added. Their principal use was in grinding circuits, but since the mid-1950s, they have been largely replaced by hydrocyclones.

Spiral or Rake Mechanical Classifiers. Spiral or rake classifiers are semirectangular tanks with parallel sides (sides may flare somewhat toward the overflow end) and a sloped bottom. Inside the tank, a rake or a spiral mechanism conveys coarse material upward to a sands return chute. Figure 4.23 shows schematics of a rake classifier and the more modern spiral classifier (Hitzrot and Meisel 1985).

General Characteristics. Take L to be the length of the classifier. Feed enters at a point that is about 0.6 L (high weir type) or 0.5 L (overflow end of spiral submerged) or 0.3 L (low weir type) from the

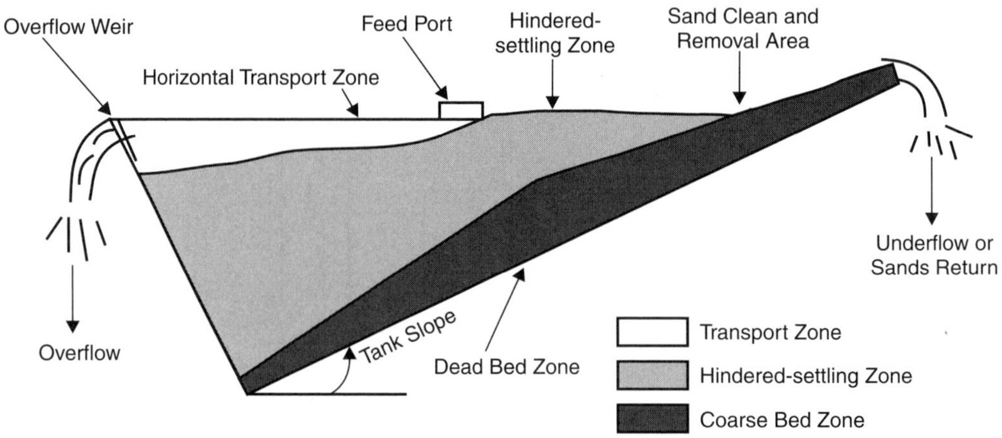

FIGURE 4.23 Classifier zones

overflow weir (the overflow weir is a movable baffle plate at the overflow end, the height of which can be adjusted to control pool area). Spirals are preferred to rakes because spirals cost less to maintain.

Either rakes or spirals move coarse sand out of the tank. Rakes employ a repetitive rectangular trajectory whose long dimension is parallel to the bottom. Rakes in their down-and-forward position move parallel to the bottom, thereby dragging coarse solids up the slope. At a certain point, the blades lift and then reverse their direction of parallel motion while in the up position (up-and-reverse stroke). Again, at a certain point, the blades drop back to the bottom and begin their down-and-forward stroke to push sand up the slope. A rake may complete up to 30 down-and-forward strokes per minute, depending on the classifier's areal efficiency (the ratio of effective pool area to actual pool area).

The spiral has a pitch of 50%–75% of its diameter, although 50% is recommended (Hill 1982), where pitch is the distance between the helix flights (the spiral arms). The axis of the spiral is parallel to the bottom and rotates at speeds of 2–10 rpm in a direction that conveys sand up-slope. In "duplex" (twin spirals) versions of the classifier (Figure 4.23), weir height can be automatically adjusted.

Classifier Zones. The diagram in Figure 4.24 shows general zones that exist in horizontal classifiers.

Separation size and overflow capacity depend on several design and operating variables that influence settling in the classifier pool, which contains zones as shown in Figure 4.24. In the horizontal flow transport zone, which is relatively dilute, the bulk of the water and the lighter particles are transferred to the overflow. Heaviest particles work their way down through the hindered-settling zone and enter the sands removal zone to be conveyed up-slope by the mechanism (e.g., spirals). A dead bed zone accumulates more or less permanently between the outer edge of the spirals and the tank bottom.

Interparticle transfer between zones is always taking place. The top of the hindered-settling zone has a lower pulp density than the bottom of the zone. Spiral motion agitates the hindered-settling zone so it behaves somewhat like a heavy-medium suspension. Particles intermediate between light and heavy are sensitive to the suspension density and viscosity of this zone. Water is often added with feed slurry by a separate line controlled by an operator. Sprays are employed to clean the mechanism.

Variables That Influence Separation Size and Capacity. Design variables of importance include degree of end flare, number of spiral flights (or, for rakes, rake blades), point of feed entry, tank slope, and spiral (or rake) speed; operating variables are feed size distribution, feed rate, mineralogical composition, weir height, and total water added to classifier.

Degree of End Flare. When the classifier overflow end is flared out, pool area, A, increases. The width increases 20%–130% of the helix diameter. Because the net upward velocity of particles is $V_n = V_{su}$

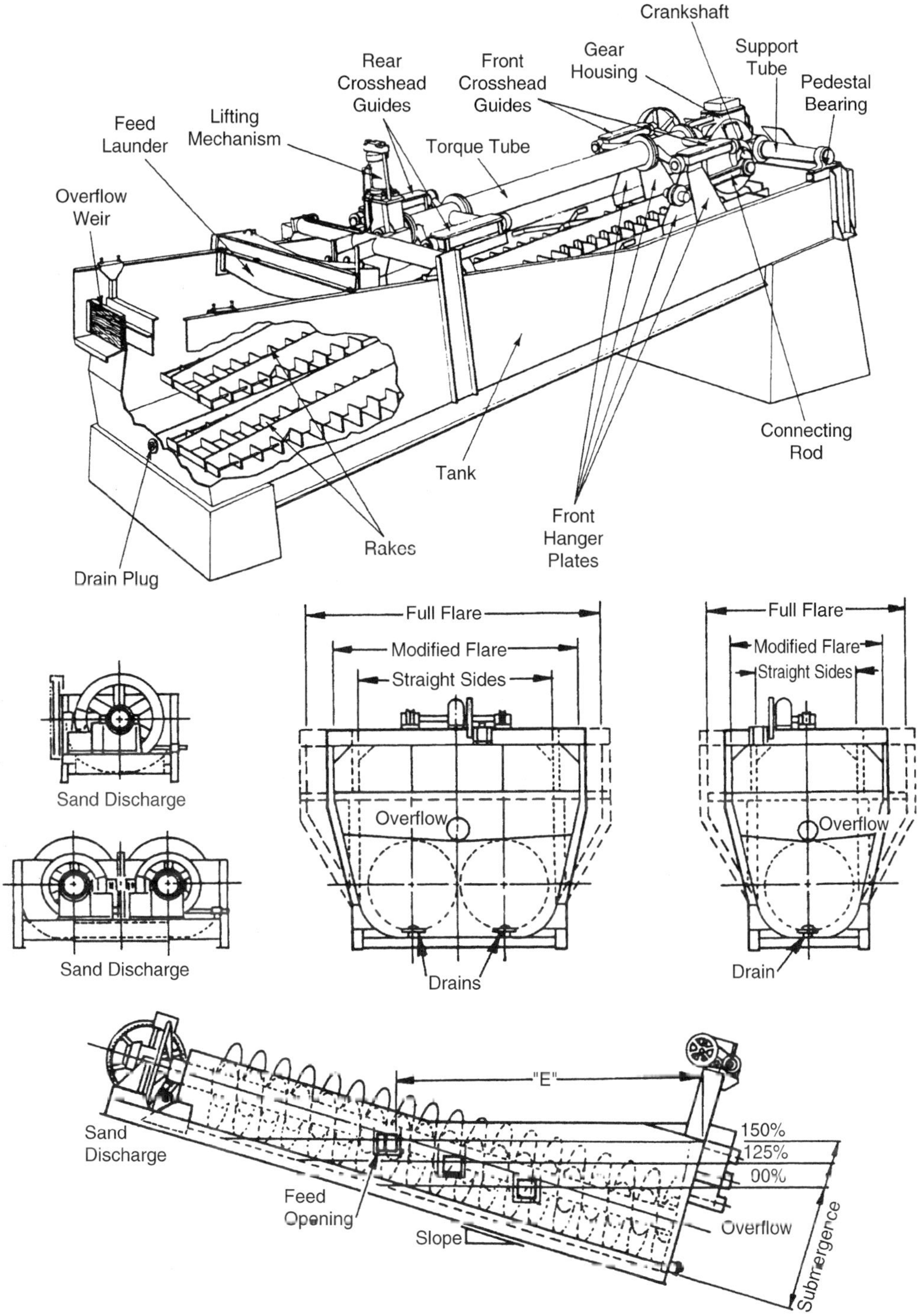

Source: Hitzrot and Meisel 1985.

FIGURE 4.24 Rake and spiral classifiers

$- V_p = (Q/A) - V_p$, where V_{su} is the upward velocity of the suspension and V_p is the settling velocity of particles in the downward direction; if A has increased, the term Q/A becomes smaller. Hence, V_n must decrease. This relationship means that the larger area, other things being equal, will cause the separation size to decrease. At the same time, the capacity is reduced.

Number of Spiral Flights. The capacity of a spiral will increase with the number of flights on the shaft, relative to a single helix. Thus, a double helix treats more, and a triple helix even more, solids. However, crowding occasioned by triple helix shafts may be detrimental to sizing.

Point of Feed Entry. Feed may enter the classifier from both sides or from one side only. By changing the location along the classifier length at which feed enters, the effective pool area can be either increased or decreased. The retention time available for settling in the hindered-settling zone, T_{pool}, will change, because $T_{pool} = V/Q$ where V is the effective pool volume and Q is the volumetric flow rate of overflow. The effect of point of addition of feed is thus explained in either of two ways: as a change in T_{pool} or a change in $V_n = (Q/A) - V_p$. Thus, when feed enters at a point closer to the overflow weir, effective pool area and volume will decrease. This decrease in turn increases V_n (or a decrease in T_{pool}). As a result, the overflow becomes coarser.

Tank Slope. Increasing the slope has two effects: it will decrease the pool area, so that V_n, which is proportional to $1/A$, will increase, and it decreases retention time. The net result is a coarser overflow (increase in separation size). In addition, the raking (sands return) capacity is reduced. At some critical slope, the sands will slough back into the pool, which causes the overflow to become coarser.

Spiral (or Rake) Speed. An increase in spiral or rake speed has somewhat the same effect as a decrease in pool area. The overflow becomes coarser because of better mixing and agitation caused by the mechanism. The lower the speed, the finer the split, provided that the loss in capacity can be tolerated.

Feed Size Distribution, Mineralogical Composition, and Feed Rate. A change in the composition of the ore may signal a change in ore hardness or a change in proportions of more dense or less dense minerals. Changes in ore hardness influence the feed size distribution; changes in composition influence the density of the hindered-settling zone. If the classifier is closed with grinding mills, changes in the fresh feed rate to the grinding circuit or changes in the size consist to the mill likewise influence the characteristics of the pool. Except for fresh feed rate these changes are not generally controllable. Hence, they are viewed as disturbance variables whose effects on separation size and capacity must be minimized. An increase in solids feed rate decreases the retention time of suspended pool solids, so that the overflow coarsens. A similar result follows when an increase in fresh feed rate of slurry causes V_n to increase.

Weir Height. Weir height is an operating variable on a long-term basis. Lowering the weir decreases pool area; raising it increases pool area. The effect is either to increase (coarsen) or to decrease (make finer) separation size.

Total Water to Classifier. The total water to the classifier is composed of feed slurry water, hydraulic water added to the feed box, and spray water that washes slime from exposed mechanisms to increase efficiency. If water is added at an increased rate, several reactions occur rapidly.

Because most of the water goes to overflow (the sands percent solids is relatively unaffected by changing water addition), $V_{su} = Q_O/A$ increases. At the same time, particles settle at a velocity given approximately by

$$V_p = K\sqrt{\left(\frac{\rho_s}{\rho_{su}} - 1\right)\frac{gd}{C_D}} \quad \textbf{(Eq. 4.30)}$$

where C_D may be a function of suspension viscosity, μ_{su}, and K probably depends on the voidage. Adding water immediately decreases the suspension density, ρ_{su}, so that the settling velocity of a particle of size d in the downward direction is increased. As more water is added, V_{su} continues to increase in direct proportion, whereas V_p reaches a constant (because ρ_{su} approaches the specific gravity of water). On the other hand, the net particle velocity, V_n, is large at the start, then goes through a minimum, and finally increases again (Figure 4.25). In Figure 4.25, note that the usual operating point (Q in the figure) is to the left of the minimum in the curve of d versus water addition, where

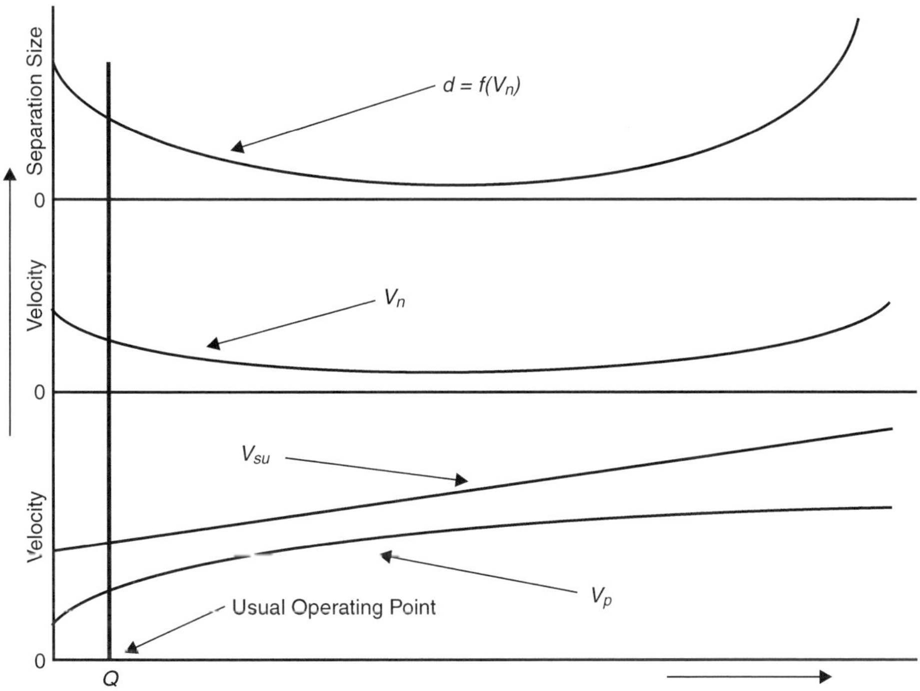

FIGURE 4.25 Effect of total water on separation size

d is proportional to net velocity, V_n, raised to some power. This relationship means that adding hydraulic water to the classifier produces a finer overflow, and cutting back on water causes the overflow to coarsen.

Overflow pulp density responds in the same manner, so that a finer overflow means a lower pulp density and a coarser overflow means a higher pulp density. Consequently, the most important variable that controls separation size is hydraulic water. Overflow pulp density can be monitored and maintained constant. A constant density maintains a relatively constant separation size.

Selection of Spiral or Rake Classifiers. For desliming operations, when overflow pulp contains about 10% solids (specific gravity = 2.65) or less, overflow capacities for open circuit operation may be estimated from the area principle with a safety factor of 1.72–2. Thus $Q = 18.06\ V_d\ WL$ (safety factor of 1.72) or $Q = 15.6\ V_d\ WL$ (safety factor of 2), where Q is gpm, V_d is given in units of in./s and W and L are, respectively, the width and length of the pool in feet.

For closed-circuit grinding operations, when overflow pulps contain 20%–45% solids, a graphical estimation method is recommended (Hitzrot and Meisel 1985). Roughly, for solids of specific gravity = 2.65, the percent solids in the overflow, P_o, is found from $P_o = -35.82 + 11.76\ ln(d_s)$, where d_s is the separation size (μm). This size is related to basic capacity, T (tons/24 h/ft^2 of pool area) by

$$T = -\ 11.97 + 3.23\ ln(d_s) \qquad \textbf{(Eq. 4.31)}$$

If C is the desired capacity in tons/24 h, $C/T = A$ (the required pool area in square feet), and $A = WL$.

If D (ft) is helix diameter, recommended speeds, S (rpm), are $S = 19.91/D$. Nominal raking capacity C (stph) at speed S is estimated from $C = 25.28D/S$. Tank slopes are 3–4 in./ft and helices are double pitch of 2–8 ft in diameter that vary in off-the-shelf increments of 0.5 ft. Horsepower can be approximated as hp = $97/S^{1.82}$. The sands raking capacity is inversely proportional to the recommended slope, which depends on the desired size of separation (that size in the overflow that passes 95%–99% of the particles). Suggested slopes (in./ft) for tanks are 4 at 20 mesh, 3.75 at 28 and 35 mesh, 3.5 at 48 and 65 mesh, 3.25 at 100 and 150 mesh, and 3 at 200 and 325 mesh (Reithmann and Bunnell 1980).

Empirical Models. Fundamental studies of wet classification have identified the importance of turbulence, diffusion processes, and retention times for building mechanistic models. Currently, useful empirical models of rake or spiral classifiers are available (Riethmann and Bunnell 1980; Plitt and Flintoff 1985; Lynch et al. 1967; Fitch and Roberts 1985). For example, Plitt and Flintoff (1985) proposed that the x_{50c} size has a settling velocity, v, equal to Q_o/A, where Q_o is the volumetric overflow rate and A is pool area. The settling velocity is determined from

$$v = \frac{Q_o}{A} = \frac{4(\rho_s - \rho_{su}) g x_{50c}^2}{a[3.646 x_{50c}^{1.5}(g\rho_s\rho_{su})^{.5} + 20.785(b\mu_{su})]} \quad \textbf{(Eq. 4.32)}$$

For galena-like particles, $a = 1.038$ and $b = 3.01$. All terms, other than x_{50c}, are measured or estimated, so that x_{50c} can be calculated from the above equation. Measurements or estimates of the feed size distribution, the sharpness of classification, the water split, and x_{50c} are then entered into the fractional recovery equation proposed by Plitt (1976) to permit calculation of the overflow and underflow size distributions.

Lynch and colleagues (1967) analyzed the performance of a ball mill–rake classifier circuit, and then fitted their corrected fractional recovery equation to corresponding data. By means of multiple regression analysis, x_{50c}, the mass flow rate of overflow water, the mass fraction of solids in the overflow, and the mass fraction of water in the sands were related to operating variables. The fitted equations provided a means to calculate the overflow and underflow size distributions from more easily measured operating variables by their fractional recovery expression:

$$\frac{\frac{Uu_i}{Ff_i} - H_u}{1 - H_u} = \frac{\exp\left(0.5\frac{x_i}{x_{50c}}\right) - 1}{\exp\left(0.5\frac{x_i}{x_{50c}}\right) = \exp(0.5) - 2} \quad \textbf{(Eq. 4.33)}$$

where U and F are solids mass flow rates in underflow and feed, respectively; u_i and f_i are solids weight fractions of size x_i in underflow and feed, respectively; and H_u is the fraction of feed water reporting to the underflow.

Drag and Bowl Classifiers. Drag classifiers (Taggart 1945) are rectangular tanks that have somewhat of a V-shape when viewed from the sides, as shown in Figure 4.26. Feed enters at the lower end, and overflow is discharged onto pan-type launders mounted at the sides just below pool level. Underflow sands are dragged up-slope by rakes (flights) mounted on the outer side of an endless belt or link chain. These horizontal-current devices are reputedly inexpensive to build and are still in use today.

Figure 4.26 contains a schematic of a bowl classifier (Hitzrot and Meisel 1985), which is used for finer sizing operations. It is essentially a rake classifier with a cylindrical bowl attached at the overflow end. Feed enters at the center of the bowl, which is cone shaped and has scraping blades inside that revolve gently to force sand toward a central discharge slot. Rakes positioned beneath the slot transport sand up an inclined bottom to a discharge launder. Overflow spills into an annular discharge launder wrapped around the outside top of the bowl. This arrangement maximizes the length of the discharge lip. In addition, the settling area is large in a relative sense. These two features are a decided advantage in terms of the ratio of overflow capacity to raking capacity.

Hydrocyclone Classifiers

Hydrocyclones use centrifugal forces to classify particles in a fluid that experiences essentially free vortex motion inside the device. They are widely used in mineral processing plants today because of their extremely favorable capacity-to-size ratios and reasonably low maintenance.

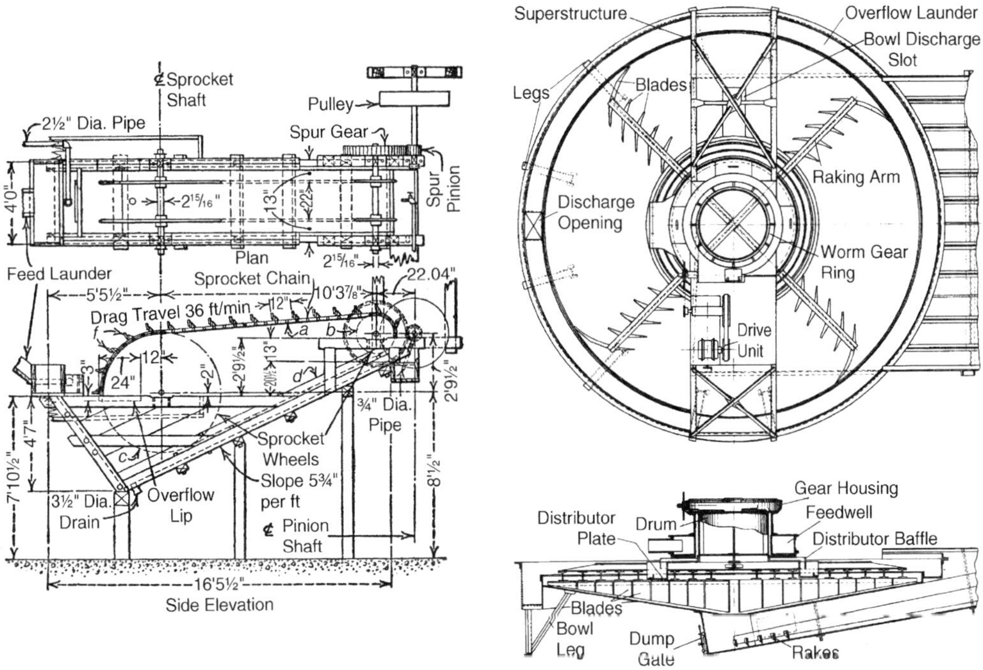

Source: Taggart 1945; Hitzrot and Meisel 1985.

FIGURE 4.26 Drag and bowl classifier

Basic Characteristics. A cutaway view of a typical hydrocyclone is shown in Figure 4.27. Feed slurry, either pumped or flowing by gravity, enters the inlet through a feed pipe and flows at a tangent to a cylindrical feed chamber under pressure. To increase retention time, a cylindrical section is often added between the upper feed chamber and the lower conical section. This section has an included angle (cyclone angle) in the range of 12° (for cyclones of 10-in. diameter or less) to 20° (for larger cyclones). Fine particles leave through the vortex finder and are directed to further processing by the overflow pipe. Coarse particles travel downward in a spiral path and discharge at atmospheric pressure through a variable apex (spigot) that connects to an underflow pipe. Cyclones are often mounted radially, with their feed pipes attached to a central vertical feed line that is capped at the top. A typical mounting assembly is called a "Cyclopac." Underflow slurry enters a circular weir trough (concentric like a doughnut) that is sloped to divert the combined underflow to a next processing step (such as the feed spout of a ball mill). Overflow lines are U-shaped at the top and discharge to an annular launder that is concentric around the central feed pipe. Standpipes that are open to atmosphere are located at the peak of each overflow line (they prevent possible siphoning if lines are below the feed line). For ease of access for maintenance and liner replacement, air-actuated valves may be installed to seal off feed pipes as desired.

Theoretical aspects of cyclones have been well developed (Kelsall 1952; Dahlstrom 1954; Lilge 1962; Rietema 1962; Bradley 1965; Tarr 1985) and have led to useful design criteria (Tarr 1985). The effects of major design and operating variables have been documented (Tarr 1985) and methods for selection (Arterburn 1982; Mular and Jull 1982; Tarr 1985) are available. Mathematical models have been proposed and improved on for selection and design (Lynch and Rao 1975; Plitt 1976; Plitt and Flintoff 1985).

Cyclone Fundamentals. Fluid motion inside a cyclone is analogous to that within a free vortex (one that persists without external energy input). Water draining from a bathtub will exhibit such motion because an air core forms as the water rotates into the drain hole. In contrast, forced vortex

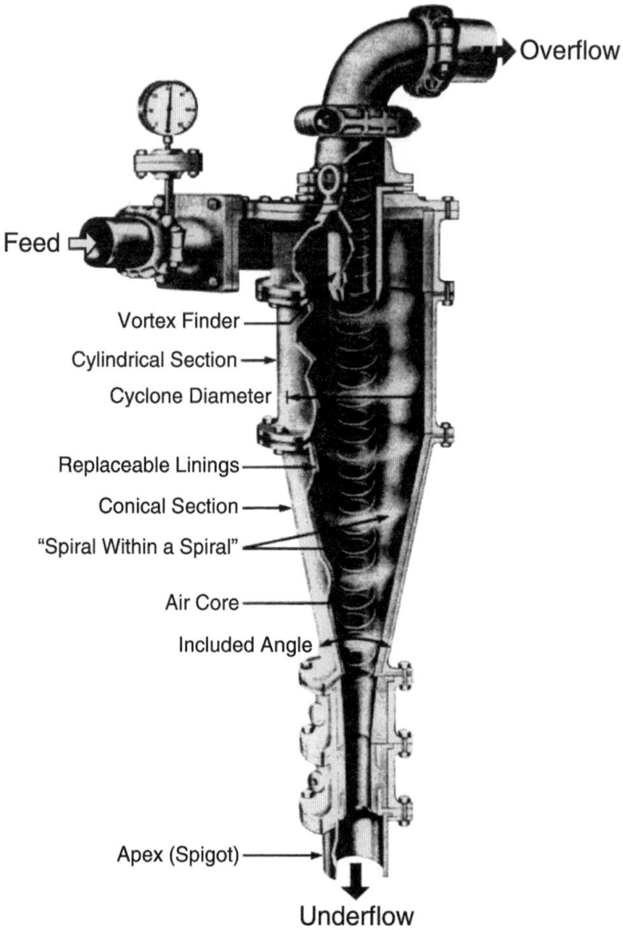

Source: Krebs Engineers.

FIGURE 4.27 Conceptual view of hydrocyclone section

motion is obtained when a body of fluid is forced to rotate by applying external energy (e.g., causing a beaker of water to rotate at angular velocity). Figure 4.28 illustrates the essential idea.

For cyclones, the tangential velocity, V_t, of an element of fluid at a horizontal distance r from the edge of the air core is given by

$$V_t = \frac{C}{r^n} \quad \textbf{(Eq. 4.34)}$$

where C is a constant and n varies from about 0.5 (turbulent flow) to 0.8 (viscous flow). Moreover, an energy balance shows that the pressure, P, at a point, q, is

$$\frac{P}{pg} = H - \frac{C^2}{2gnr^{2n}} \quad \textbf{(Eq. 4.35)}$$

with H as the total head relative to the reference plane. For forced vortex motion, as in a centrifuge, the tangential velocity at a horizontal distance r from the center is $V_t = \omega r$. At a point, q, the pressure, P, is

$$\frac{P}{\rho g} = \frac{P_o}{\rho g} + \frac{\omega^2 r^2}{2g} \quad \textbf{(Eq. 4.36)}$$

where P_o is the pressure at the reference plane with $r = 0$.

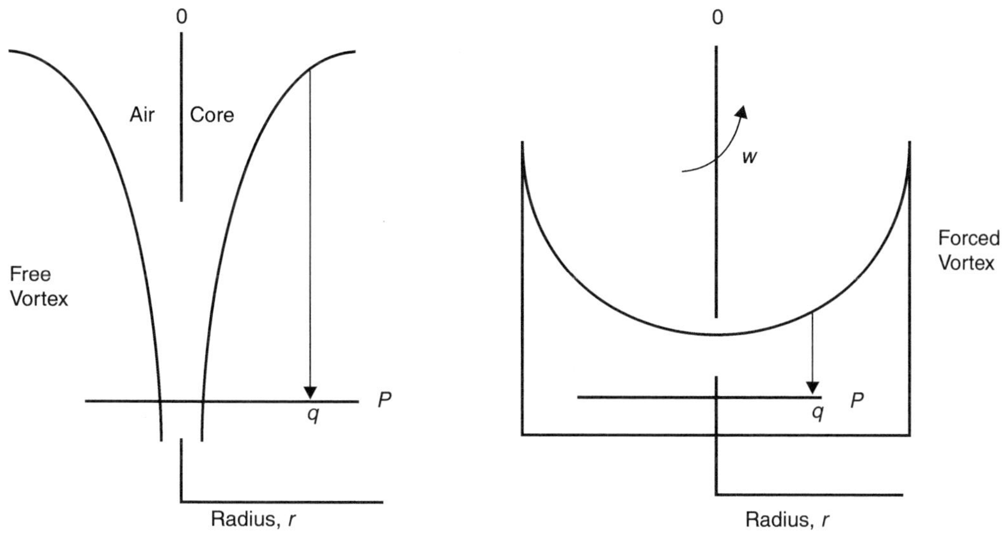

FIGURE 4.28 Free and forced vortex motion

Assuming that a particle behaves like an element of fluid, a particle of diameter d will experience a centrifugal force, F_c, equal to

$$F_c = (m - m_l)\frac{V_t^2}{r} = \frac{\pi}{6}d^3(\rho_s - \rho_l)\frac{V_t^2}{r} \qquad \textbf{(Eq. 4.37)}$$

In a centrifuge, F_c is proportional to r, whereas in a cyclone, F_c is proportional to $1/r^{2n+1}$. In a cyclone, the centrifugal force is higher near the air core than it is near the wall. Coarse particles near the core are driven outward, whereas fine particles near the wall are readily forced toward the center (they experience relatively minor centrifugal force at the wall). This feature makes the cyclone more attractive as a size-separation device.

The behavior of fluid velocities and the corresponding forces acting in a cyclone have been reported by Lilge (1962). He shows that to the left of a zero-vertical-velocity contour, fluid velocities rise sharply; to the right they decrease slowly to the wall. Radial velocities rise roughly in proportion to radius at any given level of height, whereas tangential velocities behave essentially as described previously in this chapter.

A vertical force acts downwardly on a particle to the right of a zero-vertical-velocity envelope and upwardly to the left of it. If the cyclone radius is r, the envelope trace can be initiated at a distance $r/2$ from the center and at the same level as the bottom of the vortex finder. The envelope trace extends as a cone downward to the apex and intersects at about the trace of the (spigot radius)/2. Particles to the left of the envelope tend to rise; those to the right tend to travel downward. An envelope of maximum tangential velocity lies virtually at the air core wall. The two envelopes offer insight into the resulting motion of particles (Figure 4.29).

There is always some size of particle, d_{50}, associated with the intersection of the envelopes of maximum tangential velocity and of zero vertical velocity. Half of these particles rise; the other half enter the underflow. Particles finer than this size enter the overflow; particles coarser enter the underflow. The cone section of a cyclone at steady state contains particles with a size distribution similar to that of the underflow stream. In the vicinity of the bottom edge and outer wall of the vortex finder, very fine particles predominate. Just below the vortex finder and extending a short distance into the cone section, particles of intermediate sizes are found. Near the top and inner walls of the feed chamber, the size distribution is very like that of fresh feed.

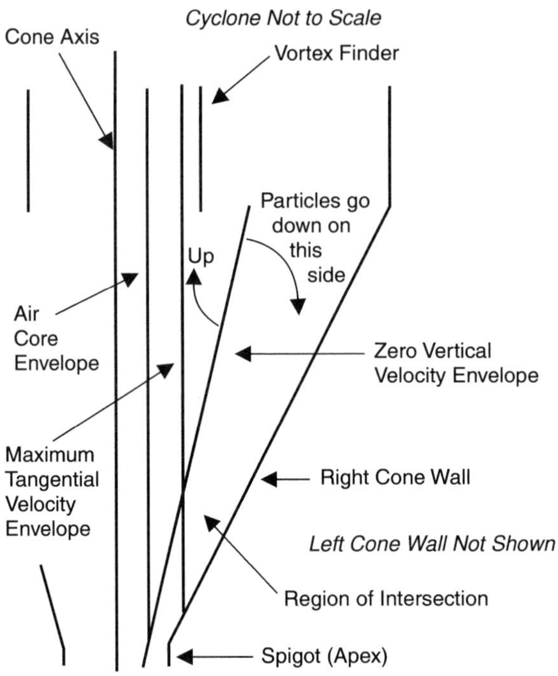

Source: Lilge 1962.

FIGURE 4.29 Maximum tangential velocity and zero-vertical-velocity contours in cyclone

Numerous studies of cyclones have dealt with single-particle behavior. Yet slurries fed to cyclone classifiers in mineral processing plants contain in excess of 55%–65% solids. Most of the feed slurry volume departs through the vortex finder, so the overflow is representative of the inside medium that "drags" particles inward and up. The underflow consists of coarse particles whose voids are filled with water and fines that have characteristics similar to those of the overflow medium. Thus, when the overflow is concentrated (or dilute), underflow voids are filled with concentrated (or dilute) overflow medium.

Cyclone Performance and Cut Size. Cut size has been usually defined as that size of overflow that passes 97%–99% of the particles. Since the early 1960s, cut size has been defined with respect to either a fractional recovery curve (d_{50}) or a corrected fractional recovery curve (d_{50c}).

For completeness, the recovery curves associated with 30-in. cyclone data have been calculated. The calculations are summarized in Table 4.9, where *F* is feed, *U* is underflow (coarse), and *O* is overflow (fines). The graphical result is shown in Figure 4.30.

Note, in Figure 4.30, that

$$R_{ic} = \frac{R_i - H_u}{1 - H_u} \quad \textbf{(Eq. 4.38)}$$

$$R_i = \frac{Uu_i}{Ff_i} \quad \textbf{(Eq. 4.39)}$$

$$H_u = \frac{W_u}{W_f} \quad \textbf{(Eq. 4.40)}$$

From the graph, the d_{50} size is estimated as 0.066 mm and the d_{50c} size as 0.145 mm. The sharpness index is approximately 0.080/0.245 = 0.33. Note that the R_i curve is displaced upwardly relative to the R_{ic} curve. This displacement is typical.

TABLE 4.9 Cyclone performance data and calculations

Mesh	Size, mm	f_i	u_i	o_i	Ff_i	Uu_i	R_i	R_{ic}
3	6.7	0.0001	0.00014	0	0.05	0.05	1	1
6	3.35	0.0029	0.0039	0	1.44	1.44	1	1
12	1.7	0.0302	0.0402	0.0016	14.97	14.73	0.9840	0.0737
20	0.85	0.0977	0.1310	0.0036	48.44	47.99	0.9907	0.9847
40	0.425	0.2129	0.2817	0.0182	105.55	103.2	0.9777	0.9634
70	0.212	0.2636	0.2919	0.1834	130.69	106.94	0.8183	0.7016
140	0.106	0.1463	0.1186	0.2246	72.54	43.45	0.5990	0.3414
270	0.053	0.0709	0.0451	0.1438	35.15	16.52	0.4700	0.1296
–270	0.053	0.1754	0.0874	0.4248	86.96	32.02	—	—
	$F = 495.79$	$W_f = 297.9$	$U = 366.37$	$W_u = 116.50$	$*H_u = 0.3911 = 116.5/297.9$			

$*H_u$ = fraction of feed water reporting to underflow = W_u/W_f.

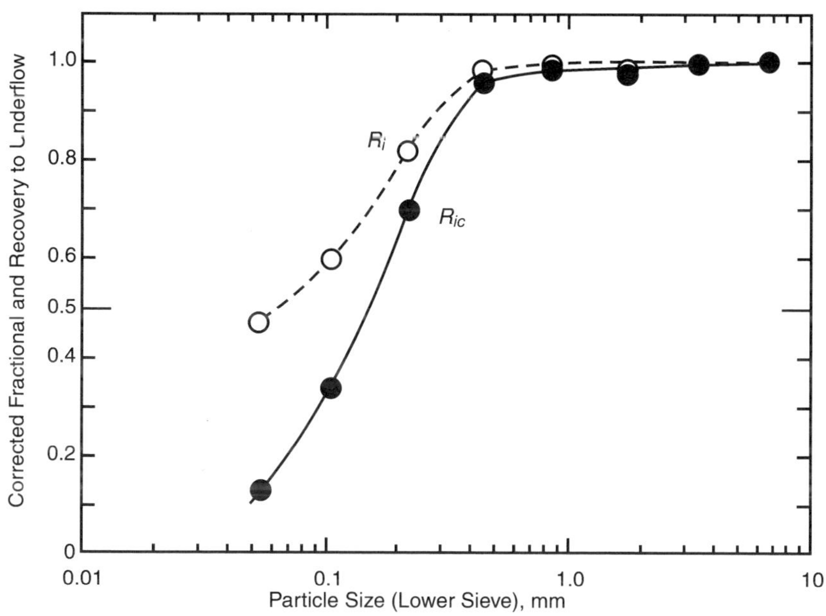

FIGURE 4.30 Corrected and uncorrected fractional recoveries to cyclone underflow

Design Variables That Influence Performance. The cyclone appears simple in design, but there is still plenty of room for improvement. The design criteria establish a standard cyclone in which definite geometric relationships are maintained among cyclone diameter, inlet area, vortex finder diameter and length, cylindrical section length, apex diameter, and included angle of the cone section. The diagram in Figure 4.31 itemizes key relations.

Because design variables interact with operating variables, a basic set of operating conditions must be employed for scale-up and selection. The base case is as follows: feed liquid is water at 20°C, feed solids are spherical particles of specific gravity 2.65, the volume of feed solids is less than 1%, and the inlet pressure is 10 psi. The influence of all variables is then relative to the base conditions as listed.

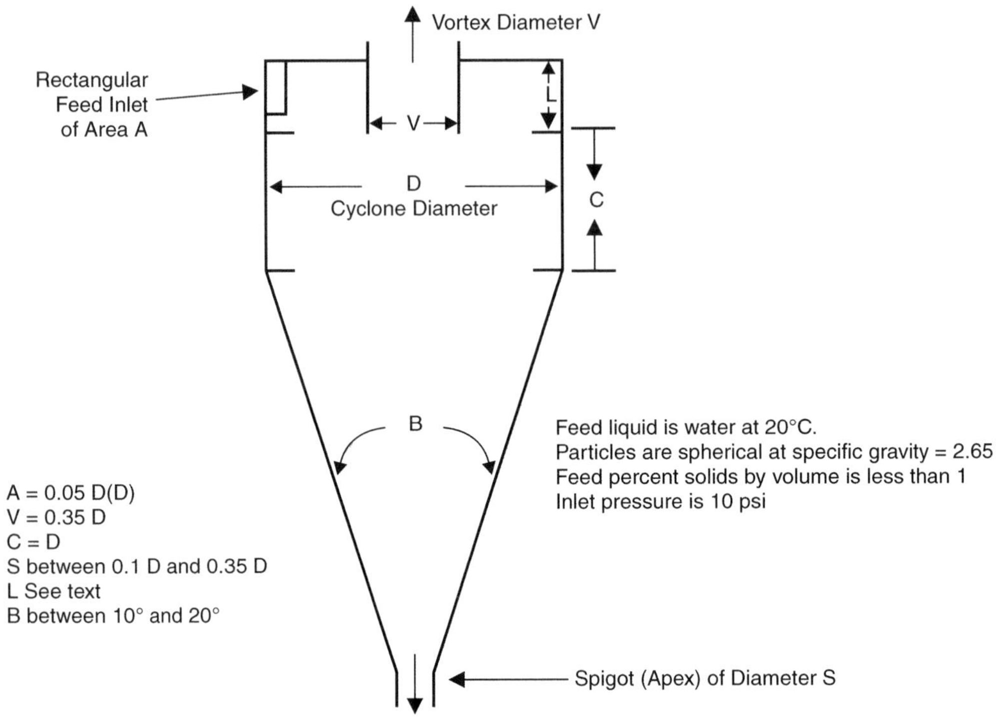

FIGURE 4.31 Scale relationships for a base cyclone classifier

Cyclone Diameter. Cyclone diameter, the inside diameter of the cylindrical feed chamber, is the most important variable governing the size split. Because centrifugal forces generated inside the cyclone vary inversely with cyclone diameter to some power, the smaller the cyclone diameter, the finer the split. In fact, it has been observed that

$$d_{50c} \alpha D^n \quad \textbf{(Eq. 4.41)}$$

where D is cyclone diameter and n has a value somewhere between 0.46 and 0.683.

Larger-diameter cyclones will have a larger solids handling capacity. Thus, the capacity, at a given inlet pressure, varies with D^2. To be more specific, if Q (gpm of water) is the capacity of a cylone of diameter D (in.)

$$Q = 0.7071 D^2 \sqrt{\Delta P} \quad \textbf{(Eq. 4.42)}$$

where ΔP is inlet pressure in psi. For greater precision, Q should be corrected when slurry is being pumped. However, by neglecting this correction, a safety factor is automatically built in.

Inlet Area. Inlet area determines the entrance velocity, which largely governs the characteristic of tangential velocity versus cyclone radius. It has been shown (Lilge 1962) that

$$\frac{V_t}{V_{in}} \approx 5\left[\frac{A_{in}}{A_c}\right]^{0.585} \quad \textbf{(Eq. 4.43)}$$

where V_t is tangential velocity, V_{in} is inlet velocity, A_{in} is inlet area, and A_c is the cross-sectional area of the cylindrical chamber just below the vortex finder. To ensure consistency in scale-up, the base inlet area of the feed nozzle is set to 0.05 D^2, so that V_t and V_{in} will be approximately equal to each other. To maintain base conditions when inlet area is increased, feed flow rate must increase. Decreasing inlet area at similar capacities increases inlet pressure slightly.

Inlet nozzle entrances should be positioned at the top of the vortex finder, well above the vortex finder bottom. They can be straight (where the outer entrance wall matches the cylindrical section wall) or involute (where the inner entrance wall matches the cylindrical section wall and the outer entrance wall pinches off gradually to match with the outer cyclone wall). In a third design, the center line of the entrance nozzle matches the cylindrical section wall. Entrance nozzles are intended to reduce turbulence in the vicinity of the vortex finder; a rectangular cross section also reduces turbulence.

Vortex Finder Diameter and Length. The vortex finder diameter is the next important design variable that governs the size split. For cyclones of fixed diameter, the vortex finder diameter affects the d_{50c} size, which is proportional to V^m, where V is the vortex finder diameter and m is a constant. The larger the vortex finder diameter, the coarser the overflow.

Most of the water in cyclone feed and about 25% of the feed solids by weight report to overflow. At base conditions, the diameter of the vortex finder is $V = 0.35\ D$.

The vortex finder must extend below the feed entrance to avoid sending feed solids directly to overflow. The bottom of the vortex finder usually terminates just below the junction of the cylindrical feed chamber and the cylindrical section. If L is the length of the vortex finder, $L \approx 0.55\ D$. A shorter vortex finder will coarsen the overflow; extending the vortex finder into the cone section will also coarsen the overflow. To decrease turbulence, the bottom edge of the vortex finder may be machined to a knife edge.

Cylindrical Section Length and Included Cone Angle. The cylindrical section's length and the cone's included angle affect the residence time in the cyclone. If C is the cylindrical section length, $C = D$ for the base condition. If C is increased (equivalent to increasing retention time), a finer separation is obtained. The zone where coarse particles are being forced toward the axis by the cone wall becomes further removed from the vortex finder.

The cone diverts coarse solids toward the center and minimizes voidage near the apex. For a fixed cyclone diameter, decreasing the cone angle will increase the length of the cone section, and hence, the retention time may increase. The d_{50c} size decreases, and the sharpness index may decrease. Increasing the cone angle at a constant cyclone diameter will decrease the length of the cone section, so that retention time may decrease. The d_{50c} size will increase, and the sharpness index may increase.

For cyclones in which D is less than 10 in., cone angles are about 12°; for larger cyclones, the cone angle is around 18°–20°.

Apex (Spigot) Diameter. The apex originates where the cone section terminates; there, apex diameter is the inside diameter at the underflow discharge point. The apex must permit classified coarse particles to exit without plugging–i.e., roping must be avoided. The central air core will become unstable and pinch shut when the cyclone ropes, a condition that arises when the apex is overloaded or is inadvertently throttled, thereby forcing coarse particles into the overflow stream. However, cyclones that operate near a rope condition (the underflow stream is cylindrical with a detectable air core) have a minimum of bypassed slurry that fills voids. Inlet pressure may be low, but efficiency is maintained. In contrast, a spray discharge indicates a more dilute underflow, which in turn suggests that a substantial amount of fines may be bypassing. A near-rope condition can be approached by reducing the apex diameter. As long as the percent of underflow solids does not exceed a critical value at a given percent of overflow solids, roping will be avoided. This relationship is illustrated in Figure 4.32 for various specific gravities (Mular and Jull 1982).

Note that roping is probable to the right of each curve and that a high underflow percent solids is possible at high overflow solids concentrations. The following equation can be used to estimate the apex diameter below which roping may occur (Mular and Jull 1982).

$$S = 4.16 - \frac{16.43}{\left(2.65 - \rho + \dfrac{100\rho}{P_u}\right)} + 1.1\left[ln\left(\frac{U}{\rho}\right)\right] \qquad \textbf{(Eq. 4.44)}$$

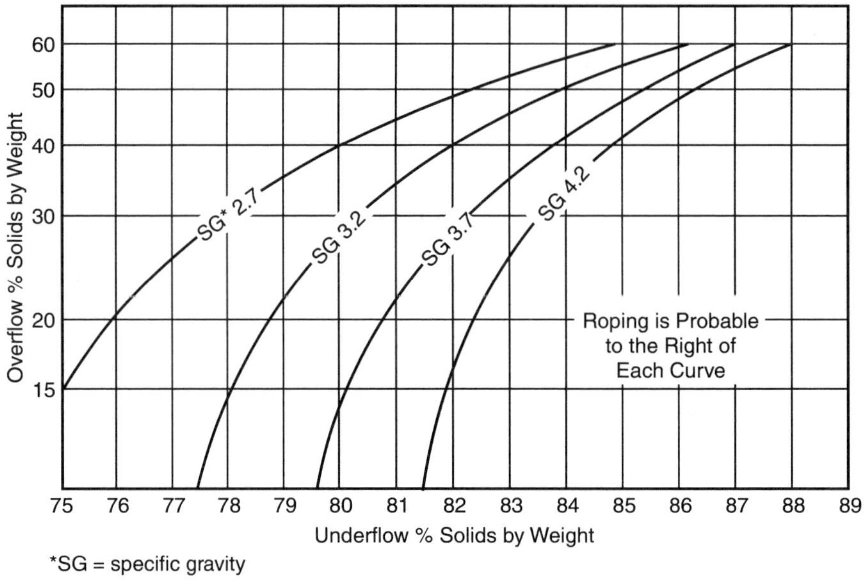

Source: Mular and Jull 1982.

FIGURE 4.32 Critical percentage solids cyclone overflow versus underflow at different specific gravities

where

S = the recommended spigot diameter (in.)

ρ = the specific gravity of the ore

P_u = the underflow percent solids by weight

U = the underflow solids tonnage (stph)

For example, if P_u = 79%, U = 150 stph, and the specific gravity of the ore = 2.65, S = about 3.7 in. A diameter less than this could create a rope condition. Conversely, if a 3.7-in. apex is employed under the same conditions,

$$\frac{U}{\rho} = \exp\left[\frac{14.94}{\left(2.65 - \rho + \dfrac{100\rho}{P_u}\right)} + 0.909S - 3.782\right] \quad \textbf{(Eq. 4.45)}$$

Roping could develop if U exceeds by any extent the value of 150 stph. Apex diameters are in the range 0.10–0.35 D.

Operating Variables That Influence Performance. Operating variables that influence cyclone performance include feed size distribution (not very controllable), the specific gravity and viscosity of the internal slurry, feed percent solids, specific gravity of solids, inlet velocity, and inlet pressure.

Feed Size Distribution. A coarse feed containing few fines will increase the separation size, while a fine feed with few coarse particles will decrease the separation size. Both the d_{50c} size and the recovery of water to the underflow are influenced.

Internal Slurry, Specific Gravity, and Viscosity. In suspensions of solids in liquids, viscosity and specific gravity are not independent of each other, and for this reason, it is difficult to isolate their effects on separation size. The separating medium inside the cyclone must strongly resemble the overflow slurry. Fine clay and slime can substantially increase viscosity with relatively minor changes in slurry specific gravity. Because internal slurry, specific gravity, and viscosity affect drag forces exerted by the medium, the split can be strongly influenced.

Feed Percent Solids. A change in the cyclone feed percent solids will change both the specific gravity and viscosity of the internal medium, which affects the d_{50c} size. Thus, the rate at which water is added is an important variable for controlling separation size. Relative to base conditions (Arterburn 1982; Mular and Jull 1982),

$$d_{50c} \propto \left[\frac{V_m}{V_m - V}\right]^{1.43} \quad \textbf{(Eq. 4.46)}$$

or

$$d_{50c} \propto \exp[-0.301 + 0.0945V - 0.00356V^2 + 0.0000684V^3] \quad \textbf{(Eq. 4.47)}$$

where V_m = 53% (an upper limit on feed percent solids by volume) and V is feed percent solids by volume. The two functions are fitted to an experimental curve, the second version of which may be employed for V greater than 53%. Note that percent solids by weight, P, is related to volume percent solids by

$$P = \frac{\rho}{\rho_m}V \quad \textbf{(Eq. 4.48)}$$

with

$$V = 100\frac{(\rho_m - \rho_l)}{(\rho - \rho_l)} \quad \textbf{(Eq. 4.49)}$$

The suspension's specific gravity, ρ_m, is calculated from

$$\rho_m = \rho_l + \frac{(\rho - \rho_l)V}{100} = \frac{\rho}{\left[\frac{P}{100} + \left(1 - \frac{P}{100}\right)\rho\right]} \quad \textbf{(Eq. 4.50)}$$

where ρ and ρ_l are specific gravities of solids and water, respectively, and V is the feed percent solids by volume.

Specific Gravity of Solids. The free-settling ratio in the Stokes' region has been experimentally observed to influence, relative to base conditions, the d_{50c} size, so that

$$d_{50c}\alpha\sqrt{\frac{\rho_{base} - \rho_l}{\rho - \rho_l}} = \sqrt{\frac{1.65}{\rho - 1}} \quad \textbf{(Eq. 4.51)}$$

where the specific gravity of water is taken as one. There are grounds for substituting the suspension specific gravity in the above expression for liquid specific gravity, because forces acting on particles depend on the internal medium. The specific gravity of the internal medium is difficult to determine at best.

Inlet Velocity and Pressure. The inlet velocity, $V_{in,}$ = Q/A_{in} (where Q is the volumetric rate of flow of feed slurry), governs the tangential velocity at any point inside the cyclone. For a given inlet, an increase in Q will increase the inlet pressure relative to overflow, because Q is proportional to the square root of the pressure drop. Increasing Q also decreases d_{50c} but the effect is weak; the pressure must drop by a factor of four to make the cyclone separate a mesh size finer, when mesh size follows the square-root-of-two ratio. The d_{50c} is related to inlet pressure by

$$d_{50c}\alpha 1.9(\Delta P)^{-0.28} \quad \textbf{(Eq. 4.52)}$$

where ΔP is in psi. Inlet pressures of 5–10 psi are recommended in grinding circuits to minimize energy requirements and reduce wear.

Selection of Hydrocyclones for Grinding Circuit. On the basis of experimental studies and field work (Arterburn 1982), the following expression has been developed from graphical data for the selection of standard cyclones relative to base conditions:

$$D = 0.02338\left(1 - \frac{V}{V_m}\right)^{2.167}(d_{50c})^{1.515}(\Delta P)^{0.4242}(\rho - \rho_l)^{0.7576} \quad \textbf{(Eq. 4.53)}$$

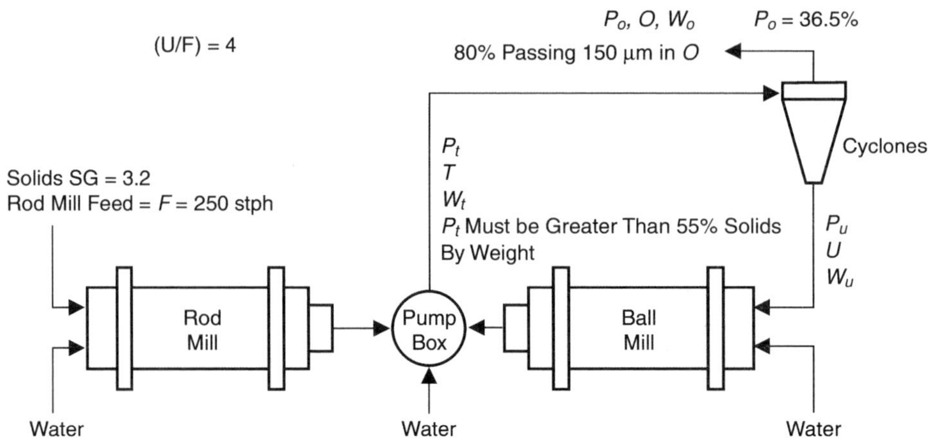

FIGURE 4.33 Grinding circuit for which cyclone classifiers are to be selected

where

D = cyclone diameter in centimeters

V = percent feed solids by volume

V_m = 53%

d_{50c} = the size (μm) at which half of the particles report to overflow and the rest to underflow after correction for bypassing

ΔP = the inlet pressure in kPa (100 kPa = 14.5 psi)

ρ = the specific gravity of the solids

ρ_m = the specific gravity of the fluid (ρ = 1 for water)

An alternative (Mular and Jull 1982) is

$$D = \frac{0.02102(d_{50c})^{1.515}(\Delta P)^{0.4242}(\rho - \rho_l)^{0.7576}}{\exp(-0.4561 + 0.1431V - 0.005394V^2 + 0.0001036V^3)} \quad \textbf{(Eq. 4.54)}$$

where D is cyclone diameter (in.) and ΔP is inlet pressure (psi).

To estimate d_{50c} use (Arterburn 1982)

$$d_{50c} = 3.14 d_y \, ln(119.12/y_d) \quad \textbf{(Eq. 4.55)}$$

where y_d is the cumulative percent finer than size d_y (μm) in cyclone overflow and d_y is the size (μm) that passes y_d percentage of the solids in the overflow. For example, suppose that the cyclone overflow is to be 80% passing 149 μm (100 mesh). The d_{50c} is

$$d_{50c} = 3.14(149) ln[119.12/(80)] = 186 \text{ μm, rounded off} \quad \textbf{(Eq. 4.56)}$$

Selecting Cyclone Classifiers: Example. Suppose that cyclone classifiers are to be selected for the grinding circuit shown in Figure 4.33. The data available in the figure are minimal. To solve the problem a water and solids balance is needed; the cyclone diameter must be determined; the number of cyclones must be estimated; and finally estimates of inlet, vortex finder, and spigot sizes are required. For the circuit shown, the circulating load required is 400%.

Obtain Circuit Mass Balance. First estimate the underflow percent solids around the cyclones from information in Figure 4.32. Because the cyclone overflow is to be 36.5% solids by weight, then at a solids specific gravity of 3.2, the underflow percent solids must not exceed 81.3% to prevent roping. Hence, 80% is considered to be safe.

Solids balance:	$F = 250$ stph (given)
	$O = F = 250$ stph (steady state)
	$U = 4F = 1{,}000$ stph (given)
	$T = F + 4F = 1{,}250$ stph (steady state)
Water balance:	$W_o = 250(100 - 36.5)/36.5 = 434.9$ stph
	$W_u = 1{,}000(100 - 80)/80 = 250$ stph
	$W_t = 250 + 434.9 = 684.9$ stph
Slurry balance:	$W_t + T = 684.9 + 1{,}250 = 1{,}934.9$ stph
	$W_o + O = 434.9 + 250 = 684.9$ stph
	$W_u + U = 250 + 1{,}000 = 1{,}250$ stph
Percent solids by weight:	$P_o = 36.5\%$ (given)
	$P_u = 80\%$ (from graph)
	$P_t = 100(1{,}250)/(684.9 + 1{,}250) = 64.6\%$
Feed percent solids (by volume):	Note: gpm = stph(4/specific gravity)
Feed slurry volume:	1,250(4/3.2) = 1,562.5 gpm solids
	684.9(4/1) = 2,739.6 gpm water
	1,562.5 + 2,739.6 = 4,302.1 gpm slurry
	100(1,562.5/4,302.1) = 36.32% solids by volume in cyclone feed
or:	$V = 100(T/\text{specific gravity})/(T/\text{specific gravity} + W_t)$
	$V = 100(1{,}250/3.2)/[(1{,}250/3.2)+684.9]$
	$V = 36.32\%$

Calculate Cyclone Diameter. First estimate d_{50c}. Thus, $d_{50c} = 3.14(150)ln[119.12/(80)] = 187.5\ \mu m$. Next find D from one of the equations that is used to calculate the cyclone diameter. Thus,

$$D = 0.02338\left(1 - \frac{V}{V_m}\right)^{2.167}(d_{50c})^{1.515}(\Delta P)^{0.4242}(\rho - \rho_l)^{0.7576} \quad \textbf{(Eq. 4.57)}$$

where $V = 36.32\%$, $V_m = 53\%$, $d_{50c} = 187.5\ \mu m$, $\rho = 3.2$, and $\rho_l = 1$. The inlet pressure is chosen to be 8 psi, which is $8(100/14.5) = 55.17$ kPa $= \Delta P$. Thus,

$$D = 0.02338\left(1 - \frac{36.32}{53}\right)^{2.167}(187.5)^{1.515}(55.17)^{0.4242}(3.2 - 1)^{0.7576} \quad \textbf{(Eq. 4.58)}$$

and $D = 52.8$ cm (20.8 in.). Hence, 20-in. cyclones should be acceptable.

Estimate Number of Cyclones Required. The cyclones must handle 1,934.9 stph of feed slurry or 4302.1 gpm at 36.32% by volume (64.6% feed solids by weight). If V is the total volume flow and if Q is the volume flow (gpm) per cyclone, the number of cyclones, N, must be

$$N = \frac{V}{Q} = \frac{V}{0.7071D^2\sqrt{\Delta P}} = \frac{4302.1}{0.7071(20)^2\sqrt{8}} = 5.3 \quad \textbf{(Eq. 4.59)}$$

$$N = 5.38 \approx 6 \text{ cyclones} \quad \textbf{(Eq. 4.60)}$$

Because Q is in terms of water, the above estimate is conservative and five cyclones will probably suffice. However, to ensure that there are enough cyclones to permit operation during maintenance, most likely seven or eight would be selected. If the inlet pressure should be raised to 10 psi, 4.86 cyclones are needed. Liner wear would then increase.

Estimate Inlet Area, Vortex Finder Diameter, and Apex Diameter. The inlet area is about 0.05 D^2 = 20 sq in., which is about 7% of the cross-sectional area of the cylindrical feed chamber.

The vortex finder diameter is equal to 0.35 D or 0.35(20) = 7 in. This diameter is normally modified during actual operation to obtain an optimum size.

To determine an apex diameter such that the cyclone does not rope, the roping expression can be used. Note that U is the underflow dry solids rate per cyclone (i.e., if five cyclones are in use, U = 1,000/5 = 200 stph).

$$S = 4.16 - \frac{16.43}{\left(2.65 - \rho + \frac{100p}{P_u}\right)} + 1.10\ ln\left(\frac{U}{\rho}\right) \quad \textbf{(Eq. 4.61)}$$

or

$$S = 4.16 - \frac{16.43}{\left(2.65 - \rho + \frac{100(3.2)}{80}\right)} + 1.10\ ln\left(\frac{200}{3.2}\right) = 3.95 \text{ in.} \quad \textbf{(Eq. 4.62)}$$

This result suggests that if a spigot of less than 4-in. diameter is installed, the cyclone is likely to enter a rope condition. Because 4 in. is at "near rope," the cyclone efficiency is likely to be fairly high, because underflow voids have a minimum of misplaced slurry.

Pneumatic (Air) Classifiers

Pneumatic classifiers effect size separations in the 0.1–1,000-µm range in air or gas, where a combination of physical forces are employed. Devices include air cyclones, expansion chambers, vane classifiers, inertial classifiers, tank through-flow classifiers, and recirculating-flow classifiers.

The bases for designing air classifiers have been summarized (Klumpar et al. 1986). Forces acting on particles entering as feed are due to gravity, aerodynamic drag, centrifugal force, and collision force. Devices employ suitable combinations of these for sizing. In the Sturtevant SD classifier shown in Figure 4.34, all forces operate. Particles are fed centrally onto a rotating distributor plate that has vertical pins (posts) mounted peripherally beneath the plate. Plate friction accelerates particles radially outward and imparts a tangential velocity component, the magnitude of which approaches that of the plate edge. Air is fed through a feed nozzle of involute design, so that a flow laden with coarse particles is generated downward toward the underflow chamber and cone, and another flow is diverted through the pins into the fine-particle chamber. Thus, the air drags the fines radially through the rotor pins, while centrifugal force acts in an opposite direction because of particle tangential velocity. For collision force to operate, particles must be captured aerodynamically by rotating pins, such as by direct interception, inertial deposition, or electrostatic precipitation (Mular and Jull 1982). Capture efficiency is defined as the ratio of the cross-sectional area of the fluid stream from which all particles are removed to the cross-sectional area, projected in the direction of flow, of the pin. It is a function of the Stokes' number. Characteristics of the feed and operating conditions determine which intermediate-size particles will hit pins and be thrown back toward the coarse-particle chamber.

If collision force is too large, particles may comminute. Spherical particles in an air classifier possess a drag force, centrifugal force, and gravitational force similar to those in hydrocyclones. The collision efficiency for direct interception thus is

$$E_k = \left(1 + \frac{d}{d_{pin}}\right) - \frac{1}{\left(1 + \frac{d}{d_{pin}}\right)} \quad \textbf{(Eq. 4.63)}$$

and depends on pin velocity, the number of pins, pin diameter, d_{pin}, particle diameter, d, and particle specific gravity.

Performance of Air Classifiers. The performance of air classifiers is measured by the same criteria applied to hydrocyclones. Thus, cut size is defined as either d_{50} or d_{50c}. The latter is determined

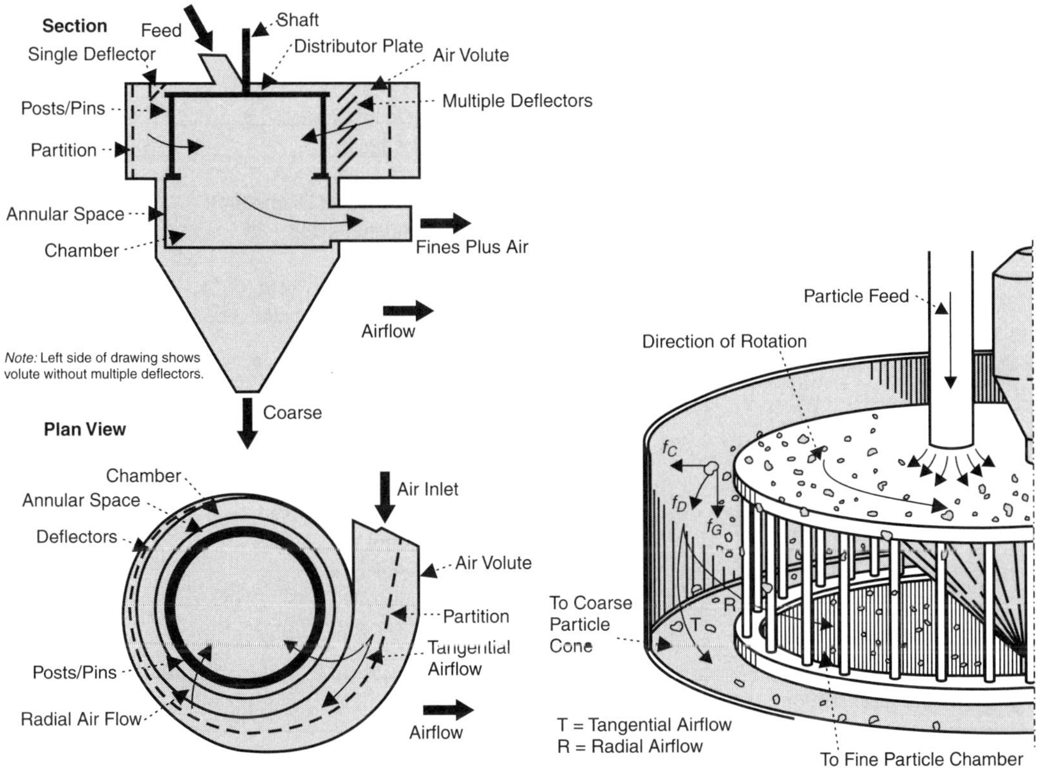

Sturtevant SD Classifier Schematic

Forces Acting on Particle in Separation Zone

Source: Klumpar et al. 1986.

FIGURE 4.34 Features of Sturtevant SD Classifier

by the fractional recovery curve (Figure 4.30), although such curves can be more complex in air classifiers. Coarse material may bypass to the fines stream, fine material may bypass to the coarse stream, or both types of bypassing may occur.

Air Cyclones. In mineral processing, air cyclones may sometimes be found near a secondary crushing/screening plant for extraction of dust. Such dust-collecting units bear strong resemblance to hydrocyclones, because physical forces act in an identical manner in air.

An empirical method for sizing air cyclones for efficiency has been formulated (Valdez et al. 1986) to account for the effect of particle specific gravity. Figure 4.35 shows relevant design parameters along with the variable nomenclature employed. Thus,

$$a = f_1 D_c,\ b = f_2 D_c,\ D_e = f_3 D_c,\ H = 4\ D_c,\ h = 1.5\ D_c,\ B = 0.375\ D_c,$$

and

$$S = 0.65\ D_c$$

where D_c, f_1, f_2, and f_3 must be chosen to give a maximum efficiency. This efficiency occurs when inlet velocity is equal to 1.25 times the saltation velocity. The latter is the minimum air velocity that prevents settling out of solid particles. The relationships below have been derived (Valdez et al. 1986):

$$D_c = \left[\left(\frac{1.085}{(f_3)^4} + \frac{0.32}{f_1(f_2)^2}\right)\frac{\rho_g Q^2}{\Delta P g_c}\right]^{\frac{1}{4}} \qquad \textbf{(Eq. 4.64)}$$

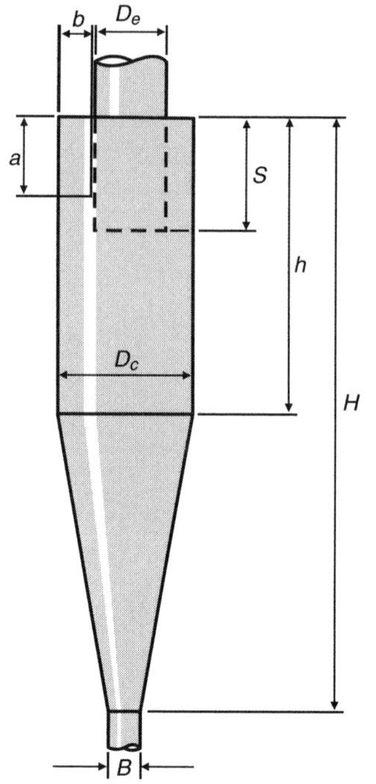

a = Inlet Height, ft
B = Dust Outlet Diameter, ft
b = Inlet Width, ft
D_c = Cyclone Diameter, ft
D_e = Overflow Outlet Diameter, ft
f_1 = Inlet Height Factor
f_2 = Width Factor
f_3 = Outlet Diameter Factor
g_c = Gravitational Constant, 32.3 ft/s^2
H = Overall Height, ft
h = Cyclinder Height, ft
ΔP = Cyclone Pressure Drop, in. H_2O
Q = Volumetric Gas Flow Rate, ft^3/s
S = Overflow Outlet Length, ft
v_i = Inlet Velocity, ft/s
v_s = Saltation Velocity, ft/s
w = Equivalent Velocity, ft/s
ρ_g = Gas Density, lb/ft^3
ρ_s = Solid Particle Density, lb/ft^3
μ = Gas Viscosity, lb/(ft)(s)

Source: Valdez et al. 1986.

FIGURE 4.35 Air cyclone design parameters and nomenclature

$$v_i = \frac{Q}{f_1 f_2 D_c^2} \qquad \textbf{(Eq. 4.65)}$$

$$v_s = 2.055w\left[\frac{f_2}{1-f_2^{\;0.333}}\right](f_2 D_c)^{0.067} v_i^{\;0.67} \qquad \textbf{(Eq. 4.66)}$$

$$w = \left[4g_c\mu\left(\frac{\rho_s - \rho_g}{3\rho_g^2}\right)\right]^{0.333} \qquad \textbf{(Eq. 4.67)}$$

Design of an Air Cyclone That Meets Given Conditions.

Example: Design an air cyclone to work under the following conditions: Q = 55.5 ft^3/s, ρ_g = 0.0554 lb(mass)/ft^3, ρ_s = lb(mass)/ft^3, ΔP = 3 in. H_2O, and $\mu = 1.44 \times 10^{-5}$ lb(mass)/(ft/s) for air. This example is taken from Valdez et al. (1986).

Start with recommended values f_1 = 0.5, f_2 = 0.2, and f_3 = 0.5. Now D_c can be calculated from ρ_g, ΔP, Q, and g_c = 32.2 ft/(s)^2. The calculated D_c is 3.06 ft. Next, calculate v_i and v_s. Thus, v_i = 59.3 ft/s and v_s = 50.7 ft/s. The ratio of v_i to v_s is too low, because it must be 1.25. To increase the ratio, try changing f_2 to 0.18 (but retain f_1 and f_3 as before). In this case, D_c = 3.17 ft, v_i = 61.5 ft/s, and v_s = 49.5 ft/s. Now the ratio of v_i to v_s becomes 1.24, which is close enough. The cyclone dimensions (see Figure 4.35) are thus

D_c = 3.17 ft	$D_e = 0.5D_c$ = 1.59 ft	$B = 0.375D_c$ = 1.19 ft
$S = 0.65D_c$ = 2.06 ft	$H = 4D_c$ = 12.7 ft	$a = 0.5D_c$ = 1.59 ft
$h = 1.5D_c$ = 4.76 ft	$b = 0.18D_c$ = 0.57 ft	

BIBLIOGRAPHY

Allen, T. 1975. *Particle Size Measurement*. London: Chapman and Hall.

Arterburn, R.A. 1982. The Sizing and Selection of Hydrocyclones. In *Design and Installation of Comminution Circuits*. Edited by A. Mular and G. Jergensen. New York: AIME.

Bergstrom, B.H. 1966. Empirical Modification of the Gaudin–Meloy Equation. *Trans. SME-AIME*, 235:45.

Bradley, D. 1965. *The Hydrocyclone*. New York: Pergamon Press.

Colman, K.G. 1963. Screening Machinery: Selection of Vibrating Screens. In *General Information PM 1.1*. Appleton, Wisc.: Allis-Chalmers.

——. 1972. *Nordberg Screening Application and Capacity Data*. Milwaukee, Wisc.: Nordberg, Inc.

——. 1980. Selection Guidelines for Size and Type of Vibrating Screens in Ore Crushing Plants. In *Mineral Processing Plant Design*. Edited by A.L. Mular and R.B. Bhappu. New York: AIME.

——. 1985. Selection Guidelines for Vibrating Screens. In *SME Mineral Processing Handbook*. Edited by N.L. Weiss. New York: AIME.

Cyclosizer Instruction Manual. 1965. Sydney, Australia: Warman Equipment Ltd.

Dahlstrom, D.A. 1954. Fundamentals and Applications of the Liquid Cyclone. *Chem. Engr. Progress Symposium*, 50(15).

Fitch, B., and E.J. Roberts. 1985. Classification Theory. In *SME Mineral Processing Handbook*. Edited by N.L. Weiss. New York: AIME.

Gates, A.O. 1915. Kick vs. Rittinger: An Experimental Investigation in Rock Crushing Performed at Purdue University, *Trans. AIME*, 51: 875.

Gaudin, A.M. 1926. An Investigation of Crushing Phenomena, *Trans. AIME*, 73:253.

——. 1939. *Principles of Mineral Dressing*. New York: McGraw-Hill.

——. 1963. Classifier Becomes a Commercial Unit. In *Chemical and Engineering News*. Nov. 18, p. 54.

Gaudin, A.M., and T.P. Meloy. 1962. Model and a Comminution Distribution Equation for Single Fracture. *Trans. SME-AIME*, 223:40.

Gilson Screen Manual. 1986. Wilmington, Calif.: Sepor Ltd.

Gluck, S.E. 1965a. Vibrating Screens: Surface Selection and Capacity Calculation. In *Chemical Engineering*. New York: McGraw-Hill.

——. 1965b. Vibrating Screens. In *Chemical Engineering*. New York: McGraw-Hill.

Harris, C.C. 1968. The Application of Size Distribution Equations to Multi-event Comminution Processes. *Trans. SME-AIME*, 241:343

Herbst, J.A., and Sepulveda, J.L. 1985. Particle Size Analysis, Section 30-3. In *SME Mineral Processing Handbook*. Edited by N.L. Weiss. New York: AIME.

Hill, R.B. 1982. Selection and Sizing of Gravity Classifiers. In *Design and Installation of Comminution Circuits*. Edited by A. Mular and G. Jergensen. New York: AIME.

Hitzrot, H.W., and G.M. Meisel. 1985. Mechanical Classifiers. In *SME Mineral Processing Handbook*. Edited by N.L. Weiss. New York: AIME.

Kelly, E.G., and D.J. Spottiswood. 1982. *Introduction to Mineral Processing*. New York: John Wiley & Sons.

Kelsall, D.F. 1952. A Study of the Motion of Solid Particles in a Hydraulic Cyclone. *Trans. Inst. Chem. Engr.*, 30(2).

Klumpar, I.V., F.N. Currier, and T.A. Ring. 1986. Air Classifiers. In *Chemical Engineering*. New York: McGraw-Hill.

Lilge, E.O. 1962. Hydrocyclone Fundamentals. *Trans. IMM, 71*, Part 6.

Luckie, P.T. 1984. Size Separation. In *Encyclopedia of Chemical Technology*. New York: John Wiley & Sons.

Lynch, A.J. et al. 1967. An Analysis of the Performance of a Ball-mill Rake Classifier Comminution Circuit. *J. Aus. IMM, 224*-Dec.

Lynch, A.J., and T.C. Rao. 1975. Modeling and Scale-up of Hydrocyclone Classifiers. *Proceedings of XI International Mineral Processing Congress*. Instituto di Arte Mineraria e Preparzione dei Minerali, Universita di Cagliari, Italy.

Malghan, S.G., and A.L. Mular. 1982. Measurement of Size Distribution and Surface Area of Granular Materials. In *Design and Installation of Comminution Circuits*. Edited by A. Mular and G. Jergensen. New York: AIME.

Matthews, C.W. 1974. *Tyler Specification Tables for Industrial Wire Cloth, Woven Wire Screens No. 74*. Mentor, Ohio: CE-Tyler Inc.

——. 1985a. General Classes of Screens. In *SME Mineral Processing Handbook*. Edited by N.L. Weiss. New York: AIME.

——. 1985b. Screening Media. In *SME Mineral Processing Handbook*. Edited by N.L. Weiss. New York: AIME.

Mular, A.L. 1966. Empirical Modeling and Optimization of Mineral Processes. *Mineral Sci. and Eng.*, 4:45.

Mular, A.L., and N.A. Jull. 1982. The Selection of Cyclone Classifiers, Pumps and Pump Boxes for Grinding Circuits. In *Mineral Processing Plant Design*. Edited by A.L. Mular and R.B. Bhappu. New York: AIME.

Nichols, J.P. 1982. Selection and Sizing of Screens. In *Design and Installation of Comminution Circuits*. Edited by A. Mular and G. Jergensen. New York: AIME.

Plitt, L.R. 1976. A Mathematical Model of the Hydrocyclone Classifier. *CIM Bulletin*, December.

Plitt, L.R., and B.C. Flintoff. 1985. Unit Models of Ore and Coal Process Equipment: Classification and Coal Processing. In *SPOC Manual*. Edited by D. Laguitton. Ottawa, Canada: CANMET.

Pryor, E.J. 1965. *Mineral Processing*, New York: Elsevier.

Rietema, K. 1962. *Cyclones In Industry*. New York: Elsevier.

Riethmann, R.E., and B.M. Bunnell. 1980. Application and Selection of Spiral Classifiers. In *Mineral Processing Plant Design*. Edited by A.L. Mular and R.B. Bhappu. New York: AIME.

Schuhmann, R., Jr. 1948. Laboratory Sizing, *Powder Metallurgy Reprint*, Cleveland, Ohio: American Society for Testing and Materials.

Taggart, A.F. 1945. *Handbook of Mineral Dressing*. New York: John Wiley & Sons.

——. 1951. *Elements of Ore Dressing*, New York: John Wiley & Sons.

Tarr, D.T. Jr. 1985. Hydrocyclones. In *SME Mineral Processing Handbook*. Edited by N.L. Weiss. New York: AIME.

Valdez, M.G., I. Garcia, and B. Beato. 1986. Sizing Gas Cyclones for Efficiency. *Chemical Engineering*. New York: McGraw-Hill.

W.S. Tyler Company. 1973. *Testing Sieves and Their Uses*. Vancouver, B.C.: W.S. Tyler Company.

CHAPTER 5

Movement of Solids in Liquids

Kenneth N. Han

INTRODUCTION

Many mineral processing operations rely on the movement of solids in fluids, especially liquids. For example, one common process is gravity separation of minerals in water. It is therefore important to understand how various solids behave in fluids and what factors govern that behavior, information that can be used to advantage in the separation of mineral particles. Understanding the movement of solids also contributes to an understanding of size reduction, size separation, various concentration technologies, dewatering, and aqueous dissolution processes.

In this chapter, the principles involved in the movement of solids in liquids are introduced.

DYNAMIC SIMILARITY

The manner in which solid particles fall in a medium strongly depends on the density and size of the particles and also on the properties of the medium. Solid particles fall or rise because of differences in density between the solid and the surrounding medium. Differences between particles in size, shape, and density cause them to fall or rise at different rates. These differences can be used to separate particles from one another.

To identify the parameters of a system that demonstrates dynamic similarity, we will perform dimensional analysis of particles settling in a fluid medium. The variables assuming important roles in the fluid/particle system include diameter of particles, d; terminal velocity of particles, v_t; density of fluid, ρ_f; viscosity of fluid, μ; and the force of gravity, represented by gravity acceleration, g. We will also define a dependent dimensionless parameter, P, consisting of one or more dimensional parameters. Therefore, a dimensional analysis yields

$$P \propto d^a v_t^b \rho_f^c \mu^d g^e \quad \text{(Eq. 5.1)}$$

where dimensions of each variable are

$$d : L$$
$$v_t : L\,\theta^{-1}$$
$$\mu : ML^{-1}\theta^{-1}$$
$$g : L\,\theta^{-2}$$
$$\rho_f : ML^{-3}$$

and where, in turn, L = length, θ = time, and M = mass.

Make $e = 1$.

Therefore,

$$L : a + b - 3c - d + 1 = 0$$
$$M : c + d = 0$$
$$\theta : -b - d - 2 = 0$$

and consequently,

$$a = c + 1$$
$$b = c - 2$$
$$d = -c$$

Finally, Eq. 5.1 can be rewritten as

$$P \propto d^{c+1} v_t^{\,c-2} \rho_f^{\,c} \mu^{-c} g = A\left(\frac{d v_t \rho_f}{\mu}\right)^c \left(\frac{dg}{v_t^{\,2}}\right) \quad \textbf{(Eq. 5.2)}$$

where A is a system constant and dimensionless.

The first dimensionless parameter, $(dv_t\rho_f/\mu)$, is referred to as the Reynolds number (N_{Re}) and is the inverse of the second dimensionless parameter, (dg/v_t^2), which is known as the Froude number (F_r). For any two systems to be dynamically similar, these two dimensionless parameters must be the same.

Example: Calculating Reynolds and Froude Numbers

Problem: A galena particle of diameter 0.5 mm falls in water at a velocity of 8.27 in./s. Calculate the Reynolds and Froude numbers for this particle.

Solution

$\rho_s = 7.5 \text{ g/cm}^3$ $\quad\quad \rho_f = 1 \text{ g/cm}^3$

$d = 0.05 \text{ cm}$ $\quad\quad \mu = 0.01 \text{ g cm}^{-1} \text{ s}^{-1}$

$g = 981 \text{ cm s}^{-2}$ $\quad\quad v_t = 8.27 \cdot 2.54 = 21 \text{ cm s}^{-1}$

$$N_{\mathrm{Re}} = \frac{0.05 \times 21 \times 1}{0.01} = 105$$

$$F_r = \frac{21^2}{0.05 \times 981} = 8.99$$

FREE SETTLING

A force balance formulated for the motion of a spherical particle in a medium is given by Eq. 5.3.

$$m'a = m'g - mg - R' \quad \textbf{(Eq. 5.3)}$$

where

m' = mass of a sphere having radius r $\left(\frac{4}{3}\pi r^3 \rho_s\right)$

m = mass of fluid having the same volume of the solid particle $\left(\frac{4}{3}\pi r^3 \rho_f\right)$

ρ_s, ρ_f = densities of solid and fluid

g = gravity acceleration (981 cm s^{-2})

R' = resistance force acting on the solid particle

a = acceleration of the motion of particle = $\frac{dv}{dt}$

The resistance force, R', has the form

$$R' = 6\pi\mu v r \quad \textbf{(Eq. 5.4)}$$

when $N_{\mathrm{Re}} < 0.1 - 1.0$

$$R' = \frac{1}{2} f \pi r^2 v^2 \rho_f \qquad \textbf{(Eq. 5.5)}$$

for all N_{Re} where

μ = viscosity of fluid
f = friction factor or drag coefficient
v = velocity of particle

The resistance force term given by Eq. 5.4 is valid for laminar flow and is strictly valid when the Reynolds number is less than 0.1. However, in practice, the equation is applicable for a Reynolds number up to 1.0 without introducing significant error.

By substituting Eq. 5.4 into Eq. 5.3 and solving for steady-state situations, where acceleration is zero, Eq. 5.6 results:

$$v_t = \frac{2r^2(\rho_s - \rho_f)g}{9\mu} \qquad \textbf{(Eq. 5.6)}$$

Note that the velocity v becomes v_t, indicating steady-state or terminal velocity. Equation 5.6 is often known as the Stokes' equation for free-settling velocity of particles. This equation cannot be used with acceptable accuracy when the Reynolds number of the system is greater than 1.

The derivation of Stokes' law contains several assumptions. Some important ones are

- The particle must be spherical, smooth, and rigid.
- The surrounding medium is of infinite extent and homogeneous.
- The particle has reached its terminal velocity, and inertia effects are negligible.

Example: Friction Factor for Laminar Flow

Problem: Show that for laminar flow the expression for the friction factor has the following form:

$$f = \frac{24}{N_{Re}}$$

Solution:

Equating Eqs. 5.4 and 5.5,

$$6\pi\mu r v = 1/2 f \pi r^2 v^2 \rho_f$$

Therefore,

$$f = \frac{12\mu}{rv\rho_f}$$

$$= 24\frac{\mu}{2rv\rho_f}$$

$$= \frac{24}{N_{Re}}$$

Example: Maximum Diameters of Particles That Exhibit Laminar Flow

Problem: The Stokes' equation for free-settling velocity can be used without introducing substantial error if the particle's Reynolds number is less than 1. Estimate the maximum diameters of silica and hematite that exhibit laminar flow in water and in air at 20°C.

Solution:

ρ_s of SiO_2 = 2.65 g/cm^3; ρ_f of water = 1.0 g/cm^3

ρ_s of Fe_2O_3 = 5.3 g/cm^3; ρ_f of air = $1.2 \cdot 10^{-3}$ g/cm^3

μ of water = 0.01 g cm^{-1} s^{-1}; μ of air = $1.81 \cdot 10^{-4}$ g cm^{-1} s^{-1}

In water,

$$\frac{dv_t\rho_f}{\mu} = 1$$

Therefore,

$$v_t = \frac{\mu}{d\rho_f} = \frac{d^2(\rho_s - \rho_f)980}{18\mu}$$

By rearranging the above equation,

$$d^3 = \frac{18\mu^2}{(\rho_s - \rho_f)980\rho_f}$$

Therefore,

$$d = \left[\frac{18(0.01)^2}{980 \times 1}\right]^{1/3} \frac{1}{(\rho_s - \rho_f)^{1/3}}$$

$$= 0.0122\frac{1}{(\rho_s - \rho_f)^{1/3}}$$

$$= 0.01032 \text{ cm or } 103.2\ \mu\text{m for } SiO_2$$

$$= 0.0075 \text{ cm or } 75.0\ \mu\text{m for } Fe_2O_3$$

and in air,

$$d = \left[\frac{18(1.81 \times 10^{-4})^2}{980 \times 1.2 \times 10^{-3}}\right]^{1/3} \frac{1}{\rho_s^{1/3}}$$

$$= 7.95 \cdot 10^{-3}\frac{1}{\rho_s^{1/3}}$$

$$= 0.00575 \text{ cm or } 57.5\ \mu\text{m for } SiO_2$$

$$= 0.00456 \text{ cm or } 45.6\ \mu\text{m for } Fe_2O_3$$

When Eq. 5.5 is substituted into Eq. 5.3 and solved for v_t, Eq. 5.7 is obtained:

$$v_t = \sqrt{\frac{8r(\rho_s - \rho_f)g}{3f\rho_f}} \quad \textbf{(Eq. 5.7)}$$

Using Eq. 5.7 to calculate the terminal velocity is impossible unless the friction factor, f, is known. The friction factor is empirical and must be found experimentally. Figure 5.1 shows a plot of friction factor versus Reynolds number. If the Reynolds number is identified for a given particle, the terminal velocity can be calculated by Eq. 5.8:

$$v_t = \frac{N_{Re}\mu}{d\rho_f} \quad \textbf{(Eq. 5.8)}$$

However, because the relationship between the friction factor and the Reynolds number is empirical, identifying the Reynolds number for any given particle is tedious and time consuming.

Equation 5.7 can be rearranged for the friction factor as given by Eq. 5.9.

$$f = \left[\frac{4}{3}\frac{gd^3\rho_f}{\mu^2}(\rho_s - \rho_f)\right]\frac{1}{N_{Re}^2}$$

$$= \frac{4}{3}Ga\frac{1}{N_{Re}^2} \quad \textbf{(Eq. 5.9)}$$

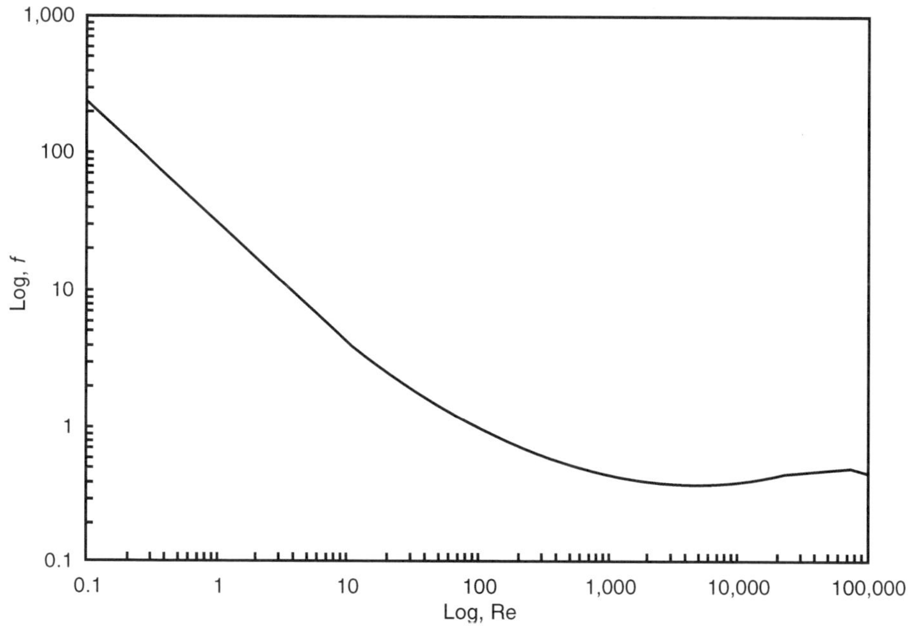

FIGURE 5.1 Friction factor versus Reynolds number for spheres

where Ga is the Galileo number, $\frac{gd^3\rho_f(\rho_s - \rho_f)}{\mu^2}$. Taking the logarithm of both sides of Eq. 5.9 yields

$$\log f = \log\,(4/3\ \mathrm{Ga}) - 2\log N_{\mathrm{Re}} \qquad \textbf{(Eq. 5.10)}$$

Therefore, for a given particle, the terminal velocity can be found by identifying the Reynolds number that satisfies both Eq. 5.10 and the empirical relationship shown in Figure 5.1.

Many investigators (Dallavalle 1948; Lapple 1951; Torobin and Gauvin 1959a,b; Olson 1961; Concha and Almendra 1979a) have attempted to describe mathematically the friction factor–Reynolds number relationship. Concha and his coworkers (Concha and Almendra 1979a,b; Concha and Barrientos 1982) have derived the expression for the terminal velocity of particles by establishing a fifth-order polynomial and solving the resulting equation and Eq. 5.10 simultaneously. The result, for the terminal velocity, v_t, is

$$v_t = \frac{20.52}{d^*}\left[(1 + 0.0921 d^{*3/2})^{1/2} - 1\right]^2 Q \qquad \textbf{(Eq. 5.11)}$$

where $$Q = \left[\frac{4}{3}\frac{(\rho_s - \rho_f)\mu g}{\rho_f^2}\right]^{1/3}$$

$$d^* = d/\left[\frac{3}{4}\frac{\mu^2}{(\rho_s - \rho_f)\rho_f g}\right]^{1/3}$$

A similar approach (Han 1984) used a quadratic expression instead of a fifth-order polynomial, and the resulting terminal velocity expression is given by Eq. 5.12.

TABLE 5.1 Free-settling velocities of spherical particles that have various densities and sizes

Density of Solids, g/cm^3	Diameter of Solids, cm	v_t, cm/s Fig. 5.1	Eq. 5.11	Eq. 5.12
2.65	0.01	0.921	0.86	0.90
	0.03	4.50	4.61	4.39
	0.05	8.35	8.49	7.92
	0.10	16.30	16.71	15.85
4.70	0.01	1.86	1.77	1.82
	0.03	8.22	8.43	7.92
	0.05	14.30	14.72	13.76
	0.10	27.10	27.41	26.40
7.50	0.01	2.88	2.89	2.93
	0.03	12.10	12.64	11.80
	0.05	20.70	21.32	20.01
	0.10	38.30	38.37	27.40

$$v_t = \frac{\mu}{\rho_f d} 10^A \quad \textbf{(Eq. 5.12)}$$

where $A = 5\ [0.66 + 0.4 \log (4/3\ \text{Ga})]^{1/2} - 5.55$.

Table 5.1 tabulates v_t values for four sizes of three different particles using Eqs. 5.11 and 5.12. The resulting values are compared with values derived from Figure 5.1.

Example: Free-Settling Velocity of Spherical Magnetite

Problem: Calculate the free-settling velocity of spherical magnetite ($\rho_s = 5.2\ g/cm^3$) of diameter 0.1 cm in water and in air at 20°C.

Solution:

In water, using Eq. 5.11,

$$d^* = 0.1 \Big/ \left[\frac{3}{4}\frac{(0.01)^2}{(5.2-1)981}\right]^{1/3} = 37.99$$

$$Q = \left[\frac{4}{3}\frac{(5.2-1)0.01\ 981}{1}\right]^{1/3} = 3.80$$

$$v_t = \frac{20.52}{37.87}\left[(1 + 0.0921 \times 37.99^{3/2})^{1/2} - 1\right]^2\ 3.8 = 28.87\ \text{cm/s}$$

and using Eq. 5.12,

$$\text{Ga} = \frac{981(0.1)^3(5.2-1)}{(0.01)^2} = 41{,}202$$

$$A = \left[0.66 + 0.4 \log (4/3 \cdot 41{,}202)\right]^{1/2} - 5.55 = 2.444$$

$$v_t = \frac{0.01}{0.1 \times 1} = 10^{2.444} = 28.87\ \text{cm/s}$$

In air, using Eq. 5.11,

$$d^* = 0.1 \Big/ \left[\frac{3}{4}\frac{(1.81 \times 10^{-4})^2}{(5.2 - 1.2 \times 10^{-3})981 \times 1.2 \times 10^{-3}}\right]^{1/3} = 62.88$$

$$Q = \left[\frac{4}{3}\frac{(5.2 - 1.2 \times 10^{-3})1.81 \times 10^{-4}981}{(1.2 \times 10^{-3})^2}\right] = 94.87$$

$$v_t = \frac{20.52}{62.88}\,[(1 + 0.0921 \times 62.88^{3/2})^{1/2} - 1]^2 \;\; 94.87 = 1{,}059.5 \text{ cm/s}$$

Using Eq. 5.12,

$$\text{Ga} = \frac{981(0.1)^3(5.2 - 1.2 \times 10^{-3})1.2 \times 10^{-3}}{(1.81 \times 10^{-4})^2} = 1.868 \times 10^5$$

$$A = [0.66 + 0.4 \log (4/3 \cdot 1.868 \cdot 10^5)]^{1/2} - 5.55 = 2.844$$

$$v_t = \frac{1.81 \times 10^{-4}}{0.1 \times 1.2 \times 10^{-3}} = 10^{2.844} = 1{,}053.2 \text{ cm/s}$$

The two methods, Eqs. 5.11 and 5.12, yield very similar results, but Eq. 5.12 is simpler to use. Therefore, in subsequent treatments that require such calculations, Eq. 5.12 is used instead of Eq. 5.11.

PARTICLE ACCELERATION

In particle settling analysis, it is commonly assumed that the terminal velocity is established instantaneously. However, this assumption is not necessarily valid in practice, especially when particles move continuously and change direction of motion in a suspension—as when the system is stirred. As the frequency of the change increases, the effects of the initial acceleration on the overall motion of particles increase and can be very large. In systems such as jigs, for example, this initial acceleration is believed to help separate the particles.

For the Stokes' regime, Eqs. 5.3 and 5.4 can be combined to yield

$$\frac{4}{3}\pi r^3 \rho_s \frac{dv}{dt} = \frac{4}{3}\pi r^3 (\rho_s - \rho_f)g - 6\pi\mu r v$$

and

$$\frac{dv}{dt} = \left(\frac{\rho_s - \rho_f}{\rho_s}\right)g - \left(\frac{9\mu}{2r^2\rho_s}\right)v$$

Therefore,

$$\int_0^v \frac{dv}{\left(\frac{\rho_s - \rho_f}{\rho_s}\right)g - \left(\frac{9\mu}{2r^2\rho_s}\right)v} = \int_0^t dt$$

By integration,

$$v = \frac{2r^2(\rho_s - \rho_f)g}{9\mu}\left[1 - \exp\left(-\frac{9\mu}{2r^2\rho_s}t\right)\right]$$

or

$$v = v_t\,[1 - \exp(-kt)] \qquad \textbf{(Eq. 5.13)}$$

where $k = \dfrac{9\mu}{2r^2\rho_s}$

Equation 5.13 demonstrates that $v = v_t$ when $t = \infty$. According to this equation, the time required to obtain v_t depends strongly on the k value, which, in turn, depends on the size and density of the particle in question and the viscosity of the medium surrounding the particle.

Example: Time Required for Two Particles to Reach 99% of Terminal Velocity

Problem: Calculate the time required for silica and galena of 70 μm radius to reach 99% of v_t in water and in air at 20°C.

Solution:

First, t for SiO_2 in water,

$$k = \frac{9 \times 0.01}{2 \times (0.007)^2 \times 2.65} = 346.5$$

Therefore,

$$v/v_t = 0.99 = 1 - e^{-kt} \text{ and } t = 0.0133 \text{ s}$$

Similar calculations for other conditions produce the following results:

	Silica		Galena	
	Air	Water	Air	Water
k (s^{-1})	6.272	346.5	2.216	122.5
t (s)	0.734	0.0133	2.08	0.0376

Suppose that two mineral particles of different densities are separated in water because of differences in their initial velocities. Assume, further, that these two particles are galena (specific gravity 7.5) and silica (specific gravity 2.65) and that they have different sizes but the same terminal velocity. Therefore, under laminar conditions,

$$v_t\,(SiO_2) = \frac{2r_s^{\,2}(\rho_s - \rho_f)g}{9\mu} \quad \textbf{(Eq. 5.14)}$$

$$v_t\,(PbS) = \frac{2r_p^{\,2}(\rho_s - \rho_f)g}{9\mu} \quad \textbf{(Eq. 5.15)}$$

Because $v_t\,(SiO_2) = v_t\,(PbS)$, from Eqs. 5.14 and 5.15,

$$\frac{r_p}{r_s} = \sqrt{\frac{2.65 - 1}{7.5 - 1}} = 0.504$$

Therefore,

$$r_p = 0.504\, r_s \quad \textbf{(Eq. 5.16)}$$

The radius of galena particles having the same terminal velocity as silica particles is about half of the radius of silica particles. If the radius of the silica particles is 70 μm, the corresponding radius of galena particles will be 35.3 μm, and they will exhibit the same terminal velocity. Table 5.2 shows the settling velocities of these two particles as a function of time. Galena particles reach terminal velocity faster than silica particles because of galena's greater k value. Although the terminal velocity of these two particles is the same, their settling velocities during the transient period are significantly different. For example, at 1 ms, the settling velocity of galena particles is about 30% greater than that of silica particles.

Such behavior has practical implications in systems such as jigs and those in which particle velocity plays an important role. In other words, the difference in the initial velocity (not the terminal

TABLE 5.2 Transient- (unsteady-state) settling velocities of galena and silica as a function of time

Time, s	v, cm/s		v (Galena)/v (Silica)
	Galena	Silica	
10^{-4}	0.0877	0.060	1.38
10^{-3}	0.726	0.515	1.31
2×10^{-3}	1.088	0.880	1.24
5×10^{-3}	1.602	1.449	1.11
8×10^{-3}	1.723	1.650	1.04
10^{-2}	1.746	1.705	1.02
$t = \infty$	1.76	1.76	1.00

TABLE 5.3 Sphericity of various shapes

Shape	φ
Sphere	1
Octahedron	0.847
Prisms	
a × a × 2a	0.767
a × 2a × 2a	0.761
a × 2a × 3a	0.725
Cylinders	
h = 3r	0.860
h = 20r	0.580
Disks	
h = r	0.827
h = r/3	0.594
h = r/15	0.254

velocity) can be used to increase the relative velocity of solids against the fluid. For example, high-frequency vibration, such as ultrasonic waves applied to a leaching system, will cause an unexpectedly high rate of dissolution, partly because the vibration accelerates the velocity of the dissolving solid particles. Such improvement in leaching is possible when mass transfer is an important factor in the overall process, because mass transfer is directly influenced by the velocity of solids relative to that of the liquid phase.

PARTICLE SHAPE

The foregoing discussions assumed that particles were spherical. In nature, however, mineral particles rarely approach spherical. To examine the effect of shape of particles on the settling velocity, we define the sphericity, φ.

$$\varphi = \frac{\text{surface area of sphere having same volume as particle}}{\text{surface area of particle}}$$

Therefore, the sphericity of a sphere is 1.

The effect of sphericity on the motion of particles can be substantial. For example, the friction factor of a disk having a sphericity value of 0.25 (Table 5.3) is about 20 times that of a sphere with the same Reynolds number. Table 5.3 lists the sphericity of various shapes.

Example: Sphericity of a Cube

Problem: Calculate the sphericity of a cube with dimensions of $d \times d \times d$ and a prism with dimensions of $d \times 3d \times 1/3\, d$.

Solution:

For the cube,

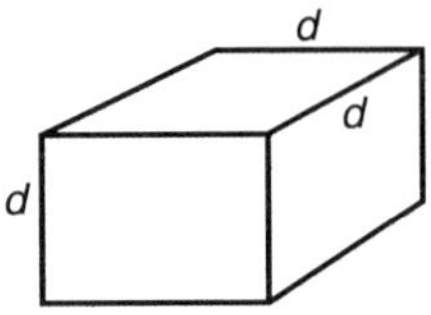

Volume of cube = volume of sphere

$$d^3 = (4/3)\,\pi\, r^3$$

Therefore,

$$r = d/[(4/3)\,\pi]^{1/3}$$

$$\varphi = \frac{4\pi r^2}{6d^2} = \frac{d^2(36\pi)^{1/3}}{6d^2} = 0.806$$

Similarly for the prism,

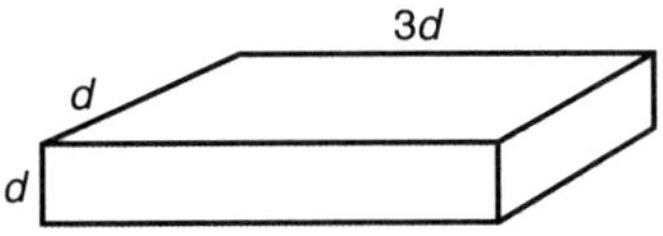

$$\varphi = \frac{\pi\left(\frac{6}{\pi}\right)^{2/3} d^2}{\left(\frac{2}{3} + 6 + 2\right) d^2} = 0.558$$

The diameter of particles used in all flow equations is defined as the diameter of a sphere having the same volume as the particle. This diameter is related to the average screen diameter, d_s, by the following relationship:

$$d = \frac{d_s}{\varphi}\frac{1}{q} \qquad \textbf{(Eq. 5.17)}$$

where

$$q = \frac{\text{specific surface area of particle}}{\text{specific surface area determined by } d_s}$$

$$= \frac{S_p}{\frac{6}{\rho_s} d_s} \qquad \textbf{(Eq. 5.18)}$$

The specific surface area, S_p, can be calculated if the geometry of the particle is known. When S_p is measured, however, the area is referred to as the "external area" and does not include the area inside pores.

Example: Calculate *q* and d/d_s for a Cube

Calculate q and the ratio of d/d_s for a cube with dimensions of $a \times a \times a$ and a prism with $a \times a \times 2a$.
Solution:
For the cube,

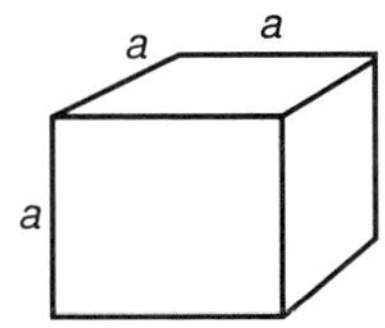

$$d_s = a;\ q = \frac{6\frac{a^2}{\rho_s}a^3}{\frac{6}{\rho_s}a} = 1$$

$$\frac{d}{d_s} = \frac{1}{\varphi}\frac{1}{q} = \frac{1}{0.806} = 1.24$$

For the prism,

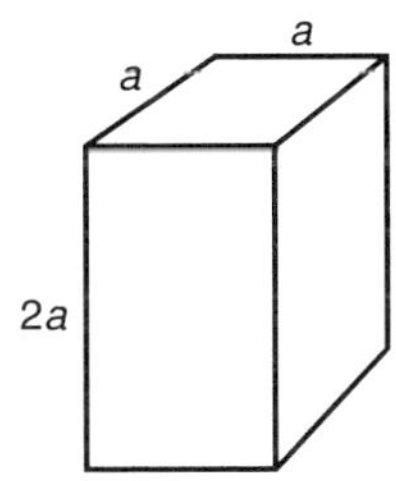

$$\frac{d}{d_s} = \frac{6}{5}\frac{1}{0.767} = 1.564$$

Example: Terminal Velocity of Galena Particles

Problem: Estimate the terminal velocity of galena particles (specific gravity 7.5) of diameter 0.1 cm in water at 20°C. Assume that $\varphi = 1$.

Solution:
When $\varphi = 1$ (sphere),

using Eq. 5.11, $d^* = 43.70$

$Q = 4.40$

$v_t = 37.40$ cm/s

and using Eq. 5.12, $v_t = 36.43$ cm/s

HINDERED SETTLING

When particles that fall in a vessel are influenced by the wall of the vessel or by neighboring particles, the flow pattern of these particles will be distorted. Therefore, the settling velocity of concentrated particles is quite different from the velocity of free-settling particles. Lorentz (1907) suggested that the drag force on a sphere given by Eq. 5.4 should be modified to be

$$R' = 6\pi r\mu v\left(1 + k'\frac{r}{L}\right) \quad \textbf{(Eq. 5.19)}$$

where L is the distance between the particle and the wall or the neighboring particle and k' is a constant for the system.

Other factors affect the settling velocity when the concentration of solids is high. For example, small particles in the system can be considered as a part of the fluid that surrounds larger particles. Therefore, as far as large particles are concerned, the density and viscosity of the medium is quite different from that of the pure fluid. As a rule of thumb, when the concentration of all solid particles in a fluid is less than 0.01 or 1% by volume, the particle will be expected to settle freely. However, when the volume of the particles in the system is >1%, hindered settling will prevail and the motion of the settling particles will be retarded.

The hindered-settling velocity is usually a fraction of the free-settling velocity. Therefore, this velocity is evaluated using a correction factor, C_F: $V_H = C_F \times V_t$. Here, C_F, V_H, and V_t are, respectively, the correction factor, the hindered-settling velocity, and the free-settling velocity. The correction factor can be obtained using the following equations (Gaudin 1939; Steinour 1944a,b,c; Brown 1950; Taggart 1951).

$$C_F = (1-\gamma^{2/3})(1-\gamma)(1-2.5\gamma) \qquad \textbf{(Eq. 5.20)}$$

or

$$C_F = \frac{(1-\gamma)^2}{10^{1.82\gamma}} \qquad \textbf{(Eq. 5.21)}$$

when $\gamma < 0.3$

$$C_F = 0.123\frac{(1-\gamma)^3}{\gamma} \qquad \textbf{(Eq. 5.22)}$$

when $0.3 < \gamma \le 0.7$.

In the above equations, γ represents the volume fraction of all the solids present in the system. Equations 5.20 and 5.21 are more specific, and the refined equations used for the specified concentration ranges and Eq. 5.21 are more general.

REFERENCES

Brown, G.G. 1950. *Unit Operations*. New York: John Wiley & Sons.

Concha, R., and E.R. Almendra. 1979a. Settling Velocities of Particulate Systems, 1. Settling Velocities of Individual Spherical Particles. *Int. J. Miner. Process.*, 5:349–367.

——. 1979b. Settling Velocities of Particulate Systems, 2. Settling Velocities of Spherical Particles. *Int. J. Miner. Process.*, 6:31–41.

Concha, R., and A. Barrientos. 1982. Settling Velocities of Particulate Systems, 3. Power Series Expansion for the Drag Coefficient of a Process. *Int. J. Miner. Process.*, 9:167–172.

Dallavalle, J.M. 1948. *Micrometrics*. New York: Pitman.

Gaudin, A.M. 1939. *Principles of Mineral Dressing*. New York: McGraw-Hill.

Han, K.N. 1984. A Simple and Accurate Method of Determining the Free Settling Velocity. *J. Korean Inst. Mineral and Mining Eng.*, 21:237–240.

Lapple, C.E. 1951. *Fluid and Particle Mechanics*. Newark, Del.: University of Delaware.

Lorentz, H.A. 1907. Uber die Entstehung turbulenter Flussigkeitsbewegungen und uber den Einfluss dieser Bewegungen bei der Stromung durch Rohren. *Abh. u. Th. Phys*, Leipzig. 1:23–33.

Olson, R. 1961. *Essentials of Engineering Fluid Mechanics*. Scranton, Pa.: International Textbook.

Steinour, H.H. 1944a. Rate of Sedimentation. Nonflocculated Suspensions of Uniform Spheres. *Ind. Eng. Chem.*, 36:618–624.

——. 1944b. Rate of Sedimentation. Suspensions of Uniform-size Angular Particles.. *Ind. Eng. Chem.*, 36:840–847.

——. 1944c. Rate of Sedimentation. Concentrated Flocculated Suspensions of Powders. *Ind. Eng. Chem.*, 36:901–907.

Taggart, A.F. 1951. *Elements of Ore Dressing*. London: John Wiley & Sons.

Torobin, L.B., and W.H. Gauvin. 1959a. Fundamental Aspects of Solid–Gas Flow, Part 1. *Can. J. Chem. Eng.*, 37:129–141.

——. 1959b. Fundamental Aspects of Solid–Gas Flow, Part 2. *Can. J. Chem. Eng.*, 37:167–176.

CHAPTER 6

Gravity Concentration

Frank F. Aplan

INTRODUCTION

Gravity concentration is a process in which particles of mixed sizes, shapes, and specific gravities are separated from each other in a fluid by the force of gravity or by centrifugal force. The process is designed to separate particles by specific gravity, but to a certain extent it also separates particles on the basis of size and shape.

Historically, the process has been used to separate ore minerals or coal from their associated gangue (refuse) on the basis of mineral density (Table 6.1). Gravity concentration is often equally applicable to other common industrial processes, such as degritting food grains, paper pulp, and chemical raw materials; recycling municipal solid waste; recovering and recycling spills, splatters, skimmings, skulls, turnings, and grindings from metal production and fabrication; and remediating toxic waste piles.

Early Use and Development of Gravity Concentration

Gravity concentration of heavy minerals is a natural geological process, and Mother Nature has concentrated minerals, such as gold, cassiterite, ilmenite, and diamonds, into natural placer (alluvial or glacial) deposits. Humans have used gravity concentration processes for thousands of years. Egyptian monuments of about 3000 BCE depict the washing of gold ores (Anon. 1970) and the Athenians undoubtedly used flowing film concentration to process ores from their mines at Laurium before the birth of Christ (Gaudin 1939). In the sixteenth century, Agricola (1556) in *De Re Metallica* described several gravity concentration devices used in Europe, and seventeenth-century Chinese concentration technology is described in *T'ien-kung K'ai-Wu* (Sung 1637). In the nineteenth century, Rittinger in Europe performed theoretical and practical studies, and in the later part of that century, Richards in the United States did much to establish the basic principles of gravity concentration that were published in his classic four-volume treatise (Richards 1906–1909). In the 1920s, Finkey (1924) established many of the mathematical relationships describing the process, and Gaudin (1939) and Taggart (1945, 1951) extended and codified the principles on which gravity concentration is based. Other valuable references that describe either processes or devices are Richards and Locke (1940), Mills (1978), Burt and Mills (1984), Aplan (1985a), Weiss (1985), Osborne (1988), and Leonard and Hardinge (1991).

Importance of Gravity Concentration in Minerals Processing

A glance at the literature reveals that gravity concentration has been studied much less than its more glamorous counterpart, flotation. Yet, in terms of commercial use, about 25% more coal and ore tonnage is treated by gravity concentration in the United States than is treated by flotation. It has been estimated that in 1988, 529 million metric tons was treated by gravity concentration, 529 million by flotation, and 153 million by magnetic separation (Aplan 1989). Of the gravity concentration tonnage,

TABLE 6.1 Densities of several common organic and inorganic minerals

Density	Mineral	Composition	Density	Mineral	Composition
1.07	Gilsonite	Asphalt	~4.0	Sphalerite	ZnS
1.10	Amber	Fossil resin	4.1–4.9	Chromite	$(Mg, Fe)\ Cr_2O_4$
1.2–1.7	Coal	Metamorphosed plant matter	~4.2	Chalcopyrite	$CuFeS_2$
1.99	Sylvite	KCl	4.25	Rutile	TiO_2
2.16	Halite	$NaCl$	~4.25	Barite	$BaSO_4$
2.32	Gypsum	$CaSO_4 \cdot 2H_2O$	4.5–5.0	Ilmenite	$FeTiO_3$
2.56	Feldspar (orthoclase)	$KAlSi_3O_8$	4.6	Zircon	$ZrSiO_3$
2.65	Quartz	SiO_2	4.9–5.3	Hematite	Fe_2O_3
2.71	Calcite	$CaCO_3$	4.9–5.3	Monazite	$(Ce, La)\ PO_4$
2.85	Dolomite	$CaMg\ (CO_3)_2$	~5.0	Pyrite	FeS_2
3.00	Magnesite	$MgCO_3$	5.2	Magnetite	Fe_3O_4
3.1–3.2	Apatite	$Ca_5\ (PO_4)_3\ (F,OH)$	5.3–7.3	Columbite-Tantalite	$(Fe,Mn)(Nb,Ta)_2O_6$
3.18	Fluorite	CaF_2	6.8–7.1	Cassiterite	SnO_2
3.50	Diamond	C	7.1–7.5	Wolframite	$(Fe,Mn)\ WO_4$
3.95	Garnet (almandite)	$FeAl_2\ (SiO_4)_3$	7.5	Galena	PbS
~4.0	Corundum	Al_2O_3	8.94	Copper	Cu
			15.6–19.3	Gold (+ some silver)	Au

about 500 million tons was raw coal treated in coal preparation plants. Although these tonnage estimates were made in 1988, the values probably are not much different today (in 1999). Of the raw coal sent to preparation plants, about 94% is cleaned by gravity methods as compared with only 6% cleaned by flotation (Aplan 1989).

Over the years a bewildering number of gravity concentration devices have been developed. Concentrating the finer sizes is difficult, and as the particle size of the material to be treated decreases, the number of devices invented to capture the fines seems to increase exponentially. Some of the more important devices are listed in Table 6.2. For details on these and similar devices, the older, more recent, and current literature should be consulted (see previous citations).

Applicability to Concentration Processes

Particles respond differently to various concentrating devices depending on the fluid, the force field, and specific properties of the particles, such as density, size, shape, chemistry, surface chemistry, magnetism, conductivity, color, and porosity. Various concentration devices are applicable to particles in various size ranges (Figure 6.1), and for any given size range, several processes or devices might be used. Gravity concentration works best in the 130-mm (about 5-in.) to 74-μm (200-mesh) range. Below about 74 μm (200 mesh), separation of particles by specific-gravity differences is increasingly difficult, and it is generally inapplicable below about 15 μm except in special circumstances. Of the processes listed in Figure 6.1, only flotation and wet magnetic separation effectively separate −10-μm particles. If minerals are liberated at a coarse size, gravity concentration is often an inexpensive and effective way to separate them from their associated gangue minerals; if not, other methods may be more attractive.

TABLE 6.2 Size ranges treated by typical gravity concentration devices

- Coarse concentration (+¼ in. [+6.4 mm])
  - Sorting with hand (held) devices
  - Jigs
    - Pulsion-suction
    - Pulsator
    - Baum
  - Heavy media
  - Heavy media hydrocyclone
  - Hindered-settling classifiers
  - Pneumatic jigs, tables, launders
- Intermediate concentration (¼ in.–100 mesh [6.4–0.150 mm])
  - Jigs
  - Heavy media hydrocyclone
  - Water-only cyclone
  - Hindered-settling classifiers
  - Kelsey jig
  - Pneumatic jigs and tables
  - Sluices
  - Shaking tables
  - Humphreys-type spirals
  - Pinched sluices
  - Cannon concentrator
  - Reichert cone
  - Wright impact tray
  - Falcon concentrator
  - Knelson concentrator
  - Burch shaken helicoid
  - Bartles-Mozley concentrator
  - Mozley multigravity separator
  - Buddles
    - Planilla
    - Lanchute
- Fines concentration (–100 mesh [–0.150 mm])
  - Jigs that emphasize suction
  - Kelsey jig
  - Shaking table
  - Humphreys-type spirals
  - Pinched sluices
  - Cannon concentrator
  - Reichert cone
  - Wright impact tray
  - Falcon concentrator
  - Knelson concentrator
  - Burch shaken helicoid
  - Bartles-Mozley concentrator
  - Mozley multigravity separator
  - Buddles
    - Planilla
    - Lanchute
  - Round table
  - Tilting frames (e.g., Denver–Buckman)
  - Vanners
    - Frue
    - Bartles crossbelt concentrator
  - Strakes
    - Blanket and corduroy tables
    - Johnson concentrator
    - Endless belt
  - Plane table

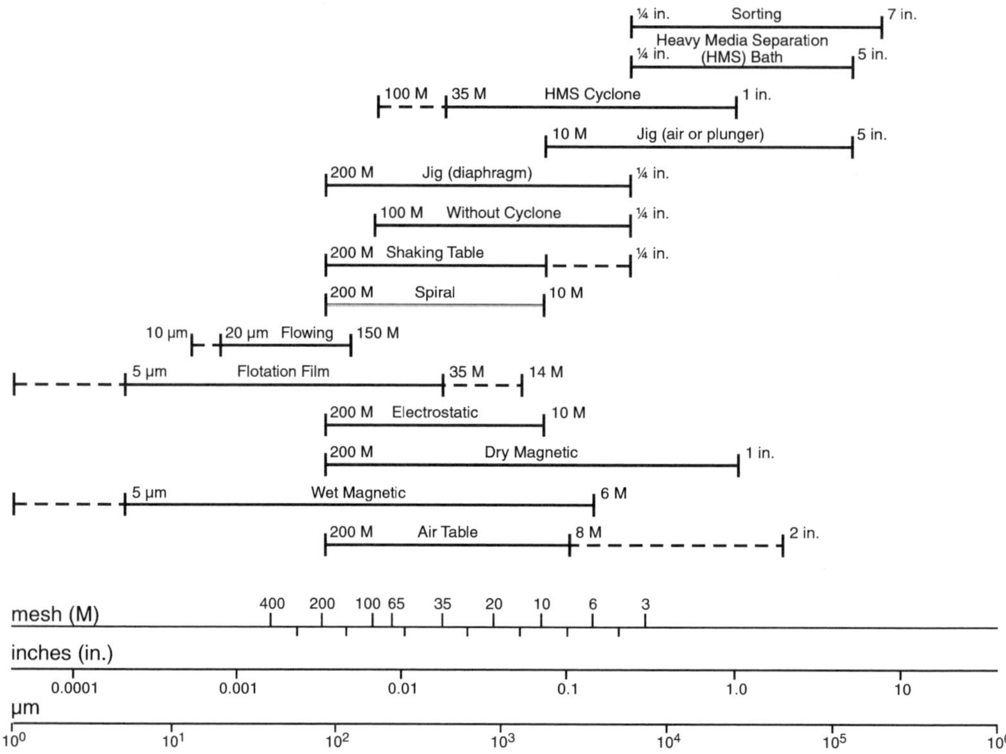

FIGURE 6.1 Approximate range of applicability of various concentrating devices (M = mesh, Tyler Standard)

THE BASICS OF GRAVITY SEPARATION

Settling phenomena, especially hindered-settling, underlie all gravity concentration processes.

Free Settling

Free settling may be defined as that process in which individual particles fall freely in a fluid without being hindered by other particles (paraphrased after Richards and Locke [1940]). The settling of these particles, which are assumed to be spheres, can be calculated from the equations of Newton and Stokes and for the Allen range from measurements or approximations, as outlined in the previous chapter. The terminal settling velocity of spheres of various densities as a function of particle diameter and the density and nature of the fluid (water or air) is given in Figure 6.2 (as modified from Lapple et al. [1956]). Above about 2,000 µm the slope of the curves is 0.5, corresponding to $\sqrt{d}$ in the Newtonian equation:

$$V_m = \sqrt{\frac{4}{3f}\,\frac{\rho-\rho'}{\rho'}\,dg} \qquad \textbf{(Eq. 6.1)}$$

where

V_m = the terminal settling velocity of the particle
f = the friction factor or coefficient of resistance (f is ~ 0.4 for spheres)
ρ and ρ' = the densities of the solid and fluid, respectively
d = particle diameter
g = the force of gravity

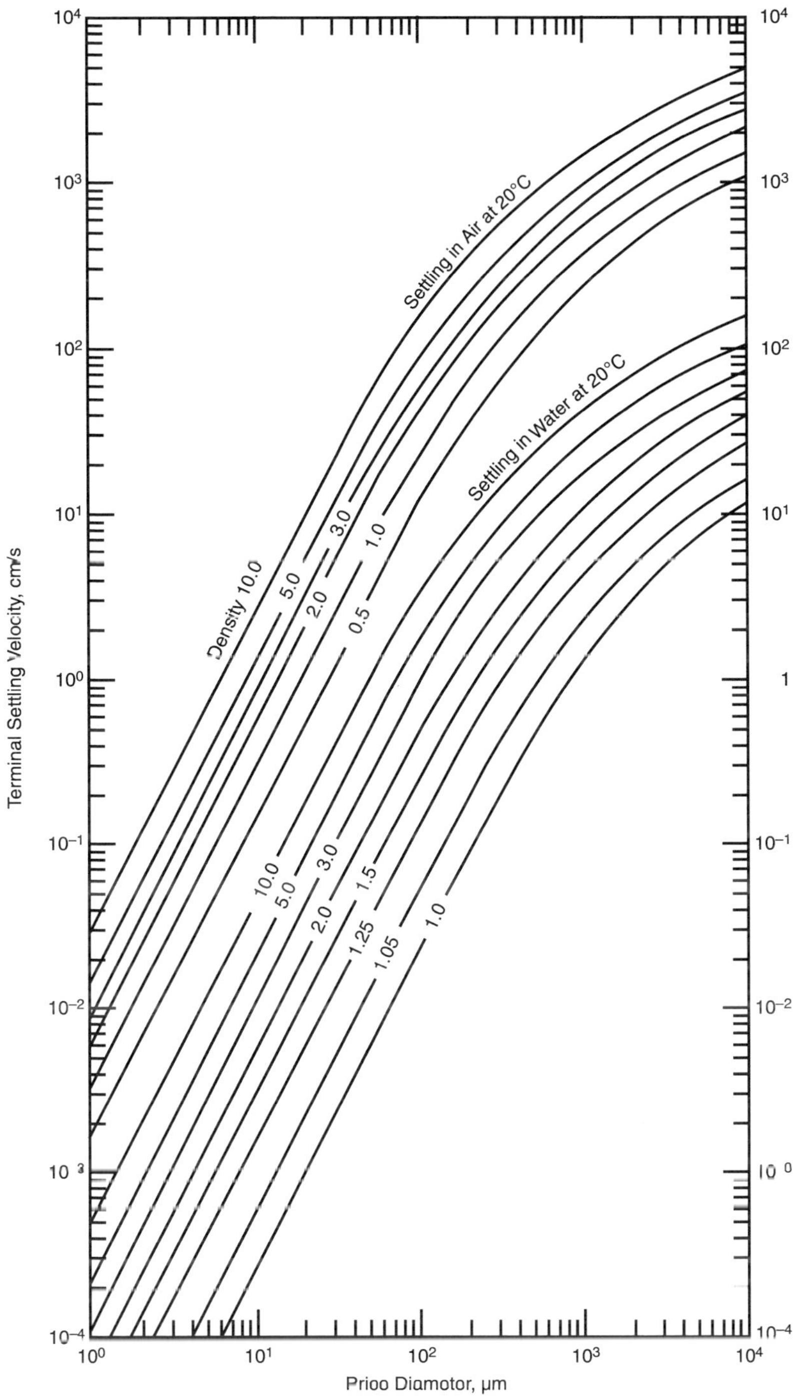

Source: Modified from Lapple et al. 1956.

FIGURE 6.2 The terminal settling velocities of spheres of various sizes and densities settling in water and in air

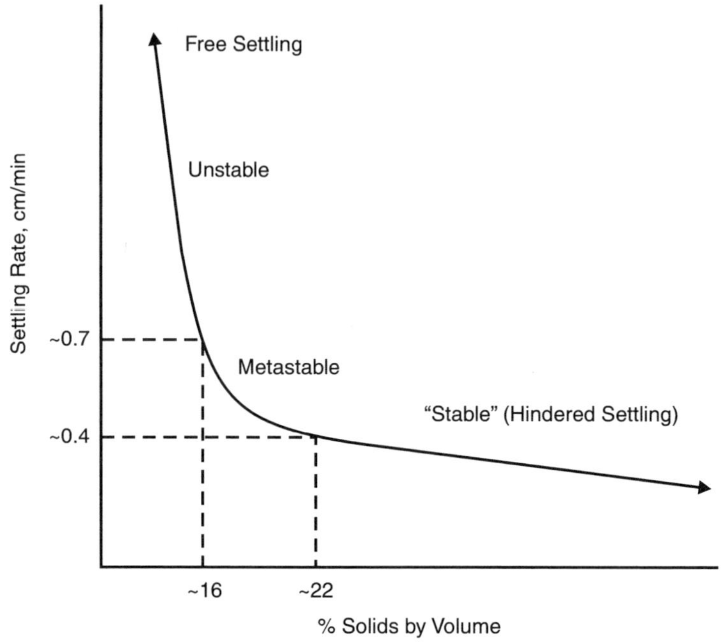

Source: Based on studies of Dalta 1977.

FIGURE 6.3 Generalized effect of percentage of solids on the settling rate of ground mineral suspensions in water. Approximate values given for coal or limestone ground to nominal –100 mesh

Below about 100 μm the slope of the curve is 2, as dictated by the Stokes' equation:

$$V_m = \frac{1}{18} \frac{\rho - \rho'}{\mu} d^2 g \qquad \textbf{(Eq. 6.2)}$$

where

V_m = terminal settling velocity of the particle
ρ and ρ' = densities of the solid and fluid, respectively
μ = fluid viscosity
d = particle diameter
g = the force of gravity

Hindered Settling

Hindered settling describes that process in which particles of mixed sizes, shapes, and specific gravities, in a crowded mass yet free to move among themselves, are sorted in a rising fluid current (paraphrased after Richards and Locke [1940]). Collisions between particles are continuous, and the assemblage will settle much more slowly than the freely settling individual particles. Hindered settling may, most conveniently, be promoted by agitation of the suspension (stirring, or the use of jets or a rapidly rising fluid current) or by the introduction of a constriction (such as a punched plate, screen, or Venturi).

The generalized curve for the settling of common mineral suspensions (typically −10 mesh) is given in Figure 6.3. The approximate numerical values for the three settling regions (unstable, metastable, and stable) are shown for the case of coal or limestone ground to a nominal −100 mesh (150 μm; Datta 1977). The settling rate values of 0.4 and 0.7 cm/min are generally applicable to ground ores except for closely sized or high-density particles. However, the values for the percent solids by volume (% S_V)

required for stabilization depend not simply on the shape and density of the particles; they are acutely sensitive to the size consist (size distribution) of the feed (e.g., the size parameter, K, and the distribution parameter, a, in the Gaudin–Schuhmann equation for ground ores [Eq. 2.55, Particle Characterization chapter]). To stabilize suspensions of most other ores, or of more coarsely ground material, will require a substantially higher percent S_V than that shown in Figure 6.3, while material of a finer size consist can be stabilized at a lower percent S_V. As more than a few particles are settled together in water, the settling rate decreases linearly as percent S_V increases, up to the onset of metastability.

Equal Settling Particles

If a heavy (H) and a light (L) particle are settled under the same free-settling conditions, some smaller dense particles will settle at the same rate as some larger light particles, and under Newtonian conditions (Eq. 6.1), the ratio of their diameters at the same settling rate is

$$\left.\frac{d_L}{d_H}\right]_{\text{Newt, }FS} = \frac{(\rho_H - \rho')^1}{(\rho_L - \rho')^1} \qquad \textbf{(Eq. 6.3)}$$

and using the same approach for Stokesian conditions (Eq. 6.2), the exponent n will be 0.5. Allen-range particles will have an exponent between 0.5 and 1.0. Typically, the ratio of diameters is small, and particles settle in about the same settling regime, so the equation may be approximated as

$$\left.\frac{d_L}{d_H}\right]_{FS} = \left(\frac{\rho_H - \rho'}{\rho_L - \rho'}\right)^{0.5 \le n \le 1} \qquad \textbf{(Eq. 6.4)}$$

and under hindered-settling conditions, the equation becomes

$$\left.\frac{d_L}{d_H}\right]_{HS} = \left(\frac{\rho_H - \rho''}{\rho_L - \rho''}\right)^{0.5 \le n \le 1} \qquad \textbf{(Eq. 6.5)}$$

where ρ' is replaced by the apparent specific gravity, ρ'', of the suspension.

Knowing the density of the solid in the hindered-settling zone and its volumetric percent solids, γ_s, the ρ'' of an aqueous suspension may be approximated by the formula

$$\rho'' = \gamma_s\rho + (1 - \gamma_s)\rho' \qquad \textbf{(Eq. 6.6)}$$

Table 6.3 indicates that galena (ρ_H) of 2-mm diameter, assumed spherical, with a free-settling ratio of 3.94, will settle at about the same rate as 8-mm-diameter quartz (ρ_L). Thus, all particles of galena greater than 2 mm will be separated from all quartz particles less than 8-mm diameter. However, in a 45% quartz–water suspension, +2-mm galena can be separated from all quartz particles less than 12.7-mm diameter. Fine particles are separated less effectively, although their separation is better in a suspension than in a true fluid.

TABLE 6.3 Free- and hindered-settling ratios for spheres of the mineral pair galena (ρ_H = 7.5)/ quartz (ρ_L = 2.65) settling in water and in quartz/water suspensions

System	Settling of Large Particles (+10 mesh)	Settling of Small Particles (–150 mesh)
Free-settling ratio, water	3.94	1.98
Hindered-settling ratio		
25% quartz suspension = 1.41	4.91	2.22
45% quartz suspension = 1.74	6.33	2.52

The best condition for separating particles during hindered-settling (Table 6.3 and Eqs. 6.5 and 6.6) is favored by

- Large particles
- A large density difference between the particles
- A large apparent density difference between the heavier particles and the suspension
- A small apparent density difference between the lighter particles and the suspension
- A high volumetric percent solids in the hindered-settling zone
- A high volumetric percent solids achieved with a dense solid

The hindered-settling zone is called a "teeter" zone. Eventually the large, denser particles will displace other particles in the teeter zone, raising ρ'', and, within limits, improving the separation. A mechanism needs to be provided that removes some of the heavier particles, either periodically or continually, from the teeter zone.

Suspension Stability

Closely allied with the hindered-settling phenomenon is suspension stability. The onset of hindered-settling improves suspension stability (Figure 6.3).

Because mineral processing plants must minimize sanding in equipment, such as pipes, pumps, sumps, launders, process vessels, and tailings lines, and in pumped products, the mineral process engineer must design for slurry stability under all conditions. The start-up and shutdown of a plant are the most unpredictable stages, because the very high water-to-solids ratios at those times encourage suspension instability. Consequently, allowance must be made to thwart the problem of suspension instability in any solids–liquid system. Certain factors favor suspension stability and minimize the problem (Figure 6.3 and Table 6.4). Many of the factors favoring suspension stability cannot be changed in a given plant (e.g., particle density, size, and shape), so greater reliance must be placed on those factors that can be easily altered.

Sorting Classifiers

Sorting classifiers not only classify particles by size but also permit a crude gravity separation to be made based on the hindered-settling principle. Perhaps the first sorting classifier was the Spitzkasten, dating back to the early nineteenth century or before. It is a series of inverted pyramidal or conical vessels of increasing cross-sectional area and, hence, of decreasing flow velocity. The heaviest and coarsest particles are removed first, and smaller and lighter particles are removed sequentially. The lower part of each vessel is a pocket region that constitutes a hindered-settling zone. An orifice or siphon device can be used to remove the particles from the successive pockets.

Following the development of the Spitzkasten came the Evans, the Richards, and the Fahrenwald–Dorrco hindered-settling classifiers. (Following common technical usage, these and other trade, inventor's, or manufacturer's names are used throughout this chapter. Such terms are used to identify the general type of equipment only, and they do not imply endorsement of the equipment of any particular manufacturer.) These devices had the same size pockets but supplied differing amounts of water to each pocket to achieve a different rising current velocity in each. The current velocity was greatest in the first pocket to facilitate removal of the heaviest and coarsest particles; current velocity was successively lower in each of the subsequent pockets. The heavy concentrate was removed by ingenious sensing and orifice, siphon, or intermittent flow devices. In the coal industry, a similar device was the Rheolaveur launder separator, now essentially obsolete. That separator consisted of a series of hindered-settling boxes fitted in a launder. The heavier refuse settled against the rising water current in the pocket, while this same current tended to thwart entry of the lighter coal. An orifice continuously discharged refuse from the pocket. The heavy product from the primary pockets in the launder was cleaned several times in another Rheolaveur launder to remove additional coal (Leonard and Hardinge 1991).

TABLE 6.4 Achieving suspension stability in a solids–liquid system

Suspension stability favored by

- Low density solids (coal, ρ = 1.4, better than galena, ρ = 7.5)
- Small particle size
- A broad range of particles (poly-sized better than mono-sized)
- Irregular shape of particles
- No excess of large, oversized particles
- High percent solids in suspension
- Turbulent flow
- Relatively high suspension viscosity
  - — Nature of the particles (e.g., clays)
  - — Percent solids
  - — Particle size and size distribution
  - — Externally added material:
    - Clays and xanthan gums to increase pulp viscosity

Simple cures for suspension instability

Elimination of tramp (oversize) particles:

- Protective screens (fail-safe)
- Classifiers—mechanical, hydrocyclone

Turbulence to prevent stratification by particle size and specific gravity created by

- Vigorous pumping
- Constrictions—Venturi principle (usually not practical with abrasive slurries)
- Baffles in an open launder
- Jumps (a short vertical drop in a gravity line to improve mixing and minimize stratification)
- Dump into pond and repump

A high percent solids

- Minimize water during:
  - — Comminution
  - — Classification
  - — Concentration
- Reduce the water content of the plant process stream by dewatering:
  - — Classifier
  - — Thickener
  - — Fine-screening devices
  - — Centrifuge

Later, Fahrenwald-type pocket sizers, siphon sizers, and hydrosizers were developed. In favorable conditions, such as a closely sized feed, sorting classifiers can produce a final product typified by the production of pebble phosphate. These sorting devices are also used to make an intermediate concentrate for subsequent final separation by another device, such as a shaking table.

Hindered-settling classifiers have recently been reborn as large-capacity (up to several hundred tons per hour) bin-like units, such as those made by Linatex (the Hydrosizer) and CFS, Inc. (the Density Separator; Figure 6.4). Numerous jets of water issue from perforated pipes to fluidize particles. The device may be used for both sorting and sizing in operations, such as eliminating iron-containing particles and

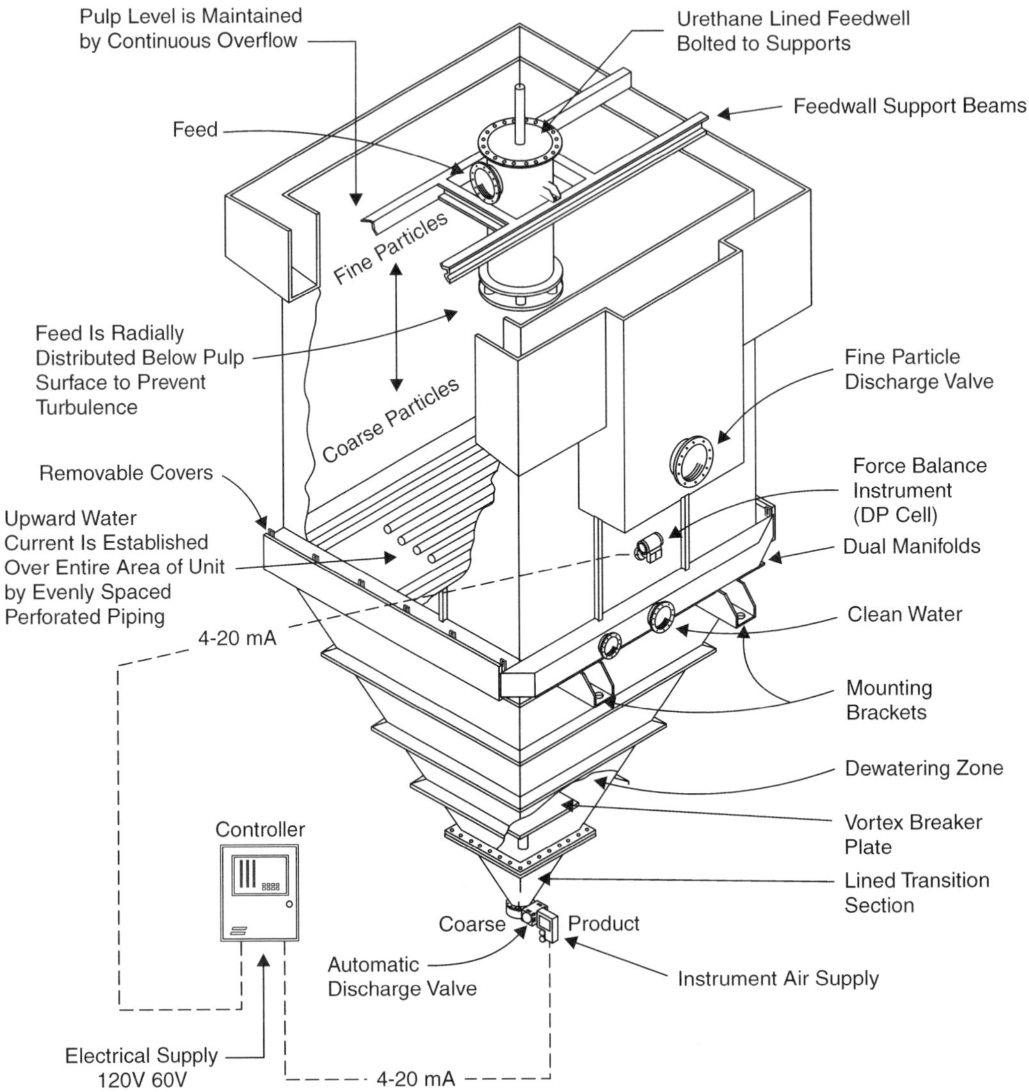

Source: Classification and Flotation Systems, Inc.

FIGURE 6.4 The CFS Density Separator, a hindered-settling, sorting, and sizing classifier

clays from glass sand, rejecting lower-density contaminants (such as coal or charcoal) or chemicals from particulate suspensions, and removing shale and other undesirable particles from construction sands. Both the lower density and the platey nature of shale facilitates its removal from the sand. The removal of impurities from sand is usually done on particles in the 20–200 mesh-size range. The usefulness of a hindered-settling classifier in washing soil for environmental remediation is obvious.

In practice, many sizing classifiers, such as the Allen cone and mechanical classifiers of the Akins, Dorr, Esperanza Drag, and Hardinge types, use the hindered-settling principle even though their primary function is sizing. Sometimes the hindered-settling phenomenon causes heavier constituents to accumulate in the circulating load of a ball mill-classifier circuit, which leads to overgrinding of the heavier constituents.

FLOAT–SINK SEPARATION

Rather than rely on the hindered-settling phenomenon and the hindered-settling ratio, a plant engineer might choose to use a liquid of density intermediate between those of the minerals to be separated. Heavy solutions, heavy liquids, semistable suspensions, and ferro fluids have all been used, but the most enduring has been an aqueous suspension of fine particles of a magnetic solid, such as magnetite or ferrosilicon.

The current process of heavy media separation is also sometimes called dense media separation; it is the most important float–sink separation process in use today. Its applicaction for both coal and ore separations began to increase significantly in about 1940, and use of the process in the last 35 years, primarily for coal, has increased very rapidly. Today, more than 40% of the coal sent to preparation plants in the United States is cleaned by some variation of this process. Those interested in greater detail should consult Burt and Mills (1984); Aplan (1985b); Miller, De Mull, and Matoney (1985); Osborne (1988); and Leonard and Hardinge (1991).

Early Development

In 1858, Bessemer patented the use of calcium chloride solutions to separate the lighter coal from refuse. Much later, in the 1930s, several commercial processes of this type were used in the United States to make a raw coal separation at about 1.5 specific gravity (Aplan 1985b). These processes are obsolete today. Heavy liquids are typified by the halogenated hydrocarbons, and only pilot-plant separations have been made on coals (using halogenated hydrocarbons of $\rho = \sim 1.5$) and ores (using 1,1,2,2, tetrabromoethane, $\rho = 2.96$). Several problems are associated with these processes—toxicity, evaporative losses, "drag out" losses related to the surface area and porosity of some of the materials treated, and high capital and operational costs. Heavy liquid separations have not been a commercial success.

A heavy medium technique now in the laboratory stage of development uses a ferro fluid. Very fine magnetite is stabilized (for example, with a fatty acid) in kerosene. The specific gravity of the ferro fluid is then adjusted by an external magnetic field.

Heavy Media Separation

Heavy media separation (HMS) has been used industrially to

- Produce a finished concentrate and a rejectable waste in one operation
- Reject a relatively coarse waste leaving an enriched product ready for further processing (a step that can greatly reduce expensive grinding costs)
- Produce a finished concentrate and a lower-grade product ready for further processing
- Produce two finished products of differing composition

Ordinarily, HMS is used for the first two of these functions—the first for coal, the second for ores.

The most successful way to achieve a float–sink separation has been to use a quasi-stable suspension of a solid that is appreciably heavier than the mineral to be floated. Various solids have been employed to make an aqueous heavy media suspension (Table 6.5). Silica sand in an inverted conical vessel is used in the Chance Cone system, which dates from 1917. Because this suspension is relatively unstable, particular care must be used to keep the sand in suspension (Leonard and Hardinge 1991). The sand is separated from the coal and refuse by water sprays and screening. Although this process is not used in newer plants, several Chance Cone plants still exist today. Barite finely ground to facilitate suspension stability was formerly used in a few plants. Loess in a hydrocyclone separatory vessel has been used to separate coal from refuse. Today, finely ground magnetite is the medium of choice for coal cleaning. Its fine particle size gives good suspension stability, and its excellent magnetic properties greatly facilitate its removal from the separated products.

Ores, such as those of the base metals of copper, lead, and zinc, were first separated from their associated gangue minerals using a readily available galena ($\rho = 7.5$) flotation concentrate (–65 mesh)

TABLE 6.5 Solids used for heavy medium

Material Treated and Heavy Medium	Density, g/cm^3	Typical Mesh Size
Coal		~–65
Loess*	~2.6	35 × 100
Quartz	2.65	–200
Barite	4.5	–200
Magnetite	5.2	
Ore		
Ferrosilicon	6.8	–65
Galena	7.5	–65

*Usually windblown soil; density and size vary.

as the heavy medium. Galena has been largely supplanted by 15% Si-ferrosilicon (ρ = 6.8), easily made in an electric furnace. The cold-furnace product is ground to about –65 mesh, or, for higher gravity separations, the molten metal is steam shotted to produce spherical particles. It is strongly ferromagnetic. An idiosyncracy of magnetite and ferrosilicon is the variation in the size consist of these media, because ferrosilicon (either ground or shotted) contains only a small proportion of fines. For example, a nominal –65-mesh (–208-µm) ground magnetite may contain about half –325-mesh (–44-µm) material, whereas a ground ferrosilicon of the same nominal top size will contain only about a quarter –325-mesh material (Aplan 1985b). At –10 µm, ground magnetite may contain 20% or more fines, whereas ferrosilicon will contain only about 5% fines. The finer the size consist, the more stable the suspension.

A typical flowsheet for the treatment of coal or ore in a magnetic medium is given as Figure 6.5. The principal features of this process are

- Preparation of feed
- Separation in heavy medium
- Removal of medium from products
- Reclamation and recycle of medium

Preparation of Feed. Raw coal or ore feed is typically prepared by wet screening (Figure 6.5). The purpose is twofold: to prepare a feed size range that is compatible with the separatory vessel to be used and to remove fine particles that would otherwise contaminate the medium suspension and thereby lower its specific gravity and increase its viscosity.

Separation in a Heavy Medium. A variety of separatory vessels has been used for HMS: inverted cones and pyramids, Akins spiral classifiers, trough-type vessels (also called drag tanks) (Figure 6.6), rotating drums (Figure 6.7), and hydrocyclones. The choice of the vessel is related to the nature of the feed to be treated, the medium and its inherent stability at the suspension specific gravity to be used, and, to some extent, the wishes of the plant operators and design engineers. Generally, the coarsest fractions, such as 5½ in. (127–12.7 mm), are treated in a pseudostatic bath in a drag tank or in a drum vessel. The capacity of heavy media vessels that treat coarse materials is highly variable and depends on the vessel type and size, the nature and size of the feed, and the amount of float or sink products to be removed. Cone vessels can handle up to 300 tph, whereas drums and trough or drag tank vessels can treat tonnages as high as 700–800 tph. Very small tonnages can be treated in small, commercially available vessels.

The heavy media hydrocyclone usually treats particles in the size range 38 mm–0.5 mm (1½ in.–1⁄50 in.), although under certain conditions it can treat material as fine as 100 mesh (150 µm). Coal preparation plants, especially, will commonly have both a heavy media drag tank and hydrocyclone separating vessels. A heavy media hydrocyclone is similar in design to a classifying hydrocyclone, and it

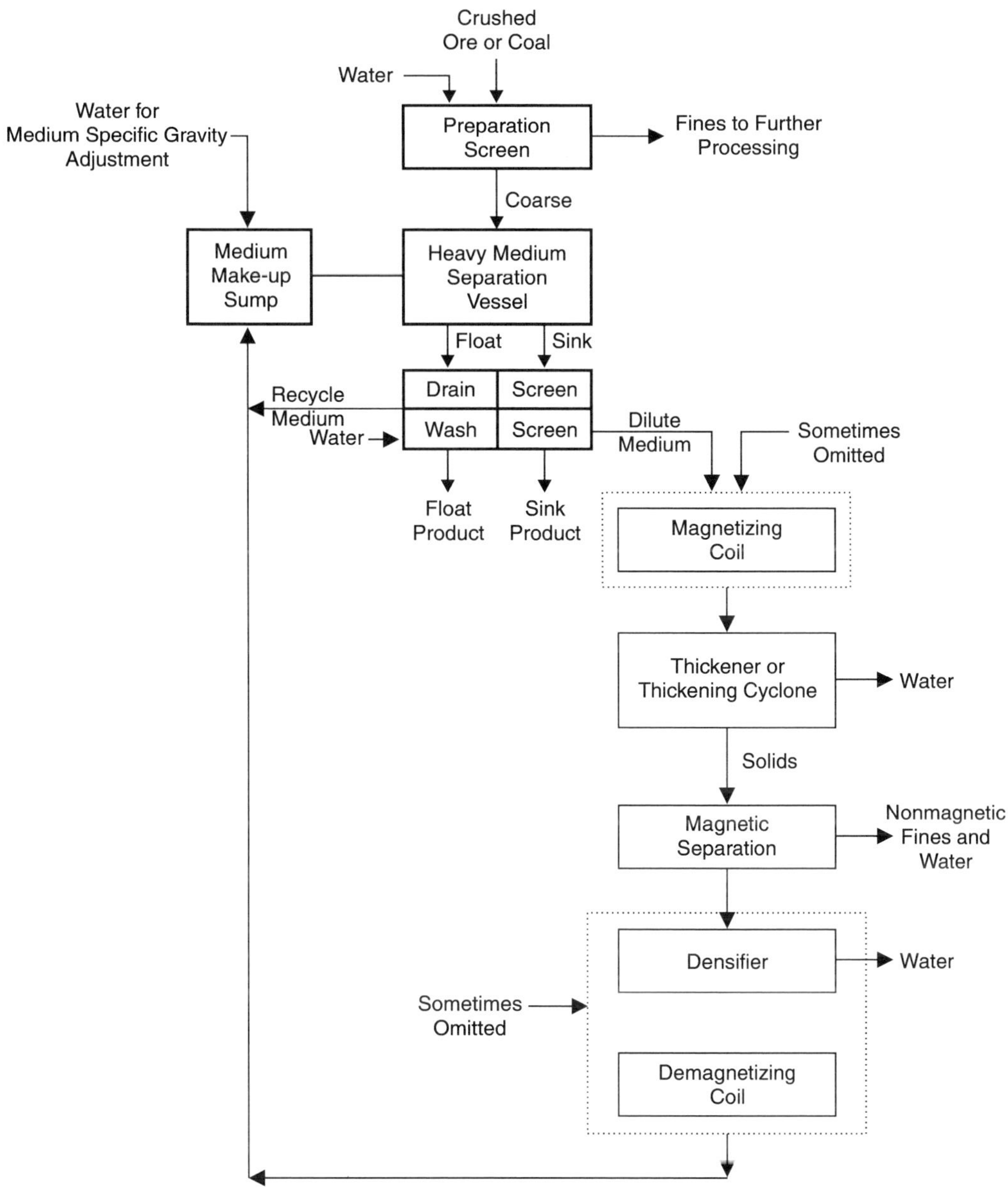

Source: Aplan 1985.

FIGURE 6.5 Typical heavy media flowsheet

can also process large volumes. A 26-in. (660-mm) diameter cyclone can accommodate upward of 2,000 gpm (about 7,600 L/min) while treating 135 mtph coal (Table 6.6). The heavy media cyclone is usually installed at an angle of about 20° from the horizontal. This angle allows the sink and the float products to be discharged at roughly the same elevation, and thus it allows the product drain and rinse screens to be installed at the same level in the plant. A similar device is the cylindrical Dynawhirlpool Separator (Miller, De Mull, and Matoney 1985). It is used much less frequently than the heavy media hydrocyclone.

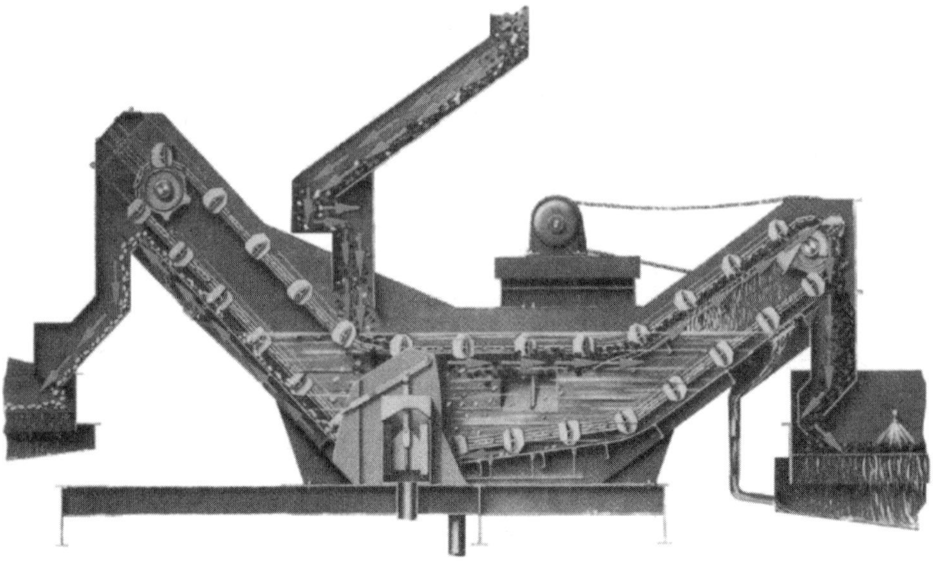

Source: Metso Minerals.

FIGURE 6.6 Trough- or drag tank-type heavy media separatory vessel; McNally Lo-Flo Vessel

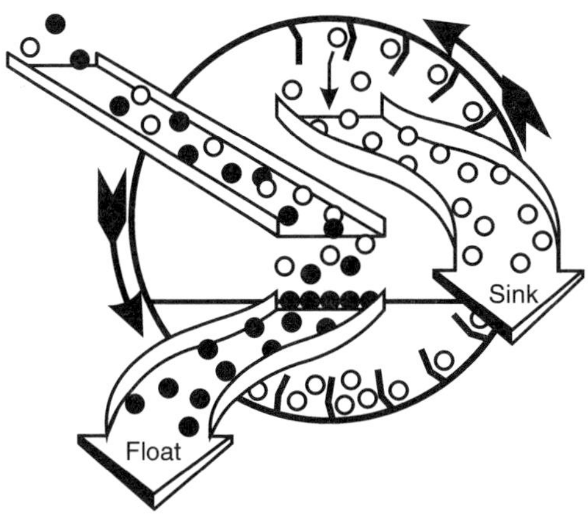

Source: Dorr-Oliver Eimco USA Inc. 2003. All Rights Reserved.

FIGURE 6.7 Schematic drawing of WEMCO drum-type heavy media separatory vessel

TABLE 6.6 Approximate capacities for heavy media hydrocyclones operating at 10–15 psi (69–103 kPa) and treating coal with a magnetite medium/coal ratio ~4:1

Cyclone Diameter		Top Feed Size,	Dry Feed,	Pulp Flow,
mm	in.	mm	mtph	L/m
380	15	9.4	40	2,100
500	20	38	75	4,200
660	26	51	135	7,600

Source: Hopwood 2000.

Removal of Medium from Products. The bulk of the medium is removed from the float and the sink products on separate wedge-wire vibrating screens of about 35 mesh. The first part of the screen, which has its own sump underneath, is called the drain section. Medium removed here is still at the separation specific gravity and is returned to the separatory vessel. The second section, which contains overhead sprays, removes any remaining medium still clinging to the particles. It is placed directly over a dilute-medium sump, and the medium is recovered from the bulk of the water and from undersize float–sink particles by magnetic separation.

Reclamation and Recycle of Medium. Usually two stages of wet magnetic drum separation are used to ensure nearly complete capture of the magnetite or ferrosilicon. As initially developed, the magnetic separators were preceded by a magnetizing coil to help flocculate the medium for easy settling and thickening. Currently, the magnetizing coil is commonly omitted. The thickener is also sometimes omitted, but it is more commonly retained not only to eliminate water but also to store medium during a plant shutdown or upset. A simple, annular alternating-current step coil surrounds the exit pipe that carries medium coming from magnetite separation. The coil facilitates deaggregation of the medium particles, as does shear during pumping. For ores, which require media of much higher specific gravity (2.6–3.85), a spiral classifier or similar device is used to thicken the medium after magnetic separation. For coal, which is cleaned in a much lower specific-gravity (about 1.5) suspension, this step is invariably omitted. Loss of medium during processing is generally in the range of 0.1–0.5 kg/t treated.

Suspension Rheology

Previously, attention has been called to the importance of hindered-settling and of agitation on the suspension of solids in a fluid. Another important factor is the suspension viscosity (or, for a non-Newtonian suspension, its apparent viscosity). For dense solids, such as heavy media, and for the pumping of ores and concentrates, attention to slurry viscosity is crucial.

The standard shear curve for Newtonian and non-Newtonian suspensions (Figure 6.8), reflects that in a Newtonian fluid (e.g., water or oil), the viscosity, μ, is the proportionality constant between the shearing stress (*F/A*, force per unit area) and the rate of shear (*dv/dx*) according to the formula

$$\frac{F}{A} = \mu\frac{dv}{dx} \qquad \textbf{(Eq. 6.7)}$$

However, in practice, a dilatant suspension, such as sand or ground ore particles in water, or a Bingham plastic, such as a relatively thick clay or slime-containing suspension, are more common. For the latter suspension, the formula for shearing stresses is

$$\frac{F}{A} = \eta\frac{dv}{dx} + \tau_y \qquad \textbf{(Eq. 6.8)}$$

where η is the plastic viscosity and τ_y is the yield stress, which is the theoretical intercept on the shearing stress axis.

Although the viscosity term, as defined in Eq. 6.7, holds for a Newtonian fluid, an infinite number of lines may be drawn from the origin to these non-Newtonian shear curves (Figure 6.8) for the dilatant suspension, Bingham plastic, or other non-Newtonian mineral suspensions commonly found in practice. To circumvent this problem, the term "apparent viscosity," μ_{app}, is defined as the slope of the line drawn from the origin to a point on the curve at some fixed rate of shear (see dashed lines in Figure 6.8). This term describes a reasonable way of distinguishing the relative viscosities of two different slurries at some constant shear rate. The μ_{app} values for several heavy media suspensions as a function of the pulp specific gravity are given in Figure 6.9 (Aplan 1985b). These suspensions are relatively usable for heavy media separations up to about the point at which the μ_{app} begins to rise rapidly. Thus, magnetite can be used up to about 2.5 (although lower for more finely ground material), ground ferrosilicon up to about 3.2 (about 3.4 for "run-in" particles whose sharp edges have been abraded),

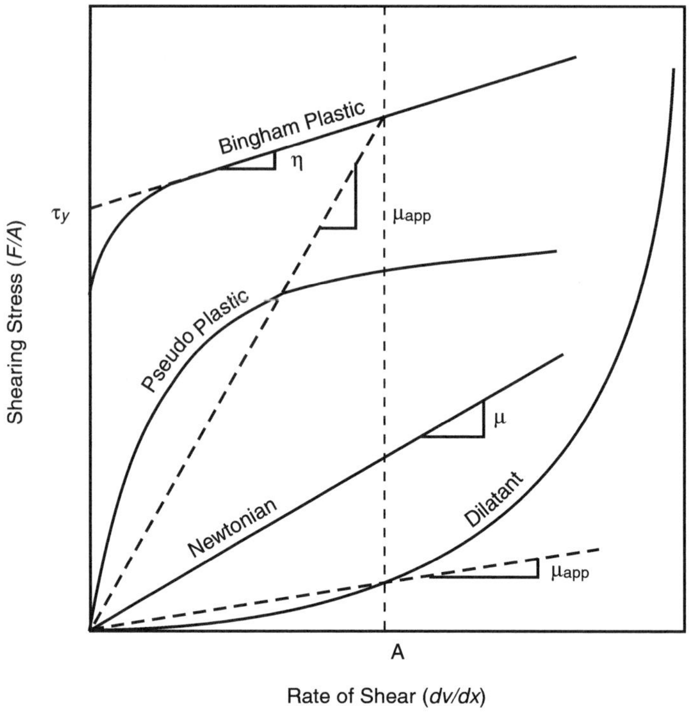

FIGURE 6.8 Shear diagram for various liquids and suspensoids. The apparent viscosity, μ_{app}, for non-Newtonian suspensions is the slope of the dashed line drawn from the origin to the curve at some rate of shear, such as at shear rate A.

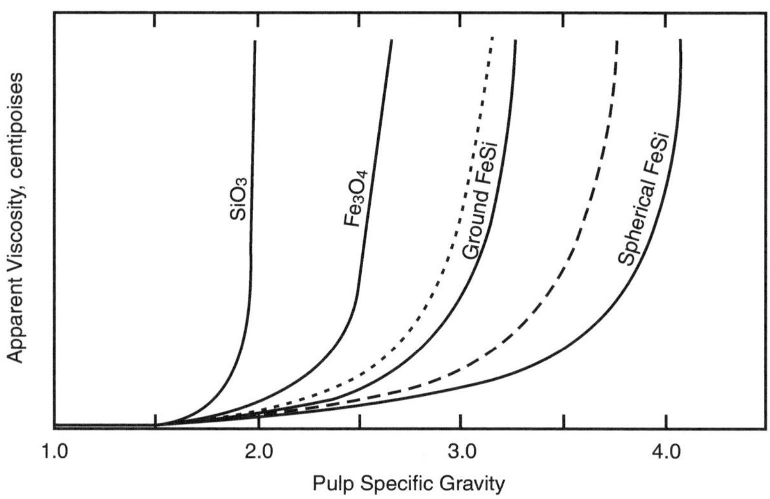

Source: Aplan, unpublished data; Aplan 1985b.

FIGURE 6.9 Apparent viscosities of suspensoids of various heavy media solids; all solids –65 mesh except quartz, 35–100 mesh. Dotted and dashed lines refer to the addition of 5% slimes to ground and spherical ferrosilicon, respectively.

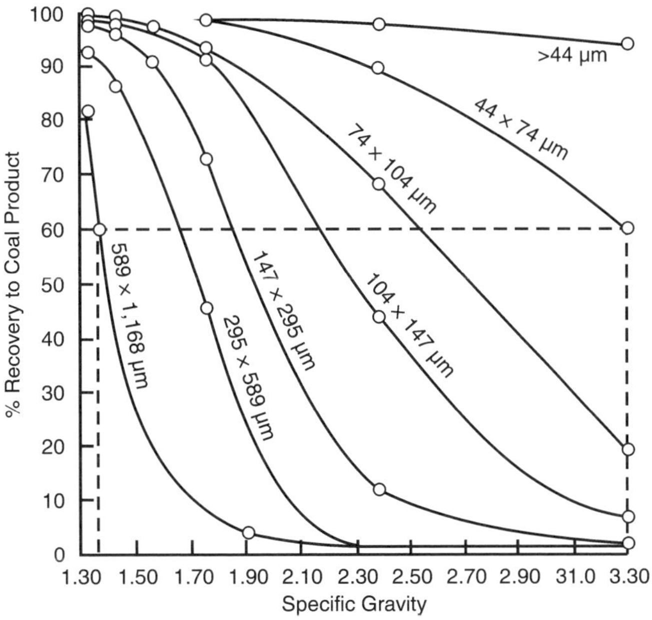

Source: Miller, De Mull, and Matoney 1985.

FIGURE 6.10 Recovery of various sizes of bituminous coal using a water-only cyclone. Note the recovery of 589- to 1,168-µm particles at 1.35 specific gravity is the same as the recovery of 44- to 74-µm particles at 3.3 specific gravity.

and spherical (shotted) ferrosilicon can be used up to about 3.85. When contaminated by very fine particles, the suspension will increase in apparent viscosity, μ_{app} (dashed lines in Figure 6.9). Excessive slurry viscosity can seriously diminish the effectiveness of, especially, a noncyclonic heavy media separator (Aplan 1985b), although a dispersant for the slime particles can mitigate the problem somewhat (Aplan 1980). The problem is most serious in a heavy media unit operated near the upper end of its practical specific gravity or capacity range. The high shear in a heavy media hydrocyclone makes this device less sensitive to viscosity than a vessel using a relatively static suspension.

Water-only Cyclone

Closely allied to a heavy media hydrocyclone using a loess medium is the water-only hydrocyclone that uses no external media. Its design differs from that of classifying or heavy media hydrocyclones. The water-only hydrocyclone is stubby, it uses a long vortex finder that typically extends downward to near the top of the conical section; and the conical section has a large included angle (≤120°). It is substantially less effective than a heavy media cyclone, but it is simple and inexpensive to operate. The separation is very sensitive to particle size (Figure 6.10) (Miller, De Mull, and Matoney 1985). It does a reasonably good job on 14 to 28 mesh particles (about 1.0–0.6 mm), a fair job down to 100 mesh (150 µm), and is generally ineffective below that. However, the separation specific gravity (d_{50}) becomes progressively higher as the particle size is reduced (Figure 6.10). This form of concentration works best for a closely sized feed and when a different-sized cyclone is used for each feed size (smaller cyclones for the smaller sizes of feed particles). It is usually not convenient to feed several different size fractions, so a compromise is usually reached between cyclone size, feed size, $d_{50,}$ and coal recovery. This compromise is typically weighted toward the coarsest particle sizes because they contain the greatest weight of particles and, usually, the highest potential product recovery. Feed solids of 10%–15%

TABLE 6.7 Approximate capacities of water-only cyclones operating at 15–20 psi (103–138 kPa) and treating coal at ~15% solids

Cyclone Diameter		Top Feed Size,	Finest Effective Size,	Feed Dry,	Pulp Flow,
mm	in.	mm	mesh	tph	L/m
250	10	0.6	100 × 150	5	900
380	15	6.0	65 × 100	18	2,700
500	20	12.7	48 × 65	30	3,500
660	26	19.0	35 × 48	60	7,700

Source: Hopwood 2000.

by weight can be used at operating pressures of 15–20 psi (103–138 kPa), and feed sizes up to ¾ in. (19 mm) can be accumulated. Table 6.7 lists typical capacities of various water-only cyclones used for cleaning coal.

JIGS

The history of jigging likely goes back to antiquity, and the phenomenon was undoubtedly known in Grecian times. As rich, easy-to-mine ores became exhausted and selective mining no longer produced a smelting-grade product, hand sorting became necessary. Very quickly humans learned that sizing and washing particles of ores such as silver, lead, copper, and tin greatly facilitated the sorting process. Soon thereafter it was likely learned that if a wicker basket containing particles to be sized and washed was jogged up and down in water, the heavy particles soon congregated at the bottom and the light particles at the top. This act of alternately fluidizing and collapsing a bed of particles to concentrate the denser mineral on the bottom is the essence of the jigging process. By the middle to late nineteenth century, coarse ore jigging was well developed, and by the first few decades of the twentieth century, jigs that recovered at least some the fine particles (~100 mesh) were developed.

The essence of a jig is captured in Figure 6.11, which shows a Harz Jig. The downward movement (called the pulsion stroke) of a plunger fluidizes the bed of particles on a sieve plate so that heavy particles move to the bottom and light particles to the top of the bed. As the plunger then moves upward, it creates a suction stroke that collapses the jig bed.

The capacity of jigs, usually given as tons per square meter of bed area per hour ($t/m^2/h$), is so variable as to be meaningless unless the jig type and conditions are closely specified. Capacity is a function of the jig type and size, the particle size and nature of the feed, the amount of the various products to be removed, and the desired quality and recovery of concentrate.

The Jigging Process

Gaudin (1939) defined the three principles of jigging as hindered-settling, differential acceleration, and consolidation trickling.

Hindered Settling. Hindered settling was described in detail earlier in this chapter, and the formula for the equal-hindered-settling ratio (Eq. 6.5) captures the essence of the process. In a 45% solids (by volume) suspension of quartz particles, a ratio of about 6 to 1 can be achieved between galena ($\rho = 7.5$) and quartz ($\rho = 2.65$) for the settling of large particles; less than half that is possible for small particles (Table 6.3). Perhaps a ratio as high as 30 to 1 could be achieved with a hindered-settling suspension under ideal conditions (Taggart 1951). However, if some of the particles are very fine, about 200 mesh (74 μm), and the exponent in Eq. 6.5 is only 0.5, a very much smaller ratio would pertain. Furthermore, in the commercial recovery of placer cassiterite by jigging, cassiterite ($\rho = 7.0$) as fine as 200 mesh (74 μm) is separated from 12.7-mm quartz ($\rho = 2.65$). The jigging size ratio for the extreme particle sizes is thus 172 to 1. A jig is obviously much more than a mechanized hindered-settling classifier.

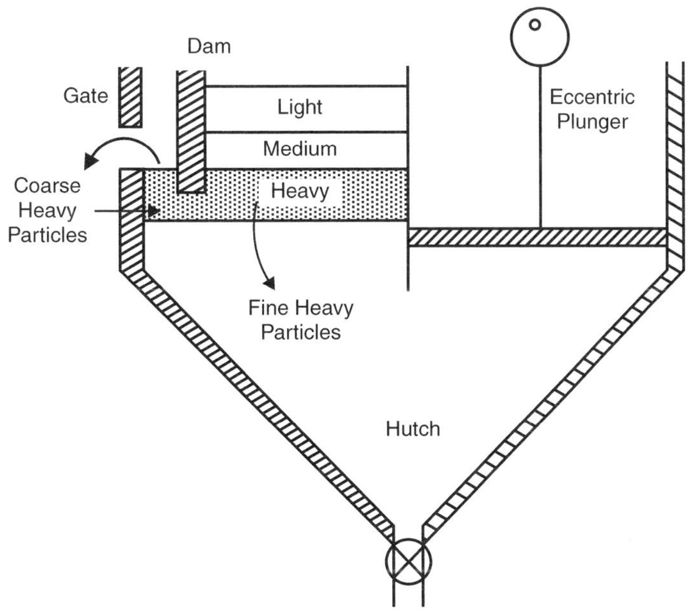

Source: Aplan 1980.

FIGURE 6.11 Schematic diagram of a Harz Jig

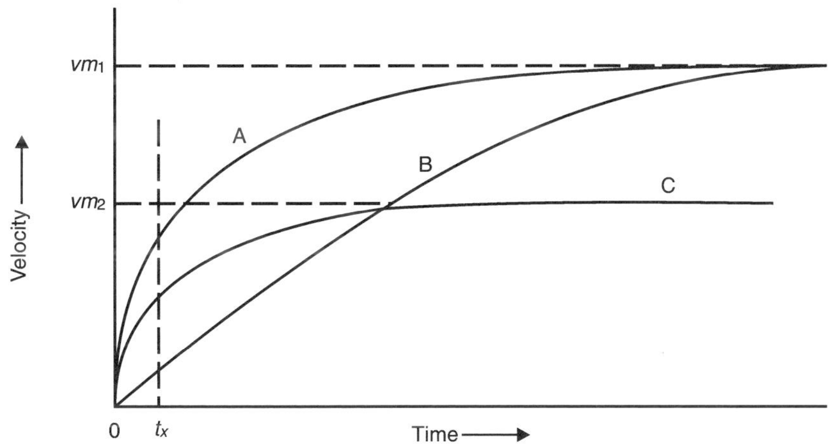

Source: Gaudin 1939.

FIGURE 6.12 Velocity–time relationships for various particles. A—a heavy particle, B—a larger light particle of the same terminal settling velocity (vm_1), and C—a smaller heavy particle of lower terminal settling velocity (vm_2).

Differential Acceleration. Particles that settle in a fluid accelerate until they reach their terminal velocities. Using the particle acceleration equations developed in chapter 5, Movement of Solids in Liquids, we can calculate that a 2-µm particle of quartz settling in water achieves half its terminal velocity (t_{50}) in 0.00003 s and 99.9% ($t_{99.9}$) in 0.0004 s; that is, for all practical purposes a small particle achieves its terminal velocity almost instantaneously. By way of contrast, a 2-cm spherical particle of quartz takes 0.1 s to achieve t_{50} and 0.65 s to achieve $t_{99.9}$. The acceleration period for coarse particles, in the time frame of seconds or major fractions thereof, is thus very significant. We can expect denser particles to achieve their terminal velocity more quickly. A large, low-density particle and a smaller, high-density particle can be selected so as to fall at the same terminal velocity (vm_1), but their acceleration profiles will differ markedly (Figure 6.12). Particle A, the denser and smaller particle, will approach its terminal velocity quickly, whereas the large, lighter particle B will accelerate more slowly. Some smaller high-density particle C will follow a pathway to a lower terminal settling velocity (vm_2). However, at time t_x both high-density particles (A and C) could be separated from the low-density particle (B). The time of fluidization in a jig can thus be controlled in such a way as to make maximum use of this differential in acceleration. Depending on the type, a jig will run at a speed of 60–400 strokes/min, well within the region of coarse particle acceleration. Both hindered-settling and differential acceleration occur during jig bed fluidization; that is, during the pulsion stroke.

Consolidation Trickling. Consolidation trickling occurs as a jig bed collapses. The movement of coarse particles has been interrupted, but fine particles can still move in the interstices of the just-collapsing bed of larger particles. The suction stroke will greatly encourage the downward movement of fine, heavy particles into the lower portion of the jig bed and also into the hutch region below the sieve plate. Higher-density particles are greatly favored in their downward movement because the preceding pulsion stroke tends to move the fine, light particles upward where they are removed from the bed by the horizontal flow of surface water.

Jig Types

Jigs were formerly classified as movable (where the bed of particles moved up and down in a tank of static water) and fixed sieve (where the bed is fixed and the water is moved). With but few important commercial exceptions (principally the Remer Jig), nearly all jigs in use today are of the fixed sieve type. The Remer Jig is composed of an oblong box with a sieve plate bottom that supports the bed of particles being jigged. It is surrounded by a rubber diaphragm that is also attached to the enclosed vat of water in which it sits. One of two longitudinal, eccentric shafts located below the jig compartment supplies the main jigging action; the second supplies a higher frequency action to keep the bed in motion.

Although jigs are not as efficient as HMS, they have much lower capital and operating costs that make them ideal for temporary use or to treat smaller tonnages of relatively coarse material. It is convenient to classify jigs as plunger, pulsator, or full-suction jigs.

Plunger Jigs. Plunger jigs are typified by variations of the Harz Jig (Figure 6.11) that provide both a pulsion and a suction stroke. Although many types of these jigs were once used to concentrate ores (Taggart 1945), they are used today only in special circumstances largely because today's ores, if coarse, may be more effectively concentrated by HMS. Or, if the ore mineral is disseminated, as is usually the case today, a full-suction jig or some other concentration device will commonly do a superior job. Harz-type jigs operate at 100–300 strokes/min and use a 0.4- to 10-cm stroke length. Today, such jigs are not commonly used, although one specific type, the Denver Mineral Jig, is used in special circumstances. This jig admits water during the upstroke, which lessens the suction somewhat but gives superior capacity.

Pulsion Jigs. Pulsion jigs were used extensively in the days before the advent of HMS. They resemble a small Harz Jig except that the plunger was replaced by a standpipe about 10-m high with a rotating valve that alternately admitted and shut off water. This action created a pulsion that cyclically fluidized the jig bed while the suction cycle was essentially nonexistent and greatly increased the capacity of the jig. The Richards Pulsator was reported to have a capacity as high as 65 t/m^2/h (Richards and Locke 1940) for relatively coarse sulfide ore. Although these examples illustrate the advantages of pulsion, the devices are essentially obsolete for treating ores because of the efficiency of HMS and the dearth of ores liberated at a relatively coarse size.

A closely allied jig used almost exclusively for cleaning coal, the Baum Jig (Figure 6.13A), also greatly emphasizes the pulsion stroke. It has a U-shaped tank, and pulsion is created by alternately admitting and exhausting compressed air in one leg of the U. This sequence pulses the water and fluidizes the

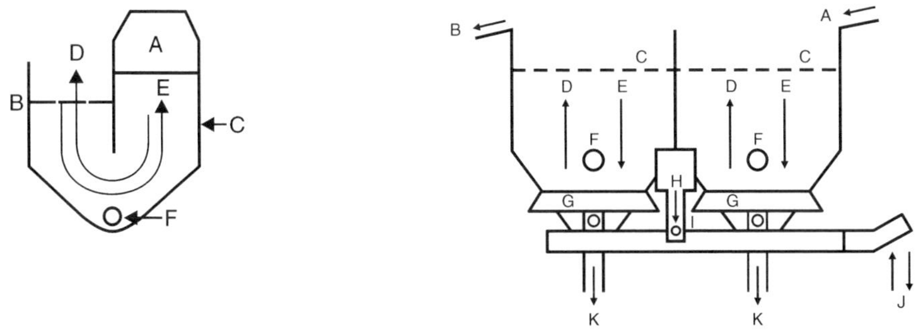

A–Compressed Air Source
B–Jig Screen
C–Hutch Compartment
D–Pulsion Phase
E–Suction Phase
F–Screw Conveyor

(A)

A–Feed End
B–Discharge End
C–Screen
D–Pulsion Phase
E–Suction Phase
F–Hutch Water Inlet
G–Diaphragm
H–Central Pivot
I–Rocker Arm
J–Eccentric Motion
K–Hutch Product Discharge Port

(B)

Source: Thomson, Laros, and Aplan 1985.

FIGURE 6.13 Schematic diagrams of (A) a Baum Jig and (B) a Pan-American Placer Jig

bed (Leonard and Hardinge 1991). Any trace of suction is represented by the absence of pulsion. The water is typically pulsed 60–80 pulses/min. Baum jigs can handle coal up to 130 mm, and at that size they have capacities of about 40 $t/m^2/h$. Jigs with capacities of 270–700 tph are not uncommon. Such jigs are not very efficient for concentrating particles below about 6 mm, but circumstances generally allow the device to work well in practice anyway. Although fines, low ash or not, tend to report to the clean coal product, the fine sizes in many raw coals are mostly particles of coal. Thus, the weight of noncoal fines reporting to the clean coal is typically small. Further, fine clays reporting to the coal product are easily removed when the coal is dewatered, as by screening. As a consequence, a Baum Jig can often be used on a broad feed size without adding much ash to the clean coal product. Next to HMS processes, jigging is the most widely used means of cleaning coal.

Full-suction Jigs. Full-suction jigs are ordinarily used in treating ores such as placer gravels for the recovery of fine gold, cassiterite, and diamonds. Rather than a separate pulsion compartment, they have a rubber diaphragm in the hutch area, which is located a short distance below the sieve that holds the jig bed (Figure 6.13B). In this way, the full-suction stroke is transmitted directly to the bed, which greatly facilitates consolidation trickling. The jig thus has both a full-pulsion and a full-suction stroke. The location of the diaphragm underneath the jig bed favors the use of these jigs on dredges, where space is at a premium, because they require only about half the floor space of a Harz-type jig. Typical jigs of this type are the Cleaveland, IHC-Holland, Pan-American Placer, Kuoss, and Yuba. When treating −13-mm placer gravel, they have a capacity of about 5 $t/m^2/h$. This lower capacity in comparison with some other jigs is related to the full-suction stroke and the small particle size of the valuable mineral to be recovered. Further, the valuable cassiterite contained in the −13 mm placer gravel to be jigged is typically much less than 10 mesh (1.7 mm) in size. Placer jigs do a good job of recovering particles greater than 200 mesh (74 µm), a less effective job from 200 to 400 mesh, and a poor job below 400 mesh (37 µm). The high specific gravity of gold makes its recovery somewhat better than that of other heavy minerals, unless the gold is in flakes. These jigs are typically used to treat lean ore; placer cassiterite ore will contain only about 0.04% cassiterite, and gold and diamond placer gravels will contain very much less.

Placer jigs are rarely used to make a final concentrate in one step. In the recovery of placer cassiterite, for example, rougher, cleaner, and recleaner jigs are used to produce a final heavy-mineral wet concentrate at an overall ratio of concentration of about 1,000:1. The cassiterite is then removed from the other heavy minerals in a "dry" plant using magnetic, electrostatic, and gravity methods.

Ragging. Where other jigs typically remove most of the heavy particles from the bed, usually by a downcomer and dam arrangement (see Figure 6.11), the placer jig recovers most of the valuable material through the jig sieve and into the hutch. Accordingly, to prevent much of the fine material from flowing directly into the hutch, ragging must be used. The ragging consists of about three layers of particles having a specific gravity similar to that of the heaviest particles being jigged. Hence, steel shot ($\rho = 7.8$) is used for many ores, hematite ($\rho = 5.0$) for cassiterite, and feldspar ($\rho = 2.56$) for raw coal. In many jigging operations, the heavy constituent in the ore being jigged will quickly accumulate during operation and serve as the ragging. However, when the material to be jigged lacks relatively coarse heavy minerals, artificial ragging is required.

Jig Cycles

The standard Harz-type jig cycle impressed on the system by the plunger takes a sinusoidal form. However, research by Thomson, Laros, and Aplan (1985), who used pressure transducers in the hutch just below the jig sieve plate, indicates that the actual situation is quite different. In the Denver, Pan-Am Placer, and Baum jigs, the wave form at the bed only remotely follows the sinusoidal form impressed on the system and does so only at relatively high speeds. Two pressure transducers were used in a Baum Jig, the front one below the jig bed and the rear one in the hutch water near the air cylinder. The rear pressure trace showed a strong pressure spike of very short duration when air was admitted to the jig. The front trace was greatly muted and showed only a small pressure increase during bed fluidization. In all of these studies, the closest approach to the sinusoidal form by a Harz, Denver Mineral, or a Pan-Am Placer jig occurred only at excessive speed. A bed could not be maintained, water flew from the jig bed, and the jigging process was out of control.

Just as a strong pulsion stroke favors the jigging of coarse particles, so should a strong suction stroke and consolidation trickling favor the recovery of fine, heavy particles. The use of a full-suction jig to recover fine particles illustrates the case. Another way to accomplish the same end is to use a sawtooth cycle, in which a short-duration pulsion stroke is followed by a prolonged suction phase. This cycle has been suggested in the literature, and studies by Laros and Aplan (1991) showed its effectiveness in recovering fine, heavy particles using either a Harz or a Baum feldspar jig. Greater details on jigs and jigging are available elsewhere (Taggart 1945; Burt and Mills 1984; Pickett and Riley 1985; Leonard and Hardinge 1991).

FLOWING FILM CONCENTRATORS, SLUICES, AND SHAKING TABLES

The processes described in this section are used to treat intermediate- and fine-size particles in the range of, roughly, ¼ in. (6.4 mm) to about 15 µm and include many of the intermediate-particle-size and nearly all of the fine-particle-size concentrating devices listed in Table 6.2. No one device is fully effective in concentrating this entire feed size range. Furthermore, it is rarely possible to simultaneously achieve a high recovery and a high concentrate grade, so the rougher concentrate is cleaned, often several times, on a similar or a different device to achieve an acceptable final concentrate grade.

Flowing Film Concentration

In a flowing film concentration process, a thin layer of a slurry of fine particles in water flows down a slight incline and is subsequently washed with a gentle flow of water. The particles on the incline will then distribute themselves in the sequence of fine, heavy particles highest upslope, coarser heavy particles and fine light particles in between, and coarser light particles farthest downslope. The extremely fine particles, the slimes, are lost to the water discharge.

This process dates from antiquity, certainly several centuries BCE. It is so elementary (an inclined flat rock probably was used initially) that it was likely developed independently at several locations around the globe. By the sixteenth century in Europe, Agricola in his treatise *De Re Metallica* (1556) described the use of a buddle, a form of flowing film concentrator in which the deposited heavy particles are repeatedly pushed back uphill with a hoe to facilitate the further removal of light particles by flowing water. Subsequently, similar devices—such as the strake—were developed. This device used materials such as canvas, hides, and blankets to achieve a slightly roughened surface that recovered particles not readily collected on a smooth surface. Flowing film devices, although largely made obsolete by the flotation processes (especially for sulfides), are still used today in certain situations: to treat (1) those minerals not effectively concentrated by the flotation process; (2) large tonnages of presized material containing only a few percent of a heavy mineral (typified by beach sands); or (3) minerals in primitive or very small-scale operations, such as those in Southeast Asia, where the lanchut is used to remove impurities from cassiterite concentrate.

Because flowing film concentrators have historically been used to treat fines and slimes, it is necessary to define these terms. Unfortunately, the terms are moving targets whose definitions have changed with the material being treated, the devices available, and time. Fines, before the advent of flotation, were often defined as −10 mesh (1.7 mm), and slimes were those particles below about 150 mesh (100 μm). Today, fines are considered to be particles below either −10 (165 μm) or −100 mesh (150 μm), and slimes are often defined as about −15 μm, although these definitions may vary depending on the mineral assemblage being treated. Flowing film concentrators, like all gravity concentration devices, cause minerals to be recovered or rejected on a particle-by-particle basis. The major defect of flowing film concentrators is that for fine particles, innumerable decisions must be made to recover even a small weight of concentrate. Because the layer of particles in the flowing film is only one to a few particles thick, the goal of equipment designers has been to design a device with a large flowing film surface area that occupies a small floor area.

Flowing Film Concentration Principles

The thin film of water that flows down the incline shows a vertical, half parabolic-like flow pattern ranging from near zero at the surface of the deck to a maximum near the top surface of the flowing film. The size of the particles to be treated will influence the depth of film required—the coarser the particle, the thicker the flowing film. The push of the fluid will obviously be greater on the larger particles in the fluid film, assuming that they are totally incorporated into the flowing film. The fluid velocity v', at any distance in the film from the top surface, may be calculated (Gaudin 1939; Michell 1985) by

$$v' = \frac{\rho' g \sin \sigma}{2\mu}(2\theta - z)y \quad \textbf{(Eq. 6.9)}$$

where

ρ' = the fluid density
g = the acceleration caused by gravity
σ = the angle of the incline from the horizontal
μ = the fluid viscosity
θ = the film thickness

The feed is invariably in the form of a wet slurry, but penetration through the flowing film will be a function of the size, shape, and density of the particles; the pulp density and viscosity; and the thickness and velocity of the film. Based on Stokes' law (Eq. 6.2) and Eq. 6.10, the distance traveled by a spherical particle from the top to the bottom of the film, z, is (Michell 1985)

$$z = \frac{18\mu Q}{2(\rho - \rho')d^2 g \cos \sigma} \quad \textbf{(Eq. 6.10)}$$

where Q is the flow rate per unit time and width.

Gaudin (1939) summarized the behavior of particles to be deposited at the bottom of the flowing film as influenced by

- Specific gravity of the particles
- Shape of the particles
- Effective coefficient of friction between particle and deck
- Deck roughness
- Deck slope
- Fluid film thickness (which is influenced by rate of flow)
- Viscosity of the fluid

Flowing Film Concentration Devices

Many flowing film concentration devices have been developed to concentrate fine particles (Table 6.2). Excellent summaries have been given by Burt and Mills (1984) and Michell (1985). A brief discussion of some of the more important ones follows.

Tilting Frames. Tilting frames were derived from an elementary flowing film device. The frame is in inclined plane, and in the modern version, such as the Denver–Buckman Tilting Concentrator, a half-dozen or so decks at 5°–10° from the horizontal are stacked, with space between, in a frame assembly. The whole device is mechanized. The feed slurry flows over each surface for a fixed time, the feed is then stopped, and the frames are tilted to a steeply reverse slope and washed with water to remove the concentrate. The frames are then tilted forward again and the feed is reintroduced. Feed size is typically −65 mesh or so, and a material, such as corduroy, fiber matting, or indented rubber sheeting, is used to create a slightly rough surface. The recovery of heavy particles below about 15 μm is low.

A recent variant of the tilting frame is the Bartles–Mozley Concentrator. It consists of two assemblages of 20 fiberglass decks, each with a surface area of about 1.8 m^2 suspended in frame at an angle of about 2° from the horizontal. An unbalanced weight imparts an orbital shear, allowing the heavy particles in the flowing film to settle while the suspended light particles pass to the tailings. The frame is then briefly tilted to 45° and wash water is used to remove the concentrate. Treating −40-μm feed, this apparatus effectively removes particles down to about 10 μm and less efficiently down to about 5 μm. It has a capacity of about 1 t/m^2/day and is usually used as a roughing device.

Vanners. A vanner is a continuous moving belt separator with a gentle downward slope from feed end to tailings discharge. The fine, heavy minerals deposited on the belt are conveyed over the head end pulley by the slow up-slope movement of the belt; the light minerals are carried to the tail end by the forward flow of pulp and wash water. One of the more popular variations, the Frue Vanner, used a side shake to facilitate particle separation. The device has been obsolete for decades. More recently, the vanner has been reintroduced as the Bartles Crossbelt Concentrator. Like the Bartles–Mozley Concentrator, it uses an orbital motion. The feed is introduced over the upper half of a slight central, longitudinal ridge, and light particles are discharged over the two longitudinal edges. A slow, forward-running belt transfers the heavy particles to a washing zone and then to discharge. The device recovers particles in the same size range as the Bartles–Mozley Separator belt but is used as a cleaning device, whereas the Bartles–Mozley Concentrator is used primarily as a roughing device. This crossbelt separator has a capacity of about 1.5 t/m^2/day.

Spiral Separators. The Humphreys Spiral was the first spiral separator developed. Floor space is conserved by hanging a spiral trough (Figure 6.14) around a central post. The original Humphreys Spiral was made of cast iron, was about 60 cm in diameter, and had five turns. More modern versions, made of 2- to 3-m diameter fiberglass spirals, were developed for the Australian beach sand industry. Further, a second or third spiral wrap just below the turns of the first spiral (called multiple starts) can double or triple the capacity per floor area. The device has been used for both roughing and cleaning. The typical feed size of ores is about 10–200 mesh (1.65 mm–74 μm), but for coal about 20–65 mesh (833–208 μm) is preferred. Typically, wash water is supplied from an inner spiral to wash the heavy concentrates, which

Source: Roche Mining (MT).

FIGURE 6.14 Photo of a spiral concentrator treating beach sand. The heavy dark minerals are represented by the dark inside zone, whereas the lighter sand particles are shown as the white outside zone.

are removed through a series of ports on the inner spiral surface. The older, smaller cast-iron spirals had a capacity of only about 3 tph of 20–200 mesh (833–74 μm) feed, but large modern spirals have capacities 10 times as great per spiral start.

Another flowing film device, also largely improved in Australia, is the pinched sluice (Figure 6.15). It is another gently sloping flowing film device that is pinched in plan view. The wide feed end facilitates low-velocity deposition of fine, heavy particles while lighter particles, still in the fluid, are accelerated toward the pinched end. A small transverse slot near the discharge allows the heavier minerals clinging to the sluice surface to exit, while the light particles in the bulk of the water cross the slot and are discharged as tailings. Many units—both in parallel and in series—are used, and some other device is typically used for final cleaning. One popular device of this type is the Reichert Cone System. It consists of a series of about 2-m or 3.5-m inverted, gently sloping conical surfaces, fed from the center. This conical surface may be alternated with pinched sluices just below and flowing inward (Figure 6.16). Individual concentrating units are stacked in a vertical assembly as many as a dozen high.

Another variation is the Wright Impact Tray. This device is similar to the pinched sluices except that near the discharge, the pulp strikes a transverse plate causing the heavy-mineral-rich and the light-mineral-rich fractions to split hydrodynamically on impact. The plate angle is set so that the lower stream contains most of the heavies (although still in dilute form), whereas the upper part contains mostly barren sand. Multiple stages are required to achieve an effective separation.

Sluices

A sluice is an inclined trough that typically has a series of transverse ridges, called riffles, placed in the bottom. Behind these riffles a turbulence or boil is created, which, in effect, is a hindered-settling zone that helps to concentrate heavy minerals behind the riffles. The device has been known since antiquity and has been in continual use since. Today, sluices are used frequently in small gold and cassiterite concentration operations. In Southeast Asia, they are called palongs and are common in small

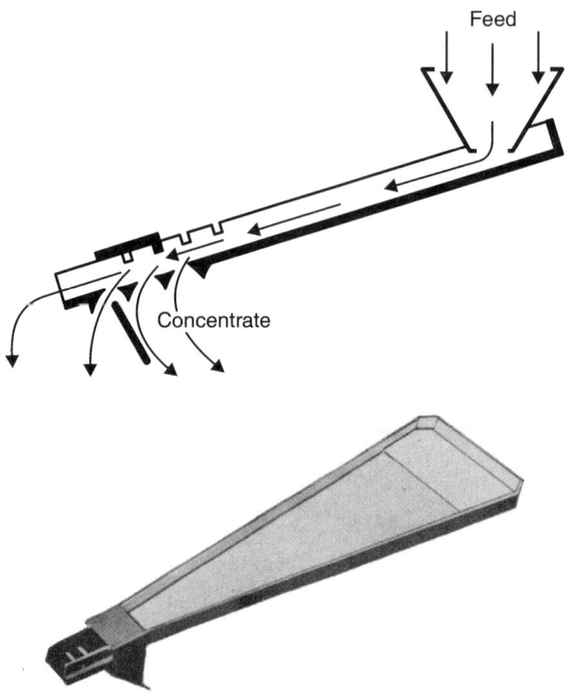

Source: Roche Mining (MT).

FIGURE 6.15 Pinched sluice

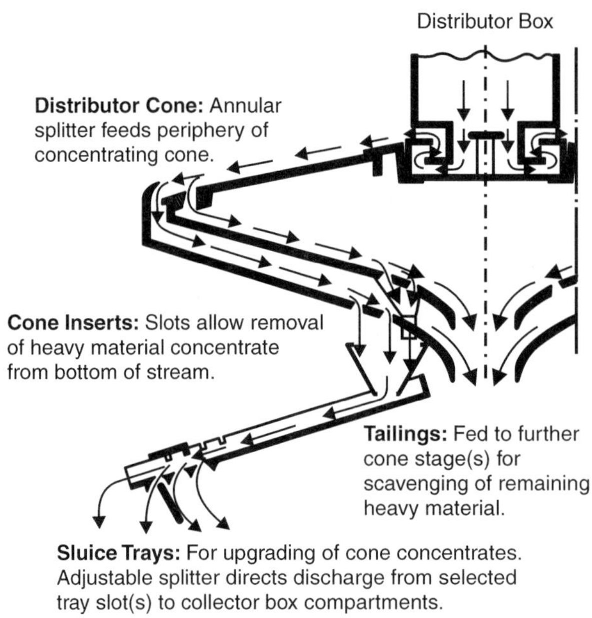

Source: Roche Mining (MT).

FIGURE 6.16 Schematic cross section of a Reichert Cone System

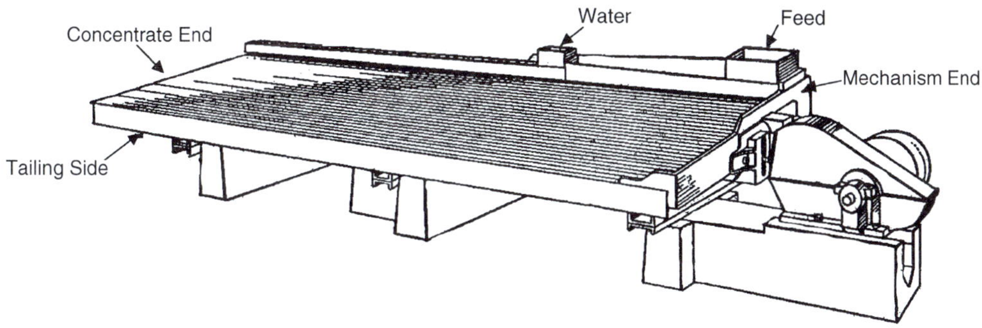

Source: Taggart 1951.

FIGURE 6.17 Wilfley Shaking Table

cassiterite operations. Periodically, the sluice is shut down, and the heavy concentrate behind the riffles is upgraded by use of the gold pan (miners' or prospectors' pan). This circular pan is 30–45 cm in diameter and has sloping sides truncated by a flat bottom. Material (water and sluice concentrate) placed in the pan is given both a swirling and a horizontal bumping action. Initially, a hindered-settling zone separates the heavy minerals from the light minerals, and the latter are largely spilled out of the pan. Final separation is made by a gentle, swirling, flowing film concentration action.

If the ore lacks nuggets of gold or other coarse heavy minerals, the riffles in a sluice may be small or nonexistent. Rough surfaces, such as hides, sod, and blankets, were often used to create miniature riffles. In a common version used today, AstroTurf is placed in the bottom of the sluice covered by a layer of expanded metal. The latter prevents coarser gangue particles from scouring the deposited gold-enriched upper layer and assists in recovering gold particles by acting as a small riffle or a catch basin. The AstroTurf is then removed periodically and washed vigorously to yield its gold.

Shaking Table

The most popular of several similar devices, the Wilfley Shaking Table was developed in the 1890s in Kokomo, Colo. It has been in use since and is a common device for concentrating particles in the intermediate range, such as 10–200 mesh (1.65 mm–74 µm) particles for ore and 3–100 mesh (6.7 mm–150 µm) for coal. It is an oblong, shaken deck, typically 1.8- to 4.5-m wide; the deck is partially covered with riffles that taper from right to left as in Figure 6.17. The deck is gently sloped downward in the transverse direction. Feed enters at the upper right and flows over the riffled area, which is continually washed from a water trough along the upper edge of the deck. Heavy particles are concentrated behind the riffles and are transported by a bumping action (of 12–25 mm throw at 200–300 strokes/min) to the left end of the table where flowing film concentration takes place.

Principles of Operation. Gaudin (1939) identified three principles of operation: hindered-settling, asymmetrical acceleration, and flowing film concentration. The hindered-settling action takes place in the boil behind the riffle. Asymmetrical acceleration, from a spring and from the bumping action supplied by a pitman and toggle arrangement (not unlike that in a Blake-type jaw crusher), not only transports the material behind the riffles but also helps to separate heavy from light materials. Heavy minerals are influenced less by the bumping action than are light ones, and thus the heavier particles have much longer residence times on the deck than do light ones. The bumping action also keeps particles in motion and allows the wash water to remove light particles more thoroughly. Final particle separation is made on the flowing film part of the deck, which produces a superior heavy mineral concentrate.

The final slope sequence is fine-to-intermediate heavy particles highest upslope, fine light and intermediate-to-coarse heavy particles in between, and coarse light particles furthest downslope. This sequence differs from that of a hindered-settling classifier. Accordingly, it is common practice to have separate shaking tables, each with different settings, treat the various spigot products from a hindered-settling (sorting) classifier.

Industrial Applications. Whether shaking tables or some other intermediate-to-fine gravity concentration devices are used for roughing, shaking tables are commonly used for cleaning to produce an acceptable concentrate. They are frequently used for upgrading heavy minerals that are not well floated, such as −¼-in. (6.4-mm) coal particles (as coarse as −10 mm in some instances), small middling streams, and heavy particles to be removed during environmental remediation. The latter is typified by such procedures as removing metal splatter from foundry sands and separating metal grindings from abrasives. In treating relatively coarse −8-mesh (2.4-mm) ore particles, the tables can handle several tons per hour, but for much finer, 150- to 400-mesh (100- to 37-µm) particles, their capacity may drop to about 0.3 tph. If −10-mm coal is treated, a capacity of 12 tph per deck may be achieved. To clean coal, shaking tables are often stacked two or three decks high, all controlled by the same shaking mechanism. For further information on shaking tables, consult the literature (Gaudin 1939; Taggart 1945; Mills 1978; Burt and Mills 1984; Deurbrouck and Agey 1985; Leonard and Hardinge 1991).

CENTRIFUGAL DEVICES

Stokes' law (Eq. 6.2) demonstrates that as a particle becomes finer, its settling time increases as the square of the particle diameter. Furthermore, Klima and Luckie (1989) have shown that for small particles, even as coarse as 500 µm, the quality of the fractional recovery curve is adversely affected unless long settling times (several minutes) are used. The use of centrifugal force substantially decreases particle settling time. Centrifugal devices can improve overall and fine-particle recovery, increase throughput, reduce water usage, and perform the work of other intermediate- and fine-particle concentrating devices.

A relatively new centrifugal device is the Falcon Concentrator. It is a smooth-surface, truncated cone with a superimposed cylindrical upper annular riffled section that is rotated at high speed. A relatively thick feed (up to 45% solids) is introduced into the bottom, moves up the bowl in a thin layer, and is accelerated at up to 300 times gravity. Feed size ranges from 45 µm to 6.25 mm, and capacities of 0.1–200 tph can be achieved depending on feed size and density and the size of the machine. Recovery of gold particles as fine as 10 µm has been reported (Falcon 1999).

The Knelson Concentrator is a similar device except that the rotated, truncated conical bowl bears a series of ring riffles, and, in addition, water is forced through perforations in the bowl. The added water both fluidizes the bed of particles and serves as wash water. The heavy minerals are collected behind the ring-like riffles, whereas the light particles overflow the bowl. Several hundred of these devices have been sold worldwide, and they are widely used for the recovery of gold. The Kelsey Jig and the Mozley Multi-Gravity Separator are other devices that use centrifugal force.

PNEUMATIC DEVICES

The use of air to separate materials of differing density has long been known and is typified by the winnowing of grain using an air current to remove the chaff. Over the years various hindered-settling classifiers, dry pans, and rockers were developed for use in arid regions (Taggart 1945). Dry concentrators make use of particle density, size, and shape, but in some cases the bulk or apparent density is the major criterion of merit, as in separating exfoliated vermiculite from gangue. Although pneumatic devices have a decided value in arid regions or where water cannot be tolerated in feed, product, or process, they suffer from two major impediments: a poor equal setting ratio and the difficulty and expense of dust containment.

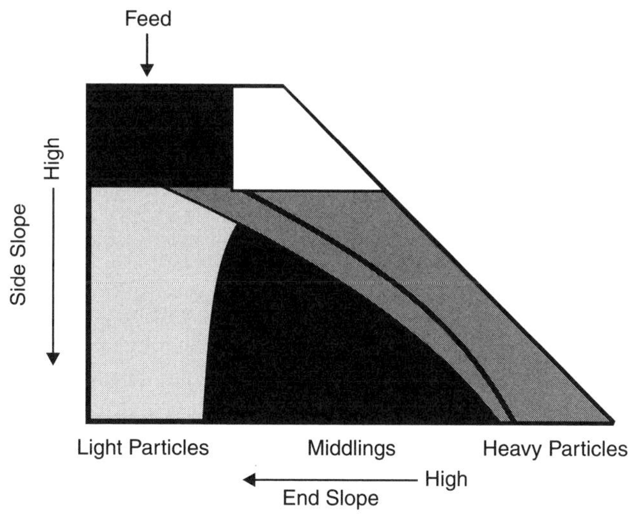

Source: Jarman 1985.

FIGURE 6.18 Schematic diagram of an air table

Application of Eqs. 6.3 through 6.6 in which the density of water (= 1.0) is replaced by that of air (= 0.001) shows that using air will lead to lower ratios of free or hindered-settling. Air devices are thus inherently inferior to their water counterparts, except where water cannot be tolerated or is not available. Dust containment in a large-tonnage, low-cost processing plant is always difficult. Further, dust is commonly controlled with a wet scrubber, which, in turn, transfers an air-pollution-control problem to a water-pollution-control problem that must also be solved.

One of the largest users of pneumatic processes was the coal industry. Here an air flow jig, such as the Stump Air Flow Jig, was used. This jig was a sloped device in which pulsating air was admitted through a porous oscillating deck to fluidize and stratify the coal and refuse. The device could treat feed particles up to about 50 mm. Although 15% of the bituminous coal cleaned in the United States was treated by this process in 1940 (Arnold, Hervol, and Leonard 1991), few if any air flow jigs survive today. Their demise was likely hastened by the difficulty in satisfying today's dust-control standards. Further, these separations were much inferior to those of most common wet methods (they had higher probable errors, E_p).

Another formerly popular pneumatic device was the air-aspirated screen used to remove asbestos from gangue. The process went into eclipse after the asbestos market collapsed for environmental reasons.

One pneumatic concentration device, the Sutton, Steele, and Steele (Triple S) Air Table is still in use today (although sparingly). It is, however, more properly called a jig. It is essentially a porous vibrated surface that slopes in both the forward and cross directions (Figure 6.18). Air admitted below the porous surface fluidizes the particles, and concentrate, middlings, and tailings are produced. To achieve a good separation, close sizing of the feed is necessary. Feed sizing by $\sqrt{2}$ sizes is common and sometimes $\sqrt[4]{2}$ sizing is employed. A common mineral-separation use of the device is the dry processing of dredged ilmenite, rutile, or cassiterite. The heavy-mineral gravity concentrates from a dredge may contain a dozen or more mineral species, a few of them in large amounts and many of them in small to trace amounts. The species are largely separated from one another by magnetic and electrostatic means, but small quantities of middling particles often remain to be separated by air tabling. Dust is usually not a great problem because the extreme fines were previously excluded during the wet concentration process used on the dredge or elsewhere.

Air tables have also been used to eliminate a host of small problems in the food industry and in applications such as separating abrasive grains in the cleaning of foundry sand and removing metal from crushed slag. Another air device, the zig-zag classifier, is used to separate paper and other light materials from heavier glass, stone, and metals during the recovery of valuable constituents from municipal solid waste.

Further information about these devices is given by Taggart (1945); Burt and Mills (1984); Jarman (1985); and Arnold, Hervol, and Leonard (1991).

PROCESS SELECTION AND EVALUATION

The essence of the selection or evaluation of any process is to be able to quantify the results expected or achieved. Techniques that allow this quantification are invaluable to the process engineer.

Preliminary Evaluation

A preliminary evaluation must precede the selection of an appropriate gravity concentration technique, or of any beneficiation method for that matter. This evaluation starts with a crude identification of the ore and gangue minerals, or, for coal, the coal macerals and refuse minerals. These tentative mineral identifications are followed by a much more detailed evaluation technique called quantitative mineragraphy. This study requires careful sampling, size separation of the comminuted product, specific-gravity fractionation of selected products, a careful petrographic study using reflected- or transmitted-light microscopy, and the use of other identification instruments and techniques as needed. Studying the fractionated particles in a polished section with reflected light is invaluable for determining the mineral type, purity, texture, and association, and for estimating the size of grind needed to liberate the desired species.

Heavy liquids or heavy solutions are used for specific-gravity fractionation. For coal and low-density ore minerals ($\rho < 2$), a series of plentiful and low-cost halogenated hydrocarbon liquids, or heavy solutions of calcium or zinc chlorides, make the task relatively simple. Intermediate-density ores (one or more species of $\rho < 3.3$) may require much more costly halogenated hydrocarbon liquids, whose cost generally restricts the separations to small samples. Higher-density materials may require a variety of gravity-, magnetic-, or conductivity-based laboratory devices, or handpicking visually or under a low-power microscope. The procedures to be used in this evaluation are well detailed in the literature (Aplan 1973; Mills 1978, 1980; Burt and Mills 1984; Osborne 1988; Leonard and Hardinge 1991; and the *Process Mineralogy* volumes published by The Minerals, Metals, and Materials Society and the Society for Mining, Metallurgy, and Exploration).

For coal, relatively large samples (several kilograms) of sized material are selected for a washability test. The size range is selected based on the capability of the various devices the process engineer feels may be appropriate for the separation in question. For determining the washability of coal (and certain other light minerals), the procedure of Coe (1938) is appropriate (Figure 6.19). This figure shows the weight percent to either the float or the sink product at any specific gravity of separation, as well as the elementary (incremental) and cumulative ash at any specific gravity of separation. One curve shows the weight percent of particles within ±0.1 specific-gravity units of the separation specific gravity. Similar curves may be constructed that use sulfur or pyritic sulfur as the criterion of merit, rather than ash.

The ±0.1 specific gravity indicates the difficulty of separation. Obviously, if only a few particles are near the gravity of separation, it is a simple separation to make, but an efficient separation becomes increasingly difficult as the percentage of near-separation-gravity particles increases. Table 6.8, based on Coe (1938), Zimmerman (1950), and Mills (1980), compares the difficulty of separation and methods that can be used to treat particles with various percentages of near-gravity material.

Another method of evaluating the effectiveness of gravity separation is to use the equal setting ratio (Eq. 6.3) and Table 6.9 (Arbiter 1955). The special circumstances in which certain devices may be used to recover particles as fine as 10–20 μm are discussed in the section on flowing film devices.

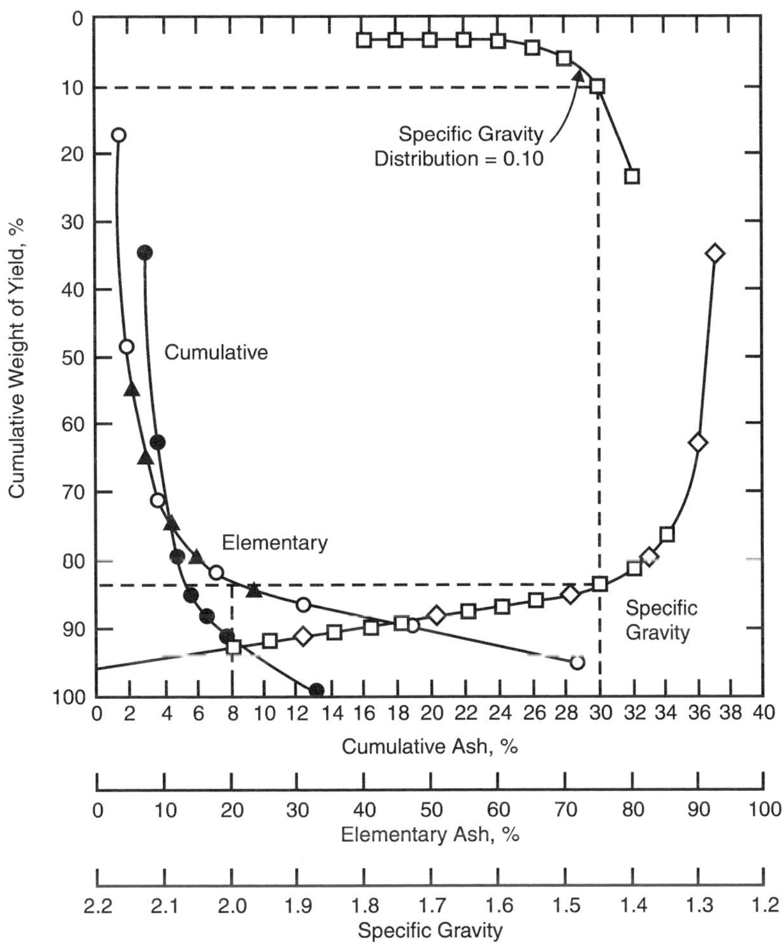

Source: Coe 1938.

FIGURE 6.19 Washability curves for coal. Dashed lines illustrate the results obtained with a bituminous coal in which a 1.4 specific-gravity separation produces an 83.5% coal yield at 8% ash and 10% near-gravity (±0.1 specific gravity) material

TABLE 6.8 Influence of near gravity material on the difficulty of separation and on process selection

Wt% ± 0.1 Specific Gravity	Degree of Difficulty	Suggested Gravity Process	Suggested Devices
0–7	Simple	Almost any	Jigs, HMS, tables, spirals, sluices, vanners
7–14	Moderately difficult	Efficient process	
10–15	Difficult	Efficient process, good operation	
15–20	Very difficult	Very efficient process, expert operation	Heavy media separation
20–25	Exceedingly difficult		
>25	Formidable	Exceptionally efficient process, expert operation	Heavy media separation, close control

Source: Modified from Coe 1938; Zimmerman 1950; and Mills 1980.

TABLE 6.9 Gravity separation effectiveness based on the equal setting ratio (the concentration criterion)

Equal Setting Ratio in Water	Separation Effectiveness
>2.5	Down to ~200 mesh*
2.7–1.75	Down to ~100 mesh
1.75–1.50	Possible to ~10 mesh, but difficult
1.50–1.25	Possible to 6.4 mm, but difficult
<1.25	HMS, or another process (e.g., flotation)

Source: Modified from Arbiter 1955.
*Can be extended down to 10–20 μm in special circumstances.

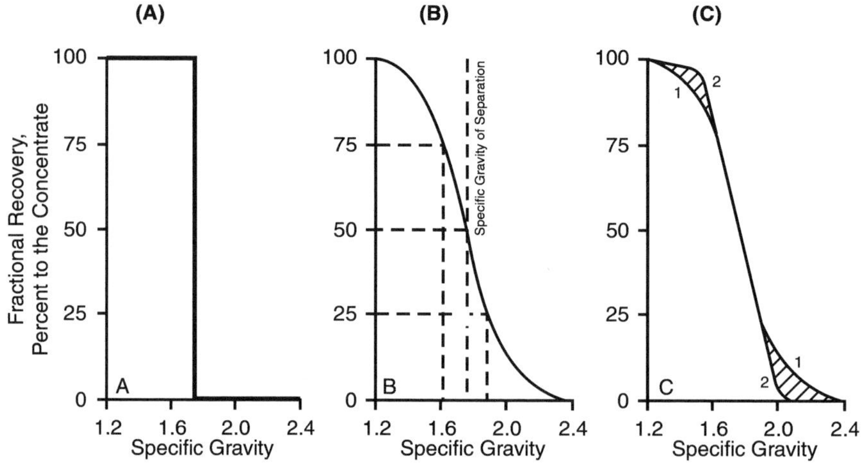

Source: Aplan 1989.

FIGURE 6.20 Partition curves: (A) perfect separation, (B) actual separation (curve 1), and (C) same E_p as for (B) but with superior recovery of misplaced particles (shaded area between curves 1 and 2)

The washability techniques so effective for coal are partially or wholly inadequate for most ores, largely because of the lack of appropriate heavy liquids. In these cases, test runs on laboratory- or pilot-sized equipment (such as jigs, shaking tables, spirals, and pinched sluices) must be employed.

Process Evaluation

In addition to the standard techniques for evaluating process effectiveness, such as concentrate grade and recovery, ratio of concentration, and the metallurgical balance, the partition (or fractional recovery) curve (Figure 6.20) is often used as a measure of the effectiveness of gravity concentration. The partition curve is used to evaluate coal cleaning effectiveness.

The fractional recovery to the concentrate (or to the float in the case of coal) is plotted at the midpoint of the specific-gravity interval used (Figure 6.20). A perfect separation (Figure 6.20A) would produce a vertical line at the specific gravity of separation (d_{50}), in this case, $\rho = 1.75$. In practice, such sharp separations are impossible, and a curve such as that in Figure 6.20B is typical. For convenience, the 50%, 75%, and 25% weight recovery values defined as d_{50}, d_{75}, and d_{25} are used. From these values, the probable error, E_p, is defined as follows:

TABLE 6.10 Approximate E_p values for coal cleaning devices at 1.5 specific gravity

Coal Size and Appropriate Cleaning Device	E_p at Stated Size				Broad Size Range of Feed	
	+½ in. (+12.7 mm)	½ × ¼ in. (12.7 × 6.4 mm)	14 × 28 M* (1.7 × 0.6 mm)	20 × 200 M (830 × 74 mm)	Range	E_p
Coarse coal						
Baum Jig	0.06	0.16	0.30	—	6 in. × 48 M	0.12
HMS, static bath	0.03	—	—	—	6 in. × ¼ in.	0.03
Intermediate to fine coal						
HMS, cyclone	0.02	0.03	0.05	—	¾ in. × 28 M	0.03
Shaking tables	—	0.07†	0.10	0.20	⅜ in. × 200 M	0.09
Water-only cyclone	—	0.15†	0.20	—	¼ in. × 200 M	0.28

Source: Gottfried and Jacobson 1977, modified by Aplan 1989.
*All mesh sizes (in.) in Tyler Standard mesh.
†¼ in. × 14 M.

$$E_p = \frac{d_{25} - d_{75}}{2} \tag{Eq. 6.11}$$

or in this specific case as

$$E_p = \frac{1.88 - 1.62}{2} = 0.13 \tag{Eq. 6.12}$$

Another index often used is the sharpness index, SI:

$$\text{SI} = \frac{d_{75}}{d_{25}} \text{ or } \frac{1.62}{1.88} = 0.86 \tag{Eq. 6.13}$$

If the percent of the material reporting to the sink is used, a mirror-image curve results. In this event, an absolute value sign should be used in Eq. 6.10, and Eq. 6.11 should be adjusted to give a number less than 1.0.

The E_p is a function of the separating device used, how it is operated, the content of ±0.10 specific-gravity particles, and the particle size in question. The treating of large particles typically gives better E_p values than does the processing of small particles, and HMS devices are superior to most other gravity separation devices (Table 6.10; Gottfried and Jacobsen 1977; Aplan 1989).

The main defect with the E_p approach is that it is based only on the data between d_{75} and d_{25}. The potential difference between two separations of the same E_p is shown as the shaded area in Figure 6.20C. These misplaced particles must be taken into account in the final evaluation of machine efficiency.

An approach similar to the partition curve described here can also be used with nongravity separating devices by substituting the appropriate physical parameter instead of specific gravity. Partition curves are widely used to evaluate particle size separations, such as for classifiers.

Equipment Selection

Armed with appropriate data, the experience of a process engineer, and a list of the devices potentially available to make the separation (such as those given in Figure 6.1 and Table 6.2), one can select an appropriate gravity separation device. Rarely will just one device suffice, and two to four devices, each treating different size ranges, are often used. Sometimes an alternative means of concentration, such as flotation, is blended into the mix, especially for the separation of fine particles for which gravity concentration devices have a severely limited capacity.

BIBLIOGRAPHY

Agricola, G. 1556. *De Re Metallica*. Translated from the first Latin edition by H.C. and L.H. Hoover, 1950. New York: Dover Publications.

Anonymous. 1970. Gold. In *Encyclopaedia Britannica*. Edited by W.E. Preece. Chicago.

Aplan, F.F. 1973. Evaluation to Indicate Processing Approach. In *SME Mining Engineering Handbook,* Section 27.3. Edited by A.B. Cummins and I.A. Given. New York: AIME.

——. 1980. Gravity Concentration. In *Kirk-Othmer Encyclopedia of Chemical Technology,* Vol. 12. 3rd ed. New York: John Wiley & Sons.

——. 1985a. Gravity Concentration. In *SME Mineral Processing Handbook*. Edited by N.L. Weiss. New York: AIME.

——. 1985b. Heavy Media Separation. In *SME Mineral Processing Handbook*. Edited by N.L. Weiss. New York: AIME.

——. 1989. Whither Gravity, Magnetic and Electrostatic Separations. In *Challenges in Mineral Processing*. Edited by K.V.S. Sastry and M.C. Fuerstenau. Littleton, Colo.: SME.

Arbiter, N. 1955. Solids Concentration. *Chem. Eng.,* 62(8):164–177.

Arnold, B.J., J.D. Hervol, and J.W. Leonard. 1991. Dry Particle Concentration. In *Coal Preparation*. 5th ed. Edited by J.W. Leonard and B.C. Hardinge. Littleton, Colo.: SME.

Burt, R.O., and C. Mills. 1984. *Gravity Concentration Technology*. Amsterdam: Elsevier.

Coe, G.D. 1938. *An Explanation of Washability Curves*. U.S. Bureau of Mines IC 7045, Washington, D.C.

Datta, R.S. 1977. *Rheology and Stability of Mineral Suspensions*. Ph.D. diss. Department of Mineral Processing, The Pennsylvania State University, University Park, Pa.

Deurbrouck, A.W., and W.W. Agey. 1985. Wet Concentrating Tables. In *SME Mineral Processing Handbook*. Edited by N.L. Weiss. New York: AIME.

Falcon. 1999. Bulletin.

Finkey, J. 1924. *The Scientific Fundamentals of Gravity Concentration*. Translated from Hungarian and edited by J. Pocsubay, C.O. Anderson, and M.H. Griffitts. 1930. Missouri School of Mines and Metallurgy Bulletin, 2(1).

Gaudin, A.M. 1939. *Principles of Mineral Dressing*. New York: McGraw-Hill.

Gottfried, B.S., and P.S. Jacobsen. 1977. *Generalized Distribution Curve for Characterizing the Performance of Coal Cleaning Equipment*. U.S. Bureau of Mines RI 8328. Washington, D.C.

Hopwood, J.W. 2000. Krebs Engineers, Tucson, Ariz., Personal communication.

Jarman, W. 1985. Pneumatic Concentration. In *SME Mineral Processing Handbook*. Edited by N.L. Weiss. New York: AIME.

Klima, M.S., and P.T. Luckie. 1989. Application of an Unsteady-State Pulp-Partition Model to Dense-Medium Separations. *Coal Preparation*, 6:227–240.

Lapple, C.E. et al. 1956. *Fluid and Particle Mechanics*. Newark, Del.: University of Delaware.

Laros, T.J. and F.F. Aplan. 1991. Comparative Jig Performance Using Standard and Saw-Tooth Jig Cycles. Abstracts of the SME Annual Meeting, Feb. 28 in Denver, Colo.

Leonard, J.W. and B.C. Hardinge. 1991. *Coal Preparation*. 5th ed. Littleton Colo.: SME. See also earlier editions: 1st ed., 1943, and 2nd ed., 1950 (both edited by D.R. Mitchell); 3rd ed., 1968 (edited by J.W. Leonard and D.R. Mitchell); 4th ed., 1979 (edited by J.W. Leonard). New York: AIME.

Michell, F.B. 1985. Flowing Film Concentration. In *SME Mineral Processing Handbook*. Edited by N.L. Weiss. New York: AIME.

Miller, F.G., T.J. De Mull, and J.P. Matoney. 1985. Centrifugal Specific Gravity Separation. In *SME Mineral Processing Handbook*. Edited by N.L. Weiss. New York: AIME.

Miller, F.G., J.M. Podgursky, and R.P. Aikman. 1977. Study of the Mechanism of Coal Flotation and Its Role in a System for Processing Fine Coal. *Trans. AIME*, 238:276–281.

Mills, C. 1978. *Gravity Concentration*. Short Course Oct. 14–16 at Mackay School of Mines, University of Nevada, Reno, Nev.

——. 1980. Process Design, Scale-Up and Plant Design for Gravity Concentration. *Mineral Processing Plant Design*. 2nd ed. Edited by A.L. Mular and R.B. Bhappu. New York: AIME.

Osborne, D.G. 1988. *Coal Preparation Technology*. London: Graham and Trotman.

Pickett, D.E., and G.W. Riley. 1985. Hindered Settling and Jigging. In *SME Mineral Processing Handbook*. Edited by N.L. Weiss. New York: AIME.

Richards, R.L. 1906–1909. *Ore Dressing*, Vol. 1, 1906; Vol. 2, 1908; Vols. 3 and 4, 1909. New York: McGraw-Hill.

Richards, R.L., and S.B. Locke. 1940. *Textbook of Ore Dressing*. 3rd ed. New York: McGraw-Hill.

Sokaski, M., P.F. Sands, and W.L. McMorris. 1991. Wet Fine Particle Concentration: Dense Media. In *Coal Preparation*. 5th ed. Littleton, Colo.: SME.

Sung, Y.H. 1637. *T'ien-Kung K'ai-Wu, Chinese Technology in the Seventeenth Century*. Translated by E.Z. and S.C. Sun. 1966. University Park, Pa.: The Pennsylvania State University Press.

Taggart, A.F. 1945. *Handbook of Mineral Dressing*. (See also earlier edition: 1927, *Handbook of Ore Dressing*.) New York: John Wiley & Sons.

——. 1951. *Elements of Ore Dressing*. New York: John Wiley & Sons.

Thomson, R.S., T.J. Laros, and F.F. Aplan. 1985. A Study of Jig Cycles in the Fine Jigging of Pyrite from Coal. In *Proceedings of the XV International Mineral Processing Congress*, Vol. 1, in Cannes, France.

Weiss, N.L., ed. 1985. *SME Mineral Processing Handbook*. New York: AIME.

Zimmerman, R.E. 1950. Plant Control and Efficiencies. In *Coal Preparation*. 2nd ed. Edited by D.R. Mitchell. New York: AIME.

CHAPTER 7

Magnetic and Electrostatic Separation

Partha Venkatraman, Frank S. Knoll, and James E. Lawver

INTRODUCTION

Knowledge of magnetic and electrostatic forces dates back at least to the Greek philosopher Thales of Miletus, who lived about 600 B.C. Thales knew some of the magnetic properties of the mineral lodestone, and he was also aware that when amber was rubbed with animal fur, the electrostatic charge produced on the amber (or fur) would attract light, nonconducting particles. The first record of magnetic separation of minerals appears to be a patent issued in 1792 to an English experimenter, William Fularton, that described the concentration of iron ore. About a century later, in 1886, F.R. Carpenter obtained a U.S. patent for electrostatic concentration of ore.

The application of magnetic separators has extended well beyond removing tramp iron. The latest developments in material science and magnet technology have allowed high-intensity and high-gradient industrial magnetic separators with field strengths as high as 6 tesla (60,000 gauss) to be developed. Development of permanent rare-earth magnets and superconducting magnets has opened new markets for magnetic separators.

The electrostatic separator is still the most reliable and economic unit operation for processing beach sand deposits rich in minerals such as ilmenite, rutile, leucoxene, zircon, and garnet. Increased environmental awareness has promoted the demand for unit operations that process secondary materials. A classic example is the successful use of electrostatic separators to remove plastics from metals. Triboelectrostatic separators, which can successfully separate two nonconductors, are being used in minerals and plastics separation.

REVIEW OF MAGNETIC THEORY

All materials can be classified based on their magnetic properties. "Paramagnetic" minerals are attracted along the lines of magnetic force to points of greater field intensity. "Diamagnetic" minerals are repelled along the lines of magnetic force to a point of lesser field intensity. "Ferromagnetic" minerals, a special category of paramagnetic materials, have a very high susceptibility to magnetic forces and retain some magnetism (remanent magnetism) after removal from the magnetic field. The magnetic susceptibility and electrostatic response of minerals are provided in Table 7.1. In this section, the fundamentals of magnetic separation are introduced without detailed explanations of the physics involved.

Magnetic Force or Flux Density

In an electromagnet, an electric charge in motion sets up a magnetic field in the space surrounding it, and a magnetic field exerts a force on an electric charge moving through it. In fact, all magnetic

TABLE 7.1 Minerals and their magnetic and electrostatic response

Mineral	Composition	Specific Gravity	Magnetic Response: Ferromagnetic	Magnetic Response: Paramagnetic	Magnetic Response: Nonmagnetic	Electrostatic Response: Conductive	Electrostatic Response: Nonconductive
Actinolite	$Ca_2(Mg,Fe)_5(Si_4O_{11})_2(OH)_2$	3.0–3.2		X			X
Albite	$Na(AlSi_3O_8)$	2.6			X		X
Almandine	$Fe_3Al_2(SiO_4)_3$	4.3		X			X
Amphibole	$(Fe,Mg,Ca)_xSiO_3$	2.9–3.5		X			X
Anatase	TiO_2	3.9			X	X	
Andalusite	Al_2SiO_5	3.2			X		X
Andradite	$3CaO \cdot Fe_2O_3 \cdot SiO_2$	3.8		X		(2) ←	X
Anhydrite	$CaSO_4$	3.0			X		X
Ankerite	$Ca(Mg,Fe)(CO_3)_2$	2.9–3.1		X			X
Apatite	$(F_1Cl_1OH)Ca_5(PO_4)_3$	3.2			X		X
Aragonite	$CaCO_3$	3.0			X		X
Arsenopyrite	FeAsS	5.9–6.2		X →	(1)	X	
Asbestos	$Mg_3[Si_2O_5](OH)_4$	2.4–2.5			X		X
Augite	$Ca(Mg,Fe,Al)[(Si,Al)_2O_6]$	3.2–3.5		X		X →	(1)
Azurite	$Cu_3[CO_3]_2(OH)_2$	3.8			X		X
Baddeleyite	ZrO_2	5.6			X		X
Barite	$BaSO_4$	4.5			X		X
Bastnaesite	$(Ce,La,F)CO_3$	5.0		X			X
Bauxite	$Al_2O_3 \cdot 2H_2O$	2.6			X		X
Beryl	$Be_3Al_2[Si_6O_{18}]$	2.7–2.8			X		X
Biotite	$K(Mg,Fe)_3[Si_3AlO_{10}](OH,F)_2$	3.0–3.1		X			(4)
Bismuth	Bi	9.8			X	X	
Borax	$Na_2B_4O_7 \cdot 10H_2O$	1.7			X		X
Bornite	Cu,FeS_4	4.9–5.0		(1) ←	X	X	
Brannerite	$(UO,TiO,UO_2)TiO_3$	4.5–5.4		X		X	
Brookite	TiO_2	4.1			X	X	
Calcite	$CaCO_3$	2.7			X		X
Carnotite	$K_2(UO_2)_2V_2O_8 \cdot 2H_2O$	5.0			X	(2) ←	X
Cassiterite	SnO_2	7.0			X	X	
Celestite	$SrSO_4$	4.0			X		X
Cerussite	$PbCO_3$	6.6			X	(2)	X

(Table continues on next page)

TABLE 7.1 Minerals and their magnetic and electrostatic response (continued)

Mineral	Composition	Specific Gravity	Magnetic Response			Electrostatic Response	
			Ferromagnetic	Paramagnetic	Nonmagnetic	Conductive	Nonconductive
Chalcocite	Cu_2S	5.5–5.8			X	X	
Chalcopyrite	$CuFeS_2$	4.1–4.3		(1) ←	X	X	
Chlorite	$(Mg,Al,Fe)_{12}[(Si,Al)_8O_{20}](OH)_{16}$	2.6–3.2		X			X
Chromite	$(Fe,Mg)(Cr,Al)_2O_4$	4.6		X		X	
Chrysocolla	$CuSiO_3 \cdot nH_2O$	2.0–2.3			X		X
Cinnabar	HgS	8.1			X		X
Cobalitite	$(Co,Fe)AsS$	6.0–6.3		X		X	
Colemanite	$Ca_2B_6O_{11} \cdot 5H_2O$	2.4			X		X
Collophanite	$Ca_3P_2O_8 \cdot H_2O$	2.6–2.9			X		(3)
Columbite	$(Fe,Mn)(Ta,Nb)_2O_6$	5.2–8.2		X		X	
Copper	Cu	8.9			X	X	
Corumdum	Al_2O_3	3.9–4.1			X		X
Covellite	CuS	4.7			X	X	
Cryolite	Na_3AlF_6	3.0			X	(2) ←	X
Cuprite	Cu_2O	5.8–6.2			X		X
Diamond (natural)	C	3.5			X		X
Diamond (synthetic)	C	3.5		X			X
Diopside	$CaMg[Si_2O_6]$	3.3–3.4		X →	(1)		X
Dolomite	$CaMg(CO_3)_2$	1.8–2.9			X		X
Epidote	$Ca_2(Al,Fe)_3Si_3O_{12}(OH)$	3.4		X			X
Euxenite	$(Y,Er,Ce,La,U)(Nb,Ti,Ta)_2(O,OH)_6U_3O_8$	4.7–5.2		X		X	
Feldspar group	$(K,Na,Ca..)_x(Al,Si)_3O_8$	2.6–2.8			X		X
Ferberite	$FeWO_4$	7.5	(1) ←	X		X	
Flint	SiO_2	2.6			X		X
Fluorite	CaF_2	3.2			X		X
Franklinite	$(Zn,Mn)Fe_2O_4$	5.1–5.2	X			X	
Gahnite	$ZnAl_2O_4$	4.6			X		X
Galena	PbS	7.5			X	X	
Garnet complex	Ca,Mg,Fe,Mn silicates	3.4–4.3		X →	(1)	(2) ←	X
Gibbsite	$Al(OH)_3$	2.4			X		X
Geothite	$FeO(OH)$	4.3		X		(2) ←	X

(Table continues on next page)

TABLE 7.1 Minerals and their magnetic and electrostatic response (continued)

Mineral	Composition	Specific Gravity	Magnetic Response			Electrostatic Response	
			Ferromagnetic	Paramagnetic	Nonmagnetic	Conductive	Nonconductive
Gold	Au	15.6–19.3			X	X	
Graphite	C	2.1–2.2			X	X	
Grossularite	$Ca_3Al_2(SiO_4)_3$	3.5			X	(2)	← X
Gypsum	$CaSO_4{\cdot}2H_2O$	2.3			X		X
Halite	NaCl	2.5			X	(2)	← X
Hematite	Fe_2O_3	5.2		X		X	
Hornblende	$Ca_2Na(Mg,Fe^{2+})_4(Al,Fe^{3+})[(Si,Al)_4O_{11}](OH)_2$	3.1–3.3		X		(2)	← X
Huebnerite	$MnWO_4$	6.7–7.5		X →	(1)	X	
Hypersthene	$(Mg,Fe)SiO_3$	3.4		X			X
Ilmenite	$FeTiO_3$	4.7		X		X	
Ilmenorutile	$(Nb_2O_5,Ta_2O_5)_xTiO_2$	5.1		X		X	
Ilvaite	$CaFe_2(FeOH)(SiO_4)_2$	4.0		X		X →	(1)
Kaolinite	$Al_2Si_2O_5(OH)_4$	2.6			X		X
Kyanite	$Al_2O[SiO_4]$	3.6–3.7			X		X
Lepidolite	$[OH,F]_2KliAl_2Si_3O_{10}$	2.8–2.9			X		X
Leucoxene	$FeTiO_3$ TiO_2(alteration product)	3.6–4.3		X →	(1)	X	
Limonite	$HFeO_2{\cdot}nH_2O$	2.2–2.4		X →	(1)	(2)	← X
Magnesite	$MgCO_3$	3.0			X		X
Magnetite	Fe_3O_4	5.2	X			X	
Malachite	$Cu_2CO_3(OH)_2$	4.0			X		X
Manganite	$MnO(OH)$	4.3		X →	(1)	X	
Marcasite	FeS_2	4.6–4.9			X	X	
Martite (see hematite)							
Microline	$KAlSi_3O_8$	2.6			X		X
Microlite	$Ca_2Ta_2O_7$ (see Pyrochlore)	5.5			X		X
Millerite	NiS	5.2–5.6		X		X	
Molybdenite	MoS_2	4.7–5.0			X	X	
Monazite	$(Ce,La,Y,Th)PO_4$	4.9–5.5		X			X
Mullite	$Al_6Si_2O_{13}$	3.2			X		X
Muscovite	$KAl_2[AlSi_3O_{10}][F,OH]_2$	2.8–3.0			X		(4)

(Table continues on next page)

TABLE 7.1 Minerals and their magnetic and electrostatic response (continued)

Mineral	Composition	Specific Gravity	Magnetic Response			Electrostatic Response	
			Ferromagnetic	Paramagnetic	Nonmagnetic	Conductive	Nonconductive
Nahcolite	$NaHCO_3$	2.2			X		X
Nepheline Syenite	$(Na,K)(AlSi)_2O_4$	2.6			X		X
Niccolite	NiAs	7.6–7.8		X		X	
Olivine	$(Mg,Fe)_2[SiO_4]$	3.3–3.5		X			X
Orpiment	As_2S_3	3.4–3.5			X	X	
Orthoclase	$K[Al,Si_3O_8]$	2.5–2.6			X		X
Periclase	MgO	3.6			X		X
Perovskite	$CaTiO_3$	4.0			X		X
Petalite	$LiAl(Si_2O_5)_2$	2.4			X		X
Phosphate (pebble)	(see Collophanite)						
Platinum	Pt	14.0–21.5		(1) ←	X	X	
Pyrite	FeS_2	5.0		(1) ←	X	X	
Pyrochlore	$(Na,Ca...)_2(Nb,Ta...)_2O_6[F,OH]$	4.2–4.4			X	X	
Pyrolusite	MnO_2	4.7–5.0		(1) ←	X		X
Pyrope	$Mg_3Al_2(SiO_4)_3$	3.5			X	(2) ←	X
Pyroxene	$(Ca,Mg,Fe,Al)_2Si_2O_6$	3.1–3.6		X →	(1)	(2) ←	X
Pyrrhotite	$Fe_{1-x}S_x$	4.6–4.7	X			X	
Quartz	SiO_2	2.7			X		(3)
Realgar	AsS	3.6			X	X	
Rhodochrosite	$MnCO_3$	3.7			X	(2) ←	X
Rhodonite	$MnSiO_3$	3.6–3.7			X	(2) ←	X
Rutile	TiO_2	4.2–4.3			X	(2)	
Samarskite	$(Y,Er...)_4[(Nb,Ta)_2O_7]_3$	5.6–5.8	(1) ←	X		X	
Scheelite	$CaWO_4$	6.1			X		X
Serpertine	$Mg_6[Si_4O_{10}](OH)_8$	2.5–2.7		X			X
Siderite	$FeCO_3$	3.9		X		(2) ←	X
Sillmanite	$Al_2O[SiO_4]$	3.2			X		X
Silver	Ag	10.1–11.1			X	X	
Smithsonite	$ZnCO_3$	4.1–4.5			X		X
Sodalite	$Na_8[Al_6Si_6O_{24}]Cl_2$	2.1–2.3			X		X
Spessarite	$Mn_3Al_2[SiO_4]_3$	4.2			X		X

(Table continues on next page)

TABLE 7.1 Minerals and their magnetic and electrostatic response (continued)

Mineral	Composition	Specific Gravity	Magnetic Response			Electrostatic Response	
			Ferromagnetic	Paramagnetic	Nonmagnetic	Conductive	Nonconductive
Sphalerite	ZnS	3.9–4.0		X →	(1)	X →	(1)
Sphene	$CaTi[SiO_4](F_3OH)$	3.3–3.6			X	(2) ←	X
Spinel	$MgAl_2O_4$	3.6		(1) ←	X	X	(1)
Spodumene	$LiAl(SiO_3)_2$	3.1–3.2			X		X
Stannite	Cu_2FeSnS_4	4.3–4.5			X	X	
Staurolite	$Fe^{2+}Al_4[Si_4O_{11}]_2O_2(OH)_2$	3.6–3.8		X		(2) ←	X
Stibnite (Antimonite)	Sb_2S_3	4.6			X	X	
Struverite	$(Ta_2O_5,Nb_2O_5)_xTiO_2$	5.1		X		X	
Sulpher	S	2.1			X		X
Sylvite	KCl	2.0			X		X
Talc	$Mg_3Si_4O_{10}(OH)_2$	2.7–2.8			X		X
Tantalite	$(Fe,Mn)(Ta,Nb)_2O_6$	5.2–8.2		X		X	
Tapiolite	$Fe(Ta,Nb)_2O_6$	7.3–7.8		X		X	
Tetrahedrite	$(Cu,Fe)_{12}Sb_4S_{13}$	5.0		X		X	
Thorianite	ThO_2	9.7			X		X
Thorite	$ThSiO_4$	4.5–5.4			X		X
Topaz	$Al_2SiO_4(F,OH)_2$	3.5–3.6			X		X
Tourmaline	$(Na,Ca)(Mg,Fe^{2+},Fe^{3+},Al,Li)_3Al_6(BO_3)_3Si_6O_{18}(OH)_4$	2.9–3.2		X →	(1)	(1,2) ←	X
Uraninite	UO_2	11.0		X			X
Vermiculite	$Mg_3[Al,Si_3O_{10}](OH)_2 \cdot nH_2O$	2.4–2.7			X		X
Wolframite	$(Fe,Mn)WO_4$	6.7–7.5		X		X	
Wollastonite	$CaSiO_3$	2.8–2.9			X		X
Wulfenite	$PbMoO_4$	6.7–7.0			X	X	
Xenotime	YPO_4	4.4–5.1		X			X
Zeolite	Hydrous alumino-silicate usually of Ca and Na	2.0–2.5			X		X
Zincite	ZnO	5.7			X	(1) ←	X
Zircon	$ZrSiO_4$	4.7			X		X

phenomena arise from forces acting between electric charges in motion. Therefore, the flux density B, at a point P, resulting from a current element I, of length dl, is as specified by Ampere's law:

$$dB = KI\left[\frac{dl(\sin\theta)}{r^2}\right] \quad \textbf{(Eq. 7.1)}$$

where

B = magnetic flux density, newtons/ampere-meter
K = constant of proportionality
I = current element, amperes
l = length of current element, meters
θ = angle between current element and radius vector to point P
r = distance to P, meters

The magnetic flux density B, or force per pole, is defined as the number of magnetic flux lines per unit area normal to the lines. The unit for magnetic flux is weber. The term B has units of webers per square meter called tesla. The largest values of magnetic induction that can be produced in the laboratory are about 50 to 60 tesla.

Industrial high-intensity superconducting separators can reach fields of about 5 tesla. High-intensity induced-roll dry separators reach values as high as 2 tesla, and low-intensity wet drum separators for concentration of iron ore have drum-surface field strengths on the order of 0.1 tesla. However, the magnetic force acting on particles depends not only on the magnetic field B, but also on its gradient dB/dz, where z is the direction of the changing field.

Magnetization

A ferromagnetic material can be magnetized simply by being brought close to a permanent magnet or by passing current through a wire winding around the material. The magnetic state of a body can be defined by (1) stating the magnetization of all points within the body, (2) defining the strength of the magnetic poles, or (3) defining the magnitude of equivalent surface currents.

Coulomb's Law for Magnets. Coulomb's law for magnets is similar to that for electric charges; that is,

$$F = \frac{1}{4\pi\mu_o} * \frac{m_1 m_2}{r^2} \quad \textbf{(Eq. 7.2)}$$

where

F = force, newtons
μ_o = permeability of a vacuum ($4\pi \times 10^{-7}$ henry/meter)
m_1 and m_2 = pole strength, amperes per meter
r = distance, meters

Magnetization (or more completely, the intensity of magnetization M) is the total magnetic moment of dipoles per unit volume in units of amperes per meter, or pole strength m per unit area A.

$$M = \frac{m}{A} \quad \textbf{(Eq. 7.3)}$$

Magnetic Field

The strength of the magnetic field H has the same units as M (amperes/meter) and can be thought of as the cause of magnetization. It is defined as

$$H = \frac{B}{\mu} \quad \textbf{(Eq. 7.4)}$$

where

μ = absolute permeability

In free space, a magnetic field produces a magnetic force given by $B = \mu_o H$. However, in most applications the space is filled with some magnetic substance that causes an induced magnetization, $\mu_o M$. Therefore, the total magnetic flux density B is the vector sum of the flux caused by the magnetic field H and the flux resulting from the magnetization M of the material. For ferromagnetic materials, however, the contribution of M usually dominates B.

$$B = \mu_o(H + M) \quad \text{(Eq. 7.5)}$$

Permeability and Susceptibility

The magnetic flux density, B, the strength of the magnetic field, H, and the magnetization, M, can be used to compare the magnetic response of various materials. The ratio M/H is a dimensionless quantity called volume susceptibility or simply magnetic susceptibility, χ. Similarly, the ratio B/H is called absolute permeability, μ. Relative permeability μ_r is defined as

$$\mu_r = \frac{\mu}{\mu_o} \quad \text{(Eq. 7.6)}$$

where

μ_r = relatively permeability
μ = absolute permeability
μ_o = permeability in a vacuum ($4\pi \times 10^{-7}$ henry/meter)

The relative permeability of diamagnetic materials is slightly less than one, and that of paramagnetic materials is slightly greater than one. In the case of ferromagnetic materials, relatively permeability is very high (for example, μ_r for iron with 0.2% impurities is about 5,000).

Summary

Paramagnetic minerals have higher magnetic permeabilities than the surrounding medium, usually air or water, and they concentrate the lines of force of an external magnetic field. The higher the magnetic susceptibility, the higher the field intensity in the particle and the greater the attraction up the field gradient toward increasing field strength. Diamagnetic minerals, on the other hand, have lower magnetic permeabilities than the surrounding medium, usually air or water, and they repel the lines of force of an external magnetic field. These characteristics cause the expulsion of diamagnetic minerals down the gradient of the field toward decreasing field strength. This negative diamagnetic effect is usually orders of magnitude smaller than the positive paramagnetic attraction. Thus, a magnetic circuit can be designed to produce higher field intensity or higher field gradient, or both, to achieve effective separation.

CONVENTIONAL MAGNETS

Magnets are used in the mineral industry to remove tramp iron that might damage equipment and to separate minerals according to their magnetic susceptibility.

Low-intensity Magnetic Separators

Low-intensity magnetic separators have flux densities up to 2,000 gauss. These separators are mainly used to remove ferromagnetic materials, such as iron, to protect downstream unit operations, such as conveyor belts, or to scalp ferromagnetic materials to improve the performance of permanent or electromagnetic separators used to separate weakly magnetic materials. Low-intensity separators can treat wet slurry or dry solids.

Protective Magnets. The device most widely used to protect downstream operations from tramp iron is a magnetic pulley installed in the head of the conveyor (Figure 7.1). These devices remove tramp metals from dry solids. They contain either a permanent magnet or an electromagnet. Many types of magnets can be used—for example, plate magnets, cross-belt magnets, cobbing magnets,

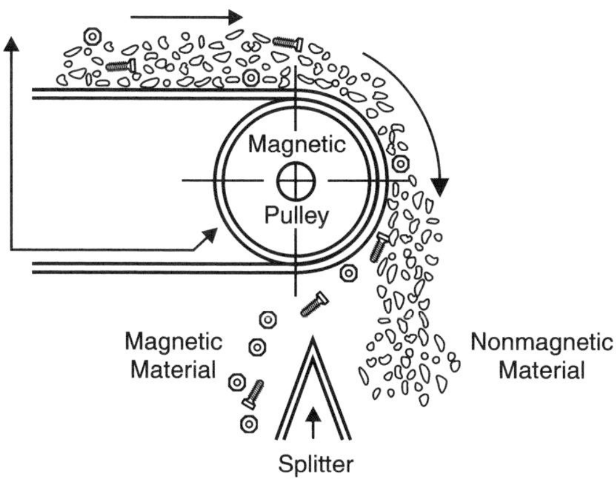

FIGURE 7.1 Typical magnetic pulley

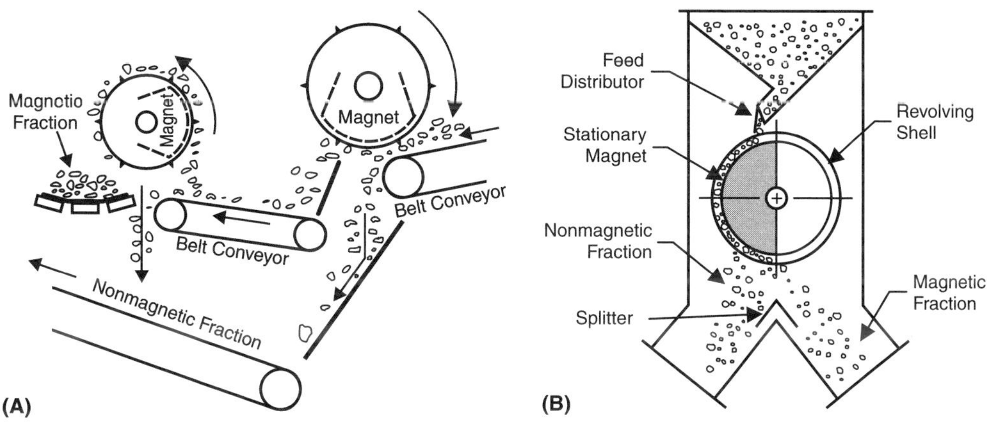

Source: Eriez Magnetics.

FIGURE 7.2 Magnetic drum operating as a lifting magnet

grate magnets, magnetic humps, and magnetic filters. The arrangement of magnetic drum separators is shown in Figures 7.2A and 7.2B.

Wet Magnetic Separators. Low intensity wet magnetic separators have been the workhorse of the iron ore industry for several decades. Iron ore rich in magnetite has traditionally been enriched by these magnets. The coal industry uses these magnets to recover magnetite or ferrosilicon in a media recovery circuit. Several types of separators work on the same principle but have different design features. The common types are counter-rotation drum separators and concurrent-rotation drum separators (Figures 7.3A and 7.3B).

High-intensity Magnetic Separators

Separating paramagnetic or weakly magnetic particles requires a higher flux density. This higher density is achieved by designing electromagnetic circuitry that can generate a magnetic force of up to 2 tesla. For example, in a silica sand processing plant, these separators are used to remove weakly magnetic iron-bearing particles.

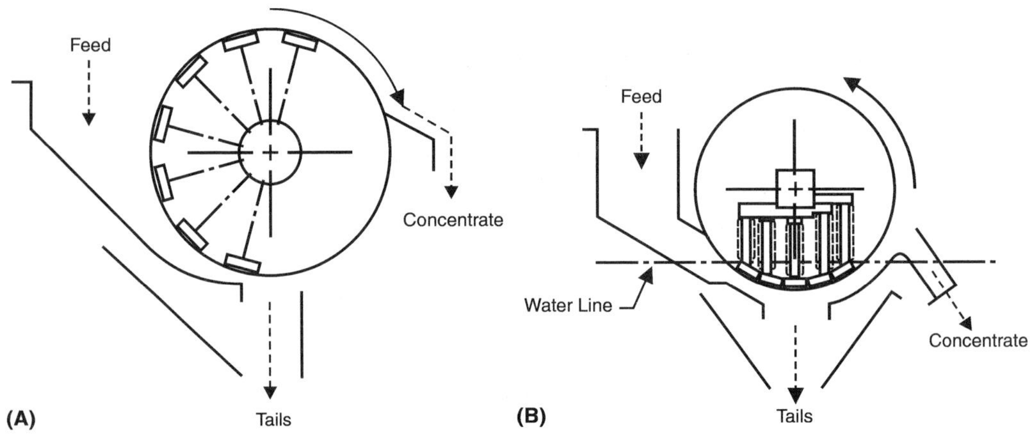

FIGURE 7.3 (A) Counter-rotation drum magnetic separator, nonsubmerged magnetic field; (B) Concurrent-rotation drum magnetic separator, submerged magnetic field

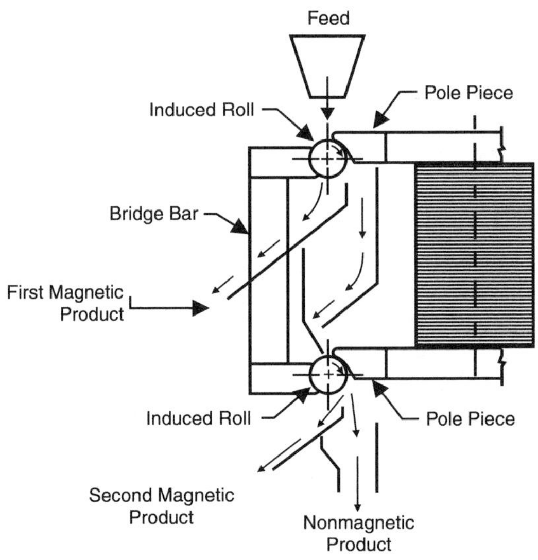

FIGURE 7.4 Trajectory of particles in an induced-roll dry magnetic separator

Induced-roll Magnetic Separator. Induced-roll dry magnetic separators are widely used to remove trace impurities of paramagnetic substances from feedstocks such as quartz, feldspar, and calcite. The machine contains laminated rolls of alternating magnetic and nonmagnetic discs. A magnetic flux on the order of 2 tesla is obtained, and very high gradients are obtained where the flux converges on the sharp edges of the magnetic laminations. A thin stream of granular material is fed to the top of the first roll. The magnetic particles are attracted to the roll and are deflected out of their natural trajectory (Figure 7.4). Selectivity is obtained by varying roll speed and magnetic flux. A rather closely sized material must be treated if high selectivity is required. An industrial induced-roll magnetic separator consists of several rolls and can treat up to 10 tph (Figure 7.5).

Lift-type magnetic separators are used on granular and powdered material that is dry and free flowing. This type of separator produces a clean magnetic product because the magnetic particles are

Source: Outokumpu Technology Inc., Physical Separation Division.

FIGURE 7.5 Industrial induced-roll magnetic separators treating silica sand

lifted out of the stream against the force of gravity, which minimizes entrapped particles (Figure 7.6). The selectivity of the lift-type separator is superior to that of induced-roll separators. Their main limitation is lower capacity. The cross-belt separator, a type of lift magnetic separator, has been used to some extent in processing ilmenite, garnet, and monazite in beach sands.

Jones Separator. The Jones separator is a wet high-intensity separator built on a strong main frame made of structural steel (Figure 7.7). The magnet yokes are welded to this frame, and the electromagnetic coils are enclosed in air-cooled cases. The actual separation takes place in the plate boxes that are on the periphery of the one or two rotors attached to the central shaft. The feed, which is a

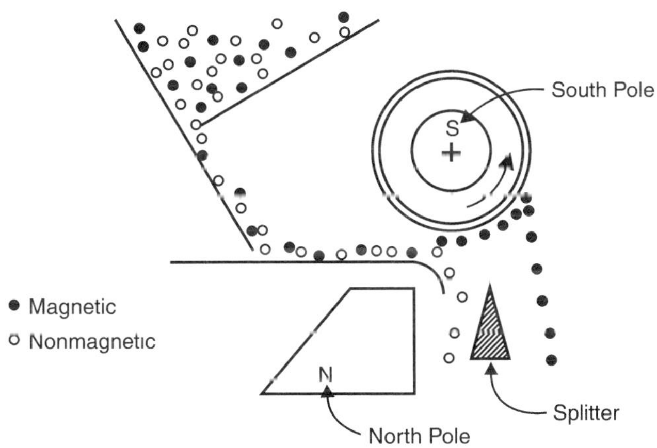

FIGURE 7.6 Trajectory of particles in lift-type magnetic separators

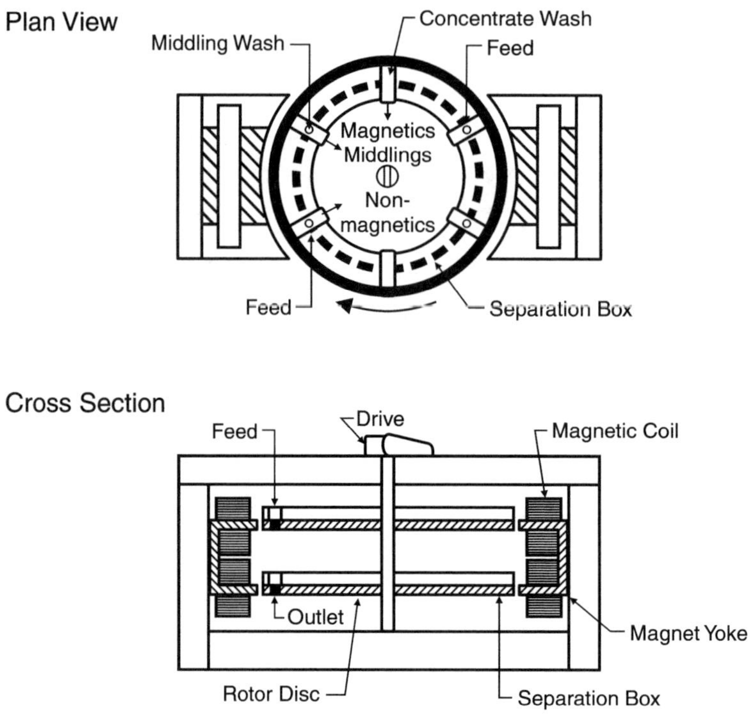

FIGURE 7.7 Jones high-intensity wet magnetic separator

thoroughly mixed slurry, flows through the separator by means of fitted pipes and launders and into the plate boxes. The plate boxes are grooved to concentrate the magnetic field at the tips of the ridges. Feeding is continuous as a result of the rotation of plate boxes and the rotors, and the feed points are at the leading edges of the magnetic fields. Each rotor has two symmetrically placed feed points.

The feebly magnetic particles are held by the plates, whereas the remaining nonmagnetic slurry passes straight through the plate boxes and is collected in a launder. Before leaving the field, the entrained nonmagnetic particles are washed by low-pressure water and are collected as a "middlings product." When the plate boxes reach a point midway between the magnetic poles, where the magnetic field is essentially zero, the magnetic particles are washed out under high-pressure scour water sprays of up to 5 bars of pressure. Field intensities greater than 2 tesla can be produced in these machines. They are widely used to recover iron minerals from low-grade hematite ore. Some other common applications include removing magnetic impurities from cassiterite concentrate, removing fine magnetics from asbestos, and purifying talc.

Frantz Isodynamic Separator. The Frantz Isodynamic Separator, introduced in the early 1930s, is the most efficient magnetic separator for separating minerals with field-independent magnetic susceptibilities. The isodynamic field, generated by a bipolar magnet with special pole tip profiles, provides constancy of the product of the field and the field gradient. However, mineral separation in an isodynamic magnetic field is limited to minerals that have a constant susceptibility at the laboratory scale. Only this category of mineral then experiences a constant force throughout the isodynamic area.

PERMANENT MAGNETS

Most of the weakly magnetic minerals, such as garnet, ilmenite, and magnetic impurities in silica sand, can be effectively separated with a magnetic separator that has a flux density greater than 6,000 gauss.

For nearly a century, induced-roll magnetic separators were the only economically viable unit operation in these applications. In spite of their considerable success, induced-roll separators have certain limitations in their selectivity and application. The development of permanent-magnet technology during the last two decades has reestablished the importance of magnetic separation and has increased the efficiency of fine-particle separations that were not successful with induced-roll magnets.

Principle and Design

In the last decade, magnetic separation technology has undergone a revolution. Research in material science and ceramic technology has culminated in the development of new permanent rare-earth magnets and superconducting alloys that can be used to build high-gradient magnetic separators. Successful adaptation of these new magnetic materials combined with the knowledge of magnet geometry has led to the design and development of a number of new magnetic separators. These separators have opened niche markets that were previously considered beyond the realm of magnetic separation. These new separators are capable of

- Effectively removing magnetic impurities or reducing their concentration (even to ppm levels)
- Producing high-grade mineral separates
- Operating on virtually no energy, which makes them economical
- Generating higher magnetic flux levels up to 21,000 gauss or 2.1 tesla

Dry Permanent Magnetic Separator

Recent improvements in magnet composition and design have led to the development of permanent magnetic separators. These improved rare-earth permanent magnets (e.g., NdFeB magnets) have a magnetic attractive force an order of magnitude greater than that of conventional permanent magnetic circuits. The two main types of dry permanent magnetic separators that have found wide industrial applications are the rare-earth drum (RED) and the rare-earth roll (RER). They are widely used to separate weakly magnetic materials, such as garnet, ilmenite, and chromite, and also to separate magnetic impurities present in low concentrations in silica sand.

Rare-earth Drum Separator. In an RED separator, the NdFeB magnets are uniquely arranged to provide an intense (up to 9,000 gauss) and "deep" magnetic field perpendicular to the drum surface (Figure 7.8). Once the particles are on the drum surface, they experience uniform flux density that minimizes the misplacement of pinned particles to the middlings. The weakly magnetic particles pinned to the drum are carried to the region of no magnetic intensity and are released as magnetics. The centrifugal force of the rotating drum throws those particles not influenced by the magnetic field into the nonmagnetic hopper.

An industrial-scale RED separator usually has three drums (Figure 7.9). In general, the top drum is a low-intensity (up to 2,000 gauss) scalper magnet to remove ferromagnetic particles, and the nonmagnetic fraction is subsequently treated on the REDs. The main purpose of the scalper is to protect the bottom two REDs, as well as to increase their capacity. Some separators have a built-in internal air-cooling system to protect the magnets from overheating when the feedstock is preheated, as in plants that process beach and silica sand.

Rare-earth Roll Separator. The feed is fed onto a thin belt (usually 7.6×10^{-3} to 5.1×10^{-2} cm) that travels at a very high velocity. The unique aspect of these separators is the way in which the magnetic separator is configured as a head pulley. The feed material is passed through the magnetic field, and the magnetic (or weakly magnetic) particles are attached to the roll and separated from the nonmagnetic stream (Figure 7.10).

Drum separators can effectively handle coarse particles (12.5–0.075 mm), whereas roll separators are very effective in treating fine particles (<1 mm). The capacity of the deeper-field drum separators is generally higher than that of the roll separators at 400–500 lb/h/in. for drum separators but

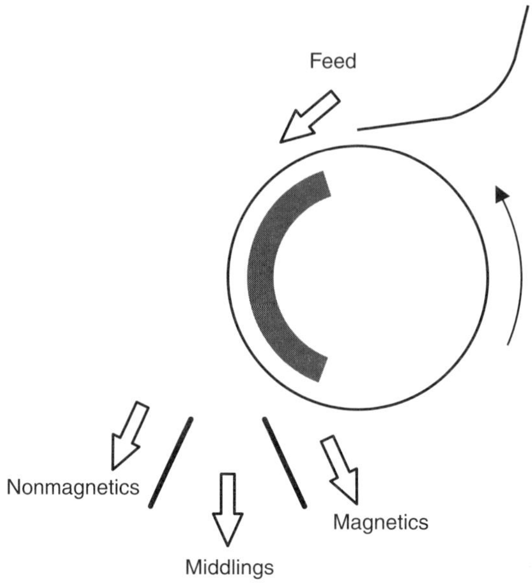

FIGURE 7.8 Operating principle of a RED separator

Source: Outokumpu Technology Inc., Physical Separation Division.

FIGURE 7.9 Industrial-scale RED separator

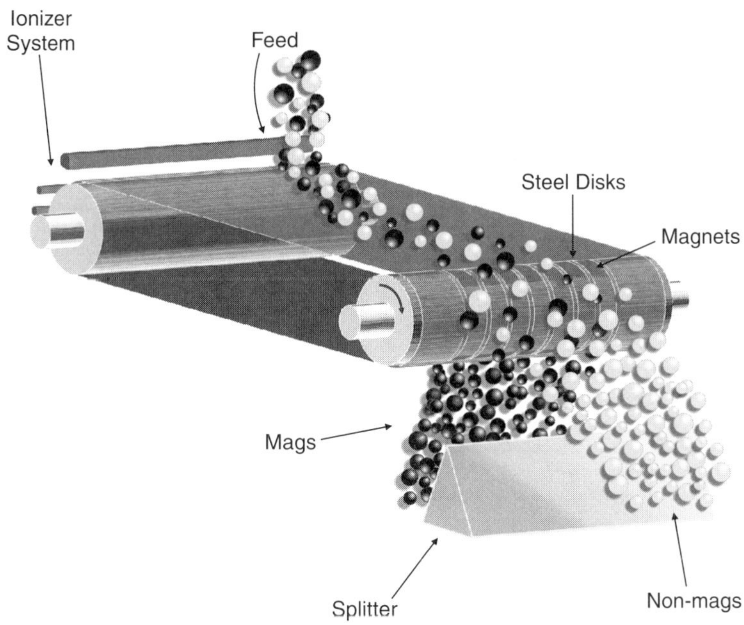

FIGURE 7.10 Operating principle of typical RER magnetic separator

100–300 lb/h/in. for roll separators. The major advantage of the drum separator is its low maintenance cost, because it does not contain a belt that must be replaced frequently. However, both separators have their own niche markets. Drum separators can treat coarser particles, such as garnet, ilmenite, and iron ore, at a higher throughput. Roll separators can be used in producing high-grade, high-purity glass sand products when the feed material is not preheated.

Case Study I: Recovering titanium minerals as a nonmagnetic product using a RED separator. A series of tests used a RED separator and an induced-roll magnetic separator to process a titanium-rich magnetic feed. The feed contained 76% TiO_2 and 1% Al_2O_3; it was obtained from a beach-sand processing plant. This test work compared the performance of the RED and the induced-roll magnets, widely used in the plant, in producing a high-grade nonmagnetic TiO_2 product (Table 7.2). The product of the RED separator contained nearly twice the TiO_2, 45% rather than 24%, at a comparable product grade (TiO_2 grade of +90% and Al_2O_3 content of about 1%).

Case Study II: Performance evaluation of the RED and RER separators. A detailed study of the effect of feed rate on the RED and the RER separators used a garnet-rich heavy mineral sample (Figure 7.11). The effect of feed rate on the RED separator was minimal. The garnet recovery decreased very marginally with increase in feed rate; it was 96.5% at a 2.6-tph feed rate and 95.1% at a 7.8-tph feed rate.

TABLE 7.2 TiO_2 recovery and composition of nonmagnetic product of an induced-roll magnetic separator* and a RED-7000 gauss drum separator

Unit Operation	TiO_2 Recovery, %	TiO_2, %	Al_2O_3, %
Induced-roll magnet	24.2	92	0.96
RED-7000 gauss separator	44.7	92	1.03

*Feed contained 76% TiO_2 and 1% Al_2O_3; two passes; nonmagnetic retreat.

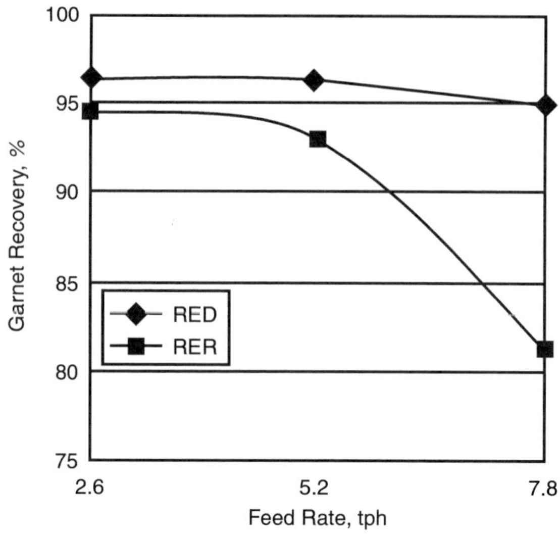

FIGURE 7.11 Performance of RED and RER separators

However, the RER separator seemed to be very sensitive to feed rate. Garnet recovery was 94.6% at a 2.6-tph feed rate, but it fell to 81.3% at a 7.8-tph feed rate.

Eddy Current Separator. In the case of the RED and RER magnetic separators, the outer drum or the belt rotates, while inside the shell, rare-earth magnets are mounted on a stator. However, in an eddy current separator not only does the nonmetallic outer drum rotate but also the inside shell, which is a faster-moving rotor containing rare-earth magnets arranged in alternating polarity to produce induced eddy currents. The induced eddy current sets up a repulsive force in good conductors and thus separates nonferrous electrically conductive metals, such as copper and aluminum, from nonconducting materials.

Wet Permanent Magnetic Separator

A wet permanent magnetic separator is shown in Figure 7.12. The permanent-magnet (NdFeB) bars are positioned more or less horizontally inside a revolving drum made of stainless steel. This separator provides a field strength of 0.7 tesla on the drum surface. The pulp tank is made of stainless steel with concurrent or semicountercurrent flow tank design and an adjustable discharge gap at the magnetics discharge end. The separator is equipped with an adjustable valve on the nonmagnetics discharge pipe. This valve helps to control the flow rate and overflow level. It has found industrial applications in processing low-grade (martitic) iron ore.

SUPERCONDUCTING MAGNETS

High magnetic fields (up to 2 tesla) are generated by passing current through a resistive coil or by permanent magnets. The development, through the use of finite element analysis techniques, of newer computer models has helped to achieve higher magnetic force. However, there is a logical maximum magnetic field for both the resistive coil and permanent magnet. Resistive coils are limited by the intrinsic resistance applied by the windings; the field strength of existing permanent magnets can be increased only marginally by modifying the magnet geometry. In the future, new magnetic materials may help to overcome this limitation.

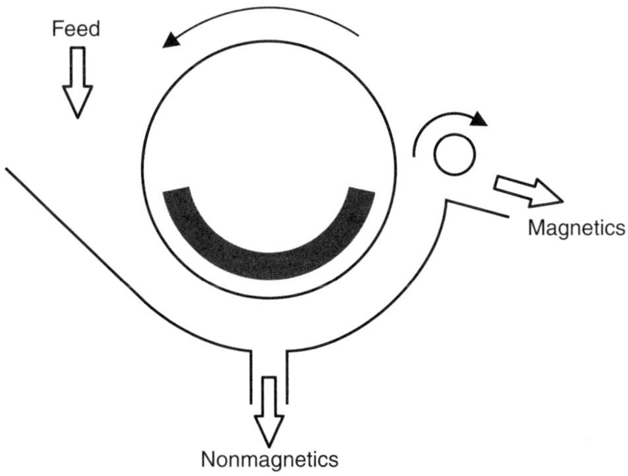

FIGURE 7.12 Operating principle of a wet drum RED separator

Principle and Design

Currently, superconducting magnets are the only economically and technically viable way to achieve field strengths as high as 5 tesla. Fundamental requirements of superconducting magnets are a suitable conductor and a cryogenic system. During the last decade, extensive research in material science has resulted in new alloys that are suitable candidates for superconducting magnets. Because of its reliability and favorable economics, a niobium and titanium alloy is the most suitable for low-temperature (about 4 K) industrial applications. High-temperature (about 20–30 K) superconducting magnetic separators have yet to be developed.

The cryogenic system is the most expensive component of the superconducting magnet, and it determines the economic viability and practicality of these machines. To date, three cryogenic systems have been successfully applied.

Closed-cycle Liquefier System. In a closed-cycle liquefier system, the superconductor resides in a bath of liquid helium, and boil-off gas is recirculated through a helium liquefier. Although the installation of such a system is quite complex, its performance has been good and reliable, provided there are no long-term interruptions in the supply of electrical power and cooling water.

Low-loss System. In a low-loss system the superconductor windings reside in a reservoir of liquid helium. A very efficient insulation system enables the magnet to operate for long periods, typically 1 year or more, between liquid helium refills. The salient feature of this system is its relative immunity to short-term electrical failures. This feature has allowed this technology to be used where equipment is operated under difficult conditions.

Indirect Cooling. The advent of heat engines based on the Gifford McMahon cycle, which generate temperatures of 4 K or less, has made it possible to cool superconducting windings without the need for liquid helium. This technique offers great potential for small-scale systems in which the economics of helium supply or the cost of a liquefier cannot be justified. However, a constant power supply is essential for reliable operation.

In summary, the superconducting magnets have two main advantages:

- Low power consumption resulting from zero resistance of the magnet winding
- Generation of much higher magnetic fields

Superconducting High-gradient Wet Magnetic Separator (HGMS)

In an HGMS, the magnetic particles are captured on a stainless steel–wool matrix contained within the bore of a high-intensity magnet. The high intensity is generated using a superconducting coil. Because these coils have essentially zero resistance, little electrical power is required to energize the magnet. Furthermore, once the magnet is energized, the coil ends can be shorted, leaving the magnet in a fully energized state without any additional power supply. This practice is called operating the magnets in persistent mode. Unloading the trapped magnetic particles from the matrix is an essential step that determines the separation efficiency and the capacity of the unit operation. The demagnetization is achieved either by de-energizing the magnet (a state commonly referred to as switch-mode) or by moving the matrix canister (referred to as a reciprocating canister HGMS). In reciprocating technology, captured magnetic particles are flushed using a ram to remove the trapping zone from the magnetic field regions. The ram operates on a magnetically balanced canister that houses a multisection separation region with unique and separate trapping zones (Figure 7.13). Units that combine reciprocating canister technology and a low-loss cryogenic system have been used in kaolin processing throughout the world. Figure 7.14 shows the installation of a typical large-scale reciprocating canister HGMS.

Superconducting Open-gradient Dry Magnetic Separator (OGMS)

In a conventional OGMS, the magnet structure is arranged to provide a region in open space with a highly divergent field. Thus, the magnet geometry supplies both the magnetic field and the field gradient. Any paramagnetic material passing through this region will experience a force directly proportional to the field intensity and the magnitude of the field gradient. However, a superconducting OGMS offers not only higher magnetic force but also a deeper magnetic field, which in turn translates to larger separation volume than that obtained by conventional electromagnets and permanent magnets.

A novel dry OGMS device is being developed by Outokumpu Technology, Inc. In this separator, the magnet is inclined, and feed is allowed to fall through the magnetic region. This separator is capable of

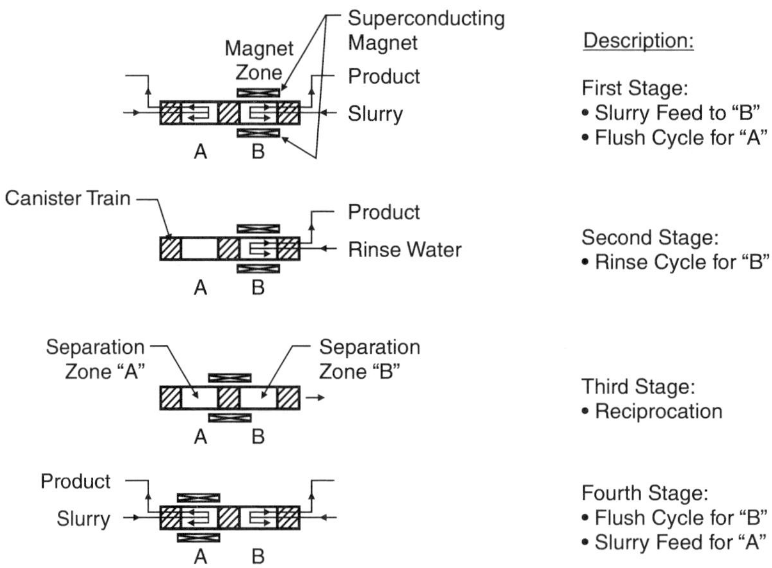

Source: Outokumpu Technology Inc., Physical Separation Division.

FIGURE 7.13 Basic process cycle of a reciprocating canister superconducting magnetic separator

Source: Outokumpu Technology Inc., Physical Separation Division.

FIGURE 7.14 Superconducting magnetic separator widely used in kaolin processing

treating 25- × 0.5-mm particles. The winding structure and the overall geometry of this system make it an ideal candidate for indirect cooling, thereby providing both overall simplicity and economic benefits. It operates with a peak field of 4 tesla (T), and a magnetic force of 250 T^2/m allows separation of minerals with magnetic susceptibilities in the order of 10^6 emu/g. In essence, this unit will be an industrial-scale version of the well-known laboratory-scale Frantz Isodynamic Separator.

ELECTROSTATIC SEPARATION

Almost all minerals show some degree of conductivity. An electrostatic separation process uses the difference in the electrical conductivity or surface charge of the mineral species of interest. The electrostatic separation process has generally been confined to recovering valuable heavy minerals from beach-sand deposits. However, the growing interest in plastic and metal recycling has opened up new applications in secondary material recovery.

When particles come under the influence of an electrical field, depending on their conductivity, they accumulate a charge that depends directly on the maximum achievable charge density and on the surface area of the particle. These charged particles can be separated by differential attraction or repulsion. Therefore, the important first step in electrostatic separation is to impart an electrostatic charge to the particles. The three main types of charging mechanisms are contact electrification or triboelectrification, conductive induction, and ion bombardment. Once the particles are charged, the separation can be achieved by equipment with various electrode configurations.

Triboelectrification

Triboelectrification is a type of electrostatic separation in which two nonconductive mineral species acquire opposite charges by contact with each other. The oppositely charged particles can then be separated under the influence of an electric field. This process uses the difference in the electronic surface structure of the particles involved. A good example is the strong negative surface charge that silica acquires when it touches carbonates and phosphates.

The surface phenomenon that comes into play is the work function, which may be defined as the energy required to remove electrons from any surface (Figure 7.15). The particle that is charged positively after particle–particle charging has a lower work function than the particle that is charged negatively.

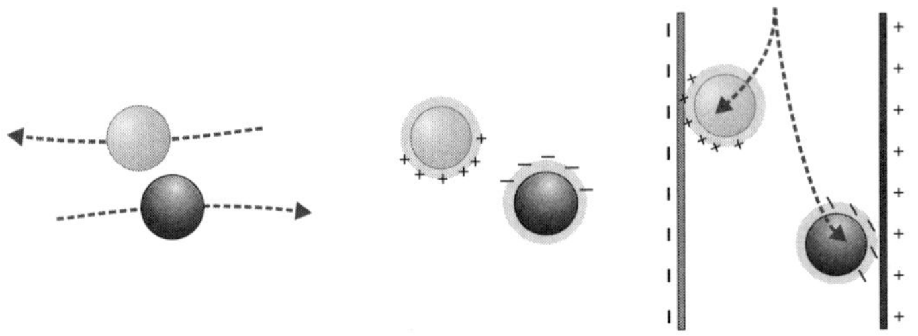

FIGURE 7.15 Particle charging mechanism; the particle charged positively has a lower work function and the particle charged negatively has a higher work function

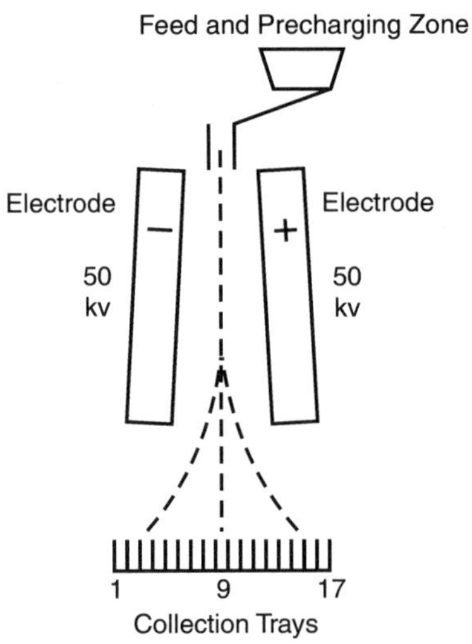

Source: Outokumpu Technology Inc., Physical Separation Division.

FIGURE 7.16 Operating principle of a V-Stat separator

Tube-type Separator. In a tube-type separator, the precharging zone and the separation zone are integral parts of the machine (Figure 7.16). The precharging zone, or triboelectrification process, exploits the difference in the electronic appearance of the particles involved. The particles become charged by particle–particle contact, particle–wall contact, or both. Particle–particle contact between two dissimilar particles results in the transfer electrons (charges) from the surface of one particle to the surface of the other. After this transfer, one of the particles is positively charged and the other is negatively charged.

The separation zone consists of two vertical walls of rotating tubes that oppose each other. Each tube "wall" is electrified with opposite potential. As the charged particles enter the separation zone, they are attracted toward oppositely charged electrodes. The separated products are collected at the base of the separator. This separator very effectively removes silica from other nonconductive minerals, such as calcium carbonate, phosphate, and talc. A typical grade–recovery curve obtained on treating limestone on the V-Stat Separator is shown in Figure 7.17. An industrial-scale triboelectrostatic separator capable of treating up to 20 tph is shown in Figure 7.18.

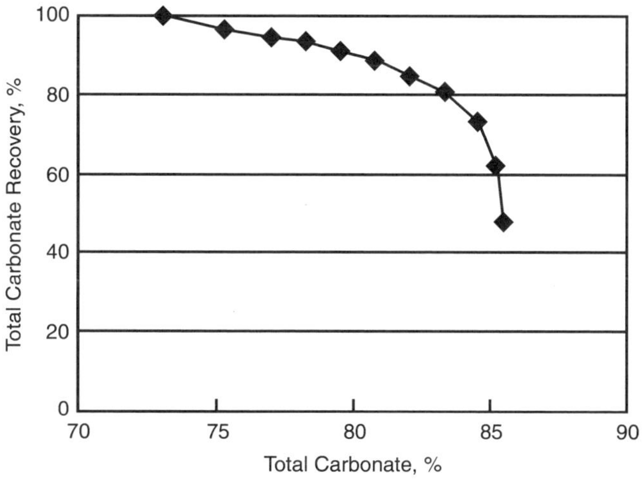

Source: Outokumpu Technology Inc., Physical Separation Division.

FIGURE 7.17 Performance of a V-Stat Separator

Source: Outokumpu Technology Inc., Physical Separation Division.

FIGURE 7.18 Industrial V-Stat Triboelectrostatic Separator

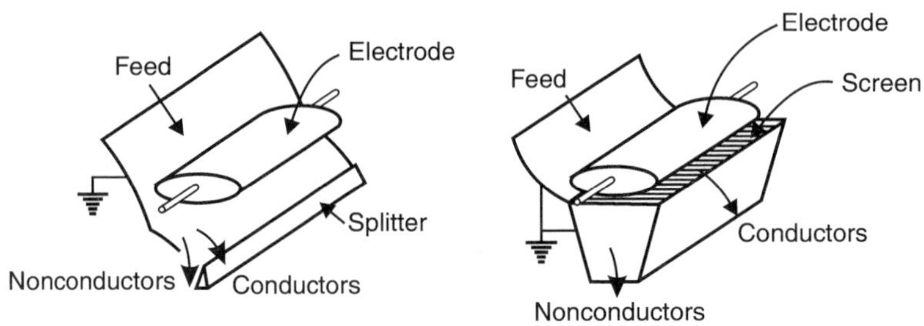

Source: Integrated Mineral Technology, Ltd.

FIGURE 7.19 Operating principle of a plate-type electrostatic separator

Belt-type Separator. In a horizontal belt-type separator, fast-moving belts travel in opposite directions adjacent to suitably placed plate electrodes of the opposite polarity. Material is fed into a narrow gap between two parallel electrodes. The particles are swept upward by a moving open-mesh belt and conveyed in opposite directions, thus facilitating the particles' charging by contact with other particles. The electric field attracts particles up or down depending on their charge. The moving belts transport the particles adjacent to each electrode toward opposite ends of the separator.

Conductive Induction

When uncharged particles, conductors or nonconductors, contact a charged surface, the particles assume polarity and the potential of the surface. The electrically conductive minerals will rapidly assume the polarity and the potential of the surface. However, in the case of nonconductors, the side away from the charged surface will more slowly acquire the same polarity as the surface. Hence, if both the conductor and nonconductor particles are just separated from contact with a charged plate (Figure 7.19), the conductor particles will be repelled by the charged plate, and the nonconductor particles will be unaffected by the charged plate–they will be neither attracted nor repelled.

The most common industrial separators working on this principle are plate- and screen-type separators. The feed particles fall under gravity onto an inclined, grounded plate and into an electrostatic field induced by a high-voltage electrode. These electrodes are generally oval. Here, the conductor particles acquire an induced charge from the grounded plate and move toward the oppositely charged electrode; that is, the particles experience a "lifting effect." The nonconductor particles are generally not affected by this field. Because the lifting effect depends on the surface charge as well as the mass of the particle, fine conductors are effectively separated from coarse nonconductors.

Ion Bombardment

When conductor and nonconductor particles placed on a grounded conducting surface are bombarded with ions of atmospheric gases generated by an electrical corona discharge from a high-voltage electrode, both the conductor and nonconductor particles acquire a charge. When ion bombardment ceases, conductor particles rapidly lose their acquired charge to the grounded surface. However, nonconductor particles react differently. The nonconductor particle surface that faces away from the grounded conducting plate is coated with ions of charge opposite in electrical polarity to that of the grounded conducting plate. Therefore, nonconductor particles remain "pinned" to the grounded plate because of electrostatic force (Figure 7.20).

An industrial high-tension electrostatic separator using the pinning effect is shown in Figure 7.21. This separator consists of a rotating roll made from mild steel that is grounded through its supporting bearing. The electrode assembly consists mainly of two types of electrodes, a beam or corona electrode or a static-type electrode. The beam electrode, usually connected to a d-c supply of up to 50 kv

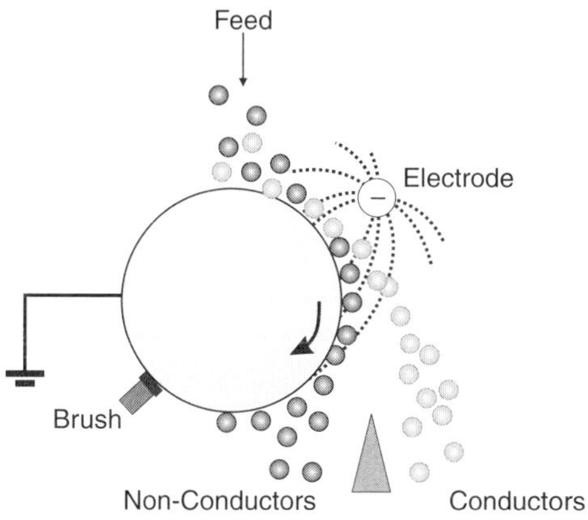

FIGURE 7.20 Operating principle of a roll-type electrostatic separator

Source: Outokumpu Technology Inc., Physical Separation Division

FIGURE 7.21 High-tension electrostatic separator used in processing plastics and metal scrap

of negative polarity, is used to charge all particles and pin the nonconductors to the roll. The feed is presented uniformly to the rotating roll surface by a velocity feed system. Both conductor and nonconductor particles are sprayed with ions. Conductor particles rapidly lose their charge to the grounded roll surface and are thrown off by centrifugal force. Nonconductor particles are pinned to the rolled surface and are brushed off that surface. Both conductor and nonconductor particles are collected in a partitioned product hopper at the bottom of the unit. The operating variables–roll speed, applied voltage, feed rate, splitter position, and the electrode combination and position–are adjusted to achieve effective separation.

BIBLIOGRAPHY

Alfano, G., P. Carbinj, R. Ciccu, M. Ghani, R. Peretti, and A. Zucca. 1988. Progress in Triboelectric Separation of Minerals. In *XVI International Mineral Processing Congress*. Edited by K.S.E. Forssberg. Amsterdam: Elsevier.

Andres, U., and W. O'Reilly. 1992. Separation of Minerals by Selective Magnetic Fluidization. *Powder Technol.,* 69:279–284.

Carpco SMS Ltd. 1993. *Specification for Cryofilter HGMS*. Model No. HGMS 5/460/10/S. Jacksonville, Fla.

D'Assumpção, L.F.G., J.D. Neto, J.S. Oliveira, and A.K.L. Resende. 1995. High Gradient Magnetic Separation of Kaolin Clay. Preprint 95–119. Littleton, Colo.: SME.

Dingwu, F., S. Jin, S. Zhougyuan, and P. Shiying. 1997. Technical Innovation and Theoretical Approach of a New Type of Permanent High Gradient Magnetic Separators (PHGMS). In *XIX International Mineral Processing Congress, San Francisco, 1995*. Littleton, Colo.: SME.

Falconer, T.H. 1992. Magnetic Separation Techniques. *Plant Engineering*, File 4599, February:85–87.

Gaudin, A.M. 1939. *Principles of Mineral Processing*. New York: McGraw-Hill.

Knoll, F.S. 1997. *Solid–Solid Operations and Equipment. Perry's Chemical Engineer's Handbook*. Edited by R.H. Perry and D.W. Green. New York: McGraw-Hill.

Knoll, F.S. et al. 1997. *Superconducting Magnetic Separators in Mineral Dressing*. Preprint 97–150. Littleton, Colo.: SME.

Knoll, F.S., and J.B. Taylor. 1985. Advances in Electrostatic Separation. *Minerals and Metallurgical Processing*. New York: AIME.

Oberteuffer, J.A. 1974. Magnetic Separation: A Review of Principles, Devices and Applications. *IEEE Trans. on Magnetics,* 10(2):223–234.

Prabhu, C. 1999. Design and Testing of a Triboelectrostatic Separator for Cleaning Coal. Master's thesis. Department of Mining and Minerals Engineering, Virginia Polytechnic Institute and State University, Blacksburg, Va.

Svboda, J. 1987. Magnetic Methods for the Treatment of Minerals. In *Developments in Mineral Processing*. Edited by D.W. Fuerstenau. Amsterdam: Elsevier.

Wasmuth, H.D., and E. Mertins. 1997. A New Medium-Intensity Drum Type Permanent Magnetic Separator and Its Practical Application for Processing Ores and Minerals in Wet and Dry Modes. In *XIX International Mineral Processing Congress, San Francisco, 1995*. Littleton, Colo.: SME.

Wills, B.A. 1992. *Mineral Processing Technology*. 5th ed. New York: Pergamon Press.

Zdenko, C. 1988. WHIMS Improves Recovery of Fine-grained Limonite at the Omarska Iron Mine. *Eng. & Mining J.*, October: 28–33.

CHAPTER 8

Flotation

Maurice C. Fuerstenau and Ponisseril Somasundaran

SURFACE PHENOMENA

Separation of minerals by froth flotation techniques depends primarily on the differences in the wettability of particles. Particles to be floated must selectively attach to air bubbles, so they must be hydrophobic. A few minerals, like sulfur, are naturally hydrophobic, so they can be floated directly, but most minerals are hydrophilic and have to be made hydrophobic by adding selected surface-active chemicals called *collectors*. These chemicals selectively coat or adsorb on the desired minerals, usually assisted by any of a number of auxiliary reagents. Most of the auxiliary reagents assist flotation by adsorbing selectively on the particles or by complexing with the many chemical species that interfere with adsorption of collector on minerals.

Wettability of the mineral particles and changes in their wettability caused by adsorption are usually expressed in terms of contact angle, defined as the angle, θ, subtended by a bubble on the solid immersed in the liquid and measured through the liquid (Figure 8.1).

Contact angle is related to interfacial tension, γ, between the gas (*G*), solid (*S*), and liquid (*L*) by the Young–Dupre equation:

$$\gamma_{SG} = \gamma_{SL} + \gamma_{SG} \cos \theta \quad \textbf{(Eq. 8.1)}$$

$$\cos \theta = \gamma_{SL} - \gamma_{SG} \quad \textbf{(Eq. 8.2)}$$

Free energy change on particle-bubble contact results from creation of a solid–gas interface and destruction of an equivalent area of solid–liquid and liquid–gas interfaces. The relationship can be written as

$$\Delta G_{ad} = \gamma_{SG} - \gamma_{SL} - \gamma_{LG} \quad \textbf{(Eq. 8.3)}$$

Combining Eqs. 8.2 and 8.3 yields

$$\Delta G_{ad} = \gamma_{LG}(\cos \theta - 1) \quad \textbf{(Eq. 8.4)}$$

Particle-bubble contact is established if θ is greater than zero and, thus, Δ*G* is negative. Collector species can be adsorbed on all three interfaces and can reduce their tensions. It is clear, then, that adsorption on all these interfaces can be important in determining particle-bubble attachment. The extent of collector adsorption and reduction in surface tension required depends on the nature of the solid and its original surface tension in the liquid. A highly hydrophilic solid (one with low γ_{SL}) also requires a low γ_{SG} or γ_{LG} for effective flotation, which could be obtained with high collector adsorption. Conversely, a less hydrophilic mineral requires minimum collector adsorption for particle-bubble attachment to occur. Direct application of Eq. 8.4 to an actual flotation process may not be appropriate, however, because the particle-bubble aggregate may be far from equilibrium under the dynamic and turbulent conditions in a flotation cell. Nevertheless, particles can be made selectively hydrophobic by altering the appropriate solid interfacial tensions through adsorption of surfactant species.

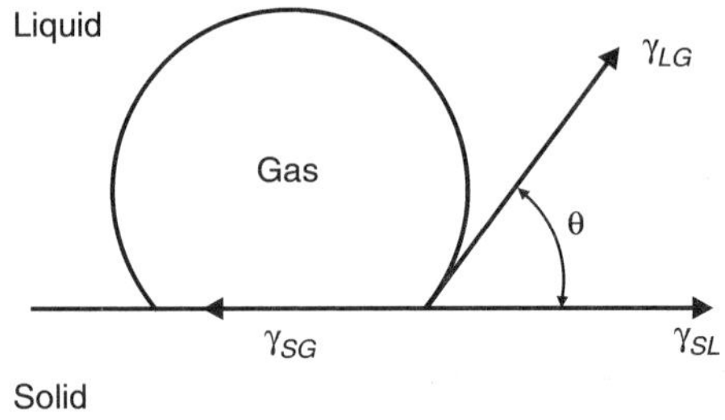

FIGURE 8.1 Schematic of the equilibrium contact between an air bubble and a solid immersed in a liquid

Adsorption

Adsorption can be considered as preferential partitioning of the adsorbents (surfactants and inorganic species) in the interfacial region resulting from favorable energy changes. Thus, when a surfactant species adsorbs on a bubble or solid, removal of hydrocarbon chains from water permits water dipoles and dissolved ions to interact with each other without an interfering nonpolar species separating them. Such adsorption results in favorable flotation conditions with respect to both the bubble and the solid particles. Adsorbed surfactants on the bubble surface provide stability to the bubbles in the froth, and they also act as a reservoir of surfactants that can migrate to the solid–gas interface when particles and bubbles collide to provide stability to the particle-bubble aggregate. In all cases, adsorption leads to a lower interfacial tension, γ, as dictated by the Gibbs adsorption equation:

$$\Gamma_A = -\frac{1}{nRT}\frac{d\gamma}{d\ln C_A} \quad \text{(Eq. 8.5)}$$

where n is the equivalents produced when one mole of the salt is dissolved in water, Γ is the adsorption density in mol/cm^2, γ is the interfacial tension in ergs/cm^2, C is the bulk concentration in mol/L, n is the number of equivalents, R is the gas constant in cal/deg·mol, and T is the absolute temperature. Thus, for dodecylammonium chloride, $n = 2$. Γ_A and C_A are adsorption density and bulk concentration, respectively, of the surfactant ion A. Equation 8.5 indicates that when adsorption takes place and Γ is positive, a Γ versus log concentration plot will have a negative slope; that is, interfacial tension will decrease with the addition of surfactant. The actual decrease in surface tension depends on the type of hydrophobic and hydrophilic groups on the surfactant molecule. Surfactants produce marked decreases in surface tension because a large number of hydrophobic aliphatic and aromatic groups are present (Figure 8.2). The surfactant's hydrocarbon chain length has the most significant effect on surface tension, because an increase in chain length by one CH_2 or CH_3 group can reduce by one-third the dosage required to produce a given reduction in surface tension.

Surfactant layers at liquid surfaces can also strengthen the interfacial film between bubbles by making them more elastic and viscous, thus producing a stable froth. The elasticity is the direct result of the decrease in surface tension on adsorption of the surfactant.

When a mechanical shock leads to thinning of the liquid film between the bubbles, the film can rupture, destroying some bubbles and producing a larger, coalesced bubble. A series of such mergers of bubbles naturally results in the destruction of most bubbles in the froth, making it unstable. Surfactant layers make the film stronger, however, because the extension of the film during its thinning produces a decrease in adsorption density of the surfactant and a corresponding increase in surface tension from γ_1 to γ_2 in the thinned area (Figures 8.2 and 8.3). This local increase in surface tension has the effect of pulling surfactant species to the damaged area. As the surfactant film flows to the damaged area, it pulls along with it some water from the surrounding area, restoring the thickness of the film. In this

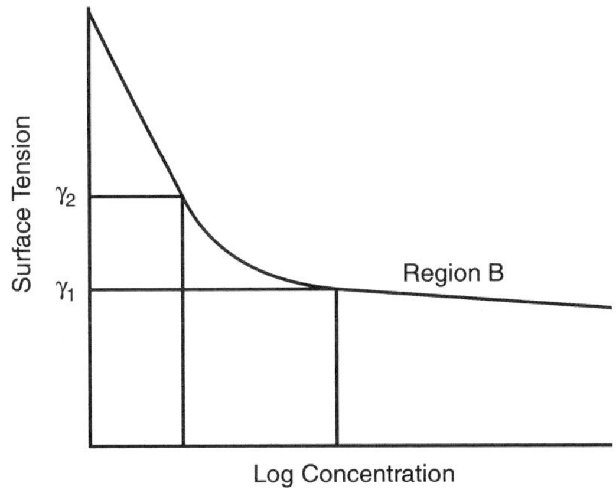

FIGURE 8.2 Variation of surface tension with concentration of surfactant

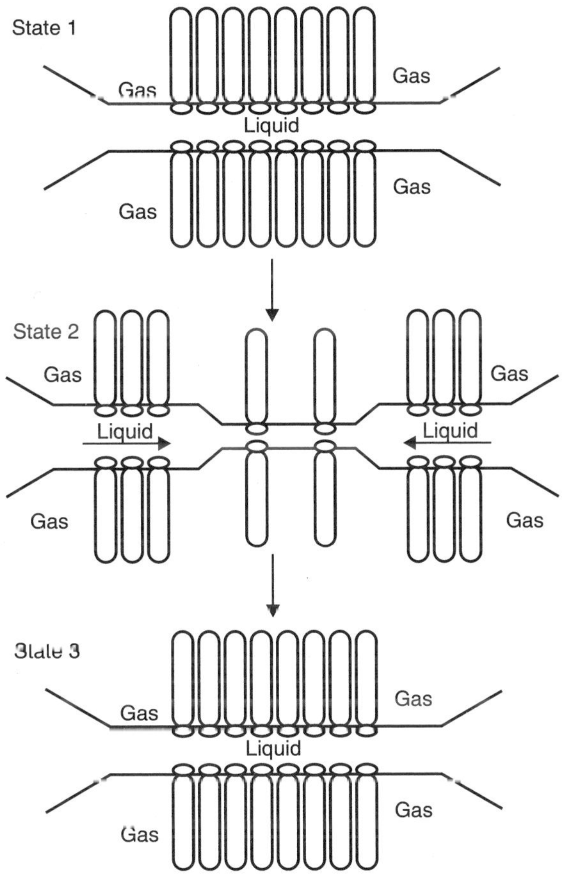

FIGURE 8.3 Diagram illustrating the repair of a stretched film for froth stability

way, the elastic nature of the film contributes to its stability. Also, it is well known that the surfactant film along with the bound water exhibits a higher viscosity than the bulk water. This increased viscosity of the film makes it difficult to rupture.

The film can be elastic only if the decrease in adsorption leads to a reduction in surface tension. In Region B of Figure 8.2, the film cannot be expected to be elastic, even though it can be extremely strong because of the high interfacial viscosity possible in concentrated solutions of certain surfactants. Also, note that if the surfactant species diffuses from the bulk water to the interface at a faster rate than from the nearby region, the surface flow will not occur, allowing rupture caused by film thinning. Manipulating both the surfactant's molecular structure and its concentration should result in the optimum combination of relative surface and diffusion rate, thus achieving the desired elasticity and viscosity.

Films can be strengthened further by the fine particles that collect at bubble surfaces. Hydrophobic particles can adsorb at the interface with the surface that contains the most hydrophobic regions in the gaseous phase and that which is essentially hydrophilic immersed in the water. Such orientation of the particles lowers the free energy of the froth system, helping to stabilize the froth. Thermodynamic conditions at the interface allow flotation if $\gamma_{LG} > \gamma_{SG}$ or if none of surface tensions of the interfaces, γ_{LG}, γ_{SG}, and γ_{LS}, is greater than the sum of the other two.

Adsorption on solids can take place because of a number of interactive forces between the adsorptive reagent and the adsorbate (solid). This phenomenon has been traditionally distinguished as physical adsorption and chemical adsorption, depending on the forces involved. Adsorption caused by weak forces such as van der Waals forces and hydrogen bonding has been called "physical adsorption," and that resulting from stronger bonding, particularly covalent bonding, has been called "chemisorption." Adsorption caused by electrostatic and hydrophobic bonding has also been labeled physical adsorption even though strong forces may be involved. A knowledge of the governing force for each mineral in a system is vital, because that understanding allows the system to be manipulated for increased selectivity. Forces normally involved in reagent adsorption on minerals include electrostatic attraction of charged surfactant or inorganic species to oppositely charged mineral surface sites, covalent bonding between the adsorptive species and surface species, cohesive lateral hydrocarbon chain–chain interaction among long-chained adsorbed surfactant species, nonpolar interaction between hydrocarbon chains and hydrophobic regions of the solid particle, hydrogen bonding, and hydration or dehydration of species because of the adsorption process. For each mineral-reagent system, one or more of these forces can be responsible for adsorption, depending on the mineral, the nature of the surfactant and its concentration, pH, temperature, ionic strength, and the nature of the dissolved species. For example, for oxides such as quartz and clays, electrostatic and hydrocarbon chain–chain interaction forces can be important, while for sulfide minerals such as galena, the covalent term is usually more important. Because of the critical role these forces play in flotation, they will be discussed in detail in the following sections.

Electrostatic Forces

Electrostatic adsorption occurs because most minerals are charged in water because of preferential dissolution of lattice ions or to hydrolysis of the surface species with pH-dependent dissociation of the hydroxyls. The latter process, which invariably occurs in mineral systems, can be described for silica as shown in Figure 8.4. At low pH values, an excess of positive sites is left on the surface, and at high pH values, an excess of negative sites results. The positive surface at low pH adsorbs additional negative ions, and the negative surface at high pH adsorbs positive ions. The pH at which the net charge of the surface is zero is a useful indicator of a mineral's electrostatic properties and is called the "point-of-zero" charge or "PZC." The ions, H^+ and OH^-, that govern the surface charge are called "potential determining" ions. The PZC of a mineral can be determined experimentally by monitoring the mobility of particles toward electrodes under the desired solution conditions, such as pH. Table 8.1 gives the PZC for a number of oxide minerals. The location of a mineral's PZC is directly related to the acidity or alkalinity of the

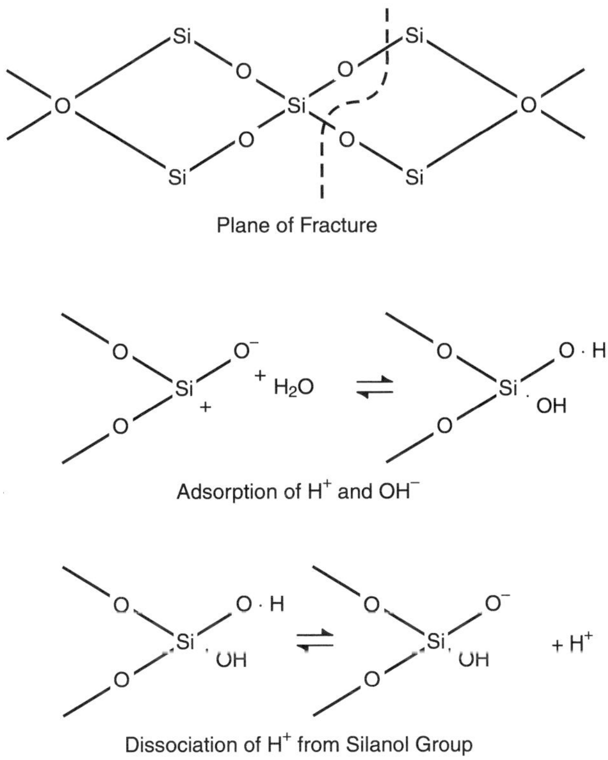

Source: Yopps and Fuerstenau 1964.

FIGURE 8.4 Schematic of surface charge development on quartz

TABLE 8.1 PZC for selected oxide minerals

Mineral	pH
Quartz, SiO_2	2
Rutile, TiO_2	6
Hematite, Fe_2O_3	7
Corundum, Al_2O_3	9
Magnesia, MgO	12

mineral itself. Silica, which forms silicic acid in water, yields an acidic PZC, while magnesia, which forms magnesium hydroxide, yields a basic PZC.

Salt-type minerals, such as calcite and apatite, can have surface charges that result from either preferential dissolution or preferential reaction of species with dissolved species in water, or both, leading to the subsequent adsorption of these minerals. Calcite, for example, undergoes various reactions on contact with water (Table 8.2).

When calcite approaches equilibrium with water at high pH values, excess negative HCO_3^-, OH^-, and CO_3^{2-} exist; at low pH values, excess positive Ca^{2+}, $CaOH^+$, and H^+ occur in solution. The total activity of the negative ions is equal to that of the positive ions at pH 8.2, the PZC.

Minerals can also become charged as a result of isomorphous substitution in the lattice. Thus, clays develop residual charge when Si^{4+} is substituted with Al^{3+} or when Al^{3+} is substituted with Mg^{2+}.

TABLE 8.2 Calcite equilibria

$CaCO_{3(s)} \Leftrightarrow CaCO_{3(aq)}$	$K_1 = 10^{-5.09}$
$CaCO_{3(aq)} \Leftrightarrow Ca^{2+} + CO_3^{2-}$	$K_2 = 10^{-3.25}$
$CO_3^{2-} + H_2O \Leftrightarrow HCO_3^- + OH^-$	$K_3 = 10^{-3.67}$
$HCO_3^- + H_2O \Leftrightarrow H_2CO_3 + OH^-$	$K_4 = 10^{-7.65}$
$H_2CO_3 \Leftrightarrow CO_{2(g)} + H_2O$	$K_5 = 10^{1.47}$
$Ca^{2+} + HCO_3^- \Leftrightarrow CaHCO_3^+$	$K_6 = 10^{0.82}$
$CaHCO_3^+ \Leftrightarrow H^+ + CaCO_{3(aq)}$	$K_7 = 10^{-7.90}$
$Ca^{2+} + OH^- \Leftrightarrow CaOH^+$	$K_8 = 10^{1.47}$
$CaOH^+ + OH^- \Leftrightarrow Ca(OH)_{2(aq)}$	$K_9 = 10^{1.37}$
$Ca(OH)_{2(aq)} \Leftrightarrow Ca(OH)_{2(s)}$	$K_{10} = 10^{2.45}$

Source: Hanna and Somasundaran 1976.

Talc, with no such substitution, exhibits a nonpolar surface with pH-dependent charge at the edge of its sheets, which is caused by broken Si–O and Al–O bonds.

Kaolinite, on the other hand, exhibits negative charge at the face of its platelets and positive or negative charge at the edges, depending on the pH.

Electrical Double Layer

Calculation of adsorption caused by electrostatic forces requires a knowledge of the electrical potential at the surface of the particle and the changes in it that result from changes in solution composition. Electrochemical potential of the surface of particles, ψ_0, is given by

$$\psi_0 = \frac{RT}{zF} \ln \frac{a^+}{a_+^0} = \frac{RT}{zF} \ln \frac{a^-}{a^0} \qquad \text{(Eq. 8.6)}$$

where R is the gas constant, F is the Faraday constant, T is the absolute temperature, and a^+ and a^- are the activities of the positive and negative potential determining ions with valences z^+ and z^- (inclusive of sign). The subscript, 0, of ψ_0 refers to the zero distance from the surface of the particle; the superscript, 0, of a^0 refers to activities of potential determining ions at the PZC.

A suspension that contains such charged particles must be electrically neutral, so it must contain an equivalent amount of oppositely charged ions, called *counter ions*. Because of the attraction by the charged surface sites, these counter ions, instead of being uniformly distributed in the solution phase, are located closer to the surface, as shown in Figure 8.5. The figure also illustrates the potential decrease that can result from this adsorption. Potential at the plane, δ, of the closest distance of approach by counter ions determines maximum adsorption and is known as the "Stern potential." Although it is not possible to determine the Stern potential experimentally, it is possible to measure the zeta potential, ζ, at the plane of shear where the liquid will move past the solid when forced using electrokinetic methods. The techniques used widely for measuring the zeta potential of minerals are electrophoresis and streaming potential.

Electrophoresis and Streaming Potential

This technique is based on the fact that when an electric field is applied to a solution containing charged particles, the particles move with a speed and direction indicative of the magnitudes of their charges and the signs of charges, respectively. Electrophoresis can be used reliably only for micron-size particles. For coarser, flotation-size particles, zeta potential can be measured using the streaming potential technique. In this technique, the desired solution is streamed through a plug of mineral particles, and the potential

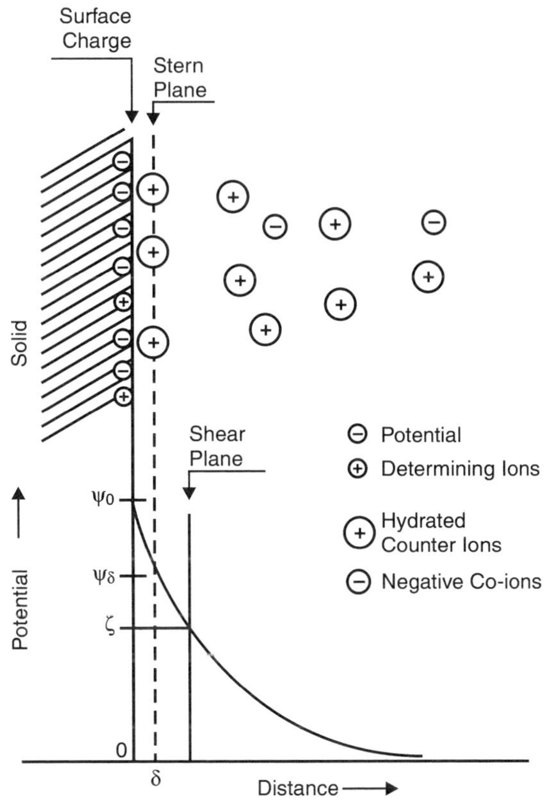

Source: Somasundaran 1975.

FIGURE 8.5 Schematic of electrical double layer

developed across the plug is measured. This potential, called the "streaming potential," E, is related to the zeta potential, ζ, by

$$\zeta = \frac{4\pi\eta}{\varepsilon}\frac{E}{P}\lambda \tag{Eq. 8.7}$$

where η is viscosity, ε is the dielectric constant, P is the driving pressure across the bed, and λ is the specific conductivity.

Note that the absolute value of the zeta potential itself might be subject to question because of uncertainties in the values used for various parameters and assumptions involved in the derivation of the equation. Nevertheless, it is the *change* in the zeta potential resulting from adding reagents that is helpful in understanding the mechanisms of adsorption of these reagents. For example, Figure 8.6 shows that a change in pH of a goethite suspension from, for example, pH 4 to pH 7, will reverse the surface potential from positive to negative. The figure also illustrates that adding salt can lead to a reduced zeta potential caused by the crowding of the counter ions in the interfacial region (compression of the double layer). This phenomenon can be expected to reduce the electrostatic adsorption of the oppositely charged surfactant ions. Note, however, that adding salt can lead to increased salting out of the surfactant from the bulk solution, and the resulting decreased solubility will increase the driving force for adsorption. The net effect of salts on the zeta potential and the surfactant solubility will determine whether surfactant adsorption and mineral flotation are decreased or increased.

The zeta potential of the mineral particle also determines the extent of flocculation and dispersion between particles of different minerals in the suspension. Because flocculation between particles

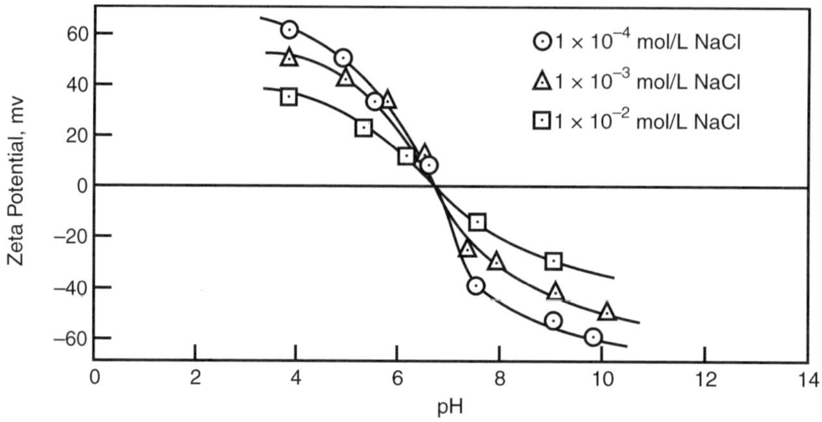

FIGURE 8.6 Zeta potential of goethite as a function of pH and ionic strength

naturally leads to poor selectivity in flotation, it must be prevented by adjusting the pH or adding reagents that can suitably modify the zeta potential of one or more particle types. This precaution becomes particularly important when the slurry contains very fine particles (slimes) that can coat the valuable coarse mineral particles and destroy their distinctive surface properties. Such coating of minerals by clay slimes is the culprit in many flotation problems. Manipulation of the zeta potential of various particles should prevent slime coating and even flocculate the slime particles for subsequent removal by gravity techniques.

FLOTATION REAGENTS

A number of organic and inorganic reagents are used in flotation and auxiliary processes to achieve separation, including collectors, frothers, extenders, activators, depressants, deactivators, flocculants, and dispersants. Collectors, frothers, and extenders are surfactants added to impart hydrophobicity to the minerals and to make *selective* adsorption of the collector possible or to eliminate interference to flotation by various dissolved or colloidal species.

Collectors

The primary role of the collector is to adsorb selectively, which imparts hydrophobicity to particles of the mineral to be floated. If it is to have the dual ability to adsorb and to impart hydrophobicity, the collector molecule must contain at least two functional parts, a nonpolar group of sufficient hydrophobicity and a polar or ionic group that will be electrostatically or chemically reactive toward species on the mineral surface. The nonpolar part of a collector used for flotation of oxides is usually a long-chained hydrocarbon (10 to 18 CH, CH_2, and CH_3 groups); short-chained hydrocarbons (2 to 5 CH_2 or CH_3 groups) are used for flotation of sulfides. The polar group is usually anionic sulfate, sulfonate, phosphate, carboxylate, oxime or thiocarbonate (xanthate), cationic amine, or nonionic oximes.

Examples of collectors used in froth flotation are given in Tables 8.3 through 8.6. Ethyl xanthate is used for flotation of galena, sphalerite, and pyrite; oleic acid is used for flotation of phosphates and hematite; and dodecylamine is used for flotation of quartz, potash, and feldspars. Collection by these surfactants depends on their properties, such as ionization constant, solubility, critical micelle concentration, and emulsifying power. Any surfactant must be solubilized or dispersed properly so that it can distribute itself in the pulp and contact the mineral surface with minimum use of mechanical energy. However, note that a highly soluble surfactant has a low tendency to depart from the solution and

TABLE 8.3 Formulas for cationic collectors

Amine	Formula
n-amyl amine	$C_5H_{11}NH_2$
n-dodecylamine	$C_{12}H_{25}NH_2$
Di-n-amylamine	$(C_5H_{11})_2NH$
Tri-n-amylamine	$(C_5H_{11})_3N$
Tetramethylammonium chloride	$[(CH_3)_4N]^+Cl^-$
Tallow amine acetate	RNH_3Ac (96% C_{18})

TABLE 8.4 Structural formulas of sodium salts of various anionic collectors

Collector	Structural Formula	
Carboxylate	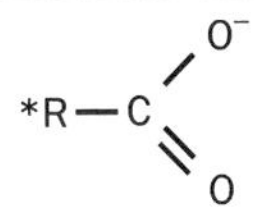	......Na^+
Sulfonate	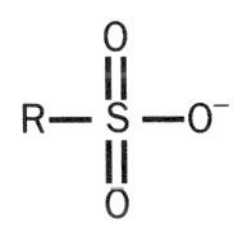	......Na^+
Alkyl sulfate	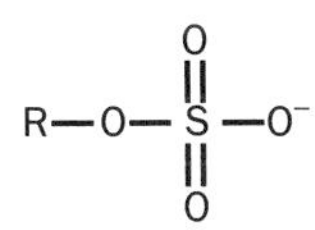	......Na^+
Hydroxamate	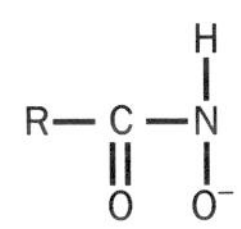	......Na^+

*R represents the hydrocarbon chain.

TABLE 8.5 Structure and solubility of selected fatty acids

Fatty Acid	Formula	Solubility of Undissociated Molecule (mol/L), 20°C
Capric	$CH_3(CH_2)_8COOH$	3.0×10^{-4}
Lauric	$CH_3(CH_2)_{10}COOH$	1.2×10^{-5}
Myristic	$CH_3(CH_2)_{12}COOH$	1.0×10^{-6}
Palmitic	$CH_3(CH_2)_{14}COOH$	6.0×10^{-7}
Stearic	$CH_3(CH_2)_{16}COOH$	3.0×10^{-7}
Oleic	$CH_3(CH_2)_7CH = CH(CH_2)_7COOH$	
Linoleic	$CH_3(CH_2)_4CH = CHCH_2CH = CH(CH_2)_7COOH$	
Linolenic	$CH_3CH_2CH = CHCH_2CH = CHCH_2CH = CH(CH_2)_7COOH$	
Abiotic	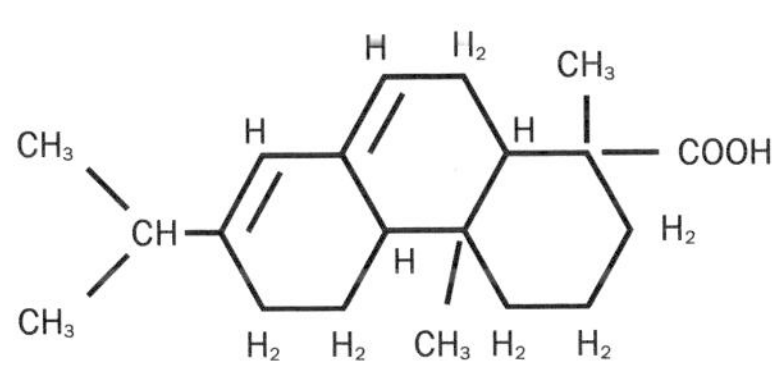	

Source: Schubert 1967.

TABLE 8.6 Various anionic sulfhydryl collectors

Xanthate	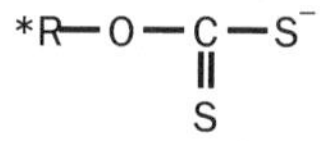	......Na^+
Thiophosphate	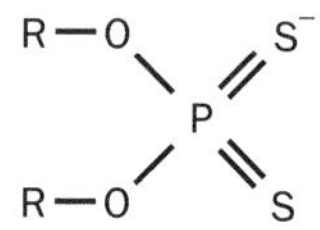	......Na^+
Thiocarbamate	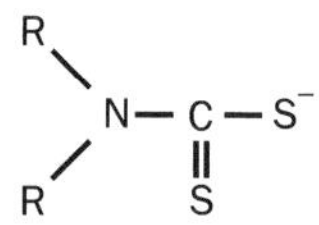	......Na^+
Mercaptan	$R-S^-$	......Na^+
Thiourea	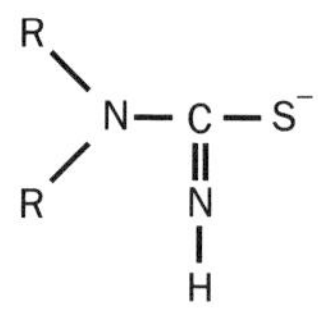	......Na^+
Mercaptobenzothiazole	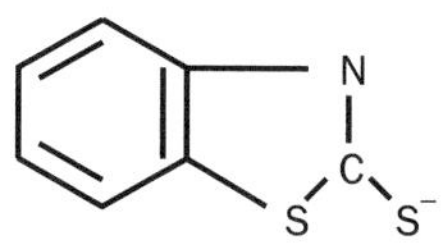	......Na^+

*R represents the hydrocarbon chain.

TABLE 8.7 Solubility of undissociated molecules of various amines (mol/L)

Decylamine	5×10^{-4}
Dodecylamine	2×10^{-5}
Tetradecylamine	1×10^{-6}

Source: Aplan and Fuerstenau 1962.

adsorb on interfaces. The tendency to form micelles also influences the utility of the surfactant for flotation. Micelles are aggregates of surfactants with hydrocarbon chains oriented toward the interior of the aggregates and the polar or ionic part oriented to be in contact with the water (Figure 8.7). Each surfactant forms micelles when its bulk concentration reaches a particular value known as the "critical micelle concentration" (CMC). Above the CMC, important properties of the surfactant solutions undergo a marked change. For example, surface tension of water decreases with the addition of a surfactant but only up to the CMC. Above the CMC (the point of onset of Region B in Figure 8.2), surface tension remains approximately constant, suggesting that the activity or concentration of the surface-active monomer species is constant above the CMC and that the micelles themselves are not surface-active. Solubility and CMCs of the most commonly used surfactants are given in Tables 8.7, 8.8, and 8.9.

Surfactants can form salts with the dissolved species of the mineral and other additives, and solubility of these salts can also have a major influence on the extent of flotation obtained. Solubility products of various metal carboxylates and xanthates are given in Tables 8.10 and 8.11. Good correlation exists between the flotation and precipitation properties of surfactants. In many systems, precipitation can also be expected to occur on the mineral surface and lead to good flotation. An example is shown

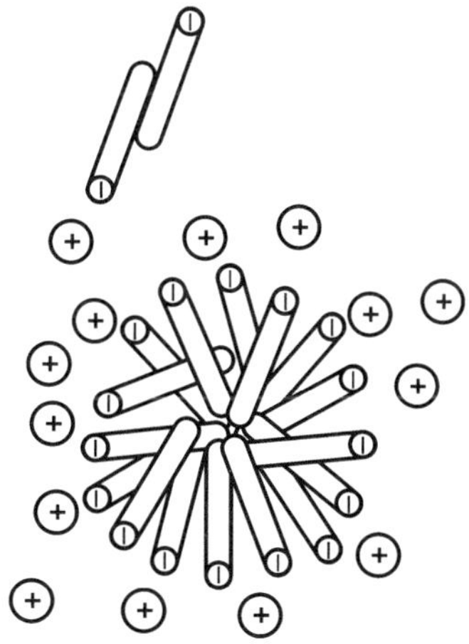

FIGURE 8.7 Schematic of a micelle

TABLE 8.8 Critical micelle concentrations of various amines (mol/L)

Decylamine	3.2×10^{-2}
Dodecylamine	1.3×10^{-2}
Tetradecylamine	4.1×10^{-3}
Hexadecylamine	8.3×10^{-4}
Octadecylamine	4.0×10^{-4}

Source: Aplan and Fuerstenau 1962.

TABLE 8.9 Critical micelle concentrations of various carboxylates, sulfonates, and alkyl sulfates

	CMC, mol/L		
Chain Length	**Carboxylate**	**Sulfonate**	**Alkyl Sulfate**
C_{12}	2.6×10^{-2}	9.8×10^{-3}	8.2×10^{-3}
C_{14}	6.9×10^{-3}	2.5×10^{-3}	2.0×10^{-3}
C_{16}	2.1×10^{-3}	7.0×10^{-4}	2.1×10^{-4}
C_{18}*	1.8×10^{-3}	7.5×10^{-4}	3.0×10^{-4}

* Temperature, 50°C; other determinations at room temperature.

TABLE 8.10 Solubility products of various metal carboxylates

	Ag^+	Pb^{2+}	Cu^{2+}	Zn^{2+}	Cd^{2+}	Fe^{2+}	Ni^{2+}	Mn^{2+}	Ca^{2+}	Ba^{2+}	Mg^{2+}	Al^{2+}	Fe^{3+}
Palmitate	12.2	22.9	21.6	20.7	20.2	17.8	18.3	18.4	18.0	17.6	16.5	31.2	34.3
Stearate	13.1	24.4	23.0	22.2	—	19.6	19.4	19.7	19.6	19.1	17.7	33.6	—
Oleate	10.9	19.8	19.4	18.1	17.3	15.4	15.7	15.3	15.4	14.9	13.8	30.0	34.2

Source: Du Reitz 1975.

TABLE 8.11 Solubility products of various metal xanthates

	Ag^+	Pb^{2+}	Cu^{2+}	Ni^{2+}	Co^{2+}	Fe^{2+}	Zn^{2+}	Mn^{2+}
Ethyl	18.6	16.7	24.2	12.5	—	—	8.2	—
Isopropyl	18.6	17.8	24.7	13.4	—	—	—	—
Butyl	19.5	18.0	26.2	—	—	—	—	—
i-Butyl	19.2	17.3	26.3	—	—	—	—	—
Amyl (*i*-)	19.7	17.6	27.0	14.5	—	—	—	—
Hexyl (*n*-)	20.8	20.3	29.0	16.5	14.3	—	—	—
Octyl (*i*-)	20.4	21.3	—	17.7	—	—	—	—
n-Nonyl	22.6	24.0	30.0	22.3	21.3	11.0	16.2	9.9

Source: Du Reitz 1975.

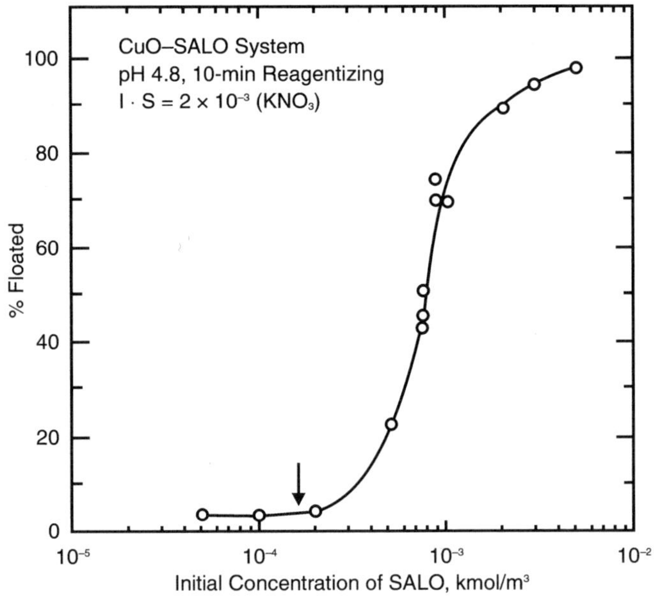

Source: Nagaraj and Somasundaran 1981.

FIGURE 8.8 Dependence of CuO flotation on concentration of salicylaldoxime (SALO). Arrow indicates onset of precipitation of Cu–SALO complex.

in Figure 8.8, where the onset of flotation can be seen to correlate well with the onset of precipitation calculated using data for bulk-solution chemical equilibria. Note that excessive collector loss is possible if collector precipitation occurs exclusively in the bulk (because of a rate of metal ion dissolution and diffusion through the interface that is faster than the rate of diffusion of collector to the particle), because bulk precipitates are not potent collectors.

Oxime-type reagents can act as very good collectors for problematic minerals because of their ability to chelate with the metallic surface species. Thus, hydroxamate and salicylaldoxime can adsorb on chrysocolla or tenorite (CuO) and result in their flotation. Potential for the use of oximes is related directly to their solubility. Also, bulk precipitation as well as detachment of the surface chelate from the particle can interfere with flotation, as bulk chelates are incapable of causing collection.

TABLE 8.12 Common frothers

Frother	Constituent	
Cresylic acid	Xylenol	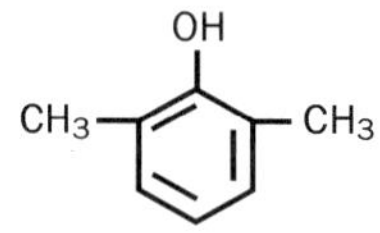
MIBC	Methyl isobutyl carbinol	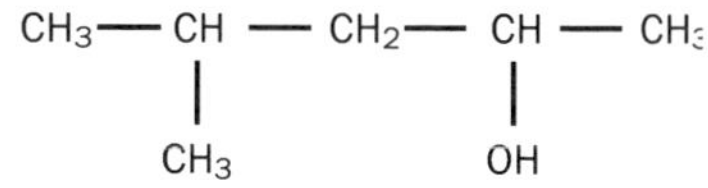
Polyglycols	Polypropylene	CH_3—$(O$—$C_3H_6)_nOH$
Pine oil	Terpineol	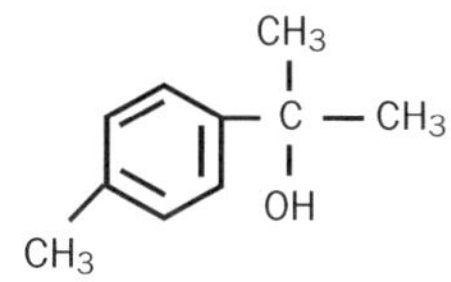

Frothers

The bubbles that rise to the top of the flotation cell must not break until they are skimmed off to collect the floated particles. To produce the desired stability of the froth that forms in the cell, nonionic surfactants such as monohydroxylated cresols are usually added unless the collector itself can act as a frother. When long-chained collectors such as oleic acid are used, they will adsorb also at the bubble surface in sufficient amounts to achieve an elastic surface and stable bubbles. However, when short-chained chemicals such as ethyl xanthate are used as collectors, additional reagents must be added for froth stability. Table 8.12 lists some commonly used frothers.

Along with the benefits of froth stability, frother species can co-adsorb with the collector on the particle. Also, the frother on the bubble surface can migrate to the particle–gas interface during the time of contact to arrive at the desired equilibrium adsorption density at that interface on the solid-bubble contact perimeter. Such migration and co-adsorption of the frother can anchor the bubble onto the particle and cause it to adhere as desired.

Extenders

In addition to frothers and collectors, nonionic and nonpolar surface-active agents are used in many flotation schemes simply to enhance the hydrophobicity of the particles and the resultant flotation recovery. Kerosene and fuel oils are used in the flotation of phosphates and coal, for example. These reagents are thought to act by forming a multilayer coating on the already partly hydrophobic surfaces. They can also act like frothers by co-adsorbing with collectors. Intensive agitation is required in some cases to disperse and "smear" these reagents onto particle surfaces.

Activators

Many minerals do not adsorb collectors, so they do not float unless special reagents are added to activate adsorption. For example, oleate will not float quartz on its own, but flotation will occur when calcium salts are added to the pulp at high values of pH where hydrolysis of Ca^{2+} has occurred. Similarly, copper sulfate acts as an activator for the flotation of sphalerite using xanthate as a collector at relatively low concentration. An activator normally acts by adsorbing on the mineral, providing sites for adsorption of the collector species. Copper ion exchanges for zinc ion of the mineral surface, and the sphalerite particle then behaves in flotation like a copper sulfide particle.

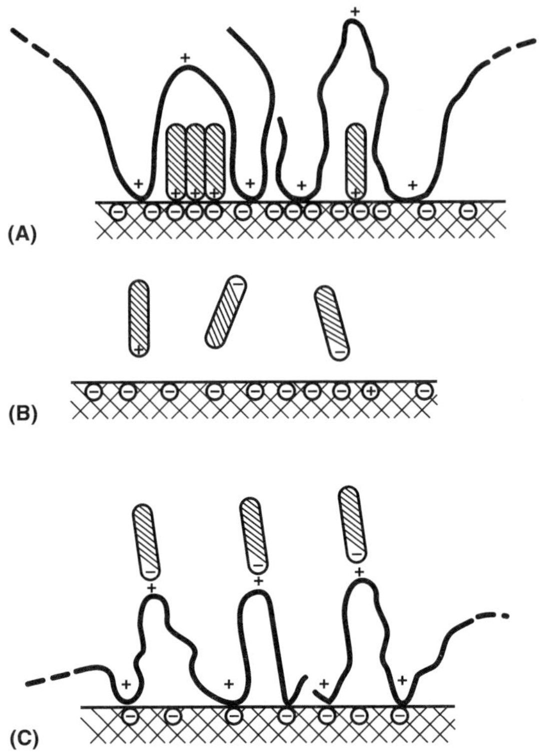

Source: Somasundaran and Cleverdon 1985.

FIGURE 8.9 (A) Schematic of the cationic polymer PAMA and dodecylamine co-adsorption on quartz particles resulting in their flotation depression, (B) schematic representation of quartz/dodecyl sulfonate system, (C) schematic representation of the cationic polymer PAMA and dodecyl sulfonate co-adsorption on quartz particles resulting in their flotation activation

Multivalent ions can adsorb on oppositely charged particles and reverse their zeta potential, causing adsorption of collectors that have a charge of the same sign as that of the mineral. An example is sulfate activation of alumina.

Depressants

Depressants retard or inhibit flotation of a desired solid. The action of a depressing agent is often a result of its adsorption on the particle surface, which preempts the collector from adsorbing and masks the adsorbed collector from the bulk solution so that the particle does not exhibit a hydrophobic exterior. For example, multivalent ions, such as phosphate, can prevent oleate adsorption on apatite because of charge reversal by the phosphate species. Multivalent ions can also act by depleting the collector through precipitation; that is, calcium can depress flotation of apatite by removing oleate from the solution as calcium oleate precipitate. Other chemicals used as depressants include silicates, chromates, dichromates, and aluminum salts. Organics are also used as depressants. Common examples include starch, tannin, quebracho, and dextrin. These massive molecules probably act by adsorbing on the mineral surface, sometimes even with the collector species, and then masking the collectors' hydrophobic tails with their own large size. Figure 8.9 illustrates depression of flotation of quartz by amine through action of a cationic polymer. In this case amine does adsorb on quartz, even in the presence of the polymer, but flotation is prevented. Note that the same polymer can activate quartz flotation using an anionic collector such as dodecyl sulfonate.

Deactivators

Deactivators are chemicals that react with activators to form inert species, thus preventing flotation. For example, activation of sphalerite with copper using xanthate as a collector is prevented by adding cyanide, which complexes with copper.

Dispersants and Flocculants

Flotation is often hampered by the presence of fine particles called slimes, which can coat the coarser mineral particles and consume excessive amounts of reagents because of their large specific surface areas. When slimes are a problem, chemicals such as silicates, phosphates, and carbonates are usually added to disperse them. Some of these chemicals also influence flotation, because they can complex with deleterious chemical species. Oxalic acid, tartaric acid, and ethylenediaminetetraacetic acid (EDTA) are often used for this purpose.

Systems for beneficiation and effluent treatment often deal with fines by flocculation using polymers. The polymers used include starch and its derivatives, polyacrylamides, and polyethylene oxide. Polymers flocculate particles into larger aggregates (flocs) by forming bridges between them. Adsorption of polymers on the mineral particles is attributed to hydrogen bonding between functional groups, such as OH and $-NH_2$ and surface –OH on the mineral particles, or chemical or electrostatic bonding between polymer functional groups and surface sites. In addition to the mineral and polymer properties, the extent of flocculation also depends on variables such as mode of polymer addition, dosage, and agitation. Polymers that can selectively adsorb on mineral fines are used also to selectively flocculate them, followed by separation of the flocs from gangue using elutriation or flotation. For example, starch is used to selectively flocculate hematite from fine taconite ore, which is then separated by floating the coarse quartz using amine. Another example is hydroxamated polyacrylamide, which is strongly adsorbed on iron oxide in "red mud" effluents from the Bayer process. Polymers added for flocculation should not interfere with downstream processes such as flotation, filtration, or effluent treatment. Note also that many low-molecular-weight polymers can act as dispersants.

pH as Modifier

The pH of the pulp must be carefully controlled to maximize recovery and selectivity. Sodium hydroxide, lime, sodium carbonate, ammonia, hydrochloric acid, and sulfuric acid are used to control pH.

CHEMISTRY OF FLOTATION

Minerals fall conveniently into five categories of flotation systems: naturally floatable, sulfides, insoluble oxides and silicates, semisoluble salts, and soluble salts. Each of these systems is treated separately in the sections that follow.

Natural Floatability

Natural hydrophibicity of solids results principally from structural and bonding phenomena (Gaudin, Miaw, and Spedden 1957; Chander, Wie, and Fuerstenau 1975). Gaudin and colleagues (1957) stated that native floatability results when at least some fracture or cleavage surfaces form without rupture of interatomic bonds other than residual bonds. As an example, molybdenite, one of a number of solids displaying natural hydrophobicity, is composed of electrically neutral layers of molybdenum sulfide. These layers, in turn, are held together by weak residual forces (i.e., van der Waals bonds). This structure results in a preferential cleavage along the (0001) basal planes. Some minerals that exhibit natural hydrophobicity are listed in Table 8.13.

TABLE 8.13 Naturally hydrophobic minerals and their respective contact angles

Mineral	Composition	Surface Plane	Contact Angle, degree
Graphite	C	0001	86
Coal	Complex HC		20–60
Sulfur	S		85
Molybdenite	MoS_2	0001	75
Stibnite	Sb_2S_3	010	
Pyrophyllite	$Al_2(Si_4O_{10})(OH)_2$	001	
Talc	$Mg_3(Si_4O_{10})(OH)_2$	001	88
Iodyrite	AgI		20

Source: Gaudin, Miaw, and Spedden 1957; Derjaguin and Shukakidse 1960–1961; Arbiter et al. 1975; Chander, Wie, and Fuerstenau 1975.

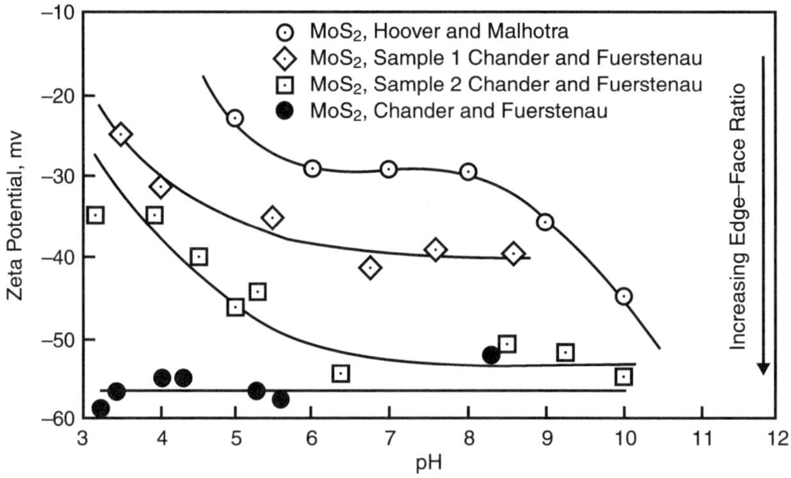

Source: Chander and Fuerstenau 1972; Hoover and Malhotka 1976.

FIGURE 8.10 Zeta potentials of various samples of molybdenite and molybdic oxide as a function of pH

Although the selected surfaces of these solids have a net hydrophobic character, two additional factors must be considered. First, the overall behavior of the surface is classified as having a net hydrophobic character, but a significant number of hydrophilic sites may exist on the surface. As a result, a hydrophobic solid may still exhibit a surface charge and may have an adsorption potential for solutes arising from coulombic forces, chemisorption forces, hydrogen bonding forces, or all three. As might be expected, the degree of hydrophobicity is greatest when the surface potential exhibits a minimum.

These phenomena are revealed in the electrokinetic responses of various molybdenite samples given in Figure 8.10. More negative potentials are realized as the ratio of edge to face of the molybdenite is increased. The greater the contribution of the edges to the total area, the greater is the number of hydrophilic sites arising from oxidation and formation of thiomolybdate anion.

Furthermore, during size reduction, crystal planes other than those exhibiting a net hydrophobic character may be exposed. As a consequence, particles might be considered to exhibit native floatability when, in fact, a significant fraction of their surfaces consist of other cleavage planes that do not exhibit hydrophobic character.

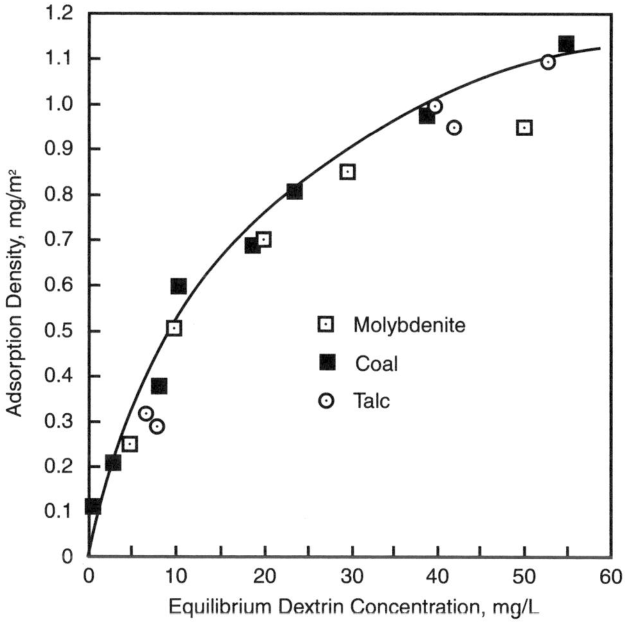

Source: Miller, Laskowski, and Change 1983.

FIGURE 8.11 Adsorption isotherm of dextrin on various naturally hydrophobic minerals

Flotation of various sulfides in oxygen-deficient systems has shown the natural floatability of these minerals under these conditions (Ravitz 1940; Lepetic 1974; Finkelstein and Allison 1976; Yoon 1981; Fuerstenau and Sabacky 1981; Heyes and Traher 1984; Luttrell and Yoon 1983; Buckley and Woods 1984). It is postulated that the presence of elemental sulfur or surface polysulfides formed under slightly oxidizing conditions is responsible for the natural floatability observed under these conditions. The absence of hydrogen bonding of water molecules to surface sulfide atoms may also influence this natural hydrophobicity.

Depression of solids that exhibit natural floatability can be achieved by chemically altering their surfaces. Usually, extensive oxidation is required. Alternatively, depression can be achieved by the adsorption of organic colloids, specifically derivatives of starch (Figure 8.11). Apparently, these polymers bond hydrophobically to the mineral surface and extend their polar hydroxyl groups to the aqueous phase so that water molecules are oriented in the polar force field. The naturally hydrophobic mineral then becomes hydrophilic. The nonspecific nature of this bonding is revealed by the fact that the same dextrin adsorption isotherm fits three naturally hydrophobic minerals of quite different chemical composition. Finally, the heat of adsorption is the same in each case at about −0.5 kcal/mole of monomeric unit (Miller, Laskowski, and Chang 1983).

Flotation of Sulfide Minerals

Electrochemical Phenomena. Sulfide minerals are semiconductors, enabling electrochemical phenomena to occur in these systems. These minerals, therefore, develop a potential, termed the *rest potential,* when placed in an aqueous solution. Rest potentials of various sulfide minerals have been established under flotation conditions, as Table 8.14 shows for a solution containing 6.25×10^{-4} mol/L ethyl xanthate at pH 7.

TABLE 8.14 Rest potentials and products of interaction of sulfide minerals with 6.25 × 10^{-4} mol/L ethyl xanthate at pH 7

Mineral	Rest Potential, v*	Product
Pyrite	0.22	Dixanthogen
Arsenopyrite	0.22	Dixanthogen
Pyrrhotite	0.21	Dixanthogen
Chalcopyrite	0.14	Dixanthogen
Bornite	0.06	Metal xanthate
Galena	0.06	Metal xanthate

Source: Allison et al. 1972.

*Reference, S.C.E., standard calomel electrode.

These values should be compared with the reversible potential for xanthate oxidation to dixanthogen in this system, which is 0.13 volt (IUPAC). This value is obtained as follows:

$$X_2 + 2e \leftrightarrow 2X^- \qquad E^\circ = -0.06 \text{ v} \tag{Eq. 8.8}$$

$$E_{rev} = E^\circ - \frac{RT}{nF} \ln \frac{(X^-)^2}{(X_2)_{(l)}} \tag{Eq. 8.9}$$

$$E_{rev} = -0.06 - \frac{1.98 \times 298}{2 \times 23{,}060} \ln \frac{(6.25 \times 10^{-4})^2}{1} \tag{Eq. 8.10}$$

$$E_{rev} = -0.06 + 0.19 = +0.13 \text{ v} \tag{Eq. 8.11}$$

Because dixanthogen, $X_{2(l)}$, is a pure liquid, its activity is unity.

When the rest potential is anodic, or larger than the reversible or Nernst potential, oxidation of xanthate to dixanthogen occurs. Referring to Table 8.14, dixanthogen is the reaction product found on the various mineral surfaces with rest potentials greater than +0.13 v.

When the rest potential is cathodic, or less than the reversible xanthate/dixanthogen potential, oxidation of xanthate cannot occur, and only metal xanthates are observed on the sulfide surface.

Galena Flotation. The flotation response of galena with 1×10^{-5} mol/L ethyl xanthate in the presence of air is presented in Figure 8.12. The figure shows that complete flotation is effected in the range from pH 2 to pH 10.

Xanthate is present in two forms on a galena surface under conditions in which flotation occurs. One form is xanthate chemisorbed at monolayer coverage; the other is bulk-precipitated lead xanthate adsorbed at multilayer coverage (Taylor and Knoll 1934; Leja, Little, and Poling 1963). Dixanthogen does not form under these conditions and is not observed (Allison et al. 1972; Kuhn, 1968; see Table 8.14). Confirmation of the presence of bulk lead xanthate on the galena surface has been provided by infrared spectrometry (Leja, Little, and Poling 1963).

The multilayers of lead ethyl xanthate are held together by van der Waals bonding of the hydrocarbon chains of the xanthate, and these layers can be dissolved with organic reagents such as acetone. However, the xanthate chemisorbed at monolayer coverage cannot be leached from the surface.

In the chemisorption of xanthate at monolayer coverage, one xanthate ion adsorbs on each surface lead ion to form an unleachable phase of lead xanthate. Electrochemical measurements suggest that monolayer adsorption involves charge transfer, with the discharged xanthate ion being fixed at the galena surface and hydroxyl ion being formed (Woods 1976). Ion exchange of xanthate ion for hydroxyl ion may also occur under these conditions.

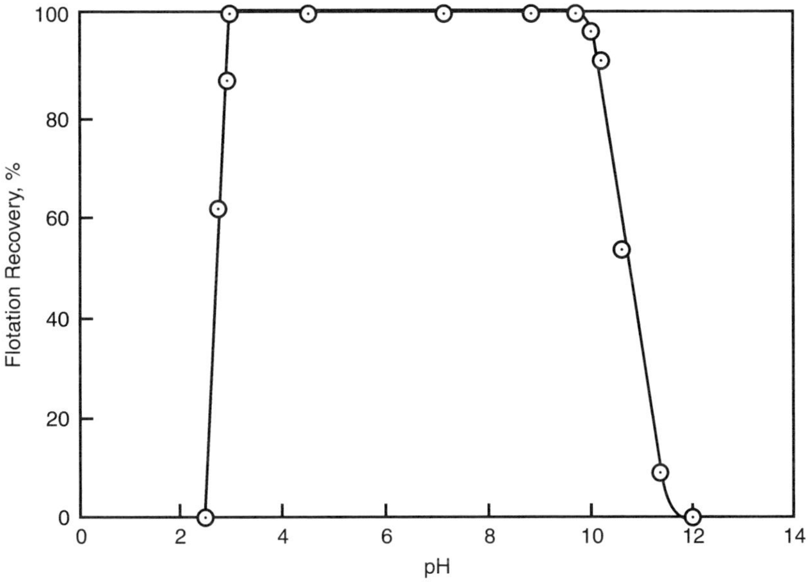

Source: Fuerstenau, Miller, and Kuhn 1985.

FIGURE 8.12 Flotation recovery of galena as a function of pH with 1 × 10^{-5} mol/L ethyl xanthate in the presence of air

Multilayer coverage occurs from the following sequence of events:

1. *Oxidation of surface sulfide to thiosulfate and sulfate.* In the presence of oxygen, galena is oxidized to lead sulfate according to Eq. 8.12:

$$PbS_{(s)} + 2O_{2(g)} \Leftrightarrow PbSO_{4(s)} \qquad K = 10^{126} \qquad \textbf{(Eq. 8.12)}$$

with the following mass action expression:

$$\frac{(a_{PbSO_{4s}})}{(a_{PbS_s})(P_{O_2})^2} = 10^{126}$$

At equilibrium, $P_{O2} = 10^{-63}$ atm. Because the oxygen tension in air is 0.2 atm, oxidation of surface sulfide to thiosulfate and sulfate occurs spontaneously in galena systems open to the air.

2. *Metathetic replacement of surface thiosulfate and sulfate by carbonate.* Because the CO_2 content of air is 300 ppm by volume in systems open to the air, carbonate ion will be present in solution, and the galena surface will carbonate at the expense of thiosulfate and sulfate.

$$PbSO_{4(s)} + CO_3^{2-} \Leftrightarrow PbCO_{3(s)} + SO_4^{2-} \qquad \textbf{(Eq. 8.13)}$$

3. *Metathetic replacement of surface carbonate, sulfate, and thiosulfate by xanthate.* At the usual flotation pH of 8 to 9, lead xanthates are more stable than lead carbonate, sulfate, or thiosulfate, and lead xanthate will form by metathetic replacement of these lead salts.

$$PbCO_{3(s)} + 2X^- \Leftrightarrow PbX_{2(s)} + CO_3^{2-} \qquad \textbf{(Eq. 8.14)}$$

$$PbSO_{4(s)} + 2X^- \Leftrightarrow PbX_{2(s)} + SO_4^{2-} \qquad \textbf{(Eq. 8.15)}$$

The exchange between xanthate ion abstracted and reduced sulfur-oxy, sulfate, and carbonate ions released to solution is stoichiometric (Taylor and Knoll 1934). Further evidence supporting metathetic replacement of surface anions by xanthate has been provided by Mellgren and Rao (1963), who used calorimetry to obtain the data.

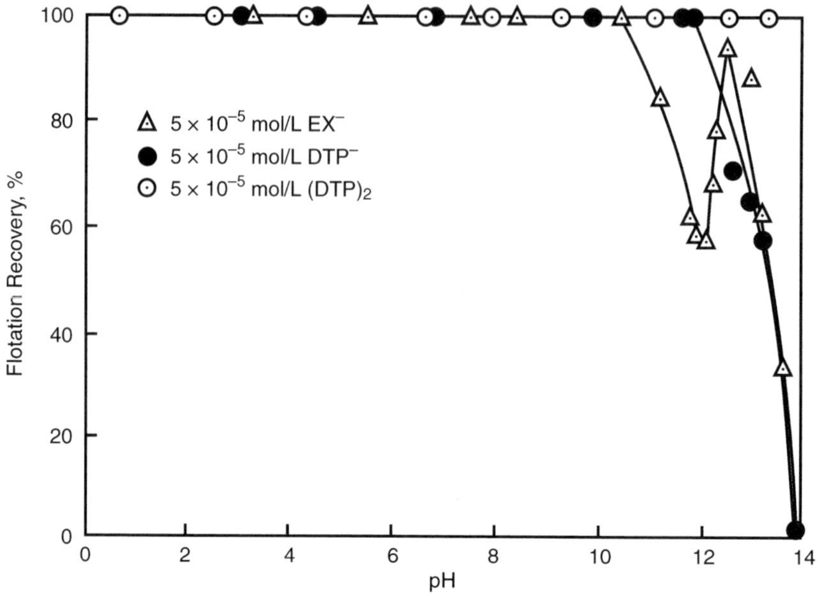

Source: Fuerstenau, Huiatt, and Kuhn 1971.

FIGURE 8.13 Flotation recovery of chalcocite as a function of pH with various additions of ethyl xanthate, diethyl dithiophosphate, and diethyl dithiophosphatogen

Adsorption of xanthate on galena, then, apparently occurs in two stages. The first stage comprises chemisorption of one xanthate ion on each surface lead ion. The second stage comprises the formation and adsorption of bulk precipitated lead xanthate formed by metathetic replacement of sulfur-oxy species and carbonate on the surface.

Copper Sulfide Flotation. Chalcocite (Cu_2S) and chalcopyrite ($CuFeS_2$) are the two most commonly floated copper sulfide minerals. Bornite (Cu_5FeS_4), covellite (CuS), and enargite (Cu_3AsS_4) are normally present in smaller quantities. Both chalcocite and chalcopyrite are floated readily with common sulfhydryl collectors. Figures 8.13 and 8.14 show the responses of chalcocite and chalcopyrite to flotation with ethyl xanthate.

The active species of collector when xanthate is added to the chalcocite system is xanthate ion. Dixanthogen does not form on the chalcocite surface (Allison et al. 1972; Kuhn 1968; see Table 8.14).

Xanthate adsorption on chalcocite is a two-stage process similar to that for galena. The presence of an unleachable xanthate species on the chalcocite surface following xanthate adsorption was demonstrated by Gaudin and Schuhmann (1936). These authors also demonstrated that following the formation of an unleachable chemisorbed layer, multilayers of cuprous xanthate form and adsorb on the surface.

Ion exchange experiments similar to those in the galena-ethyl xanthate system were also conducted by Dewey (1933) in the chalcocite-amyl xanthate system. The two principal anions exchanged when xanthate chemisorbs on chalcocite are hydroxyl and carbonate.

The rest potential of chalcopyrite is so close to the reversible xanthate-dixanthogen potential that xanthate has been shown to chemisorb on this mineral (Kuhn 1968) or be floated by the formation of dixanthogen (Allison et al. 1972). Whether xanthate or dixanthogen is the active species is obviously sample-dependent.

Sphalerite Flotation. The flotation characteristics of sphalerite have received considerable attention, both in the absence and presence of activating ions. Some investigators have observed flotation

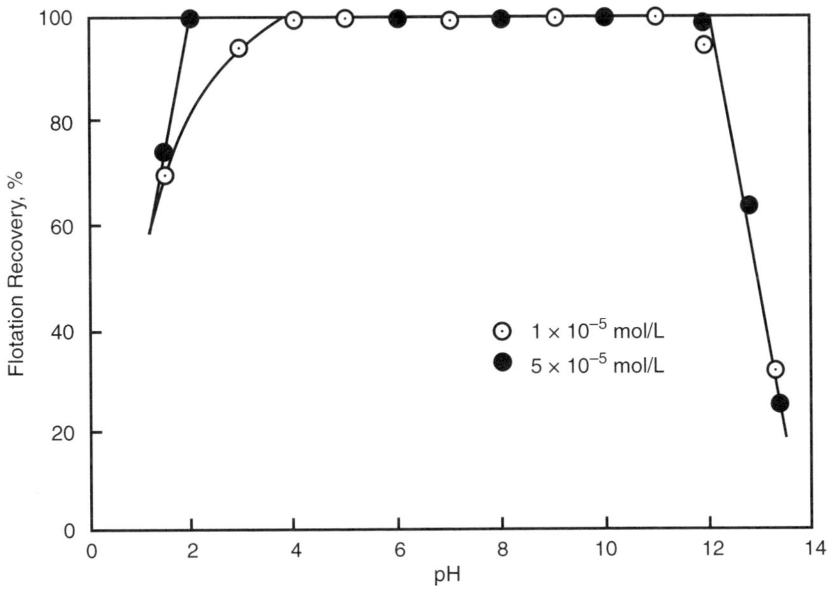

Source: Fuerstenau, Miller, and Kuhn 1985.

FIGURE 8.14 Flotation recovery of chalcopyrite as a function of pH with two additions of ethyl xanthate

with ethyl and amyl xanthates in the absence of activators; others have not (Steininger, 1967; Girczys and Laskowski 1972; Fuerstenau, Clifford, and Kuhn 1974; Harris and Finkelstein 1975; Finkelstein and Allison 1976). These differences in response may have been caused by differences in the oxidation characteristics of the sphalerites involved. In other words, the formation and adsorption of bulk precipitates of zinc xanthates on sphalerite have been shown to be necessary for flotation in the absence of activators (Fuerstenau, Clifford, and Kuhn 1974). With sphalerites that are refractory to oxidation, only a limited quantity of Zn^{2+} will be available for the formation of multilayers of zinc xanthate on the surface. Flotation recovery of sphalerite as a function of xanthate concentration and hydrocarbon chain length is shown in Figure 8.15. Note that ethyl xanthate floats unactivated sphalerite but a high concentration is required.

In this regard, xanthate adsorption on sphalerite is similar to that on chalcocite and galena in that xanthate appears to adsorb via two stages. The first stage involves chemisorption of an initial layer of xanthate at 1:1 coordination. The second apparently involves the formation and adsorption of bulk-precipitated zinc xanthate on the sphalerite surface.

After exposure to xanthate, the collector species on the surface, readily identifiable with infrared analysis, is bulk-precipitated zinc xanthate (Yamasaki and Usui 1965; Fuerstenau, Clifford, and Kuhn 1974). Dixanthogen is not present on the surface.

These observations are in agreement with those of other investigators. Plaksin and Anfimova (1954) concluded that two forms of adsorption occur in this system. Weakly attached xanthate is removed by water washing and firmly attached xanthate is dissolved with pyridine. Shvedov and Andreeva (1938) showed that under flotation conditions, four to five times monolayer coverage is adsorbed.

Pyrite Flotation. The mechanisms by which pyrite is floated with sulfhydryl collectors are well-understood. The species of xanthate responsible for flotation in the presence of short-chained xanthates is dixanthogen. This conclusion has been drawn from electrochemical, electrokinetic, flotation,

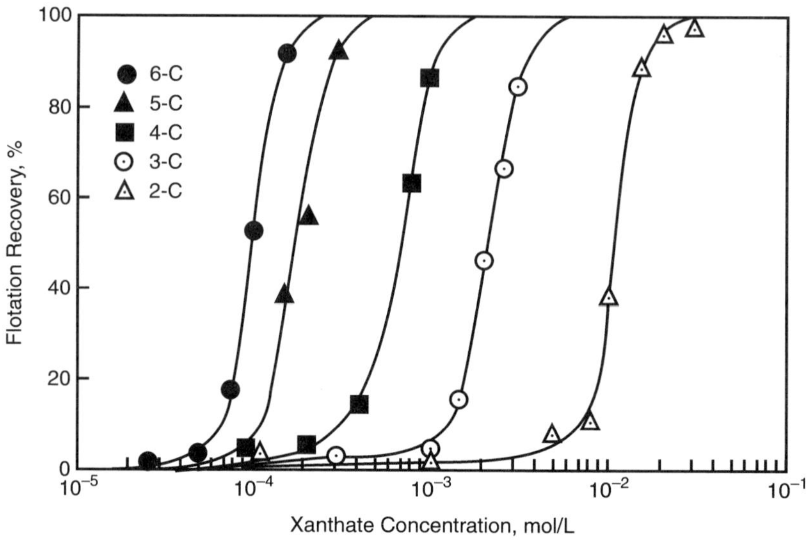

Source: Fuerstenau, Clifford, and Kuhn 1974.

FIGURE 8.15 Flotation recovery of sphalerite as a function of xanthate concentration and hydrocarbon chain length at pH 3.5

spectroscopic, and thermochemical data (Mellgren 1966; Fuerstenau, Kuhn, and Elgillani 1968; Fuerstenau, Miller, and Kuhn 1985; Majima and Takeda 1968; Usul and Tolun 1974; Woods 1976). In the case of electrokinetic experiments, the zeta potential of pyrite is the same in the absence and presence of ethyl xanthate, indicating that an electrically neutral species is adsorbed on the surface (Fuerstenau, Kuhn, and Elgillani 1968; see Figure 8.16).

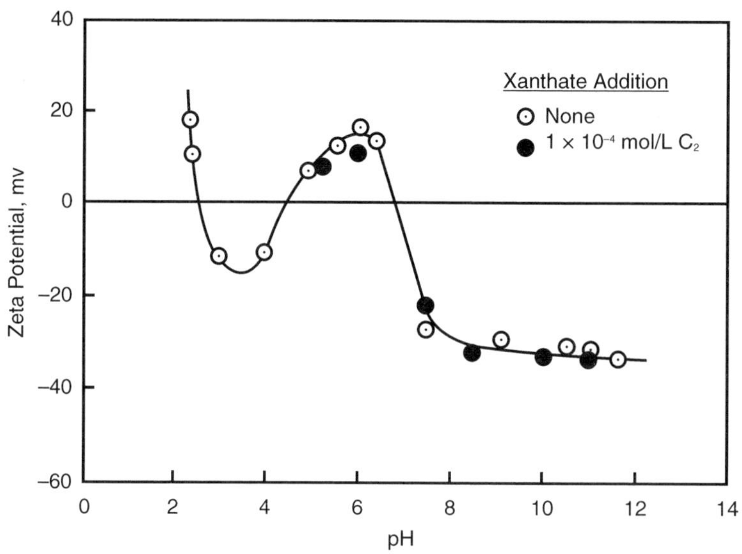

Source: Fuerstenau, Kuhn, and Elgillani 1968.

FIGURE 8.16 Zeta potential of pyrite as a function of pH in the absence and presence of ethyl xanthate in the presence of air

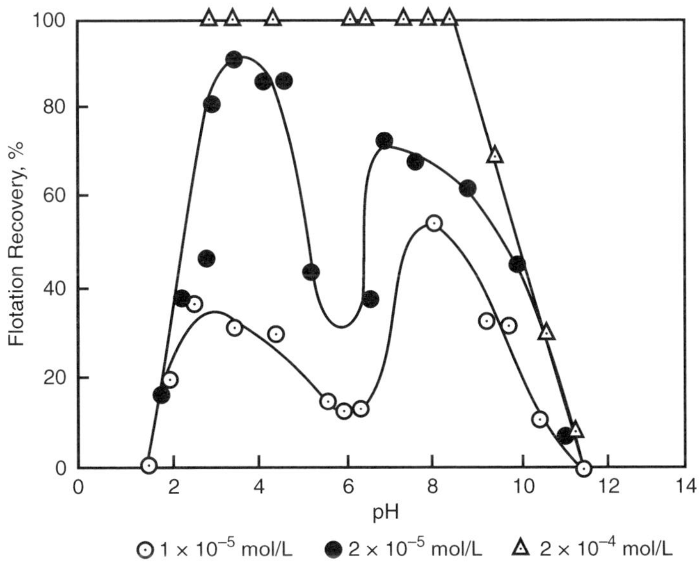

Source: Fuerstenau, Kuhn, and Elgillani 1968.

FIGURE 8.17 Flotation recovery of pyrite as a function of pH with various additions of ethyl xanthate

Dixanthogen forms by anodic oxidation of xanthate ion on the surface of pyrite coupled with cathodic reduction of adsorbed oxygen (Majima and Takeda 1968; Woods 1976; Usul and Tolun 1974). That is,

$$2X^- \Leftrightarrow X_2 + 2e^- \quad \text{Anodic} \qquad \textbf{(Eq. 8.16)}$$

$$\tfrac{1}{2}O_{2(ads)} + H_2O + 2e^- \Leftrightarrow 2OH^- \quad \text{Cathodic} \qquad \textbf{(Eq. 8.17)}$$

where X^- represents xanthate ion and X_2 represents dixanthogen. Because sulfides are electronic conductors, electron transfer occurs through the solid. Schematically,

$$FeS_2 \;\; 2e^- \begin{cases} 2e^- + \tfrac{1}{2}O_2 + H_2O \rightarrow 2OH^- \\ 2e^- + X_2 \leftarrow 2X^- \end{cases}$$

The overall reaction is

$$2X + \tfrac{1}{2}O_2 + H_2O \Leftrightarrow X_2 + 2OH^- \qquad \textbf{(Eq. 8.18)}$$

This reaction occurs up to about pH 11; above this pH, xanthate ion is the stable species of xanthate.

Flotation of pyrite, then, is possible below pH 11 with short-chained xanthates, but it is depressed above about pH 11. These phenomena are shown clearly in Figure 8.17. Addition of low levels of ethyl xanthate gives two regions of flotation from about pH 3 to pH 9. The intermediate region of depression is not related to a lack of dixanthogen, however. This phenomenon has been ascribed to the formation of basic ferric xanthate under these conditions (Wang, Forssberg, and Bolin 1989).

Another collector, dithiophosphate, has been shown to function similarly in the pyrite system. Dithiophosphate, DTP, is more difficult to oxidize to its dimer, dithiophosphatogen, however, than xanthate is to dixanthogen (Woods 1976). That is,

$$X_{2(l)} + 2e^- \Leftrightarrow 2X^- \quad E^o = -0.06 \text{ v} \qquad \textbf{(Eq. 8.19)}$$

$$(DTP)_{2(l)} + 2e^- \Leftrightarrow 2DTP^- \quad E^o = 0.25 \text{ v} \qquad \textbf{(Eq. 8.20)}$$

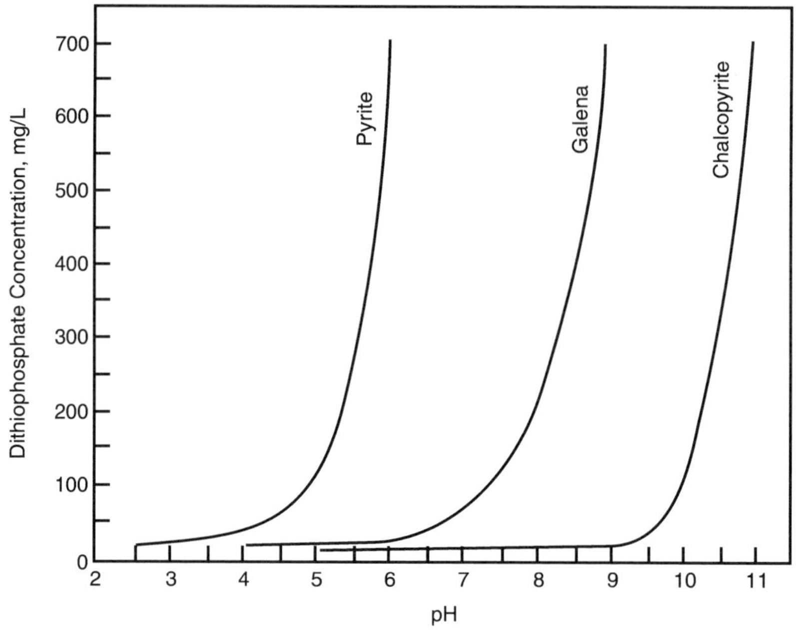

Source: Wark and Cox 1934.

FIGURE 8.18 Bubble contact curves for several sulfide minerals as a function of diethyl dithiophosphate concentration and pH

Diethyl dithiophosphatogen has been found experimentally to form at pH 4 and below in the presence of pyrite but not at pH 6 and above, and flotation of pyrite with dithiophosphate does not occur above about pH 6 (see Figure 8.18).

Modulation of Flotation of Sulfide Minerals. Selectivity in sulfide flotation systems requires the addition of specific reagents (depressants) capable of modifying surfaces selectively or complexing ions in solution. Reagents that fall into this category are hydroxyl, cyanide, chromate, sulfide, and sulfite ions.

Hydroxyl. The role that hydroxyl assumes can be seen in Figures 8.12 through 8.14 and Figures 8.17 through 8.19. Above pH 10.5, lead hydroxide [$Pb(OH)_{2(s)}$] and plumbite [$HPbO_2^-$] are more stable than lead ethyl xanthate, and flotation of galena is not possible above this pH. Chalcocite, on the other hand, is still floated at high values of pH because of the stability of cuprous ethyl xanthate relative to cuprous hydroxide, their solubility products being 5.2×10^{-20} and 2×10^{-15}, respectively. Depression occurs at about pH 14 in the presence of this collector, and calculations show that cuprous hydroxide becomes stable with respect to cuprous ethyl xanthate at about this pH.

In the case of pyrite, electrochemical oxidation of xanthate to dixanthogen on the pyrite surface does not occur above about pH 10.5, indicated in Eq. 8.18. Because dixanthogen is the species actively responsible for flotation of pyrite, depression occurs above this pH.

Cyanide. Cyanide is also an extensively used depressant in the selective flotation of sulfides. Cyanide is especially effective in depressing iron-bearing sulfides (e.g., pyrite, marcasite, and chalcopyrite; see Figure 8.20). The complex ion ferrocyanide is formed upon addition of cyanide in the presence of ferrous iron. The formation of ferric ferrocyanide on the surface of pyrite has been proposed by Elgillani and Fuerstenau (1968) to occur according to the following half cell:

$$7\ Fe^{2+} + 18\ HCN \Leftrightarrow Fe_4[Fe(CN)_6]_3 + 18\ H^+ + 4e^- \quad \textbf{(Eq. 8.21)}$$

This half cell is part of the overall reaction that must occur in several steps: dissolution of pyrite to produce dissolved Fe^{2+}, formation of $Fe(CN)_6^{-4}$, and formation of surface $Fe_4[Fe(CN)_6]_3$.

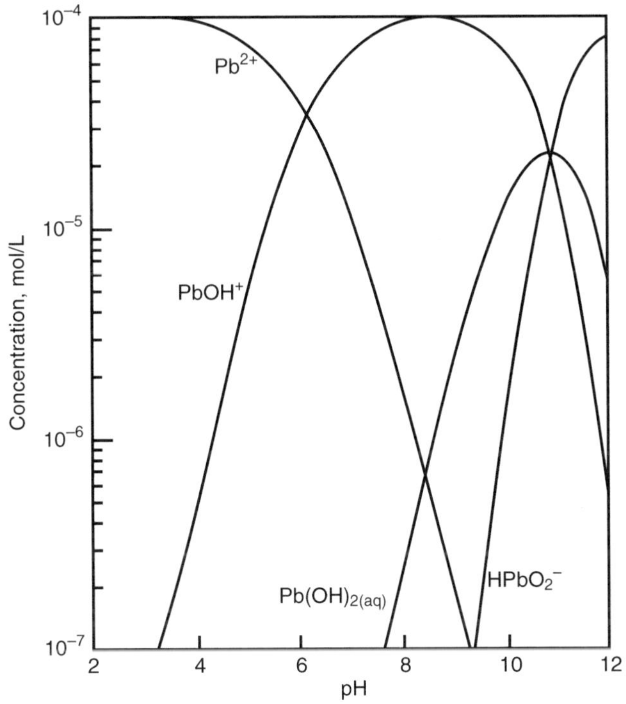

Source: Latimer 1952.

FIGURE 8.19 Speciation diagram for 1 × 10^{-4} mol/L Pb^{2+}

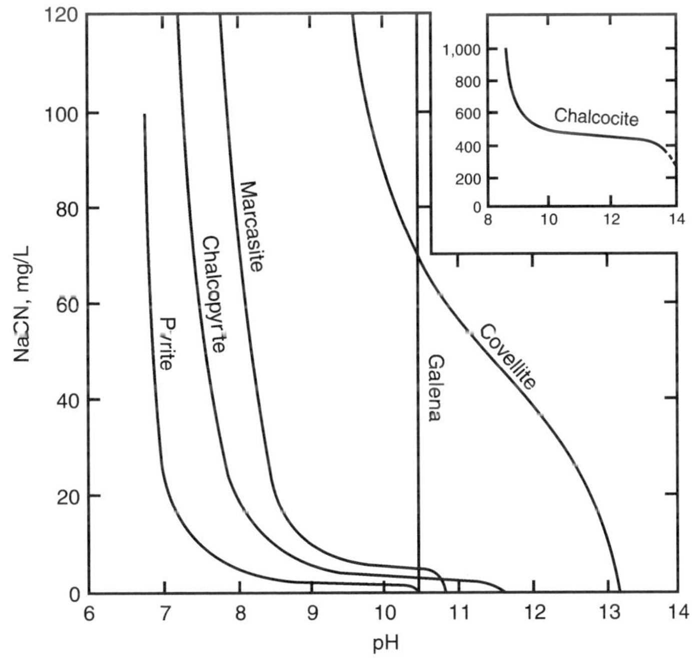

Source: Sutherland and Wark 1955.

FIGURE 8.20 Bubble contact curves for several sulfide minerals

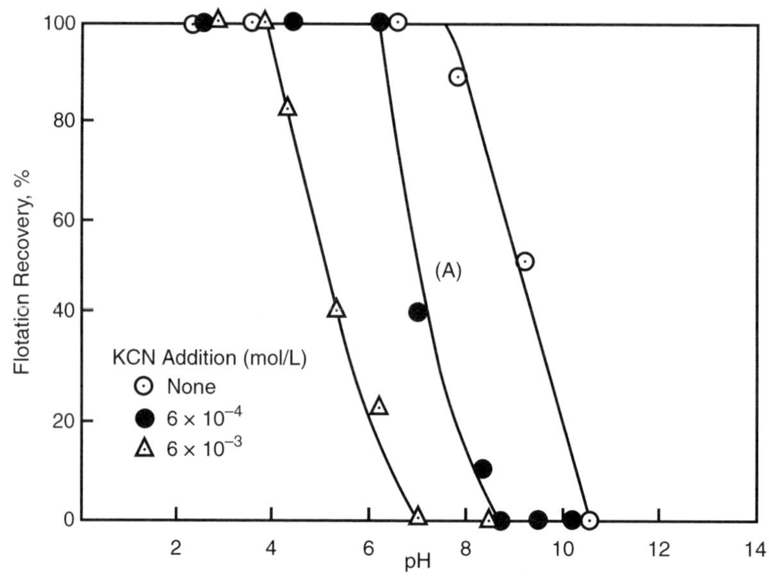

Source: Elgillani and Fuerstenau 1968.

FIGURE 8.21 Flotation recovery of pyrite as a function of pH with 5 × 10^{-4} mol/L ethyl xanthate in the absence and presence of cyanide

The flotation response of pyrite in the absence and presence of cyanide and in the presence of ethyl xanthate as collector is given in Figure 8.21. Complete flotation is effected up to about pH 6 when 1×10^{-4} mol/L ethyl xanthate and 6×10^{-4} mol/L cyanide are used. System depression occurs above this pH value.

Comparison of these data with the Eh–pH diagram for the pyrite-cyanide-xanthate system shows that $Fe_4[Fe(CN)_6]_3$ is stable under those conditions (Figure 8.22). The presence of this compound would block anodic oxidation of xanthate to dixanthogen.

Activation. In practice, flotation systems take advantage of the fact that sphalerite does not float in alkaline medium with modest levels of ethyl xanthate. Prevention of activation is ensured by suitable reagent schedules and other sulfides, such as chalcocite, chalcopyrite, and galena, are floated selectively from sphalerite.

To float sphalerite, activation is accomplished by adding a metal ion whose metal sulfide is more stable than ZnS. A number of metal ions possess this property, notably cuprous, cupric, mercurous, mercuric, silver, lead, cadmium, and antimony (Sutherland and Wark 1955; Gaudin 1957).

The most commonly added activator is copper sulfate. In a study using radioactive copper, Gaudin, Fuerstenau, and Mao (1959) showed that Cu^{2+} displaces Zn^{2+} from the sphalerite lattice. Exchange is quite rapid until three layers of zinc are replaced. Additional exchange follows a parabolic rate law typical of diffusion-controlled systems.

The activation reaction with Cu^{2+} is represented by Eq. 8.22:

$$ZnS_{(s)} + Cu^{2+}{}_{(aq)} \Leftrightarrow CuS_{(s)} + Zn^{2+}{}_{(aq)} \quad K = 9 \times 10^{10} \qquad \textbf{(Eq. 8.22)}$$

Cupric ion will replace zinc until the activity of Zn^{2+} is 9×10^{10} that of Cu^{2+} in solution. Following activation with Cu^{2+}, the flotation response of sphalerite is similar to that of copper sulfide minerals.

Other activation reactions are

$$ZnS_{(s)} + 2Ag^{+} \Leftrightarrow Ag_2S_{(s)} + Zn^{2+} \quad K = 10^{26} \qquad \textbf{(Eq. 8.23)}$$

$$ZnS_{(s)} + Pb^{2+} \Leftrightarrow PbS_{(s)} + Zn^{2+} \quad K = 10^{3} \qquad \textbf{(Eq. 8.24)}$$

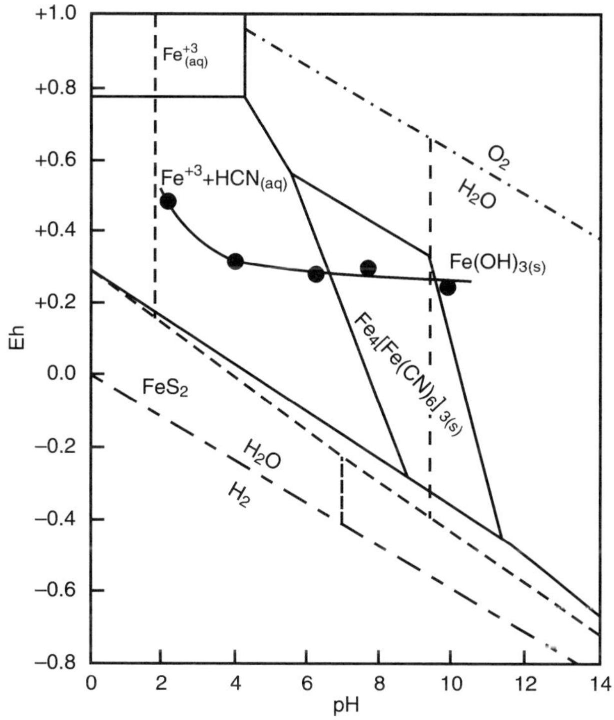

Source: Elgillani and Fuerstenau 1968.

FIGURE 8.22 Stability of FeS_2, $Fe(OH)_3$, and $Fe_4[Fe(CN)_6]_3$ at 3×10^{-4} mol/L total dissolved sulfur, 5×10^{-5} mol/L total dissolved iron, and 6×10^{-4} mol/L cyanide addition. Black circles indicate Eh values corresponding to Curve A in Figure 8.21.

Gaudin, Fuerstenau, and Turkanis (1952) showed that the abstraction of Ag^+ by sphalerite is rapid and that the sphalerite sample turned black in a short period of time. These authors also showed that the uptake of silver occurs indefinitely and that after 2 months of reaction, a +325-mesh sample of sphalerite contained 18% silver. Exactly 2 moles of Ag^+ are exchanged for 1 mole of Zn^{2+} during this process.

Prevention of Activation. Unintentional activation of sphalerite in ores is most commonly the result of Cu^{2+} and Pb^{2+} in solution. In the case of Cu^{2+}, cyanide is most commonly added to prevent activation. The stability of the cupro-cyanide complex, $Cu(CN)^-_2$, relative to $Zn(CN)_4^{2-}$ results in ratios of dissolved copper to zinc such that activation cannot occur. Relevant equilibria are as follows (Vladimirova and Kakavskii 1950; Latimer 1952; Gaudin 1957):

$$Cu^{2+} + HCN_{(aq)} \Leftrightarrow Cu^+ + H^+ + \tfrac{1}{2}(C_2N_2)_{(g)} \qquad K = 2.1 \times 10^{-4}$$

$$Cu(CN)^-_2 \Leftrightarrow Cu^+ + 2CN^- \qquad K = 2 \times 10^{-24}$$

$$Zn(CN)_4^{-2} \Leftrightarrow Zn^{2+} + 4CN^- \qquad K = 1.2 \times 10^{-18}$$

$$HCN_{(aq)} \Leftrightarrow H^+ + CN^- \qquad K = 4 \times 10^{-10}$$

In the case of Pb^{2+}, activation can be prevented when the activity of Zn^{2+} is 1,000 times that of Pb^{2+} in solution. Because sphalerite commonly resists oxidation, very little Zn^{2+} dissolves from this mineral. Zinc sulfate is therefore added, and by virtue of the equilibria involving basic lead carbonate and zinc hydroxide, the activity ratio of Zn^{2+}/Pb^{2+} is higher than the equilibrium ratio, and activation cannot occur.

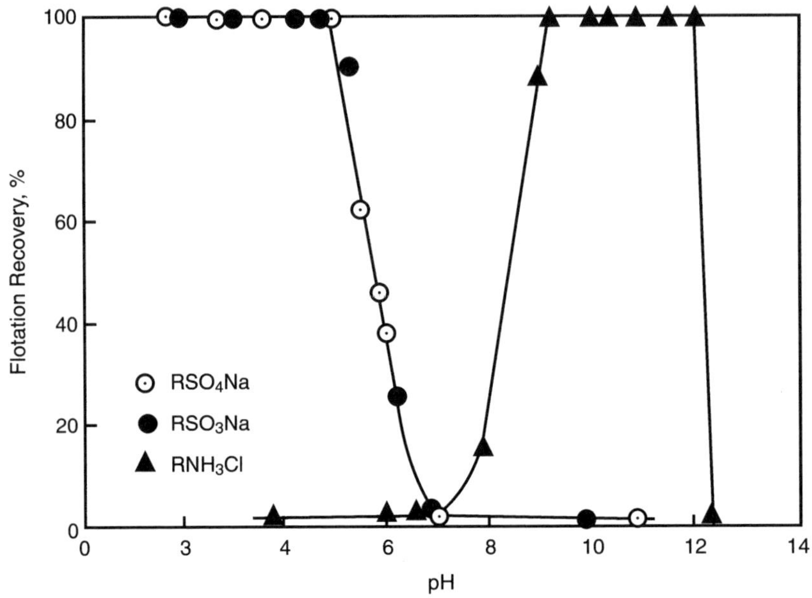

Source: Iwasaki, Cooke, and Colombo 1960.

FIGURE 8.23 Flotation recovery of goethite as a function of pH with 1×10^{-3} mol/L additions of cationic and anionic collectors

In addition to these thermodynamic considerations, considerable evidence indicates that the colloids of zinc salts, formed under conditions in which precipitation occurs, function as depressants for sphalerite. These include precipitates of zinc hydroxide, zinc carbonate, zinc sulfite, and zinc cyanide.

The depressant role of zinc hydroxide colloids was first presented by Malinovsky (1946). His observations were later confirmed by Livshitz and Idelson (1953), who also demonstrated that the extent of depression of sphalerite and the concentration of colloidal zinc hydroxide occurring in the pulp are directly related.

Grosman and Khadzhiev (1966) confirmed this finding. Their results showed that both sphalerite and chalcopyrite take up the equivalent of many tens of monolayers of zinc colloids. Chalcopyrite retains its hydrophobic character under these conditions; sphalerite is hydrophilic. As the adhering colloids are successively removed from sphalerite, floatability is increased until the equivalent of three monolayers remain, at which point complete flotation is obtained. These authors concluded that basic zinc carbonate was formed under the conditions of sphalerite depression.

Insoluble Oxide and Silicate Flotation

A large number of minerals fall into this category, and whether a particular mineral can be floated with a particular collector depends on the electrical properties of the mineral surface, the electrical charge of the collector, the molecular weight of the collector, the solubility of the mineral, and the stability of the metal-collector salt. Depending on these phenomena, adsorption of collector may occur either by electrostatic interaction with the surface (physical adsorption) or by specific chemical interaction with surface species (chemisorption).

Flotation by Physical Adsorption. Many collectors achieve adsorption by electrostatic interaction with oxide and silicate surfaces. Such collectors can be used only with knowledge of the PZC values for the minerals in question.

Figure 8.23 clearly shows the dependence of geothite flotation on electrostatic phenomena when either amines or certain anionic collectors are used. These anionic collectors do not form insoluble

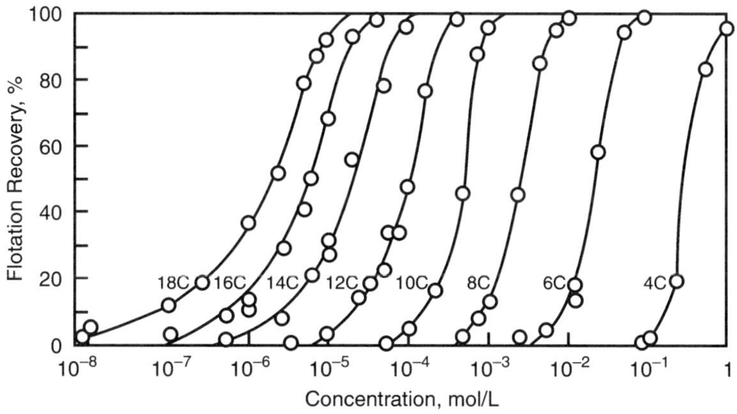

Source: Fuerstenau, Healy, and Somasundaran 1964.

FIGURE 8.24 The effect of hydrocarbon chain length on relative flotation response of quartz in the presence of various ammonium acetates at neutral pH

metal-collector salts in this system. Below the PZC, the surface is positively charged; negatively charged sulfonate ions are adsorbed in this region, and complete flotation is effected. Above the PZC, the surface is negatively charged, and sulfonate ions are repelled from the surface. On the other hand, aminium ions, which are positively charged, are adsorbed on the negatively charged surface.

As discussed previously, extensive hydrogen bonding of water molecules occurs on oxide and silicate surfaces. As a result, the presence of hemimicelles of collector or precipitates of collector appear to be necessary to render these surfaces sufficiently hydrophobic for flotation to occur. As shown in Figure 8.24, the concentration associated with a rapid rise in flotation recovery of quartz provides evidence of these phenomena. Verification of this premise has been provided by Fuerstenau, Healy, and Somasundaran (1964). These authors showed that a plot of the logarithm of collector concentration required for a rapid rise in flotation recovery as a function of the number of carbon atoms in the hydrocarbon chain yields a straight line with a slope corresponding to a specific adsorption potential of −0.62 kcal/mole CH_2 group. This is the free energy decrease associated with the removal of hydrocarbon chains from solution by either hemimicelle formation or precipitate formation. The association of hydrocarbon chains (hemimicelles) at the solid–liquid interface is shown in Figure 8.25.

These phenomena are shown very clearly with the adsorption isotherm of dodecyl sulfonate on alumina at neutral pH, as shown in Figure 8.26. Three distinct changes in slope of the isotherm can be noted. At low concentration of collector, adsorption of individual ions occurs, and the zeta potential remains constant. With increasing additions of dodecyl sulfonate, hemimicelles form, adsorption density increases markedly, and zeta potential decreases drastically with concentration. At even higher concentrations, a third change in slope occurs which probably marks the formation of a bilayer of sufonate ions at the interface.

Laskowski, Vurdela, and Liu (1988) suggested that the colloids of precipitated amine formed in alkaline solution may be responsible for quartz flotation under these conditions (see Figure 8.27). The amine colloids are charged in solution, and they have a PZC at relatively high pH; for example, dodecylamine has a PZC at pH 11. The pH region in which the amine colloids form is the same as that in which optimal flotation of quartz is obtained (see Figure 8.28; Fuerstenau 1957).

The association of hydrocarbon chains either as hemimicelles or as a precipitate of collector salt on the mineral surface is desirable for flotation. Interactions in the interfacial region depend essentially on the relative concentrations of surfactant required to form hemimicelles and to precipitate the surfactant

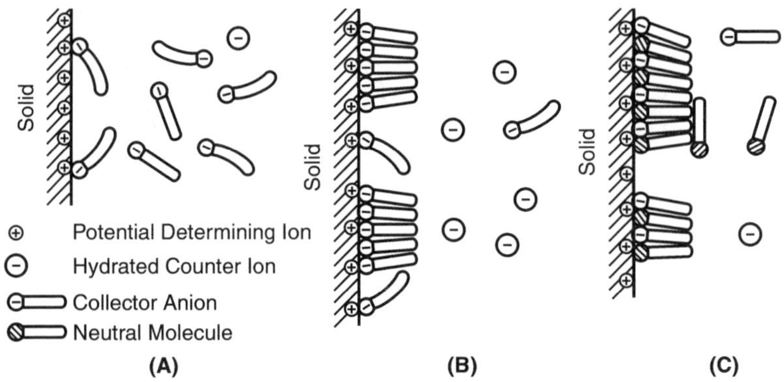

Source: Aplan and Fuerstenau 1962.

FIGURE 8.25 Schematic representation of the electrical double layer in the presence of surface-active organic compounds: (A) adsorption as single ions at low collector concentration, (B) hemimicelle formation at higher concentration, and (C) co-adsorption of collector ions and neutral molecules

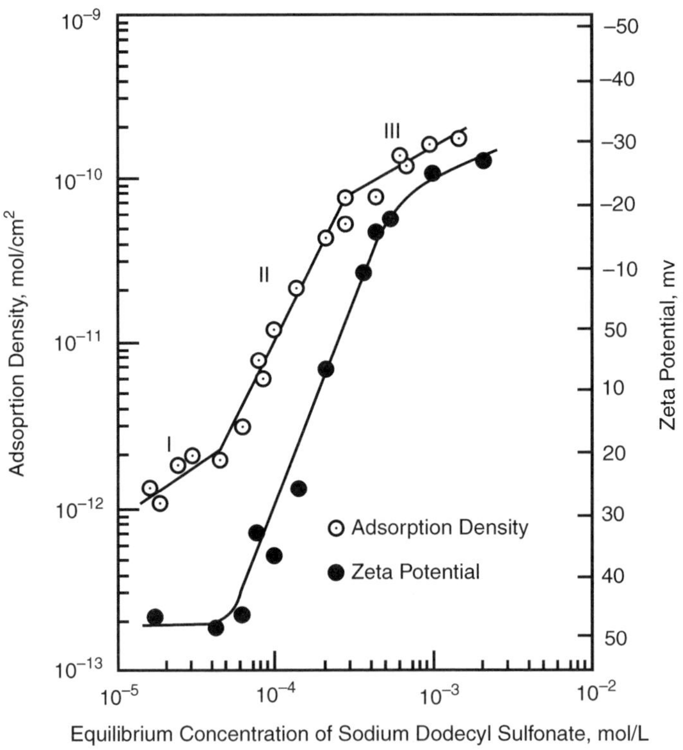

Source: Wakamatsu and Fuerstenau 1973.

FIGURE 8.26 Adsorption density and zeta potential of alumina as a function of dodecyl sulfonate concentration at pH 7.2 and 2 × 10^{-3} mol/L ionic strength

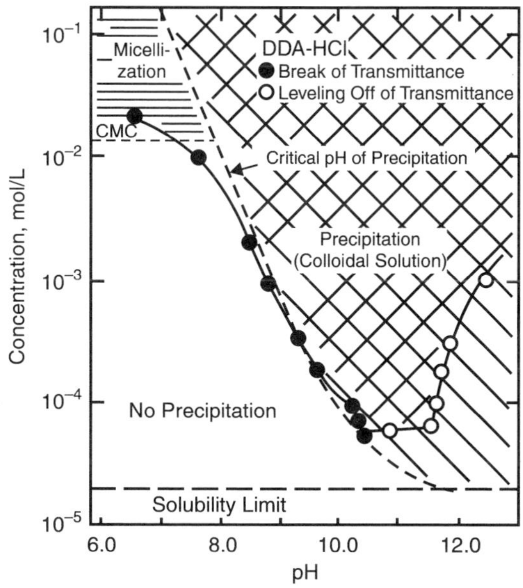

Source: Laskowski, Vurdela, and Liu 1988.

FIGURE 8.27 Thermodynamic equilibrium diagram showing the CMC (30°C) and the lines of critical pH of precipitation and solubility limit at 25°C. The points shown are experimental solubility (•) and redispersion (o) determined from the transmittance curves.

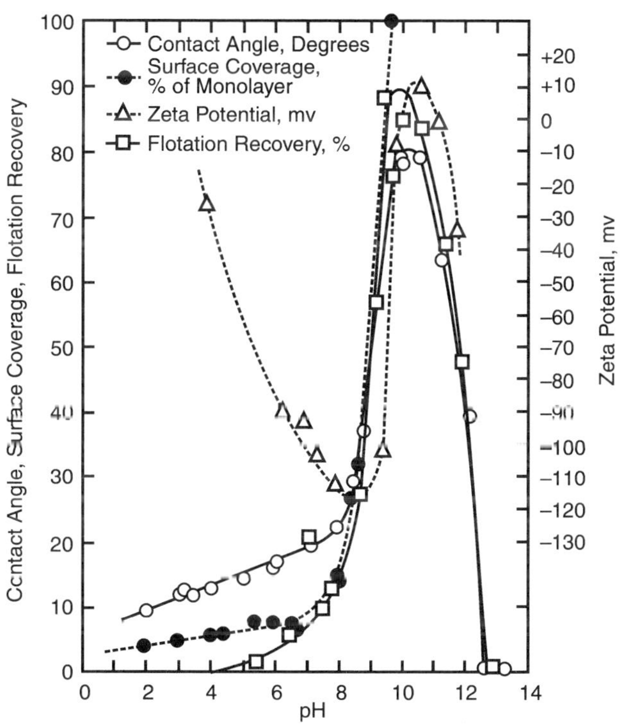

Source: Fuerstenau, Elgillani, and Miller 1957.

FIGURE 8.28 Correlation of adsorption density, contact angle, and zeta potential with flotation of quartz with 4 × 10^{-5} mol/L dodecylammonium acetate additions

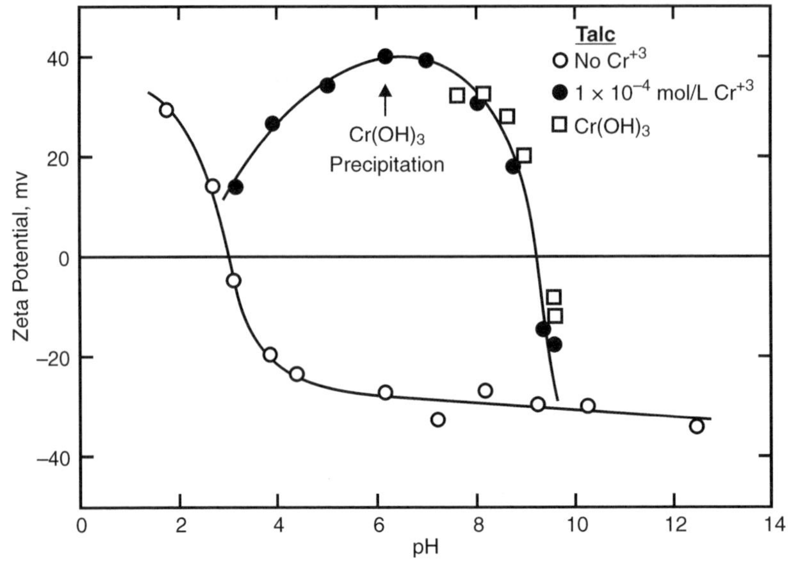

Source: Fuerstenau, Valdivieso, Fuerstenau 1988.

FIGURE 8.29 Zeta potential of talc and precipitated $Cr(OH)_3$ as a function of pH

salt. If the HMC (concentration of hemimicelle formation) is the lower of the two, the formation of hemimicelles would be preferred over salt precipitation.

Flotation by Chemisorption. In many cases, chemisorption of high-molecular-weight collectors on oxides and silicates appears to involve hydrolysis of cations comprising these minerals. The hydroxy complexes thus formed are very surface active; they adsorb strongly on solid surfaces and reverse the sign of the zeta potential if their concentration is sufficiently high (Matijvic and Tezak 1953; Matijvic et al. 1961; Fuerstenau, Elgillani, and Miller 1970; Fuerstenau and Palmer 1976). In fact, hydroxy complexes even adsorb on positively charged surfaces (Fuerstenau, Elgillani, and Miller 1970).

As the pH increases, further hydrolysis of the hydrolyzed species to the metal hydroxide occurs. These phenomena may be seen in Figure 8.29, which shows the zeta potential of talc in the absence and presence of chromium species. With the formation and adsorption of hydroxy complexes of chromium at around pH 4, the zeta potential changes sign from negative to positive. At higher values of pH, the zeta potential of talc in the presence of 1×10^{-4} mol/L Cr^{3+} is the same as that of precipitated $Cr(OH)_3$. Under these conditions, the surface of talc is that of chromium hydroxide.

James and Healy (1972) examined the adsorption of hydrolyzable metal ions at the oxide interface with adsorption and electrokinetic studies of cobalt on SiO_2 and TiO_2. They presented a thermodynamic model of adsorption of hydrolyzed species on these surfaces.

These authors provided an excellent analysis of these systems in terms of competing energy changes as an ion approaches an interface. The attractive energy is the electrostatic free energy, possibly supplemented by short-range forces. The opposing energy involves the secondary solvation energy changes as parts of the solvation sheath are rearranged or replaced. Their analysis shows that solvation energy change is much more favorable for a hydroxy complex than for a hydrated divalent ion with a solid of low dielectric, such as quartz. Hence, the overall free energy of adsorption will be much more favorable.

Studies on the adsorption of Ca^{2+} species on quartz are supportive of these concepts (Figure 8.30). $CaOH^+$ and $Ca(OH)_2$ are formed at high values of pH, the region in which extensive adsorption of Ca^{2+} species is shown to occur.

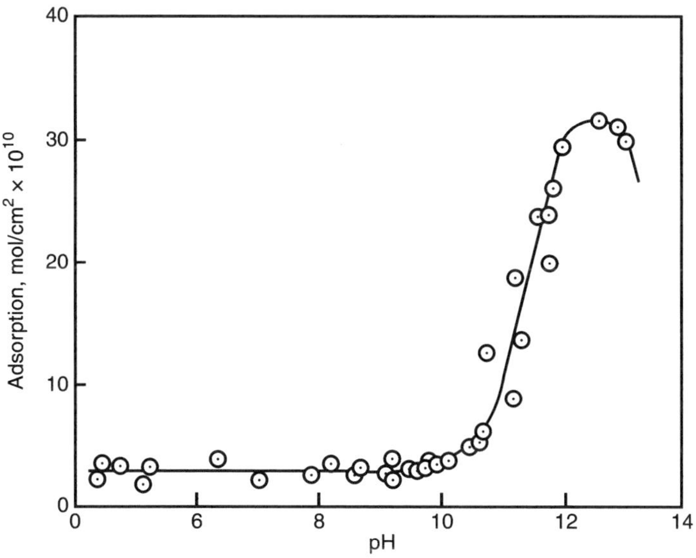

Source: Clark and Cooke 1968.

FIGURE 8.30 Adsorption of calcium species on quartz as a function of pH from solutions containing 100 ppm in Ca^{2+}

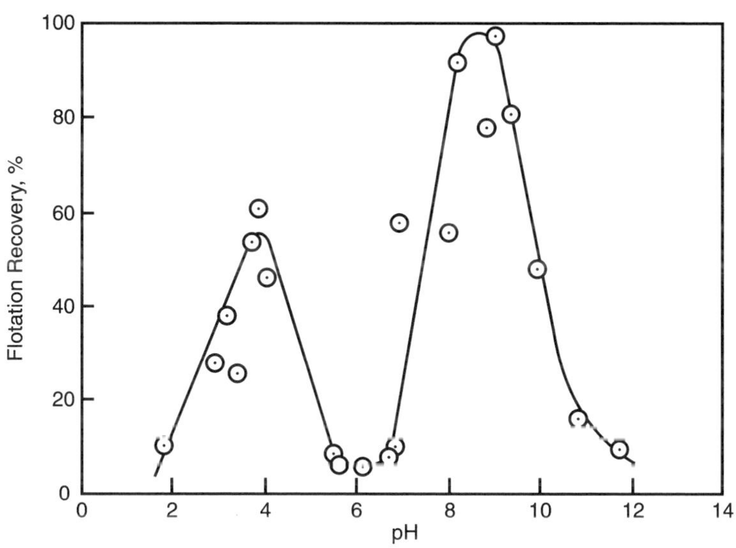

Source: Fuerstenau and Rice 1968.

FIGURE 8.31 Flotation recovery of pyrolusite as a function of pH with 1×10^{-4} mol/L oleate

The relationship of these phenomena to flotation can be seen in the pyrolusite-oleate system (Figure 8.31). The PZC of this mineral is pH 7.4. The flotation response at pH 4 can be attributed to physical adsorption of oleate ion.

The response observed at pH 8.5 is intimately involved with hydrolysis of Mn^{2+}, confirmed by the speciation diagram for Mn^{2+} shown in Figure 8.32 and the electrokinetic data in Figure 8.33. $MnOH^{+}$ is present maximally, and $Mn(OH)_2$ is known to precipitate at about this pH. The zeta potential of

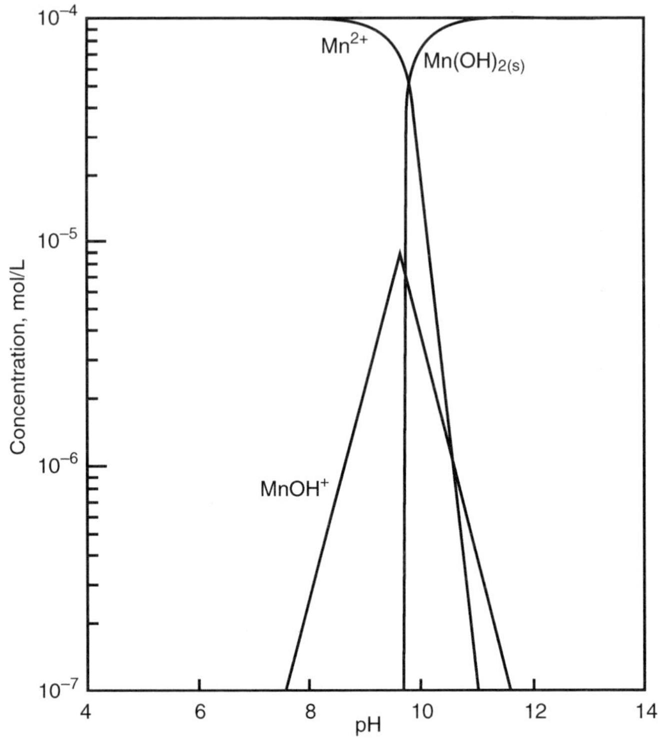

Source: Butler 1964.

FIGURE 8.32 Speciation diagram for 1 × 10^{-4} mol/L Mn^{2+}

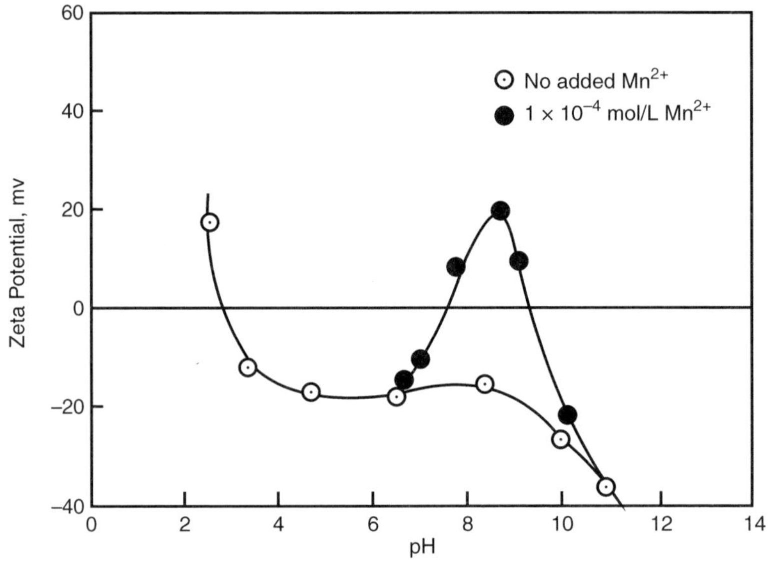

Source: Palmer, Gutierrez, and Fuerstenau 1975.

FIGURE 8.33 Zeta potential of rhodonite as a function of pH in the absence and presence of Mn^{2+}

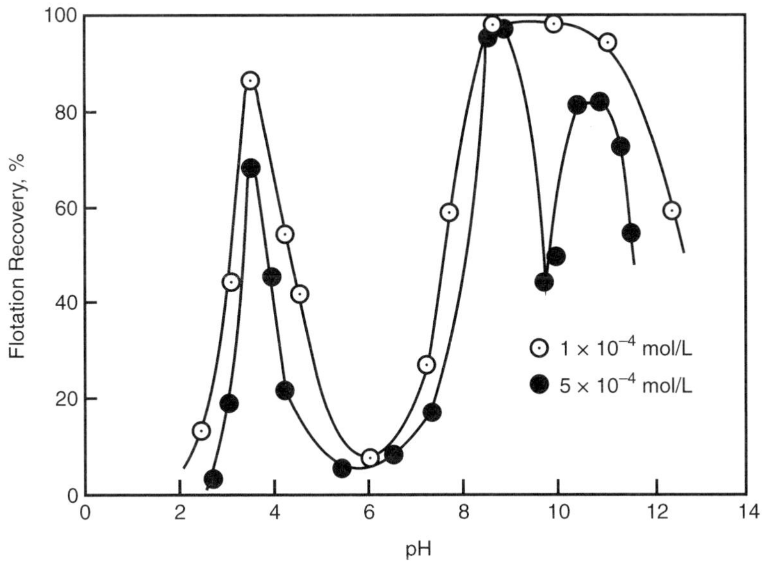

Source: Palmer, Fuerstenau, and Aplan 1975.

FIGURE 8.34 Flotation recovery of chromite as a function of pH and oleate concentration

rhodonite ($MnSiO_3$) is also noted to change sign when Mn^{2+} is added in this pH range. As will be demonstrated later, this is the same pH range in which activation of quartz occurs with Mn^{2+}.

Pyrolusite is manganic oxide, and for the hydrolyzed species of manganous ion to be controlling in this system, the following phenomena must have occurred: (1) slight mineral dissolution, (2) oxidation-reduction between manganic ion and water, (3) formation and adsorption of hydroxy complexes, and (4) formation of metal hydroxide on the surface. In practice, dissolution of pyrolusite has been enhanced by the addition of sulfur dioxide to ore systems (McCarroll 1954).

The role of metal ion hydrolysis in chemisorption within anionic flotation systems is also clearly illustrated in the chromite-oleate system. The theoretical composition of chromite is $FeO{\cdot}Cr_2O_3$. As found in nature, however, Mg^{2+} is frequently substituted for Fe^{2+}, and Fe^{3+} and Al^{3+} are substituted for Cr^{3+}. Flotation response with oleate as collector is given in Figure 8.34. The response in the vicinity of pH 11 is attributed to $MgOH^+$ and $Mg(OH)_{2(s)}$; that in the vicinity of pH 8 is attributed, in part, to $FeOH^+$ and $Fe(OH)_{2(s)}$. The PZC of this mineral is pH 7.0, and the response at about pH 4 is attributed to physical adsorption of oleate. Similar phenomena were noted to occur in the pyrolusite-oleate system (Figure 8.31).

Because this chromite contains Al^{3+}, we would suspect that hydroxy complexes of aluminum and aluminum hydroxide should be involved at around pH 4 in this system. This is apparently not the case, however. Similarly, the hydroxy complexes of chromium are apparently not involved in chromite flotation. The pH range in which $CrOH^{2+}$ is abundant is the pH range in which depression is observed. Aluminum and chromium are coordinated octahedrally with oxygen, whereas the divalent cations are coordinated tetrahedrally with oxygen. As a result the divalent cations would be expected to dissolve much more readily than the trivalent cations. Interestingly, then, the divalent ions comprising this and similar minerals control flotation response.

Additional phenomena may be involved in the flotation of oxides and silicates with oleate at pH 8. Flotation of ilmentite, monazite, and zircon by Dixit and Biswas (1969) is presented in Figure 8.35. Ilmenite, of course, is iron-bearing, and zircon is known to frequently contain small quantities of iron oxide. Hematite has been shown to float maximally at around pH 8 by Peck, Raby, and Wadsworth (1966) and by Fuerstenau, Harper, and Miller (1970). Further, Kulkarni and Somasundaran (1975)

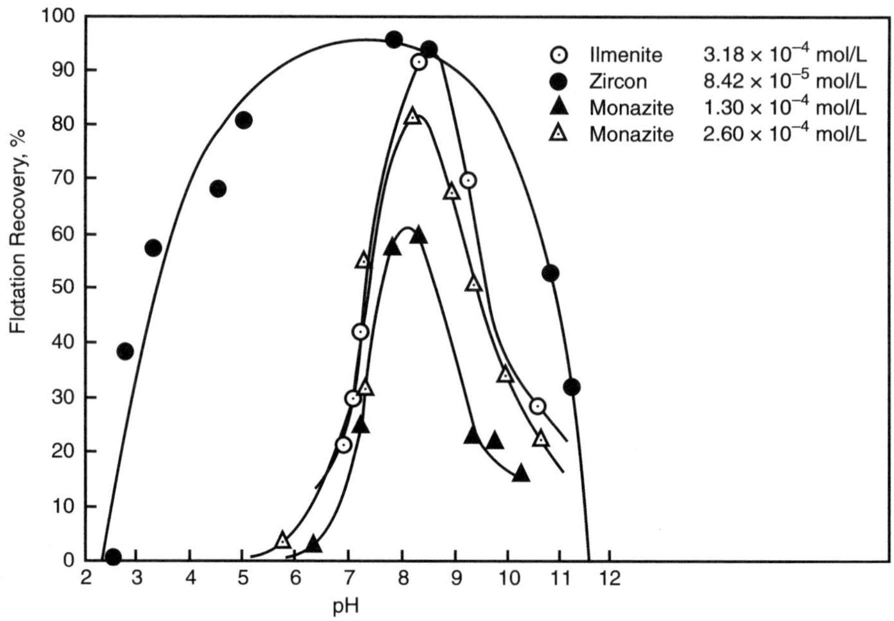

Source: Dixit and Biswas 1969.

FIGURE 8.35 Flotation recovery of beach-sand minerals with oleate as a function of pH

showed that for hematite flotation with oleate, collector adsorption at the air–water interface and subsequent lowering of surface tension may also be involved in maximal flotation observed at this pH. The dynamic surface tension in such systems appears to be related to acid soap formation.

Quartz Activation. Solubility of quartz is very limited, and because the only cation comprising the mineral is silicon, anionic flotation is obtained only when metal ions are contained in the system or are added to it in a pH range where hydrolysis occurs. Quartz activation can be achieved with most anionic collectors, including xanthate, providing certain conditions are satisfied (Kraeber and Boppel 1934; Gaudin and Rizo-Patron 1942; Clemmer et al. 1945; Cooke and Digre 1949; Schuhmann and Prakash 1950; Fuerstenau, Martin, and Bhappu 1963; Fuerstenau, Pray, and Miller 1965; Fuerstenau and Cummins 1967; and Fuerstenau and Miller 1967).

In sulfonate flotation, Figure 8.36 shows the edges outlining minimum values of pH at which flotation is obtained in the presence of 1×10^{-4} mol/L sulfonate and 1×10^{-4} mol/L of various metal ions. Also shown are the pH values at which hydroxy complexes of these metal ions are formed in significant concentration. The formation and adsorption of the metal hydroxide is also necessary for flotation, and these precipitates would be present at about the same values of pH.

The pH ranges in which flotation occurs with three of these activators are shown in Figure 8.37. Competition between the metal hydroxide and metal oleate will be in effect. At some value of pH, the metal hydroxide is more stable than the metal oleate, and flotation depression will occur. This is shown by the maximum flotation edges for each activator.

Precipitation of the metal collector has been shown to be involved in some of these systems. As shown in Figure 8.38, flotation of quartz occurred only after precipitation of calcium laurate occurred in solution. Arrows indicate the activity of laurate at which calcium laurate precipitated in each system. The critical effect of addition of collector relative to that of activator has been shown by Gaudin and Rizo-Patron (1942) and Cooke and Digre (1949). In systems where precipitation of the metal collector has occurred, at constant collector addition, increasing the metal ion addition by an order of magnitude reduced the minimum pH at which flotation is possible by 1 unit.

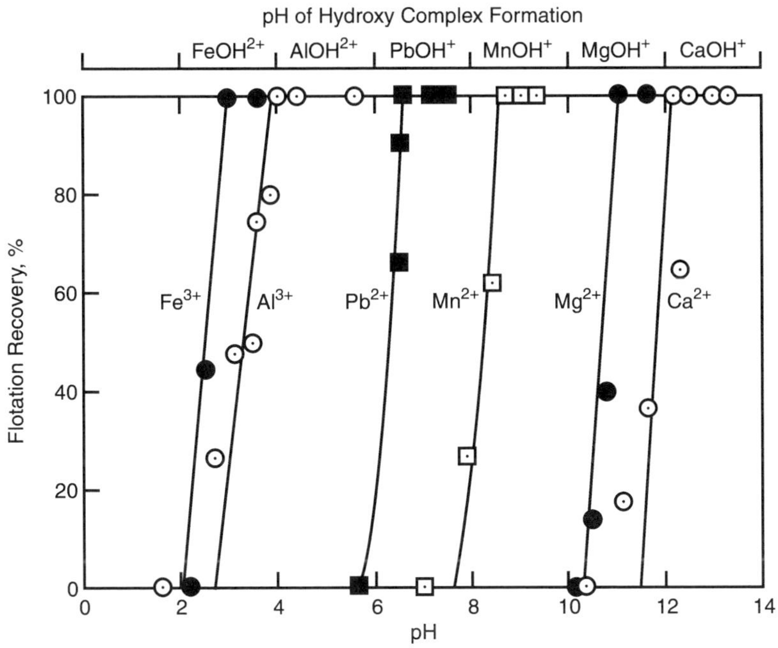

Source: Fuerstenau and Palmer 1976.

FIGURE 8.36 Minimum flotation edges of quartz—Conditions: 1 × 10^{-4} mol/L sulfonate, 1 × 10^{-4} mol/L metal ion

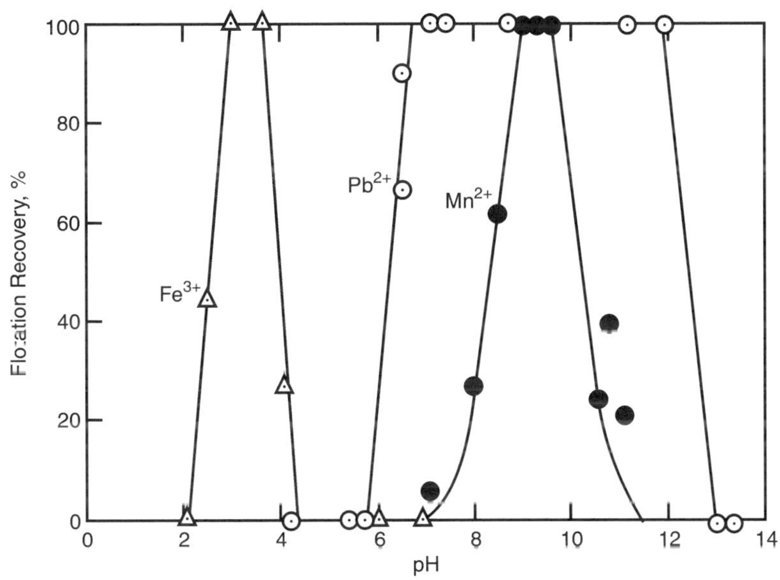

Source: Fuerstenau and Palmer 1976.

FIGURE 8.37 Flotation recovery of quartz as a function of pH—Conditions: 1 × 10^{-4} mol/L sulfonate, 1 × 10^{-4} mol/L metal ion

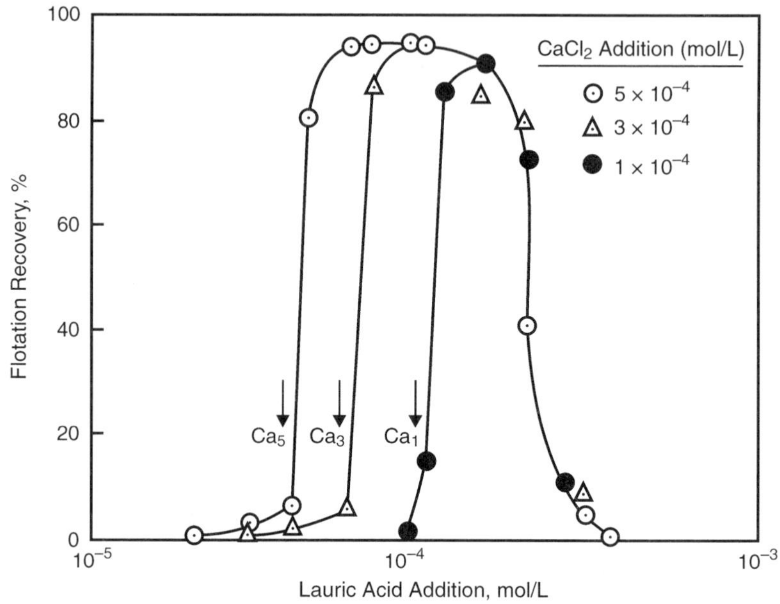

Source: Fuerstenau and Cummins 1967.

FIGURE 8.38 Flotation recovery of quartz as a function of lauric acid and calcium chloride additions at pH 11.5

Fluoride Activation. Manser (1975) obtained flotation data with minerals contained in the five groups of silicates:

- Orthosilicates (andalusite, beryl, tourmaline)
- Pyroxenes (augite, diopside, spodumene)
- Amphiboles (hornblende, tremolite, actinolite)
- Sheet silicates (muscovite, biotite, chlorite)
- Framework silicates (quartz, feldspar, nepheline)

The orthosilicates were found to be sensitive to fluoride addition, whereas the pyroxenes and amphiboles are scarcely affected by fluoride. Sheet silicates are activated by fluoride; framework silicates are activated to a lesser extent with the exception of quartz.

Quartz and feldspar were studied by Smith (1963) with dodecylamine in the absence and presence of sodium fluoride. These results are presented in Figure 8.39. Maximum contact angle on microcline and, hence, maximum flotation are obtained at approximately pH 2 in the presence of 10^{-2} molar fluoride.

A number of theories have been presented as to the possible role of fluoride ion under these conditions (Smith 1963; Warren and Kitchener 1972). Smith and Smolik (1965) suggested the following:

- HF attack of the surface silicic acid to form SiF_6^{2-}
- Fluosilicate ion adsorption on aluminum sites:

$$Al{\cdot}OH + SiF_6^{2-} \Leftrightarrow Al{\cdot}SiF_6^{2-} + OH- \quad \textbf{(Eq. 8.25)}$$

- Adsorption of aminium ions on the alumino-fluosilicate sites:

$$Al{\cdot}SiF_6^{-} + RNH_3^{+} \Leftrightarrow Al{\cdot}SiF_6{\cdot}NH_3R \quad \textbf{(Eq. 8.26)}$$

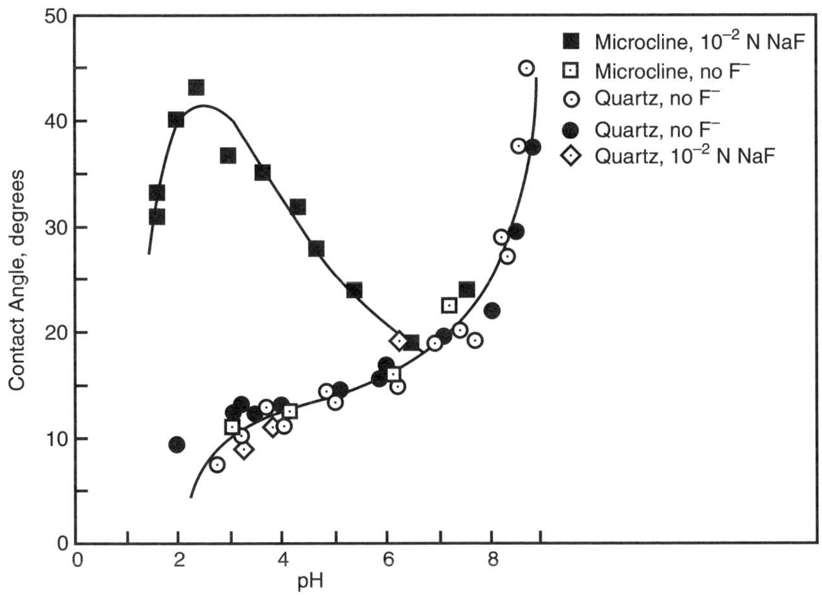

Source: Smith 1963.

FIGURE 8.39 Contact angle on microcline and quartz as a function of pH with 4 × 10^{-5} mol/L dodecylamine in the absense and presence of fluoride

Semisoluble Salt Flotation

The semisoluble salt minerals include carbonate, phosphate, sulfate, tungstate, and some halide minerals. These minerals are characterized principally by their ionic bonding and their moderate solubility in water. Many of these minerals have solubilities on the order of 10^{-4} mol/L.

The surface charge of these minerals in water would be expected to be a function of the concentration of the ions of which their lattices are composed. Hydrogen ion activity also plays a major role in establishing surface charge. The dependency arises when the surface anion of the salt acts as a weak acid. Note, for example, that the PZC of calcite (pH 10.8) corresponds to the second dissociation constant of carbonic acid.

$$H_2CO_3 \Leftrightarrow H^+ + HCO_3^- \quad K = 4.16 \times 10^{-7} \qquad \textbf{(Eq. 8.27)}$$

$$HCO_3^- \Leftrightarrow H^+ + CO_3^{2-} \quad K = 4.84 \times 10^{-11} \qquad \textbf{(Eq. 8.28)}$$

As demonstrated in the case of oxides, the PZC for semisoluble salts can also be estimated from solubility data for saturated solutions as shown for the calcite system in Figure 8.40. Notice that the solution isoelectric point (IEP), calculated from thermodynamic data, is pH 8.2, whereas the PZC of calcite is pH 10.8. When equilibrium is attained (after days of aging), however, the PZC is reported to approach pH 8.2, the value predicted from solubility calculations (Somasundaran and Agar 1967). Anionic collectors are most frequently used in the flotation of semisoluble salt minerals. In particular, carboxylic acids—unsaturated alkyl fatty acids and resin acids—are used extensively for the alkaline earth minerals, and shorter chained (C-8) saturated alkyl fatty acids–coconut oil derivatives–are used for the metallic semisoluble salt minerals. Other anionic collectors, such as sulfonate, have found limited application in these systems.

Also of importance in the flotation of metallic semisoluble salt minerals are the sulfhydryl type collectors, such as the longer chained xanthates. In many instances, their use requires sulfidization or else collector consumption becomes prohibitive.

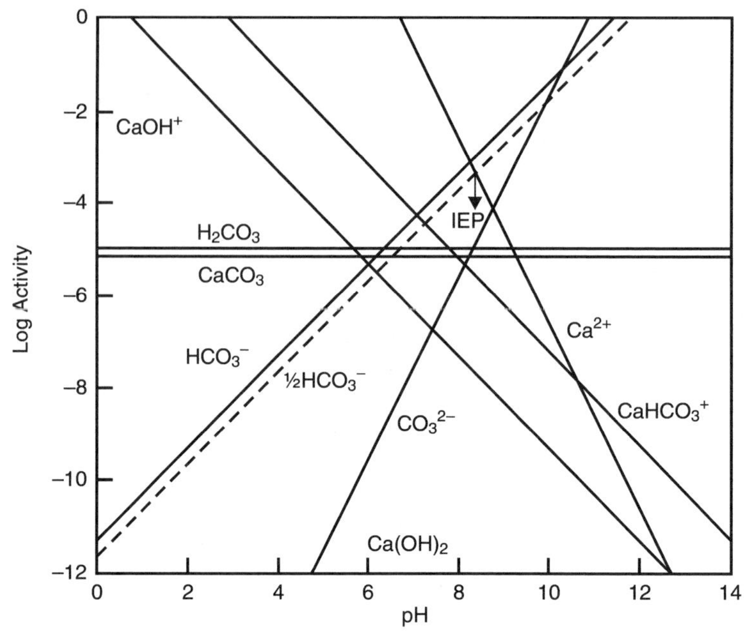

Source: Somasundaran and Agar 1967.

FIGURE 8.40 Isoelectric point of calcite established from thermodynamic data

In most systems, it is difficult to achieve high selectivity, and desliming is frequently required. Flotation is generally accomplished in alkaline media. To avoid calcium activation of gangue silicates and bulk phase precipitation of the calcium salt of the collector, soda ash is used almost exclusively for pH control (instead of lime).

In most instances, collector adsorption in these systems involves chemisorption. This phenomenon results from the stability of most multivalent cation-carboxylate salts and the moderate solubility of the semisoluble salt minerals. Dodecyl sulfate, on the other hand, has been shown to adsorb on calcite by physisorption when relatively low concentration of collector is involved (Somasundaran and Agar 1967). With higher additions of collector, however, chemisorption would likely occur in this system.

Calcite Flotation. Because calcite is ubiquitous in ore deposits, the surface properties, adsorption characteristics of various collectors, and flotation response of this mineral have received considerable study. Peck (1964) showed with infrared absorption spectroscopy that oleate chemisorbs on calcite. The researchers found no evidence of physically adsorbed oleic acid on calcite, however. Fuerstenau and Fuerstenau and Miller (1967) showed that calcite flotation occurs with shorter chained carboxylates (e.g., lauric acid) after the formation and adsorption of calcium carboxylate has occurred. Szczypa and Kuspit (1979) observed a strongly bound layer of laurate with the calcite surface, on which multilayers of calcium laurate were observed. Somasundaran (1969) observed a new phase of calcium oleate on calcite after contact with oleate. In the case of dolomite and magnesite, Predali (1969) noted two types of anion collector action, physical adsorption in acid medium and chemisorption of collector in basic medium.

Apatite Flotation. Calcium phosphate occurs either as the mineral apatite or as collophane, a substituted cryptocrystalline calcium phosphate, chemically similar to apatite but quite dissimilar in appearance, surface area, and porosity.

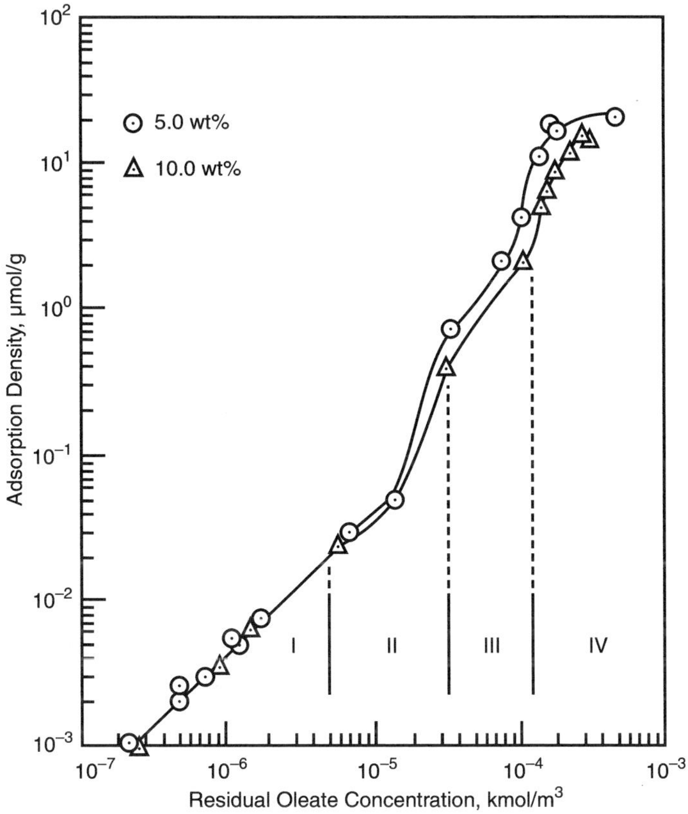

Source: Moudgil, Vasundevan, Blaakmeer 1987.

FIGURE 8.41 Adsorption isotherm of oleate on apatite

Moudgil, Vasudevan, and Blaakmeer (1987), Cases et al. (1988), and Rao, Antti, and Forssberg (1990) all conducted extensive studies on the adsorption of oleate on apatite. Distinct regimes are found in the adsorption isotherm of oleate on apatite. Moudgil and colleagues (1987) found adsorption of oleate ion at low oleate concentration (Figure 8.41, Region I). From solution conditions, calculation indicates that precipitation of calcium oleate was possible in Region II. In Region III, bulk precipitation of calcium oleate was observed, and in Region IV surface saturation with calcium oleate precipitate is approached. Correlation of the adsorption results with flotation response suggests that surface precipitation of calcium oleate is the mechanism predominantly responsible for flotation.

Fluorite Flotation. Fluorite has also received considerable study, and similar adsorption phenomena have been noted. Chemisorption of oleate on the fluorite surface occurs followed by the formation and adsorption of calcium oleate on the chemisorbed collector layer. Bahr, Clement, and Sarmatz (1968) and Bilsing (1969) observed stoichiometric release of fluoride ion with chemisorption of collector. Infrared spectroscopic studies by Peck (1964) showed that the predominant species in the region of optimum flotation is chemisorbed calcium oleate. The presence of adsorbed multilayers of calcium oleate on the surface has been observed. Interestingly, Free and Miller (1996) suggested that the predominant mechanism of calcium oleate adsorption is the formation of calcium oleate in solution and transportation to the fluorite surface rather than nucleation and growth of the precipitate at the surface.

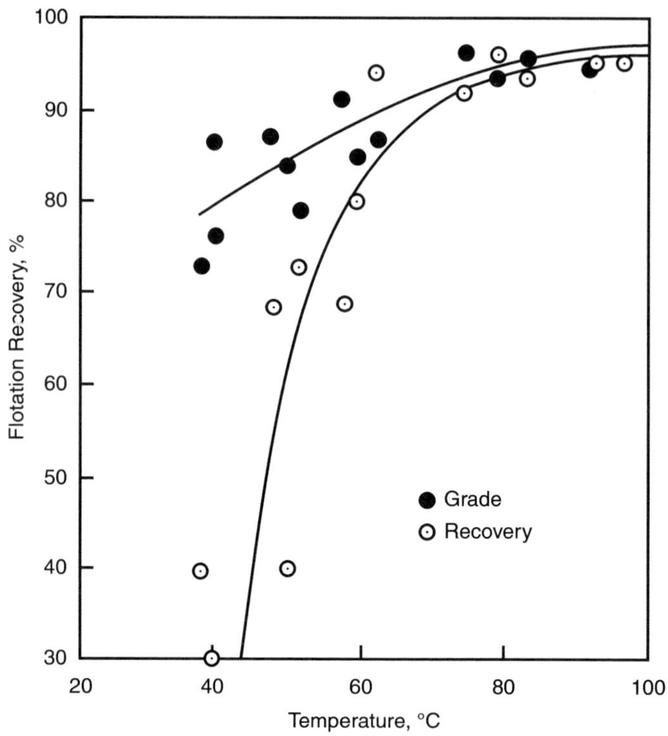

Source: Cook and Last 1950.

FIGURE 8.42 Flotation recovery and grade of fluorite from ore with 2.5 lb/ton oleic acid, 1.5 lb/ton quebracko, and 10 lb/ton soda ash

The relative proportion of tightly held chemisorbed calcium oleate to loosely held colloidal calcium oleate precipitate has been reported by Giesekke and Harris (1984) to be significantly higher for fluorite than for calcite. Such a phenomenon helps to explain the difference in the floatability of these minerals. Further, researchers have established that under certain circumstances, these films of calcium oleate are removed during bubble detachment simply by the bouyant force of the air bubble (Miller and Misra 1984).

Enhancement of fluorite flotation is obtained at elevated temperature (Cook and Last 1950). The sensitivity of flotation of fluorite from an ore is given in Figure 8.42. As the temperature increases, both grade and recovery are considerably enhanced.

In addition to the effect of temperature, oxygen partial pressure significantly increases the hyrdophobic character of the fluorite/oleate system (Plaksin 1959; Miller and Misra 1984). When the ineffectiveness of saturated fatty acids as collectors is considered, these results suggest that double bond interaction may be an important factor in this system. Cross-linking between adsorbed oleyl chains via an epoxide linkage has been suggested, leading to polymerized surface species with greater hydrophobic character (Miller and Misra 1984). Interestingly, Kellar, Young, and Miller (1992) demonstrated that only chemisorbed oleate appears to undergo oxidation; precipitated calcium oleate adsorbed on the surface does not.

Anglesite, Cerussite, and Malachite Flotation. Anglesite, cerussite, and malachite all respond well to flotation with relatively long-chained fatty acids because of the insoluble nature of heavy metal soaps and the hydrophobicity imparted by long-chained collectors. When short-chain xanthates are used as collector, however, these minerals are not floated nearly as effectively as are their sulfide counterparts. This fact is because of the extensive hydration of the carbonate and sulfate surfaces resulting from hydrogen bonding of water molecules that occurs in these systems as compared to sulfide minerals under similar circumstances.

Both anglesite and cerussite are floated completely with amyl xanthate. However, anglesite requires a tenfold increase in collector concentration over that required to float cerussite. This phenomenon is related to the fact that solubility of anglesite is significantly greater than that of cerussite.

The benefit of sulfidization of these minerals with sodium sulfide has been known for some time. After sulfidization, the surface is much less hydrophilic because chemisorbed sulfide ion is present. Under these conditions, collector requirements are reduced significantly in magnitude.

Modifiers and Depressants. Mechanisms involved in depression reactions have not been thoroughly investigated, and as a result, understanding of these systems is limited. Not only are the depressant adsorption reactions complex, but the depressants are frequently complex molecules themselves. The basic depressants used in semisoluble salt flotation systems are sodium carbonate, sodium silicate, sodium metaphosphate, lignin sulfonate, starch, quebracho, and tannin derivatives.

Sodium Carbonate. Surface carbonation is an important reaction that occurs in semisoluble salt flotation systems, resulting from CO_3^{2-} in solution derived from the CO_2 in the atmosphere or from carbonate minerals present.

In the case of fluorite, dissolved carbonate reacts with the surface to form a calcium carbonate compound. In fact, this phenomenon probably accounts for the PZC of fluorite occurring at pH 10.0 (Miller and Hiskey 1972). Calculation shows that this reaction should occur at pH values greater than pH 8 for a system open to the atmosphere.

In another study by these authors, examination of the infrared spectra of barite prepared in an aqueous suspension revealed the same carbonate doublet observed with fluorite samples, whereas spectra of barite prepared in nonaqueous environments gave no indication of surface carbonation.

The practical implication of these phenomena is that in the flotation of semisoluble salts where soda ash is used for pH control, surface carbonation seems inevitable, and the fact that good selectivity is achieved seems remarkable.

Sodium Silicate. The composition of aqueous sodium silicates is expressed by the general formula, $m\text{Na}_2\text{O}–n\text{SiO}_2$, where the ratio n/m is referred to as the "modulus" of sodium silicate. Commercial sodium silicate is available with ratios of SiO_2 to Na_2O of 1.6, 2.75, 3.22, and 3.75. The commercial sodium silicate most widely used industrially as a dispersant or depressant is Type N (modulus 3.22).

The critical importance of this reagent in achieving selectivity in nonmetallic flotation systems has been presented by Sollenberger and Greenwalt (1958), Joy and Robinson (1964), and Fuerstenau and Gutierrez (1968). In their study on apatite, calcite, and fluorite, Fuerstenau and Gutierrez (1968) showed that calcite is by far the most sensitive to sodium silicate additions. Flotation recovery of calcite with 5×10^{-4} mol/L oleate in the presence of 5×10^{-4} mol/L sodium silicate is given in Figure 8.43. At pH 6, essentially complete flotation is obtained. Under acidic conditions, protonation of silicate anions occurs. The charge on these species is reduced (perhaps even to a neutral aqueous species) under these conditions, and the effectiveness of these species as depressants is reduced.

Above pH 10, complete flotation is again obtained in the presence of sodium silicate. The natural domain of sodium silicate is pH 7 to 10; at relatively high pH values, different equilibria involving dissolved species can be expected.

The difference in bond strengths of oleate on calcite and fluorite can be seen by comparing their flotation responses in Figure 8.43 and 8.44. Fluorite is floated completely in the pH domain in which calcite is depressed in the presence of sodium silicate. The effectiveness of sodium silicate for separation of these two minerals is apparent.

Starch. Starch was recommended as a depressant as early as 1931. Industrial applications have been extensive since that time, but basic, systematic work dealing with mechanisms governing the depressant action has been limited.

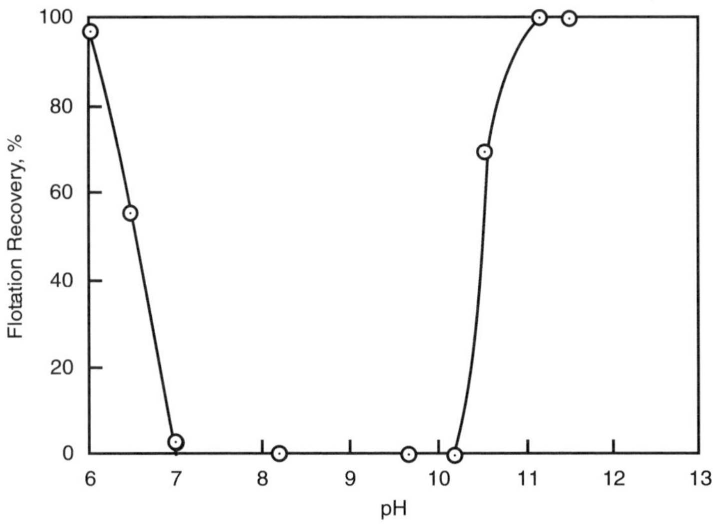

Source: Fuerstenau, Gutierrez, and Elgillani 1968.

FIGURE 8.43 Flotation recovery of calcite as a function of pH with 5 × 10^{-4} mol/L oleate and 5 × 10^{-4} mol/L sodium silicate (Type N)

The basic component of starch and dextrin is the dextrose molecule (Conn and Stumpf 1980):

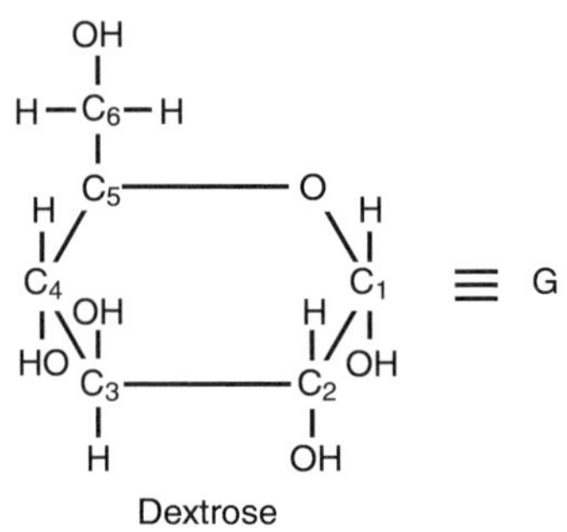

In starches, the linear chain (amylose) and branched chain (amylopectin) components have molecular weights reaching millions. In dextrin formation, these chains are fragmented and recombined to form low-molecular-weight but highly branched structures. Structural formulas of amylose, amylopectin, and dextrin are

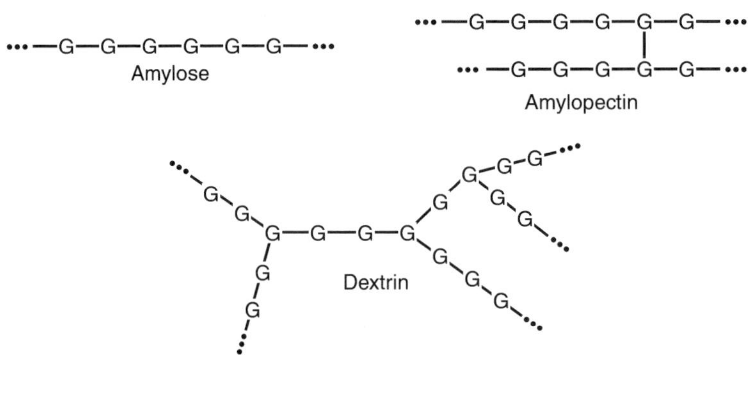

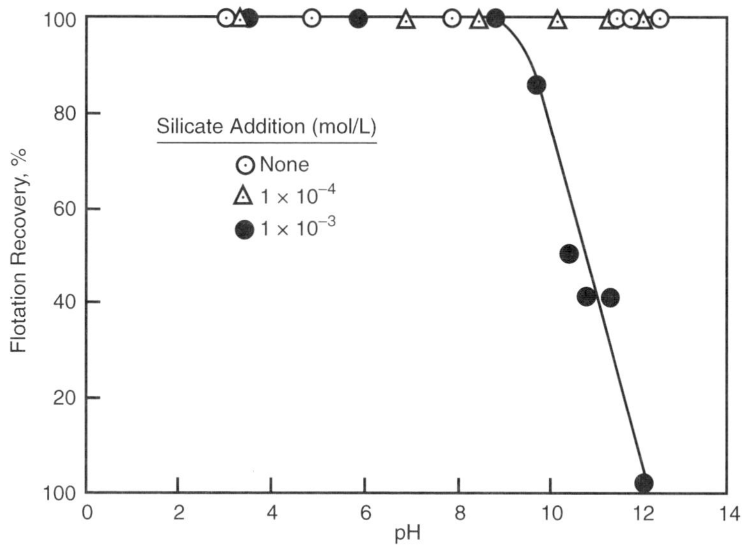

Source: Fuerstenau, Gutierrez, and Elgillani 1968.

FIGURE 8.44 Flotation recovery of fluorite as a function of pH with 1×10^{-4} mol/L oleate in the absence and presence of sodium silicate (Type N)

Starches are anionic species, and their adsorption behavior is strongly affected by their molecular weight (Iwasaki and Lai 1965). Adsorption can occur by coulombic attraction and by hydrogen bonding with surface oxygen atoms (Balajee and Iwasaki 1969). In the few systems that have been studied in detail, co-adsorption of collector and starch or starch derivatives has been reported (Somasundaran 1969; Miller, Laskowski, and Chang 1983). In the calcite/oleate/starch system, the starch actually causes an increase in oleate adsorption. The calcite surface, however, becomes hydrophilic. Such organic colloids seem able to actually blind the hydrocarbon chain of the collector and project its polar hydroxyl groups, thus creating a hydrophilic surface (Somasundaran 1969). Among semisoluble salts, this is particularly true for calcite, which may be related to the nature of the adsorbed calcium oleate at the calcite surface as previously discussed.

In a study of starch adsorption and its effect on semisoluble salt flotation with oleic acid, Hanna (1974) showed that calcite is depressed at lower concentrations of starch than is barite, which, in turn, is depressed at lower concentrations than fluorite (Figure 8.45).

Soluble Salt Flotation

Soluble salt flotation systems differ from other nonmetallic flotation systems in that ionic strengths on the order of 5*M* are typically encountered, such as in the processing of potash. Under these conditions, the zeta potential is approximately zero; the electrical double layer is essentially one ion in thickness; and the solubility of collectors is limited. These conditions result in unusual flotation phenomena.

A number of premises have been advanced to explain these phenomena: an ion exchange model (Fuerstenau and Fuerstenau 1956); a heat-of-solution model (Rogers and Schulman 1957); formation of insoluble reaction products between collector ions and surface alkali metal ions (Halbich 1933); a crystallographic properties model (Bachmann 1951); and a collector and surface hydration model along with a surface charge–ion pair model (Roman, Fuerstenau, and Seidel 1968). Each can explain some of the systems, but none can explain all observed behavior.

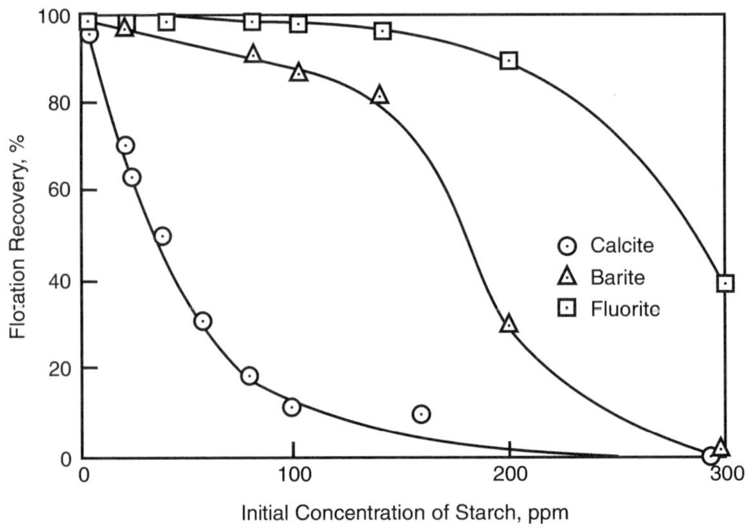

Source: Hanna 1974.

FIGURE 8.45 Flotation recovery of barite, calcite and fluorite as a function of starch concentration with 7 × 10^{-5} mol/L oleic acid at pH 8

TABLE 8.15 Sign of surface charge for selected alkali halides

	Negative Gaseous Ion Hydration Free Energies, kcal/mol			Sign of Surface Charge	
Chlorides	**Cation**	**Anion**	**ΔG***	**Predicted†**	**Experimental‡**
LiCl	112.0	82.5	29.5	–	–
NaCl	88.4	82.5	5.9	–	+
KCl	71.1	82.5	11.4	+	–
RbCl	65.9	82.5	16.6	+	+
CsCl	58.2	82.5	24.3	+	+

Source: Yalamanchili 1993.
*Difference in cation and anion gaseous hydration free energies.
†Simplified lattice ion hydration theory.
‡Electrophoretic mobility measurements by laser-Doppler electrophoresis.

Roman and colleagues (1968) showed with flocculation experiments that sylvite and halite are oppositely charged in saturated brine. These investigators also used lattice ion hydration theory to predict the surface charge on these minerals, and they predicted that sylvite would be positively charged and that halite would be negatively charged in their brines. Lattice ion hydration theory involves comparing the free energies of hydration of the corresponding gaseous ions. If the hydration free energy of the surface lattice cation is more negative than the hydration free energy of the surface lattice anion, the surface is negatively charged. The converse situation is also true.

Yalamanchili, Kellar, and Miller (1993) conducted nonequilibrium electrokinetic experiments using laser-Doppler electrophoreses on various alkali halides. A comparison between the experimentally determined and predicted surface charges of these salts is given in Table 8.15. The agreement is quite good. Note that the measured surface charges on sylvite and halite are negative and positive, respectively.

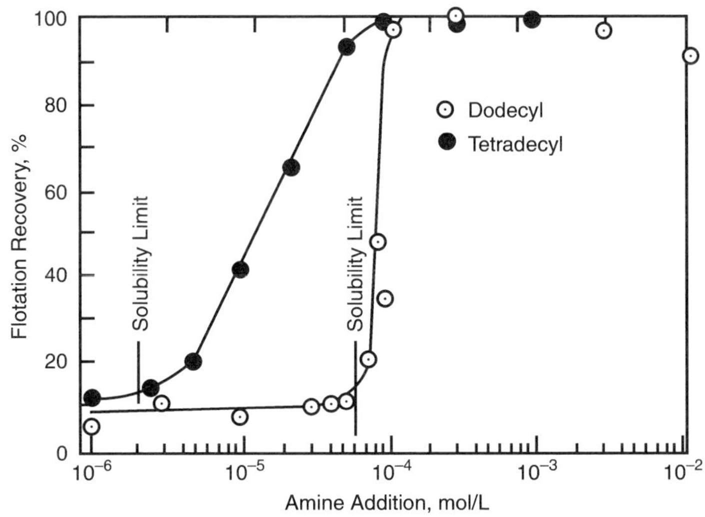

Source: Roman, Fuerstenau, and Seidel 1968.

FIGURE 8.46 Flotation recovery of sylvite as a function of amine addition

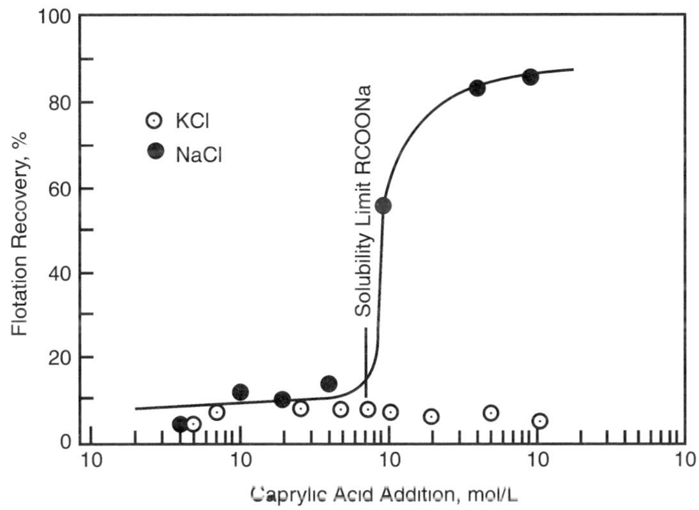

Source: Roman, Fuerstenau, and Seidel 1968.

FIGURE 8.47 Flotation recovery of KCl and NaCl as a function of caprylic acid

As shown in Figure 8.46, sylvite is floated with 12- and 14-carbon amines only after precipitation of amine chloride has occurred. With octylamine, however, flotation is achieved without precipitation of amine chloride, although relatively high additions of collector are required. Similar phenomena have also been observed for amine flotation of KNO_3 (Pizarro 1967).

In the case of halite, flotation is not achieved with amines under any circumstances. On the other hand, good recovery is obtained with a carboxylate collector after the particular sodium carboxylate has precipitated (Figure 8.47).

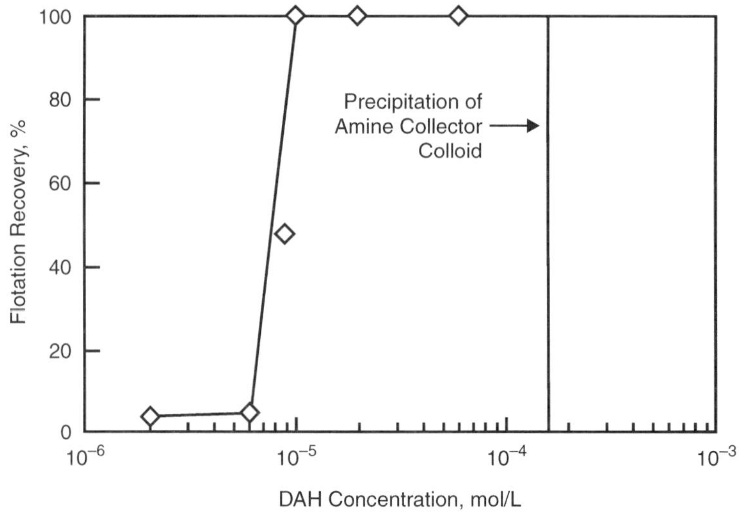

Source: Hancer et al. 1997.

FIGURE 8.48 Flotation recovery of K_2SO_4 with dodecylamine

Laskowski, Vordela, and Liu (1988) showed that precipitated amine colloids are positively charged up to about pH 11; Yalamanchili, Kellar, and Miller (1993) demonstrated that precipitated colloids of sodium laurate are negatively charged in basic medium. The positively charged amine colloids are adsorbed on the negatively charged sylvite, resulting in its flotation, but not on the positively charged halite. Conversely, the negatively charged sodium carboxylate colloids are adsorbed on halite and not on sylvite.

Sylvite has also been shown to respond to flotation with octyl and decyl sulfonate below the solubility limit of both potassium sulfonates (Roman, Fuerstenau, and Seidel 1968). Both the collector ion and the sylvite surface are negatively charged under these conditions.

Anhydrous potassium sulfate has also been shown to float with octyl sulfonate before precipitation of potassium sulfonate (Pizarro 1967). This effect is interesting, again, because both the collector ion and the K_2SO_4 surface are negatively charged under these conditions. Hancer et al. (1997) showed that K_2SO_4 is also floated with dodecylamine before precipitation of amine sulfate (Figure 8.48). It should be noted that anhydrous K_2SO_4 is stable at room temperature.

Despite the negative charge of Na_2SO_4, anhydrous Na_2SO_4 is not floated with dodecyl amine or with anionic collectors, for that matter, at room temperature. This phenomenon appears to result from the fact that anhydrous Na_2SO_4 is not stable under these conditions, and substantial hydration of the surface will occur. This transformation would create substantial instability at the surface and prevent adsorption of collector species. On the other hand, at somewhat higher temperatures, ≥32.4°C, anhydrous Na_2SO_4 is stable, and flotation is possible (Figure 8.49).

Waters of crystallization, though, apparently do not inhibit a salt's flotation. As shown in Figure 8.50, $Na_2SO_4 \cdot 10H_2O$ is floated with dodecyl amine at 26°C, under which condition this salt is stable. Anhydrous Na_2SO_4 is not stable at this temperature and is not floated. With both K_2SO_4 and Na_2SO_4 salts, then, flotation is possible when the salt is in its stable state.

Clearly, the understanding of the mechanisms of flotation occurring in this system is not at the same level as the understanding for other general mineral systems.

FLOTATION MACHINES

Flotation machines are designed to ensure flow of the pulp into good, active contact of particles with bubbles and levitation of mineral-laden air bubbles to the top of the cell, allowing entrapped particles

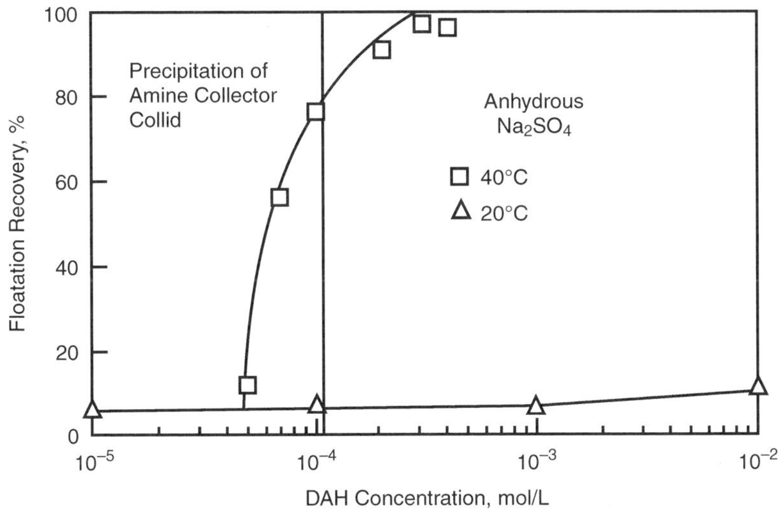

Source: Hancer et al. 1997.

FIGURE 8.49 Flotation recovery with anhydrous as a function of dodecylamine addition and temperature

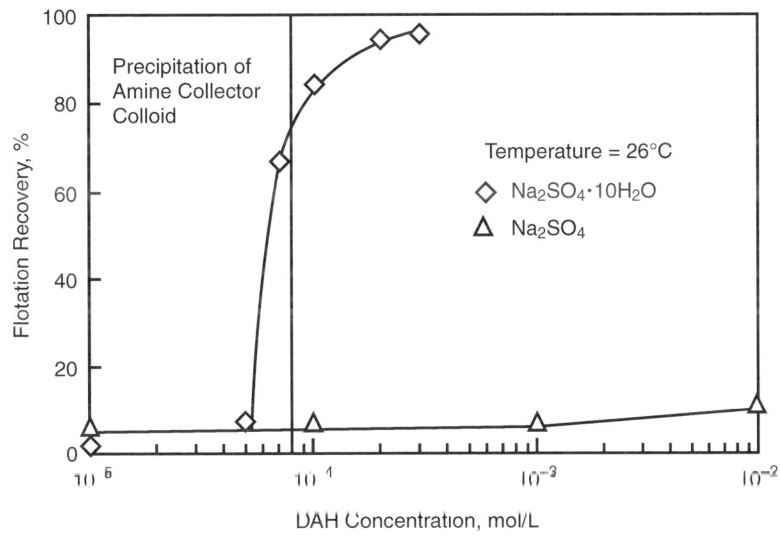

Source: Hancer et al. 1997.

FIGURE 8.50 Flotation recovery of anhydrous and hydrated Na_2SO_4 as a function of dodecylamine addition

to be removed. In addition, some laboratory machines are also designed to allow study of the physicochemical principles involved in flotation subprocesses. Different attempts to meet these requirements have resulted in many designs.

Essentially, flotation machines for production are divided into two types depending on the mechanisms by which air is introduced into the cell. Many variations of these types are seen, allowing

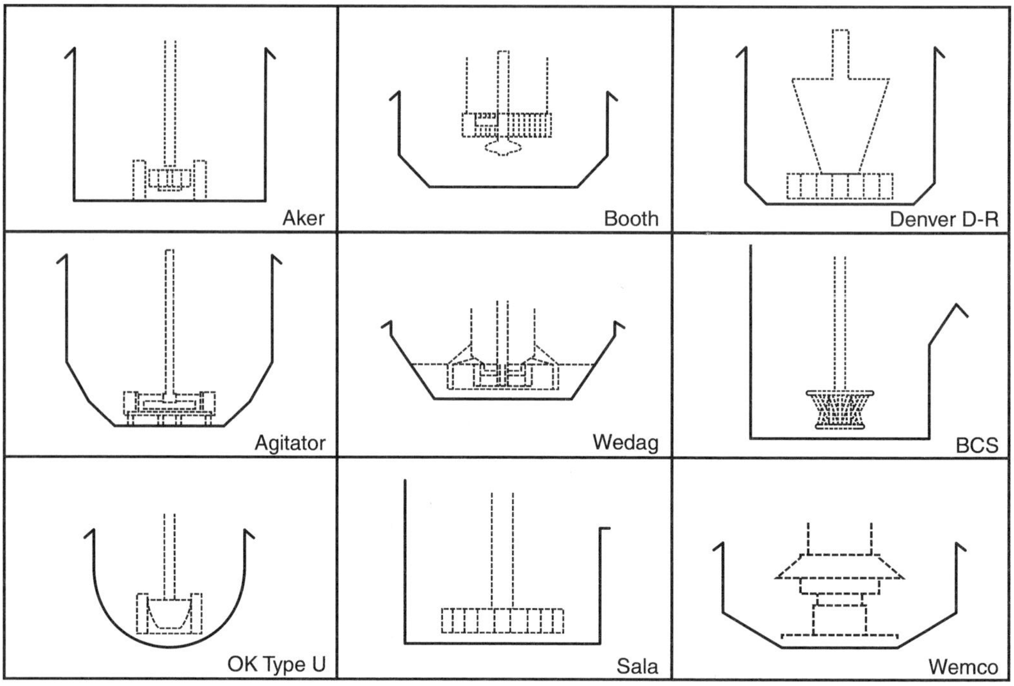

Source: Young 1982.

FIGURE 8.51 Flotation tank profiles of open-flow machines

different intensity of agitation and different flow patterns in equipment with a variety of sizes and shapes. The two main types are pneumatic and mechanical machines. In pneumatic machines, air entering the turbulent pulp is dispersed into bubbles by baffles or perforated bases, ensuring maximum opportunity for contact with the mineral particles. Dispersion of the pulp itself is achieved by agitation caused by compressed air. These cells have essentially disappeared, except for the "apatite" machine manufactured in the USSR. A variation of this machine is used in the Davcra air cell, in which feed enters through a nozzle at the bottom of the cell and impacts on a baffle where sufficient turbulence is created to provide dispersion of bubbles and contact with particles. Although these cells are mechanically simpler than other designs because they lack many moving parts, they also give less effective performance. The mechanical machines, on the other hand, achieve dispersion of the pulp through agitation by a mechanically driven impeller. Mechanical machines employ one of two types of pulp flow and aeration systems: cell-to-cell flow with adjustable wiers between cells or open flow without the wiers and air intake via suction resulting from the rotation of the impeller or an external blower.

Manufacturers offer many types of cell and impeller geometries, each with its aims and claims. Cell shapes vary from rectangular to truncated rectangular to U-shaped, and impeller components vary from simple flat turbine set to cylindrical finger set to propellers (Figures 8.51 and 8.52). The position of the impellers relative to the cell bottom and the type and placement of the stator also vary to achieve the desired pulp suspension and circulation. Major cells listed by Young (1982) are Aker, Booth, Denver, Agitair, Outokumpu (OK), Wedag, Sala, Minemet BCS, Wemco, Maxwell, and Dorr–Oliver. The heart of the flotation machine is the impeller, and the major features of these machines' impellers are described in the following paragraphs.

Aker's impeller consists of a flat turbine. Air is charged through the impeller shaft and delivered through slots behind the impeller. Booth, on the other hand, uses a truncated shallow rectangular

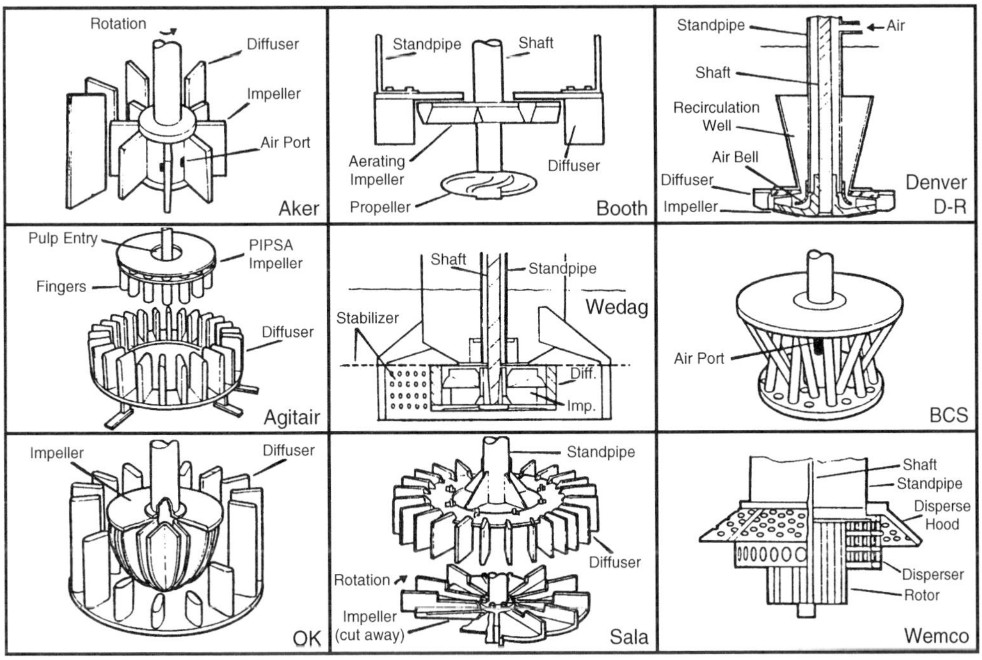

Source: Young 1982.

FIGURE 8.52 Flotation cell impeller mechanisms

tank and impellers significantly above the tank bottom. Agitation at the bottom is achieved with the help of a propeller mounted below the impeller. The Denver D-R cell is an open tank type with vertical recirculation enhanced by means of a well that directs pulp to the impeller region. In contrast to the D-R cell, the Denver subaeration cell has adjustable wiers between cells that reduce chances for back mixing and also allow control of pulp level in each cell. The Agitair cell uses an impeller consisting of a horizontal disk with cylindrical fingers extending toward the bottom. Air blown in through the hollow shaft is mixed with pulp via a secondary pump on the disk and a stator. In contrast to all these designs, the Outokumpu cell is characterized by teacup-shaped impeller blades with tapered slots between the blades for air passage from the hollow shaft. Flotation of the impeller causes pulp to be drawn from the bottom toward the slots and expelled toward the periphery. This machine claims better suspension and lower power consumption. The Wedag design for coal, on the other hand, uses a shallow self-aerating impeller with blades at the periphery under a horizontal hood with a stator projecting downward at the periphery. This design permits stratification of the pulp because of minimum vertical circulation dispersing air into finer bubbles, according to claims, and floating finer particles more efficiently than other models.

A major new addition to cell design and operation is the froth separator developed in the USSR. In this machine, the flotation feed is discharged after conditioning directly to the top of a froth bed. Hydrophobic particles are apparently retained in the froth column and get many chances to attach themselves to many bubbles, while the others sink and, thus, get separated as tailings. These types of separators are reported by Young to work well for coarse particles, with a high process rate. The main advantages of this cell are longer contact time between particles and bubbles, the possibility of contact of particles with a multiplicity of bubbles, and low tendency for detachment of particles already attached to bubbles.

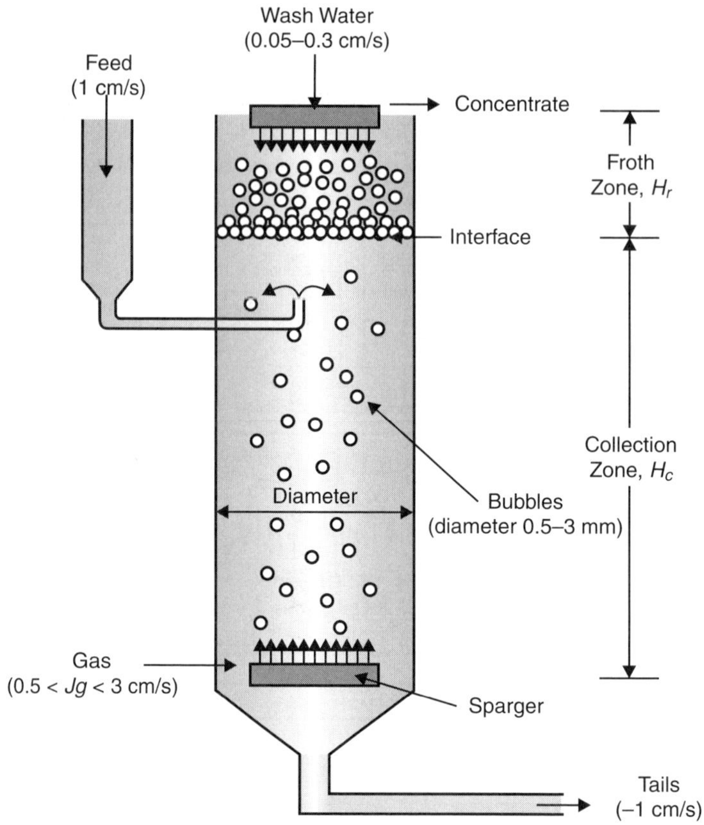

Source: Finch, Uribe-Salas, and Xu 1995.

FIGURE 8.53 Schematic of a conventional column flotation cell

The Minemet cell uses a totally different type of impeller made up of two series of circular bars forming opposite cones between a horizontal plate about one-third larger than the lower plate with perforation for recirculation of the slurry. This design yields excellent, fine dispersion of the air and good solids suspension. The Wemco impeller is composed of a rotor with blades and is located far above the tank bottom. A perforated dispenser with a hood on top acts as a stator, and the design reportedly requires less power and maintenance than others require. The Maxwell cell stands out from all the others in that it is essentially a conditioner-type tank with flat-bladed turbine and air sparged through a tube. It has no baffle and offers low installation and maintenance costs as well as low power consumption. Another machine is that offered by Dorr–Oliver. It is similar in design to that of Outokumpu. Air from a hollow shaft enters the cell from the channels between rotors and blades and the pulp around the rotor, which is surrounded by short stator vanes. This design is claimed to minimize unnecessary turbulence and, thus, power requirements.

COLUMN FLOTATION

The column flotation concept has been around for nearly 40 years, but it attracted attention with the copper mining problems of the early 1980s. The column flotation technique uses the countercurrent principle to improve separation by reducing entrapment of particles. A schematic diagram of column flotation cells is shown in Figure 8.53. The important operating difference from mechanical flotation cells is the lack of an impeller, or any other agitation mechanism, which reduces energy and maintenance costs. The other major difference is that for most ore-processing applications, wash water is sprayed into the froth at the top of the column, which is impossible to accomplish in a mechanical cell as it can kill the froth. The amount of wash water added is a major factor in determining flotation selectivity and recovery as well as column operation stability.

In column flotation, the ore is fed into the column via a distributor located at about two-thirds of the height of the column; the tails are removed from the bottom; concentrate overflows at the top; and the air bubbles are generated at the bottom of the column by a porous sparger. Three characteristic features are the use of a sparger to generate bubbles near the base, a countercurrent slurry/bubble flow in the collection zone, and a deep froth zone (0.5–2.0 m) coupled with the use of wash water to induce a cleaning action. This design was first patented in Canada in the early 1960s and is sometimes known as the "Canadian" or "conventional" column.

Industrial column equipment has a height of 9 to 14 m and a diameter of not more than 2 m without baffling. Generally, the unit is operated with enough overhead wash water to provide a net downward flow of water, a condition known as a "positive bias." Positive bias is the norm in column operation, because the froth layer in a column is then stabilized by the wash water. The greater the flow of water down the column, the greater the selectivity, and the thicker the froth layer. The froth depth in a stable operation is a little deeper than one meter. A negative bias eliminates the froth altogether, which is very deleterious for a process where the concentrate is the desired product.

The design of any ore-processing operation with columns must ensure that the rate-controlling flotation mechanism is always bubble capture of mineral particles that have been precoated with collectors in a prior flotation step.

It is customary to describe the operating conditions of flotation columns in terms of superficial velocities (*J*) in order to normalize the data for different size columns. Typical values (from Crozier 1992) are

- Gas velocity, Jg = 0.5 to 3.0 cm/s
- Pulp feed velocity, Jp = 0.7 to 2.0 cm/s
- Wash water velocity, Jw = 0.1 to 0.8 cm/s
- Bias water velocity, Jb = 0.07 to 0.3 cm/s

In addition, in scale-up equations, it is customary to normalize the gas velocity for different column heights by the pressure correction:

$$Jg = \frac{(P_c)(Jg^*)(\ln(P_s/P_c))}{P_s - P_c} \quad \textbf{(Eq. 8.29)}$$

where

Jg^* = gas velocity under standard conditions at the top of the column
P_c = absolute pressure at the top of the column
P_S = absolute pressure at the sparger

The effect of gas velocity on recovery and grade is dominated by the bubble size, which depends on the pore size of the sparger. The size of bubbles produced is also determined by the type of bubble generation system, frother type, and dosage.

Sparger

Sparging through a porous medium without high external shear is the most common approach for column flotation. Industrial sparging material is made up of either pierced rubber or fabric such as filter cloth. Pierced rubber generally generates smaller gas bubbles, but is more difficult to fabricate. Rubinstein (1995) examined the effect of filter cloth permeability on gas holdup and suggested an upper permeability level of 6 $m^3/m^2/min$. At the laboratory scale, inflexible porous materials such as porous steel, bronze, glass, or plastic are generally used. Early work showed that an inflexible medium would get plugged with solids or precipitate within several hours or days, making it unsuitable for industrial use.

Sparging through a porous medium with high external shear uses a porous sparger placed in a high-velocity slurry or wet line. In this process, bubble generation is controlled by both the nature of the porous medium and the shear action created by the flowing slurry.

Jetting is also used to generate bubbles when either a gas stream is jetted from an orifice into the liquid or when the liquid is jetted from an orifice into the pool.

Bubble Size

Bubble velocity in a flotation column is usually considerably higher than the slurry velocity. Therefore, the major hydrodynamic and flotation characteristics are determined by the airflow rate and the method of sparging. Slurry flow rate mainly influences the particle retention time. Coalescence as well as dispersion of bubbles can occur depending on the hydrodynamic and physicochemical conditions, and this effect can result in marked variation in bubble size distribution. With a limited increase in the column height, bubble size distribution in the upper portion will reach a steady state and will not depend on the sparger parameters; instead, it is determined by the condition of minimum potential energy. The time required to reach a steady-state bubble size distribution depends on aeration rate, surfactants, and solids concentrations as well as material properties. An increase in frother concentration results in lower mobility of the bubble surface and, consequently, in the reduction of the bubble rise velocity (Zhou, Egieor, and Plitt 1992). The reduction in surface tension significantly decreases coalescence intensity, which in turn causes a reduction in average bubble size. Depending on the rise velocities of small and large bubbles, their retention times in the column differ. The bubble size distribution at the sparger differs from the average distribution in the column, even in the absence of coalescence, breakage, and bubble growth caused by pressure reduction. As a result of the lower retention time of larger bubbles, mean bubble size in the slurry is lower than the initial mean size.

Mixing of Phases

Nonuniform aeration in a flotation apparatus reduces the selection efficiency significantly as a result of large-scale liquid circulation. An increase in the airflow rate results in nonuniform aeration. Such heterogeneous behavior of the column operation is unfavorable, as an increase of the air-lift effect causes particle entrainment in the froth.

Column flotation has the following advantages: low power requirement, low capital investment, large aerated space, a possibility of controlling airflow rate, and dispersion. It allows production of high-grade concentrates, reduction of consumption of depressants, and simplification of process flowsheets.

Laboratory Flotation Machines

Laboratory flotation machines employed for research purposes include the Hallimond cell, the Buchner-type cell, and their variations. A typical Hallimond tube is shown in Figure 8.54. The lower part of the cell consists of a glass chamber with a frit having pores of a uniform size not less than 40 μm. The upper part consists of a bent glass tube with a vertical stem just above the bend. The top of the tube is connected to a flowmeter to monitor the gas flow. The pressure in the reservoir can be adjusted to control the gas flow. A magnetic stirring bar inside the glass well is used to agitate the particles. Conditioning is done outside the Hallimond tube, and the pulp is then transferred into the cell and floated for a desired time. The floated particles fall into the vertical stem or stay attached to the top of the cell from where they are easily separated. Another microflotation cell consists essentially of a modified Buchner funnel with a stem parallel to the bottom and with a bent lip at the top of the cell to discharge the froth. A microscope slide suspended inside the cell acts as a baffle. Unlike the Hallimond tube, the reagentizing of the particles can be done in

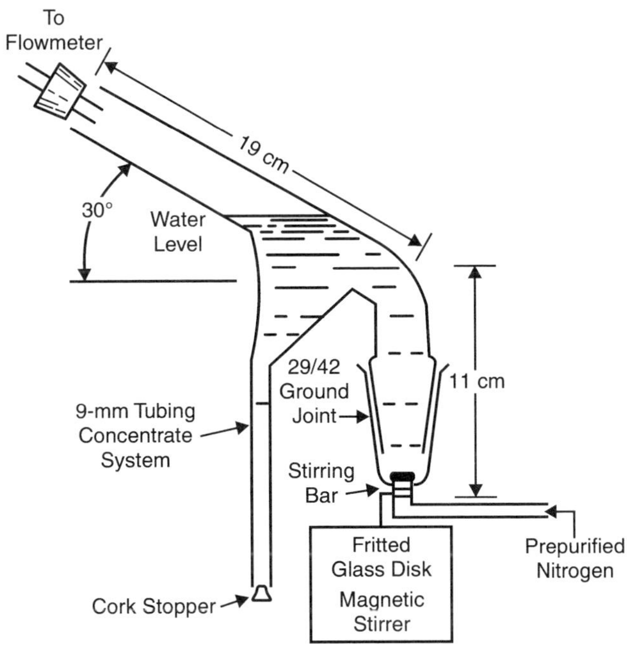

FIGURE 8.54 Schematic of a Hallimond flotation cell

the cell itself. Addition of a frother is necessary here. The gas flow system and the rest of the assembly can be similar to those described earlier for the Hallimond tube.

Chemical conditions for separation of minerals by flotation can be studied by means of microflotation experiments. The subsequent step in a process design scheme should be flotation testing using 200- to 1,000-g samples of material in laboratory-size machines. Denver, Wemco, Agitair, and Fagergren are among the machines available for laboratory testing. These replicas of commercial machines allow researchers to test the feasibility of a process. The mineral is conditioned with the collector solution for the desired time in the cell, and then other reagents, such as pH modifiers and dispersants, are added during additional conditioning. Toward the end of conditioning, frother is added, and the system is stirred for another 2 min. The impeller speed is reduced and aeration begins. Flotation often continues until completion, perhaps as long as 15 min. The floated products and tailings are analyzed for grade and recovery.

FLOTATION CIRCUITS

The elements of flotation circuits are (following Arbiter 1985):

1. *Rougher circuit.* New feed and recycled products (scavenger concentrate and cleaner tailing) are fed to this circuit.
2. *Scavenger circuit.* Rougher tailing is fed to this circuit. Scavenger concentrate and scavenger tailing are produced. Scavenger concentrate may be recycled with or without grinding to the rougher circuit or may be cleaned separately. Scavenger tailing is the final tailing.
3. *Cleaner circuit.* Rougher concentrate is fed to this circuit. Cleaner concentrate and cleaner tailing are produced. Cleaner concentrate may be used directly or cleaned additionally. Cleaner tailing is returned with or without grinding to the rougher circuit.

Cell arrangement can establish either series or parallel flow. Banks of cells are arranged in parallel when flows are too large for a single series line. Cell requirements as a function of feed rate, pulp

TABLE 8.16 Variations in cells required with cell size at different tonnages and pulp densities*

Dry Tons per Day, % solids	10,000	25,000	50,000	100,000
100 ft^3 cells				
20	77	192	383	767
30	49	122	245	489
40	32	81	161	322
500 ft^3 cells				
20	16	40	80	160
30	10	25	50	100
40	6	16	32	64
1,000 ft^3 cells				
20	8	20	40	80
30	5	12	24	50
40	3	8	16	32

Source: Arbiter 1985.

*8-min flotation time; 3.0 specific-gravity ore.

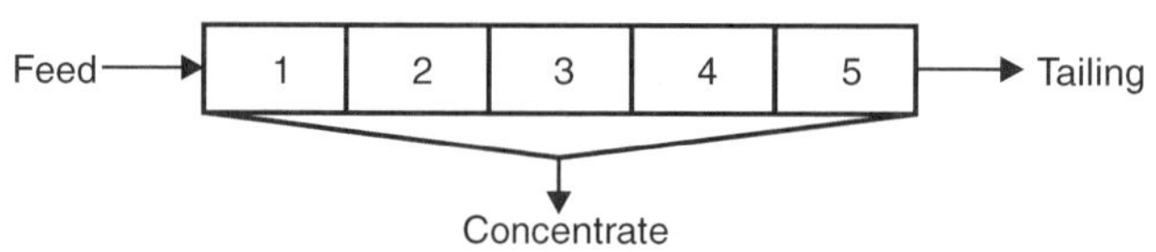

When the rougher concentrate requires cleaning, the following circuits may be involved:

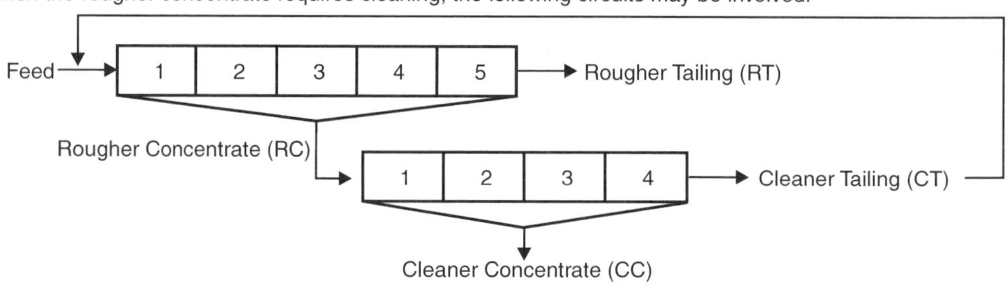

The following flowsheet shows the presence of middlings in the circuit. Regrinding of cleaner tailing (CT) and scavenger concentrate (SC) is needed.

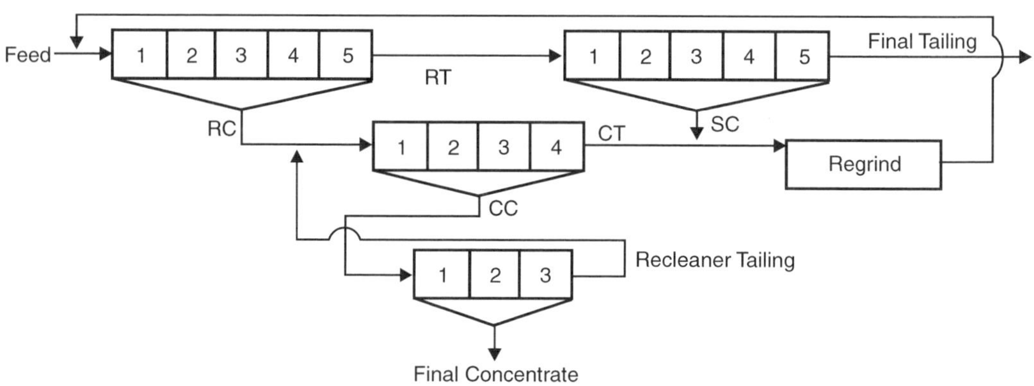

Source: Gaudin 1939.

FIGURE 8.55 Typical flotation flowsheets

density, and cell size are presented in Table 8.16. As can be noted, the use of 500 and 1,000 ft^3 cells reduces the number required very dramatically. Cells with volumes of 5,000 ft^3 are currently in use.

The complexity of flotation circuits is a function of the complexity of the ore being processed, as Figure 8.55 shows. For single value ores, relatively simple circuits are involved. When concentrate cleaning is not necessary to produce a satisfactory grade, a more elaborate circuit can be employed.

BIBLIOGRAPHY

Allison, S.A., L.A. Goold, M.J. Nicol, and A. Granville, 1972. A Determination of the Products of Reaction between Various Sulfide Minerals and Aqueous Xanthate Solution and a Correlation of the Products with Electrode Rest Potentials. *Met. Trans.*, 3:2613.

Anonymous. 1965. Flotation Column Due for Mill Scale Test, *E&MJ*, 166, No. 1.

Aplan, F.F., and D.W. Fuerstenau, 1962. Principles of Nonmetallic Flotation. In *Froth Flotation*. Edited by D.W. Fuerstenau, New York: AIME.

Arbiter, N. 1985. Flotation. In *SME Mineral Processing Handbook*. Edited by N.L. Weiss. Littleton, Colo.: SME.

Arbiter, N., U. Fujii, B. Hansen, and A. Raja, 1975. Surface Properties of Hydrophobic Solids. In *Advances in Interfacial Phenomena. AIChE Symposium Series*, 71:176.

Bachman, R. 1951. *Erzmetall*, 4:316.

Bahr, A., M. Clement, and H. Surmatz. 1968. On the Influence of Organic and Nonorganic Substances on Flotation of Non-sulfide Minerals with Fatty Acid Collectors. *Proceedings of the VIII International Mineral Processing Congress*. Mehkandobr, Leningrad. p. 337.

Balajee, S.R., and I. Iwasaki. 1969. Adsorption Mechanism of Starches in Flotation and Flocculation of Iron Ores. *Trans. AIME*, 244:401.

Bilsing, U. 1969. Ph.D. diss. Bergakademie, Freiberg.

Buckley, A.N., and R. Woods, 1984. An X-Ray Photoelectron Spectroscopic Investigation of the Surface Oxidation of Sulfide Minerals. In *Proceedings of the International Symposium on Electrochemistry in Mineral and Metal Processing*, Edited by P.E. Richardson and S. Srinivasan. Cincinnati: 84:10.

Butler, J.N, 1964. *Ionic Equilibrium*. Reading, Mass.: Addison-Wesley.

Cases, J.M., P. Jacquier, S. Smani, J.E. Poiorier, and J.Y. Bottero. 1988. Revue de L'Industri Minerale. Congress de Marrakech, Marrakech.

Chander, S., and D.W. Fuerstenau. 1972. On the Natural Floatability of Molybdenite. *Trans. AIME*, 252:62.

Chander, S., J.M. Wie, and D.W. Fuerstenau, 1975. On the Native Floatability and the Surface Properties of Naturally Hydrophobic Solids. Advances in Interfacial Phenomena. Edited by P. Somasundaran and R.B. Grieves. *AIChE Symposium Series*, 71:183.

Clark, S.W., and S.R.B. Cooke. 1968. Adsorption of Calcium, Magnesium and Sodium Ions by Quartz. *Trans. AIME*, 241:334.

Clemmer, J.B., B.H. Clemmons, C. Rampacek, M.F. Williams, and R.H. Stacey. 1945. U.S. Bureau of Mines, Report of Investigation (RI 3799).

Conn, E.E., and P.K. Stumpf. 1980. *Outlines of Biochemistry*. New York: John Wiley & Sons.

Cook, M.A., and W.A. Last. 1950. Utah Engineering Experiment Station. Bulletin No. 47.

Cooke, S.R.B., and M. Digre. 1949. Studies on the Activation of Quartz with Calcium Ion. *Trans. AIME*, 184:299.

Crozier, R.D. 1992. *Flotation Theory, Reagents, Designs and Practices*. New York: Pergamon Press.

Debrivnaya, L.B. 1962. Tr. Vses. Nauch.–Issled, I Proekt. Dust. Mekh. Obrab. Polez. *Iskop*, 131:147.

Derjaguin, B.V., and V.D. Shukakidse, 1960–61. Dependence of the Floatability of Antimonite on the Value of Zeta Potential. *Trans. IMM*, 70:569.

Dewey, F. 1933. Master's thesis, Montana School of Mines, Butte, MT.

Dixit, S.G., and A.K. Biswas, 1969. pH-Dependence on the Flotation and Adsorption Properties of Some Beach Sand Minerals. *Trans. AIME*, 244:173.

Du Reitz, C. 1975. *Proceedings of the 11th International Mineral Processing Congress*. Cagliari, Italy.

Elgillani, D.A., and M.C. Fuerstenau, 1968. Mechanisms of Cyanide Depression of Pyrite. *Trans. AIME*, 241:437.

Finch, J.S., A. Uribe-Salas, and M. Xu. 1995. Column Flotation. In *Flotation Science and Engineering*. Edited by K.A. Matis. New York: Marcel Dekker.

Finkelstein, N.P., and S.A. Allison. 1976. The Chemistry of Activation, Deactivation and Depression in the Flotation of Zinc Sulfide. In *Flotation*. Edited by M.C. Fuerstenau. New York: AIME.

Free, M.L., and J.D. Miller. 1996. The Significance of Collector Colloid Adsorption Phenomena in the Fluorite/Oleate Flotation System as Revealed by FTIR/IRS and Solution Chemistry Analysis. *Int. J. of Min. Proc.*, 48:197.

Fuerstenau, D.W. 1957. Correlation of Adsorption Density, Contact Angle and Zeta Potential with Flotation of Quartz. *Trans. AIME*, 208:1365.

Fuerstenau, D.W., and M.C. Fuerstenau. 1956. Ionic Size in the Flotation Collection of Alkali Halides. *Min. Eng.*, 8:302.

Fuerstenau, D.W., T.W. Healy, and P. Somasundaran. 1964. The Role of the Hydrocarbon Chain of Alkyl Collectors in Flotation. *Trans. AIME*, 229:321.

Fuerstenau, M.C., K.L. Clifford, and M.C. Kuhn. 1974. The Role of Zinc Xanthate Precipitation in Sphalerite Flotation. *Int. J. Miner. Process.*, 1:307.

Fuerstenau, M.C., and W.F. Cummins. 1967. The Role of Basic Aqueous Complexes in Anionic Flotation of Quartz. *Trans. AIME*, 238:196.

Fuerstenau, M.C., D.A. Elgillani, and J.D. Miller. 1970. Adsorption Mechanisms in Nonmetallic Activation. *Trans. AIME*, 247:11.

Fuerstenau, M.C., G. Gutierrez, and D.A. Elgillani. 1968. The Influence of Sodium Silicate in Nonmetallic Flotation Systems. *Trans. AIME*, 241:319.

Fuerstenau, M.C., R.W. Harper, and J.D. Miller. 1970. Hydroxamate vs. Fatty Acid Flotation of Iron Ore. *Trans. AIME*, 247:69.

Fuerstenau, M.C., J. Huiatt, and M.C. Kuhn. 1971. Dithiophosphate vs. Xanthate Flotation of Chalcoite and Pyrite. *Trans. AIME*, 250:227.

Fuerstenau, M.C., M.C. Kuhn, and D.A. Elgillani. 1968. The Role of Dixanthogen in Xanthate Flotation of Pyrite, *Trans. AIME*, 241:148–156.

Fuerstenau, M.C., C.C. Martin, and R.B. Bhappu. 1963. The Role of Metal Ion Hydrolysis in Sulfonate Flotation of Quartz. *Trans. AIME*, 226:449.

Fuerstenau, M.C., and J.D. Miller. 1967. The Role of the Hydrocarbon Chain in Anionic Flotation of Calcite. *Trans. AIME*, 238:153–160.

Fuerstenau, M.C., J.D. Miller, and M.C. Kuhn. 1985. *Chemistry of Flotation*. New York: AIME.

Fuerstenau, M.C., and B.R. Palmer. 1976. Anionic Flotation of Oxides and Silicates. In *Flotation*. Edited by M.C. Fuerstenau. New York: AIME.

Fuerstenau, M.C., R.E. Pray, and J.D. Miller. 1965. Metal Ion Activation in Xanthate Flotation of Quartz. *Trans. AIME*, 232:359.

Fuerstenau, M.C., and D.A. Rice. 1968. Flotation Characteristics of Pyrolusite. *Trans. AIME*, 241:453–457.

Fuerstenau, M.C., and B.J. Sabacky. 1981. On the Natural Flotability of Sulfides. *Int. J. Miner. Process.*, 8:79.

Fuerstenau, M.C., A. Valdivieso, and D.W. Fuerstenau. 1988. Role of Hydrolyzed Cations in the Natural Hydrophobicity of Talc. *Int. J. Miner. Process.*, 23:161.

Gaudin, A.M. 1939. *Principles of Mineral Processing*. New York: McGraw-Hill.

——. 1957. *Flotation*. 2nd ed. New York: McGraw-Hill.

Gaudin, A.M., D.W. Fuerstenau, and G.W. Mao. 1959. Activation and Deactivation Studies with Copper on Sphalerite. *Trans. AIME*, 214:430–436.

Gaudin, A.M., D.W. Fuerstenau, and M.M. Turkanis. 1957. Activation and Deactivation Studies with Copper on Sphalerite. *Trans. AIME*, 208:65.

Gaudin, A.M., H.L. Miaw, and H.R. Spedden. 1957. *Proc. II Int. Cong. Of Surf. Activity*. London: Butterworths.

Gaudin, A.M., and A. Rizo-Patron. 1942. The Mechanism of Activation in Flotation. *Trans. AIME*, 153:462.

Gaudin, A.M., and R. Schuhmann. 1936. The Action of Potassium Amyl Xanthate on Chalcocite. *J. Phys. Chem.*, 40:259.

Giesekke, E.W., and P.J. Harris. 1984. *Proceedings, MINTEK 50 International Conference on Recent Advances in Mineral Science and Technology*, Johannesburg.

Girczys, J., and J. Laskowski. 1972. Mechanism of Flotation of Unactivated Sphalerite with Xanthates. *Trans. IMM*, 81:C118.

Grosman, L.B., and P.G. Khadzhiev. 1966. Izv. Vysshikh Uchebn, Zavedenii. *Tevetn. Met.*, 9:25.

Halbich, W. 1933. *Metall. und Erz.*, 30:431.

Hancer, M., H. Yuehua, M.C. Fuerstenau, and J.D. Miller. 1997. Amine Flotation of Soluble Salts. *Proceedings of the XX International Mineral Processing Congress*, Aachen: Germany.

Hanna, H.S. 1974. *Recent Advances in Science and Technology*. Edited by A. Bashay. New York: Plenum Press.

Hanna, H.S., and P. Somasundaran. 1976. Flotation of Salt-type Minerals. In *Flotation*. Edited by M.C. Fuerstenau. New York: AIME.

Harris, C.C. 1976. Flotation Machines. In *Flotation*. Edited by M.C. Fuerstenau. New York: AIME.

Harris, P.J., and N.P. Finkelstein. 1975. *Proceedings of the 11th International Mineral Processing Congress*, Calgiari, Italy.

Heyes, G.W., and W.J. Traher. 1984. *Proceedings of the 165th Meeting of Electrochemical Society*, Cincinnati.

Hoover, R.M., and D. Malhotra. 1976. Emulsion Flotation of Molybdenite. In *Flotation*. Edited by M.C. Fuerstenau. New York: AIME.

Iwasaki, I., S.R.B. Cooke, and A.F. Colombo. 1960. U.S. Bureau of Mines, Report of Investigation (RI 5593).

Iwasaki, I., and R.W. Lai. 1965. Starches and Starch Products as Depressants in Soap Flotation of Activated Silica from Iron Ores. *Trans. AIME*, 232:364.

James, R.O., and T.W. Healy. 1972. Adsorption of Hydrolyzable Metal Ions at the Oxide-water Interface, Parts I, II, and III. *J. Coll. Int. Sci.*, 40:42–81.

Joy, A.S., and A.J. Robinson. 1964. *Recent Progress in Surface Science*. Edited by J.F. Danielli. New York: Academic Press.

Kakovsky, I.A. 1957. *Proceedings of the 2nd International Congress of Surface Activity*. Edited by J.H. Schulman. London: Butterworths.

Kellar, J.J., C.A. Young, and J.D. Miller. 1992. In-situ FTP-IR/IRS Investigation of Double Bond Reactions of Adsorbed Oleate at a Fluorite Surface. *Int. J. Miner. Process.*, 35:239.

Kraeber, L., and A. Boppel. 1934. *Metall. und Erz.*, 31:417.

Kuhn, M.C. 1968. *A Mechanism of Dixanthogen Adsorption on Sulfide Minerals*. Ph.D. diss., Colorado School of Mines, Golden.

Kulkarni, R.D., and P. Somasundaran. 1975. Kinetics of Adsorption at the Liquid/Air Interface and Its Role in Hematite Flotation. *AIChE Symposium Series*, 71:124.

Laskowski, J., R.M. Vurdela, and Q. Liu. 1988. The Colloid Chemistry of Weak Electrolyte Collector Flotation. *Proceedings of 16th International Mineral Processing Congress*. Edited by K.S. Eric Forssberg. Stockholm.

Latimer, W.M. 1952. *Oxidation Potentials*. Englewood Cliffs, N.J.: Prentice-Hall.

Leja, J., L.H. Little, and G.W. Poling. 1963. Xanthate Adsorption Studies Using Infra-red Spectroscopy, Evaporated Lead Sulfide, Galena and Metallic Lead Substrate. *Trans. IMM*, 72:414.

Lepetic, V.M. 1974. *CIM Bulletin*, 71.

Livshitz, A.K., and E.M. Idelson. 1953. Concentration and Metallurgy of Non-ferrous Metals (papers), Metallurg. Zdat.

Luttrell, G.H., and R.H. Yoon. 1983. Paper presented at 112th Annual AIME Meeting, Atlanta, GA.

Majima, H., and M. Takeda. 1968. Electrochemical Studies of the Xanthate-Dixanthogen System on Pyrite. *Trans. AIME*. 241:431.

Malinovsky, V.A. 1946. *Nonferrous Metals*, Moscow, No. 1.

Manser, R.M. 1975. *Handbook of Silicate Flotation*. Stevenage, Herts, England: Warren Springs Laboratory.

Matijvic, E., K.G. Mathai, R.H. Ottewell, and M. Kerker. 1961. Detection of Metal Ion Hydrolysis by Coagulation. III. Aluminum. *J. Phys. Chem.*, 65:826.

Matijvic, E., and B. Tezak. 1953. Coagulation Effects of Aluminum Nitrate and Aluminum Sulfate on Aqueous Sols of Silver Halides in Statu Nascendi Detection of Polynuclear Complex Aluminum Ions by Means of Coagulation. *J. Phys. Chem.*, 57:951.

McCarroll, S.J. 1954. Upgrading Manganese Ore. *Min. Eng.*, 6:289.

Mellgren, O. 1966. Heat of Adsorption and Surface Reactions of Potassium Ethyl Xanthate on Galena. *Trans. AIME*, 235:46.

Mellgren, O., and M.G.S. Rao. 1963. *Trans. IMM*, 72:673.

Miller, J.D., and J.B. Hiskey. 1972. Electrokinetic Behavior of Fluorite as Influenced by Surface Carbonation. *J. Coll. Inter. Sci.*, 41:567.

Miller, J.D., J. Laskowski, and S.S. Chang. 1983. Dextrin Adsorption by Oxidized Coal. *Coll. and Surf.*, 8:137.

Miller, J.D., and M. Misra. 1984. *Proceedings, MINTEK 50 International Conference on Recent Advances in Mineral Science and Technology*, Johannesburg.

Moudgal, B.M., T.V. Vasudevan, and J. Blaakmeer. 1987. Adsorption of Oleate on Apatite. *Trans. AIME*, 282:50.

Mukerjee, P., and K.J. Mysels. 1971. *Critical Micelle Concentrations of Aqueous Surfactant Systems*. National Bureau of Standards. U.S. Gov. Printing Office.

Nagaraj, D.R., and P. Somasundaran. 1981. Chelating Agents as Collectors in Flotation. *Trans. AIME*, 270:1351.

Palmer, B.R., M.C. Fuerstenau, and F.F. Aplan. 1975. Mechanisms Involved in the Flotation of Oxides and Silicates with Anionic Collectors. Part 2. *Trans. AIME*, 258:261.

Palmer, B.R., G.B. Gutierrez, and M.C. Fuerstenau. 1975. Mechanisms Involved in the Flotation of Oxides and Silicates with Anionic Collectors. Part 1. *Trans. AIME*, 258:257.

Peck, A.S. 1964. U.S. Bureau of Mines, Report of Investigation (RI 6202).

Peck, A.S., L.H. Raby, and M.E. Wadsworth. 1966. An Infrared Study of the Flotation of Hematite with Olei Acid and Sodium Oleate. *Trans. AIME*, 235:301.

Pizarro, R.S. 1967. Master's thesis. Colorado School of Mines, Golden, Colo.

Plaksin, I. 1959. Interaction of Mineral with Gasses and Reagents in Flotation. *Min. Eng.*, 11:319.

Plaksin, I.N., and E.A. Anfimova. 1954. *Proc. Min Inst.*, U.S.S.R.: Acad. Sci.

Predali, J.J. 1969. Flotation of Carbonates with Salts of Fatty Acids: Role of pH and Alkyl Chain. *Trans. IMM*, 78:C140–C147.

Rao, K.H., B. Antti. and E. Forssberg E. 1990. Mechanism of Oleate Interaction of the Salt-type Minerals. Part II. Adsorption and Electrokinetic Studies of Apatite in Presence of Sodium Oleate and Sodium Metasilicate., *Int. J. Min. Proc.*, 28:59–79.

Ravitz, S.F. 1940. *AIME Tech. Pub. 1147.* New York: AIME.

Rogers, J., and J.H. Schulman. 1957. *Second International Congress of Surface Activity*, III:243. London: Butterworths.

Roman, R.J., M.C. Fuerstenau, and D.C. Seidel. 1968. Mechanismd of Soluble Salt Flotation. Part 1. *Trans. AIME*, 241:56.

Rubinstein, J.B. 1995. *Column Flotation Process, Designs and Practices*. Switzerland: Gordon and Breach Science Publishers S.A.

Schubert, A.W. 1967. Aufhereitung Fester Mineralischer Rohstaffe, Band II VEB Deutsher Verlag fur Girundstaffindustrie Leipzig, 311.

Schuhmann, R., and B. Prakash. 1950. Effect of $BaCl_2$ and Other Activators on Soap Flotation of Quartz. *Trans. AIME*, 187:591.

Shvedov, D.A., and A.O. Andreeva. 1938. Sci. Invest. Work on Theory and Practice of Flotation C papers, No. 4, ONTI.

Smith, R.W. 1963. Effects of Fluoride on Contact Angle in the System, Microcline–Dodecylamine Solution-Nitrogen. *Proceedings of the South Dakota Academy of Science*. 42:60.

Smith, R.W., and T.J. Smolik. 1965. Infrared and X-ray Diffraction Study of the Activation of Beryl and Feldspar by Fluorides in Cationic Collector Systems, *Trans. AIME*, 232:196.

Sollenberger, C.L., and R.B. Greenwalt. 1958. Relative Effectiveness of Sodium Silicates of Different Silica Soda Ratios as Gangue Depressants in Nonmetallic Flotation. *Min. Eng.*, 10:691.

Somasundaran, P. 1969. Adsorption of Starch and Oleate and Interaction Between Them on Calcite in Aqueous Solutions. *J. Coll. Int. Sci.*, 31(4):557–565.

Somasundaran, P. 1975. Interfacial Chemistry of Particulate Flotation. *Advances in Interfacial Phenomena*. Edited by P. Somasundaran and R.B. Grieves. *AIChE Symposium Series*, 71, No. 150:1.

Somasundaran, P., and G.E. Agar. 1967. The Zero Point of Charge of Calcite. *J. Coll. Inter. Sci.*, 24:433.

Somasundaran, P., and J. Cleverdon. 1985. A Study of Polymer/Surfactant Interaction at the Mineral/Solution Interface. *Coll. and Surf.*, 13:73.

Steininger, J. 1967. Collector Ionization in Sphalerite Flotation with Sulfhydryl Compounds. *Trans. AIME*, 238:251.

Sutherland, K.L., and I.W. Work. 1955. *Principles of Flotation*. Melbourne: Aus. AusIMM.

Szczypa, J., K. and Kuspit. 1979. Mechanism of Adsorption of Sodium Laurate by Calcium Carbonate. *Trans. IMM*, 88:C11–C13.

Taggart, A.F. 1967. *Handbook of Mineral Processing*. New York: John Wiley & Sons.

Taylor, T., and A.F. Knoll. 1934. Action of Xanthates on Galena. *Trans. AIME*, 112:382.

Usul, A.H., and R. Tolun. 1974. Electrochemical Study of the Pyrite–O_2–Xanthate System. *Int. J. of Miner. Process.*, 1:135.

Vladimirova, M.G., and I.A. Kakavskii. 1950. Zhur. Priklad. *Khim.*, 23:580.

Wakamatsu, T., and D.W. Fuerstenau. 1973. Effect of Alkyl Sulfonates on the Wettability of Alumina. *Trans. AIME*, 254:123.

Wang, X., K.S. Forssberg, and N.J. Bolin. 1989. Thermodynamic Calculations on Iron-containing Sulfide Mineral Flotation Systems. I. The Stability of Iron Xanthates. *Int. J. Miner. Process.*, 27:1.

Wark, I.W., and A.B. Cox. 1934. Principles of Flotation, II—An Experimental Study of the Influence of Cyanide, Alkalis and $CuSO_4$ on the Effect of KEX at Mineral Surfaces. In *Milling Methods, Trans. AIME, Vol. 112*. New York: AIME.

Warren, L.J., and J.A. Kitchener. 1972. Role of Fluoride in the Flotation of Feldspar: Adsorption on Quartz, Corundum and Potassium Feldspar. *Trans IMM*, 81:C137–C147.

Wheeler, D.A. 1966. Big Flotation Mill Tested. *E&MJ*, Nov.

Woods, R. 1976. Electrochemistry of Sulfide Flotation. In *Flotation*. Edited by M.C. Fuerstenau. New York: AIME.

Yalamanchili, M.R. 1993. Ph.D. Diss. University of Utah, Salt Lake City, Utah.

Yalamanchili, M.R., J.J. Kellar, and J.D. Miller. 1993. Adsorption of Collector Colloids in the Flotation of Alkali Halide Particles. *Int. J. Miner. Process.*, 39:137.

Yamasaki, T., and S. Usui. 1965. Infrared Spectroscopic Studies of Xanthate Adsorbed on Zinc Sulfide. *Trans. AIME*, 232:36

Yoon, R.H. 1981. Collectorless Flotation of Chalcopyrite and Sphalerite Ores by Using Sodium Sulfide. *Int. J. Miner. Process.*, 8:31.

Yopps, J.A., and D.W. Fuerstenau. 1964. J. The Zero Point of Charge of Alpha Alumina. *Coll. Sci.*, 19:61.

Young, P. 1982. Flotation Machines. *Mining Magazine*, 146:1.

Zhou, Z.A., N.O. Egieor, and L.R. Plitt. 1992. *CIM Bulletin*, 82:926.

CHAPTER 9

Liquid–Solid Separation

Donald A. Dahlstrom

INTRODUCTION

Water is used in most steps in the beneficiation of minerals and coal, and water is used to process most—approximately 80%–90%—of the tonnage of minerals and coal. (A small and decreasing percentage of coal is crushed and screened dry, and industrial minerals such as diatomaceous earth and bentonite can be processed dry.) Beneficiation processes usually use water because it allows greater efficiency, higher recovery, and lower cost per unit of valuable product. In addition, it eliminates air pollution.

Costs of Liquid–Solid Separation

The use of water then necessitates that solids be separated from the liquid. In general, as the particles to be separated decrease in size, cost increases and capacity per unit area decreases. When the solids are colloids (generally considered to be −10 μm or less), costs increase even faster. They are difficult to remove by filtration or centrifugation. Usually, a flocculant is added to the mixture to cause the colloids to form larger flocculi or agglomerates; otherwise the colloids remain in suspension because of Brownian movement. Accordingly, liquid–solid separation is a major cost in mineral processing, probably exceeded only by the cost of comminution, flotation, and endothermic reactions. For example, the capital cost of a coal preparation plant increases by about 30%–40% if the −28 mesh coal is processed instead of discarded, and the process water is recovered (by liquid–solid separation) for recycle and reuse. Operating costs per ton of −28 mesh coal also increase substantially as compared with coarser coals. These costs are due primarily to the use of flotation and the liquid–solid separation steps involved.

At the same time, liquid–solid separation by mechanical means (i.e., sedimentation, filtration, and centrifugation) is much less costly than thermal drying, primarily because those means consume less energy. Furthermore, thermal drying usually requires higher-cost fuels, such as gas or oil, whereas mechanical methods can use electrical energy generated by lower-cost fuels.

To illustrate some highly efficient liquid–solid separations, consider this example. A 100-ft-diameter conventional gravitational thickener (at a normal design rate for tailings concentration and water reclamation of 3 ft^2 per short ton of dry solids per day) will process more than 2,600 short tons of solids per day. With a feed of 15 wt% solids, an underflow concentration of 50 wt% solids or higher can usually be achieved if the solids contain 50%–55% particles that are −200 mesh or coarser. This size consist means that 4.67 lb of water per pound of solids has been eliminated and that more than 82% of the water will report to the thickener overflow for reuse. The thickener drive head will be equipped with a only 5- or 7½-hp motor. The only other energy-consuming process, pumping, is merely an incremental cost as compared with the alternative—normally a large tailing pond at some distance from the plant.

In the processing of magnetite concentrates derived from the beneficiation of taconite, disk filters are used to dewater magnetite concentrates before the balling step. For an 1,800 cm^2/g Blaine concentrate (approximately 85%–90% at 325 mesh), a filtration rate of 230 lb of dry solids/h/ft^2 is used as a design basis (Wolf et al. 1971). The feed concentration would be maintained at 65 wt%, and the vacuum level should be at 24 in. of mercury by using a vacuum pump capacity of about 6 cfm/ft^2 of filtration area. Power consumption will be very high because of the high vacuum and flow rate. Considering also the filter drive, compressed air requirements, filtrate pump, and agitator, a total of 36 hp would be required for 100 ft^2 of filtration area. This power requirement is equivalent to 91,620 Btu/100 ft^2 of filtration area. However, 23,000 lb of dry magnetite solids per hour are dewatered to the 9 wt% moisture required for hauling. At the same time, 10,120 lb of water are extracted. Thus, only 9.05 Btu are required per pound of water eliminated.

Thermal drying, on the other hand, requires around 1,800 Btu per pound of water evaporated. Obviously, mechanical methods of dewatering have relatively low energy consumption per unit of liquid removed. As energy costs continue to increase, other mechanical methods will be developed to further decrease energy consumption.

Liquid–solid separation is also critical in water reclamation and closed water circuits (Wolf et al. 1971). Water that has been used for processing and beneficiating minerals and coals will contain solids that can range in size from a fraction of a micrometer to ¼ in. or more. Some streams will contain the valuable solids and others the refuse or tailings. In both cases, the solids must be separated out if the water is to be reused. Furthermore, the concentration of suspended solids in recycled water must be low enough so that the water does not contaminate the next product. In the case of iron ore processing, the return water must contain 100–150 mg/L or less to minimize the percentage of silica in the final pellet. Coal requires a suspended solids concentration of less than 1 wt%.

This brief discussion shows why liquid–solid separation is a prerequisite for a closed water circuit. It will also be required before any effluent is disposed of in lakes, streams, or other public water sources. State and federal regulations generally require that such effluents contain no more than 10–50 mg/L of suspended solids.

Steps in Liquid–Solid Separation

Separating liquid and solids requires many steps (Hassett 1969; Kynch 1952; Hitzrot and Meisel 1985). In a coal-washing plant, clean coal coarser than ½–1 in. will normally be dewatered by screening, probably the simplest and lowest-cost method of liquid–solid separation. Dewatering increasingly finer particles may require centrifugation, filtration, expression (mechanically squeezing water from feed slurries), or batch filters operating at pressures of 250 psig (Sandy and Matoney 1979).

Medium-size clean coal (½ in. to +28 mesh) is dewatered by centrifuges and sometimes screens (Anon. 1963). The −28 mesh clean coal is dewatered either by continuous filters (normally disk type) or solid bowl scroll discharge centrifuges. Gravitational thickeners are used to concentrate the refuse tailings and reclaim water for reuse. Refuse tailings are dewatered by disk or drum filters, belt presses, or pressure filters.

In minerals beneficiation, ore is usually ground much finer to achieve liberation and produce both desired grade and recovery (Henderson et al. 1957; Kobler and Dahlstrom 1979). Thus, base metals and iron ore concentrates, large-tonnage minerals, will usually be rated according to the percentage of −325 mesh or even −400 mesh solids. Because of their abrasive character, these and other minerals will be dewatered on filters. Both continuous filters and centrifuges are used with crystalline solids such as potash or sodium sulphate. Gravity thickeners are used for both concentrate and tailings (Coe and Clevenger 1916; Roberts 1949; Terchick et al. 1975). The latter are generally sent to tailings ponds as their tonnage and volume can be very large. However, more stringent regulation of the construction of tailings ponds ensures that mechanical dewatering methods will be increasingly used in the future (Chironis 1976).

Hydrometallurgical processing always creates abundant colloidal solids during the leach step. Proper liquid–solid separation enables recovering the maximum amount of pregnant liquor while minimizing its dilution. Thus, multistage countercurrent sedimentation, countercurrent washing filtration on a single filter, or two- or three-stage filtration with single washes per stage are practiced. The mineral or metal is commonly precipitated from solution and washed to maximize purity. Thus, both continuous vacuum filters and pressure filters are involved (Kobler and Dahlstrom 1979; Nelson and Dahlstrom 1957; Osborne 1975).

Because of today's emphasis on water reuse, by-products of beneficiation—dissolved salts and metals, pH, suspended solids, and temperature—must be controlled. If effluent is disposed of in a tailings pond, that effluent must be satisfactory for reuse or disposal to a watercourse, and it must not contaminate ground water or aquifers if it migrates out of the pond. If solids are disposed of in a landfill, the discharged solids must be low in moisture and must not rewet with rain. In either case, tailings ponds and landfills must be carefully monitored to ensure their stability and thus to avoid ground water pollution.

These paragraphs very briefly discuss only the major liquid–solid separation steps. However, it should be apparent that liquid–solid separation is critical to efficient and low-cost processing and probably substantially affects capital and operating costs.

MAJOR INFLUENCES ON LIQUID–SOLID SEPARATION

Many factors external to the liquid–solid separation equipment itself influence its performance and productivity. The most common of these follow (Cross 1963; Dahlstrom 1978; Weber 1977; Hsia and Reinmiller 1977; Robins 1964; Wetzel 1974; Sakiadis 1984; Rushton 1978; Bosley 1974; Scott 1970; Silverblatt et al. 1974):

1. Particle size and shape
2. Weight and volume percentage of solids
3. Fluid viscosity and temperature
4. pH and chemical composition of the feed
5. Variation and range in feed quality (items 1–4)
6. Specific gravity of solids and liquid
7. Quality requirements of discharge streams from liquid–solid separation steps, particularly as they influence results upstream and downstream

Particle Size and Shape

Size distribution greatly affects liquid–solid separation rates. Stokes' law can be used to illustrate this fact. This law permits the terminal settling velocity (maximum velocity achieved during free fall) to be determined as follows:

$$v_t = \frac{(\rho_s - \rho)gD^2}{18\mu} \quad \textbf{(Eq. 9.1)}$$

where

v_t = terminal velocity, ft/s
ρ_s = particle density, lb/ft^3
ρ = liquid density, lb/ft^3
g = gravitational acceleration, (ft/s^2) × (lb mass/lb force) = 32.17, at sea level at 45° latitude. This value is used for g, the standard gravitational acceleration on this planet, and is usually symbolized by gc.
D = particle diameter, ft
μ = fluid viscosity, lb/ft × s

Viscosity is normally measured in centipoises. One centipoise = 6.72×10^{-4} lb/ft-s. (All nomenclature has been given in English units but metric [Système International] units can be used as long as they are used consistently.)

The equation has some limitations. It assumes laminar flow (i.e., a Reynolds number of less than 0.1) and nonhindered settling. In nonhindered settling, a particle is not influenced by the presence of other particles or by the slurry's specific gravity, conditions that require very dilute slurries.

If we assume a 20-µm particle and an 80-µm particle of the same density and in the same fluid, from Eq. 9.1 the terminal velocity is found to be directly proportional to the square of the diameter. In this case, the 80-µm particle will fall 16 times as fast as the 20-µm particle.

Fine particles, particularly under laminar flow, achieve terminal settling velocity almost immediately. This rapidity is caused by the drag force, which resists settling and quickly equals the buoyancy force; the drag force acts on particles through the difference in specific gravity between liquid and solid. This behavior is generally true of most particles that are settled or otherwise classified in the mineral industry. The drag force is given by Eq. 9.2:

$$F_d = C\,A_p\,\rho\,v^2/2g_c \qquad \textbf{(Eq. 9.2)}$$

where

F_d = drag force, lb force
C = drag coefficient, dimensionless
A_p = projected area of particle in direction of motion, ft^2
ρ = density of liquid, lb/ft^3
v = velocity of particle, ft/s
g_c = conversion factor = 32.17

The buoyancy force is given by

$$F_p = \frac{m_s}{\rho_s}\left(\rho_s - \rho\frac{g}{g_c}\right) \qquad \textbf{(Eq. 9.3)}$$

where

F_p = buoyancy force, lb force
m_s = mass of particle, lb mass
ρ_s = density of solid, lb/ft^3
ρ = density of liquid, lb/ft^3
g = acceleration resulting from gravity ft/s^2

At terminal velocity, drag force equals the buoyancy force or

$$C\,A_p\,\rho\,v^2/2g_c \qquad \textbf{(Eq. 9.4)}$$

Solving for v_t and for a spherical particle,

$$v_t = \left[\frac{4gd(\rho_s - \rho)}{3\rho c}\right]^{1/2} \qquad \textbf{(Eq. 9.5)}$$

Figure 9.1 is a plot of $C = (F_d/A_p)/(\rho v^2/2g_c)$ (from Eq. 9.2) as a function of the Reynolds number. The Reynolds number is dimensionless and for spheres, disks, and cylinders, it equals $\rho vd/\mu$. Laminar flow exists for the constant slope value up to Re = 0.1. The actual value of the coefficient C is 24/Re for laminar flow. Substituting this value of C in Eq. 9.5 will yield Stokes' law. Thus, Eq. 9.5 and Figure 9.1 can be used to solve for v_t, knowing d, ρ_s, ρ, and μ, although it becomes a trial-and-error solution.

The trial-and-error solution can be eliminated by using of the terms CN_{Re}^2 and C/N_{Re}. This substitution works because v_t is not in CN_{Re}^2 and d is not in C/N_{Re}. Thus, the following equations are written:

$$CN_{\text{Re}}^2 = 4g\rho d^3(\rho_s - \rho)/3\mu^2 \qquad \textbf{(Eq. 9.6)}$$

$$C/N_{\text{Re}} = \frac{4g\mu(\rho_s - \rho)/3m^2}{3\rho^2 v_t^3} \qquad \textbf{(Eq. 9.7)}$$

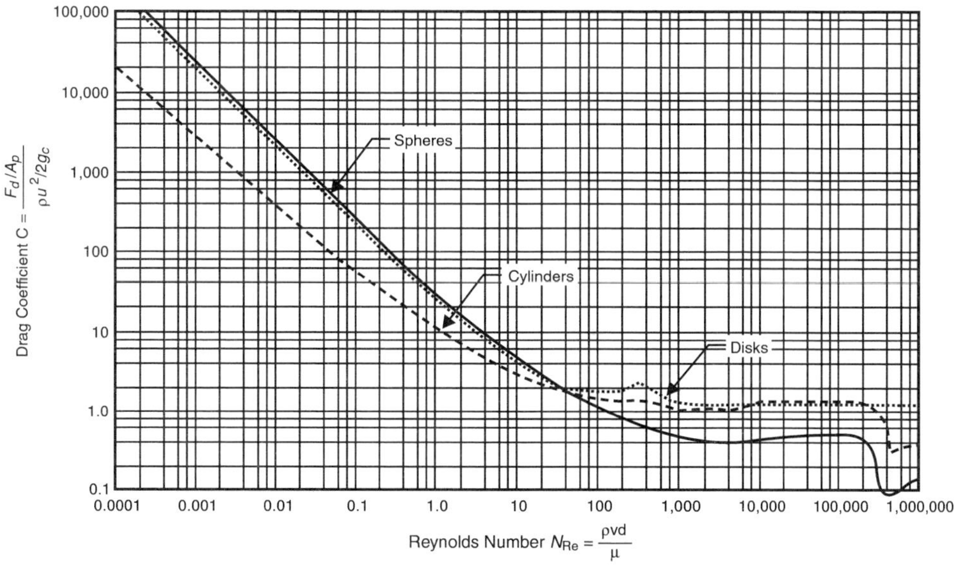

FIGURE 9.1 Drag coefficients for spheres, disks, and cylinders

Table 9.1 can be used to plot CN_{Re}^2 versus C/N_{Re} by using various values of v_t. Thus, at the actual value of CN_{Re}^2, the value of C/N_{Re} can be obtained and used to solve for v_t^3.

Figure 9.2 is a plot of the terminal velocity of spheres settling in air and settling in water, both at 70°F, as a function of the particles' spherical diameter in micrometers and their specific gravity. This figure can be used for quick approximation of nonhindered settling.

As Figure 9.1 shows, particle shape can greatly influence the drag coefficient and therefore the terminal velocity. The term "sphericity" incorporates the shape factor. The symbol ψ = sphericity, the surface area of a sphere of same volume as the particle divided by the surface area of the particle; the quantity is dimensionless. Refer to the literature for detailed values and other empirical approaches to drag coefficient as a function of particle shape (Sakiadis 1984).

Another common method of determining settling velocity is to use the equivalent spherical diameter. This unit is the diameter of a sphere whose terminal velocity is equal to that of the particle in question that has the same specific gravity. This unit can be particularly useful in the −200 mesh (−74 μm) range (see Figure 9.2). In water, particles of these dimensions usually are in the laminar flow range and thus follow Stokes' law for nonhindered settling. As Figure 9.1 illustrates, sphericity values will change depending on the Reynolds number.

As particle sizes become smaller, capillary passages in filters also become smaller. In addition, a wide particle size distribution tends to block capillaries partially or totally. Accordingly, filtration rates almost always decrease for finer particle sizes. However, this decrease will largely depend on the width of the size distribution. A very narrow range near −200 mesh will still yield a good filtration rate as long as particles are larger than colloids. An example was given earlier of magnetite concentrates containing 85%–90% of −325-mesh (44-μm) particles that filtered at a rate of 230 lb/h/ft². The colloidal range of extreme fines had been almost entirely eliminated by the preceding beneficiation methods.

Centrifuges will also lose capacity and recover fewer solids as the particle size decreases. This response can be seen from Stokes' law when centrifugal force is substituted for gravity. As terminal velocity decreases to prevent the loss of solids, capacity may have to be decreased (detention time increased).

TABLE 9.1 Drag coefficient and related functions for spherical particles

Re†	C	CN_{Re}^2	C/N_{Re}
0.1	244	2.44	2440
0.2	124	4.96	620
0.3	83.8	7.54	279
0.5	51.5	12.9	103
0.7	37.6	18.4	53.8
1	27.2	27.2	27.2
2	14.8	59.0	7.38
3	10.5	94.7	3.51
5	7.03	176	IAI
7	5.48	268	0.782
10	4.26	426	OA26
20	2.72	$(1.09)(10^3)$	0.136
30	2.12	$(1.91)(10^3)$	0.0707
50	1.57	$(3.94)(10^3)$	0.0315
70	1.31	$(6.42)(10^3)$	0.0187
100	1.09	$(1.09)(10^4)$	0.0109
200	0.776	$(3.10)(10^4)$	$(3.88)(10^{-3})$
300	0.653	$(5.87)(10^4)$	$(2.18)(10^{-3})$
500	0.555	$(1.39)(10^5)$	$(1.11)(10^{-3})$
700	0.508	(2A9) (10^5)	$(7.26)(10^{-4})$
(1×10^3)	0.471	$(4.71)(10^5)$	$(4.71)(10^{-4})$
(2×10^3)	0.421	$(1.68)(10^6)$	$(2.11)(10^{-4})$
(3×10^3)	0.400	$(3.60)(10^6)$	$(1.33)(10^{-4})$
(5×10^3)	0.387	$(9.68)(10^6)$	$(7.75)(10^{-5})$
(7×10^3)	0.390	$(1.91)(10^7)$	$(5.57)(10^{-5})$
(1×10^4)	0.405	$(4.05)(10^7)$	$(4.05)(10^{-5})$
(2×10^4)	0.442	$(1.77)(10^8)$	$(2.21)(10^{-5})$
(3×10^4)	0.456	$(4.10)(10^8)$	$(1.52)(10^{-5})$
(5×10^4)	0.474	$(1.19)(10^9)$	$(9.48)(10^{-6})$
(7×10^4)	0.491	$(2.41)(10^9)$	$(7.02)(10^{-6})$
(1×10^5)	0.502	$(5.02)(10^9)$	$(5.02)(10^{-6})$
(2×10^5)	0.498	$(1.99)(10^{10})$	$(2.49)(10^{-6})$
(3×10^5)	0.481	$(4.33)(10^{10})$	$(1.60)(10^{-6})$
(3.5×10^5)	0.396	$(4.86)(10^{10})$	$(1.13)(10^{-6})$
(3.75×10^5)	0.238	$(3.34)(10^{10})$	$(6.34)(10^{-7})$
(4×10^5)	0.0891	$(1.43)(10^{10})$	$(2.23)(10^{-7})$
(4.25×10^5)	0.0728	$(1.32)(10^{10})$	$(1.71)(10^{-7})$
(4.5×10^5)	0.0753	$(1.53)(10^{10})$	$(1.67)(10^{-7})$
(5×10^5)	0.0799	$(2.00)(10^{10})$	$(1.60)(10^{-7})$
(7×10^5)	0.0945	$(4.63)(10^{10})$	$(1.35)(10^{-7})$
(1 x 106)	0.110	$(1.10)(10^{10})$	$(1.10)(10^{-7})$
(2×10^6)	0.150	$(6.00)(10^{11})$	$(7.50)(10^{-8})$
(3×10^6)	0.163	(I.47) (10^{12})	$(5.44)(10^{-8})$
(5×10^6)	0.174	$(4.35)(10^{12})$	$(3.48)(10^{-8})$
(7×10^6)	0.179	$(8.75)(10^{12})$	$(2.55)(10^{-8})$
(1×10^7)	0.182	$(1.82)(10^{13})$	$(1.82)(10^{-8})$

Source: Perry's Chemical Engineer's Handbook, 6th Ed., New York: McGraw-Hill. 1980.

†For values of Re less than 0.1, $C = 24/\text{Re}$.

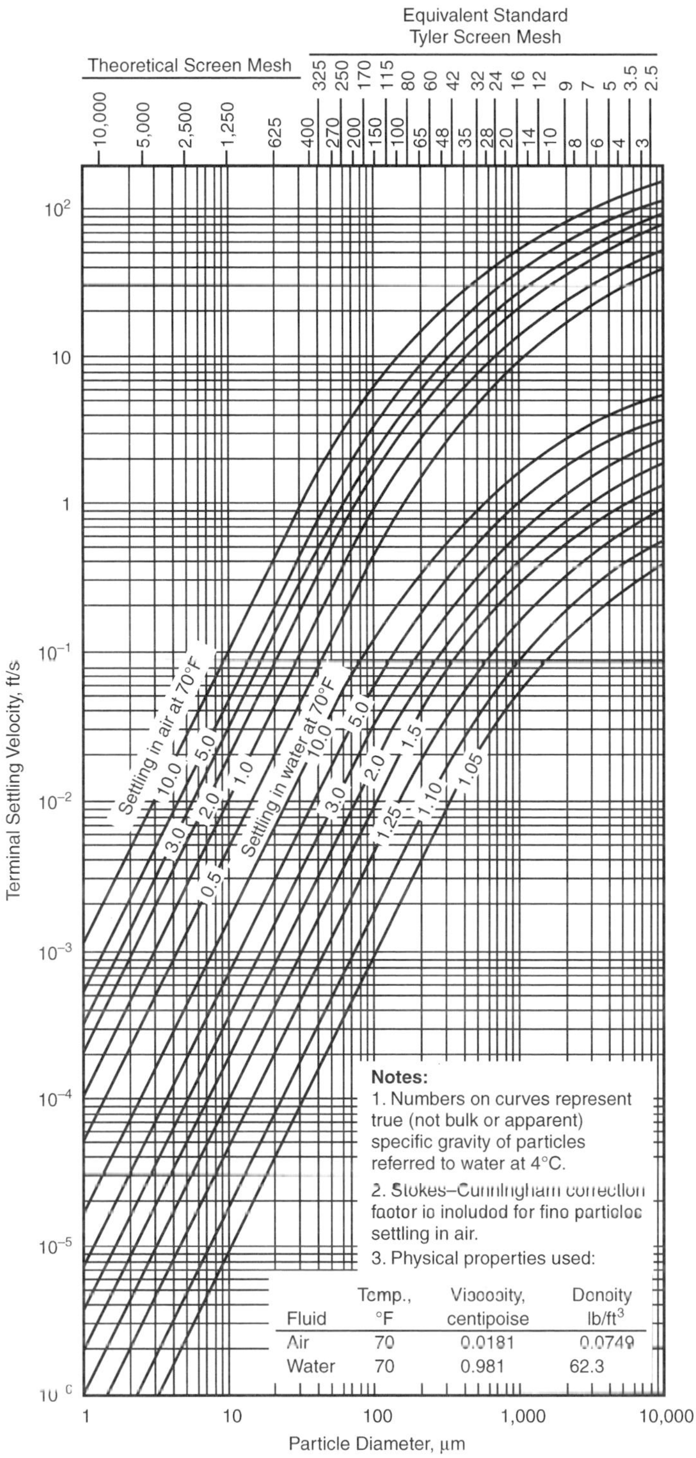

FIGURE 9.2 Terminal velocity of spheres as a function of diameter in air and water

In both filtration and centrifugation, liquid content of the discharged solids will increase as particle size decreases because of the increase in specific surface area. For solid spheres, these relationships can be shown as follows:

$$\text{Surface area of a sphere} = \pi D^2 \quad \textbf{(Eq. 9.8)}$$

$$\text{Volume of a sphere} = \pi D^3/6 \quad \textbf{(Eq. 9.9)}$$

$$\text{Weight of a sphere} = \pi D3\rho_s/6 \quad \textbf{(Eq. 9.10)}$$

$$\text{Specific surface area/unit weight, spheres} = 6/D\rho_s \quad \textbf{(Eq. 9.11)}$$

Ten-micrometer solid spheres of 2.7 specific gravity have a specific surface area of 2,222 cm^2/g, whereas the surface area of a 100-µm particle is one tenth of this value. Because liquid forms a film on particle surfaces, the liquid content per unit weight of solids will increase as the particles reduce in size.

Weight and Volume Percent Solids

The concentration of solids also greatly affects particle dynamics. For instance, settling velocity decreases as the solid concentration increases. This phenomenon is caused by two factors. First, specific gravity of the slurry increases so that the buoyancy force (Eq. 9.3) should be rewritten as

$$F_p = \frac{m_s}{\rho_s}(\rho_s - \rho_{s_1})\frac{g}{g_c} \quad \textbf{(Eq. 9.12)}$$

where ρ_{s_1} = slurry density, lb/ft^3. Second, as the concentration increases, particles are more abundant and thus more likely to collide or impede each other's fall. Slurry viscosity will also increase, which influences the Reynolds number.

Attempts have been made to predict this influence by the following equation:

$$V_{ts} = V_t\,(1\text{-}\beta)^n \quad \textbf{(Eq. 9.13)}$$

where

V_{ts} = terminal velocity of a particle in a suspension with other particles, ft/s
β = volumetric fraction of solids in the slurry, dimensionless
n = exponent, which is a function of the Reynolds number

Figure 9.3 is a plot of n as a function of the Reynolds number. The term V_t is employed in the Reynolds number value and ρ and μ apply to the liquid.

Although the terminal velocity of all the particles will be reduced as the concentration of solids increases, mineral processing capacity in terms of tons of solids per hour per square foot may still be equal to or greater than its capacity at more dilute concentrations. This topic will be discussed further under sedimentation.

An increase in solids concentration benefits both filtration and centrifugation because less liquid must be removed per unit of solid. If 20 wt% solids were concentrated to 40 wt% and then filtered to a final cake moisture content of 20 wt%, only 1.25 lb of water per pound of solids would need to be removed by filtration. If a 20 wt% solids slurry were filtered directly to a final moisture content of 20 wt%, then 3.75 lb of water per pound of solids would have to be removed—three times as much. In many cases, using filtration or centrifugation after gravitational thickening can be economically justified on both capital and operating-cost bases.

Viscosity

The equations developed to this point show that an increase in liquid viscosity will decrease settling, filtration, and centrifugation rates. Furthermore, it has been shown many times that mineral

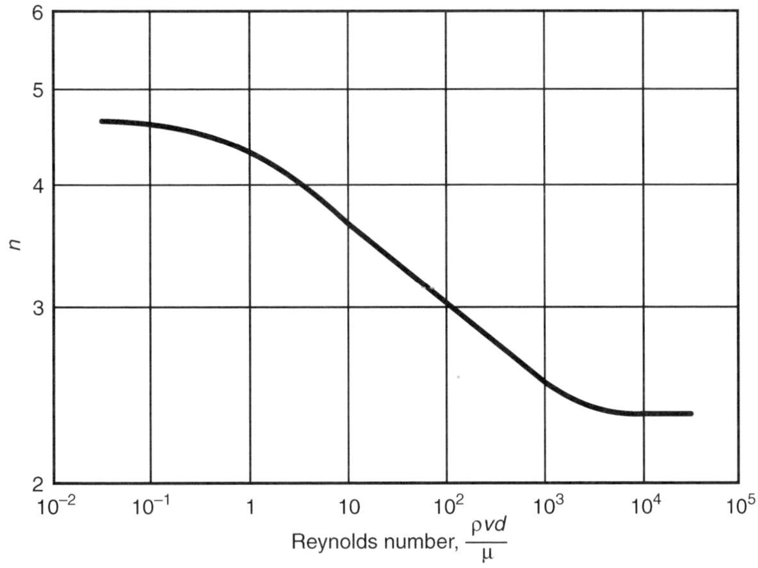

FIGURE 9.3 Values of exponent *n* as function of Reynolds number

processing at an industrial scale operates in agreement with the theoretical influence of viscosity discussed previously.

Slurry viscosity can also be an important factor (Whitmore 1957), particularly with extreme fines or colloidal solids. A slurry of 20 wt% is composed of 1-μm solids, as compared with a slurry of 20-μm solids, which contains 20^3 as many 1-μm particles as 20-μm particles in the same volume. Obviously, the viscosity of the 1-μm slurry could be much higher.

Chemical Conditions

The acidity or alkalinity of a solution (pH) can affect liquid–solid separation in several ways. For example, a highly acid or basic pH may reflect large amounts of colloidal solids if the slurry has come from a leaching step. Extremes of pH are commonly associated with elevated temperatures, mechanical agitation, detention times measured in hours, and in some cases elevated pressures, all of which generate colloidal solids. These solids must be flocculated if sedimentation, filtration, or centrifugation is to be economically practiced. It is difficult to think of a single hydrometallurgical plant that does not employ a flocculant in its flowsheet.

Sodium hydroxide or sodium carbonate may have been used in alkaline leaches. The sodium ion tends to disperse solids into individual particles (as it does with soap made of sodium-type stearates) and thus will make the colloidal solids even more difficult to separate. Again, flocculants will most likely be required.

Finally, alkaline pHs or acid slurries that are neutralized may contain metal hydroxides, such as the hydroxides of iron, calcium, magnesium, or nickel. Metal hydroxides are almost always difficult to separate, but the oxides (if the hydroxides can be converted) separate much more easily. Accordingly, it is usually wise to develop methods that precipitate metal oxide or promote self-flocculation. Even if some added flocculant is required, the dosage should be much smaller.

Specific Gravity of Solid, Liquid, and Slurry

The difference in specific gravity between solid and liquid or solid and slurry drives sedimentation and, in conjunction with centrifugal force, centrifugation. The importance of differences in specific gravity may make it desirable to consider volume percent solids in certain research work. For example, in a 45-vol% slurry of magnetite (ρ_s = 5.0), sand (ρ_s = 2.7), and coal (ρ_s = 1.5), coal would have a wt% of 55.1, sand of 68.8, and magnetite of 80.4.

All of these concentrations could be achieved for the three cases if the extreme fines are not excessive and if the slurries can be pumped by a centrifugal pump. Forty-five volume percent may be close to the maximum concentration that will be pumpable.

Quality Requirements of Discharge Streams

In today's modern processing plant, the products from a liquid–solid separation step may go on to further processing or be recirculated upstream. An example of the former is the filtration of alumina trihydrate ($Al_2O_3 \cdot 3H_2O$), which is filtered and washed to recover sodium hydroxide before calcining to Al_2O_3 for eventual conversion to aluminum.

The moisture content of a filter cake will influence the amount of energy consumed in the kiln. At the same time, Na_2O content on a dry solids basis usually must be 0.02–0.04 wt%. Finally, the filtrate must have a low solids concentration (usually 20 mg/L), because this stream returns to digestion after evaporation. A filtrate too dilute or too high in suspended solids will require excessive amounts of energy.

The performance of each liquid–solid separation step must be considered, because each influences the efficiency and operating cost of the unit operation steps to which the product streams report. This consideration may eliminate certain types of equipment or require additional liquid–solid separation steps.

Where tailings or effluent are discharged to water bodies or to water treatment plants, governmental regulations must be met. Several characteristics of effluents normally must conform to some standard:

- Suspended solids, mg/L or ppm
- pH, normally 6–9 or 5.5–8.5
- Biological oxygen demand (BOD), mg/L
- Chemical oxygen demand (COD), mg/L
- Total dissolved solids (TDS), mg/L
- Heavy metal concentrations (dissolved or total), mg/L or ppm
- Hazardous wastes and chemicals, mg/L or ppm, or µg/L or ppb
- Toxic compounds, mg/L or ppm, or µg/L or ppb
- Oil and grease, mg/L or ppm

Regulations may require that tailings possess a certain minimum bearing load, moisture content, and stability, and meet requirements that will avoid ground water contamination. If tailings are determined to contain hazardous or toxic waste, even more stringent requirements will be mandated. The state will usually specify the required standards (which normally can be no more lenient than the federal requirement). Accordingly, when an expansion or a new plant is considered, the state must be consulted well before design work begins.

Special mention should be made of water reclamation and recycling. Because modern plants usually find it economical to reuse a high percentage of process water, the water's content of suspended solids, dissolved solids, and pH will be important. Recycling water with too high a suspended solids concentration may markedly reduce product quality or mineral recovery. Excessive dissolved salts or other contaminants together with pH may cause rapid corrosion within the plant. Conservation of process heat may help to reduce energy consumption or promote performance. Again, the design of such steps must be carefully considered. For instance, a flocculant may be required (with proper mixing and control) to achieve clarity, pH may have to be brought to neutral to prevent sulfides in the ore or raw coal from generating acid, or some dissolved salts may need to be removed to obtain proper flotation performance.

LIQUID–SOLID SEPARATION EQUIPMENT

Because ores and coal are processed almost exclusively in aqueous liquids, liquid–solid separation is an important technology. The large tonnages of solids and large volumes of fluid involved require equipment of high productivity that is largely automated or is highly self-regulating. Thus, this section will discuss equipment used in three unit operations that are the most widely practiced in the mineral industry for these separations: gravitational sedimentation, filtration, and centrifugation. The basic types of equipment will be briefly described to indicate the method or concept employed, and their principal advantages and limitations will be considered.

GRAVITATIONAL SEDIMENTATION

The force of gravity can be used to concentrate suspended solids. Thus, both particle size and specific gravity of the solids will be important. Where the solids are colloidal, flocculants will improve operation by causing the colloids to form agglomerates or flocculi of much larger size (but that still contain probably 95% or more liquid) that will settle at reasonable rates.

The principle of gravitational sedimentation is employed in classifiers, thickeners, and clarifiers. Thickeners concentrate the solids to the thickener underflow (and produce an overflow acceptable for recycling); clarifiers produce a higher-quality overflow that may meet special reuse requirements for or be fit for direct disposal to public water bodies. Normally the solids concentration in the underflow is less than the maximum possible concentration obtainable by sedimentation.

Classifiers

Two basic types of gravitational classifiers employed today are the spiral classifiers and the rake classifier (Hitzrot and Meisel 1985). They both introduce the feed submerged into a pool area and subject the solids to an upflow current. The overflow is obtained from a peripheral weir that provides the vertical velocity. A solid whose terminal settling velocity is high enough will settle against this current and report to the base of the pool. At this point, a rake or screw conveys the solids out of the pool and up the beach to drain the "sands" product before discharging it over the end of the slope. Figure 9.4 is a schematic of a spiral classifier. The rake classifier has a series of rakes (actually blades) that operate in a reciprocating fashion to move the sands up the beach.

A third type is termed a "hydroseparator." The feed again enters submerged through a circular centralized feedwell, and the peripheral overflow weir causes an upward velocity. The underflow is raked toward the central outlet and is pumped out. The unit looks much like a conventional gravitational thickener or clarifier except that the unit is shallower and the bottom slope is usually steeper. The speed of a rake mechanism is normally two or more times that of a thickener, because of the more granular nature of the solids and the higher solids rate per unit area.

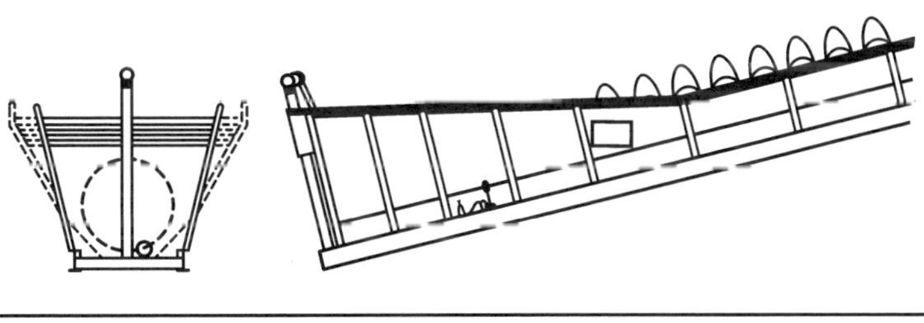

FIGURE 9.4 Spiral classifier

The main advantage of the rake and spiral classifiers is their higher percent solids of the sands product, because the product is drained. The hydroseparator has the advantage of higher capacity from a single unit as it uses a much larger pool area.

The liquid cyclone (also called a hydrocyclone) is a widely used classifier that employs centrifugal force. It is covered in another chapter.

Thickeners

A conventional thickener consists of a circular tank with a central feedwell and peripheral weir overflow (Dahlstrom 1985). Bottom slope will have a ratio of 1:12 up to 3:12 depending primarily on tank diameter and particle size distribution. For large-diameter units or for feeds containing a high percentage of coarser, fast-settling solids, a double slope is used. The inner third (approximately) of the diameter might have a 2:12 or 3:12 slope and the outer portion a 1:12 slope. (The use of rectangular and square tanks in coal processing is gradually decreasing.)

Although the overflow usually must be of reasonable clarity, the emphasis is primarily on underflow concentration. Accordingly, higher-torque driveheads are used so that the rake arms can move the higher concentration and more viscous solids. The following equation is used to determine the drivehead torque:

$$\text{Drivehead torque} = K'B^2 \tag{Eq. 9.14}$$

where

K' = constant, lb force/ft
B = rake diameter, ft

Drivehead torque is measured in ft-lb. Duty is normally specified as standard, heavy, and extra heavy; extra heavy duty is required in most mineral processing applications. Thus, K values will generally be 5–10 for standard, 10–20 for heavy, and 20–30 for extra heavy duty. In special cases, the K value may be over 100 because of very fast settling of coarse solids without extreme fines or a high solids concentration, either of which may lead to an extremely viscous compacted mud.

Most thickeners will run at 0%–20% of rated torque and as such will have a service life of 20 years or more. The great amount of extra torque is called into play in an upset condition (such as excessively coarse solids, underflow pump shutdown, excessive tonnages, or foreign object) to prevent a shutdown and the need to dig out the unit. The overflow is usually recycled back to processing for reuse. If it is to be disposed of in a public water body, additional clarification and other treatment may be required to meet regulations.

Several auxiliary devices may be used on these units. A rake-lifting device may be used to elevate the rake if torque exceeds a certain percentage (approximately 50%) of rated torque. Lifting protects the mechanism and also prevents a shutdown. When the torque drops below 50%, the mechanism gradually lowers until it reaches its minimum elevation.

Other auxiliaries are underflow pumps (centrifugal or positive displacement), flocculation systems, skimmers that remove floating material, such as solids or oils, special feedwells for feed distribution, a "killing" inlet velocity and elevation head, and various rake mechanism designs depending on the rheology of the thickened sludge. In addition, various control devices are possible depending on the application.

There are three basic types of thickeners: bridge, center pier, and traction.

The bridge thickener supports the drivehead, centershaft, and rake mechanism from a bridge across the tank diameter. Normally, the bridge thickener is the most economical only up to tank diameters of 100 ft, but bridges have been built as long as 140 ft. For tanks larger than 100 ft in diameter, a center pier usually is less expensive. In a center pier thickener, a reinforced concrete or steel center pier supports the drivehead and rake mechanism, and a cage connects the two items (Figure 9.5). The

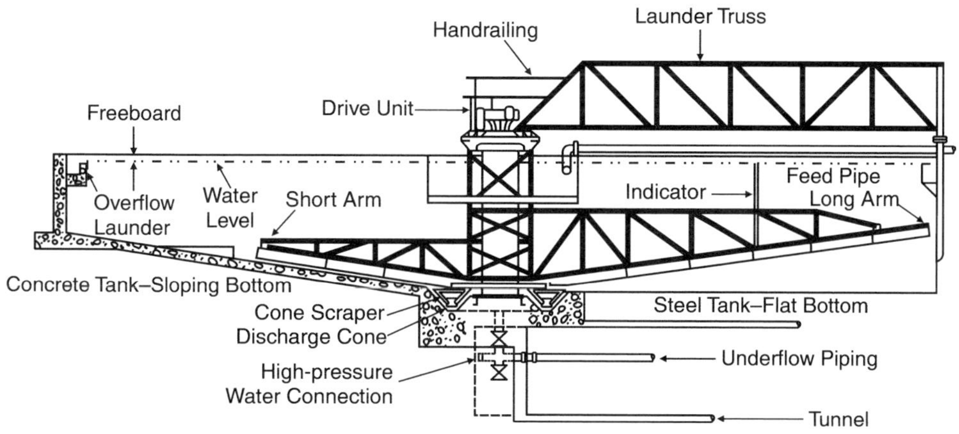

FIGURE 9.5 Center pier thickener

underflow collects in a circular trough around the center pier, and trough scrapers move the sludge. Normally, two or more ports are used to connect to the pump suction manifold.

Both of these units can be fitted with a lifting device, and the units can be covered if required for process reasons. They normally have two long arms and an additional two short arms if the raking load is high.

The traction thickener employs a wheel drive that normally rides on a rail on the side wall of the thickener and pulls one long arm. Usually, three short arms are also used. Because no lift mechanism can be used, the installed torque is normally higher than in other types of thickeners. Power enters the center of the machine and connects through a commutator to the peripheral drive. This type of unit is not widely employed but finds its greatest application in milder climates. The rail surface must be at a constant elevation, and the unit is more expensive to cover.

A thickener may be modified to produce a high-rate unit. When flocculants are added, a relatively small amount of high-viscosity, low-specific-gravity fluid (the flocculant) must be mixed with a large amount of feed slurry of higher specific gravity that can also have a high viscosity. As the flocculant adsorbs onto "what it sees," slurry particles can be unevenly flocculated unless mixing is relatively quick and thorough. Inserting flocculant into a launder with baffles or a pipe in turbulent flow does not guarantee complete mixing. Some slurries will require more shear or even more detention time than others to obtain best results, although exposing the slurry to excessive shear or pumping after flocculation will normally degrade the floc.

In one type of high-rate unit, a relatively small diameter feedwell with internal annular baffles may have up to three compartments (Figure 9.6). Various agitation methods in the three compartments rapidly mix flocculant and slurry with minimal hydraulic shear. A feed pipe permits flocculant to enter each compartment on a controlled basis. About half usually enters in the top compartment, and the rest is divided equally in the remaining ones. Baffling also permits plug flow through the feedwell, and after exiting from the bottom, the flocculated pulp falls to its specific gravity without further shear. Thus, no floc degradation occurs.

When flocculation is optimized, the area required by a thickener can be reduced to only one-third to one-tenth of that required by conventional units. At the same time, the unit must be more highly instrumented. These thickeners are used only if flocculation is practiced.

Thickeners have the double advantage of requiring relatively low horsepower and allowing the construction of very large units (units of 750 ft in diameter have been constructed), meaning that they have a large capacity. Very little operating labor is required and maintenance is low.

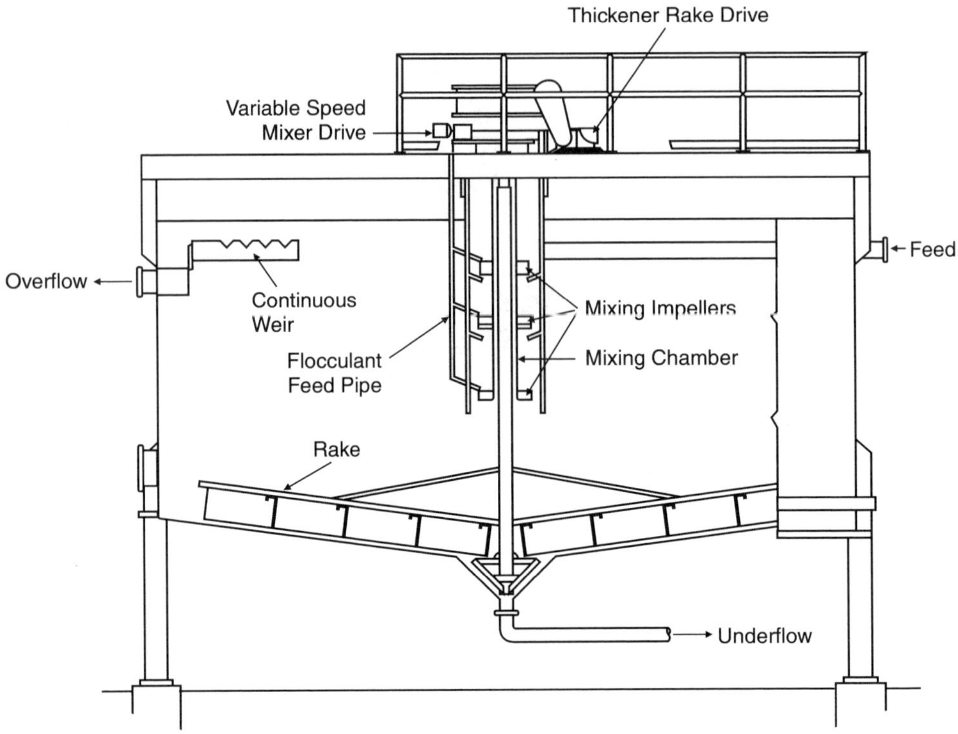

FIGURE 9.6 High-rate thickener

Clarifiers

The conventional gravitational clarifier looks very similar to a thickener, in that circular tanks are normally employed in the mineral industry. They also use a feedwell, underflow pumps, and a peripheral weir overflow. However, several features may indicate the difference. First, the underflow normally has a low solids concentration, so K' in Eq. 9.14 (Wolf et al. 1971) has a value of 5–10. The feedwell design should achieve low solids concentration in the overflow (under 100 mg/L and normally less than 10 mg/L). These feedwells will usually be deeper than normal and may be so constructed to dissipate inlet energy in the feedwell. Radial launders, in addition to the peripheral ones, may be added to improve the efficiency with which the tank handles overflow. Flocculation, precipitation, or the addition of alum [$Al_2(SO_4)_3 \cdot 14\ H_2O$] to cause an “alum floc” will almost always be employed. The latter traps fine solids within the floc structure, which produces a relatively low suspended-solids concentration in the overflow.

A second type is the solids-contact clarifier (Figure 9.7 illustrates one design). The feed enters through a pipe into a central draft tube and is immediately mixed with partially thickened solids slurry picked up from near the base of the unit. Any chemicals used for flocculation or precipitation purposes are added in the feed pipe. A high-volumetric-rate turbine at the top of the draft tube furnishes the pumping action for the draft tube. The mixture then flows into the inverted conical section, which normally has a detention time of around 10–15 minutes. The flow exits this zone at the base and enters the clarification zone, normally into a floc bed that filters out more of the unflocculated fine solids. The settling solids move to the center with the raking action and, on average, recirculate through the unit about seven times.

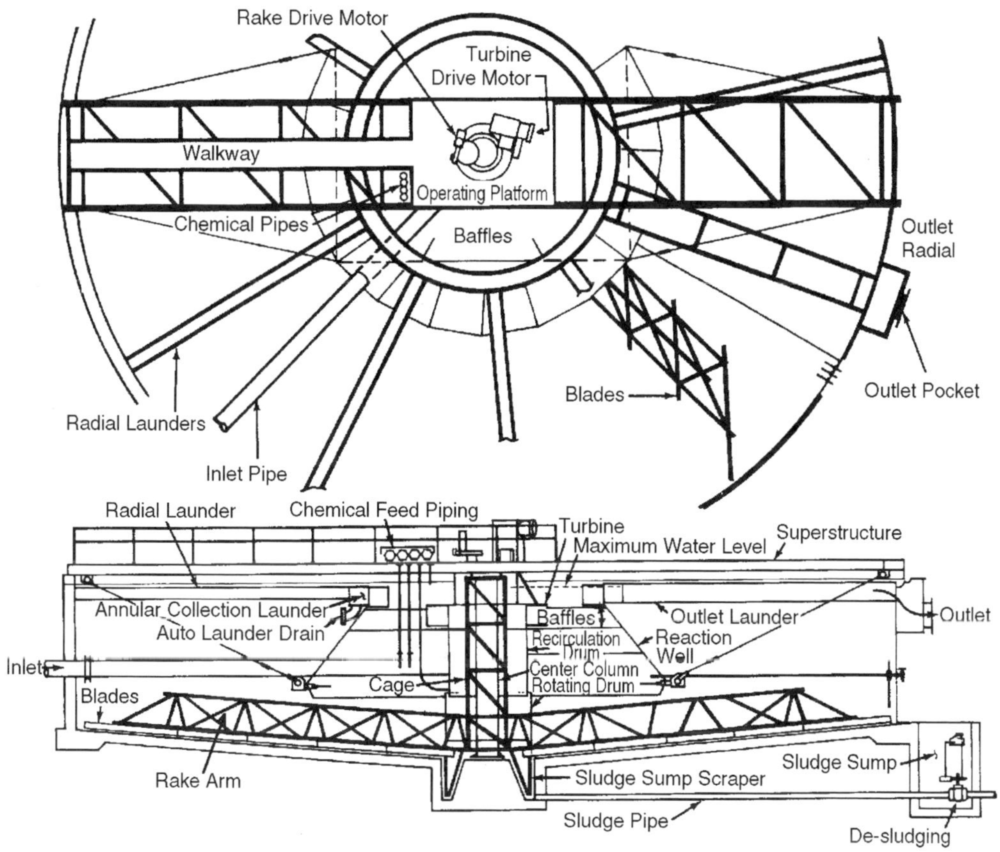

FIGURE 9.7 High-rate solids-contact clarifier

This system promotes the growth of flocculi or crystals that settle faster, and in most cases the overflow solids concentration is 10 mg/L or less. Also, upflow rates can be as high as 2 gpm/ft^2 of clarification surface area so that very large flow rates are handled in a single unit. Average upflow rates would probably be around 1 gpm/ft^2. A classifier based on another basic concept is the "lamella" type clarifier. A series of inclined planes (usually made of plastic) are arrayed within a special housing normally inclined at the same angle. The separation between planes is usually 2–3 in. and the angle of inclination is 45°–60°; 55°–60° is the most common. The following equation describes the basic concept.

$$A_s = \cos\gamma/X \tag{Eq. 9.15}$$

where

A_s = specific surface area per unit volume projected on a horizontal plane normally measured as ft^2/ft^3

γ = angle of inclination to the horizontal

X = perpendicular distance between planes, ft

Feed enters toward the sludge discharge point of each inclined plane. Overflow occurs by flow along the roof made by an inclined plane. Thus, solids or flocculi have only a short distance to fall before reaching the floor caused by an inclined plane. The angle α must be greater than the angle of repose of the solids, and this is why the 55°–60° value for α is common. It is apparent that as α and X

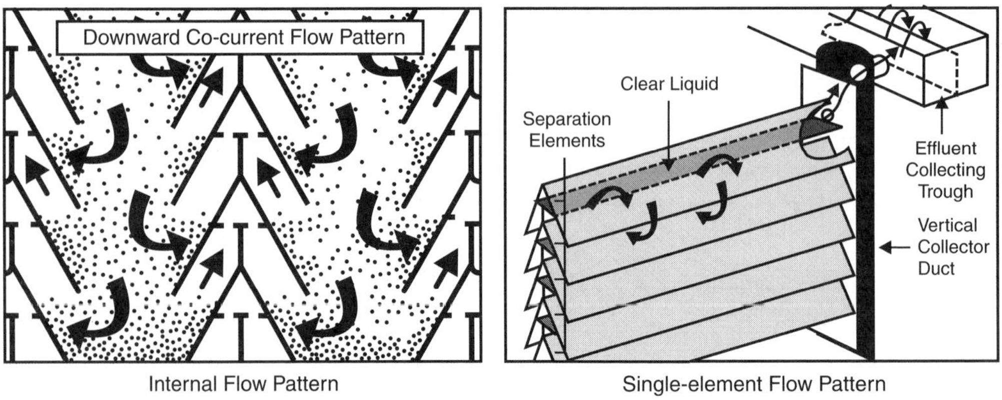

FIGURE 9.8 Chevron clarifier

decrease, *A* increases. However, if *X* is less than 45°, the settled solids may not slide down the incline properly. Operating rates range from 0.41 to 1.23 gpm/ft^2 of projected horizontal area.

A solids-contact clarifier operates more as a clarifier because the lack of compression zone volume and rake action means that the underflow does not concentrate as much. Its advantage is the higher flow rates normally possible per unit area.

Another type of clarifier uses a special "chevron" principle (Figure 9.8). The vertical series of chevrons are stacked parallel to each other. A slot between adjacent chevrons feeds into a horizontal central pipe at the top of each pair of chevrons. Feed enters the slot and the solids that fall faster than the upflow velocity in the slot will settle out and report to the next lower slot. This sequence is repeated until solids drop into the area of thickened underflow.

At one end of each pipe, an orifice controls the flow rate. As the solids fall down the vertical stack of chevrons, the solids concentration increases, which means that the upflow velocity must decrease. The upflow velocity and the size of the orifice in each pipe can be calculated from a plot of height versus time on a graduated cylinder sedimentation test. The overflow from each vertical collection riser at the end of each stack must flow over a weir so that the elevation drop across the unit is very small.

In essence, each chevron acts as an individual clarifier. By stacking 6–12 chevron units, each of which is only 4–6 in. wide, flow rates per unit of cross-sectional area are greater than those that can be obtained by other methods. The units can occupy square or rectangular tanks, and rake mechanisms are possible but rarely used. Thus, this unit is used more as a clarifier. Its chief advantages are a high feed rate per unit area and the use of square or rectangular tanks for better use of floor space.

Flocculants may be used with lamella and chevron units. However, because the feed must be flocculated outside the unit, flocculi can degrade if excessive hydraulic shear or detention time is permitted.

FILTRATION

As indicated earlier, filtration can be divided into three modes: continuous, batch and semicontinuous, and clarifying. Each of these modes can be further subdivided. The discussion that follows is limited to those units with significant application in mineral and coal processing.

Continuous Filters

Continuous filters may be divided into those forming their filter cake against gravity and those forming their filter cake with gravity (Dahlstrom 1985).

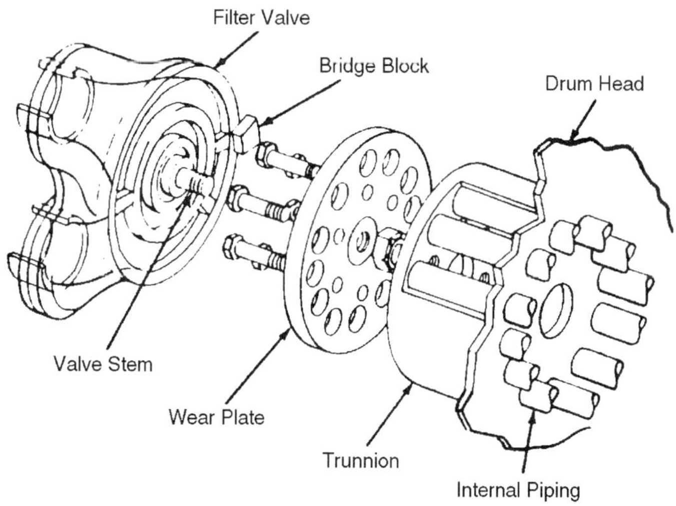

FIGURE 9.9 Arrangement of components in a continuous-filter valve

Filters Forming Cake Against Gravity. Disk and drum filters all form cake against gravity. The latter can be further divided into scraper discharge, roller discharge, and continuous-belt drum filters.

A disk-type filter contains a series of individual disks mounted on a center barrel. The barrel is held in trunion bearings mounted on either end of the filter tank. The disks are partially submerged in the feed slurry to a standard apparent submergence of about 35%. A higher submergence would require stuffing boxes around the center barrel, a procedure that is very seldom used because of the large diameter required of the stuffing boxes and the abrasiveness of the solids generally processed.

Each disk is divided into 8 to 12 pie-shaped disk sectors depending on the disk diameter. A filter bag covers the sectors' filtration area, and filtration occurs on both sides of the disk sector. Each sector is held in by radial rods between sectors that attach to the center barrel. A bag clamp that covers half of the end of the adjacent sectors holds the sector in place after a nut is applied to the end of the radial rod. At the narrow end of the sector, a pipe outlet connects to the ferrule socket with proper gasketing, and it delivers filtrate and air pulled through the cake to a port within the center barrel. The number of port filtrate channel equals the number of sectors per disk. The filter bag is tied around the filtrate outlet of the sector and nailed, stapled, or clamped to the top of the sector. Channels within the center barrel end in a wear plate (which attaches to the pipe plate) with the port openings in a circle. A stationary face of the filter valve is held to the rotating wear plate by a centering pin. Figure 9.9 is an exploded view of a typical valve. The stationary valve portion has a bridge ring so that bridge blocks can be inserted to separate the various phases of the filter cycle. For example, a bridge block whose width covers the port width would be placed before and after that portion of the filter cycle when compressed air is blown through the ports, sector, and filter media to dislodge the cake. A scraper blade also assists in the cake discharge by riding on either side of the disk. By pivoting the blade in the rear of the sector and hanging the front end so that the shoe rides on the periphery of either side of the disk, the blade can be automatically separated from the face of the sector by 1/4–3/8 in. and thereby conform to any vertical variation in each disk.

The exit from the stationary filter valve face (normally one large or two smaller exits) connects to a cylindrical receiver. Here the liquid filtrate separates from the gas pulled through the cake, and the overhead line connects to the vacuum pump. Thus the pressure differential of the vacuum pump provides the driving force for filtration. The filtrate is either pumped from the receiver or discharged from a vertical barometric leg, which is usually at least 5 ft higher than the maximum vacuum that can be applied (measured in feet of water).

The cake is discharged by a blow-back of compressed air through the valve, port, sector, and filter media. A steady low-pressure blow of 3–5 psig, which continues until the trailing edge of the sector passes the scraper blade, is commonly used. Higher pressure blows of 10–30 psig use a solenoid valve triggered by a cam rider on the trunion. A full blow for only a very few seconds tends to shock the cake from the filtering surface. In this way, the filter cloth is not inflated when it passes the scraper blade.

Because the disk is applied to high-permeability filter cakes, the filtration rate is generally high—25–700 lb of dry solids/h/ft^2 of filtration area. Therefore, particles are generally coarser than normal and must be agitated to be kept in suspension. At the base of the filter tank, a shaft with paddles between each disk and on either end maintains the solids in suspension. The shaft has outboard bearings and drive, and it usually runs between 60–120 rpm.

The cake discharges into chutes between each disk and on each end to a belt conveyor on the lower floor.

Disk filters are made in five diameters: 6 ft; 6 ft, 9 in.; 8 ft, 10 in.; 10 ft, 6 in.; and 12 ft, 6 in. Respective filtration areas, in ft^2/disk, are 40, 50, 110, 160, and 220. If more than seven disks are used, a valve on either end of the center barrel should be employed. The maximum number of disks per filter is 15.

This filter costs the least per unit area of filtration and needs the smallest amount of floor space on the same basis.

The drum filter consists of a cylinder with peripheral sections parallel to the central axis. Each section is connected by tubing to the pipe plate, as in the disk filter, and a wear plate matches the tube diameters and location. A filter disk, normally of plastic grids of polyethylene or polypropylene, is contained between the wings of the leading and trailing division strips. The filter cloth is caulked into each division strip so that each section can be isolated from the adjacent ones by appropriate bridge blocks in the valve. Either a caulking groove or a flat strip is applied on both drum ends for sealing this portion of the filter cloth.

Another method of applying the filter cloth to the drum is by wire winding. In this method a caulking rope or elastomer is inserted in each division strip so that the edge protrudes through the caulking grove. Thus by wire winding the cloth over these seals, the section can be isolated from adjacent ones as required during each revolution. Wire winding is normally spaced at ½–2-in. intervals, and stainless-steel wire is most commonly used. In both systems, the edges of either drum end are sealed with wire winding or plastic strapping.

Tubing connections to the leading and trailing edges of a section are normally joined to a single manifold pipe that in turn connects to the wear plate. The size and number of leading and trailing edge connections should cause the minimum hydraulic restrictions. Pressure drop at maximum flow should not exceed 2 in. of mercury between the filter media and the suction side of the vacuum pump. If high temperatures and vacuum levels cause the filtrate to flash, resulting in two-phase flow, the pressure drop is even more important in designing the filter drainage network.

The type of drum filter is determined by the way in which cake is discharged. The most common types in the mineral industry are scraper discharge, roller discharge, and continuous belt.

In a scraper discharge drum filter (Figure 9.10), the cake is removed by a scraper blade, which is assisted by a blow-back of pressurized air. The scraper blade should not contact the filter media during blow-back, so a ¼-in. separation is usually required. Scraper filters are probably the most commonly used in continuous service.

The second type, a roller discharge drum filter, contains a small-diameter roll that moves in a direction opposite to that of the drum. Because the roll's peripheral speed is normally slightly faster than the drum's, a pool of cake forms between the drum and the roll. The surface of the drum is covered at discharge so that very thin cakes (1/32–1/16 in.) can be discharged completely. Probably the most common roll is fabricated of a plain steel or an alloy steel. The cake sticks best to itself, and therefore, a "heel" about 1 in. deep is plastered onto the roll. The cake stuck to the roll is then cut off by a knife at 90° or 180° from this point. Figure 9.11 is a schematic of a roller discharge system for the

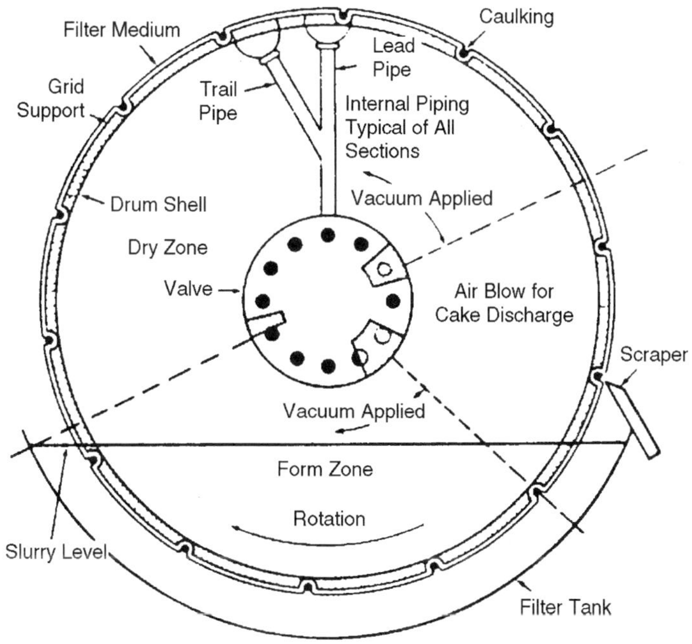

Source: Schweitzer 1979.

FIGURE 9.10 Rotary drum vacuum filter

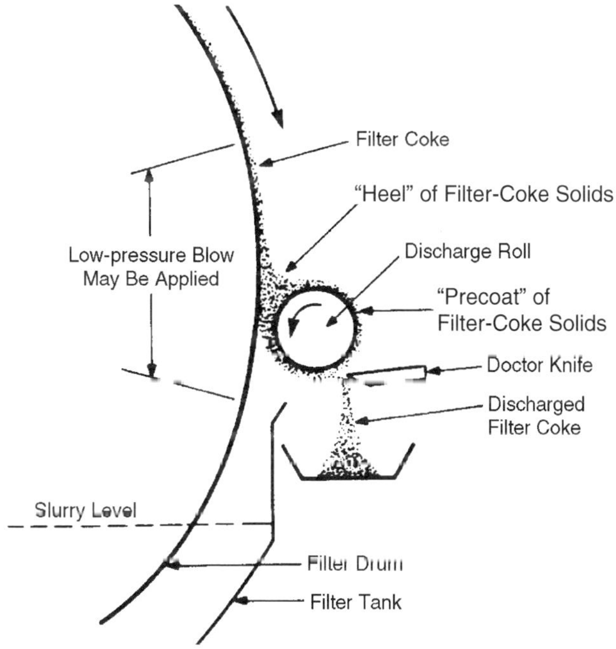

FIGURE 9.11 Roller discharge drum filter

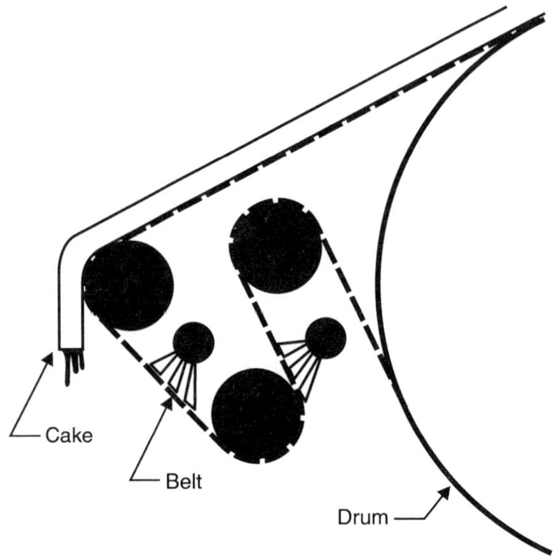

FIGURE 9.12 Cake discharge and medium washing on a continuous-belt drum filter

filtration of kaolin clay that has been leached with sulfuric acid to remove iron. The roll has a heel of clay, and the drum rotates as fast as once every 20 rpm.

This system has also been used to treat alumina red mud (bauxite residue after leaching in caustic solution) and extremely fine tailings. It can discharge very thin cakes at high drum speeds, and colloids compose 80%–100% of the total suspended solids in the cakes.

The third type of drum filter, the continuous-belt drum filter, discharges cake by continuously removing the cloth (Figure 9.12). The cloth is carried over a small-diameter discharge roll where the large difference in the radius of curvature tends to release the cake from the cloth. A deflector blade completes the discharge. To maintain a clean and unblinded cloth, spray nozzles can then be used to wash the cloth on one or both sides. This wash water is collected separately, and the cloth travels around the wash roll and then around a return roll to be placed back on the drum to begin the cycle again.

A continuous filter can discharge cakes as thin as $\frac{1}{16}$–$\frac{1}{8}$ in., a thickness that maximizes the filtration rate. Furthermore, maintaining the cloth free of blinding again increases filtration rate, which is normally 20%–50% higher than that of a scraper discharge filter. This difference persists when it is measured over the life of the filter cloth.

Because of the dimensional instability of textiles, continuous-belt filters must always control four aspects of filter media alignment at all times: the cloth must be centered across the face of the drum; one edge must not lead or trail the other; the center must not lead or trail the edges; and the cloth must be free of wrinkles.

This type of filter finds wide application where solids that cause blinding are encountered or where compounds can chemically precipitate within the filter cloth. Its higher price per unit area is more than offset by its high capacity per unit area.

Filtration area is measured by the surface area of the drum or π (diameter $\times$ length). Drums are usually 4–12 ft in diameter, at 2-ft increments. A 14-ft–diameter drum would have to be shipped separately from the filter tank to clear bridges and tunnels and generally represents the largest size that can be shipped.

Face widths of 40 ft are the largest used in scraper discharge and roller discharge drums; widths of 20–24 ft are about the maximum in continuous-belt drum filters.

Another type of drum filter, the continuous drum precoat filter, is used to produce clear liquids. The machine employs a drum similar to the ones previously described, but a microadvance knife cuts off a very thin layer of the precoat bed on each revolution, normally 0.0015–0.006 in. per revolution. The precoat bed consists of diatomaceous earth (fossil remains of diatoms), expanded perlite, or other precoat material that will filter out the usually very fine particles that must be removed from the liquid.

The precoat bed is applied by filtering it on the filter media to a thickness of 3–6 in., depending on the application and the filter design. The feed is then applied to the tank and the precoat knife cut and the filter cycle time adjusted to the optimum value for the feed. The normal precoat cut will be about 0.003 in. per revolution. The solids filtered out must be cut off during each revolution; otherwise, the bed may blind. If penetration is too great, it will be more economical to use a "tighter" grade of precoat material.

After the precoat thickness is shaved down to approximately ¼ in., the knife is retracted, the bed is sliced off, and the process is repeated.

Feeds for this type of filter are normally less than 5 wt% suspended solids, and, in most cases, less than 2 wt% suspended solids. The unit finds its widest application in such areas as hydrometallurgy, where a clear filtrate must be produced. Thus, it would be used on gravitational thickener overflows or continuous filter filtrates. It is the only continuous filter that produces a clear filtrate.

Filters Forming a Cake with Gravity. The scroll discharge horizontal table filter and the continuous horizontal belt filter are filters that form a cake with gravity. The former consists of a circular disk with filter media on the top side. The table is divided into pie-shaped sections, and a gridwork supports the filter media in each section. The filtration area is calculated as the annular area between the outside and inside diameters.

At the discharge point a scroll cuts off the cake and drops it over the side to a conveyor belt or other type of transfer unit. At the inner radius of the section, the filtrate pipe connects to the wear plate of the filter valve. The valve is mounted underneath the filter in the center and the outlets point downward. Otherwise, the valve is similar to a drum and disk filter valve.

Because of the scroll discharge, a heel of cake must be left on the filter because otherwise the filter media would wear rapidly. Normally, the heel is ½–¾ in. thick. The cake can be reoriented by blowing back with compressed air during the initial portion of the feed phase of the filter cycle to retard blinding.

The filter cake can also be washed to recover soluble constituents of interest. Although the drum filter can be washed only by a single stage, a countercurrent wash can be employed on this unit. Thus the wash fluid is applied first to the last wash, and this filtrate is used as wash for the preceding stage. Two stages usually suffice, but three have been used.

This type of filter has been used to treat granular solids with a high cake permeability, such as certain crystalline solids, particularly coarser ones whose solids should be washed to remove impurities or to recover brines or dissolved valuable salts. Sand or other granular solids of 20–100 mesh size are also dewatered on this type of machine because high solids capacities and low moistures can be achieved.

The horizontal belt filter was developed to permit washing of the cloth and thus to prevent blinding similar to that common in the continuous-belt drum filter. At the same time, it is possible to use any number of countercurrent washes on the individual machine as long as they are incorporated into the design.

Figure 9.13 is a schematic of a continuous horizontal belt filter that illustrates major construction features. Two large-diameter major pulleys are employed and a special grooved endless elastomer belt rides over the pulleys. The head pulley (cake discharge end) is driven and normally the molded belt has a full-length rib that is accommodated by a circumferential slot in each of the rubber-covered pulleys. The drainage grooves, which are perpendicular to the direction of motion of the elastomer belt, are

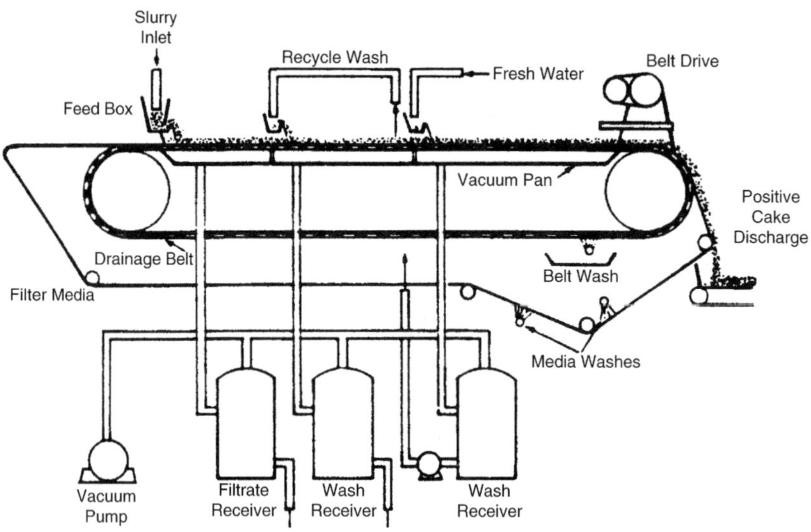

FIGURE 9.13 Horizontal belt filter

such that the two adjacent grooves drain to the same circular hole. The hole is drilled through the elastomer belt and the center line of the rib (more than one drainage hole across the width may be used, depending on the width of the belt and the required hydraulics). The belt must incorporate enough plies of fabric to give it sufficient strength in tension. The belt also employs side flaps or flanges that contain the feed slurry and wash fluids. Metal deckle sides may be used to contain feed or wash water, and hoods may be used to contain steam or hot gases that might be applied for maximum dewatering.

The filter media is made of plastic material such as polyethylene, polypropylene, nylon, or polyvinyl chloride. It rides on top of the elastomer belt and is held in place by the pressure differential across the cake and the filter cloth. The media is separated from the rubber belt after the vacuum has been terminated, and cake is discharged over a small-diameter roll. The roll's small radius of curvature at discharge helps separate cake from the filter cloth. The cloth is then washed to prevent blinding, in a manner similar to that used for the continuous-belt drum filter, and returned underneath the filter to the head pulley for a repeat of the cycle.

The cloth is kept in alignment by control systems such as those used on the continuous-belt drum filter. In addition, a take-up system for the cloth must be employed during the return of the cloth underneath the filter to take care of any stretching or shrinking. The rubber belt is maintained in alignment by the individual takeups on the tail pulley.

The rubber belt rides over the vacuum box (or boxes, depending on width or hydraulic requirements) and a support table. The vacuum box serves as a "valve" on the filter because a seal must be made between the stationary face of the vacuum box and the moving face of the elastomer belt. Because of the pressure across these faces, low-friction surfaces on the vacuum box, such as fluorocarbon plastics, must be used. These faces can be lubricated by water or a clear filtrate delivered through pressure lines into the faces. In the case of heavy cakes or long filters, it is desirable to use a low-pressure fan blowing into a plenum with entrance ports into the support table under the belt.

Dividers in the vacuum box are used to separate the various filtrates or washes as desired. If different vacuum levels are employed within the cycle, the dividers must ride against the belt for a width of one diameter of a drainage hole. The gas and liquid filtrate are carried by pipes to the appropriate receiver where the filtrate is separated from the gas. Flexible connectors allow the vacuum box to be dropped for maintenance. Where scaling occurs, such as in phosphoric acid manufacture, the vacuum box can be dropped by a gear motor or hand crank to expose the box for easier descaling. The overhead of the receiver passes to the vacuum pump to supply the driving force for filtration.

Wash boxes are normally used to keep spray nozzles from plugging; spray headers may also be used, particularly in single-stage washing or the last stage of countercurrent washing. As many as five stages of countercurrent washing have been employed to minimize the consumption of wash fluid. Major advantages of this filter are

- It can be employed with as many countercurrent wash stages as desired at a high wash efficiency.
- Cloth-washing systems eliminate cloth blinding without diluting the feed.
- Coarse, fast-settling solids can be filtered because the cake forms with gravity.
- Very high belt speeds of 200 ft/min (61 m/min) or more can be used to yield very high capacities per unit area.
- The rectangular structure of the filter and its basic concept use floor space efficiently, and all auxiliaries can be installed on the same floor as the filter.

A disadvantage is a higher price per unit area because of the rubber covering and special elastomer belts that must be employed. However, capital costs should be viewed on bases such as cost per unit of production or improved product quality.

Batch and Semicontinuous Filters

Continuous filters tend to be more widely used in the mineral and coal processing field, particularly where large tonnages are involved. This preference reflects the lower capacities of batch or semicontinuous units and the increased labor requirements, both of which result in higher operating costs. However, at low-tonnage plants and under special conditions, these filters can have distinct advantages. Also, where pressure drops must be used that are higher than those obtained by continuous vacuum filters (because of the low cake permeability), the pressure filter may be applicable. For instance, tailings may be filtered to recover water or to dewater them to a high enough solids concentration to allow land disposal. The mineral and coal processing industry use batch and semicontinuous filters of four types: plate and frame filters, recessed plate filters, vertical disk pressure filters with or without sluice discharge, and automatic discharge plate and frame-type filters. All four employ pressure filtration.

Plate and Frame Filter. The plate and frame filter uses a plate that has a grooved or other type of drainage system supporting the filter cloth. Both sides of the plate have grooved patterns so that filtration occurs on either side. The frames will contain the feed and filter cake and seal against the plate. Usually there are connecting ports through the four corners of both plates and frames so that either feed or filtrates can be accommodated. The feed may enter through one or more of the corners where the ducts are accommodated into the frame interior but not into the plate. The plates contain ducts in other corners for conducting the filtrate. The filter cloths are draped over the plates and holes of the ports in the corners are matched. A seal is made by the filter cloth between the plates and the frames, although slurry can leak if the cloth is wrinkled or if a piece of cake sticks between the plate and the frame. This possibility can usually be prevented by using a gasketed plate.

Figure 9.14 shows a typical plate and frame unit and also shows the hydraulic closing system. One head is stationary and the other can be moved by a double-acting hydraulic cylinder. Thus, relatively high hydraulic pressures are employed to close the press so that normal operating pressures up to 125 to 250 psig can be obtained.

Also shown in Figure 9.14 is a shifter mechanism, which discharges cake by mechanical movement of the plates and frames. An operator should make sure that the cake discharges and that pieces do not hang up on the sealing surfaces; such pieces could cause a very leaky joint or even break the plate or frame on closure. The operator usually has a wooden paddle to take care of these instances, and as

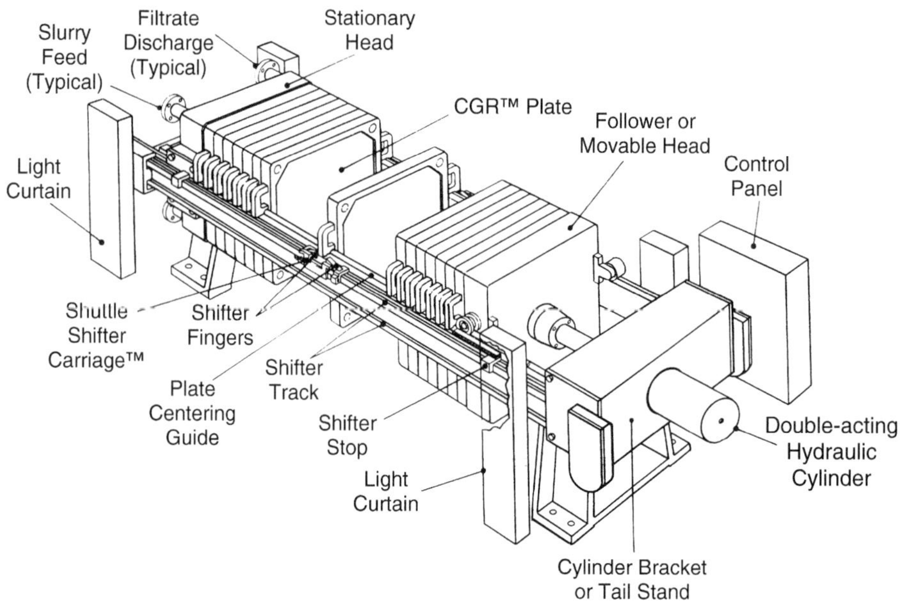

FIGURE 9.14 Plate and frame filter

soon as the paddle breaks a photocell path, the mechanical operation stops. After the operator has corrected the situation, he restarts operation by pressing a button that moves with him.

In smaller filters, a hand crank can be used in place of the hydraulic cylinder at a lower cost. The plates and frames can also be moved manually, but doing so usually requires two operators.

The filter cakes can be washed by using the feed lines for introducing wash fluid. Because the cake is being washed from a point source, the fluid tends to follow shorter paths to the filtrate side and the wash is not as effective. The wash is more efficient if every other plate is a washing plate. The wash fluid enters this plate behind the filter media on both sides and passes through the cloth to the opposite filter media and plate. Thus, a more consistent short flow path is obtained across the cake area.

Many different filter media can be used, ranging from canvas to synthetic woven fibers to nonwoven synthetics. To obtain very clear filtrates, special papers are also employed either as the sole media or over a backing cloth.

Plates and frames were formerly made of wood, cast iron or other metals, and rubber-covered steel. Currently, thermoplastics, such as polyethylene and polypropylene, have largely supplanted the earlier materials. These newer materials not only reduce costs but also greatly reduce weight. They are used at the normal operating pressures of 125–250 psig.

Frames generally are 1–2 in. deep depending on the specific cake permeability. Plates are usually square and are 12–48 in. on each side. The frame depth should be carefully determined, because the plate and frame filter work best when the frame is entirely full of cake at the end of the filtration cycle. If it is not, the frame may contain too much fluid and produce a high-moisture cake. In addition, a thicker final cake may result in an appreciably lower filtration rate in terms of pounds of dry solids per hour from the press.

The feed pump may be either a centrifugal pump or a positive displacement pump, depending on the type of solids in the feed and the filtration pressures. If final filtration pressures permit and the solids or flocculi are not injured by the pump, single-stage or multistage centrifugal pumps can be used. At higher pressures or where the solids or flocculi are fragile, the positive displacement type pump is used. Controls should be used to prevent excessive pressures that could injure the filter press.

A common auxiliary is an air or gas compressor that blows the cake at the end of the filtration cycle for further dewatering. Where tailings or refuse are filtered, the solids must be flocculated because they contain large amounts of colloids. Flocculation requires a mix tank preceding the feed pump and complementary equipment to prepare and dilute flocculant. Bench-scale investigations will be required to determine the type and dosage of flocculant, the mixing power, and the duration of mixing. Because flocculi generally deteriorate with time, the length of time that flocculated pulp is stored will be important, as the filtration rate is not constant at all times with batch equipment.

Recessed Plate Pressure Filter. A similar type of press but one that eliminates the frame is the recessed plate pressure filter. The plate has a center feed and all feed enters through this port. The filter cloth must be sewn or a fixture employed to seal the cloth on both sides of the plate at the feed port. In addition, the plate is recessed to allow for cake buildup. This recess is usually ½–1 in. deep, yielding a cake thickness of 1–2 in. Filtrate is collected at any one or more of the four corners; filtrate ports are cast in the plates as with the plate and frame filter. Cake washing, if necessary, is best practiced by using every other plate as a washing plate with wash fluid entering at the top behind the filter media on both sides of the plate. The wash fluid passes more evenly through the cake to the opposite plate, where the filtrate is collected at the bottom corners.

The usual plate sizes vary from 12-in. square to as much as 6 × 9 ft. As many as 175 plates may be incorporated into one press; the maximum plate size yields a filtration area of 18,585 ft^2. Plates are available in a wide range of materials but molded plastic dominates, particularly in the large sizes.

Auxiliary equipment is similar to that discussed under plate and frame filters. However, with the larger units, the feed ports may be doubled to achieve the proper hydraulics and feed distribution. In addition, the press is constructed so that the feed ports may be blown out by compressed air through the follower end (a movable closure head); the compressed air removes the higher moisture core and also further dewaters the cake.

Mechanical plate shifters are also employed on recessed plate filters. Because of the large size of these filters, large plate shifters that move as many as 12 plates at a time for cake discharge can be used. This device reduces the time for cake discharge to a very few minutes and increases the overall filtration rate.

Figure 9.15 is a picture of a typical recessed plate with a gasketed construction that eliminates leakage and reduces filter-media wear.

Vertical Disk Filter. A vertical disk filter contains a series of disks mounted on a central pipe that also serves as a filtrate conduit. Because it is a pressure filter, it must be contained in a pressure shell; this shell is almost always a horizontal cylinder with a vertical flange at one end. The flange is typically a quick-opening type to minimize downtime. Normal operating pressures can be up to 125 psig.

If a dry cake discharge is required, the disks are pulled out by supporting the near end from a monorail or other mechanism. The cake can then be discharged to a container or conveyor and can be shocked off using a rubber mallet.

A second method is to discharge the cake by blow-back of gas or air through the filter media that drops the cake to the base of the shell. A mechanical plow is then used to remove the cake. If a slurry discharge can be used, a sluice header for each disk surface is mounted between the disks. At the discharge portion of the cycle, the spray headers are turned on and the disks rotated a few times to sluice off the cake, then the resultant slurry is drained from the pressure shell. This latter method is easier and requires less labor, but it is restricted to a slurry discharge.

Entry to the tank and the filtrate outlet is usually on the fixed head or on the walls. An air or gas purge valve is needed during the initial filling of the tank. Some gas is left at the top of the tank to reduce or eliminate pressure fluctuations.

The disks do not have to be divided into sections as with a continuous disk filter. Usually the disks are made in two halves or as a single piece, depending on the method of construction. In many of the applications, the filter media is stainless-steel screen soldered to the drainage frame for long service. Textiles can be employed if the design permits.

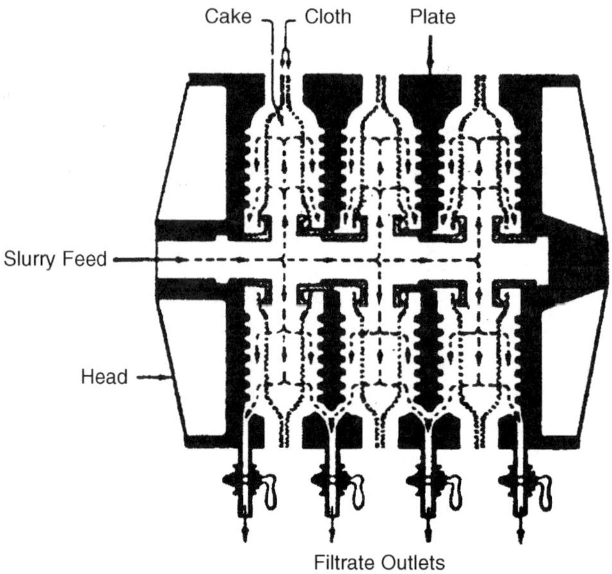

FIGURE 9.15 Typical recessed plate filter

In operation, the pressure shell should be filled quickly to maximize the filtration rate during the cycle. Shell volume will be large, so for slow-filtering solids, it may be desirable to use a higher capacity pump to fill the shell followed by a low-volume pump.

When the maximum filtration pressure has been obtained or filtration rate has dropped to a minimum value, the shell is drained of feed.

Some pressure should be maintained in the shell to hold the cake on the filter surface and simultaneously drain the unit faster. If necessary, the cake is further dewatered by blowing with compressed air or gas. However, if the cake is to be washed, the filter must be refilled with washing fluid while a pressure drop is maintained across the cake. Cake is discharged as described earlier.

Horizontal Leaf Filter. The horizontal leaf filter is highly automated and operates on a relatively short cycle of 2–5 min.

The filter consists of a series of plates, stacked one on top of the other, whose vertical movement and position can be controlled. The upper side of the plate contains a drainage system and a filtrate outlet; they are connected by a flexible connector to a common manifold for all of the filtrate outlets. The bottom side of the plate contains a flexible, impervious elastomer membrane with a connection through the side of the frame (or by other means, depending on design) for the feed. The filter media zigzags through the filter controlled by rollers and rides atop the upper surface of the plate.

The cycle begins with closing the filter by sealing the filter cloth against a gasket on the plate surface, and then sealing the frame against the cloth. A hydraulic system is used to open and close the filter; it operates against the filter pressure, which may be 60–125 psig.

Feed is pumped into the frame. When the cycle indicates the end of filtration, the diaphragms are pneumatically or hydraulically squeezed to compress the cake. If further dewatering is required, the cake can be blown using compressed gas. Thus, the frame does not need to be full of cake, an important consideration in obtaining a flexible operation. If the cake is to be washed, wash fluid is pumped in for the required time and the dewatering step is again performed. The plates are then lowered to leave a gap between the filter cloth and the frame that is slightly larger than the thickness of the frame. The filter media is prompted to travel quickly to discharge the cake over the side of the rollers on either side of the filter. If necessary, the cloth can travel one revolution and be washed on both sides of the media.

The filter is then closed and the cycle repeated. The diaphragm membrane yields an important advantage to this filter and to the plate and frame or recessed plate filter—it is not necessary to fill the cake chamber completely, because the membrane will push out the remaining feed and compress the solids. However, experience has shown that, at the end, the cake should be blown using compressed gas to minimize moisture content. Thus a dry cake is ensured regardless of cake thickness.

The filter can be washed efficiently because of the horizontal position of the cake; the wash fluid travels down through the filter cake rather than originating from a point source. Low moisture content of the discharged cake is also claimed. A disadvantage is the high cost per unit area and the usually higher maintenance costs.

Clarifying Filters

Many process liquors must have a very low suspended-solids concentration, and others must have a sparkling clarity. This requirement applies particularly to hydrometallurgy where elements or compounds derived from precipitation must have a high purity. Plant effluents being discharged to natural streams or lakes must usually contain no more than 10–50 mg/L of suspended solids. Also, many solids (e.g., gold, uranium, and silver) have a high dollar value and extremely high recoveries are justified. These streams, then, must be clarified.

As indicated earlier, the solids contact type clarifier is used for clarification purposes. Overflows of less than 10–30 mg/L can be produced, especially if flocculation or precipitation can be practiced. They have been used in many hydrometallurgy processes.

Gravitational thickeners are not normally used to produce high-quality overflows. They are extremely important in water reclamation where the thickener overflow is recirculated back to the plant for reuse. Recycled water usually does not have to be as low in solids as does a plant effluent.

The continuous filter used for clarification is the continuous vacuum precoat filter described earlier. Clarification is its major application, and it produces a filtrate essentially free of solids. It is used in many applications, such as clarifying phosphoric acid and clarifying cobalt liquors. In metallurgical applications, it is usually employed on feeds up to 2 wt% suspended solids. Above this value, it is usually more economical to employ a thickener before the precoat filter. The precoat filter clarifies the thickener overflow, which has a low solids concentration, at a much higher rate.

Several of the pressure filters are used as clarifiers—the plate and frame, recessed plate, and vertical disk pressure filters discussed earlier. If necessary, they can be precoated with a skin of diatomaceous earth, perlite, or other filter aid. Both the plate and frame and the recessed plate filters can be equipped with petcocks that allow filtrate to be discharged from each plate and delivered to a launder. Thus visual samples can be taken of each discharge to ensure proper quality. If a plate has a tear in the filter cloth, the valve can be shut without taking the filter off line. These filters are widely used in zinc-dust precipitation of gold. The zinc cements out gold very rapidly so even a pipeline reactor preceding the filter can be used.

Another type of clarifying filter is the tubular or candle filter. A number of tubes 2–4 in. in diameter are connected to a tube sheet at the top horizontal flange. The tube bundle is in a pressure shell with a drain at the base of the tank. A small chamber is placed above the tube flange, and the filtrate issues to this chamber.

A shock method may be used to discharge cake from this type of filter. The filtrate line is closed and the feed continues, so that pressure builds up in the discharge chamber. Air is also compressed in the chamber. When the drain is opened through a quick-opening valve, the very large pressure drop shocks the cake from the filter surface. A clean-water backflush may also be used.

These filters all use a slurry discharge that can usually be recirculated upstream so as not to lose any valuable constituents. The filter may be precoated. Various types of filter media are employed (e.g., fabric bags, wire cloth, and precision-wound wire that maintains a fixed distance between adjacent wires).

BASIC GUIDELINES FOR APPLICATION

Many factors influence the flowsheet and the equipment used in any liquid–solid separation application. The following factors usually must be considered (Bosley 1974; Scott 1970; Silverblatt et al. 1974):

1. Required performance:
   a. Moisture content of solids product
   b. Suspended solids and dissolved solids content of liquid or effluent product
   c. Energy requirements
   d. Average and peak production rates
   e. Availability of equipment (percentage of downtime)
   f. Recovery of dissolved solids or their elimination from plant product
2. Capital cost
3. Operating cost
4. Energy consumption
5. Maintenance costs and history for the specific application
6. Requirements governing plant effluent
7. Regulatory requirements
8. Other

Other factors apply in many cases but those listed are most important. Several flowsheets and types of equipment can be used for any particular liquid–solid separation application.

Some guidelines can be used to simplify the investigation of a liquid–solid separation. However, the use of flocculants to agglomerate colloidal solids has broadened the application of many individual types of separation equipment. Only those generalities that are widely accepted are considered here.

The first guideline concerns the state of dissolution of the feed to a liquid–solid separation step. If gravity sedimentation can remove about one-third of the water or fluid associated with a slurry, a thickener can probably be economically used before a final dewatering step. Consider the following example.

A slurry of 40 wt% solids (specific gravity of solids = 2.7) will contain 60/40 = 1.5 pounds of water per pound of solids. Removing one-third of the water by thickening would yield one pound of water per pound of solids or 50 wt% solids. This slurry might even thicken to 55 wt% solids (depending on the particle size distribution), in which case only 45/55 = 0.818 pounds of water per pound of solids would remain; 45.5 wt% of the water would be removed.

These solids would normally be filtered to 18–20 wt% moisture. If the moisture content is 20%, 0.25 pounds of water per pound of solid are present in the final filter cake. Thus, 83.3% of the water originally in the slurry has been removed, and 40.0–54.6 wt% of the water has been removed by the thickener.

As will be shown later, the filtration rate is ideally proportional to the square root of the weight of dry solids per unit volume of filtrate in the feed slurry. Thus, at 40% solid in the feed and 20% moisture in the final cake, the value is approximately 0.8 pounds of solids per unit volume of filtrate. On the other hand, if the feed to the filter contains 50% solids (or one pound of solids per pound of water) the value becomes 1.333 pounds of solids per unit volume of filtrate. The square root of the ratio of these values equals 1.29, which represents a 29% higher filtration rate by concentrating the feed to 50% solids. Actually, the filtration rate will be even higher because the calculation just made assumed no change in the filter cycle time. At the same cycle time, a much thinner cake would be produced by the more dilute slurry.

To estimate the filtration rate at the same cake thickness, it will be shown that cake weight of the dry solids per unit area per revolution is ideally proportional to the square of the filter cycle time (at constant effective submergence). Thus, the filter cycle time for the dilute feed would be increased by the square of the ratio, and it would be 2.776 times longer than the cycle time at 50% solids. However, cake weight increased from 0.8 to 1.333 or 1.666 times. Thus, the new rate ideally is now 1.666/2.776 or 60% of the rate at 50% solids. In other words, a 50% solids slurry ideally has a two-thirds higher filtration rate than the dilute slurry at the same cake thickness. Under actual conditions of higher feed solids concentration, the filtration rate would probably be even higher.

There are several reasons to consider a flowsheet that uses a thickener for final dewatering, particularly when the slurry can be reduced in fluid content by one-third or more.

1. A thickener has a low operating cost and requires relatively little energy.
2. Thickening will usually reduce both capital and operating costs.
3. The liquid product of a thickener contains a low concentration of suspended solids; the product of final dewatering can contain a high solids content. Thus, solids and liquid recovery in separate streams is very high if filtrate is recirculated to the thickener.
4. A thickener also supplies surge capacity.
5. The flowsheet is easier to operate and control.

Accordingly, consider whether solids should be concentrated ahead of final dewatering and whether means of concentration other than gravitational thickening (e.g., hydrocyclones or classifiers) should be used.

Final dewatering devices are primarily continuous filters and batch filters, or pressure filters and centrifuges. Let's first make some generalizations about continuous filters. The disk filter is used primarily on granular, relatively easily filtered slurries, such as fine clean coal (–28 mesh), base metal concentrates, and iron ore concentrates. The slurries usually must form at least a 3/8-in. cake in 30 s or less and require no cake washing. This behavior generally yields the lowest cost operation.

Drum filters are used for more difficult filtration and where cake may need to be washed. Media blinding should be slow. If blinding is a problem, the continuous-belt drum filter is probably a better unit.

The horizontal scroll–discharge table filter is widely used on coarser solids such as 20 × 200 mesh solids that are difficult to maintain in suspension. Cake can also be washed although washing may be limited to two or at most three countercurrent stages. A filter cake of at least 1/2-in. thickness must form in 10–15 s.

The horizontal belt filter is widely used as a cake-washing filter and has definite advantages in countercurrent washing of any number of stages. Because of its filter media cleaning principle, it can prevent blinding and is therefore used on straight dewatering operations even with slow-filtering slurries.

Pressure filters are generally used for more difficult slurries and where pressures well above 15 psig must be employed to develop sufficient cake thickness. They also will be used for relatively small volumes or for clarification operations. The recessed plate, the plate and frame, and the semicontinuous horizontal plate filters can all deliver a dewatered solid cake, which many other types of pressure filters cannot do. They also permit cake washing although at lower efficiencies than obtained with washing by continuous filters.

Centrifuges are used with less abrasive solids such as fine coal and crystalline solids. They can also beneficiate coal as the centrate solids are usually appreciably higher in ash content than the feed.

Membrane filtration (ultrafiltration or reverse osmosis) has not been used much in the mineral industry as yet. However, it should find future use in such areas as hydrometallurgy, recovery and separation of heavy metals, and control of toxic and hazardous materials in waste effluent.

Expression or expelling will undoubtedly grow in application where flocculants can be employed. Thus, the practice of dewatering tailings so they can qualify for land disposal should become more widespread in the future.

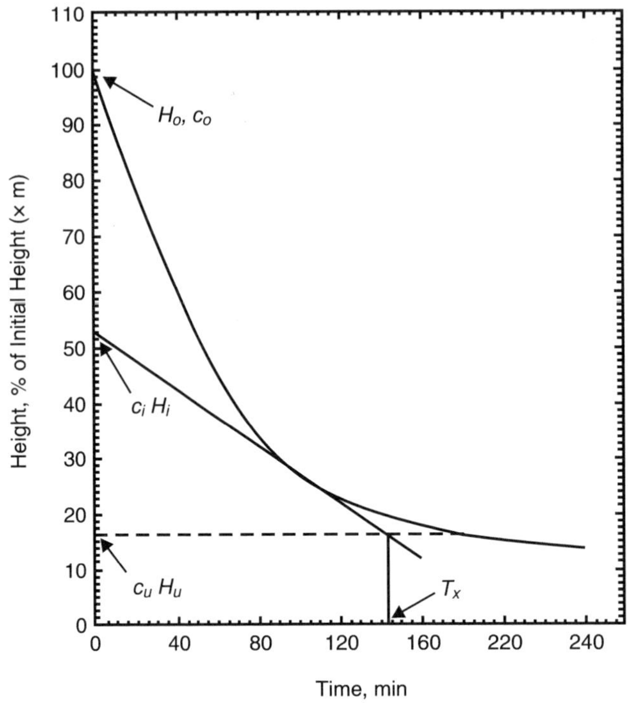

FIGURE 9.16 Gravitational sedimentation plot showing height of interface as a function of settling time

GRAVITY SEDIMENTATION APPLICATIONS

To properly apply gravitational sedimentation, both basic and applied theory must be considered. Bench-scale testing must be used to determine the type and size of most thickener applications. If we are to design such equipment based on performance, data must be correlated on the basis of applied theory. Applied theory will predict performance over a range of conditions so that requirements for quality and capacity will be met. (Knowledge of particle dynamics in a liquid is also desirable.)

Basic Theory

In most thickener applications, solids do not settle freely. Instead, there is a mass settling of the particles, and essentially all solids settle at the same velocity if the feed concentration is high enough. However, solids coarser than 60 mesh (specific gravity 2.7) will normally settle rapidly and in a sense will fall out of mass settling.

If a 15 wt% solids slurry of a typical −60 mesh size distribution (specific gravity 2.7) is placed in a 2-L graduated cylinder, mass subsidence will normally result. Thus, there will be an interface between the supernatant fluid and the solids. At the same time, if a substantial amount of −10-μm solids is present in a colloidal state, and self-flocculation does not occur, they will remain in suspension and leave a very dirty supernatant.

Bench-scale tests for gravitational sedimentation should be performed in 2-L (or larger) graduated cylinders to avoid wall effects. If the height of the interface is recorded as a function of time, a typical plot will look like that shown in Figure 9.16. An initial straight line passes through a "knee" and becomes asymptotic to some minimum value. The straight-line portion settles at a constant velocity until it reaches the knee of the curve. In this phase, the solids fall through the liquid. However, at some concentration, the weight of the settling solids above squeezes or compresses water out of the settling mass of solids below. The region in which this squeezing happens is termed the "compression zone."

Kynch showed that tangents at any point on the settling curve extrapolated back to the vertical axis yield the weight percent solids for a layer at the point of tangency (Kynch 1952). Thus, by knowing the amount of solids in the original slurry and thereby the amount of water present, the weight percent solids is equal to the weight of the original solids divided by the original slurry weight minus the water eliminated. This rule assumes that the amount of solids in the supernatant is negligible (which is almost always a good assumption). Furthermore, Kynch proved mathematically that the required thickening area for any particular underflow solids concentration could be determined from a single test (Kynch 1952).

In Figure 9.16, the initial concentration of the slurry allowed to settle in 2-L graduated cylinders is indicated as c_o. At some intermediate point on the settling curve past the straight line or constant velocity portion, a tangential line has been extrapolated back to the vertical axis, indicating c_i at H_i. Finally, the underflow concentration has been extrapolated back to the same axis by a horizontal line to indicate c_u and H_u. Using the English system, solids concentration c is given in terms of tons solids per cubic foot, and height is measured in feet. To determine thickener area requirements, the unit area is calculated. Unit area is expressed as the cross-sectional area per ton of dry solids per day. It can be shown that the following equation can be used to determine the unit area:

$$\text{U.A.} = \frac{\frac{1}{c_i} - \frac{1}{c_u}}{v} \qquad \textbf{(Eq. 9.16)}$$

where

U.A. = unit area, ft^2/ton solids/day

c = solids concentration, tons/ft^3

v = settling rate at solids concentration c_i, ft/day

But,

$$R = \frac{H_i - H_u}{T_x} \qquad \textbf{(Eq. 9.17)}$$

where

R = settling rate, ft/day

H = height of supernatant in graduated cylinder, ft

T_x = time, days

From Kynch's original proof,

$$c_o H_o A = c_i H_i A = c_u H_u A \qquad \textbf{(Eq. 9.18)}$$

where

A = area, ft^2

Then,

$$c_i = \frac{c_o - H_o}{H_i} \text{ and } c_u = \frac{c_o - H_o}{H_u} \qquad \textbf{(Eq. 9.19)}$$

Substituting Eqs. 9.17 and 9.19 into Eq. 9.16 obtains

$$\text{U.A.} = \frac{\frac{H_i}{c_o H_o} - \frac{H_u}{c_o H_o}}{\frac{H_i - H_u}{T_x}} \qquad \textbf{(Eq. 9.20)}$$

and

$$\text{U.A.} = \frac{T_x}{c_o H_o} \qquad \textbf{(Eq. 9.21)}$$

In addition, because flocculants are so widely used today, any test work must simulate the degree of mixing and other hydraulic shear that occurs in the full-scale application. The work must also simulate the influence of detention time between the start of flocculation and the beginning of sedimentation. Hydraulic shear and lengthy detention time tend to degrade the flocculi and therefore reduce settling rate. The factor T_x should also be determined. In any test work, the sample employed must represent the actual feed and its solids concentration.

Applied Theory

Coe and Clevenger (1916) were the first to develop an applied theory of thickening. They reasoned that all feed solids concentrations existed within the thickener, but the one with the lowest solids flux rate per hour per square foot would be controlling. Therefore, they developed the following relationship, which is still employed today in some applications.

They hypothesized that the critical flux rate existed when the upflow velocity of the liquor equaled the settling velocity of the mass of solids at a particular critical solids concentration. This hypothesis assumes that a negligible amount of solids are lost to the thickener overflow, which is normally reasonable. Thus, the following equation sets the sedimentation velocity equal to the upflow liquid velocity.

$$R(24)\rho\ (62.4)\ A = (F - U)S(2000) \qquad \textbf{(Eq. 9.22)}$$

where

F = dilution of solids, pounds of liquid per pound of solids with settling rate $R\rho$ = liquid specific gravity

U = underflow dilution, pounds of liquid per pound of solids

S = solids rate, tons per day

and

$$\frac{A}{S} = \text{U.A.} = \frac{1.335(F-U)}{R\rho} \qquad \textbf{(Eq. 9.23)}$$

In each of a series of tests performed in a 2-L graduated cylinder, the unit area is calculated using Eq. 9.23. Normally, the initial solids concentration is high, about 5 or 10 percentage points above the desired underflow concentration. The initial settling rate obtained serves as the value of v. Some of the solids in the original slurry are then withdrawn, and the remaining slurry is diluted back to the original 2 L. This sequence is repeated at several solids concentrations; the last one is the predicted solids concentration. The unit area as a function of the individual feed solids concentration is then plotted, and the highest or critical value is used as a design value.

The thickener rake mechanism should also be simulated, because the test mechanism will usually yield an underflow solids concentration that is a minimum of two to four percentage points higher than that in the full-scale unit. This difference is caused by the channels it forms to permit escape of fluid that has been squeezed out by the weight of the compressing solids. Usually, the test raking mechanism consists of three or four vertical rods in a cage-like construction. The cage is usually rotated by a clock motor that produces 6 to 10 revolutions per hour for a 2-L cylinder.

As will be shown later, the Coe–Clevenger test and analysis method should not be used if a flocculant has been employed, because flocculi structure changes as the solids concentration is changed.

Although the Kynch method appears very easy to use, the calculation of the time in days for the solids to thicken from the feed to the underflow concentration—a critical calculation—can be in error if not done correctly. The recommended method is the Oltmann procedure (Baczek et al. 1997).

Figure 9.16 illustrates the design of the thickener area proposed by Oltmann, using Eq. 9.24.

$$\text{Unit area of thickener} = t_u/C_oH_o \qquad \textbf{(Eq. 9.24)}$$

where

t_u = settling time, days

C_o = test or feed solids concentration in kg/L or ton/ft^3

H_o = initital height of pulp in the test, m or ft

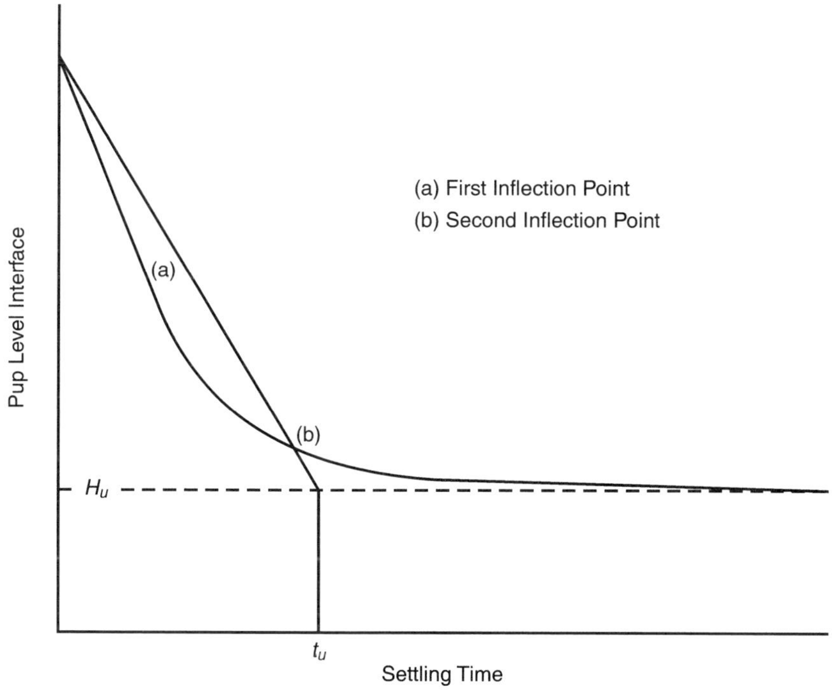

FIGURE 9.17 Typical sedimentation plot—Kynch analysis

The procedure originally derived by Kynch can be used to determine the unit area in square feet per ton per day. The settling test carried out using a measuring cylinder is used to plot a chart, such as Figure 9.17. The height of the interface is plotted as a function of time. Normally, there will be two discontinuity points: the first inflection point (a), and the second inflection point (b), at which point compression is believed to begin. Oltmann proposed to draw a line from the start of settling to the second inflection point (b) of Figure 9.17. The extension of this line to the underflow line (H_u) gives t_u.

The Talmadge and Fitch method, which was developed first, employed a line tangent to the initiation of compression, which point was determined by empirical means (Talmadge and Fitch 1955). However, the method was not precise and yielded a greater unit area than the Oltmann construction.

The modified Kynch method usually gives a larger unit area than the Coe–Clevenger method, and the latter method may even yield an undersized thickener (Talmadge and Fitch 1955). The modified Kynch method can also be used when a flocculant is added, because only one test is required. However, it is essential that the flocculation test uses the actual feed solids concentration.

A third analysis, termed the Wilhelm–Naide method, starts with the solids flux rate in a continuous thickener as follows (Wilhelm and Naide 1981):

$$G_i = x_i R_i + x_i R_u \quad \text{(Eq. 9.25)}$$

where

G_i = solids flux rate in layer i, lb/day/ft^2

x_i = concentration of solids in layer i, lb/ft^3

R_i = settling rate in layer i, ft/day

R_u = velocity downward caused by underflow removal, ft/day

The limiting solids flux G_L can be described by the same equation, as follows:

$$G_L = x_L R_L + x_L R_u \quad \textbf{(Eq. 9.26)}$$

where

G_L = limiting solids flux rate, lb/day/ft^2
x_L = limiting solids concentration, lb/ft^3
R_L = limiting settling rate, ft/day

At the limiting flux rate the following must occur:

$$\left(\frac{dG_i}{dx_i}\right)_{i=L} = 0 \quad \textbf{(Eq. 9.27)}$$

Substituting Eq. 9.27 into Eq. 9.25, the following is obtained:

$$R_u = \left[\frac{-d(x_i R_i)}{dx_i}\right]_{i=L} \quad \textbf{(Eq. 9.28)}$$

Wilhelm and Naide found that x_i was a function of the solids concentration in the hindered-settling and compression zone, and thus the following equation could be used:

$$R_i = a x_i^{-b} \quad \textbf{(Eq. 9.29)}$$

where

a = coefficient defined by Eq. 9.29
b = exponent defined by Eq. 9.29

The values of a and b will change between the zones. However, Eq. 9.29 can be substituted into Eq. 9.28 to obtain the following:

$$R_u = a(b-1)x_L^{-b} \quad \textbf{(Eq. 9.30)}$$

Substituting Eqs. 29 and 30 into Eq. 9.26 yields

$$G_L = x_L\,(a x_L^{-b}) + x_L[a(b-1)]\,x_L^{-b} = abx_L^{-b} \quad \textbf{(Eq. 9.31)}$$

Also, another equation can be written for G_L:

$$G_L = x_u R_u \quad \textbf{(Eq. 9.32)}$$

where

x_u = underflow solids concentration, lb/ft^3

Substituting Eq. 9.30 into Eq. 9.32,

$$G_L = a\,(b-1)x_u x_L^{-b} \quad \textbf{(Eq. 9.33)}$$

From Eq. 9.31 and 9.33,

$$x_L = \frac{b-1}{b}x_u \quad \textbf{(Eq. 9.34)}$$

Substituting Eq. 9.34 into 9.31,

$$G_L = ab\left(\frac{b-1}{b}x_u\right)^{b-1} \quad \textbf{(Eq. 9.35)}$$

The unit area is simply the reciprocal of G, and thus Eq. 9.35 becomes the form used for U.A.:

$$\frac{1}{G_L} = \text{U.A.} = \frac{\left(\frac{b-1}{b}\right)^{b-1}}{ab}x_u^{\,b-1} \quad \textbf{(Eq. 9.36)}$$

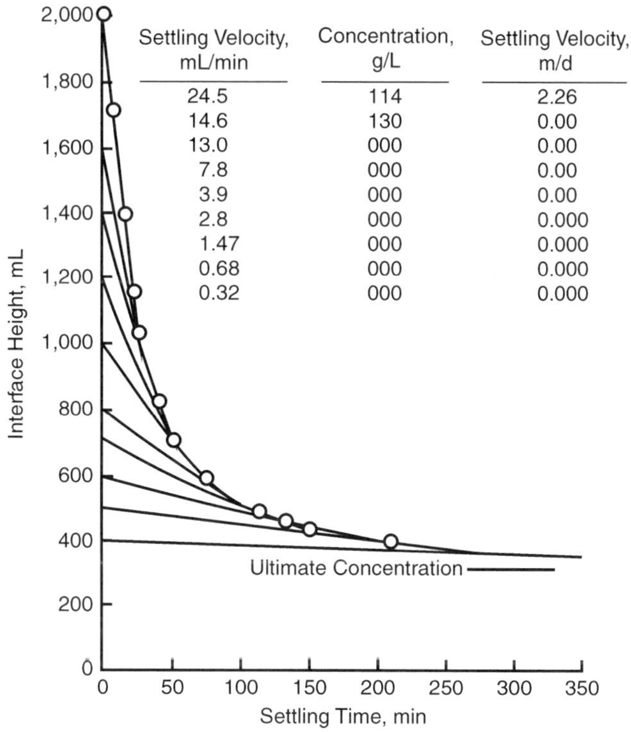

Source: Wilheim and Naide 1981.

FIGURE 9.18 Determination of settling velocity versus concentration relationships for coal refuse C_o = 90 g/L.

To determine values of *a* and *b*, Wilheim and Naide plotted the interface height against settling time in a 2-L cylinder test (as shown in Figure 9.18 for a coal refuse). The tangent lines drawn permit the calculation of the settling velocity of the interface at the point of tangency as well as the solids concentration.

A log-log plot of settling velocity versus concentration was then developed (see Figure 9.19). There are three distinct straight lines of different slopes and, accordingly, two discontinuities. The values of *a* and *b* have also been calculated for each line. Finally, predicted unit area as a function of underflow concentration from Eq. 9.36 is shown in the log-log plot of Figure 9.20. Thickening data from a continuous pilot thickener on the same sludge were compared to predicted values and results were very close.

To relate 2-L graduated cylinder tests to full-scale results where sludge depths will be much higher, the following equation is used (Dahlstrom and Emmett 1983):

$$(\text{U.A.})\text{actual} = (\text{U.A.})\text{test}\left(\frac{h_T}{H_a}\right)^N \qquad \textbf{(Eq. 9.37)}$$

where

h_T = average height of pulp during settling test, ft

H_a = design height of pulp in full-scale thickener, ft

N = empirical exponent, less than 1.0

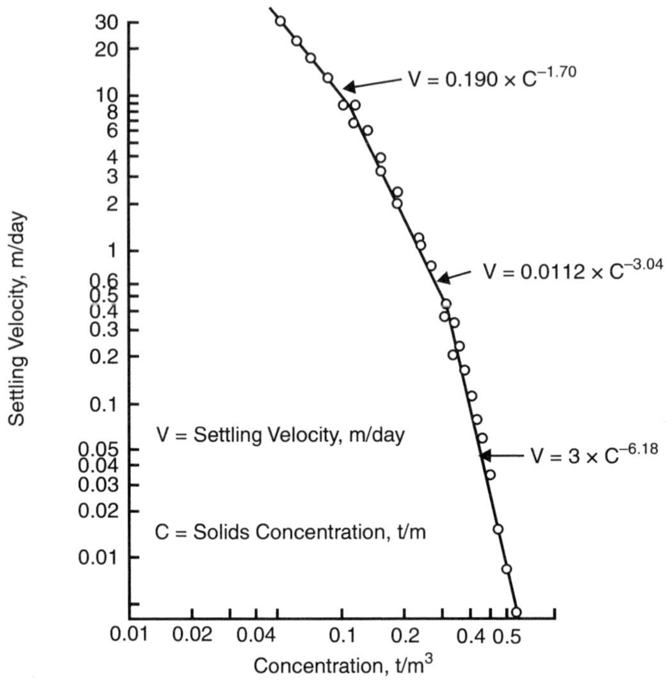

FIGURE 9.19 Log-log plot of settling velocity versus solids concentration

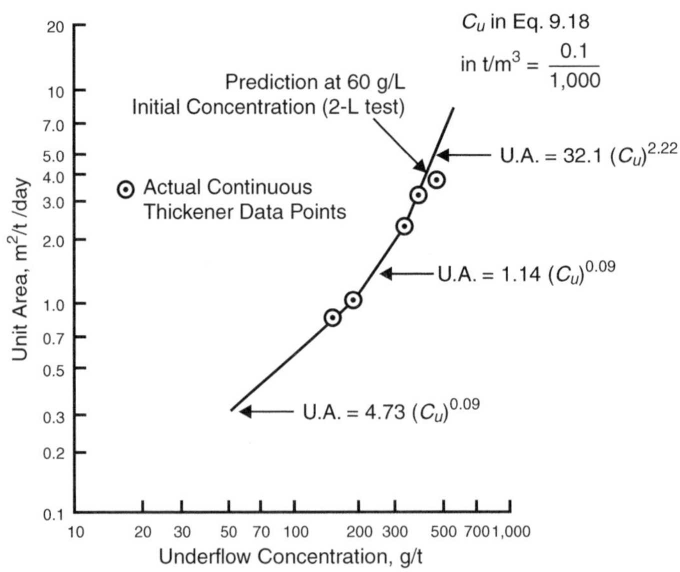

Source: Wilheim and Naide 1981.

FIGURE 9.20 Thickener operating prediction for coal refuse

Source: Dahlstrom and Emmett 1983.

FIGURE 9.21 Empirical exponent, *N*, as function of settling rate at underflow concentration

The average height in the test is obtained by integrating the area under the curve of Figure 9.18 to the desired underflow concentration and dividing by the time. The design height is equal to the side water depth plus one-third of the conical section. Figure 9.21 is a plot of *N* versus a function of the settling rate at the underflow concentration (Dahlstrom and Emmett 1983).

The Wilhelm–Naide method appears to approximate actual results to a much closer extent and should prevent overdesign.

Scale-up Factor

All thickeners should be designed with scale-up factors to account for tank inefficiency. Upflow velocity across the diameter of the thickener is uneven because of the basic concept. An additional scale-up amount should normally be used to account for fluctuations in factors such as particle size distribution, feed solids concentration, pH, and temperature. Thus, the unit area, U.A., is multiplied by 1.2 for thickeners of 100 ft in diameter or larger, and by 1.5 for tanks of 15 ft in diameter or smaller. In between, the multiplier is proportional.

Operating Variables and Their Influence

Operating variables that greatly affect thickening results are particle size distribution, feed and underflow solids concentration, temperature, and flocculation. The particle size of most concern are colloids, which are usually finer than 10 μm. Appreciable changes in the concentration of this fraction can substantially change settling rates and thus unit areas. Moreover, the colloidal fraction may change from one clay type to another if the ore body changes. A reasonably consistent feed should be maintained by blending if ores vary a great deal.

The feed solids concentration affects performance and should be stabilized. The same is true of the underflow concentration. Many times, underflow concentration is lowered by pumping at too high a rate. Although this pumping may make it easier to operate the thickener, it will increase the difficulty of operating the final dewatering device. Thus, it is desirable to have a solids concentration indicator on the underflow discharge so that operations can be achieved at a reasonably high value. If feed quality changes, the maximum underflow concentration and the proper control value will also change.

Many thickeners are installed outside, even in very cold regions, because of the great expense of housing such units. With proper design, thickeners can be installed outside where temperatures can be down to −40°F, and they can operate continuously even with an ice cap on the surface (except in the feedwell area). Viscosity near freezing temperatures is approximately 1.5 centipoise, a value that can greatly influence settling and compression rates. Accordingly, design should be based on the cold fluid temperature that will occur during winter.

General Installation Requirements

The thickener tank is usually the most expensive item in an installation. It may be constructed of materials such as earthen basins, wood, steel, concrete, acid brick, rubber-covered steel, and stainless steel.

An earthen basin may be lower in cost for large units (200 ft in diameter and larger). However, cost will depend on the type of soil, local water table, amount of excavation, and other factors, and it will usually require chemicals for hardening the base. Wood can be used for units of 60–100 ft diameter. A wooden tank will take some time to swell after water is added, and it should remain full at all times. A tank with a concrete bottom and sidewalls of steel is very common and is reasonable in cost. Total concrete construction is also common. Other materials are used primarily for acidic conditions. A membrane liner with acid brick is most commonly used, but rubber-covered steel is also employed.

A tunnel may lie below a thickener if the underflow lines must be accessible in an emergency and short suction lines are desirable. Tunnels increase the costs substantially, and U.S. Occupational Safety and Health Administration (OSHA) regulations normally require a full tunnel with "light at both ends." Alternatively, the thickener tank can be put on piers; this solution is not uncommon but it is even more costly.

The feed lines and feedwell strongly influence the quality of the overflow. Normally a pipeline or launder is hung from the bridge. Feed lines should be designed to minimize drop into the feedwell, to reduce the influence of excessive turbulence. The feedwell may be specially designed if very good overflow quality is necessary.

Peripheral launders that withdraw overflow from thickeners are almost always used. Weir rates range from 5 to 25 m^3/h/m (9,660 to 48,310 gal/day/ft). The higher rates are used with well-flocculated and easily settled solids. By setting the launder away from the wall so that both sides flow freely, the rate per foot of launder is doubled. Radial launders delivering into the peripheral launder are also used for high upflow rates on solids contact clarifiers.

The underflow should be connected to the suction side of the underflow pump with a minimum of directional changes. This pump should always be spared, because most thickeners can allow the underflow pump to be shut down only for a brief period. For applications that treat an underflow with a high solids concentration or large amounts of +60 mesh solids of 2.7 specific gravity, high-pressure water or air connections (or both) should be placed at intervals on the suction and discharge lines to blow out any plug and prevent a shutdown.

Flocculation

Flocculants are used in most thickeners today to obtain concentrations of overflow solids that will allow water to be reused or to abide by government regulations if the overflow is to be discharged. Governmental regulation of effluent commonly makes it less expensive to recycle water back to the processing plant. Reclaimed water that contains from 200 mg/L to 1% solids (depending on the processing plant requirements) is generally acceptable. Most thickeners can achieve this concentration by using a flocculant. In hydrometallurgical operations, more colloidal solids are produced because of the chemical conditions, temperatures, and detention times employed. Flocculants are required in essentially all hydrometallurgical thickeners.

Several aspects of flocculant usage should be considered. First, there are three basic types: anionic, cationic, and nonionic. Although flocculants are used only in small quantities (0.05–0.25 lb/ton of solids), they are expensive ($1.50–$3.00/lb), and the most economical one should be employed.

Bench-scale tests must be performed to determine which is best in terms of economics, followed by full-scale plant tests.

Second, the flocculants used in thickeners should be added at concentrations of 0.025–0.05 wt% dry substance, so that the volume is sufficient to improve mixing. If necessary, thickener overflow can be used to dilute the stock solution. Because flocculant adsorbs onto "what it sees" and does not desorb, good mixing is essential. Mixing can be promoted by using several input points on the feed line and proper baffling. To produce "tough" flocculi, the flocculant may be added before the centrifugal feed pump.

High-rate thickeners obtain their high rate by optimizing flocculation, so flocculation methods are extremely important. Flocculant is added at several points and baffled small-diameter feedwells are used. In this way, a high degree of plug flow can be achieved, and rapid mixing and minimum shear will prevent flocculi degradation. The flocculated solids then issue from the feedwell and fall to their own specific gravity level.

Flocculants can also produce agglomerated solids that stick together on the raking mechanism. They can grow and form an island or even a complete donut. These formations will increase torque even though the underflow may be coming out at a low solids concentration (usually caused by "rat holing"). This phenomenon can be caused by too much flocculant; if allowed to continue, it may damage or completely shut down the thickener. It usually can be prevented by controlling the flocculant-to-solids ratio or by periodic lifting and lowering of the rakes (once a shift to once a day) through the lift device.

The structure of the floc will depend on the solids concentration at which it formed. A large voluminous floc, commonly formed at a higher concentration, has a slow settling rate. Diluting the feed with thickener overflow may achieve a lower unit area and a higher solids concentration underflow. Bench-scale tests can easily ascertain this outcome. As is now readily apparent, Coe–Clevenger tests cannot be conducted with flocculated pulp. Furthermore, high-rate thickeners should be used only if a flocculant is employed. Otherwise, no advantage will be seen.

Countercurrent Decantation

A series of thickeners may be used to recover soluble values in situations where the liquid is flowing in one direction and the solids in the opposite direction. This practice, termed "countercurrent decantation" (CCD), may use as few as three stages or as many as nine. Water or some other wash fluid is added to the last stage feed and contacts and mixes with the underflow from the previous stage. The overflow of this final stage then passes to the previous stage feed and mixes with the underflow of that preceding stage. This sequence is repeated as many times as there are thickener stages. A most common method is for the leached pulp to be used as feed to the first stage; it then contacts the overflow of the second stage. This sequence is pictured in Figure 9.22 as a three-stage CCD. Another common flowsheet that can produce higher strength pregnant liquor and sometimes save some acid or base chemical illustrates adding the second-stage overflow to the incoming solids or slurry for leach feed along with make up chemicals. This leached slurry then becomes the feed to the first stage. A disadvantage may be that the more dilute leach stage that results will require more agitation and larger leach tanks.

To illustrate the calculations for CCD, an example is given in Figure 9.22. The example is a gold ore, grade 0.0006 wt% gold/ton of ore, which is processed at the rate of 2,500 tons per day. Other information is (1) all of the gold is dissolved in the agitator, (2) the dissolution of non-gold-bearing minerals is negligible, (3) agitator discharge contains 35 wt% solids, (4) thickener underflows contain 55 wt% solids, (5) 5,000 tons per day pregnant liquor overflow the first thickener and are sent to gold recovery, (6) underflow from the third thickener is discarded, (7) part of the barren solution from the gold recovery circuit is added as wash water to the third thickener, while part is returned to the agitator, and (8) make-up water is added to the agitator. Solid and solution flows are calculated from the mass balances and are given in Figure 9.22. The recovery of gold is calculated as follows.

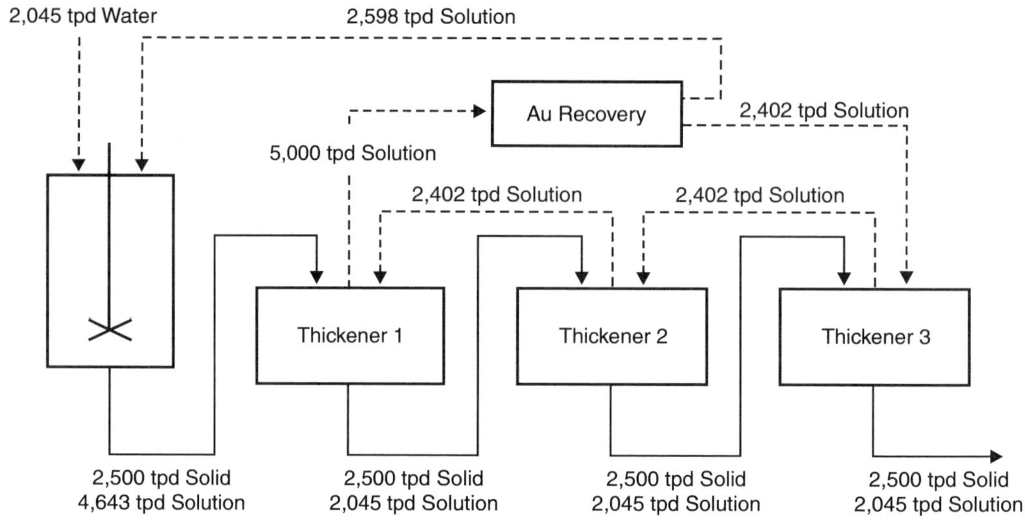

FIGURE 9.22 Three-stage countercurrent decantation circuit for gold ore

Let x = concentration of gold in solution in thickener 1 (ton gold/ton solution)

y = concentration of gold in solution in thickener 2 (ton gold/ton solution)

z = concentration of gold in solution in thickener 3 (ton gold/ton solution)

Gold balances around each of the thickeners are

Thickener 1

$$2{,}500\,(0.000006) + 2{,}402\,(y) = 5{,}000\,(x) + 2{,}045\,(x)$$

Thickener 2

$$2{,}045\,(x) + 2{,}402\,(z) = 2{,}045\,(y) + 2{,}402\,(y)$$

Thickener 3

$$2{,}045\,(y) + 2{,}402\,(0) = 2{,}045\,(z) + 2{,}402\,(z)$$

Solving these simultaneous equations yields the concentration of gold in the solutions of each of the thickeners. These concentrations are

$x = 2.66 \times 10^{-6}$ ton gold/ton solution

$y = 1.63 \times 10^{-6}$ ton gold/ton solution

$z = 7.46 \times 10^{-7}$ ton gold/ton solution

Recovery can be calculated from

$$\text{Recovery (100)} = \frac{(5{,}000)(2.66 \times 10^{-6})(100)}{(2{,}500)(0.00006)} = 88.6\%$$

CONTINUOUS VACUUM FILTRATION

Continuous vacuum filtration is actually a series of batch filtrations with a relatively short cycle that are continuously repeated to make the sequence continuous. Drum filters, for example, are divided into a number of sections. In an 8-ft diameter × 8-ft wide drum, there are usually about 16 sections around the periphery. Each section would be 8 ft long and $8\pi/16$ (1.57) ft high. In each section, cake will first form under a pressure differential. Cake thus formed is then dewatered by air that flows through the pores in response to a pressure differential. While the drum revolves each section is exposed to air. When the section reaches the scraper blade, and after the vacuum in the section has been neutralized, a reverse flow of air above atmospheric pressure helps to discharge the cake. Finally, the cycle is repeated when the section is again submerged in the feed in the filter tank.

The cake can also be washed on a drum by a wash fluid in part of the third and fourth quadrants of the cycle. The wash fluid is drawn through the cake by a pressure differential and tends to force another liquor ahead of it, although complete plug flow cannot be achieved. Those filters forming their cake with gravity can also employ countercurrent washing, which will increase recovery or reduce the amount of wash fluid required.

Basic Theory

Much of the theory of filtration comes from the flow of fluids through capillaries, such as in ground water movement or oil reservoir production. Darcy's law for flow through capillaries is given as follows:

$$v = \frac{J(\Delta p)}{\mu l} \quad \textbf{(Eq. 9.38)}$$

where

v = flow rate, ft/s
J = permeability, ft^2
Δp = pressure drop, lb/ft^2
l = length of capillary or cake, ft
μ = viscosity, lb/ft-s

Poiseuille's law, in which D_i = average capillary diameter, is very similar:

$$v = \frac{d_i \Delta \rho g}{32\ \mu l} \quad \textbf{(Eq. 9.39)}$$

Both equations involve a driving force divided by a resistance. The latter is the reciprocal of permeability. The following substitutions can be made in Poiseuille's law:

$$v = \frac{1}{A}\frac{dv}{d\theta} \quad \textbf{(Eq. 9.40)}$$

$$\frac{d_i g}{32} \cong J = 1/\alpha \quad \textbf{(Eq. 9.41)}$$

$$l \cong \frac{wV}{A} \quad \textbf{(Eq. 9.42)}$$

$$-\frac{dV}{Ad\theta} = \Delta\rho/((\mu\alpha wV)/A) \quad \textbf{(Eq. 9.43)}$$

where

θ = time, normally min
V = volume of filtrate, ft^3
w = lb dry solids filter cake/ft^3 filtrate
α = specific cake resistance (s^2/ft^2)

Because of the change in the dimensions of wV/A compared to l, the specific cake resistance, α, now has dimensions of foot per pound. This expression is the basic one that is used in filtration, but it is modified by adding a term for the resistance of the filter media and drainage network down to the point where the pressure is known. Thus, the basic expression is as follows:

$$\frac{dV}{Ad\theta} = \frac{\Delta p}{\mu\left(\alpha w\frac{V}{A} + r\right)} \quad \textbf{(Eq. 9.44)}$$

where r = resistance of the cloth and the drainage network in consistent units. This equation is most commonly employed in cake deposition.

Another basic expression of dewatering of a filter cake was developed by Brownell and Gudz as follows (Nelson and Dahlstrom 1957):

$$d\theta_d = \frac{-\phi l^2 \mu[(1-s_r)/(1-s_r s_e)]^2}{J\Delta p s_e^{\,y}} ds_e \qquad \textbf{(Eq. 9.45)}$$

where

θ_d = dewatering time/cycle, s

ϕ = cake porosity, void fraction

s_r = residual saturation at equilibrium

s_e = saturation, voids containing wetting fluid in flow, and voids containing both liquid and gas in flow

y = exponent determined by particle size

A final basic expression was developed by Rhodes as a washing equation (Choudhury and Dahlstrom 1957):

$$\frac{z}{z_o} = e^{-kft/l} \qquad \textbf{(Eq. 9.46)}$$

where

z = concentration of solute in the filtrate leaving the cake at any time after washing begins

z_o = concentration of the solute in the original liquor

k = constant

f = wash flow rate/unit area

t = time

All these expressions are basic to the applied theory.

Applied Theory

Previous discussions have shown that various rates occur within the filter cycle. First, cake forms at some rate in every continuous-filter application. Cake is dewatered at some rate in almost every continuous-filter application. Some exceptions do exist. In a few installations that use continuous-belt drum filters, cake is washed until the filter media leave the drum surface. In such cases, recovery of soluble values such as gold is very important, and the cake is to be repulped for disposal to a tailings pond. These are exceptions, however, and a cake dewatering rate must usually be considered.

Finally, if cake is washed, it will be washed at some rate. Washing actually involves two rate functions: the rate of penetration of the cake washing fluid and the rate of displacement of the washing fluid. In summary, the rate functions in continuous filtration are as follows (Dahlstrom 1980):

1. Cake formation rate (always a factor)
2. Cake dewatering rate (almost always a factor)
3. Cake washing rate (a factor only if cake is washed; actually two rate functions):
   a. Washing-fluid penetration rate
   b. Washing-fluid displacement rate

As with other process operations that involve more than one rate function (e.g., heat transfer), the slowest rate function will usually control. However, all should be determined to properly incorporate them into the filter cycle. By using rate functions, the filtration step can be designed on the basis of performance, solids handling rate, desired moisture content, and recovery or elimination of soluble constituents.

Cake Formation Rate

Because continuous filtration is a cyclic process that is continuously repeated, the Poiseuille equation must be integrated over the time of cake formation in the filter cycle. Accordingly, Eq. 9.44 is arranged as follows:

$$\int_0^{V_f} \frac{\mu\alpha w V dv}{A^2} + \int_0^{V_f} \frac{\mu r dV}{A} = \int_0^{\theta_f} \Delta p d\theta \quad \textbf{(Eq. 9.47)}$$

The term r describes the resistance of the filter and drainage network downstream of the media. However, the cake resistance, $\alpha wv/A$, is almost always much larger than the filter-media resistance. Additionally, the filter drainage network should be designed hydraulically so that even at maximum rates, the maximum pressure drop from the bottom of the filter media to the suction side of the vacuum pump should be no more than 2 in. of mercury and normally less. Thus the term r can be considered negligible. The equation can now be integrated between limits of 0 and V_f and 0 and θ_f and rearranged as follows (Dahlstrom 1978):

$$\frac{1}{A^2}\int_0^{V_f} V dV = \frac{\Delta p}{\mu\alpha w}\int_0^{\theta_f} d\theta \quad \textbf{(Eq. 9.48)}$$

where

V_f = volume of filtrate per cycle, ft^3

θ_f = cake formation time per cycle, min

Five different terms that have been assumed constant are here considered individually.

Filtration area, A, should not change, and even if it does on a normal-size machine because of a "mushroom effect," the change will be small.

Pressure differential should not change during the cake formation portion of the cycle, because the filter design allows vacuum level to be achieved very rapidly.

Viscosity of the liquid, μ, will stay constant unless operation occurs at or near the boiling point of the liquid. Such applications usually require an internal drainage network design that can accommodate two-phase flow. Such a design can offset the viscosity effect downstream where a problem would occur.

The term w will not change appreciably unless the feed solids concentration changes. Dilution of feed will always affect cake formation rates and should be controlled at optimum values.

Finally, specific cake resistance, α, might change. Any change will not be caused by cake compressibility factors, because the pressure drop (Δp) is held constant. Rather, it will be because of the migration of extreme fines toward the filtrate side after the bulk of the solids are initially deposited, thereby changing the specific resistance of the cake downstream. This change can be handled by data correlation. Accordingly, Eq. 9.48 is integrated to the following:

$$\frac{V^2}{A^2} = \frac{2\Delta p\theta_f}{\mu\alpha w} \quad \textbf{(Eq. 9.49)}$$

Taking the square root of both sides and then dividing by θ_f,

$$\frac{V_f}{A\theta_f} = \left[\frac{2\Delta p}{\mu\alpha\theta_f}\right]^{1/2} \quad \textbf{(Eq. 9.50)}$$

This equation expresses the volume of filtrate per unit area per unit time of cake formation. Because most industrial filtration is concerned with the solids handling rate, both sides can be multiplied by w (Dahlstrom 1978).

$$\frac{wV_f}{A\theta_f} = Z_f = \left[\frac{2\Delta p w}{\mu\alpha\theta_f}\right]^{1/2} \quad \textbf{(Eq. 9.51)}$$

where Z_f = form filtration rate, weight of dry solids per unit area per unit time of cake formation.

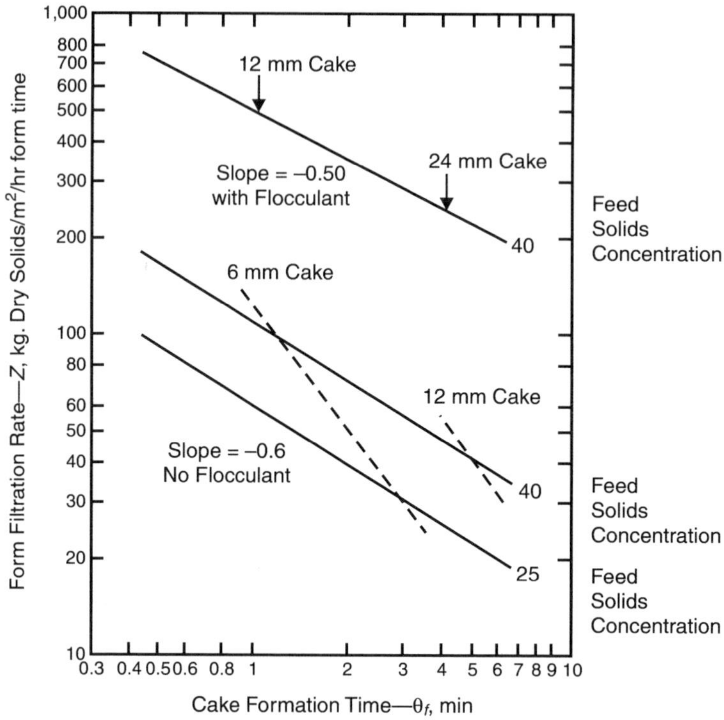

FIGURE 9.23 Form filtration rate as function of time

To obtain the full-scale rate, the value of Z_f would be multiplied by the fraction of time during the total filter cycle devoted to cake formation. If θ_f is given in minutes, it should be changed to hours because full-scale rates are usually given in pounds of dry solids per hour per foot squared.

Thus, bench-scale test work, the normal method of predicting full-scale results, should be run at constant pressure. The pressure generally depends on the air- or gas-handling requirements of the vacuum pump. Twenty inches of mercury vacuum is the most common operating-pressure drop. Where minimum moisture content is important, as in a balling operation, vacuums as high as 26 in. of mercury have been used. With highly permeable filter cakes, vacuum levels of 12 in. to as low as 4 in. of mercury are used. The vacuum pump capacities range from about 15–200 ft^3 of air per square foot of area based on the vacuum pump's suction-side conditions.

The term α can change with time during the cake formation portion of the cycle. This is apparent on some feeds when Z_f is plotted as a function of θ_f at constant Δp. According to Eq. 9.51, the ideal slope of a log-log plot of Z_f as a function of θ_f should be −0.5. If the slope is more negative, extreme fines are migrating with time and increasing the specific resistance in layers of filter cake downstream (Dahlstrom 1978). In any case, it can be shown that the exponent can be no smaller than −1.0. Figure 9.23 is a typical log-log plot of form filtration rate as a function of the cake formation time. For most applications, the slope will range from −0.5 to −0.65. An increasingly negative slope calls attention to the need to maintain a minimum cake formation time that will produce a dischargeable cake. Appreciably higher filtration rates can be achieved with flocculation, particularly where excessive amounts of extreme fines are migrating. It becomes strictly a matter of economics.

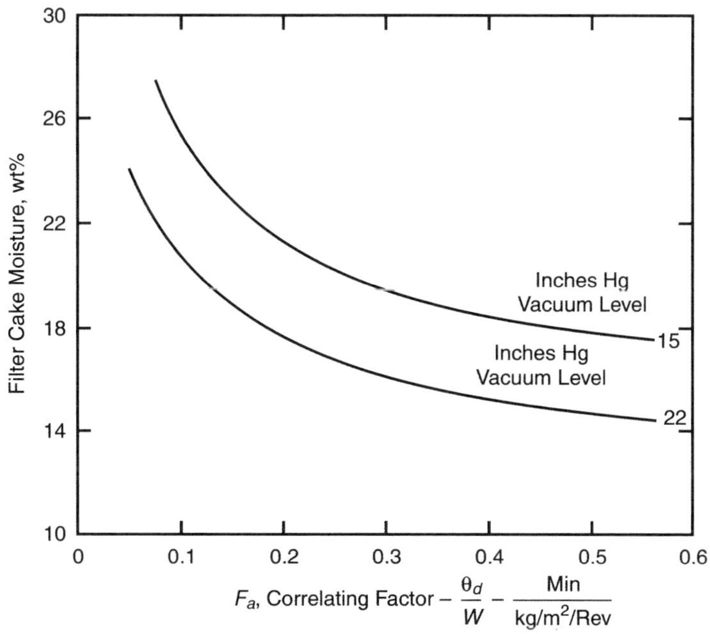

FIGURE 9.24 Filter cake moisture as a function of correlating factor (θ_d/W)

Cake Dewatering Rate

The Brownell equation (Eq. 9.45) can be simplified to the following expression (Nelson and Dahlstrom 1957).

$$\%M = f(F_a, \%M_r, Y) \qquad \textbf{(Eq. 9.52)}$$

where

$\%M$ = moisture content of discharged cake, wt%

F_a = factor which indicates the approach to $\%M_r$

$\%Mr$ = residual moisture content. This term is equal to the moisture obtained at infinite time with the same cake thickness, pressure drop, size distribution, temperature, etc., when 100%-humidity air is pulled through the cake.

Y = factor depending on particle size distribution and shape factor

$$F_a = \frac{\theta_d Z_f}{W} \frac{\Delta p_d}{\Delta p_f}$$

θ_d = dewatering time/cycle, normally min

Z_f = Form filtration rate, volume per unit area per unit time of cake formation

W = weight of dry cake solids per unit area per cycle

Δp_d = cake dewatering pressure drop

Δp_f = cake formation pressure drop

In most cases, Δp_d and Δp_f are equal or nearly so. Thus, unless the air rate through the cake during dewatering reaches 20 cfm/ft^2, it is more convenient to use the term F_a as $(\theta_d/W)_{\Delta p}$. Thus, correlating %M as function of (θ_d/W) with parameters of Δp should yield an initially sharply descending curve that passes through a "knee" and becomes asymptotic to some minimum value of moisture percent. Figure 9.24 shows a typical plot of moisture content as a function of the correlating factor (θ_d/W) for copper concentrates at two different pressure drops during dewatering. This plot shows that a higher pressure drop appreciably lowers moisture content. The higher vacuum level will undoubtedly produce a net energy savings, because less moisture will need to be evaporated in the smelter.

The full-scale filtration rate is determined by selecting the moisture content desired from the curve, which yields the required correlating factor. A given cake thickness (selected to permit good cake discharge) will automatically determine the cake dewatering time, θ_d. The full-scale rate based on cake dewatering only will be determined by the following equation:

$$\text{Full-scale rate} = W\left(\frac{60}{\theta_d}\right)q_d \quad \textbf{(Eq. 9.53)}$$

where q_d = fraction of cycle time devoted to dewatering.

The full-scale rate is given in terms of pounds of dry solids per hour per square foot. Where moisture must be minimized, the filtration rate will be lower than when it is determined only by cake formation. Accordingly, the time devoted to cake formation must be reduced. The time can be reduced by changing the bridge block setting on the filter valve for filters forming their cake against gravity, or by properly incorporating the rate functions into the cycle for those filters forming their cake with gravity.

Filter Cake Washing Rates

The cake washing penetration rate can be predicted by the Poiseuille equation, assuming that the viscosity of the washing fluid is equal to that of the liquor in the cake. Because most applications involve aqueous solutions, this assumption is generally satisfactory. The following equation should predict this rate (Choudhury and Dahlstrom 1957):

$$\frac{dV}{Ad\theta_d} = \frac{\Delta p}{\mu \alpha w \frac{V_f}{A}} = \text{constant rate} \quad \textbf{(Eq. 9.54)}$$

This equation can be simplified to

$$\theta_w = k'(\theta_f)N \quad \textbf{(Eq. 9.55)}$$

where

θ_w = cake washing time, normally min

k′ = a constant

N = wash ratio, i.e., the volume of wash fluid/volume of liquor in the cake before washing

The term N in most cases can be reduced to pounds of wash fluid per pound of liquor in the cake before washing, when the specific gravities are essentially the same.

A coordinate plot of wash time θ_w as a function of wash ratio N should yield a series of straight lines passing through the origin with parameters θ_f. Furthermore, the parameter spacing should be directly proportional to the cake formation time.

A single straight line should be achieved if the horizontal axis is NW^2 or WV_w, where V_w equals the volume of cake wash per unit area per cycle (Dahlstrom and Silverblatt 1977). A typical plot is shown in Figure 9.25. Here, the straight-line relationship exists in the lefthand portion of the graph, but then the line shifts to a lesser slope. The line shifts because the liquor in the cake has a higher viscosity than the wash fluid; after a wash ratio of around 0.75 is achieved, the bulk of the original liquor has been displaced and a higher wash rate is possible.

Thus, by selecting a wash ratio that will achieve the desired recovery of soluble constituents, the full-scale rate is determined. The following equation should be used:

$$\text{Full-scale rate} = W\left(\frac{60}{\theta_w}\right)q_w \quad \textbf{(Eq. 9.56)}$$

where q_w = fraction of cycle time devoted to cake washing. This equation will determine the full-scale rate on the basis of weight of dry solids per hour per unit area. The rate of cake washing usually is

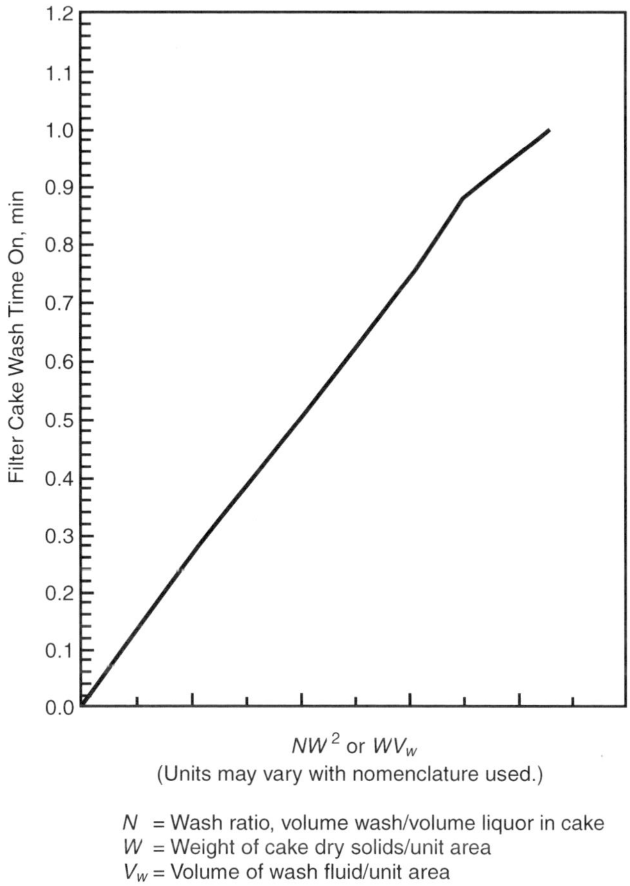

FIGURE 9.25 Filter cake wash time as function of NW^2

slower than the rates of cake formation and cake dewatering and, accordingly, must be properly designed into the cycle.

The cake washing fluid displacement rate can be determined from the Rhodes cake-washing equation. This equation can be simplified to the following logarithmic decrement function (Choudhury and Dahlstrom 1957):

$$\frac{\Sigma}{100}\left(1-\frac{E}{100}\right)^{N} \quad \text{(Eq. 9.57)}$$

where

Σ = percent of solubles remaining after cake washing based on 100% prior to wash

E = 100 − Σ at N = 1.0 (this value is termed the "wash displacement efficiency")

Equation 9.57 indicates a semi-log plot of the log of Σ as a function of N. Figure 9.26 is a typical plot; it contains a straight-line relationship down to an N value of around 1.0–1.5 for the three results shown. It then tends to become asymptotic to some minimum value. The right-hand portion of the curve likely results from blocked capillaries or channeling. These problems are common and indicate that wash rates above the range of 2.5 to 3.0 are influenced by the law of diminishing returns. Curve 1 in Figure 9.26 illustrates a low permeability cake that will usually have a high displacement efficiency. Curve 2 is typical of a majority of filter cakes. Curve 3 illustrates results for many high-permeability filter cakes.

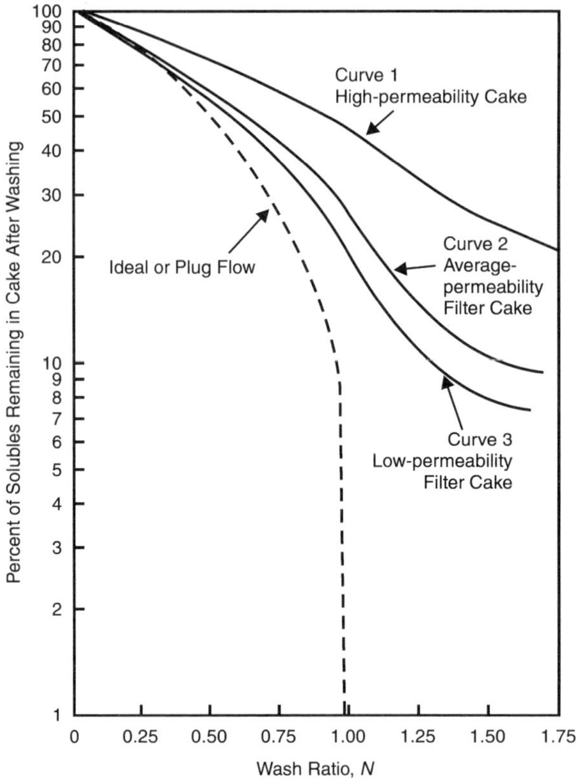

FIGURE 9.26 Percent of solubles remaining after cake washing as function of wash ratio

Values of E generally range from about 45% to 85%; most are in the range between 65% and 80%. Lower values tend to be associated with high-permeability cakes, which may tend to channel excessively.

Normally, a safety factor is built into an E determined from bench-scale tests. This safety factor allows for a somewhat uneven cake thickness (the greater amount of liquid passing through the thinner cake does not offset the lesser amount passing through the thicker cake) and any uneven distribution of wash fluid across the cake. Thus, five percentage points are subtracted from the bench-scale value of E.

The value of Σ can be determined from experimental data by the following equation (Dahlstrom and Silverblatt 1977):

$$\Sigma = \frac{z_1 - z_w}{z_0 - z_w}(100) \qquad \textbf{(Eq. 9.58)}$$

where

z_1 = solute concentration in washed cake liquor

z_w = solute concentration in washing fluid

It is now possible to make a complete material balance around a continuous filter using the Σ values and other data obtained during bench-scale testing. An example will be given to illustrate.

In the processing of bauxite for the manufacture of alumina (Al_2O_3), which is used to produce aluminum by electrolysis, $A1_2O_3 \cdot 3H_2O$ is produced by crystallization from liquor containing essential caustic constituents and dissolved alumina trihydrate. The alumina trihydrate must be recovered by filtering the feed at 80°C and washing the cake before calcining the alumina trihydrate to alumina in a rotary kiln. The feed to the filter contains 45 wt% suspended solids and has a concentration of 7.0% NaOH in the liquor. Before washing, the filter cake contains 15% moisture by weight. The liquor in the filter cake contains 5,400 mg/L NaOH after washing with 0.2 lb of wash fluid per pound of liquor in the cake, and the washing fluid contains 300 mg/L of NaOH. The final moisture content of the discharge cake is 9.5%. Determine the percentage of Na_2O in the final cake on a dry basis.

Basis: 1,000 lb of feed slurry

Pounds $Al_2O_3 \cdot 3H_2O$ = 0.45 (1,000) = 450

Pounds NaOH = 550 × 0.07 = 38.5

Pounds liquor/pounds $A1_2O_3 \cdot 3H_2O$ suspended solids in feed 550/450 = 1.222

$$\text{Pounds liquor/pounds } A1_2O_3 \cdot 3H_2O \text{ solids before wash} = \frac{15/0.93}{85 - 15(0.07/0.93)} = 0.192$$

$$\text{Percent recovery of NaOH by filtration alone} = \frac{1.222 - 0.192}{1.222} \times 100 = 84.29\%$$

$$\text{Pounds NaOH remaining in the cake before wash} = \frac{38.5(100 - 84.29)}{100} = 6.048$$

$$z_l = 5{,}400 \text{ mg/L NaOH}$$

$$z_o = 70{,}000 \text{ mg/L NaOH}$$

$$z_w = 300 \text{ mg/L NaOH}$$

$$\Sigma = \frac{5{,}400 - 300}{70{,}000 - 300}(100) = 6.40\%$$

Specific gravity 7% NaOH at 80°C = 1.046

$$N = \frac{0.20(450)}{0.15(450)/1.046} = 1,395$$

$$\frac{\Sigma}{100} = \left(1 - \frac{E}{100}\right)^N$$

$$E = 86.06\%$$

$$\text{Scale-up } E = 86.05 - 5 = 81.06\%$$

$$\Sigma = 9.81\%$$

But additional dewatering occurs from 15.0% to 9.5% final moisture:

$$\frac{\text{pounds liquor}}{\text{pounds solid}} \text{ at } 15.0\% = \frac{15}{85} = 0.1765$$

$$\frac{\text{pounds liquor}}{\text{pounds solid}} \text{ at } 9.5\% = \frac{9.5}{90.5} = 0.105$$

Percent remaining from dewatering = 0.105/0.1765 × 100 = 59%–49%

Final NaOH weight = 6.048 × 0.0981 × 0.5949 = 0.35 lb

Figure 9.27 is a plot of filtration rate as a function of filter cycle time for a continuous-vacuum drum filter with cake washing. The wash ratio and cake thickness are specified for this acid-leached uranium ore. It is immediately apparent that increasing the wash ratio for a given cake thickness reduces the filtration rate markedly. However, the increased recovery of soluble constituents more than offsets the increased equipment cost. Most drum filters with cake washing that treat uranium-leached slurries were designed with an approximate 1.5 wash ratio; this ratio then usually controls the filtration rate.

Scale-up

As we can see from the previous discussion of applied theory, it is possible to determine performance results over a range of operating filtration rates and operating conditions identified in bench-scale testing. However, bench-scale predictions must still be scaled up to full-scale operations that allow for

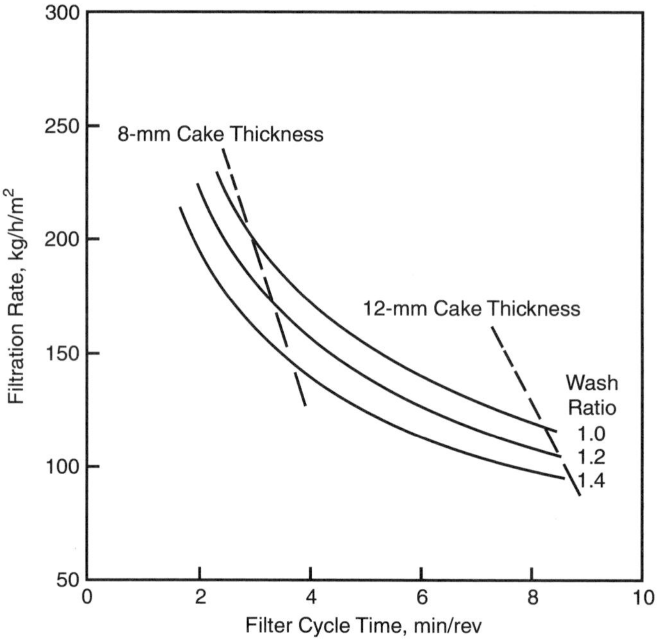

FIGURE 9.27 Filtration rate as function of filter cycle time for a continuous-vacuum drum filter

normal feed quality fluctuations as well as the approximate ideal situation of the bench-scale tests. It is safe to use an 80% scale-up factor for filtration rate. However, this factor assumes that the media blinds very slowly and that the quality of the feed changes by no more than 10%.

Bench-scale tests can be run so as to yield results at typical operating conditions. These tests permit a much wider investigation, at lower cost than a pilot-plant test, because a much broader suite of operating conditions can be investigated. For example, the ratio of θ_d/θ_f and θ_w/θ_f can be varied a wide range, as is also true of N.

Limitations of the various types of filters must be carefully considered. Table 9.2 lists the basic types of continuous filters; the percent of cycle time that can be devoted to cake formation, washing, and dewatering; and the minimum cake thickness that allows good cake discharge. Filters that form their cake with gravity are more flexibile in the use of the cycle time for the various phases of cake formation, dewatering, and washing so that the area is used more effectively. The filtration rate of a horizontal belt filter is based only on active area (area under vacuum). In other filters, it is based on total area and includes cake discharge and dead time.

General Installation Requirements

The filters must be installed in sufficient area to permit normal maintenance for the type of machine used. The equipment should allow the ratio of cake formation, dewatering, and washing to be changed in a reasonable amount of time. The time required to change filter media should be minimized. Connections from the filter valve to the filtrate receiver should be maintained separately, if possible, to minimize pressure drop. Moisture traps should normally be employed for the receiver overhead to the vacuum pump, particularly if the filtrate is valuable or is high in acid or alkali. The traps also protect the vacuum pump. Any centrifugal filtrate pumps should be installed to eliminate air binding. Consideration must also be given to sparing of filtrate and vacuum pumps and to OSHA requirements.

TABLE 9.2 Typical equipment factors for continuous vacuum filters, standard designs

Filter Type	Submergence %* Apparent	Submergence %* Effective	Total % Under Vacuum	Max. % Cake Wash	Max. % Dewatering Only	% Required for Cake Discharge	Minimum Cake Thickness in.	Minimum Cake Thickness mm
			Filters Forming Cake Against Gravity					
Drum								
Scraper	35	30	80	29	50–60	20	1/4	6.4
Roll 1	35	30	80	29	50–60	20	1/32	0.8
Belt	35	30	75	29	45–55	25	1/8–3/16	3.2–4.8
Precoat	35	35	100	29	50	0	0	0
Disk	35	28	75	NA†	45–50	25	3/8–1/2	9.5–12.7
			Filters Forming Cake with Gravity					
Horizontal belt as required			100*		As required	0	1/8–3/16	3.2–4.8
Horizontal table as required			80		As required	20	3/4	19.1

*Horizontal belt filter is based only on effective area (area under vacuum).
†NA = not available.

BATCH PRESSURE FILTERS

It is considerably more difficult to develop an applied theory of batch pressure filters that fits operational results. The major problem is the lack of homogeneity in the filter cake. Consider a feed that contains solids that can settle. In a vertical filtration area, some of these solids will settle within the filter. In a plate frame or recessed plate filter, these solids can settle when the upflow velocity is less than the settling velocity. On a recessed plate below the center feed inlet, settling can occur at the very start of filtration. With the plate and frame filter, settling may start very early also in the central area and at the base of the frame. Thus, it is possible to have a cake-specific resistance that is different–and changing–at all points within the filter.

This situation is further complicated by the fact that most of the solids filtered in pressure filters are compressible, meaning that the specific resistance of a homogeneous cake is still changing with pressure. Most theories of batch pressure filters are based on forming a homogeneous cake, but such a cake may not form in feed slurries containing solids that can settle. Still, this situation can be reasonably approximated unless very serious settling tendencies are encountered. This qualification emphasizes that such filters are best operated with high concentrations of feed solids such as those obtained from thickener underflows.

Basic Theory

The Poiseuille equation is also employed in pressure filtration. In this case, it is usually rearranged to the following form:

$$\frac{d\theta}{dv} = \frac{\mu\alpha wV}{A^2\Delta p} + \frac{\mu r}{\Delta p} \quad \textbf{(Eq. 9.59)}$$

or it may be used in the differential form:

$$\frac{\mu\alpha wVdV}{A^2\Delta p} + \frac{\mu r dv}{\Delta pA} = d\theta \quad \textbf{(Eq. 9.60)}$$

Applied Theory

Application of the theory is normally studied from several operational vantage points: constant-pressure operation; constant-rate operation; and variable-pressure, variable-rate operation. Equation and correlation methods for each operation follow.

Constant-pressure Operation

Equation 9.60 can be integrated for constant-pressure operation if we assume constant Δp. This assumption theoretically makes μ, α, w, A, and r constant, as long as the feed quality and the temperature are constant. Accordingly, integrating yields

$$\frac{\mu \alpha w V^2}{2A^2 \Delta p} + \frac{\mu r V}{\Delta p A} = \theta \qquad \textbf{(Eq. 9.61)}$$

Dividing through by V and multiplying by A, the following equation is obtained:

$$\frac{\theta}{V/A} = \frac{\mu r w V}{2\Delta p A} + \frac{\mu r}{\Delta p} \qquad \textbf{(Eq. 9.62)}$$

A plot of $\theta V/A$ against V/A should yield a straight-line relationship whose intercepts should yield the following:

$$\text{slope} = \frac{\mu \alpha w}{2\Delta p} \qquad \textbf{(Eq. 9.63)}$$

$$\text{intercept} = \frac{\mu r}{\Delta p} \qquad \textbf{(Eq. 9.64)}$$

As μ, w, and Δp are easily measured, it is then possible to calculate μI and r. However, this calculation yields the value of α only at a specific pressure drop. A strictly empirical relationship that has been assumed by many investigators is

$$\alpha = \alpha_o (\Delta p)^n \qquad \textbf{(Eq. 9.65)}$$

where

α_o = specific resistance at $\Delta p = 1$
n = exponent whose value is between zero and one

Thus, if several constant-pressure runs are made, it may be possible to develop an equation for α that would be reasonably representative.

Another way to employ α over a range of pressure is to use an average value of α_{av}. This value can be determined by the following equation (Osborne 1981):

$$\frac{1}{\alpha_{av}} = \frac{1}{\Delta p}\int_0^{\Delta p} \frac{d\Delta p}{\alpha} = \int_0^{\Delta p} \frac{d\Delta p}{\alpha_o (\Delta p)^n} \qquad \textbf{(Eq. 9.66)}$$

This expression yields

$$\alpha_{av} = (1-n)\alpha_o \Delta p^n \qquad \textbf{(Eq. 9.67)}$$

Constant-rate Operation

For an incompressible cake, Eq. 9.44 can be modified to

$$\frac{dv}{d\theta} = \text{constant} = Q = \frac{\Delta p A}{\mu \alpha w \frac{V}{A} + \mu r} \qquad \textbf{(Eq. 9.68)}$$

where Q = a constant, underpressure filtration when constant rate filtration is employed, volume filtrate per unit time.

Because $V = Q\theta$, $Q\theta$ can be substituted into Eq. 9.68 as follows:

$$\Delta p = \frac{\mu \alpha w Q^2 \theta}{A^2} + \frac{\mu r Q}{A} \qquad \textbf{(Eq. 9.69)}$$

Thus, if Δp is plotted as a function of θ, a straight line would result.

$$\text{slope} = \mu\alpha p\frac{Q^2}{A^2} \qquad \text{(Eq. 9.70)}$$

$$\text{intercept} = \mu rQ/A \qquad \text{(Eq. 9.71)}$$

Again, both r and α can be determined because μ, w, and Q/A are constant and measurable.

To apply the theory to compressible solids, the following relationships are assumed (Osborne 1981):

$$\Delta p = \Delta p_c \Delta p_m \qquad \text{(Eq. 9.72)}$$

where

Δp_c = pressure drop across the cake, lb/in^2

Δp_m = pressure drop across the media, lb/in^2

Also,

$$\Delta p_m = \frac{\mu r dV}{Ad\theta} \qquad \text{(Eq. 9.73)}$$

and

$$\Delta p_c = \frac{\mu\alpha wV}{A^2}\frac{dV}{d\theta} \qquad \text{(Eq. 9.74)}$$

Substituting Eq. 9.67 into Eq. 9.74,

$$\Delta p_c = (1-n)\alpha_o \Delta p_c^{\ n} \mu\alpha w\frac{VdV}{A^2 d\theta} \qquad \text{(Eq. 9.75)}$$

This equation can be simplified to the following:

$$\frac{\mu wVdV}{A^2 d\theta} = \frac{(\Delta p_c)^{1-n}}{(1-n)\alpha_o} \qquad \text{(Eq. 9.76)}$$

and because $dV/d\theta = Q$ and $V = Q\theta$, it can be written

$$(\Delta p_c)^{1-n} = \alpha_o(1-n)\mu w\frac{Q^2}{A^2}\theta \qquad \text{(Eq. 9.77)}$$

Accordingly, a plot of the log of Δp against logθ should yield a straight line, because

$$\log \Delta p_c = \frac{1}{1-n}\log\theta + \frac{\log[\alpha_o(1-n)\mu wQ^2/A^2]}{1-n} \qquad \text{(Eq. 9.78)}$$

where

$$\text{slope} = \frac{1}{1-n} \qquad \text{(Eq. 9.79)}$$

and

$$\text{intercept} = \frac{\log[\alpha_o(1-n)\mu wQ^2/A^2]}{1-n} \qquad \text{(Eq. 9.80)}$$

Thus, both n and α_o can be determined for compressible solids with the assumptions made.

Variable-rate, Variable-pressure Operation

Considering first the incompressible solids, Eq. 9.44 can be altered to

$$V = \frac{A}{\mu\alpha w}\left(\frac{\Delta pA}{Q} - \mu r\right) \qquad \text{(Eq. 9.81)}$$

and because $Q = dV/d\theta$, then

$$\theta = \int_0^v \frac{dV}{Q} \qquad \textbf{(Eq. 9.82)}$$

Because many pressure filters are fed by a centrifugal pump that has a characteristic curve of Δp as a function of capacity or the value Q, a plot of $1/Q$ as a function V can be developed. Various values of Q from the characteristic curve are used in Eq. 9.81, along with the known values of A, μ, and w. The term Δp is obtained from the characteristic curve at the assumed values of Q. Finally, α and r have already been determined from a test as indicated earlier. Thus, by integrating the area under the curve to the desired amount of filtrate, the time θ is determined.

Compressible cakes are considered next. The terms Δp_c, Δp_m, V, Q, and θ are all variables, and if it is assumed that Eqs. 9.72, 9.76, and $dV/d\theta = Q$ still apply, the following equation can be developed (Osborne 1981):

$$V = \frac{A^2}{(1-n)\alpha_o \mu w} \frac{(p - \Delta p_m)^{1-n}}{Q} \qquad \textbf{(Eq. 9.83)}$$

The values of n, α_o and r can be determined by methods indicated earlier. Equation 9.72 can be modified to

$$\Delta p_m = \frac{\mu r Q}{A} \qquad \textbf{(Eq. 9.84)}$$

From the centrifugal pump characteristic curve, various values of Q can be assumed that also yield values of Δp. Because g, n, α_o, and A are constant, it is possible to solve for Δp_m at various values of Q. Accordingly, $1/Q$ versus V can be plotted, and the area under the curve integrated to the desired value of V will yield the required filtration time.

One set of units must be employed consistently, either English or metric (Système International). Also, to the filtration time must be added the time for discharge of the filter plus closing of the unit and filling the unit so that the cycle can be completed.

Scale-up

Again, a scale-up factor should be applied, in this case ranging from 70% to 80%. This factor accounts for changes in feed quality and scale-up from very small-scale equipment. Binding of the filter cloth may be avoided or reduced by using high-pressure sprays to clean the media. The sprays can be purchased with the filter and the spraying done manually.

General Installation Requirements

The complexity of batch pressure filters ranges from very simple machines with all manual operations and no controls to highly automated units with controls and means to reduce variables such as time to cake discharge and blinding of media. The recessed plate or plate and frame filters may be purchased with the following options:

- Opening and closing of filter press (manual or hydraulic)
- Moving of plates and frames for discharge (manual plate shifters or multiple plate shifters)
- Filtration, cake blowing (manual or automatic)
- Cloth wash, manual or automatic

These decisions are essentially economic ones.

If flocculation is to be used, as, for example, to dewater tailings, remember that flocculi generally deteriorate with time and mixing. Also, the filtration rate may vary with time, so the flocculation rate may not equal the filtration rate. Usually, a flocculation tank precedes the pump so that flocculation can be performed in batches, or so that the flocculation rate can be matched to the filtration rate. Because the flocculant dosage is proportional to the solids concentration, batch flocculation is usually favored. Flocculation normally is promoted by a positive displacement type of pressure pump, which greatly reduces hydraulic shear. In any event, detention time after flocculation should normally be less than 30 min, and 15 min may save operating costs to offset other costs.

REFERENCES

Anonymous. 1963. *Selection of Vibrating Screens.* Milwaukee, Wisc.: Allis-Chalmers.

Baczek, F.A. et al. 1997. Sedimentation. In *Handbook of Separation Techniques for Chemical Engineers,* 3rd ed. Edited by P.A. Schweitzer. New York: McGraw-Hill.

Bosley, R. 1974. Vacuum Filtration Equipment Innovation. *Filtr. Sep.,* 11:138–149.

Chironis, N.P. 1976. New Clarifier/Thickener Boosts Output of Older Coal Preparation Plant. *Coal Age,* 81:140–145.

Choudhury, A.P., and D.A. Dahlstrom. 1957. Prediction of Cake Washing Results with Continuous Filtration Equipment. *Chem. Eng. J.,* 3:23:433–438.

Coe, H.S., and G.H. Clevenger. 1916. Methods for Dewatering the Capacities of Slime Settling Tanks. *Trans. AIME,* 55:356–384.

Cross, H.E. 1963. A New Approach to the Design and Operation of Thickeners. *J.S. Afr. MM.,* 63:271–298.

Dahlstrom, D.A. 1978. Practical Use of Applied Theory of Continuous Filtration. *Chem. Eng. Prog.,* 74 (Mar.):69.

———. 1980. *How to Select and Size Filters, Mineral Processing Plant Design*. New York: AIME.

———. 1985. Filtration. In *SME Mineral Processing Handbook*. Vol. 1. Edited by N.L. Weiss. New York: AIME.

Dahlstrom, D.A., and R.C. Emmett. 1983. Recent Developments in Gravitational Sedimentation. In *Proceedings of the Third Pacific American Chemical Engineering Congress,* Seoul, Korea.

Dahlstrom, D.A., and C.E. Silverblatt. 1977. Continuous Vacuum and Pressure Filtration. In *Solid–Liquid Separation Equipment Scale-Up.* Edited by D.B. Purchas. Croyden, England: Uplands Press.

Hassett, N.J. 1969. Thickening in Theory and Practice. *Miner. Sci. Eng.,* 1:24–40.

Henderson, A.S., C.F. Cornell, A.F. Dunyon, and D.A. Dahlstrom. 1957. Filtration and Control of Moisture Content on Taconite Concentrates. *Mining Eng.,* March.

Hitzrot, H.W., and G.M. Meisel. 1985. Mechanical Classifiers. In *SME Mineral Processing Handbook*. Vol. 1. Edited by N.L. Weiss. New York: AIME.

Hsia, E.S., and F.W. Reinmiller. 1977. How to Design and Construct Earth Bottom Thickeners. *Min. Eng.,* 29:36–39.

Kobler, R.W., and D.A. Dahlstrom. 1979. Continuous Development of Vacuum Filters for Dewatering Iron Ore Concentrates. *Trans. SME-AIME,* 266:2015–2021.

Kynch, G.J. 1952. Theory of Sedimentation. *Trans. Farady Soc.,* 48:66–176.

Nelson, P.A., and D.A. Dahlstrom. 1957. Moisture Content Correlation of Rotary Vacuum Filter Cakes. *Chemical Engineers' Progress,* 53:320–327.

Osborne, D.G. 1975. Scale-up of Rotary Vacuum Filter Capacities. *Trans. IMM,* 84:C158–C166.

———. 1981. Gravity Thickening. In *Solid–Liquid Separation*. 2nd ed. Edited by L. Svarovsky. London: Butterworths.

Perry's Chemical Engineer's Handbook. 6th ed. 1984. Edited by R.H. Perry and D. Green. New York: McGraw-Hill, p. 5-67.

Roberts, E.J. 1949. Thickening, Art or Sciences. *Trans. AIME,* 184:61.

Robins, W.H.M. 1964. The Theory of the Design and Operation of Settling Tanks. *Trans. Inst. Chem. Eng.,* 42:T158–T163.

Rushton, A. 1978. Design Throughputs in Rotary Disc Vacuum Filtration with Incompressible Cakes. *Powder Technol.,* 21:161–169.

Sandy, E.J., and J.P. Matoney. 1979. Mechanical Dewatering. In *Coal Preparation*. 4th ed. Edited by J.W. Leonard. New York: AIME.

Sakiadis, B.C. 1984. Fluid and Particle Mechanics. In *Perry's Chemical Engineering Handbook*. 6th ed. Edited by R.H. Perry and D. Green. New York: McGraw-Hill.

Scott, K.J. 1970. Continuous Thickening of Flocculated Suspensions. *Ind. Eng. Chem. Fundam.*, 9:422–427.

Silverblatt, C.E., H. Risbud, and F.M. Tiller. 1974. Batch, Continuous Processes for Cake Filtration. *Chem. Eng.*, 81:127–136.

Sweitzer, X., ed. 1979. *Handbook of Separation Techniques for Chemical Engineers*. 3rd ed. New York: McGraw-Hill.

Talmadge, W.P., and E.B. Fitch. 1955. Determine Thickener Unit Areas. *Industrial and Engineering Chemistry*, 47:38–41.

Terchick, A.A., D.T. King, and J.C. Anderson. 1975. Application and Utilization of the Enviro-Clear Thickener in a U.S. Steel Coal Preparation Plant. *Trans. SME-AIME*, 258:148–151.

Weber, F.R. 1977. How to Select the Right Thickener. *Coal Min. Process.*, 14:98–104.

Wetzel, B. 1974. Disc Filter Performance Improved by Equipment Redesign. *Filtr. Sep.*, 11:270–274.

Whitmore, R.L. 1957. The Relationship of the Viscosity to the Settling Rate of Slurries. *J. Inst. Fuel*, 30:238–242.

Wilhelm, J.H., and Y. Naide. 1981. Sizing and Operating Continuous Thickeners. *Mining Engineering*, 1710–1718.

Wolf, J., et al. 1971. Present Methods and Future Needs in Iron Concentrate Dewatering the Process Water Reclamation. Paper presented at Annual Meeting, SME-AIME. March 1–4.

CHAPTER 10

Metallurgical Balances and Efficiency

J. Mark Richardson and Robert D. Morrison

To properly design, control, and optimize mineral processing plant circuits, the mineral processing engineer must have the means to adequately measure process performance. Metallurgical balances provide an absolute measure of plant performance, while efficiencies usually provide a comparative or relative indication of performance for individual unit operations within the plant. This chapter explains metallurgical balances and efficiencies, discusses their various applications, and presents information on the basics of how to calculate them.

TERMINOLOGY

Before the main issues of this chapter are addressed, a discussion of the primary terms is in order.

Metallurgical Balance

"Metallurgical balance" is the term usually applied to the overall accounting of material and energy entering and leaving a metallurgical process. With the notable exception of hydrometallurgical processes and operations, such as ore roasting, most common mineral processing operations (comminution, classification, liquid–solid separation, concentration, etc.) are primarily physical in nature and do not require energy balances. For that reason, the discussion in this chapter will be limited to material balances.

Most metallurgical balances are performed under the assumption that the system being balanced is at steady state, where total mass input is equal to total mass output. At steady state, there is no internal accumulation of material; all unit operations are assumed to be functioning with a constant amount of material present. Typically, a metallurgical balance around a mineral process may apply to any or all of the following measurable characteristics of materials processed:

- Total material, or "pulp" (solids plus fluids)
- Individual material phases (solids, fluids, or gases)
- Individual "assayable" chemical components of the various phases
- Individual mineral components of the solid phase
- Individual particle size fractions of the solid phase
- Changes in accumulation of material (in those cases where the accumulation cannot be avoided)

Process Efficiency

There are many traditional measurements of process efficiency. For base metals, the terms "concentrate grade" and "recovery" (i.e., the percentage composition of desired metal in the concentrate and the percentage of this metal in the plant feed that reports to the concentrate, respectively) are in common use. Recovery is also favored for precious metals, but with bullion, product grade is not usually an

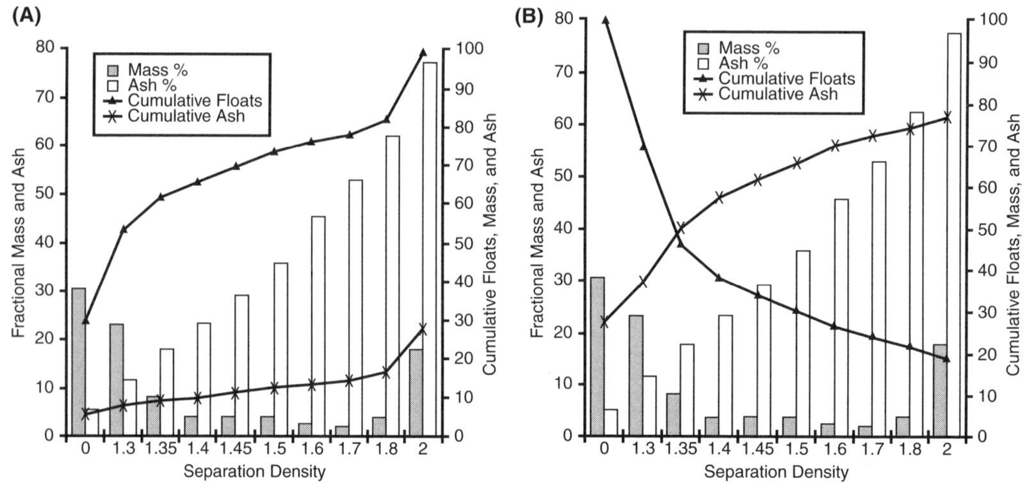

FIGURE 10.1 Cumulative "floats" curve (A); cumulative "sinks" curve (B)

issue. None of these terms address the question: How do we know that we did as well as we might have done for this particular parcel of ore?

Industrial minerals, such as iron ore and coal, offer a useful way of considering process efficiency: the "washability" curve. Coal and (some) iron ores are essentially binary mixtures of product and waste with different densities. Hence, the density of each ore particle can be measured, as well as its grade (i.e., ash percent), and sorted in terms of density. This curve can be plotted cumulatively in either separation direction ("floats" or "sinks") as shown in Figures 10.1A and 10.1B. The bars in these figures show the fractional mass and ash measured. The lines (and right axes) show cumulative values obtained by adding in each direction.

With this base information, we can carry out a "perfect" separation at any density and compare this separation with an actual process. A separation (or partition) curve can be plotted as shown in Figure 10.2. A perfect separation would provide a "Z" shape. Real separations provide an "S"-shaped curve, which is often characterized in terms of the slope of the curve at the point where 50% of the feed reports to each product. This slope is usually defined as the gradient between the points of 25% (of feed) reporting to product and 75% (of feed) reporting to product; it is usually referred to as the "Ecart probable" or, more commonly, "the error"–i.e., the density difference between the 25% and 75% points.

If there are very many particles at or close to the separation density, the ore can be considered to be difficult; that is, critically dependent on the separation accuracy of the process. If there are very few particles around the target density, the separation is said to be easy. In an example of technical irony, the "easy" separation is very difficult to assess by sampling because there are few particles in the range of interest–hence, tracers that mimic the behavior of these "rare" particles provide a practical approach (Davis, Wood, and Lyman 1987). For these reasons, no sale contract for a commodity product should ever be written without a good knowledge of ore washability. For a detailed discussion of the application of washability curves to coal, see Partridge (1994).

In the past, it has been necessary to use separation curves (Dell 1961) to assess other processes. A separation curve such as in Figure 10.2 considers recovery of desired metal against recovery of feed mass to concentrate. Such curves provide a simple, generic separator model. The area between the separation curve and the diagonal provides an indication of separation achieved by the process. These models can easily be computerized by fitting a cubic spline curve through the measured points.

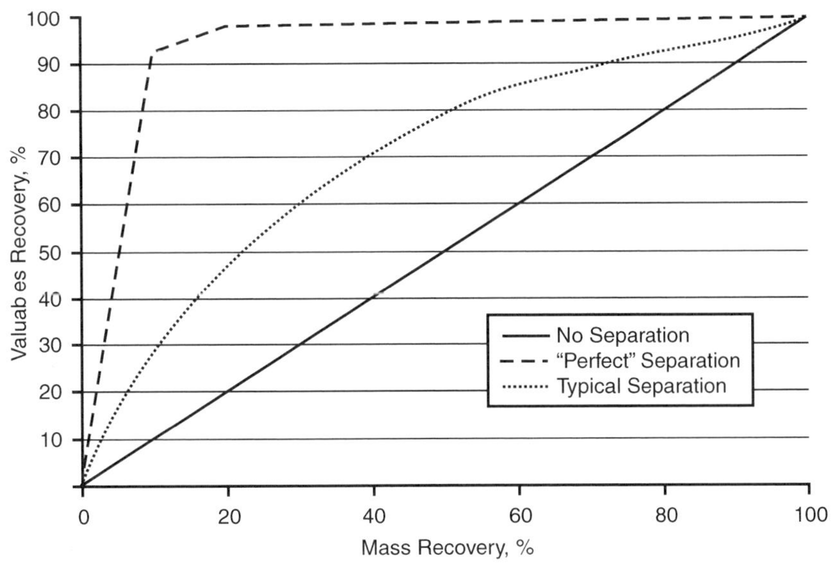

FIGURE 10.2 Separation curve

Modern image analysis techniques offer an opportunity to generalize the washability model. In other words, we may characterize particles at a particular stage of comminution (or even by texture in the ore) to develop a theoretical separation (or washability) curve that ranks the particles in terms of proportion of valuable material. The raw image data can be corrected for stereological bias to provide this curve (Gay and Lyman 1995). This means that we can start with the minerals that are completely liberated valuable material (or as close to it as possible) and then rank each of the other particles in terms of its proportions of this valuable mineral. An actual process can then be compared with this curve to provide an "absolute" measure of separation efficiency.

To consider the potential for preconcentration, we would start at the "sinks" end of the washability curve (see Figure 10.1B). An additional dimension can be considered by carrying out the same analysis for different size fractions to ascertain an optimum grind size. It should be noted that a grind for maximum gangue rejection may be very different from that required to maximize liberation of valuables.

Economic Efficiency

The term "economic efficiency" considers the economic return from each unit of ore in the deposit. In an idealized case, a high economic efficiency might correspond only to achieving the maximum recovery of metal units from a deposit. In the real world, we must also account for operating costs, smelter contracts, contaminant penalties, transport costs, royalties, and so on. Attempting to recover all of the metal in an orebody would provide a short route to bankruptcy.

In general, the broader the scope of an economic evaluation, the more difficult the task; that is, assessing the economics of an entire "orebody to market" scenario is substantially more difficult than defining the economic optimum for a concentrator treating a defined parcel of ore for sale via a particular smelter and shipping contract. If we can plot a separation curve for our concentrator, maximizing economic efficiency is equivalent to finding the operating point on this curve that provides the maximum net revenue from the ore parcel.

The presence of constraints on throughput/ore availability (i.e., having a goal of maximizing recovery) may result in quite different "best" operating points than the absence of such constraints (i.e., having a goal of maximizing throughput; Morrison 1993). The overall optimization requires a sophisticated mine model combined with a realistic process model. This technology is not yet fully developed, but good progress is being made (Whittle and Vassiliev 1998). More simply, optimizing the comminution process from blasting to grinding may produce a substantial increase in economic efficiency. Thus, estimating economic efficiency is the more complex process and it may yield misleading results if the scope is too narrow or (usually with much worse consequences) based on traditional or political perceptions of technical efficiency or resource "maximization."

APPLICATIONS

Metallurgical balances have wide application in mineral processing. In fact, the need to perform metallurgical balances is pervasive throughout the field. Among the most important applications are in

- Process design
- Metallurgical accounting
- Process optimization
- Process control

Process Design

During process design, engineers may calculate material balances for individual unit operations, specific circuits within a plant, or an entire process flowsheet. The primary focus of process design is usually to specify equipment and determine costs (both capital and operating); therefore, the main objectives of performing a design balance usually include:

1. Evaluation and comparison of alternative process flowsheets
2. Determination of individual unit operation equipment capacity
3. Specification of duty requirements for pumps, conveyors, pipes, and other material transport equipment
4. Determination of process surge and storage capacity requirements (bins and stockpiles)
5. Determination of process operating costs in the form of media consumption, water, air, reagents, power, etc.

Material balances also result in useful, indirect information, such as specific gravities of solids and liquids, solids bulk densities, and slurry densities, all of which are necessary for proper equipment design.

Unrealistic or superficial balances can result in capacity bottlenecks, as well as excess process capacity. Designers strive to maintain an equilibrium between minimizing bottlenecks (which result in lost revenue) and avoiding overcapacity (which results in excess capital expenditure).

Metallurgical Accounting

Metallurgical accounting deals with measuring the economic well-being of existing operations; it is primarily concerned with estimating total material processed and total valuable metal(s) produced, lost to waste, and held up in inventory. It may also deal with tracking (usually harmful or costly) by-products and consumables, such as media, reagents, and power. Metallurgical accounting is most often a scheduled activity, occurring at regular intervals. It is useful in the following types of economic assessments:

1. Life-of-mine estimation
2. Cash flow forecasting/budgeting
3. Return-on-investment (ROI) calculations
4. Valuation of assets

5. Depletion calculations
6. Taxation calculations

In addition, metallurgical accounting may be of use in the following technical contexts:

1. Production planning
2. Resource allocation
3. Process optimization

Process Optimization

Throughout their operating life, most mineral processing plants do not remain static. Engineers are constantly striving to improve either throughput or recovery and to reduce operating costs while working to overcome design shortcomings, take advantage of new technology, and cope with changes in ore feed characteristics and metal prices. Processes cannot be improved, however, unless they are measured and understood, and this cannot happen without first performing metallurgical balances and measuring process efficiency. Additionally, proposed process alternatives must be tested and measured to determine if they represent improvement over the existing process; once again, balances must be performed. Traditionally, testing of alternatives has taken place in pilot plants and even in the full-scale plants themselves. Recently, however, there has been an increased reliance on the use of computerized process models to compare and eliminate process alternatives before proceeding to the engineering design step with the most promising modifications. In either case, calculation of material balances and measurement of process efficiency are essential and unavoidable. The optimization process is also iterative (as discussed in more detail later in this chapter).

Process Control

Computerized process control has become an accepted fact of life in modern mineral processing, contributing to ever more stable and cost-effective operation in the face of often unpredictable changes in ore feed and other external influences on the operation. In fact, the end result of process control in many cases is to optimize process performance as far as possible without physically modifying the plant. However, a process cannot be controlled and optimized unless it is first measured. Once again, measurement means performing a metallurgical balance and determining current efficiencies. The extent of the underlying balance required depends on the nature of the control system; individual unit operations may be controlled by simple, programmable logic control (PLC) loops, or an entire plant may be under the supervision of an "expert" supervisory control system.

For well-instrumented concentrators, the use of on-line mass balancing to check on the precision and reliability of on-line measurements and to estimate unmeasurable streams is not unusual. For example, an iron ore concentrator that uses a different separation process for each size range will often have a single product belt and a single reject belt. Belt scales after each addition of product or reject can give a useful indication of mass yield from each stage of the process, provided the weigh scales are kept well calibrated.

A more interesting example is an on-line mass balance of a well-instrumented flotation circuit that has a comprehensive on-stream analysis system. For this case, a metallurgical accounting software package known as JKMetAccount (developed by JKTech, the commercial arm of the Julius Krutschnitt Mineral Research Centre of Indooroopilly, Queensland, Australia) is used to provide both on-line mass balancing and off-line metallurgical accounting.

In a more general context, mass balancing and optimization should be used to set the supervisory objectives for the process control system. Otherwise, the process control system may prevent the concentrator from ever operating at its economic optimum.

TYPES OF BALANCES

Although it is apparent that metallurgical balances are suited to a wide range of applications, only two classes of such balances actually exist: (1) nonmodel-based and (2) model-based. Engineers and operators must be sure to understand the differences between these two types of balances and when to apply each.

Nonmodel-based Balances

A nonmodel-based balance can also be called a "conservation of matter" balance (i.e., what goes out is equal to what goes in). A key property is that the process is reversible; that is, we can estimate the feed to a node from the node's products. This property does not hold true for many models used in model-based balancing (for example, models of the comminution process), which are essentially "one way," meaning that we cannot estimate the feed to a node from that node's products.

A nonmodel-based balance is also known by other expressions, including "mass balancing," "data reconciliation," and "data adjustment." The objective is to obtain a consistent material balance around a unit operation or a complete process flowsheet by statistical adjustment of measured process flowstream data. The resulting final balance hopefully represents the best estimate based on available data (which is almost never consistent) of the true process balance, which, of course, can never be exactly known.

Achieving the final balance does not require complex mathematical models of unit operations (or of the process, for that matter) that impose experience knowledge. To impose experience knowledge means that the model must respond in a specified manner to a specific set of feed and operating conditions; this is usually based on the *experience* of empirical observations of the process under varying conditions. This *knowledge* of the process response helps to determine the nature of the model. In nonmodel-based balances, unit operations are represented as nodes, connected by a network of process flowstreams. Simple equations define the balance of material around each node. There is no predictive aspect to this type of balance. The solution of the balance is not dependent in any way on depicting the nature of the physical or chemical processes occurring within the unit operations. Rather, the balance is defined by the measured data, which should have been extensive enough (or, as explained later, gathered in enough redundancy) to more than satisfy the balance equations.

Degrees of Freedom. The concepts of consistent data and redundant data are important in the context of metallurgical balances. To understand these notions, we must first have a good grasp of what the term "degrees of freedom" means. Murrill (1967) provides a particularly good discussion, along with this definition:

$$df = v - e \quad \textbf{(Eq. 10.1)}$$

where

df = number of degrees of freedom

v = number of variables that describe the system

e = number of independent relationships or equations that exist among these variables

To completely specify a system, all degrees of freedom must be removed; that is, df in Eq. 10.1 must be made equal to zero. This is the natural objective in achieving any balance. However, if too many degrees of freedom are removed (df becomes negative), the system is said to be overspecified and has no feasible solution. Conversely, if there are too many degrees of freedom (df is positive), the system is said to be underspecified and an infinite number of solutions to the balance exist.

Consider a simple balance around a rougher float cell, as shown in Figure 10.3. For the purposes of this example, assume we are dealing only with the solids phase. The total solids flows (in tons per hour) for each of the three streams are represented by M_1, M_2, and M_3, respectively. There are two assayable components of interest, A and B. The concentrations of each component in the streams, as a percentage,

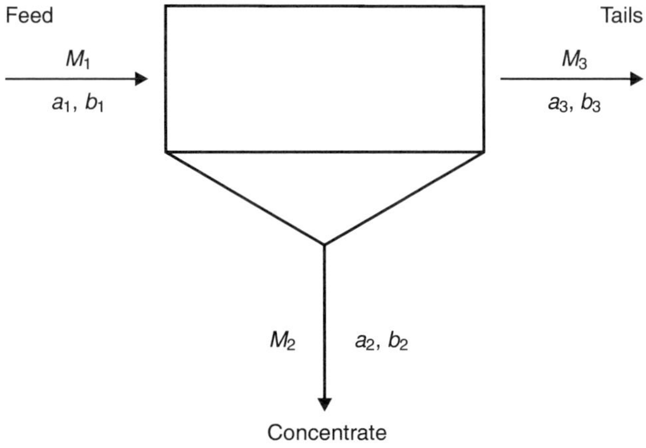

FIGURE 10.3 Rougher float cell

are given as a_1, a_2, and a_3, and as b_1, b_2, and b_3. Three independent relationships (equations) can be written to define the balances of total solids and the two components around the rougher cell:

$$M_1 = M_2 + M_3 \quad \text{(Eq. 10.2)}$$

$$a_1 \cdot M_1 = a_2 \cdot M_2 + a_3 \cdot M_3 \quad \text{(Eq. 10.3)}$$

$$b_1 \cdot M_1 = b_2 \cdot M_2 + b_3 \cdot M_3 \quad \text{(Eq. 10.4)}$$

At this stage we have three equations and nine unknowns. If we go to the plant and measure the flow rate and concentrations of A and B in the feed only, we will have three equations and six unknowns; per Eq. 10.1, the number of degrees of freedom becomes

$$df = 6 - 3 = 3 \quad \text{(Eq. 10.5)}$$

Hence, the system is underspecified, and an infinite number solutions are possible; the split of solids and components between the concentrate and tails has not been defined and could be any value. If we now go back to the plant and measure total solids flow and concentrations of A and B in one of the two product streams (such as the concentrate, for example) in addition to the feed, we will have three equations and only three unknowns, so that

$$df = 3 - 3 = 0 \quad \text{(Eq. 10.6)}$$

Now the system and the balance are exactly specified. The total solids flow and the concentrations of A and B in the tails can be solved from Eqs. 10.2, 10.3, and 10.4. However, if this is as far as we take the balance, we are assuming that the measurements taken of the feed and the concentrate streams are completely accurate and that the calculated values for the tails are correct as well. Unfortunately, in the real world, this is never true, usually because of (1) experimental error in both sampling and assaying and (2) the fact that the system is unlikely to be at perfect steady state when the measurements were taken. Hence, if we want the best possible estimate of the true values of flow rates and concentrations that satisfy the balance, we need to take an extra step. We must measure the flow and concentrations for all three streams, giving us three equations and no unknowns, so that the number of degrees of freedom becomes negative:

$$df = 0 - 3 = -3 \quad \text{(Eq. 10.7)}$$

In this case, the data are now said to be "redundant." Actually, measuring only the feed stream flow rate and the assays of A and B in all three streams would be sufficient (leaving only M_2 and M_3 to be calculated). This would likewise yield a negative value for the number of degrees of freedom:

$$df = 2 - 3 = -1 \quad \textbf{(Eq. 10.8)}$$

When we have a *complete* set of redundant data (as in the example of Eq. 10.7, where there are no unknowns because all values have been measured or can be inferred from the equations and measured values), we can then substitute our measured data into Eqs. 10.2, 10.3, and 10.4 and test to see if the equations hold. If all the equations are true, given the measured values, the data are said to be "consistent." However, in the real world, consistent data are rare, primarily because of the two factors stated above. This brings us to the objective of nonmodel-based balances, or data adjustment: *to determine, by adjusting individual data values, the best set of* consistent *estimates of the measured data that solves the balance.*

According to Richardson and White (1982), measured data cannot be adjusted arbitrarily; however, at the same time, all mass balance equations must be satisfied. These criteria should be met by adjusting the data as little as possible and by adjusting good (accurate) data less than poor data. Most modern data adjustment procedures attempt to minimize a weighted sum of squares based on the difference between measured data and adjusted data. Data are weighted according to accuracy estimates, usually based in some manner on the estimated standard deviation (S.D.) of individual data points. Therefore, the difference between an accurate data point and its corresponding adjusted value will make the same contribution to the weighted sum of squares as an inaccurate data point. This concept can be expressed by the following equation:

$$S = \sum_{i=1}^{n} \omega_i (d_i - d_i)^2 \quad \textbf{(Eq. 10.9)}$$

where

S = sum-of-squares objective function

n = number of data items

ω_i = the weighting factor associated with the ith data value (in most cases, ω_i is the inverse of the variance)

d_i = the ith measured data value

$\hat{d}_i$ = the ith adjusted data value

The most common weighting factor is the inverse of the variance (i.e., the inverse of the standard deviation squared):

$$\omega_i = \frac{1}{(\text{S.D.})^2} \quad \textbf{(Eq. 10.10)}$$

The standard deviation is defined as follows:

$$\text{S.D.} = \sqrt{\frac{\sum_{i=1}^{n} (x_i - \bar{x})^2}{n-1}} \quad \textbf{(Eq. 10.11)}$$

where

x_i = ith measured value of x

$\bar{x}$ = mean value of n measurements of x

The reason for using this weighted approach to data adjustment is that if the required adjustments to the data and the standard deviation estimates are drawn from the same experimental data population (and they should be if the circuit is indeed at steady state), then—from Eq. 10.9—on average our adjustment and standard deviation should be of similar value and the expected value of the objective function should be equal to 1. Hence, the expected value of our weighted sum of squares is simply the number of adjustments. This provides a useful statistical check on our assumptions.

Thus, our objective is to minimize the objective function (S) of Eq. 10.9 while ensuring that the solution set is consistent (i.e., that it satisfies the set of equations that define the balance around our process). As the process flowsheet (system) to be analyzed and balanced becomes more complex, so does the process of setting up and solving the equations. As we have seen in the simple example above, redundant data are necessary for successful nonmodel-based data adjustment. However, a common mistake that occurs in dealing with complex systems is underspecification through the use of "redundant equations." These equations are restatements of relationships that have already been stated in some other form and that the engineer mistakenly believes are independent equations needed to reduce the degrees of freedom. Such mistakes are fairly easy to avoid in simple flowsheets/systems involving between 5 and 10 equations and a similar number of unknowns, but with larger systems the chances of inadvertently specifying redundant equations are greatly increased. Furthermore, even if we assume that the engineer has properly defined the system with the correct number of independent equations, the task of solving those equations still remains. Thus, the mass balancing of most process flowsheets is best handled by general-purpose computerized solutions.

A general description of the basic mathematical technique used in these computerized approaches is given by Richardson and White (1982) and Richardson and Mular (1986). Richardson and Mular also give a brief review of several published balance programs available since the mid-1980s. Description of a specific algorithm for mass balance solutions is provided by Morrison (1976). A nonmodel-based balance program based on that algorithm, called JKMBal, is described by Morrison and Richardson (1991). JKMBal is a typical example of modern balance programs that allow users to graphically depict a flowsheet (see Figure 10.4) and enter measured data for stream flows and assay values, along with accuracy estimates (in the form of standard-deviation estimates). The software then converts the flowsheet connections into a node network (also depicted in Figure 10.4), from which the balance equations are automatically derived. A search algorithm, similar to that described by Morrison (1976), is used to find the most likely minimum of the weighted sum-of-squares objective function.

In practice, nonmodel-based balances are really little more than data manipulation exercises. However, such balances have found widespread application.

Metallurgical Accounting. Most metallurgical accounting exercises are concerned with accounting of totals: total material processed; total product produced; total waste; total recovery; and, of course, final grades. Metallurgical accounting is usually not directly concerned with explaining or optimizing the process; therefore, models are not essential to the balance. Only a best estimate of the true balance is required.

The day-to-day data collected from a mineral processing plant are rarely consistent and will almost always contain redundant information. In general, any two methods of calculation will yield different results. The challenge for metallurgical accounting is to produce adjusted data that are both self-consistent and as accurate a representation of plant performance as possible.

Consider a typical base metal concentrator with several products from several circuits, as shown in Figure 10.5. At each point marked with a circled X, we have Au, Cu, Fe, Pb, and Zn assays. For the feed, we have weightometer readings and load-out weights with stockpile surveys for concentrates.

If we select an accounting period that is relatively large compared with the circuit residence time, we can use a suitable software program to carry out a mass balance over this complete data set. If large adjustments are required, there may be measurement problems in sampling or assay techniques. Smaller subcircuits to mass balance should be selected to isolate these problems. Once a consistent set of adjusted data is produced for each accounting period, the sums of these sets will also be consistent. If assays and flow rates are available on a short-time scale (e.g., several times per shift), these data can be balanced for each time period, printed to an ASCII text file, and then composited by using almost any spreadsheet program that can import text files.

Process Optimization: Data Analysis and Model Building. Nonmodel-based balances are used extensively to evaluate and analyze data taken for the purpose of developing process models to be used in model-based balances. If we are to build accurate predictive models of unit operations and

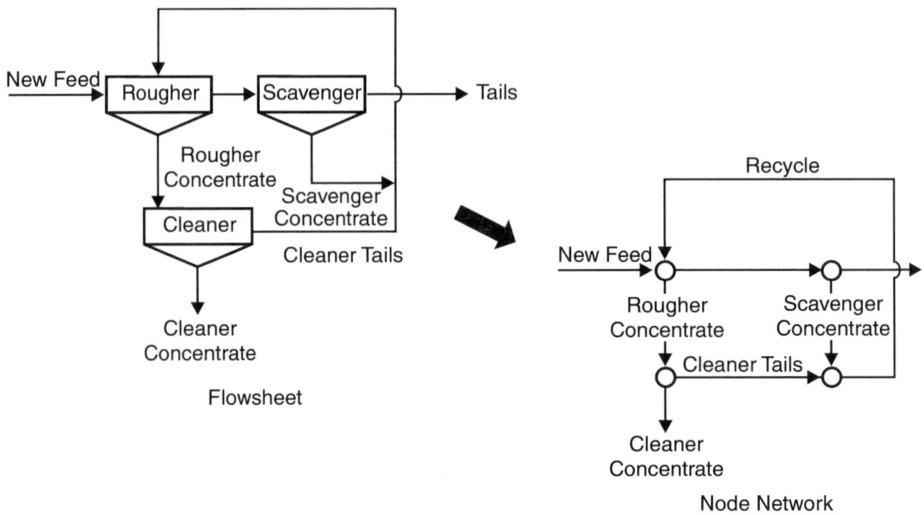

FIGURE 10.4 Converting a simple process flowsheet into a node network

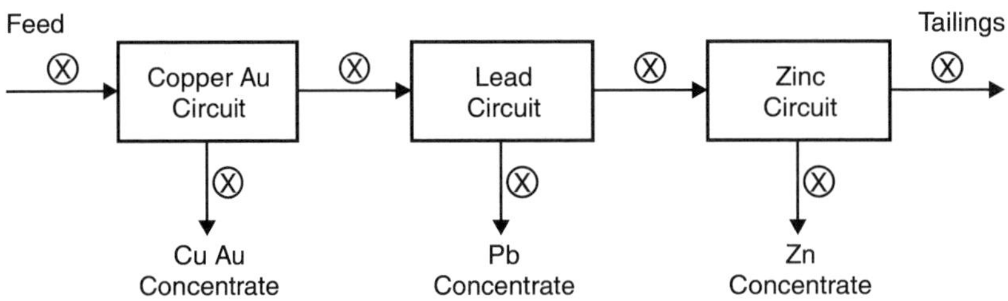

FIGURE 10.5 Metallurgical accounting: data requirements for balance of a total plant

entire processes, we must first have confidence in the underlying data that will be used to build the models and drive the balance. Hence, nonmodel-based balances play an important role in determining the validity of measured data and deciding whether to accept or reject that data.

Process Control. Effective process control requires substantial and accurate knowledge of the process being controlled. Usually the degree of self-consistency of the data used is important to the success of the control strategy, but it may only be assumed, rather than well known. Additionally, some streams that must be known for control purposes cannot be easily or reliably measured and must be estimated. Hence, both on-line and off-line nonmodel-based balances (usually performed using dedicated software programs) can yield valuable information about unmeasured or difficult-to-measure streams, and they can also provide an effective check on the self-consistency of the data. However, as demonstrated in the discussion above, good balances require large amounts of redundant information; this requirement remains important when balances are being applied to process control.

Model-based Balances

Unlike nonmodel-based balances, model-based balances are not reversible. The balance always begins from a set of known feed and process operating conditions that the engineer specifies. The feed is not estimated from the product data. In nonmodel-based balances, only the data and a few fundamental

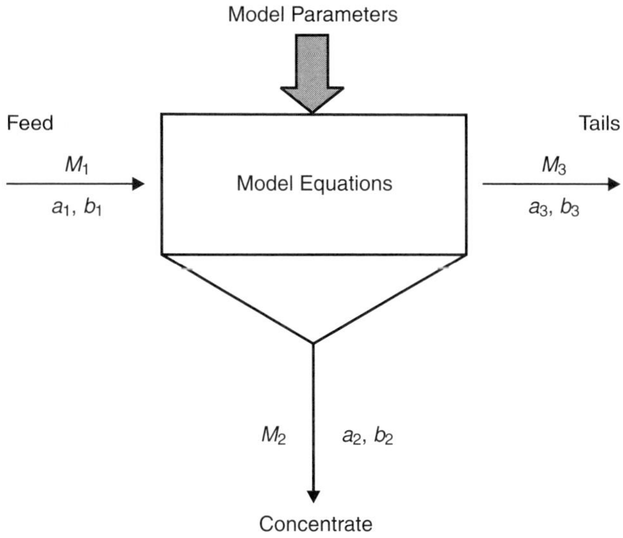

FIGURE 10.6 Rougher float cell: simple model

node-balance equations determine the result; output and input data may have equal opportunities to influence the result. In model-based balances, the feed, operating conditions, and model equations determine the result. The output data are the result.

Model-based balances are often referred to as simulation or design balances. They involve using a set of equations to represent the physical or chemical processes occurring within individual unit operations, forming a model of the unit operation. Such equations are said to impose experience knowledge on the model. These equations "predict" the values of the product streams leaving the unit on the basis of input to the unit operation. (This input includes feed and reagent streams, along with certain operating conditions and design variables that define the state of the unit, as well as model parameters that relate unit performance to operating conditions and design variables.) In this approach, there are no degrees of freedom. The input conditions are assumed to be known and fixed, as are the model equations. The result is a perfectly self-consistent balance around the unit (and ultimately around the entire process being modeled). Strictly speaking, this type of balance is determined only by the input data and is not dependent on measured data from internal or product streams. The answer is fixed by the inputs. However, as will be shown later, measured data can have an indirect effect on the balance, because most complex models rely on a process known as model fitting or parameterization to determine plant- or process-specific model parameters that affect the calculations. Model fitting is an attempt to develop the best set of model parameters that match an existing process, helping to ensure that the predictive model agrees as well as possible with the real world.

Consider now a simple model based on our previously described rougher float cell, as shown in Figure 10.6. For the purposes of this example, let us specify a very simplistic model with the following constraints:

- Each stream has two mineral components, say, chalcopyrite (component A) and gangue (component B).
- The concentrations of the two components together compose the total solid phase of each stream (i.e., concentration of A + concentration of B = 100%).
- The model has two parameters: r_A and r_B, the recoveries (as percent of mass contained in feed stream) of chalcopyrite and gangue, respectively, to the concentrate. In the real world,

specifying mineral recovery alone as a parameter is, of course, too simplistic. Mineral recovery in a more realistic flotation model would likely be based on other design variables and parameters, such as cell configuration [conventional versus column], volume/retention time, flow regime [plug flow versus perfect mixing], airflow, reagent addition rate, and individual mineral kinetics [such as fraction floating and rate of recovery].

As discussed earlier, a steady-state predictive model must be entirely self-consistent. Therefore, the number of degrees of freedom must equal zero. Because we have six unknowns (M_2, a_2, b_2, M_3, a_3, and b_3), we need six independent equations that use the known values of the feed and the model parameters:

1. *Total flow rate of concentrate stream.* From the model constraints, we know that, for each stream, the total mass of A and B must equal the total flow. Hence, for the concentrate stream, we have

$$M_2 = M_{A2} + M_{B2} \quad \textbf{(Eq. 10.12)}$$

where

M_2 = total mass of concentrate stream

M_{A2} = mass of component *A* in concentrate stream

M_{B2} = mass of component *B* in concentrate stream

We also know that the mass of *A* in the concentrate is a function of the recovery parameter r_a and mass of *A* in the feed:

$$M_{A2} = \left(\frac{r_a}{100}\right) \cdot M_{A1} \quad \textbf{(Eq. 10.13)}$$

but

$$M_{A1} = a_1 \cdot M_1$$

so

$$M_{A2} = \left(\frac{r_a}{100}\right) \cdot a_1 \cdot M_1 \quad \textbf{(Eq. 10.14)}$$

Similarly,

$$M_{B2} = \left(\frac{r_b}{100}\right) \cdot b_1 \cdot M_1$$

Therefore,

$$M_2 = M_1 \cdot \left[\frac{r_a}{100} \cdot a_1 + \frac{r_b}{100} \cdot b_1\right] \quad \textbf{(Eq. 10.15)}$$

2. *Total flow rate of tails stream.* By difference, we have

$$M_3 = M_1 - M_2 \quad \textbf{(Eq. 10.16)}$$

3. *Concentration of A in concentrate stream:*

$$a_2 = \frac{M_{A2}}{M_2} \quad \textbf{(Eq. 10.17)}$$

or

$$a_2 = \frac{\left[\left(\frac{r_a}{100}\right) \cdot a_1 \cdot M_1\right]}{M_2} \quad \textbf{(Eq. 10.18)}$$

4. *Concentration of B in concentrate stream:*

$$b_2 = 1 - a_2 \quad \textbf{(Eq. 10.19)}$$

5. *Concentration of A in tails stream:*

$$a_3 = \frac{M_{A3}}{M_3} \quad \textbf{(Eq. 10.20)}$$

or

$$a_3 = \left(\frac{M_{A1} - M_{A2}}{M_3}\right) \quad \textbf{(Eq. 10.21)}$$

$$a_3 = \left(\frac{a_1 \cdot M_1 - a_2 \cdot M_2}{M_3}\right) \quad \textbf{(Eq. 10.22)}$$

6. *Concentration of B in tails stream:*

$$b_3 = 1 - a_3 \quad \textbf{(Eq. 10.23)}$$

Example. Now consider a test of our model by assuming some values for the feed, along with some recoveries. Assume 100 tph of feed containing 2.3% chalcopyrite (about 0.6% Cu) and 97.7% gangue. Assume 90% recovery of the chalcopyrite and 1.5% recovery of the gangue. By using the six equations of the model, we can

1. Calculate concentrate total flow rate:

$$M_2 = 100 \cdot \left(\frac{90}{100} \cdot 0.023 + \frac{15}{100} \cdot 0.977\right) = 3.54 \text{ tph}$$

2. Calculate tails total flow rate:

$$M_3 = 100 - 3.54 = 96.46 \text{ tph}$$

3. Calculate assay of chalcopyrite in concentrate:

$$a_2 = \frac{\frac{90}{100} \cdot 100 \cdot 0.023}{3.54} = 0.587 \quad \text{(58.47\% chalcopyrite = 15.02\% Cu)}$$

4. Calculate assay of gangue in concentrate:

$$b_2 = 1 - 0.5847 = 0.4153 \quad \text{(41.53\% gangue)}$$

5. Calculate assay of chalcopyrite in tails:

$$a_3 = \frac{0.023 \cdot 100 - 0.5847 \cdot 3.54}{96.46} = 0.002 \quad \text{(0.2\% chalcopyrite)}$$

6. Calculate assay of gangue in tails:

$$b_3 = 1 - 0.002 = 0.998 \quad \text{(99.8\% gangue)}$$

Model-based balances have application in process design, process optimization, economic optimization, and process control.

Process Design. Model-based balances, usually determined via design simulation software packages, allow designers to evaluate (and eliminate) various process alternatives on the basis of technical feasibility, as well as to determine equipment capacities and operating costs. Unfortunately, there is no existing plant to survey and provide data for a model-fitting exercise. Therefore, model parameters must often be estimated manually or selected from a preexisting database by using parameter values that were obtained by model fitting of existing plants with similar feeds, capacities, and process conditions. Although this approach should always be combined with traditional methods for equipment sizing and process evaluation, such as laboratory and pilot-plant testing, it nonetheless allows for much more rapid evaluation of a greater number of alternatives than manual calculation of process balances.

Process Optimization. Operating companies wanting to improve the performance of existing plants can use both model fitting and model-based balances to rapidly test the technical feasibility of various process alternatives.

Economic Optimization. Even if a process is technically optimized (i.e., achieving an effective separation), it may still be operating far from its economic optimum. For a single-concentrate plant, the simple model shown in Figure 10.2 can be applied to a complete plant (contaminants can also be included). Indeed, plant records may provide us with a point on the separation curve for each day or shift of operation. Where the plant operates on this curve depends on operator-selected split points. If we have a sales contract for the product, calculating the financial return at each point on this curve is easy. For a multiproduct plant, we need a better model—one that can handle downstream implications of adjustments to the earlier stages. A general approach to this problem was described by Burns, Duke, and Williams (1982) and updated by Franzidis, Manlapig, and Morrison (1998). For a multiproduct operation, the financial benefits of getting each metal into the correct product are often substantial.

Process Control. Operation at correct process targets for multiple products is a challenging control system task. What is required is a model that can be updated on-line by current performance and then used to "look ahead" (i.e., to estimate which feasible change will yield the best increment in net smelter return). An on-line mass balance is a very useful adjunct to keeping the models current (and checking on the data). For flotation, the JK/UCT (Julius Kruttschnitt Mineral Research Centre and the University of Cape Town Department of Chemical Engineering) modeling approach (Franzidis, Manlapig, and Morrison 1998) provides a powerful technique for on-line modeling and prediction. For gravity separation processes, some knowledge of the washability of the feed is essential to such a control scheme.

CALCULATION METHODS

When it comes to actually solving a metallurgical balance—whether model-based or nonmodel-based—three types of calculation methods are generally available:

- Manual
- Spreadsheet
- Dedicated computer program

Manual Calculation of Balance

For a quick overview of the capability and performance of a complete processing plant, hand calculation is still a viable option, although the practicality of its use is heavily dependent on the size and complexity of the plant being considered.

Example. Consider the following example of a gold plant (Morrison 1991). The plant was originally designed to operate at 100 tph, with a 2-g/t feed and 90% recovery. A built-in excess capacity of 20% extra for carbon loading and a grinding circuit currently allows the plant to run at 120 tph with carbon moved as rapidly as possible. If head grade jumps to 4 g/t, with throughput maintained at 120 tph, carbon loading becomes limiting and tailings grade rises sharply. The high throughput may generate more cash flow, but a lower throughput will increase the profitability and longevity of the plant. What is the optimum throughput at design loading and a 4-g/t head grade?

First, calculate the design gold-loading capacity at increased throughput of 120 tph:

$$\text{loading capacity} = (\text{design throughput} \times \text{design grade} \times \text{design recovery}) \times \text{excess capacity}$$

$$\left(100\frac{\text{t}}{\text{h}} \cdot 2\frac{\text{g}}{\text{t}} \cdot \frac{90}{100}\right) \cdot \left(\frac{120\frac{\text{t}}{\text{h}}}{100\frac{\text{t}}{\text{h}}}\right) = 216\frac{\text{g}}{\text{h}}$$

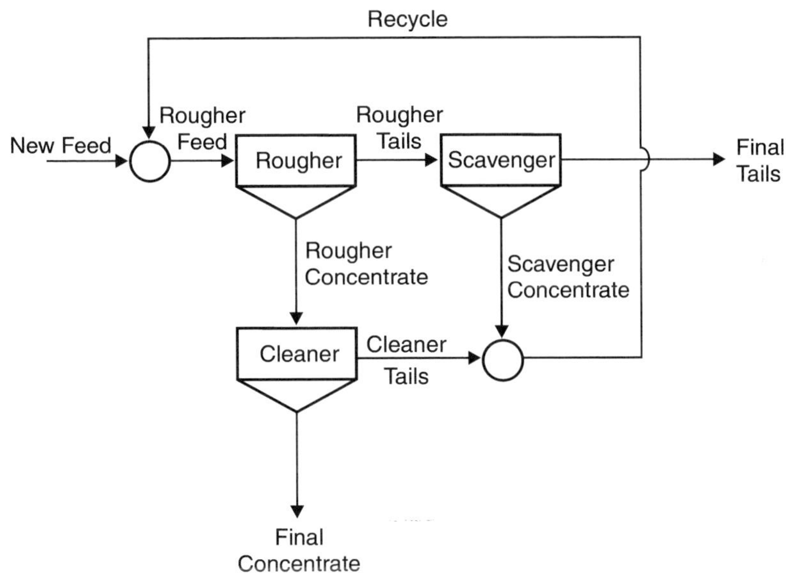

FIGURE 10.7 Copper flotation circuit

Thus, at a new grade of 4 g/t:

$$\text{optimum throughput at new grade} = \frac{\text{design loading capacity}}{\text{new grade} \times \text{design recovery}}$$

$$\text{throughput, } \frac{\text{t}}{\text{h}} = \frac{216\frac{\text{g}}{\text{h}}}{4\frac{\text{g}}{\text{t}} \cdot \frac{90}{100}} = 60\frac{\text{t}}{\text{h}}$$

Using Spreadsheet Programs to Calculate a Balance

Entire process flowsheets are modeled by combining unit models. The unit models are interconnected by the process flowstreams. However, as with nonmodel-based balances, problem complexity increases with the number of unit operations and the complexity of the process. For flowsheets of moderate complexity, the use of spreadsheet programs to solve a model-based balance can be a cost-effective method for performing repetitive, cumbersome calculations. Commercial spreadsheet programs are in wide use and well understood by most engineers.

Example. The following example expands on our simple flotation model to demonstrate the basics of setting up and solving such a problem using a spreadsheet. Consider Figure 10.7, where our original model has been expanded from a single rougher cell to represent three flotation banks (rougher, cleaner, and scavenger). Once again, for simplicity, we will concern ourselves only with the solids phase, and we will assume that two components, chalcopyrite and gangue, form the total of the solids phase.

Examine Figure 10.7 closely and note that we now have the added complication of a recycle stream that returns material from downstream back to the head of the process. Before setting up our simple spreadsheet model of the plant, we need to develop an approach for dealing with the problem of recycle streams. This approach involves arbitrarily "tearing" the recycle stream into two parts: the "guessed" part and the "calculated" part, as shown in Figure 10.8. This step allows us to successfully calculate the balance around the first node/model, which is a simple mixer that combines the recycle

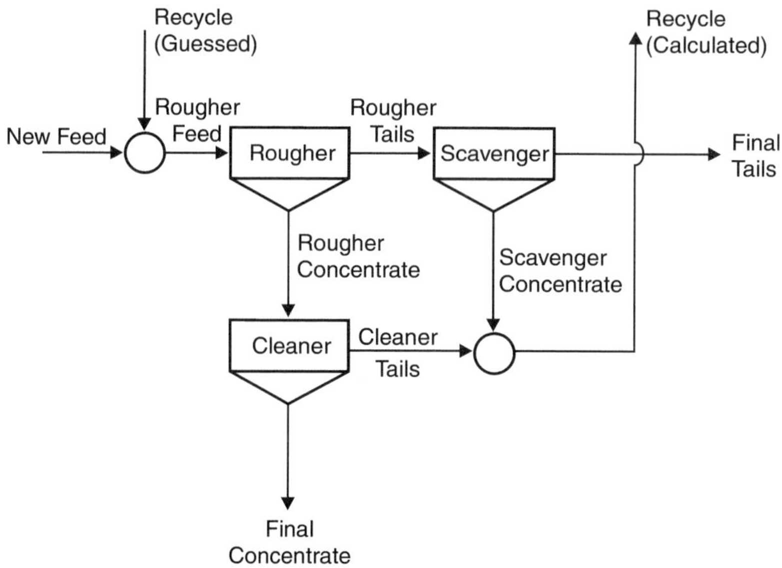

FIGURE 10.8 Copper flotation circuit: Recycle stream "torn" for iterative calculations

and new feed streams. By using guessed values for the key variables in the recycle stream, we can predict the total feed to the rougher bank. With the rougher feed known, we can then proceed to calculate our rougher products by using the six equations of our simple two-component flotation model. Once the rougher products have been predicted, the cleaner and scavenger banks can then be calculated in turn. Finally, the calculated values of the recycle stream are computed by the last node/model, another mixer that combines the scavenger concentrate and cleaner tails. Unfortunately, unless we have made a perfect starting guess, the calculated values will not match the original guessed values. Therefore, we need to replace the original guesses with better guesses and start again. The simplest form of this procedure, called successive substitution, involves simply replacing the former guessed values with the latest calculated values and recalculating. The process is repeated until the guessed and calculated values match within reason, usually within a few decimal places of accuracy; when this occurs, the model is said to be "converged." This approach usually works well enough with simple models and flowsheets; larger and more complex problems may require more sophisticated convergence strategies. This topic is discussed in some detail by Richardson and White (1982) and Richardson and Mular (1986).

Figure 10.9 shows one method for setting up and solving this circuit balance using a commercial spreadsheet program. Note that many sophisticated calculation and programming capabilities are available in modern spreadsheet programs. This example is intended to show a basic solution using manual input adjustment and recalculation; further automation of data input and iterative calculations are left to the reader. In Figure 10.9, we see the basic equations entered as formulas and the necessary inputs (feed conditions, recycle stream guesses, and model parameters) entered as numeric values. The six independent equations that form the minimum description of the simple flotation model for the rougher bank are found in cells B7-8, E7-8, and H7-8. The same six equations as written for the cleaner bank are found in cells B10-11, E10-11, and H10-11. Likewise, the independent equations for the scavenger are found in cells B13-14, E13-14, and H13-14. The rougher feed node is calculated from equations in cells B5, E5, and H5; and the calculated recycle stream node is defined from the formulas found in cells B4, E4, and H4. All other equations in Figure 10.9 can be derived from the basic balance and are not independent relationships.

	A	B	C	D	E	F	G	H	I
1	Stream Name	Solids t/h	$CuFe_2$ t/h	Gangue t/h	% $CuFe_2$	$CuFe_2$ Rec. %	% Cu	% Gangue	Gangue Rec. %
2	Feed	100	=E2/100*B2	=H2/100*B2	2.3	=100*C2/C$2	=0.257*E2	=100–E2	=100*D2/D$2
3	Recycle (Guess)	1.744	0.0915	=H3/100*B3	=100*C3/B3	=100*C3/C$2	=0.257*E3	=100–E3	=100*D3/D$2
4	Recycle (Calc)	=B10+B14	=C10+C14	=H4/100*B4	=100*C4/B4	=100*C4/C$2	=0.257*E4	=100–E4	=100*D4/D$2
5	Rougher Feed	=B2+B3	=C2+C3	=H5/100*B5	=100*C5/B5	=100*C5/C$2	=0.257*E5	=100–E5	=100*D5/D$2
6									
7	Rougher Tails	=B5–B8	=E10/100*B7	=B7–C7	=(E5*B5–E8*B8)/B7	=100*C7/C$2	=0.257*E7	=100–E7	=100*D7/D$2
8	Rougher Conc	=B5*(B19/100^2*E5+C19/100^2*H5)	=E8/100*B8	=B8–C8	=((B19/100)*E5/100*B5)*100/B8	=100*C8/C$2	=0.257*E8	=100–E8	=100*D8/D$2
9									
10	Cleaner Tails	=B8–B11	=E10/100*B10	=B10–C10	=(E8*B8–E11*B11)/B10	=100*C10/C$2	=0.257*E10	=100–E10	=100*D10/D$2
11	Final Conc	=B8*(B20/100^2*E8+C20/100^2*H8)	=E11/100*B11	=B11–C11	=((B20/100)*E8/100*B8)*100/B11	=100*C11/C$2	=0.257*E11	=100–E11	=100*D11/D$2
12									
13	Tails	=B7–B14	=E13/100*B13	=B13–C13	=(E7*B7–E14*B14)/B13	=100*C13/C$2	=0.257*E13	=100–E13	=100*D13/D$2
14	Scav Conc	=B7*(B21/100^2*E7+C21/100^2*H7)	=E14/100*B14	=B14–C14	=(B21/100^2*E7*B7)/B14	=100*C14/C$2	=0.257*E14	=100–E14	=100*D14/D$2
15									
16									
17		Input Model Parameters:							
18		$CuFeS_2$ Rec. %	Gang Rec. %						
19	Rougher	75	1						
20	Cleaner	95	1						
21	Scavenger	30	0.5						
22									
23									
24		Convergence Check:							
25									
26	Solids Conv:	=100*(B13+B11)/B2							
27	$CuFe_2$ Conv:	=100*(C13+C11)/C2							

FIGURE 10.9 Spreadsheet model of copper flotation circuit, showing formulas

	A	B	C	D	E	F	G	H	I
1	Stream Name	Solids t/h	$CuFe_2$ t/h	Gangue t/h	% $CuFeS_2$	$CuFeS_2$ Rec. %	% Cu	% Gangue	Gangue Rec. %
2	Feed	100	2.3	97.7	2.3	100	0.59	97.7	100
3	Recycle (Guess)	**10.000**	**1.5000**	8.500	15.000	65.22	3.855	85.000	8.70
4	Recycle (Calc)	2.005	0.1454	1.859	7.251	6.32	1.863	92.749	1.90
5	Rougher Feed	110.00	3.8	106.20	3.45	165.22	0.89	96.55	108.70
6									
7	Rougher Tails	**106.09**	0.95	105.14	**0.90**	41.30	0.23	**99.10**	107.61
8	Rougher Conc	**3.91**	2.85	1.06	**72.85**	123.91	18.72	**27.15**	1.09
9									
10	Cleaner Tails	**1.19**	0.14	1.05	**11.94**	6.20	3.07	**88.06**	1.08
11	Final Conc	**2.72**	2.71	0.01	**99.61**	117.72	25.60	**0.39**	0.01
12									
13	Tails	**105.28**	0.95	104.33	**0.90**	41.18	0.231	**99.10**	106.79
14	Scav Conc	**0.81**	0.0029	0.808	**0.35**	0.12	0.09	**99.65**	0.83
15									
16									
17		**Input Model Parameters:**							
18		**$CuFeS_2$ Rec. %**	**Gangue Rec. %**						
19	Rougher	**75**	**1**						
20	Cleaner	**95**	**1**						
21	Scavenger	**30**	**0.5**						
22									
23									
24		**Convergence Check:**							
25									
26	Solids Conv:	**108.00**							
27	$CuFe_2$ Conv:	**158.90**							

FIGURE 10.10 Spreadsheet model of copper flotation circuit: results of initial guesses

As for spreadsheet input data, Figure 10.9 shows that the feed solids flow rate and feed chalcopyrite concentration are found in cells B2 and E2; the guessed values for the recycle solids flow rate and chalcopyrite flow rate are found in cells B3 and C3. Note that the gangue values for all input streams can be calculated by definition from the problem statement—namely, that all remaining solids material that is not chalcopyrite is gangue. Furthermore, the chalcopyrite concentration could just as easily have been used as the starting input value for the recycle stream, instead of chalcopyrite flow rate; given the total solids flow rate, either chalcopyrite value can be calculated if the other is given. The model parameters of chalcopyrite and gangue recovery are given as inputs in cells B19–21 and C19–21, respectively. Finally, cells B26 and B27 are used to provide a convergence check; that is, if the entire model is converged for total solids, the total solids leaving the plant (final tails plus final concentrate) must equal (or approach) the total solids in the feed. Likewise, the total chalcopyrite leaving in the tails and concentrate must also equal or approach the amount of chalcopyrite in the feed. These two conditions can be expressed as percentage ratios; by watching these ratios change with each new set of guesses for the recycle stream, the user can more easily determine which direction to adjust the next set of guesses. As the ratios approach 100%, the user can decide if the overall balance is sufficiently converged.

Figure 10.10 shows the same spreadsheet as input and calculated values for an initial set of recycle stream guessed values. Note that the guessed values and calculated values for the recycle stream are not equal and that the convergence check ratios are not equal to 100%. Finally, Figure 10.11 shows the solution, with the guessed and calculated values of the recycle stream in near-perfect agreement and the convergence check ratios equal to 100%. The reader is encouraged to expand on this example by investigating other values of component recovery parameters. For example, what is the effect of better overall chalcopyrite recovery, coupled with less gangue recovery (more suppression)? What is the effect of changing recovery in the individual banks? Is this the most efficient combination of these banks? Water can also be added as a component but will definitely increase the complexity of the problem, requiring additional independent model equations.

Dedicated Computer Programs for Solving Model-based Balances

Modern process models tend to be far more complex than the very simple flotation example presented in this chapter, often with many more equations (and often more model parameters) required to accurately predict the unknown values of the product streams. This complexity soon reaches a level that requires the aid of specialized computer software (usually called process simulation software) to set up and solve the design balance around a flowsheet. Richardson and White (1982) and Richardson and Mular (1986) give detailed descriptions of the fundamentals of such software packages. Specific details of one such modern package for comminution circuits are discussed by Cameron and Morrison (1991) and Wiseman and Richardson (1991). Richardson (1990; 1992) discusses application of a modern simulator package to mineral processing plant optimization, supported by real-world case studies. Alford (1992) describes a simulator package designed specifically for solving flotation circuits. The models in that package combine cell/machine characteristics such as configuration (conventional versus column), volume (retention time), and flow regime (perfect mixing versus plug flow) with mineral charcteristics such as floatability (fraction of mineral floating, kinetic rate of flotation, etc.) to define flotation performance. Such software packages allow the user to set up and solve problems of great complexitiy, involving many individual units and flowstreams, including several levels of nested recycle loops.

Modern balance programs allow users to graphically depict a flowsheet (as shown in Figure 10.12) and enter measured data for feed stream flows and assay values, along with model design variables and parameters. The software analyzes the flowsheet connections as drawn for recycle loops and automatically reconstructs the problem so that calculations proceed sequentially, model by model. If the model parameters are known (or assumed known) in advance, the process is straightforward. The simulator will calculate the one and only unique solution to the flowsheet balance, as specified by the feed conditions,

	A	B	C	D	E	F	G	H	I
1	Stream Name	Solids t/h	$CuFeS_2$ t/h	Gangue t/h	% $CuFeS_2$	$CuFeS_2$ Rec. %	% Cu	% Gangue	Gangue Rec. %
2	Feed	100	2.3	97.7	2.3	100	0.59	97.7	100
3	Recycle (Guess)	**1.744**	**0.0915**	1.653	5.247	3.98	1.348	94.753	1.69
4	Recycle (Calc)	1.744	0.0915	1.653	5.244	3.98	1.348	94.756	1.69
5	Rougher Feed	101.74	2.3915	99.35	2.35	103.98	0.60	97.65	101.69
6									
7	Rougher Tails	**98.96**	0.60	98.36	**0.60**	25.99	0.16	**99.40**	100.67
8	Rougher Conc	**2.79**	1.79	0.99	**64.35**	77.98	16.54	**35.65**	1.02
9									
10	Cleaner Tails	**1.07**	0.09	0.98	**8.36**	3.90	2.15	**91.64**	1.01
11	Final Conc	**1.71**	1.70	0.01	**99.42**	74.08	25.55	**0.58**	0.01
12									
13	Tails	**98.29**	0.60	97.69	**0.61**	25.92	0.156	**99.39**	99.99
14	Scav Conc	**0.67**	0.0018	0.669	**0.27**	0.08	0.07	**99.73**	0.69
15									
16									
17		Input Model Parameters:							
18		$CuFeS_2$ Rec. %	Gangue Rec. %						
19	Rougher	**75**	**1**						
20	Cleaner	**95**	**1**						
21	Scavenger	**30**	**0.5**						
22									
23									
24		Convergence Check:							
25									
26	Solids Convergence:	100.00							
27	$CuFe_2$ Convergence:	100.00							

FIGURE 10.11 Spreadsheet model of copper flotation circuit: final results

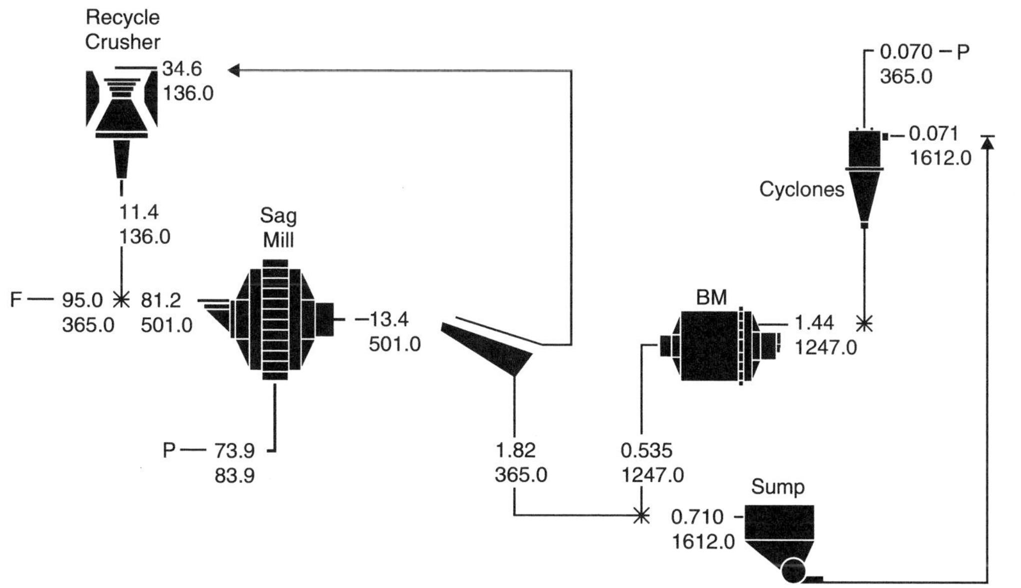

FIGURE 10.12 Flowsheet depicted graphically in mineral processing simulation software

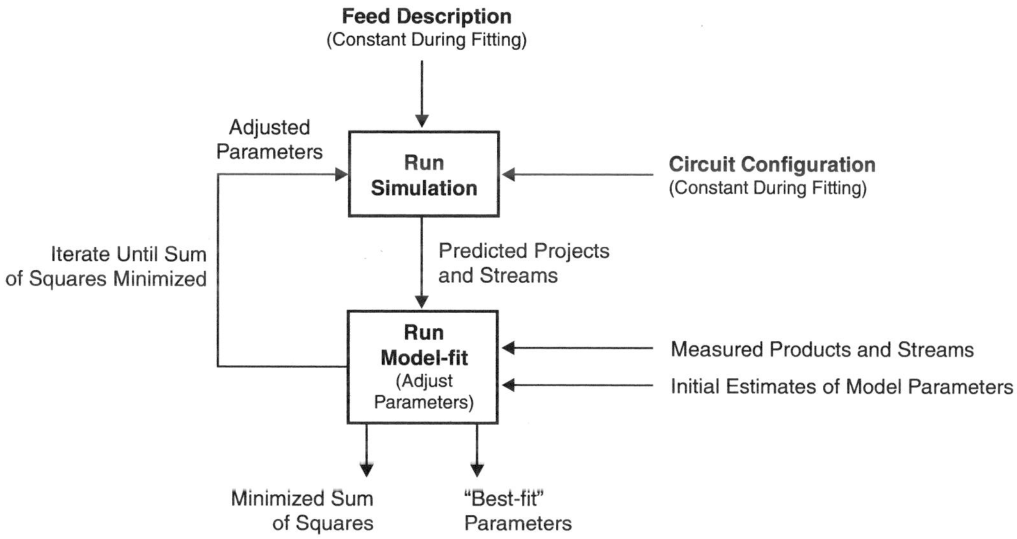

FIGURE 10.13 Diagram illustrating a model-fitting process

design variables, model parameters, and model equations. This is usually the approach taken in design circumstances, where the flowsheet being modeled does not yet exist.

In the case of existing plants in need of optimization, however, the process is slightly different and a bit more complicated. Actual plant measured data (preferably taken under near-steady-state conditions during smooth plant operation) are used in a process called model fitting to determine the most likely set of model parameters that will most accurately predict the measured plant data (Figure 10.13).

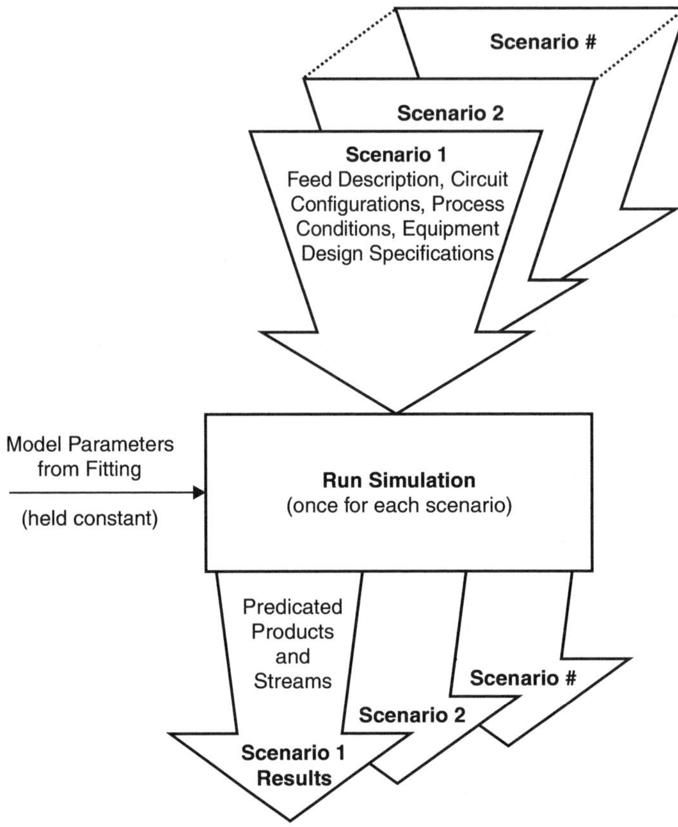

FIGURE 10.14 Diagram illustrating process simulation/optimization

During model fitting, model parameters are not considered to be constant and are adjusted in much the same way that the measured data are adjusted in nonmodel-based mass balancing. In other words, the model parameters are adjusted by using accuracy estimates to weight the error between predicted and measured values. The weighted error, in turn, is minimized by a specialized search algorithm that adjusts the model parameters. The result of this process is a base-case model that closely represents the actual plant at surveyed conditions and that can be used to more confidently predict how the plant will perform under new operating conditions and new design variable values.

After model parameters have been established by model fitting, the base-case model can be used for optimization. Model parameters are now held constant, and plant performance is predicted under different "what if" scenarios by making selected changes to the design variables, feed, and operating conditions of the base-case model. Each new, unique set of conditions results in a single and unique balance solution containing predicted product stream values. This process is commonly called process simulation/optimization and is depicted in Figure 10.14.

Model sophistication can vary widely, including

- Mechanistic models that attempt to explain the unit operation from first principles and are often highly theoretical
- Semiempirical models that have some basis in both theory and operating/experimental data
- Fully empirical models based on engineering assumptions and operating/experimental data, often relying on statistical correlations rather than theory to describe observed behavior

DATA

In closing this chapter, a brief discussion of the importance of good data as the basis of our mass-balancing efforts, regardless of application, is worthwhile. Mass balancing (nonmodel-based balancing) is perhaps most useful as a method of data assessment. However, a common–and inaccurate–perception is that mass balancing can fix bad data. Mass balancing helps the engineer to identify bad data to help refine measurement techniques and collect better data next time. The level of confidence we have in the predictions that result from our model-based balances is directly related to the quality of data used to construct these models.

Good Data Versus Bad Data: Summary Balance

The key to making mass balancing work at all is the information that the separation imposes on the streams. There is an interesting derivation based on information theory in the literature. In practice this means that the best defined balance is among feed, final products, and tailings–which are very often the best measured streams as well. The summary balance provides a very good fit with good flow definition. It is a good initial test of any data set. If an overall balance is poor, the data are probably not worth considering further.

If you want an easy way to think about this, consider the problem in terms of signal-to-noise ratio. In this analogy, the difference in assays between the streams is the signal, and the measurement error is the noise. Therefore, the feed and final products may contain substantially more information than the internal ones–especially where the internal streams are the products of "weak" separations, such as scavengers and final cleaners or (worse still) splitters. For these internal streams, mass balancing is of limited use other than to identify which flow rates should be measured.

Good Data Versus Bad Data: Complete Circuit Balance

In the case of a complete circuit balance, the product streams are well defined and flow rates (and the estimates of their accuracy) are legitimate information to use in an overall balance–especially if internal streams are poorly defined. Inside a copper flotation circuit, for example, working back from the copper concentrate does provide some reasonable balances, and these flows and accuracy estimates make "estimating" the poorly defined flows possible. In fact, there is a range of flow rates that would give otherwise very similar balances. In cases like this, even an "eyeball" estimate of flow rate will be more reliable than a mass balance "solution." The existence of multiple solutions to this type of problem is the reason for adopting a sensible approach to mass balancing. Mass balancing should be used as a kind of "sieve" to identify poor data and see which samples and flow rates need to be better measured for good definition, because the purpose of our balance is really to accept or reject data (rather than to merely adjust it) for further modeling or for accounting purposes. Ideally, we want to be able to collect data that are suitable for modeling with minimal adjustment. Therefore, balance software programs are not designed to handle poorly defined mass balances. Thus, operators of pilot plants and existing plants should concentrate on collecting very well-defined data and should not have to rely on a mass balance for flow rate estimates. The same point applies with regard to concentrator performance.

Data Collection

Napier-Munn et al. (1996) maintain that the quality and nature of data obtained from a circuit by sampling are the key to assessing circuit performance, particularly with respect to modeling and simulation studies that rely on model-based balances. Those authors define a "survey" as collecting data and samples (from a circuit over a particular operating period) that are representative of the circuit's operation during that time frame. Because our confidence in process simulation, and hence optimization, of a circuit is based on our ability to build a model that is representative of a real system, the accuracy and representative character of the survey data are very important. The engineer wishing to build an accurate model-based simulation of a circuit needs to pay particular attention to the nature of the data to be collected, the sampling points in the plant, the method of sampling, sampling equipment used, and the processing of samples. A knowledge of statistics and sampling theory is also useful (Gy 1976; Pitard 1993).

Napier-Munn et al. (1996) also maintain that although the objective of all sampling is to obtain a representative sample, that goal is rarely realized in practice. This difficulty is primarily caused by the errors and disturbances that can contribute to overall error in determining a particular data point:

- Transient nature of plant operation
- Inadequate design of sample cutter
- Subsampling of primary samples
- Analytical errors (weighing, sieving, chemical analyses)
- Propagation of error when quantities are being calculated
- Fundamental statistical uncertainty involved in choosing a small, finite sample to represent the properties of a large (effectively infinite) population

Circuit surveys should be well designed and should take these items into consideration. Before a sampling campaign is undertaken, the objectives and the list of data to be obtained should be well understood by all involved. A written plan of survey is a good idea in cases dealing with large and complex circuits involving multiple sample points. When time, economics, and plant design permit, there is no such thing as collecting "too much data." However, all too frequently, too much *bad* data are collected. The objective of careful survey design and planning should be to minimize the collection of bad data and maximize the collection of useful data that are representative of the operation at the time of survey. Ensuring the availability of sufficient human and equipment resources to adequately and accurately conduct the survey will help to minimize the necessity of repeating the survey.

The sequence of events, as recommended by Napier-Munn et al. (1996), should be as follows:

1. Define objectives of the survey and identify units to be surveyed.
2. Plan the survey, accounting for sample points, size of samples required, data to be collected, and difficulties likely to be encountered, such as accessibility, production interruptions, non-steady-state operation, and missing data.
3. Conduct the survey according to the tenets of good practice.
4. Analyze the samples with good care.
5. Analyze and mass balance (nonmodel-based) the data; reject poor or doubtful data and resurvey if necessary.
6. Use the data as defined by the objectives; for example, for parameter estimation (model fitting), as a base case for simulations, or to confirm that the expected improvements have been obtained.

Sensitivity to Data Accuracy (Standard Deviation) Estimates

Once we have collected our data, we can use nonmodel-based balancing to accept or reject the data for use in parameter estimation for our model-based simulation and optimization exercises. However, to properly analyze the data with nonmodel-based balancing and to conduct proper model fitting, we need a good estimate of the accuracy of each data point. Unfortunately, the high degree to which mass-balancing techniques are sensitive to poor estimates of data accuracy is a poorly appreciated concept. Consider the simple case of one stream, A, classified into two streams, B and C. Let us define β as the mass fraction of stream A that reports to stream B. If we know *exactly* the value of one element i in each stream, we can calculate a sum of squares of mass flow imbalance for each estimate of β from 0 to 1.0, plotting the result as shown in Figure 10.15:

$$\text{SSQ} = \sum_{i=1}^{n} (a_i - \beta b_i - (1-\beta)c_i)^2 \qquad \textbf{(Eq. 10.24)}$$

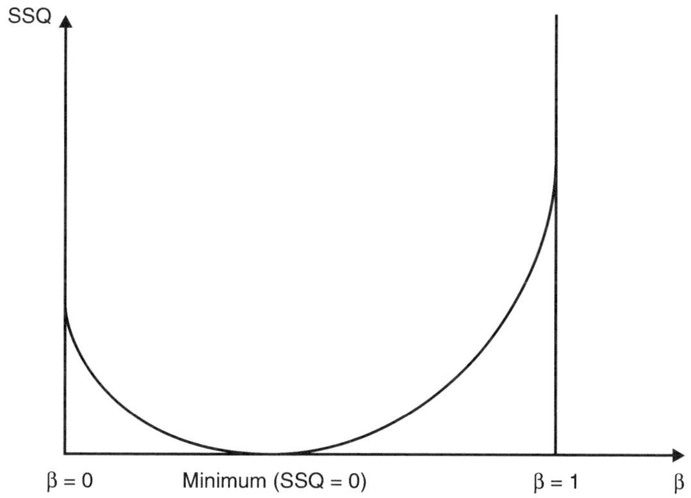

FIGURE 10.15 Sum-of-squares objective function versus mass split estimate

where

SSQ = sum of squares of mass flow imbalance
n = number of components
a_i = concentration of component i in stream A
b_i = concentration of component i in stream B
c_i = concentration of component i in stream C

Because the data are exact, the minimum of SSQ will be zero and the gradient at the minimum will also be zero.

For real data, the sum of squares of the errors will look more like the result in Figure 10.16. If the mass split (β) is well determined, the minimum will be quite sharp. If it is poorly determined, the minimum will be close to flat. Nonmodel-based balance sofware programs minimize the weighted sum of squares. It is fairly easy to see that the "flat spot" in the SSQ can be strongly affected by a change of accuracy estimates. Hence, accuracy (standard deviation) estimates should be as realistic as possible.

REFERENCES

Alford, R.A. 1992. Modeling of Single Flotation Column Stages and Column Circuits. *Int. J. Min. Process.,* 36:155–174.

Burns, C.J, P.J. Duke, and S.R. Williams. 1982. Process Development and Control at Woodlawn Mines. Paper presented at the XIV International Mineral Processing Congress/Canadian Institute of Mining annual meeting, October 17–23, at Toronto, Ontario, Canada.

Cameron, P., and R.D. Morrison. 1991. Optimisation in the Concentrator: The Practical Realities. In *Proceedings of the Mining Industry Optimisation Conference.* Sydney, Australia: Australasian Institute of Mining and Metallurgy.

Davis, J.J., C.J. Wood, and J.G. Lyman. 1987. The Use of Density Tracers for the Determination of Dense Medium Cyclone Partitioning Characteristics. *Int. J. Coal Process.,* 2(2):107–126.

Dell, C.C. 1961. Technical Efficiency of Concentration Operations. *Colorado School of Mines Quarterly,* 56(3):113–127.

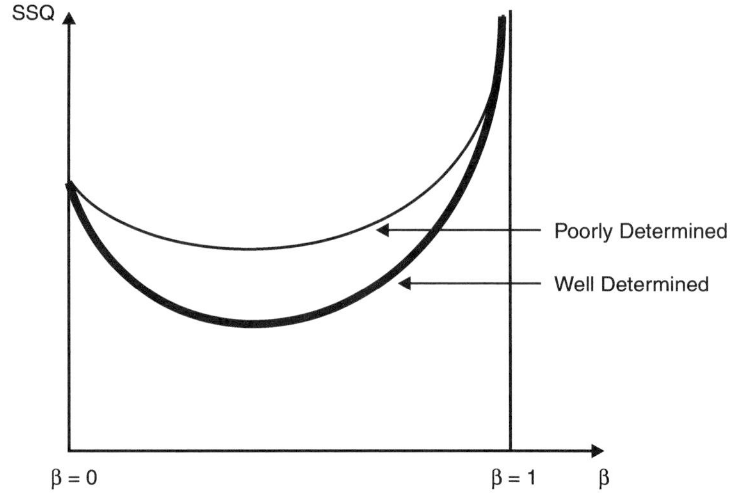

FIGURE 10.16 Sum-of-squares objective function versus mass split: actual data

Franzidis, J.P., E.V. Manlapig, and R.D. Morrison. 1998. Modeling and Control of Flotation Circuits. Paper presented at Australian Mineral Foundation–Best Practise Conference, November, at Perth, Australia.

Gay, S.L., and J.G. Lyman. 1995. Stereological Error in Particle Sections: The Answer. In *Proceedings of APCOM XXV.* Publication Series 4/95. Brisbane, Australia: Australasian Institute of Mining and Metallurgy.

Gy, P.M. 1976. *Sampling of Particulate Materials: Theory and Practise.* 2nd ed. Amsterdam: Elsevier.

Morrison, R.D. 1976. A Two-Stage Least Squares Technique for the General Material Balance Problem. Julius Krutschnitt Mineral Research Centre Internal Report No. 61 (unpublished).

——. 1991. Material Balance Techniques. In *Evaluation and Optimization of Metallurgical Performance.* Edited by D. Malhotra, R. Klimpel, and A. Mular. Littleton, Colo.: SME.

——. 1993. Concentrator Optimisation. In *Proceedings of the International Mineral Processing Congress XVII.* Publication Series 3/93. Brisbane, Australia: Australasian Institute of Mining and Metallurgy.

Morrison, R.D., and J.M. Richardson. 1991. JKMBal: The Mass Balancing System. In *Computer Applications in the Mineral Industry: 2nd Canadian Conference Proceedings.* Vancouver, B.C., Canada: Department of Mining and Mineral Process Engineering, University of British Columbia.

Murrill, P.W. 1967. *Automatic Control of Processes*. Scranton, Pa.: International Textbook.

Napier-Munn, T.J., S. Morrell, R.D. Morrison, and T. Kojovic. 1996. Surveying Comminution Circuits. In *Mineral Comminution Circuits: Their Operation and Optimisation*. Indooroopilly, QLD, Australia: Julius Kruttschnitt Mineral Research Centre.

Partridge, A.C. 1994. *Principles of Separation, Part II, Vol. I: The Advanced Coal Preparation Monograph Series.* Indooroopilly, Queensland, Australia: Australian Coal Preparation Society.

Pitard, F.F. 1993. *Pierre Gy's Sampling Theory and Practice.* 2nd ed. Boca Raton, Fla.: CRC Press.

Richardson, J.M. 1990. Computer Simulation and Optimization of Mineral Processing Plants: Three Case Studies. In *Control '90: Mineral and Metallurgical Processing.* Littleton, Colo.: SME.

——. 1992. A Rational Approach to Computerized Optimization of Mineral Processing Plants. In *Proceedings of 23rd Application of Computers and Operations Research in the Mineral Industry.* Littleton, Colo.: SME.

Richardson, J.M., and A.L. Mular. 1986. Metallurgical Balances. In *Design and Installation of Concentration and Dewatering Circuits.* New York: AIME.

Richardson, J.M., and J.W. White. 1982. Mass Balance Calculations. In *Design and Installation of Comminution Circuits.* New York: AIME.

Whittle, D., and P. Vassiliev. 1998. Synthesis of Stochastic Recovery Prediction and Cut Off Optimisation. In *Proceedings of Mine to Mill 1998 Conference.* Publication Series 4/98. Brisbane, Australia: Australasian Institute of Mining and Metallurgy.

Wiseman, D.M., and J.M. Richardson. 1991. JKSimMet: The Mineral Processing Simulator. In *Computer Applications in the Mineral Industry: 2nd Canadian Conference Proceedings.* Vancouver, B.C., Canada: Department of Mining and Mineral Process Engineering, University of British Columbia.

CHAPTER 11

Bulk Solids Handling

Hendrik Colijn

The general field of bulk solids handling may be divided into six distinct functional categories of activity. These are, in the order in which they usually occur in industry:

1. Bulk handling (dry solids and liquids)
2. Unit handling
3. Industrial packaging
4. Warehousing
5. Carrier handling
6. Handling operation analysis (industrial engineering)

This chapter's discussion is confined to granular bulk solids handling, which involves the handling and storage of all kinds of particulate matter, such as ferrous and nonferrous minerals, aggregates, cement, coal, and chemicals. In-process handling of bulk solids also involves proportioning, weighing, blending, mixing, sampling, and conveying operations.

Materials handling and storage activities in most basic industries may account for 40% to 60% of the total production cost. Therefore, close attention must be given to the engineering, design, and operations of the facilities involved. The main subjects discussed in this chapter are

- Theory of solids flow
- Design of storage silos and hoppers
- Feeders
- Mechanical conveying systems
- Pneumatic conveying systems
- Instrumentation and controls

THEORY OF SOLIDS FLOW

The theory of granular solids flow is different from that of liquid flow or hydraulics because the concept of viscosity is not applicable. In fact, the properties of solids and liquids differ so much that the mechanisms for flow in the two cases are quite different. The principal differences follow:

1. Bulk solids can transfer shearing stresses under static conditions, whereas liquids do not. Bulk solids can maintain, for instance, an angle of repose.
2. Many solids, when consolidated, possess cohesive strength and retain their shape under pressure.
3. The shearing stresses that occur in slowly deforming or flowing bulk solids can usually be considered independent of the rate of shear and dependent on the mean pressure acting within the solid. In a liquid, the situation is reversed; the stresses are dependent on the rate of shear and independent of the mean pressure.

TABLE 11.1 Sample listing of pertinent handling properties and characteristics

Physical and Mechanical Properties	Handling Characteristics
▪ Abrasiveness	▪ Aeration–fluidity
▪ External angle of friction	▪ Tendency for material to soften
▪ Angle of maximum inclination	▪ Tendency for material to build up and harden
▪ Angle of repose	▪ Corrosiveness
▪ Angle of slide	▪ Tendency to generate static electricity
▪ Angle of surcharge	▪ Degradability—size breakdown
▪ Bulk density—loose	▪ Tendency to deteriorate in storage–decomposition
▪ Bulk density—vibrated	▪ Dustiness
▪ Cohesiveness	▪ Explosiveness
▪ Elevated temperature	▪ Flammability
▪ Flowability—flow function	▪ Presence of harmful dust, toxic gas, or fumes
▪ Lumps—size and weight	▪ Hygroscopicity
▪ Specific gravity	▪ Tendency to interlock, mat, and agglomerate
▪ Moisture content	▪ Presence of oils or fats
▪ Particle hardness, size	▪ Particle shape influence

The differences from the previous page suggest that a granular bulk solid must be regarded as a plastic rather than a viscoelastic continuum.

A great many terms are used to describe the properties of bulk materials. By way of illustration, see the list in Table 11.1.

To assess a bulk solid's handleability (often referred to as flowability), a measure of the solid's shear strength must be established. The lower the resistance to internal shear within the granular material, the better the flowability. Of course, the reverse is also true; the higher the shear resistance (shear force), the worse the flowability becomes. Therefore, one of the main properties to be measured for handleability is the shear strength.

The shear strength is influenced by the type of bulk solid, degree of compaction, time of consolidation, surface moisture, ash or clay content, and particle size distribution. Other properties that play a role in flowability are bulk density, internal angle of friction, effective angle of friction, and sliding friction over specific surfaces (such as stainless steel, rusted carbon steel, plastic, or concrete).

Special testing equipment is required for measuring these flowability properties–a process that is analogous to soil testing but with further improvements and refinements. There are basically three types of shear testers: (1) linear (biaxial translatory), (2) rotational (biaxial rotational), and (3) triaxial. The most commonly used types of shear testers are in the first two categories.

Regardless of which type of shear tester is used, the test measurements are first plotted in a Mohr stress diagram, as shown in Figure 11.1. A Mohr stress circle is generally used for graphically representing combined stresses, such as normal and shear stresses (see, e.g., Merriam [1980] for more details). The resulting yield locus establishes a boundary curve for incipient failure of the test sample under a specific state of consolidation. Each Mohr diagram provides a value for the unconfined yield strength (f_c) and major principal consolidation stress (σ_1), internal angle friction (ϕ), and effective angle of friction (δ). The bulk density is also measured as part of the shear test. Angles of repose, surcharge angles, and sliding angles can also be derived.

f_c = unconfined yield strength
σ = major principal consolidation pressure
θ = internal angle of friction
φ = angle of sliding friction on wall
δ = effective angle of friction on wall

S = strength
P = pressure
P_c = pressure at which consolidation occurs

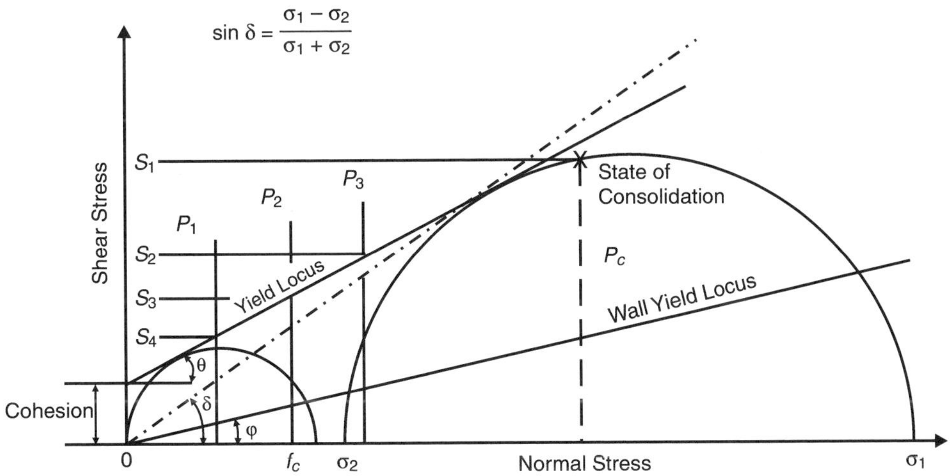

Source: Conveyor Equipment Manufacturers Association.

FIGURE 11.1 Typical plot of shear test results

DESIGN OF STORAGE SILOS AND HOPPERS

In the general field of bulk solids handling, ensuring that both the storage of materials and the movement from storage will be carried out in an effective and efficient manner is essential. However, the flow out of bins and hoppers is well known to be often unreliable; as a result, considerable costs are incurred because of consequential losses in production. Problems that commonly occur in storage bin operation include particle segregation, erratic feeding, flooding, arching, piping, and adhesion to the bin walls—all of which reduce the bin capacity below the values specified by the manufacturer. For example, a poorly flowing material may cause an arch or bridge over the hopper outlet or a stable rathole within the bin (see Figure 11.2). On the other hand, a very flowable material (dry, fine powder) may become aerated and subsequently fluidize, causing potential flooding problems.

Where flow blockages occur in practice, a common response is to resort to flow-promoting devices, which add to the expense of the installation and often result in only a marginal improvement in reliability. In most cases, the problems that occur in practice are caused by inadequate design analysis together with a lack of knowledge of the relevant flow properties of the materials.

Since 1960, significant advances have been made in the development of the theories and associated analytical procedures to describe the behavior of bulk solids under the variety of conditions encountered in materials-handling operations. Of particular note is the research associated with storage bin and discharge equipment design, for which comprehensive mathematical models and design information have been established. (See, for example, Jenike [1990].) The information enables bins to be designed to provide reliable and predictable flow under the influence of gravity.

There are basically three flow patterns in bins: mass flow, funnel flow, and expanded flow (see Figure 11.3). Each of these flow patterns has its advantages and disadvantages. Mass flow refers to a flow pattern where all the material in the bin is in a downward motion whenever the feeder is discharging. In essence, the material column slides along the hopper wall. To attain this type of flow pattern, the hopper walls must be steep and smooth. Funnel flow occurs when the material moves

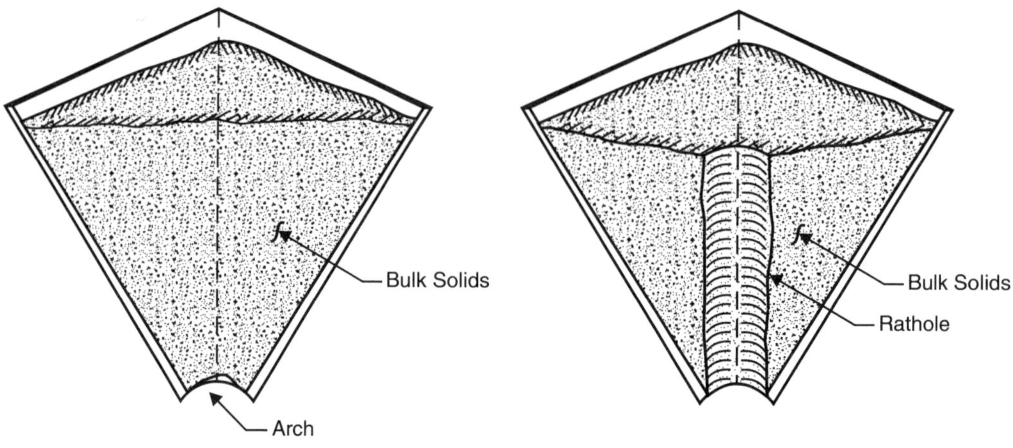

FIGURE 11.2 **Hopper flow problems: Arching and ratholes**

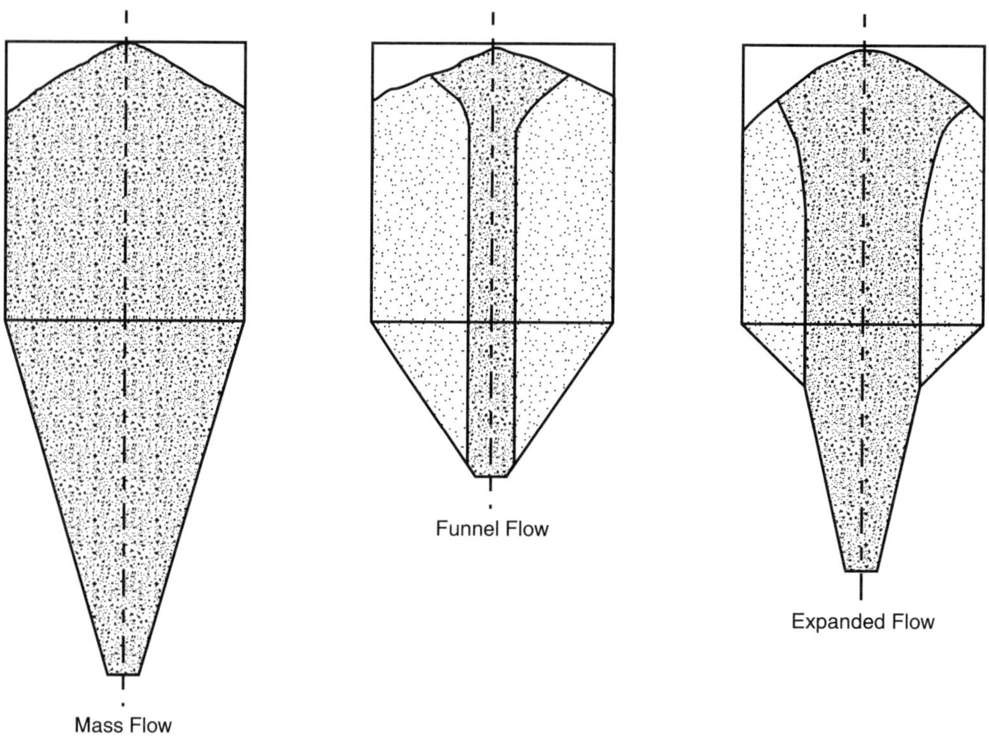

FIGURE 11.3 **Flow patterns**

strictly within a confined channel above the hopper outlet. The material outside this flow channel is at rest until the bin level drops and the material slides into the channel. The diameter of this flow channel is established essentially by the hopper outlet dimensions. However, when the cohesive strength of the material is high enough, the flow channel may possibly be emptied out without the upper layers in the bin sloughing off into the channel. In this case, a continual open channel will be formed right within

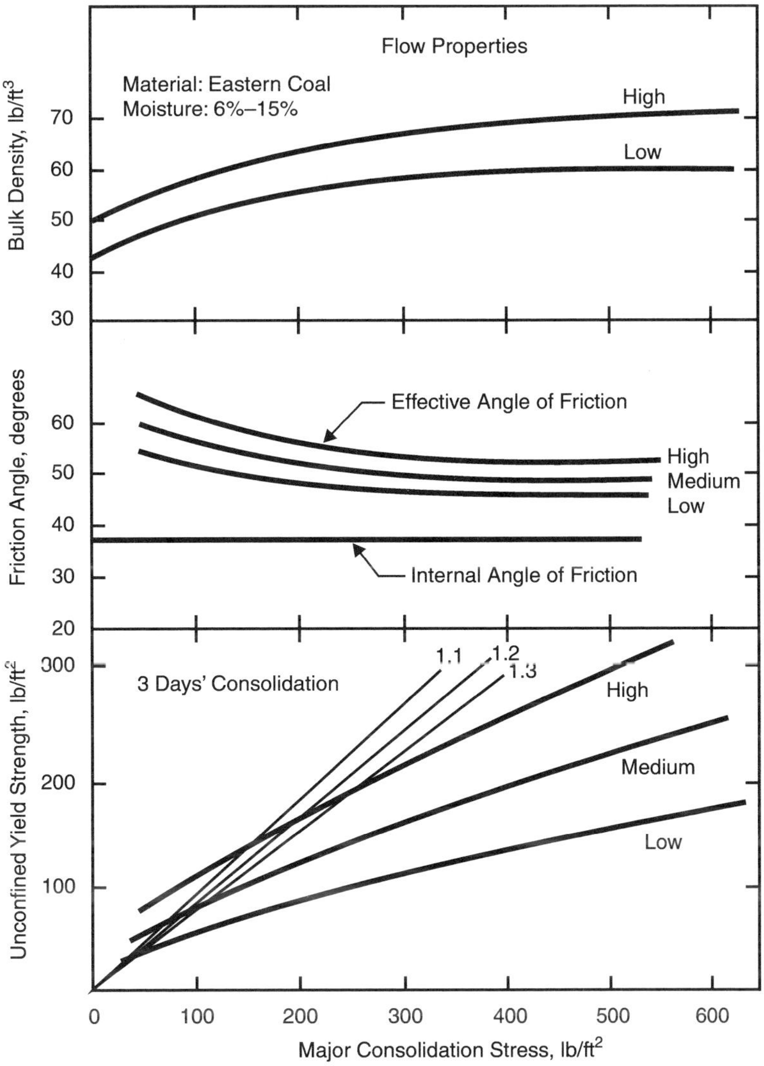

FIGURE 11.4 Typical flow-function graph for low, medium, and high strength

the bin. Such a channel is referred to as a stable rathole (see Figure 11.2). Expanded flow exhibits the mass-flow pattern in the lower hopper section up to the point where the stable rathole diameter is reached; then the flow pattern continues as funnel flow. The stable rathole diameter can be calculated when the flow properties are known.

Accurate measurement of the flow properties is essential for proper design of the storage bin and hopper. Once the shear tests have been completed, the values for unconfined yield strength (f_c) can be plotted in graphical form, as shown in Figure 11.4. The strength curves are referred to as flow functions (FF). Figure 11.4 shows three flow functions: for low-, medium-, and high-strength coals. (The lines marked 1.1, 1.2, and 1.3 represent flow factors [ff], which represent stresses in different shapes of hoppers. The intersection of FF and ff provides the critical value of the strength that is used in computing the critical arching dimension.)

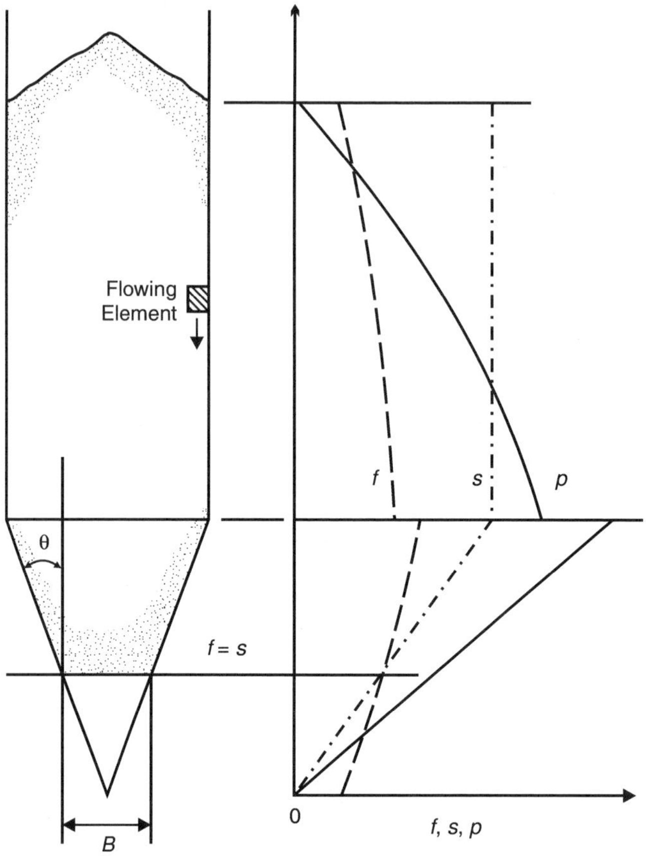

FIGURE 11.5 Flow criterion concept

Once the material strength is measured, the stresses within the granular material inside the bin can be calculated. If any arching or doming situation can develop inside the bin, the design engineer must make sure to create a geometric configuration of the bin or hopper such that the stresses in the material (s) will be larger than the strength of the material (f). The basic flow criterion requires that $f < s$ in order to maintain gravity flow.

Figure 11.5 shows a typical graphical illustration of the pressure (p), strength, and stress distributions inside a bin and hopper. The bulk solid is unconsolidated at the top of the bin because p is about zero. While the bulk solid is flowing downward, it becomes consolidated under pressure p. For each value of pressure, corresponding values exist for the material strength and stress. Close to the apex of the hopper, the f-curve and s-curve intersect. Above this point, the flow criterion $f < s$ is satisfied and gravity flow will occur. Below this intersection, we have $f > s$ and arching will occur. Therefore, this intersection identifies the critical level in the hopper and also fixes the critical opening dimension (B). A thorough engineering analysis, based on the flow functions shown in Figure 11.4, would show that the critical arching diameters for a stainless steel-lined, conical mass-flow hopper are 0.55 m (1.8 ft) for low-strength coal, 0.91 m (3.0 ft) for medium-strength coal, and 1.83 m (6.0 ft) for high-strength coal. These values represent a typical case and are intended to demonstrate the variability of coal in terms of its flowability.

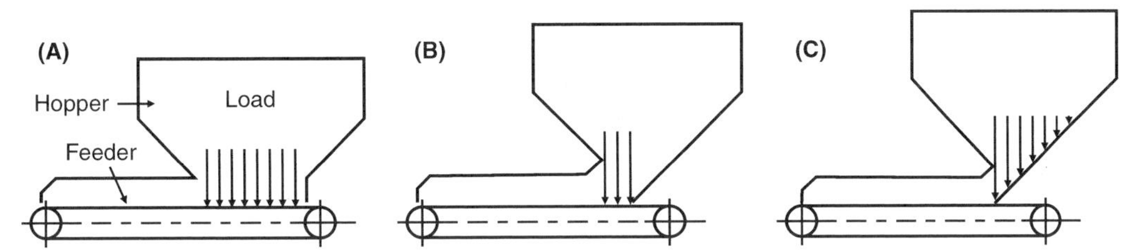

FIGURE 11.6 Feeder loads

FEEDERS

Feeders are used to provide a means of control for the withdrawal of bulk materials from storage units, such as bins, bunkers, silos, and hoppers. This control function can be performed properly only as long as the bulk materials flow by gravity to the feeder in a uniform and uninterrupted fashion. A feeder can do many things, but it should never be considered a suction pump. Many types of bulk solid feeders are on the market, but only a few will be briefly discussed in this chapter: belt feeders, apron feeders, rotary table feeders, rotary plow feeders, screw feeders, and vibratory feeders.

Feeders must be considered an integral part of the overall bin and feeder system. Improper design of either one of these parts will affect the performance of the whole system. The integral concept of bin and feeder design requires quantitative analysis of the bulk material characteristics before any attempt to design and select the components.

The design of a feeder system must start with the proper dimensioning of the hopper outlet to prevent arching, doming, or ratholing. The hopper opening size should be large enough to allow passage of the bulk solid at the required maximum discharge rate. A feeder can only throttle the flow.

Since the late 1970s, various efforts have been made to accurately determine the load or pressure on feeders mounted directly underneath the hopper opening. Many designers assume that this pressure equals the "hydrostatic" head of material above the opening (i.e., that the pressure is directly related to the head of the material, as in a water tank); they assume the pressure on a feeder to be 0.9 to 1.2 m (3 to 4 ft) of material head. Consequently, to eliminate this high pressure, the designer tends to locate the feeder in an offset position from the hopper opening and connects the two by way of a spout. However, head pressure on a feeder must be determined by using the feeder inlet dimensions and the flow properties of the bulk solids.

Figure 11.6 illustrates three examples of how the bin load may act on the feeder. In case A, the full load (which is not equal to the "hydrostatic" head of the material) acts on the feeder. In case B, the load is partly reduced by a change in the shape of the hopper. In case C, the load is completely removed from the feeder and acts only on the hopper wall. Although the advantages of cases B and C appear obvious in reducing the load on the feeder, we must consider that in these cases the effective outlet area is reduced, which may influence the flow pattern of the bulk solids. Therefore, the final choice must be related to the material characteristics. Most manufacturers consider a 0.9-m (3-ft) head load on the feeder as being equivalent to a full load, and as a result, they underestimate the head load.

Belt Feeders

A belt feeder consists of a continuous rubber belt supported by closely spaced idlers and driven by end pulleys that are generally referred to as the head and tail pulleys (see Figure 11.7). This unit is contained within a single frame; the motor can be mounted on the ground or on another frame and drives the feeder by means of V-belts. The belt feeder is usually placed under a long-slotted hopper opening feeding along the length of the hopper. Figure 11.7 shows a taper, which is an expanding dimension in the direction of feed. Usually this taper amounts to 10% expansion per 0.3 m (1 ft) length on either side

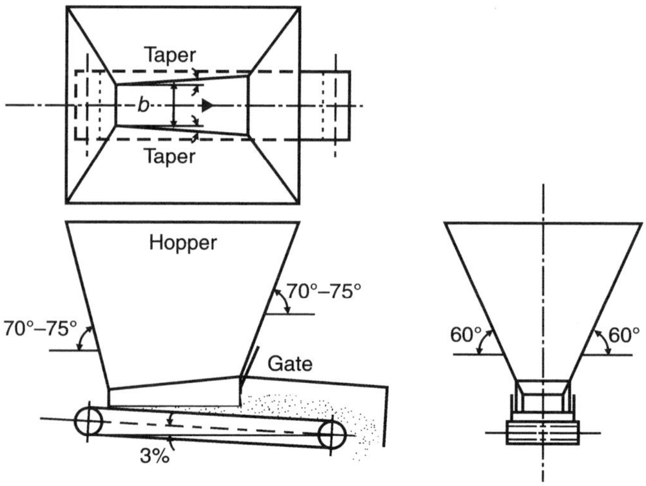

FIGURE 11.7 Belt feeder

of the belt feeder infeed. This tapering facilitates a uniform flow from a slotted hopper outlet. The slot-type belt feeder is one of the most economical feeders for bulk solids. When properly applied, the belt feeder lends itself nicely to low first cost, dependable operation, and automatic control.

Belt feeders generally range in widths of 0.6 to 1.8 m (2 to 6 ft) and have lengths of 1.5 to 4.6 m (5 to 15 ft). The capacity of the belt feeder is dependent on the width and rate of movement of the belt and is generally found to be between 4.5 and 2,270 tph (5 and 2,500 st/h).

Apron Feeders

An apron feeder consists primarily of chain-linked heavy cast manganese pans (see Figure 11.8). Usually a two-strand chain supports the feeder pan on a center rail. For very wide feeders, the use of three-strand chains is recommended. The hopper considerations for an apron feeder are, in general, the same as for a belt feeder. If the feeder is to be used under a truck dump hopper with a long hopper opening,

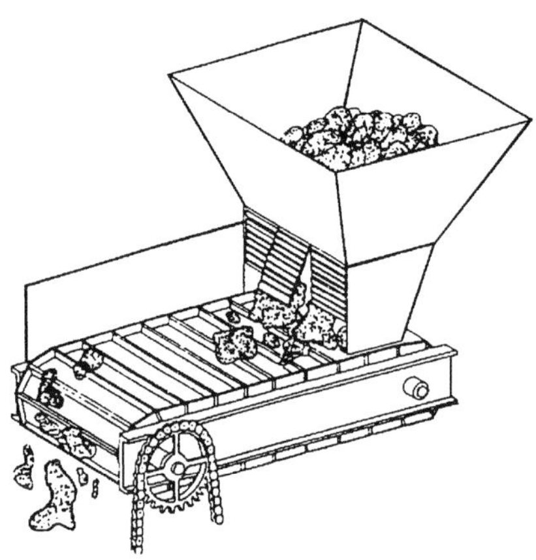

FIGURE 11.8 Apron feeder

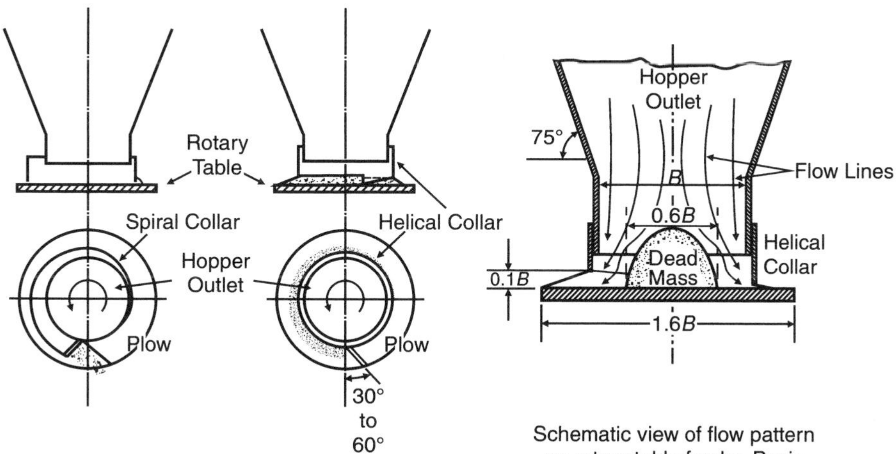

Two types of hopper-collar designs allowing a smooth flow pattern on a rotary table feeder.

Schematic view of flow pattern on rotary table feeder. Basic dimensioning is expressed in terms of hopper outlet diameter *B*.

FIGURE 11.9 Rotary table feeder

the hopper should be tapered to diverge in the direction of horizontal flow (as shown on Figure 11.7 for the belt feeder). An important point is that apron feeders are used for high-capacity, large-size materials handling, so the hopper gate must be designed to permit very large chunks to come through the hopper opening. The feeder shown in Figure 11.8 is provided with a method that overcomes potential hang-ups—caterpillar tracks are hung from the hopper outlet, thus helping to provide a flexible front wall for unrestricted flow of the large lump material. Chains are commonly used instead of caterpillar tracks for the same purpose.

Apron feeders vary in width from 0.6 to 3.0 m (2 to 10 ft) and in length from 2.4 to 30.5 m (8 to 100 ft). The lengths in excess of 4.6 m (15 ft) are used primarily for conveying material rather than as a part of the feeder itself. The capacities of apron feeders range from 91 to 2,270 tph (100 to 2,500 st/h). Power requirements for apron feeders are about twice as high as for comparable belt feeders. Apron feeders are generally used with truck dumps or in other situations where very coarse materials are handled, such as feeding primary or secondary crushers.

Rotary Table Feeders

Rotary table feeders are mostly used for cohesive materials requiring large hopper outlets, such as wet mineral concentrates, wood pulp, and wood chips, and for low feed rates (4.5 to 114 tph [5 to 125 st/h]) (see Figure 11.9). The table rotates under a stationary hopper outlet, and a fixed flow (penetrating from the side) removes the material from the table deck. This type of feeder can accommodate hopper openings up to 2.4 m (8 ft) in diameter. The table diameter is usually 50% to 60% larger than the hopper outlet diameter. Rotating speed of the table ranges from 2 to 10 rpm. The drive horsepower varies greatly from one manufacturer to another. Proper configuration of the hopper outlet, outlet collar, and plow position is essential. If the outlet collar is helical or spiral as shown in Figure 11.9, fairly uniform flow can be expected in the hopper outlet. However, a dead conical mass will still remain on the center of the table, causing most of the shearing resistance. This mass occupies a cross-sectional area of about 40% to 50% of the hopper outlet and has a height equal to about half the outlet diameter. A rotary table feeder consists primarily of a gear reducer; therefore, the cost is greatly dependent on the torque required.

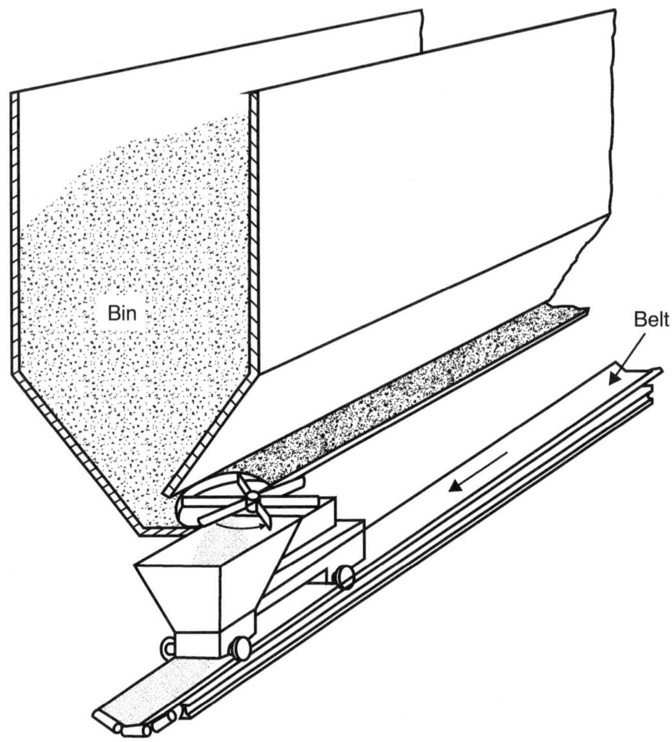

FIGURE 11.10 Rotary plow feeder

Rotary Plow Feeders

The rotary plow type of feeder has not made the same inroads in North America as it has in Europe. It was first developed in Germany in the 1930s for the feeding of lignite. Since then, it has found a wide field of application for other materials, such as sinter, coal, potash, phosphate, limestone, iron ore, and cement clinker. Rotary plow feeders are suitable for use in reclaim tunnels, under storage piles, or under long storage bins. The plowing mechanism (see Figure 11.10) consists of curved arms arranged to sweep material off a narrow shelf running the length of the storage pile or bin. The traversing and rotating plow scrapes the material from a stationary shelf. The plow machinery is attached to an independently driven carriage that contains a receiving hopper above a belt conveyor.

Screw Feeders

A screw feeder is essentially designed for very low-tonnage outputs, where positive discharge must be ensured. This type of feeder offers an advantage in that the feeder itself can be easily enclosed, making it dust-tight. Thus, it provides a closed hopper-and-chute arrangement from the hopper to the delivery point. The feeder consists primarily of a helical screw rotating beneath the hopper outlet and driven from an external source (see Figure 11.11). The screw itself can be of a fixed pitch or can have a smaller pitch spacing in the rear with gradual increases in pitch to the discharge end. This latter arrangement ensures that the material will be moving in the back portion of the hopper. Occasionally, screw feeders will be required to have a tapered screw; that is, a smaller diameter in the back that gradually increases to the largest diameter at the outlet. This taper ensures near uniform material removal from the hopper outlet.

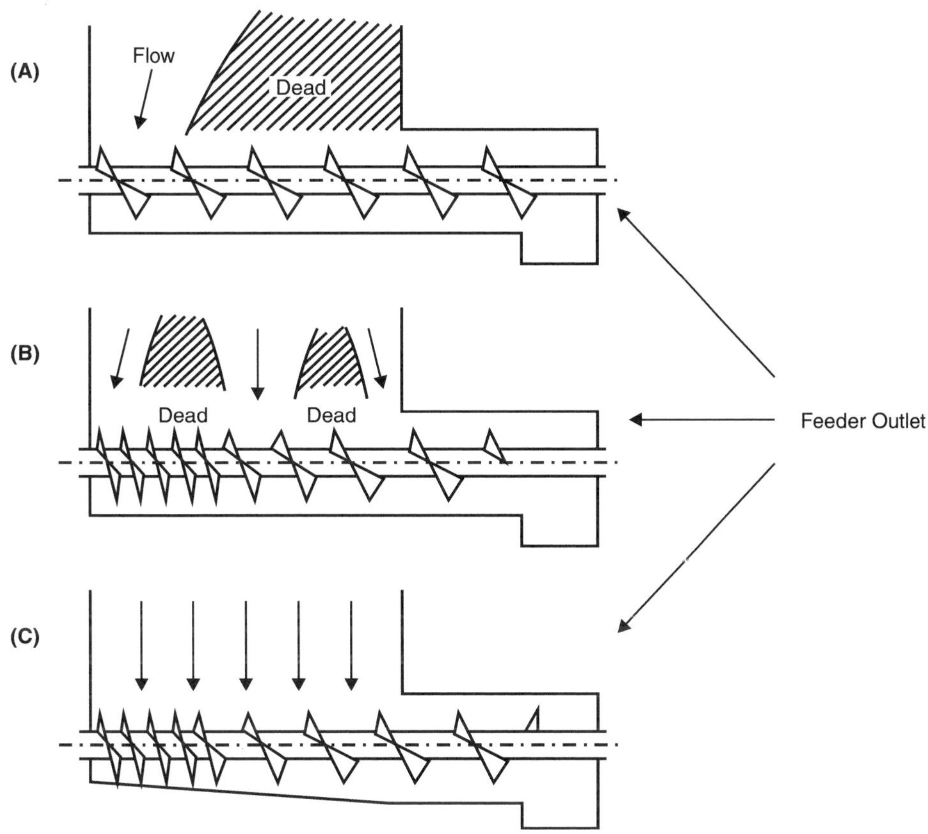

FIGURE 11.11 Screw feeder: (A) uniform pitch and uniform diameter, (B) graduated pitch and even diameter, (C) increasing pitch and increasing diameter

The entrainment pattern of the stored material in the screw is the feature that determines the pattern of flow across the hopper outlet slot. Where no inflow may take place, there will be a "dead" region in the foregoing space. Dead regions develop because the material does not feed into the screw feeder flights. Such a region does not allow a mass flow and may cause deterioration of the flow properties of the static material, along with all the other consequent disadvantages. Figure 11.11 shows typical flow patterns for various screw forms. By changing the pitch of the feed screw or changing the shaft diameter, dead regions can be minimized.

Vibratory Feeders

The process involved in determining the design parameters of a vibratory feeder—which uses vibration to induce motion of the particles that exit the bin—is rather complex. Many papers on this subject have been published since the late 1970s. Material on the feeder trough is subjected to the forces of gravity, along with normal, friction, and impact forces. Basically, the feeder trough or pan is driven by a nearly sinusoidal force at some angle θ to the trough. When the feeder is operating, the trough is oscillating along a straight line, with the amplitude and direction determined by the driving force.

The resultant linear vibration is a repetitive series of throws and catches that move the material on the trough. Figure 11.12 demonstrates the action for a single particle on the trough. The particle is in contact with the trough for approximately one-fourth of the drive cycle (shown as point A to point B in

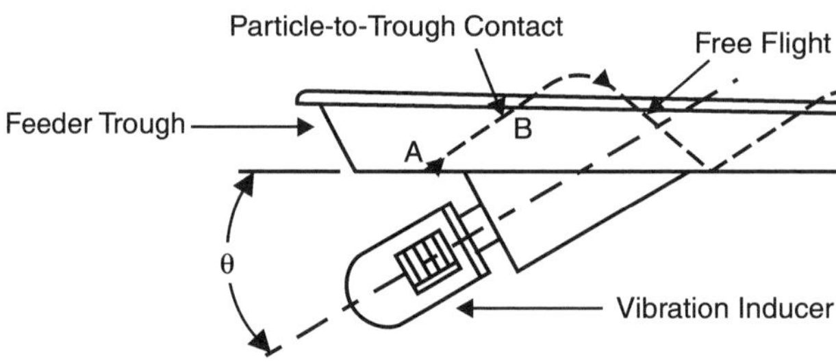

FIGURE 11.12 Movement of a single particle along a vibratory feeder

the figure). When the particle leaves the trough, it travels with a uniform horizontal velocity, but the vertical velocity gradually decreases because of gravity. At some later point, the trough again contacts the particle, and the process is repeated. This process, then, conveys material along the trough at a rate of 0 to 18 m/min (0 to 60 ft/min) depending on the combination of drive frequency, amplitude, drive angle, and feeder inclination. These parameters, as well as material flow depths and feeder trough widths, allow material to be delivered at rates ranging from several kilograms or pounds per hour to more than 1,800 tph (2,000 st/h).

Various manufacturers of vibratory feeders have selected different operating parameters for the trough movement. Operating frequencies generally vary from 600 to 3,600 vibrations per minute; amplitudes range from a few thousandths of a millimeter up to 8 mm (a few thousandths of an inch up to 1/4 in.) or more, and the drive angle ranges from 20° to 45°. For any given material, an optimum operating combination of frequency, stroke, and drive angle will exist. For vibratory feeders, subresonant tuning is mandatory.

MECHANICAL CONVEYING SYSTEMS

Manufacturers of mechanical conveyors and elevators have made available to the basic industries a wide variety of equipment for moving bulk solid materials. This section of the chapter looks closely at a number of devices, both stationary and portable, that convey bulk solids between two fixed points with a continuous drive and either a continuous or intermittent forward movement.

CEMA has defined about 80 types of conveyors, 10 types of elevators, and 50 types of feeders. Because covering each one in detail here would be impractical, this section will focus on a few of the most common types: belt, screw, chain, and vibratory conveyors, as well as bucket elevators.

Belt Conveyors

The endless moving belt, perhaps the most popular of conveyors, is widely employed to transport materials horizontally or on an incline, either up or down. Figure 11.13 shows a typical belt conveyor arrangement, identifying the five main components of the system:

1. The belt, which forms the moving and supporting surface on which the conveyed material rides
2. The idlers, which form the supports for the carrying and return strands of the belt
3. The pulleys, which support and move the belt and control its tension
4. The drive, which imparts power to one or more pulleys to move the belt and its load
5. The structure, which supports and maintains the alignment of the idlers and pulleys and supports the driving machinery

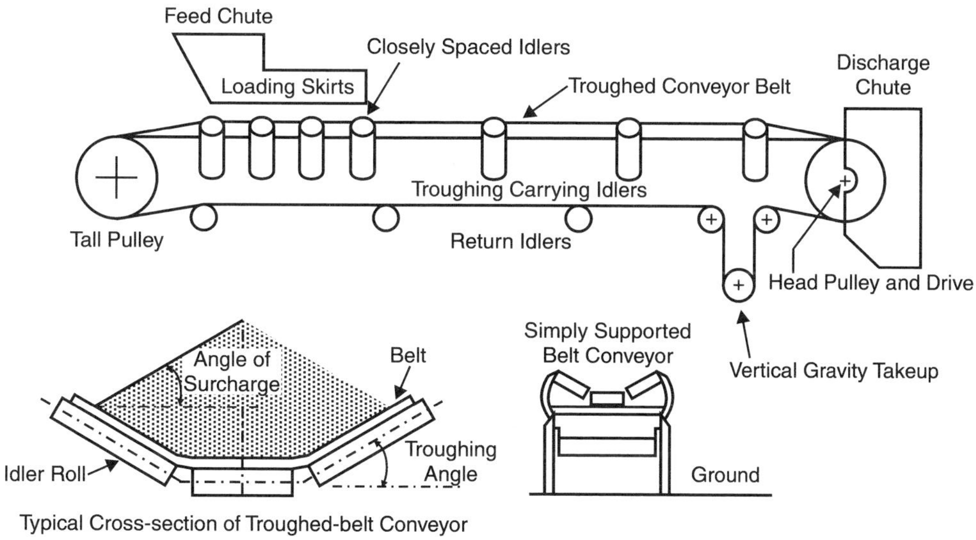

FIGURE 11.13 Schematic of belt conveyor system, showing the major components

Almost all belt conveyors for bulk solids use rubber-covered belts, the inner carcass of which provides the strength to pull and support the load. The carcass is protected from damage by rubber layers that vary in thickness for different applications.

Belt conveyors can move material at rates ranging from a few kilograms or pounds per minute to thousands of metric tons or short tons per hour. A great variety of materials can be handled. Depending on belt width, however, lump size can be a limitation, and dusty compounds can be troublesome. Wet or sticky bulk solids warrant special consideration, and temperatures higher than 66°C (150°F) should be approached with caution. Some solids react with rubber in the belt, necessitating a special covering for the belt.

The maximum slope over which a belt conveyor can operate depends, of course, on the characteristics of the product. Most conveyor manufacturers have data on the maximum suggested angles for various materials. For the average application, limiting the angle of inclination to somewhat less than the suggested maximum is a good idea.

Figure 11.13 shows a typical cross section of a troughed-belt conveyor. In North America, the standard troughing angles are 0°, 20°, 35°, and 45°. The angle of surcharge is a property of the material and can be compared with the dynamic angle of repose. Tables are available that list cross-sectional areas for different surcharge angles.

CEMA's detailed design manual for belt conveyors (CEMA 1979) is a recommended source of information. Power requirements for belt conveyors depend on many variables related to conveyor profile, the type of drive-pulley arrangement, belt tensions and belt speed, and type of idler spacing. Detailed discussions of this subject may be found in various CEMA publications. For estimating purposes, simplified methods of determining power may be used.

Screw Conveyors

A screw conveyor usually consists of a long-pitch, steel-helix flight mounted on a shaft, supported by bearings within a U-shaped trough (see Figure 11.14). As the element rotates, material fed to it is moved forward by the thrust of the lower part of the helix and is discharged through openings in the trough bottom or at the end. When properly used, this type of conveyor does a good job, and its cost

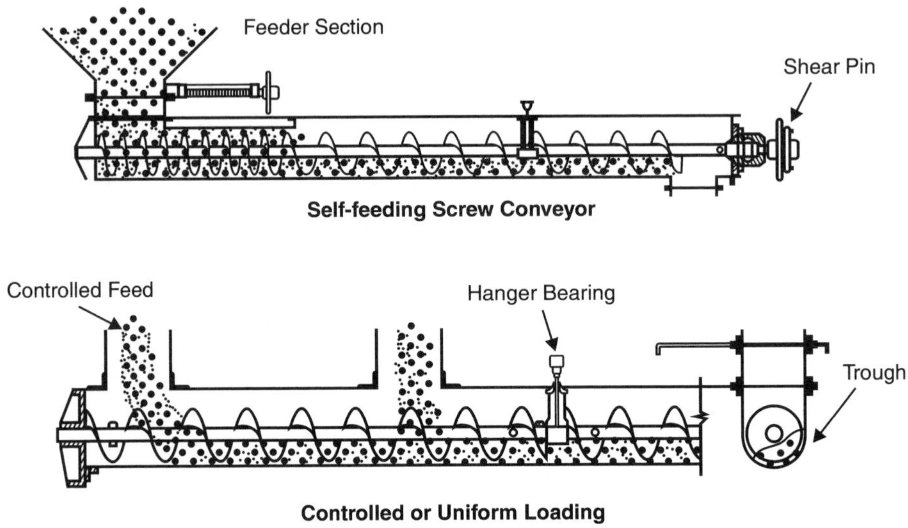

FIGURE 11.14 Arrangements by which solids enter screw conveyor

will often be only about half that of another type of conveyor. A screw conveyor is easy to maintain, inexpensive to replace, and readily made dust-tight. For many uses, it is the preferred type of conveyor.

Screw conveyors can be operated with the path inclined upward, but capacity decreases rapidly as the inclination increases. A standard-pitch screw inclined at 15° above horizontal retains 70% of its horizontal capacity. If the screw is inclined 25°, the capacity is reduced to 40%; if it is inclined 45°, the material will move along the floor of the trough at a greatly reduced rate. For steep inclines, the helix may be given a short pitch, and the trough may be made tubular to reduce the capacity loss. With a jam feed, such a conveyor can deliver about 50% of its horizontal capacity at a 45° incline.

The allowable loading and screw speed are limited by the characteristics of the material. Light, free-flowing, nonabrasive materials fill the trough deeply, permitting a higher rotating speed than with heavier and more abrasive materials. Manufacturer recommendations on screw conveyor operation should be followed.

Chain Conveyors

Chain conveyors employ continuous chains that travel the entire length of the conveyor, transmitting the pull from the driving unit and, in some cases, carrying the whole weight of the transported material. The material may be carried directly by the chains, by flights pushed or towed by the chains, or by special attachments fitted to the chains. The conveyor types derive their names from the attachment; for example, apron conveyors, flight conveyors, and drag-chain conveyors (see Figure 11.15). Chain conveyors are particularly suited for systems that require complete enclosure (for dust containment), minimal conveyor housing cross sections, the ability to load or discharge materials at different points from the same conveyor, combinations of horizontal and vertical paths, or the handling of materials at elevated temperatures.

Vibratory Conveyors

Vibratory or oscillating conveying is used widely to transfer many types of granular materials. It can be matched with such other process functions as screening, cooling, drying, and dewatering. Although construction and installation of these conveyors are relatively simple, the engineering and

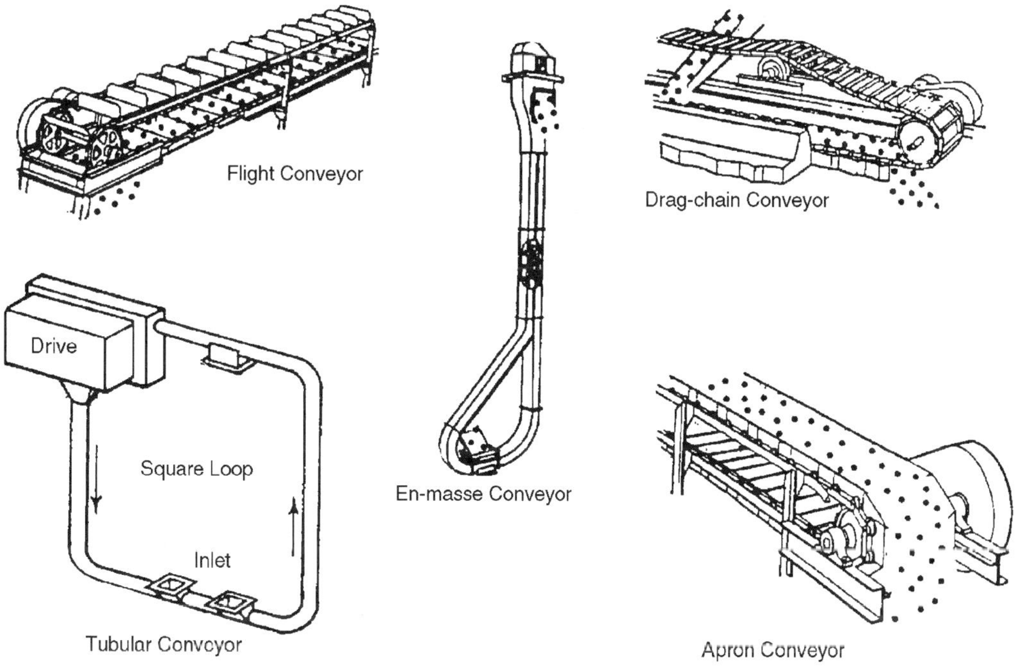

FIGURE 11.15 Variations of chain conveyors

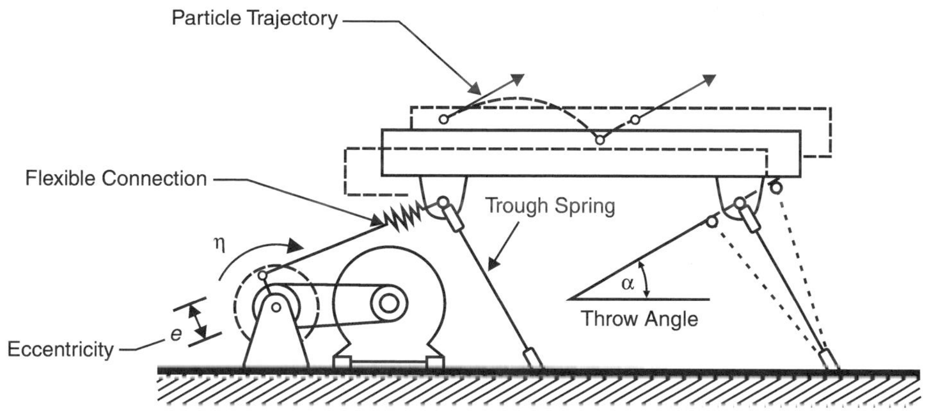

FIGURE 11.16 Simple vibratory conveyor

design analyses of the vibratory mechanics are complex, requiring a fairly high degree of mathematical understanding.

Figure 11.16 shows a typical schematic of a simple vibratory conveyor, consisting of a carrying trough, supporting legs or springs, and a drive system. The drive system imparts to the carrying trough an oscillating motion of a specific frequency and amplitude. The bulk material on the trough is moved along by the periodic trough motion. The stroke of the trough is equal to twice the amplitude of vibration. A basic distinction between vibratory and reciprocating equipment is that, in the former case, the

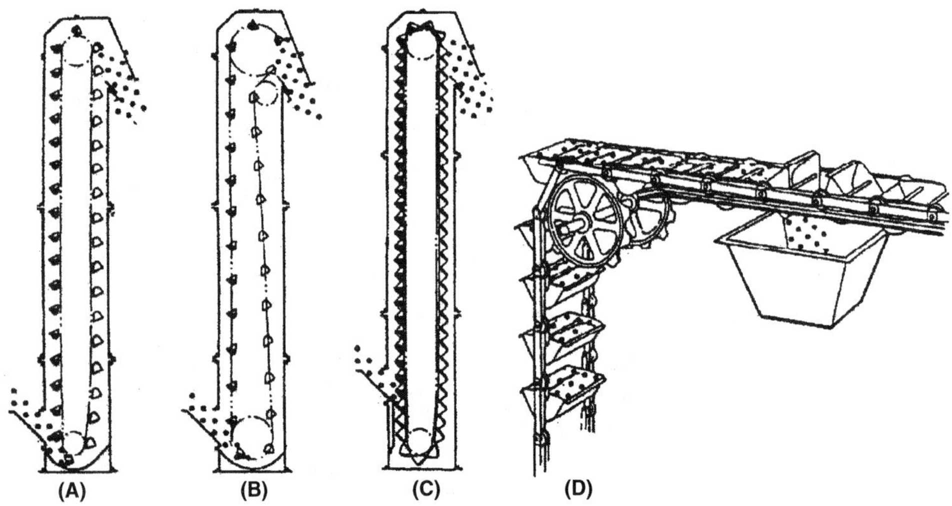

FIGURE 11.17 Design options for discharge from bucket elevators: (A) centrifugal discharge; (B) positive discharge; (C) continuous bucket; (D) pivoted-bucket conveyor/elevator

material is bounced from the conveying surface during transport, whereas in the latter case the material simply slides over the trough.

Equipment catalogs generally classify vibrating conveyors according to their ultimate application, such as foundry conveyors or grain conveyors, or by their type of duty–light, medium, heavy, and extra heavy. The design required for a specific type of service is designated by the manufacturer.

The capacity of a vibratory conveyor is determined largely by the trough cross section and the velocity at which the material is conveyed. The linear flow rate or transport velocity of the material in the trough is almost directly proportional to the product of frequency and stroke (assuming the drive angle is properly selected to provide enough, but not excessive, vertical acceleration). Longer strokes and higher frequencies are preferred. However, the combination of high frequencies and long strokes means higher structural stress and therefore more massive and costly equipment. Because the stresses are proportional to the product of the stroke and the *square* of the frequency, vibratory conveyors–which are normally fairly long pieces of equipment–are generally of the low-frequency, high-stroke design. Vibratory feeders, on the other hand, are designed as rugged, relatively small pieces of equipment with the structural integrity to withstand the high-frequency oscillation.

The power requirements of vibratory conveyors vary depending on the type of design. Quite often, power is determined solely by the start-up characteristics of the conveyor.

Bucket Elevators

CEMA (1990) has defined a bucket elevator as "a conveyor for carrying bulk materials in a vertical or inclined path, consisting of an endless belt, chain or chains to which buckets are attached, the head and boot terminal machinery, and supporting frame or casing." Because the belt or chain operates unidirectionally, this definition does not include skip hoists and freight elevators. Furthermore, the discussion here covers only vertical elevators; the use of inclined elevators is limited in the United States. In most instances, conveying horizontally, elevating, and conveying again are more economical than performing these functions simultaneously with an inclined bucket elevator.

Vertical bucket elevators can be classified into four major groups, according to the means used to convey and discharge material (see Figure 11.17). In centrifugal-discharge elevators, material is released by centrifugal action. These units, consisting of buckets mounted on a chain or belt at regular intervals, operate at a minimum rate of 76 m/min (250 ft/min). The lump size of handled material is usually no more than 50 mm (2 in.). In continuous bucket elevators, material is released by gravity. Buckets are mounted back to back on a continuous chain or belt, and the elevator operates at a rate of 36.6 to 38.1 m/min (120 to 125 ft/min). These elevators will successfully handle materials 50 to 100 mm (2 to 4 in.) in size. Positive-discharge elevators are spaced-bucket elevators in which the buckets are turned over by the idler wheels. Buckets are held over the discharge chute long enough to permit free gravity discharge. These units operate at no more than 36.6 m/min (120 ft/min) and are used to handle sticky solids. Hinged/pivoted bucket elevators are intended for a closed-circuit path in a vertical plane. They consist of a train of overlapping buckets pivotally suspended between strands of chain, with supporting rails or guides, turn wheels, dive, and tripper or dumper mechanism to up-end the buckets for discharge.

PNEUMATIC CONVEYING SYSTEMS

A pneumatic conveying system uses a flow of air as the carrying medium for transport of solids through a pipeline. The velocity of the airstream keeps the solid particles in suspension. This type of conveyance is often called "two-phase flow."

The practice of pneumatic conveying is still very empirical and is sometimes applied in inappropriate situations. Many universities around the world are conducting research in this field, but the theoretical solutions for two-phase flow are often too complex for the practicing engineer. Besides, many of these solutions require experimentally derived coefficients, which are not readily available.

Figure 11.18 shows typical layouts of a total system, which can be either a negative-pressure (vacuum) or positive-pressure system. A positive-pressure system uses an airflow with a pressure above atmospheric; a negative-pressure system uses an airflow with a pressure below atmospheric, like a vacuum cleaner.

Pneumatic conveying systems are classified into five basic categories depending on the range of velocities and pressures (Table 11.2). The high-velocity and low-pressure systems are termed "dilute-phase systems," whereas the low-velocity and high-pressure systems are known as "dense-phase systems." The air-activated gravity conveyor (sometimes referred to as an airslide) is in a separate category by itself.

The discussion here will focus primarily on dilute-phase systems because they are still the most commonly used in the industry. Dense-phase systems rely not on keeping the bulk solids in suspension in the airstream during conveyance, but rather on pushing the solids more as a plug through the pipeline—hence, the higher pressures.

Conveyance of solids suspended in an airstream through a pipeline is, in essence, similar to other hydraulic conveyances. The pressure drop along the conveying line is primarily dependent on transport velocity, pipe diameter, bends and elbows, system length, solids-to-air ratio, and types of solids handled. A few theoretical equations are available in the industry for computing the pressure drop of a pneumatic conveying system. These computations, however, are fairly complex and generally require the use of a computer.

Quite a few design combinations are possible depending on air velocity, solids flow rate, and pressure drop. Additional bends or elbows can often be simulated as "equivalent lengths." For example, for 90° bends with a bend-radius-to-pipe-diameter ratio of about 12, the equivalent length is typically about 4.6–6.1 m (15–20 ft) for air only, assuming at least 4.6–6.1 m (15–20 ft) of distance is present between elbows.

For dilute-phase systems, a general recommendation is to allow for at least 4.6 m (15 ft) horizontal run of pipeline before a bend or elbow is applied. This arrangement allows the particles in the airstream to accelerate to sufficient speed before they are slowed down again at the first bend or elbow.

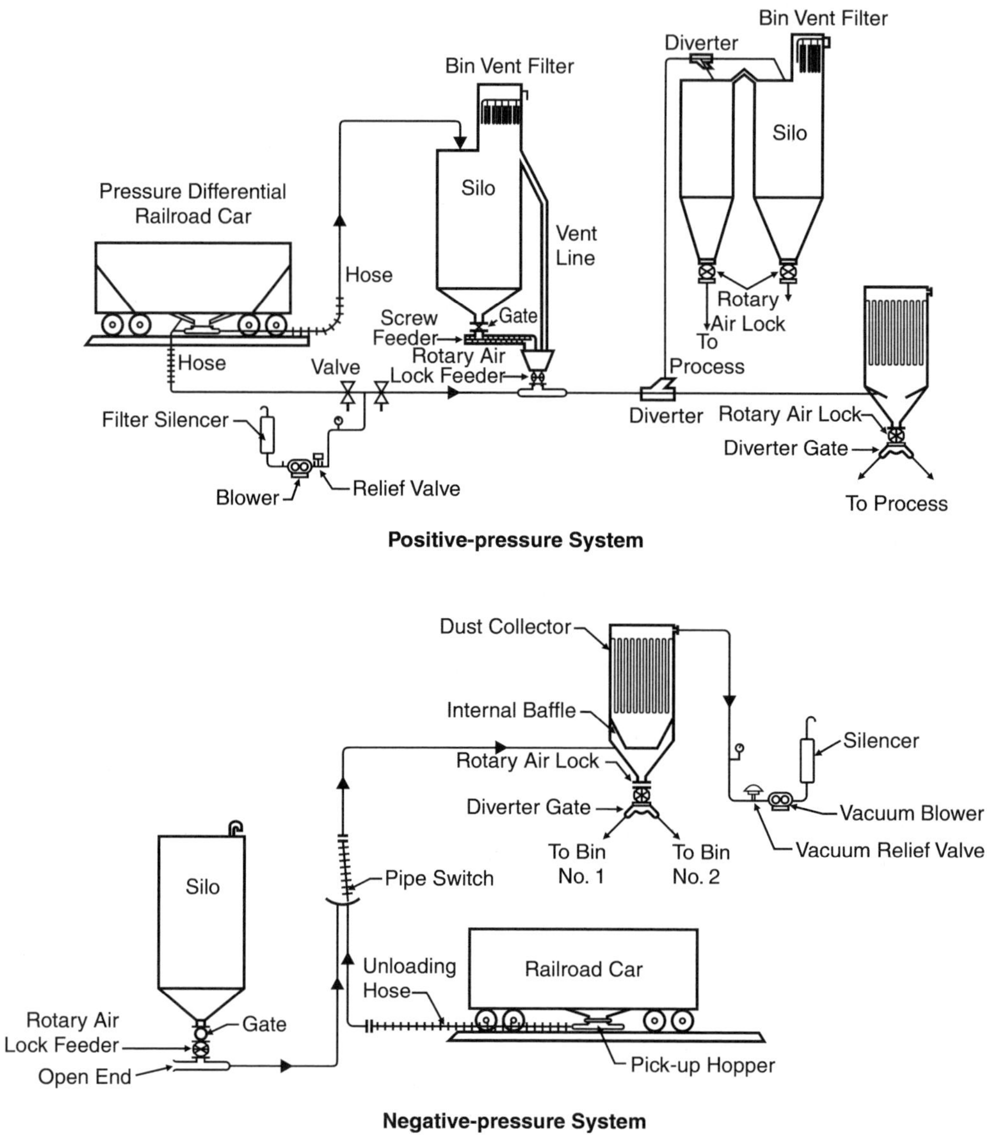

FIGURE 11.18 Pneumatic conveying systems

Routing of conveying lines is a key element in establishing a good plant layout. Short distances and a minimum number of bends are desirable. Dilute-phase conveying lines should, in general, comprise only horizontal and vertical runs. Behavior of solids in upwardly inclined dilute-phase conveying is unpredictable, so such layouts should be avoided.

INSTRUMENTATION AND CONTROL

In bulk handling systems, the subject of instrumentation and controls may refer either to the control of drive motors for conveying, stockpiling, and reclaiming systems or to the associated activities such as weighing, proportioning, and sampling. This section will deal solely with the latter aspect.

TABLE 11.2 Classification of pneumatic conveyor systems

Parameter	Dilute Phase: Fan System	Dilute Phase: Blower System	Dense Phase: Pump System (Medium Dense)	Dense Phase: Blow Tank System	Air-activated Gravity (Airslide System)
Pressure range	20 in. H_2O	7 psi	15–35 psi	30–125 psi	Fan type: 0.5–1 psi (closed); 4–5 psi (open)
Saturation, ft^3 air/lb material	Vacuum: 10–30; pressure: 4.5–13	Vacuum: 3–5; pressure: 1–3.5	0.35–0.75	0.1–0.35	3–5 $(ft^3/min)/ft^2$
Material loading, lb material/lb air	Vacuum: 1.3–0.45; pressure: 3–1	Vacuum: 4.5–2.5; pressure: 13–3.8	45–18	135–45	–
Air velocity, ft/min	6,000	4,000–8,000	1,500–3,000	200–2,000	10 through diaphragm
Maximum capacity, st/h	50	100	300	400	500
Practical distance limits, ft	Vacuum: 100; pressure: 200	Vacuum: 200; pressure: 500	2,000	5,000	100 ft in 6-ft-long sections at 3° to 10° downslope

Note: Conversions to Système International (SI) units—1 in. = 25.4 mm; 1 psi = 6.895 kPa; 1 ft = 0.305 m; 1 ft^2 = 0.093 m^2; 1 ft^3 = 0.028 m^3; 1 st = 0.907 t.

The ever-increasing price of bulk commodities forces buyers and sellers in world markets to take a more careful look at methods for obtaining accurate accounting of commodity transactions in commerce and material utilization in processes. Depending on the application of a sampling and weighing system or the use intended, inaccuracies can result in a loss of income or a loss of quality in the process. The inevitable result is a loss of control over costs. The accuracy of sampling and weighing systems is extremely important because these systems are recognized to be the key components of the overall management approach to providing both quantitative and qualitative control.

Bulk Weighing Techniques

Several types of systems are currently in use to determine the weight of bulk commodities shipped or received:

- Truck scales
- Railroad track scales
- Rotary dumper scales
- Hopper scales
- Belt conveyor scales
- Vessel drafting

Unlike static weighing devices, such as track scales and hopper scales, a belt conveyor scale is a dynamic weighing device requiring time integration. The material weight in kilograms per meter (or pounds per foot) is integrated with belt travel over a period of time. A belt scale is capable of accurate weighing (down to as low as 0.25% of the scale rating) and is the least expensive of the scale devices listed above.

For a more detailed description of bulk solids weighing systems, the published literature should be consulted (see, for example, Colijn [1983]). An important point to keep in mind is that a weighing system is not simply a scale. A scale is a manufactured piece of equipment, normally statically tested at the plant. Under actual conditions of operations, environment, and bulk solids flow, the scale may behave quite differently from what is expected.

Not all of the weighing systems listed above will be suitable for a particular application. An engineering study should be conducted for each application to evaluate all aspects of the applicable systems and to establish their cost-effectiveness. The buyer should become acquainted with the different options that are available.

The bulk weighing system selected is usually determined on the basis of several factors:

- Desired accuracy
- Capital cost of equipment
- Maintenance costs
- Customer preferences
- Regulatory requirements

If the weighing system is used for commercial payment or tariff agreements, the users should find out what regulatory agency is involved and who has jurisdiction. They should become acquainted with the specifications and requirements for the weighing system under consideration.

Particular attention should be given to the testing, scale maintenance, and certification procedures of the various weighing systems. One system can appear less expensive than another when only the initial capital cost is considered, but it may become more costly when maintenance and calibration expenses are included.

When the requirement for a weighing device is approached from a systems point of view, the feasibility of installing the device into an environment conducive to accuracy must be thoroughly examined. In other words, the features of the total materials-handling facility must be considered, such as bulk solids flow properties, flow regulation and rate of flow, potential changes in moisture, loading and unloading conditions of conveyors, spillage, structural deflections or foundation settlements, and freezing.

Since the late 1970s, there has been tremendous development of electronic equipment in the weighing industry. The point has been reached where weighing systems are now primarily thought of as being digital electronic devices controlled by microprocessors. However, this concept can—and often does—lead to problems in weighing accuracy because operators tend to forget that weight determination is still a force measurement and, therefore, subject to the basic principles of a mechanical system. The "load" to be measured—whether this measurement takes place on a belt scale or track scale—still sits or moves on a weigh bridge. This load must be transferred to the load-sensing element without the addition or subtraction of any other forces. Even a digital device will give an incorrect reading if used in an improper setting.

The use of minicomputers in weighing offers no real advantage in terms of the accuracy of weight measurement. However, it does offer distinct advantages in terms of information processing, display, data conversions, and controls, as well as self-diagnostics and troubleshooting features. A display screen may be included with a prompter to guide the operator through the selection of various options available for testing and calibration.

Microprocessors will play an invaluable role in permitting industrial users to gather data quickly—a feat that heretofore was either not available or not economically feasible. They will also permit correction of other elements within a weighing system, as well as automatic calibration to correct for recorded error (i.e., sensed but not "recorded" after calibration against a reference point).

Bulk Sampling Techniques

Over the years, bulk sampling has evolved from the use of very simple concepts to multistage sampling systems of greater and greater complexity to accommodate rapidly changing sampling requirements and increase tonnage flow rates. For example, at the time of this writing (early 1999), some installations are handling feed rates as high as 9,100 tph (10,000 st/h) with the maximum particle size sometimes exceeding 15 cm (6 in.).

The proper selection of a sample involves an extensive understanding of the physical characteristics of the material, the minimum number and mass of the increments to be taken, the lot size, flow rates, the size consist, the condition of the material (wet, dry, frozen), and the overall sampling precision that is required. The need for sampling occurs at various points from the mine face to the end user. The design requirements, however, may vary greatly as the objectives for the sampling vary. The justifications for sampling generally fall under one of the following categories:

1. To determine quality for purchase or sale
2. To control a process or operation, such as blending or combustion
3. To facilitate inventory control for the purposes of material balances, cost estimates, and taxes
4. To estimate reserves in the ground

Each of these categories will eventually influence the final design and operation of the sampling facilities. Lot size, flow rates, lump size, material properties, and variability are the basic parameters that influence the design of any sampling facility.

The designs of the majority of mechanical sampling systems are based on standards generated by the American Society for Testing and Materials (ASTM), the International Organization for Standardization (ISO), and the Japanese Standards Association. In their standards, these groups delineate methods and procedures for the collection of material samples.

The number and weight of increments required for a given degree of precision depends on the variability in the sample itself. This variability increases with the increase in free impurities. For example, an increase in ash content of a given coal usually indicates an increase in total variability.

Therefore, a mandatory requirement is that not less than a minimum specified number of increments of not less than the minimum specified mass must be collected for the total lot.

Unfortunately, the typical mechanical sampling system in use today is basically a gravity-flow-type bulk materials-handling facility, flowing at very low (frequently intermittent) mass flow rates. This fact is generally given too little recognition. In current practice, equipment is generally sized on the basis of flow rates only, without adequate consideration for the cohesive and/or adhesive properties of the sample–properties that a reduction in particle size will exacerbate tremendously. As a result, many sampling systems are seriously deficient in their performance.

REFERENCES

CEMA (Conveyor Equipment Manufacturers Association). 1979. *Belt Conveyors for Bulk Materials.* Rockville, Md.: CEMA.

——. 1980. *Classification and Definitions of Bulk Materials.* Book 550. Rockville, Md.: CEMA.

——. 1990. *Conveyor Terms and Definitions.* Book 102. Rockville, Md.: CEMA.

Colijn, H. 1983. *Weighing and Proportioning of Bulk Solids.* 2nd ed. Clausthal-Zellerfield, Germany: Trans Tech Publications.

Jenike, A.W. 1990. *Storage and Flow of Solids.* 14th printing. Bulletin 123. Salt Lake City, Utah: University of Utah, Utah Engineering Experiment Station.

Merriam, J.L. 1980. *Engineering Mechanics: Statics and Dynamics.* New York: John Wiley & Sons.

. . . .

CHAPTER 12

Hydrometallurgy and Solution Kinetics

Kenneth N. Han and Maurice C. Fuerstenau

Once desired minerals are separated from a mixture of many minerals in an ore by a physical separation process such as froth flotation, the minerals can then be subjected to chemical processes for metal extraction. There are, in general, two different ways of achieving chemical release of metals from minerals: (1) hydrometallurgical processes and (2) pyrometallurgical processes. The hydrometallurgical processes use water as the medium, whereas the pyrometallurgical processes rely on a high-temperature treatment. In this chapter, the hydrometallurgical processes of extracting metals from various sources will be studied.

INTRODUCTION

Consider an equilibrium reaction given by Eq. 12.1:

$$a<A> + b\{B\} + c(C) \rightarrow d<D> + e\{E\} \qquad \textbf{(Eq. 12.1)}$$

In general, the Gibbs free energy, ΔG, can be given in terms of activities of various species involved in the reaction:

$$\Delta G = \Delta G^o + RT\, ln \frac{a_D^d\, a_E^e}{a_A^a\, a_B^b\, a_C^c} \qquad \textbf{(Eq. 12.2)}$$

The symbols, < >, { }, and ( ) represent a solid phase, a liquid phase, and a gas phase, respectively. It should be noted that when the Gibbs free energy of reaction, ΔG, is negative, the reaction given by Eq. 12.1 will take place spontaneously. On the other hand, when ΔG is positive, the reaction will not occur.

Suppose zinc is to be extracted from three different sources—namely ZnS, ZnO, and elemental Zn—by using an acid. We could write the following stoichiometric equations representing these extracting processes.*

$$\begin{array}{lccccccccl}
 & <ZnS> & + & 2\,\{H^+\} & \rightarrow & \{Zn^{++}\} & + & (H_2S) & & \\
\Delta G^o_{f,25^\circ C} & -47.4 & & & & -35.18 & & -6.54 & \Delta G^o_{R,25^\circ C} = 5.68 \text{ kcal/mol} & \textbf{(Eq. 12.3)} \\
 & <ZnO> & + & 2\,\{H^+\} & \rightarrow & \{Zn^{++}\} & + & \{H_2O\} & & \\
\Delta G^o_{f,25^\circ C} & -75.69 & & & & -35.18 & & -56.69 & \Delta G^o_{R,25^\circ C} = -16.18 \text{ kcal/mol} & \textbf{(Eq. 12.4)} \\
 & <Zn> & + & 2\,\{H^+\} & \rightarrow & \{Zn^{++}\} & & (H_2) & & \\
\Delta G^o_{f,25^\circ C} & & & & & -35.18 & & & \Delta G^o_{R,25^\circ C} = -35.18 \text{ kcal/mol} & \textbf{(Eq. 12.5)}
\end{array}$$

As can be seen in these three equations, the dissolution of ZnS in acid is least likely from the standpoint of thermodynamics. In fact, sphalerite, ZnS, is not leached in acidic solutions. As a result, it

*The Gibbs free energy of formation of various species can be obtained from various literature sources: Weast, Astle, and Beyer (1985); Dow Chemical (1985); Kelley (1960); Kubaschewski and Evans (1979); Garrels and Christ (1965); Pourbaix (1966); Latimer (1952); FACT (1997); and Martell and Smith (1982).

is first subjected to roasting at high temperature to convert it to either ZnO or $ZnSO_4$ before being subjected to leaching.

On the other hand, the dissolution of either ZnO or Zn in acid is thermodynamically favorable. Is the dissolution of zinc metal more favorable than that of zinc oxide, as indicated in the previous equations? Unfortunately, we do not know the answer without experimental evidence. In other words, the values of the Gibbs free energy change for the reaction do not tell us how fast the reaction will take place. Rather, it is the kinetics of the reaction that determine how fast the reaction will occur, not thermodynamic considerations.

Thermodynamic calculations indicate only whether any given reaction is thermodynamically spontaneous. They do not tell us when the reaction will begin, how fast it will progress, and when it will end. However, these calculations do give the maximum extent of reaction through the equilibrium constant. For example, zinc oxide is soluble in acidic solutions, as indicated in Eq. 12.5. From the standard free energy of reaction at 25°C, it can easily be shown that $a_{Zn^{2+}} = 7.33 \cdot 10^{11} \cdot a^2_{H^+}$. At pH = 1, for example, when the activity of the hydrogen ion is 0.1, the equilibrium activity of the zinc ion, Zn^{2+}, would be 7.33×10^9. It should be noted that this numerical value does not represent the molar concentration of the zinc ion. To calculate the molar concentration from this number, it is necessary to know the activity coefficient, which is the topic of discussion in a later section. It should also be noted that the equilibrium zinc activity would be only 7.33×10^{-5} when the pH of the solution is 8. As will be seen later, this numerical value will be close to the numerical value of the molarity for such a dilution concentration.

SOLUTION CHEMISTRY

Activity Coefficient

The chemical potential of a species i, μ_i, is given by

$$\mu = \mu_i^o + RT \ ln \ a_i \qquad \textbf{(Eq. 12.6)}$$

where

μ_i^o = the standard chemical potential
R = the gas constant
T = the absolute temperature, in kelvin
a_i = the activity of species i

Note that when a_i is unity, $\mu_i = \mu_i^o$. The activity of species i, a_i, represents a thermodynamic concentration and is frequently expressed by $a_i = \gamma C_i$ for solutes, where C_i is the molar concentration of species i and γ is called the activity coefficient. Note that when γ is 1, the molar concentration and the activity become numerically the same. When the solution is very dilute—for example, when the molar concentration is far less than 10^{-3} mol/dm^3—the activity coefficient approaches unity (a condition known as the Henrian standard state).

However, for solvents, $a_i = \gamma X_i$, where X_i represents the mole fraction. In this case, the activity coefficient becomes unity when X_i approaches unity, which is often referred to as the Raoultian standard state.

It is most unfortunate that activity coefficients of any individual ions are impossible to measure. What is measured frequently, however, is the activity coefficient of a dissolved compound. For example, the measurement of the activity coefficient of {HCl} can be carried out electrochemically from a reaction:

$$1/2\ (H_2) + <AgCl> \Leftrightarrow \{HCl\} + <Ag> \qquad \textbf{(Eq. 12.7)}$$

$$\Delta G = \Delta G^o + RT \ ln \ \frac{a_{H^+} \cdot a_{Cl^-}}{P_{H_2}} = \Delta G^o + RT \ ln \ a_{\pm}^2 \qquad \textbf{(Eq. 12.8)}$$

Here, the partial pressure of hydrogen is assumed to be unity; i.e., P_{H_2} and 1 atm and $a_{\pm} = \sqrt{a_{H^+} \cdot a_{Cl^-}}$. The mean activity, $a_{\pm}$, is what is measured electrochemically. Because $\Delta G = -nFE$ (where n is the number of electrons involved, F is the Faraday constant, and E is the electrical potential), Eq. 12.8 can be written in terms of the electrical potentials:

$$E = E^o - \frac{2.303\, RT}{nF} \log \gamma_{\pm}^2 \cdot m_{\pm}^2 \qquad \textbf{(Eq. 12.9)}$$

where

$\gamma_{\pm}$ = the mean activity coefficient

$m_{\pm}$ = the mean molality of HCl

Also of note in Eq. 12.9 is the replacement of the natural logarithm, *ln*, by the base 10 logarithm, log. As can be seen in this development, what is measured is the overall activity or activity coefficient of {HCl} and not the individual activities of either H^+ or Cl^-. A number of chosen mean activity coefficient values for various strong electrolytes are given in Table 12.1.

Several observations can be made about Table 12.1. First, the mean activity coefficient decreases with the molality; however, in many cases (e.g., HCl), it increases again as the molality becomes still higher. The change of the mean activity coefficient is more pronounced with divalent ions (such as Ca^{2+} and SO_4^{2-}) as compared with monovalent ions (such as Na^+ and Cl^-), as shown in the cases of HCl and H_2SO_4. It should be noted that the activity coefficients of the cation and anion of a particular salt are not necessarily the same.

Estimation of Activity Coefficients for Ions

It is frequently desirable to estimate the activity coefficients of the individual ions given the mean activity coefficient of an electrolyte, such as those listed in Table 12.1. The MacInnes method, often referred to as the mean salt method, is used to carry out such an estimation. This method is based on an assumption that the mean activity coefficient of potassium chloride is the same as the activity coefficient of potassium ion, which is the same as that of chloride ion:

$$\gamma_{\pm KCl} = [\gamma_{K^+}\gamma_{Cl^-}]^{1/2} = \gamma_{K^+} = \gamma_{Cl^-} \qquad \textbf{(Eq. 12.10)}$$

Now, if we want to estimate the activity coefficient of M^+ in a solution containing MCl, the following relationship can be established:

$$\gamma_{\pm MCl} = [\gamma_{M^+}\gamma_{Cl^-}]^{1/2} = [\gamma_{M^+}\gamma_{\pm KCl}]^{1/2}$$

and therefore,

$$\gamma_{M^{2+}} = \frac{\gamma_{\pm MCl_2}^2}{\gamma_{\pm KCl}^2}$$

It would be good practice to repeat a similar exercise to estimate the activity coefficient for M^{2+} in an MCl_2 solution and show that the resulting equation becomes

$$\gamma_{M^{2+}} = \frac{\gamma_{\pm MCl_2}^3}{\gamma_{\pm KCl}^3} \qquad \textbf{(Eq. 12.11)}$$

The activity coefficient of a cation or an anion can also be estimated by using the Debye–Huckel method. This method is well known in the estimation of the activity coefficients for ions and compounds. Eq. 12.12 is used to estimate the coefficient for a compound, and Eq. 12.13 is for ions:

$$\log \gamma_{\pm} = -\frac{A|z_+z_-|\sqrt{I}}{1 + Ba\sqrt{I}} \qquad \textbf{(Eq. 12.12)}$$

TABLE 12.1 Mean activity coefficients of strong electrolytes

	Molality, m											
Electrolyte	**0.001**	**0.002**	**0.005**	**0.01**	**0.05**	**0.1**	**0.2**	**0.5**	**1.0**	**2.0**	**3.0**	**4.0**
$NiSO_4$	—	—	—	—	—	0.18	0.13	0.075	0.051	0.041	—	—
NH_4Cl	0.961	0.944	0.911	0.88	0.79	0.74	0.69	0.62	0.57	—	—	—
NH_4I	0.962	0.946	0.917	0.89	0.80	0.76	0.71	0.65	0.60	—	—	—
$(NH_4)_2SO_4$	0.874	0.821	0.726	0.67	0.48	0.40	0.32	0.22	0.16	—	—	—
NaCl	0.966	0.953	0.929	0.904	0.823	0.780	0.730	0.68	0.66	0.67	0.71	0.78
NaI	0.97	0.96	0.94	0.91	0.86	0.83	0.81	0.78	0.80	0.95	—	—
$NaNO_3$	0.966	0.953	0.93	0.90	0.82	0.77	0.70	0.62	0.55	0.48	0.44	0.41
Na_2SO_4	0.887	0.847	0.778	0.714	0.53	0.45	0.36	0.27	0.20	—	—	—
$PbCl_2$	0.86	0.80	0.70	0.61	—	—	—	—	—	—	—	—
$ZnCl_2$	0.88	0.84	0.77	0.71	0.56	0.50	0.45	0.38	0.33	—	—	—
$ZnSO_4$	0.70	0.61	0.48	0.39	—	0.15	0.11	0.065	0.045	0.036	0.04	—
$CuCl_2$	0.89	0.85	0.78	0.72	0.58	0.52	0.47	0.42	0.43	0.51	0.59	—
$CuSO_4$	0.74	—	0.53	0.41	0.21	0.16	0.11	0.068	0.047	—	—	—
$FeCl_2$	0.89	0.86	0.80	0.75	0.62	0.58	0.55	0.59	0.67	—	—	—
KCl	0.965	0.952	0.927	0.901	0.815	0.769	0.719	0.651	0.606	0.576	0.571	0.579
KI	0.965	0.951	0.927	0.905	0.84	0.80	0.76	0.71	0.68	0.69	0.72	0.75
$MgCl_2$	—	—	—	—	—	0.56	0.53	0.52	0.62	1.05	2.1	—
$MgSO_4$	—	—	—	0.40	0.22	0.18	0.13	0.088	0.064	0.055	0.064	—
$MnSO_4$	—	—	—	—	—	0.25	0.17	0.11	0.073	0.058	0.062	0.079
HCl	0.966	0.952	0.928	0.904	0.830	0.796	0.767	0.758	0.809	1.01	1.32	1.76
HNO_3	0.965	0.951	0.927	0.902	0.823	0.785	0.748	0.715	0.720	0.783	0.876	0.982
H_2SO_4	0.830	0.757	0.639	0.544	0.340	0.265	0.209	0.154	0.130	0.124	0.141	0.171
NaOH	—	—	—	—	0.82	—	0.73	0.69	0.68	0.70	0.77	0.89
KOH	—	—	0.92	0.90	0.82	0.80	—	0.73	0.76	0.89	1.08	1.35
$AgNO_3$	—	—	0.92	0.90	0.79	0.72	0.64	0.51	0.40	0.28	—	—
$BaCl_2$	0.88	—	0.77	0.72	0.56	0.49	0.44	0.39	0.39	0.44	—	—
$CaCl_2$	0.89	0.85	0.785	0.725	0.57	0.515	0.48	0.52	0.71	—	—	—
$Ca(NO_3)_2$	0.88	0.84	0.77	0.71	0.54	0.48	0.42	0.38	0.35	0.35	0.37	0.42

Source: Latimer, W.M., *Oxidation Potentials*, Prentice-Hall, Englewood Cliffs, NJ. 1952.

$$\log \gamma_i = -\frac{A z_i^2 \sqrt{I}}{1 + Ba\sqrt{I}} \qquad \textbf{(Eq. 12.13)}$$

where

z_+ = the valence of the cation in the compound
z_- = the valence of the anion in the compound
z_i = the valence of the ion (whether a cation or anion)
I = the ionic strength = $1/2\ \Sigma z_i^2 C_i$
C_i = the molarity of species i
A,B = constants (see Table 12.2)
a = the diameter of the compound (see Table 12.3)

TABLE 12.2 Values of constants *A* and *B* in the Debye–Huckel equation

Temperature, °C	*A*	*B* ($\times 10^{-8}$)
0	0.4883	0.3241
5	0.4921	0.3249
10	0.4960	0.3258
15	0.5000	0.3262
20	0.5042	0.3273
25	0.5085	0.3281
30	0.5130	0.3290
35	0.5175	0.3297
40	0.5221	0.3305
45	0.5271	0.3314
50	0.5319	0.3321
55	0.5371	0.3329
60	0.5425	0.3338

Source: Garrels and Christ 1965; Butler 1964; and Kielland 1937.

TABLE 12.3 Values of the term *a* in the Debye–Huckel equation

Value of *a* ($\times 10^{8}$ cm)	Inorganic and Organic Ions
2.5	Rb^+, Cs^+, NH_4^+, Tl^+, Ag^+
3.0	K^+, Cl^-, Br^-, I^-, NO_3^-, CN^-, NO_2^-, NO_3^-
3.5	OH^-, F^-, HS^-, BrO_3^-, IO_4^-, MnO_4^-, ClO_3^-, ClO_4^-, $HCOO^-$, H_2 citrate$^-$, $CH_3NH_3^-$, $(CH_3)_2NH_2^+$
4.0–4.5	Na^+, HCO_3^-, $H_2PO_4^-$, HSO_3^-, Hg_2^{2+}, SO_4^{2-}, SeO_4^{2-}, CrO_4^{2-}, HPO_4^{2-}, PO_4^{3-}, ClO_2^-, IO_3^-, HCO_3^-, $Co(NH_3)_4(NO_2)_2^+$, $S_2O_3^{2-}$, $(CH_3)_3NH^+$, $C_2H_5NH_3^+$
4.5	Pb^{2+}, CO_3^{2-}, SO_3^+,MoO_4^{2-}, $Co(NH_3)_5Cl^{2+}$, $Fe(CN)_5NO^{2-}$, CH_3COO^-, $(COO)_2^{2-}$
5.0	Sr^{2+}, Ba^{2+}, Ra^{2+}, Cd^{2+}, Hg^{2+}, S^{2-}, $WO4^{2-}$, $S2O4^{2-}$, $Fe(CN)_6^{4-}$, $CHCl_2COO^-$, $H_2C(COO)_2^{2-}$, citrate^{3-}
6.0	Li^+, Ca^{2+}, Cu^{2+}, Zn^{2+}, Sn^{2+}, Mn^{2+}, Fe^{2+}, Ni^{2+}, Co^{2+}, Co(ethylenediamine)$_3^{3+}$, $Co(S_2O_3)(CN)_6^{4-}$, $(C_3H_7)_2NH_2^+$, $C_6H_4(COO)_2^{2-}$, $(CH_2CH_2COO)_2^-$
8.0	Mg^{2+}, Be^{2+}, $(C_6H_5)_2CHCOO^-$, $(C_3H_7)_4N^+$
9.0	H^+, Fe^{3+}, Al^{3+}, Cr^{3+}, trivalent rare earths (Sc^{3+}, Y^{3+}, La^{3+}, Ce^{3+}, Pr^{3+}, Nd^{3+}, Sm^{3+}), $Co(SO_3)_2(CN)_4^{5-}$
11.0	Th^{4+}, Zr^{4+}, Ce^{4+}, Sn^{4+}

Source: Garrels and Christ 1965; Butler 1964; and Kielland 1937.

The mean activity coefficients of cupric chloride, $CuCl_2$, have been calculated for three different concentrations by using the Debye–Huckel method; the resulting values are listed below, along with those given in Table 12.1:

Mean activity coefficients for $CuCl_2$ (mol/dm^3)	0.001	0.05	0.1
Debye–Huckel method	0.896	0.575	0.51
Table 12.1	0.89	0.58	0.52

These numbers for three sets of values at three different concentrations indicate that the results obtained from the Debye–Huckel method and those measured (Table 12.1) are very comparable, with less than 5% difference.

It is important to note that all the methods described so far are applicable for rather low concentration of electrolytes, say, less than 1 mol/dm^3. When the concentration of these electrolytes is high, other methods should be used, for example, the Meissner method described by Gokcen (1982, 1979; Meissner, Kusik, and Tester 1972).

For a single electrolyte, the reduced activity coefficient, Γ^o_{12}, is defined by the following equation:

$$\Gamma^o_{12} = (\gamma^o_{12})^{1/z_1/z_2} \qquad \textbf{(Eq. 12.14)}$$

where

(γ^o_{12}) = the activity coefficient of the single electrolyte

z_1, z_2 = the valence of the cation and anion, respectively, expressed in terms of absolute value

For more than one electrolyte present in solution, the reduced activity coefficient is similarly defined:

$$\Gamma_{12} = (\gamma_{12})^{1/z_1/z_2} \qquad \textbf{(Eq. 12.15)}$$

It is interesting to note that when a single value of the reduced activity coefficient of an electrolyte for one ionic strength, I, is determined, other values at different ionic strengths can easily be found. The reason is that the logarithmic values of the reduced activity coefficients are parallel to each other for all electrolytes when these values are plotted against ionic strength. The reduced activity coefficient of a single electrolyte can be calculated by using the following equation (Gokcen 1982):

$$\Gamma^o_{12} = [1 + (0.75 + 0.065q)(1 + 0.1I)^q - (0.75 - 0.065q)]\Gamma^* \qquad \textbf{(Eq. 12.16)}$$

where Γ^* is given by

$$\log \Gamma^* = -\frac{0.5107I^{0.5}}{1 + cI^{0.5}} \qquad \textbf{(Eq. 12.17)}$$

$$c = 1 + 0.55q \cdot e^{-0.023I^3}$$

The empirically found q values for various electrolytes are tabulated in Table 12.4. These q values are all measured at 25°C. The q value at any other temperature, q_t, can be estimated by using the following equation (Gokcen 1982):

$$q_t = q_{25} + (t - 25)(aq_{25} + b^*) \qquad \textbf{(Eq. 12.18)}$$

where

q_{25} = the q value for 25°C

t = the temperature, in °C

a = −0.0079, b^* = 0.0029 for sulfates

a = −0.005, b^* = 0.0085 for all other electrolytes

TABLE 12.4 Values of *q* for various electrolytes

Electrolyte	*q*	*I* (maximum value tested)
$AgNO_3$	–2.550	6.0
HCl	6.690	16.0
$HClO_4$	9.300	16.0
HNO_3	3.660	3.0
HBr	1.150	5.5
KCl	0.920	4.5
$KClO_3$	–1.700	0.7
KI	1.620	4.5
KNO_3	–2.330	3.5
KOH	4.770	6.0
NaBr	2.980	4.0
NaCl	2.230	6.0
$NaClO_3$	0.410	3.5
$NaClO_4$	1.300	6.0
NaI	4.060	3.5
$NaNO_3$	–0.390	6.0
NaOH	3.000	6.0
NH_4Cl	0.820	6.0
$AlCl_3$	1.920	10.8
$Al_2(SO_4)_3$	0.360	15.0
$CaCl_2$	2.400	18.0
$Ca(NO_3)_2$	0.930	18.0
$CoCl_2$	2.250	12.0
$Co(NO_3)_2$	2.080	15.0
$CuCl_2$	1.400	6.0
$Cu(NO_3)_2$	1.830	18.0
$CuSO_4$	0.000	5.6
$FeCl_2$	2.160	6.0
K_2SO_4	–0.250	2.1
$MgCl_2$	2.900	15.0
$MnCl_2$	1.600	18.0
$MnSO_4$	0.140	16.0
$Na_2S_2O_3$	0.180	10.5
Na_2SO_4	–0.190	12.0
$(NH_4)_2SO_4$	–0.250	12.0
$NiCl_2$	2.330	15.0
$NiSO_4$	0.025	10.0
$Pb(ClO_4)_2$	2.250	18.0
$Pb(NO_3)_2$	–0.070	6.0
$ZnCl_2$	0.800	18.0
$Zn(ClO_4)_2$	4.300	12.0
$Zn(NO_3)_2$	2.280	18.0
$ZnSO_4$	0.050	8.0

Note: These values are all measured at 25°C.

Solubility of Gases in Aqueous Media

In hydrometallurgy, the solubility of various gases in aqueous media frequently plays an important role. For example, oxygen is used as a very important oxidant; therefore, its solubility in water is often critical in determining the overall reaction rate. Other relevant gases may include ammonia, carbon dioxide, sulfur dioxide, and hydrogen.

Let us assume that gaseous oxygen is in equilibrium with dissolved oxygen in water:

$$(O_2) \Leftrightarrow \{O_2\} \qquad \textbf{(Eq. 12.19)}$$

The chemical potential of the gaseous oxygen can be given by

$$\mu_i = \mu_i^o + RT\ ln\ \frac{p_i}{p_i^o} = \mu_i^o + RT\ ln\ p_i \qquad \textbf{(Eq. 12.20)}$$

where

μ_i^o = the standard chemical potential of gaseous species i
p_i^o = the standard partial pressure of the gaseous species i, which is usually 1 atm
R = the gas constant
T = absolute temperature

The chemical potential of the dissolved oxygen, on the other hand, is given by the following expression:

$$\mu_i = \mu_i^o + RT\ ln\ a_i = \mu_i^o + RT\ ln\ \gamma_i m_i \qquad \textbf{(Eq. 12.21)}$$

where

γ_i = the activity coefficient of oxygen in water
m_i = the molality of the dissolved oxygen

At equilibrium, Eq. 12.20 can be equated with Eq. 12.21, resulting in

$$\mu_i^o - \mu_i^o = -\Delta G_R^o = RT\ ln\ \frac{\gamma_i m_i}{p_i} \qquad \textbf{(Eq. 12.22)}$$

The standard chemical potential of the dissolved oxygen is 3,900 cal/mol, and that of gaseous oxygen is zero. Therefore, at 25°C, Eq. 12.22 gives $m_i = 0.0013 \times p_i$. Because $X_i = m_i/55.56$, we have $p_i = 4.27 \times 10^4 X_i$. It should be noted that the molality of water is approximately 55.56 and that the activity coefficient of the dissolved oxygen is assumed to be unity in this calculation. The numerical value, 4.27×10^4, is very close to the Henry's law constant for oxygen (see Table 12.5).

It is a good exercise for students to calculate the Henry's law constant for carbon dioxide given the values of the chemical potentials of gaseous and dissolved carbon dioxide: −94.26 kcal/mol and −92.31 kcal/mol, respectively. The calculated value should be 1.5×0^3 atm, which compares favorably with the measured value of 1.64×10^3 shown in Table 12.5. In these calculations, two very important assumptions have been made. First, the activity coefficient of gases in water is unity, which is reasonable

TABLE 12.5 Henry's law constants for various gases in water at 25°C

Gas	Henry's Constant, atm
Oxygen	4.38×10^4
Carbon dioxide	1.64×10^3
Carbon monoxide	5.80×10^4
Nitrogen	8.65×10^4

Source: Perry 1969.

TABLE 12.6 Solubility constants, Ki, for use in Eq. 12.23

Species *i*	K_i	Species *i*	K_i
H^+	0	NO_3^-	0.013
NH^{4+}	0.033	Cl^-	0.029
K^+	0.099	HSO_4^-	0.069
Na^+	0.107	OH^-	0.081
Zn^{2+}	0.108	HCO_3^-	0.083
Mg^{2+}	0.119	SO_4^-	0.121
NH_3	0.007		

when the solubility is so low that the Henrian standard state can be safely assumed. The other important assumption made in the calculation is that the dissolved species does not branch out to other species. Such an assumption may be acceptable when the solubility calculation is applied to gases such as oxygen or nitrogen. However, this assumption will break down when gases like sulfur dioxide or hydrogen sulfide are considered. These gases are present in solution in more than one form.

It should also be noted that all the foregoing values for the solubility of gases are valid in pure water. However, when various electrolytes are present in water, such analysis breaks down. In other words, the Henry's law constant is very much a function of ionic strength; furthermore, it very much depends on the types of electrolytes present in water. It is generally observed that the solubility of gases in water decreases as the electrolyte concentration increases. This phenomenon is often referred to as a "salting out" effect.

Narita, Han, and Lawson (1982, 1983) have found the following relationship for the solubility of oxygen in various salt solutions:

$$\log (S_o/S) = \Sigma K_i C_i \quad \textbf{(Eq. 12.23)}$$

where

S_o = the solubility of oxygen in pure water
S = the solubility of oxygen in the electrolyte solution
K_i = the constant derived semiempirically for a given ionic species i (the values for various electrolytes are tabulated in Table 12.6)
C_i = the molarity of species i

Figure 12.1 shows the solubility of oxygen in water containing selected electrolytes as a function of concentration. As we can see in this figure, the solubility of oxygen in salt solution is very much affected by the concentration of the salt. For example, the solubility of oxygen in 3 mol/dm^3 zinc sulfate is only 20% of that in pure water. In general, it is of note that the neutral molecules such as ammonia have the least effect, whereas divalent ions such as sulfate have a much greater effect on the solubility of oxygen in water.

Solubility Calculations of Compounds

The free energy of formation for a compound and those of the dissolved components should, in principle, allow us to calculate the extent of dissolution of this compound. For example, let us say that we wish to calculate the solubility of silver sulfate, Ag_2SO_4. Let us assume that when this compound is placed in water, two silver ions and one sulfate ion are the only components present in water on dissolution:

$$<Ag_2SO_4> \Leftrightarrow 2\{Ag^+\} + \{SO_4^{2-}\} \quad \textbf{(Eq. 12.24)}$$

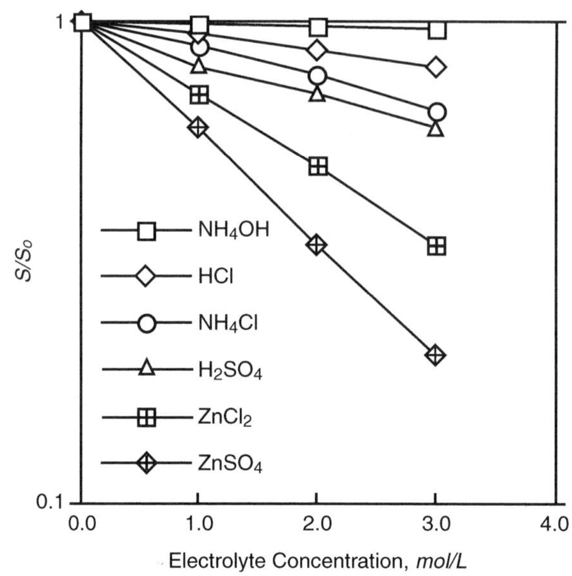

FIGURE 12.1 Values of log (S/S_0) for selected electrolytes as a function of concentration

From the standard free energy of formation values of the three components at 25°C, the equilibrium constant can easily be calculated and found to be about 6,690 cal/mol. Therefore,

$$\Delta G^o_{f,25°C} = 6{,}690 = -RT \ln K_{eq}$$

from which K_{eq} is calculated to be 1.25×10^{-5}. Because we therefore find that $\{Ag^+\} = 2.92 \times 10^{-2}$ mol/dm^3,

$$K_{eq} = \frac{\{Ag^+\}^2\{SO_4^{2-}\}}{Ag_2SO_4} = \{Ag^+\}^2\{SO_4^{2-}\} = 1.25 \times 10^{-5} = 1/2\{Ag^+\}^3$$

This simple calculation indicates that the amount of silver sulfate dissolved would be half of the concentration of silver ion; i.e., 1.46×10^{-2} mol/dm^3. However, this calculation is in error because it assumes that the activity coefficient of silver ion, as well as sulfate ion, is unity. To correct such an invalid assumption, the following analysis is carried out. Let us calculate the activity coefficients for the silver ion as well as sulfate ions by using Eq. 12.13. This step is now possible because we can calculate the ionic strength of the system based on the above calculations, although these values are inaccurate. The ionic strength for this system is calculated to be 4.38×10^{-2} mol/dm^3, which allows us to calculate the activity coefficients for silver ion and sulfate ion. These values are, respectively, 0.812 and 0.473.

Therefore, these calculations can be performed again with the activity coefficient values incorporated as follows:

$$K_{eq} = \frac{\gamma^2_{Ag^+}\{Ag^+\}^2\gamma_{SO_4^{2-}}\{SO_4^{2-}\}}{\{Ag_2SO_4\}} = \gamma^2_{Ag^+}\{Ag^+\}^2\gamma_{SO_4^{2-}}\{SO_4^{2-}\} = 1.25 \times 10^{-5}$$

$$0.812^2\{Ag^+\}^2 0.473\{SO_4^{2-}\} = 0.312\{Ag^+\}^2\{SO_4^{2-}\} = 0.156\{Ag^+\}^3$$

$$\text{since } 1/2\,\{Ag^+\} = \{SO_4^{2-}\}$$

Therefore, $\{Ag^+\} = 4.31 \times 10^{-2}$ mol/dm^3.

This process can be repeated iterratively until the final concentration of silver is merged. The results of such an iterative process are summarized below:

Iteration	I	γ_{Ag^+}	$\gamma_{SO_4^{2-}}$	$\{Ag^+\}$, mol/dm^3
First	0	1	1	2.92×10^{-2}
Second	0.0438	0.812	0.473	4.31×10^{-2}
Third	0.065	0.787	0.408	4.63×10^{-2}
Fourth	0.0694	0.776	0.411	4.68×10^{-2}
Fifth	0.0702	0.775	0.410	4.68×10^{-2}

Metal Complexation

It is frequently observed that when a metal is dissolved, more than one species can be formed. For example, when cuprous ion, Cu^+, is dissolved in water containing chloride ion, Cl^-, it may exist in water as Cu^+, CuCl (aq), $CuCl_2^-$, and $CuCl_3^{2-}$. A frequent question may be how many such complexed ions and compounds could exist and what should be the concentration of each species. The answer to how many dissolved species of these metal complexes are present in solution is usually found from thermodynamic information, such as the stability constants, which have already been identified by other investigators. Table 12.7 lists some such stability constants for various metals and complexing agents.

The thermodynamic information given in Table 12.7 is very valuable in that it specifies not only what kinds of cuprous chloride complexes are present but also the amount of each of these species present in the system, provided the total amounts of copper and chloride are known. Let us examine the case of a known amount of cuprous chloride ions present in a solution.

$$\langle CuCl \rangle \Leftrightarrow \{Cu^+\} + \{Cl^-\}$$
$$K_{s0} = \{Cu^+\}\{Cl^-\} = 10^{-6.73} \quad \textbf{(Eq. 12.25)}$$

TABLE 12.7 Equilibrium constants for ligand complexation for various metals

		Log of Equilibrium Constant				
Ligand	**Ion**	K_{s0}	K_{s1}	K_{s2}	K_{s3}	K_{s4}
Cl^-	Cu^+	−6.73	−5.0	−1.12	−1.47	
	Ag^+	−9.75	−6.70	−4.70	−4.70	−4.46
	Tl^+	−3.04	−3.15	−3.74	−4.70	
	Hg^{2+}	−13.79	−7.05	−0.57	+0.28	+1.28
Br^-	Ag^+	−12.10	−7.96	−5.00	−4.15	−3.22
	Hg^{2+}	−19.10	−10.05	−1.77	+0.64	+1.90
	Tl^+	−4.81	−4.48	−4.62	−5.10	−5.80
I^-	Ag^+	−16.35	−8.22	−5.40	−2.00	−1.90
	Hg^{2+}	−27.70	−14.83	−3.88	−0.10	+2.13
	Pb^{2+}	−8.15	−6.23	−4.47	−4.65	−3.85
CN^-	Cu^+	−19.49	−13.0	−4.23	+0.36	+2.06
	Ag^+	−15.92	−7.0	+4.62	+5.32	+4.19
	Hg^{2+}	−35.10	−17.10	−0.40	+3.43	+6.41

Source: Martell and Smith 1982; and Butler 1964.

$$\langle CuCl \rangle \Leftrightarrow \{CuCl\}$$
$$K_{s1} = \{CuCl\} = 10^{-5} \qquad \textbf{(Eq. 12.26)}$$

$$\langle CuCl \rangle + \{Cl^-\} \Leftrightarrow \{CuCl_2^-\}$$
$$K_{s2} = \frac{\{CuCl_2^-\}}{\{Cl^-\}} = 10^{-1.12} \qquad \textbf{(Eq. 12.27)}$$

$$\langle CuCl \rangle + 2\{Cl^-\} \Leftrightarrow \{CuCl_3^{2-}\}$$
$$K_{s3} = \frac{\{CuCl_3^{2-}\}}{\{Cl^-\}^2} = 10^{-1.47} \qquad \textbf{(Eq. 12.28)}$$

The preceding four equations can be solved simultaneously to establish the exact amount of each species. However, note that there are five unknowns and four relationships. Therefore, there should be at least another equation to solve these equations. The additional equations that can be considered are the mass balance on copper- and/or chloride-bearing species and the charge balance:

$$Cu_{Tot} = \{Cu^+\} + \{CuCl\} + \{CuCl_2^-\} + \{CuCl_3^{2-}\} \text{ or}$$
$$Cl_{Tot} = \{Cl^-\} + \{CuCl\} + 2\{CuCl_2^-\} + 3\{CuCl_3^{2-}\}$$

In addition to these balances, the charge balance yields

$$\{H^+\} + \{Cu^+\} = \{OH^-\} + \{Cl^-\} + \{CuCl_2^-\} + 2\{CuCl_3^{2-}\}$$

It should be noted that in these reactions, it has been assumed that chloride is added as sodium chloride and that its concentration is kept constant. It should also be noted that Eqs. 12.25 through 12.28 are valid only when solid cuprous chloride is present.

When Eq. 12.26 is subtracted from Eq. 12.25, the result is

$$\{CuCl\} = \{Cu^+\} + \{Cl^-\}$$

Therefore,

$$K_{s0}/K_{s1} = K_1 = \frac{\{Cu^+\}\{Cl^-\}}{\{CuCl\}}$$

From this relationship, the concentration of {CuCl} could be expressed in terms of $\{Cu^+\}$ and $\{Cl^-\}$, namely,

$$\{CuCl\} = \{Cu^+\}\{Cl^-\}/K_1 \qquad \textbf{(Eq. 12.29)}$$

Similarly,

$$\{CuCl_2^-\} = \{Cu^+\}\{Cl^-\}^2/K_2 \qquad \textbf{(Eq. 12.30)}$$

and

$$\{CuCl_3^{2-}\} = \{Cu^+\}\{Cl^-\}^3/K_3 \qquad \textbf{(Eq. 12.31)}$$

where

$$K_2 = K_{s0}/K_{s2}$$
$$K_3 = K_{s0}/K_{s3}$$

From the mass balance for copper-bearing species, the following equation can be established:

$$Cu_{Tot} = \{Cu^+\} + \{CuCl\} + \{CuCl_2^-\} + (CuCl_3^{2-}\}$$
$$= \{Cu^+\} + \frac{\{Cu^+\}\{Cl^-\}}{K_1} + \frac{\{Cu^+\}\{Cl^-\}^2}{K_2} + \frac{\{Cu^+\}\{Cl^-\}^3}{K_3}$$

By rearranging this relationship for an expression for the concentration of cuprous ion in terms of Cu_{Tot} and concentrations of chloride ion, the following expression is obtained:

$$\{Cu^+\} = \frac{Cu_{Tot}}{1 + \{Cl^-\}/K_1 + \{Cl^-\}^2/K_2 + \{Cl^-\}^3/K_3} \quad \text{(Eq. 12.32)}$$

The concentration of cuprous ion—and therefore those of {CuCl}, $\{CuCl_2^-\}$, and $\{CuCl_3^{2-}\}$—can be calculated provided that the total concentration of copper-bearing species, Cu_{Tot}, and the concentration of chloride ion, Cl^-, are known.

Effect of Temperature and Pressure on Equilibrium Constant

Most of the discussions so far have been concerned with room temperature, 25°C. However, when the temperature of the system is increased or decreased, the equilibrium considerations will be affected. From the second law of thermodynamics, the following equation can be derived:

$$dG = VdP - SdT \quad \text{(Eq. 12.33)}$$

where

G = the Gibbs free energy change
V = volume
P = pressure
S = entropy
T = temperature

Therefore,

$$d\Delta G^o = \Delta V^o dP - \Delta S^o dT \quad \text{(Eq. 12.34)}$$

where

ΔG^o = the standard Gibbs free energy change
ΔV^o = the partial molar volume change for a reacting system at standard state
ΔS^o = partial entropy change for a reacting system at standard state

At constant pressure,

$$d\Delta G^o = -\Delta S^o dT \quad \text{(Eq. 12.35)}$$

and since

$$\Delta G^o = \Delta H^o - T\Delta S^o \quad \text{(Eq. 12.36)}$$

where

ΔH^o = the standard enthalpy change

By rearranging Eq. 12.36, we obtain

$$\Delta S^o = \frac{\Delta H^o - \Delta G^o}{T} \quad \text{(Eq. 12.37)}$$

By combining Eqs. 12.36 and 12.37, we obtain

$$d\Delta G^o = -\left(\frac{\Delta H^o - \Delta G^o}{T}\right)dT$$

$$Td\Delta G^o - \Delta G^o dT = -\Delta H^o dT$$

Therefore, the following relationship holds:

$$\frac{d\left(\frac{\Delta G^o}{T}\right)}{dT} = -\frac{\Delta H^o}{T^2} \quad \text{(Eq. 12.38)}$$

Equation 12.38 is often referred to as Gibbs–Helmholtz equation.

Because $\Delta G^o = -RT\ ln\ K_{eq}$, we have

$$\frac{\Delta G^o}{T} = -R\ d(ln\ K_{eq})$$

and

$$d\left(\frac{\Delta G^o}{T}\right) = -R\ d(ln\ K_{eq})$$

and finally,

$$\frac{d(ln\ K_{eq})}{dT} = \frac{\Delta H_T^o}{RT^2} \qquad \textbf{(Eq. 12.39)}$$

Equation 12.39 is often known as the van't Hoff equation, and the T subscript in ΔH_T^o signifies that ΔH^o could be a function of temperature. The enthalpy change, ΔH_T^o at a temperature other than 25°C (298 K) can be identified if $\Delta H_{25°C}^o$ and the heat capacities of reactants and products (represented by R and P subscripts, respectively) are known:

$$\Delta H_T^o = \Delta H_{298}^o + \Delta H_R + \Delta H_P$$

where

$$\Delta H_R = \int_T^{298} \Sigma\ (v_i C_{p_i})_R dT$$

$$\Delta H_P = \int_{298}^{T} \Sigma\ (v_i C_{p_i})_P dT$$

where v_i is the stoichiometric coefficient for species i, and C_{p_i} is the heat capacity for species i.

Therefore, we have

$$\Delta H_T^o = \Delta H_{298}^o + \int_{298}^{T} [\Sigma\ (v_i C_{p_i})_P - \Sigma (v_i C_{p_i})_R] dt \qquad \textbf{(Eq. 12.40)}$$

By substituting Eq. 12.40 into Eq. 12.39 and rearranging the final result, we obtain

$$ln\ K_T = ln\ K_{298} + \int_{298}^{T} \left[\frac{H_{298}^o + \int_{298}^{T} [\Sigma (v_i C_{p_i})_P - \Sigma (v_i C_{p_i})_R] dT}{RT^2} \right] dT \qquad \textbf{(Eq. 12.41)}$$

Equation 12.41 allows us to calculate the equilibrium constant for a reaction at a temperature other than 25°C by knowing the value at 25°C.

Now let us examine the effect of pressure on equilibrium constant. At constant temperature, Eq. 12.34 becomes

$$d\Delta G^o = \Delta V^o dP \qquad \textbf{(Eq. 12.42)}$$

Because $\Delta G^o = -RT\ ln\ K_{eq}$, we have

$$\left(\frac{d\ ln\ L}{dP}\right)_T = -\frac{\Delta V^o}{RT} \qquad \textbf{(Eq. 12.43)}$$

Therefore, the equilibrium constant at a pressure other than 1 atm, $K_{eq,P}$, would be related to that at 1 atm, $K_{eq,1\ atm}$, by the following expression (Zena and Yeager 1967; Kestu and Pytkowicz 1970; Curthoys and Mathieson 1970; Derry 1972):

$$ln\ K_{eq,P} = ln\ K_{eq,1\ atm} - \frac{\Delta V^o}{RT}(P - 1) \qquad \textbf{(Eq. 12.44)}$$

Table 12.8 lists the partial molal volumes of a number of ionic species measured or estimated at 1 atm. Note that the partial molal volumes of the most cations are negative; when these cations are dissolved in water, their volume shrinks because of the association of the ions with water molecules. On the other hand, the partial molal volumes for anions are all positive.

TABLE 12.8 Ionic partial molal volume

Ion	V^o, cm^3	Ion	V^o, cm^3
Na^+	–6.1	Cl^-	23.7
H^+	–5.6	NO_3^-	34.8
NH_4^+	9.6	HCO_3^-	28.5
Ca^{2+}	–25.5	SO_4^{2-}	27.0
Ni^{2+}	–33.6	CO_3^{2-}	5.5
Co^{2+}	–38.4		
Al^{3+}	–46.7	Mg^{2+}	–26.5

Source: Zena and Yeager 1967; Kestu and Pytkowicz 1970.

Let us examine the effect of pressure on the dissolution of calcium carbonate into calcium ion and carbonate ion in water.

$$\langle CaCO_3 \rangle \Leftrightarrow \{Ca^{2+}\} + \{CO_3^{2-}\}$$

The partial molal volumes of calcium ion and carbonate ion are, respectively, −25.5 cm^3 and 5.5 cm^3 from Table 12.8; that of calcium carbonate can be calculated given its molecular weight (100) and its density (2.71 g/cm^3), which yields the molar volume of 36.9 cm^3. Therefore, the partial molal volume change for the above reaction becomes −56.9 cm^3. It should be noted that 1 cm^3-atm is equivalent to 0.02422 cal. The results for 500 atm and 1,000 atm are as follows:

Pressure	$K_p/K_{1\ \text{atm}}$
500 atm	3.2
1,000 atm	10.2

The effect is quite dramatic under these pressures. However, note also that pressures encountered by hydrometallurgists are not usually high. For example, the steam pressures for water at 100°C, 200°C, and 300°C are 1 atm, 15.3 atm, and 85 atm, respectively.

Correspondence Principles*

The effect of temperature on the equilibrium of ions in solution is quite different from that of neutral species. For nonionic species, the following equation is frequently applied without too much difficulty:

$$\Delta G_T^o = \Delta H_T^o - T\Delta S_T^o \qquad \textbf{(Eq. 12.45)}$$

where the T subscripts indicate that the terms vary with temperature.

However, for ionic species, Eq. 12.45 will have a more rigorous form:

$$\begin{aligned} \Delta G_T^o &= \Delta H_T^o - T\Delta S_T^o \\ &= \Delta H_{298}^o + \int_{298}^{T} \Delta C_p dT - T\Delta S_{298}^o - T\int_{298}^{T} \frac{\Delta C_p}{T}\, dT \end{aligned}$$

This equation can be rewritten as

$$\Delta G_T^o = \Delta G_{298}^o + 298\Delta S_{298}^o - T\Delta S_{298}^o + \Delta C_p]_{298}^{T} \int_{298}^{T} dT - T\Delta C_p]_{298}^{T} \int_{298}^{T} \frac{dT}{T} \qquad \textbf{(Eq. 12.46)}$$

*This section draws from Kwok and Robins (1972); Criss and Cobble (1964a, 1964b); Lowson (1971); and Macdonald (1972).

In the preceding equation $\Delta C_p]_{298}^T$ represents the mean heat capacity evaluated between the temperatures 298 and T. It should be noted that

$$\Delta S_T^o - \Delta S_{298}^o = \Delta C_p]_{298}^T ln \frac{dT}{T}$$

or

$$\Delta C_p]_{298}^T = \frac{\Delta S_T^o - \Delta S_{298}^o}{ln \frac{T}{298}}$$

According to Criss and Coble, for ionic species,

$$\Delta S_T^o = a_T + b_T \Delta S_{298}^o$$

and

$$\Delta C_p]_{298}^T = \frac{a_T + (b_T - 1)\Delta S_{298}^o}{ln \frac{T}{298}} \qquad \textbf{(Eq. 12.47)}$$

In the preceding equations, the term S_{298}^o is an adjusted value obtained by subtracting a value of $5z$ from the conventional value of S_{298}^o for nonionic species. The value z is the ion charge, including the sign. For example, for H^+, S_{298}^o would be −5 e.u., and for Ni^{2+}, it would be −25.5 − (5 × 2) = −35.5 e.u. Therefore, the mean heat capacity value for ions used in Eq. 12.46 becomes

$$\Delta C_p]_{298}^T = \alpha_T + \beta_T \Delta S_{298}^o \qquad \textbf{(Eq. 12.48)}$$

The values for a_T, b_T, α_T, and β_T for various ions at different temperatures are given in Tables 12.9 and 12.10.

TABLE 12.9 Summary of a_T and b_T values used in Eq. 12.47

Temperature, °C	Cations		Simple Anions, X^- and OH^-		Oxy-anions, XO^{-m}		Hydroxy-anions, $XO_n(OH)_f^{-m}$		H^+ Entropy
	a_T	b_T	a_T	b_T	a_T	b_T	a_T	b_T	
25	0.0	1.00	0.0	1.00	0.0	1.00	0.0	1.00	−5.0
60	3.9	0.955	−5.1	0.969	−14.0	1.217	−13.5	1.380	−2.5
100	10.3	0.876	−13.0	1.000	−31.0	1.476	−30.3	1.894	2.0
150	16.2	0.792	−21.3	0.989	−46.4	1.687	−50.0	2.381	6.5
200	23.3	0.711	−30.2	0.981	−67.0	2.020	−70.0	2.960	11.1
250	29.9	0.630	−38.7	0.978	−86.5	2.320	−90.0	3.530	16.1

TABLE 12.10 Summary of α_T and β_T values used in Eq. 12.48

Temperature, °C	Cations		Simple Anions, X^- and OH^-		Oxy-anions, XO^{-m}		Hydroxy-anions, $XO_n(OH)_f^{-m}$		H^+ $\Delta C_p]_{298}^T$
	α_T	β_T	α_T	β_T	α_T	β_T	α_T	β_T	
60	35	−0.41	−46	−0.28	−127	1.96	−122	3.44	23
100	46	−0.55	−58	0.00	−138	2.24	−135	3.97	31
150	46	−0.59	−61	−0.03	−133	2.27	−143	3.95	33
200	50	−0.63	−65	−0.04	−145	2.53	−152	4.24	35

Given ΔG_f^o at 25°C, which is −11,530 cal/mol, let us calculate ΔG_f^o for Ni^{2+} at 150°C. We also know that S_{298}^o(conventional) = −25.5 e.u. and that C_p for Ni is $4.06 + 7.04 \times 10^{-3}T$ cal/mol (where T is the temperature in kelvin). Because we are interested in the Gibbs free energy of reaction:

	<Ni>	⇔	$\{Ni^{++}\} + 2e$	
$S^o_{\text{conventional}}$	7.5		−25.5 e.u.	
S^o_{adjusted}	7.5		−35.5	$\Delta S^o_{298} = -43.0$ e.u.

From Table 12.10, for cations at 150°C (423 K), we have $a_T = 46$ and $b_T = -0.59$. Therefore, we have

For Ni^{2+}, $\quad \Delta C_p]_{298}^T = 46 - 0.59 \times (-35.5) = 66.94$ cal/mol·K

For Ni, $\quad \Delta C_p]_{298}^T = 6.6$ cal/mol·K

Finally, the Gibbs free energy of formation of Ni^{2+} at 150°C can be calculated by using Eq. 12.46:

$$\Delta G^o_{f,150°C} = -11{,}530 - (423 - 298)(-33.0) + (66.94 - 6.6)(423 - 298) - 423(66.94 - 6.6)$$
$$= -8{,}802 \text{ cal/mol}$$

Eh–pH Diagrams*

When a metal is subject to dissolution, it is important to understand the effect of the oxidation/reduction potential and the pH of the solution environment. Both the oxidation potential and the solution pH have a direct impact on how well the metal will dissolve in the solution. In a certain environment, metal may form a passive oxide film instead of dissolving to form metal ions. The phase diagram of this metal in relation to the oxidation/reduction potential and pH will serve as a guide to what the thermodynamically stable product under such a condition would be. In this section, we will examine how the phase diagram of a metal can be constructed, as well as how such a diagram can be used in the metal dissolution strategy.

It is important to understand an electrochemical cell before introducing Eh–pH diagrams. In Figure 12.2, a zinc plate is placed in the left-hand side of the electrochemical cell, which contains a

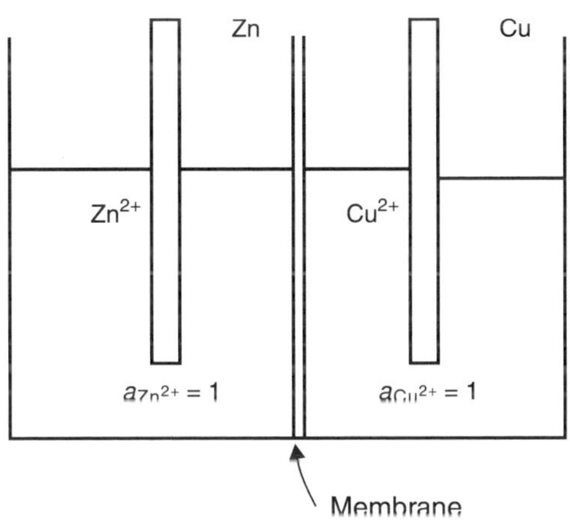

FIGURE 12.2 An electrochemical cell showing copper and zinc plates placed in a cell containing copper and zinc ions, respectively, but divided by a semipermeable membrane

*This section draws from Pourbaix (1966).

TABLE 12.11 Standard electrode potentials

Electrode	$E^o_{M/M+n}$
Au/Au^+	1.7
Au/Au^{3+}	1.50
Pt/Pt^+	1.20
Pd/Pd^{2+}	0.987
Ag/Ag^+	0.799
Hg/Hg^{2+}	0.789
Cu/Cu^+	0.521
Cu/Cu^{2+}	0.337
H_2/H^+	0.00
Fe/Fe^{3+}	–0.036
Pb/Pb^{2+}	–0.126
Sn/Sn^{2+}	–0.136
Ni/Ni^{2+}	–0.250
Co/Co^{2+}	–0.277
In/In^{3+}	–0.342
Cd/Cd^{2+}	–0.403
Fe/Fe^{2+}	–0.440
Cr/Cr^{3+}	–0.740
Zn/Zn^{2+}	–0.763
Mn/Mn^{2+}	–1.18
Zr/Zr^{4+}	–1.53
Ti/Ti^{2+}	–1.63
Al/Al^{3+}	–1.66
Be/Be^{2+}	–1.85
Mg/Mg^{2+}	–2.37
Na/Na^+	–2.714

solution of unit activity of zinc ion. A copper plate is inserted in the right-hand side of the cell containing a copper solution of unit activity. When these two plates are connected externally to each other, an electrochemical cell is formed. According to the International Union of Pure and Applied Chemistry (IUPAC), the net potential, ΔE, will be defined as the difference in potential between the two sides, i.e., $E_{right} - E_{left}$. Therefore, we have

$$\Delta E = E_{Cu/Cu^{2+}} - E_{Zn/Zn^{2+}} = 0.337 - (-0.763) = 1.1 \text{ v}$$

The electromotive force (emf) values for various metals $M/M{+}n$ are given in Table 12.11.

It should be noted that the Gibbs standard free energy of formation, given in Table 12.12, can be converted into the emf values given in Table 12.11 through the following equation:

$$E^o = \frac{\Delta G^o}{nF} \qquad \textbf{(Eq. 12.49)}$$

where

F = the Faraday constant = 23,061 cal/v·eq

n = the number of electrons involved (a negative value if the electrons appear on the right-hand side of the equation; a positive value if the electrons appear on the left-hand side of the equation)

TABLE 12.12 The Gibbs free energy formation of metal ions

Metal Ion	ΔG_f^o, kcal/mol	Metal Ion	ΔG_f^o, kcal/mol
Al^{3+}	–115.0	Cd^{2+}	–18.58
Ca^{2+}	–132.18	Cr^{2+}	–42.1
Co^{2+}	–12.8	Co^{3+}	28.9
Cu^{+}	12.0	Cu^{2+}	15.53
Au^{+}	39.0	Au^{3+}	103.6
Fe^{2+}	–20.3	Fe^{3+}	–2.52
Pb^{2+}	–5.81	Mn^{2+}	–54.4
Mn^{3+}	–19.6	Mg^{2+}	–108.99
Hg^{2+}	39.38	Ni^{2+}	–11.53
Pd^{2+}	45.5	Pt^{2+}	54.8
K^{+}	–67.466	Rb^{+}	–67.45
Ag^{+}	18.43	Na^{+}	–62.589
Sr^{2+}	–133.2	Th^{4+}	–175.2
Sn^{2+}	–6.275	Ti^{2+}	–75.1
U^{3+}	–124.4	Zn^{2+}	–35.184

For example, let us evaluate the standard emf value for the following reaction:

$$<Cu> \rightarrow \{Cu^{+2}\} + 2e \qquad \textbf{(Eq. 12.50)}$$

Because $\Delta G^o = 15{,}530$ cal/mol for Cu^{2+} (see Table 12.12), we have $E^o = -15{,}530/[(-2) \times 23{,}061] = 0.337$ v, which is in agreement with Table 12.11. On the other hand, if Eq. 12.50 is written in the opposite direction:

$$\{Cu^{+2}\} + 2e \rightarrow <Cu> \qquad \textbf{(Eq. 12.51)}$$

we have $\Delta G^o = -15{,}530$ cal/mol from Table 12.10, and $E^o = -15{,}530/(2 \times 23{,}061) = 0.337$ v. Therefore, the E^o value is affected not by how the equation is written but, rather, whether it is written as an anodic or a cathodic reaction.

It should be noted that all of the emf values for various metals given in Table 12.11 are based on the assumption that the emf of hydrogen ion discharge is zero under standard conditions, namely, at unit activity of hydrogen ion and 1 atm of hydrogen gas:

$$2\{H^+\} + 2e \rightarrow (H_2) \qquad \textbf{(Eq. 12.52)}$$

The Nernst equation for the reaction given in Eq. 12.52 can be written as

$$E = E^o - \frac{2.303RT}{nF} \log \frac{P_{H_2}}{\{H^+\}^2} \qquad \textbf{(Eq. 12.53)}$$

Note that $E = E^o = 0$ v when the partial pressure of hydrogen gas is 1 atm and the activity of hydrogen ion is unity. It should also be noted that the value $2.303RT/nF = 0.059/n$ v at 25°C. The emf values of metals listed in Table 12.11 are measured against the hydrogen electrode as given in Eq. 12.52. In practice, however, it is not convenient to carry around the hydrogen electrode; therefore, a reference electrode is used in place of the hydrogen electrode. The most commonly used reference electrode is the calomel electrode, which is a half-cell electrode whose chemical reaction is given by

$$<Hg_2Cl_2> + 2e \rightarrow 2\{Hg\} + 2\{Cl^-\} \qquad \textbf{(Eq. 12.54)}$$

It can easily be shown that the emf for this reaction becomes:

$$E = E^o - \frac{0.059}{2} \log \{Cl^-\}^2 \qquad \textbf{(Eq. 12.55)}$$

The emf of this equation will be a function of the activity of chloride in the solution. The emf values for various chloride concentrations are as follows:

KCl Concentration	*E*
0.1*N*	0.3338 v
1.0*N*	0.2800 v
Saturated	0.2415 v

The saturated calomel electrode is the most common because maintaining the saturated KCl solution is easy. When any electrode potential is measured against this calomel electrode, the measured potential has to be adjusted by adding 0.2415 v at 25°C to find the standard hydrogen electrode (SHE) potential.

Eh–pH Diagram for the Fe–H_2O System

Let us consider that an iron bar is immersed in water and that the fate of the iron under various oxidation/reduction potentials and pH values of the solution is to be observed. It could be noted that when the pH of the solution is low and the oxidation potential of the iron bar is raised, iron will dissolve into the solution in the forms of either Fe^{2+} or Fe^{3+}. As the pH of the solution is also increased, the surface iron will be subjected to oxidation to form either Fe_3O_4 or Fe_2O_3. It should be noted that the solid forms of $Fe(OH)_2$ and $Fe(OH)_3$ are thermodynamically less stable than Fe_3O_4 or Fe_2O_3. Therefore, only the following species will be considered in this study: $<Fe>$, $\{Fe^{2+}\}$, $\{Fe^{3+}\}$, $<Fe_3O_4>$, and $<Fe_2O_3>$. The values of the Gibbs standard free energy of formation for these species are, respectively, 0, −20.3, −2.5, −242.4, and −177.1 kcal/mol; the value for water is 56.69 kcal/mol.

For the following general reaction involving components A, B, C, and D:

$$a<A> + b\{B\} + m\{H^+\} + ne \rightarrow c\{C\} + d\{D\} \qquad \textbf{(Eq. 12.56)}$$

the Nernst equation is

$$E = E^o - \frac{0.059}{n}\log Q - 0.059\left(\frac{m}{n}\right)\text{pH} \qquad \textbf{(Eq. 12.57)}$$

where $Q = \dfrac{\{C\}^c\{D\}^d}{\langle A\rangle^a\{B\}^b}$ and, if the component $<A>$ is a pure solid, $<A> = 1$

The lower phase boundary of water in relation to the oxidation/reduction potential can be described by Eq. 12.52:

$$2\{H^+\} + 2e \rightarrow (H_2)$$

The corresponding Nernst equation for Eq. 12.52 is given by Eq. 12.53:

$$E = E^o - \frac{2.303RT}{nF}\log\frac{P_{H_2}}{\{H^+\}^2}$$

In terms of Eq. 12.57, Eq. 12.53 becomes

$$E = E^o - 0.059\ \text{pH} - 0.0295\log p_{H_2} \qquad \textbf{(Eq. 12.58)}$$

The upper boundary of the water stability line is presented by:

$$\tfrac{1}{2}(O_2) + 2\{H^+\} + 2e \rightarrow \{H_2O\} \qquad \textbf{(Eq. 12.59)}$$

The Nernst equation for this relationship becomes

$$E = 1.23 - 0.059\ \text{pH} + 0.0148\log p_{O_2} \qquad \textbf{(Eq. 12.60)}$$

The following equations represent the phase boundary lines for the iron-bearing species.

$$\{Fe^{3+}\} + e \rightarrow \{Fe^{2+}\}$$

$$E = 0.771 - 0.059 \log \frac{\{Fe^{2+}\}}{\{Fe^{3+}\}} \quad \textbf{(Eq. 12.61)}$$

Note that when the activity of Fe^{2+} is the same as that of Fe^{3+}, E = 0.771 v. In other words, when the oxidation potential is more than 0.771 v against the SHE, Fe^{3+} will be predominating, whereas Fe^{2+} will be more abundant when the oxidation potential is less than this value.

$$\{Fe^{2+}\} + 2e \rightarrow <Fe>$$

$$E = -0.440 + 0.0295 \log \{Fe^{2+}\} \quad \textbf{(Eq. 12.62)}$$

On the other hand, for $\{Fe^{3+}\} + 3e \rightarrow <Fe>$,

$$E = -0.0366 + 0.0197 \log \{Fe^{3+}\} \quad \textbf{(Eq. 12.63)}$$

Equation 12.62 is valid for most of $\{Fe^{2+}\}$ in view of Eq. 12.61, but Eq. 12.63 is invalid in view of Eq. 12.61 for practical concentrations of $\{Fe^{3+}\}$. It should be noted that for the copper-water system, $\{Cu^{2+}\}$ is more stable than $\{Cu^{+}\}$, which is in contrast to the case of the Fe–H_2O system.

Now the phase boundaries for $\{Fe^{3+}\}/<Fe_2O_3>$ and $\{Fe^{2+}\}/<Fe_2O_3>$ should be identified:

$$<Fe_2O_3> + 6\{H^+\} \rightarrow 2\{Fe^{3+}\} + 3\{H_2O\}$$

Because there is no electron transfer involved in this equation, Eq. 12.57 cannot be used. Therefore, from the ordinary Gibbs free energy reaction, the following equation can be derived:

$$\log K_{eq} = \log \frac{\{Fe^{3+}\}^2}{\{H^+\}^6} = -1.45 \quad \textbf{(Eq. 12.64)}$$

Because pH = $-\log \{H^+\}$, Eq. 12.64 becomes

$$\log \{Fe^{3+}\} = -0.72 - 3\,pH \quad \textbf{(Eq. 12.65)}$$

As we can see in the preceding equation, as soon as the activity of Fe^{3+} is identified, the pH of the system will be uniquely defined. For example, if $\{Fe^{3+}\} = 10^{-6}$, the pH of the phase diagram dividing the Fe^{3+} phase and the $<Fe_2O_3>$ phase will occur at pH = 1.76 and will be represented by a vertical line.

Now the phase boundary for $\{Fe^{2+}\}/<Fe_2O_3>$ should be determined:

$$<Fe_2O_3> + 6\{H^+\} + 2e \rightarrow 2\{Fe^{2+}\} + 3\{H_2O\}$$

$$E = 0.728 - 0.1773\,pH - 0.059 \log \{Fe^{2+}\} \quad \textbf{(Eq. 12.66)}$$

The phase boundary line represented by Eq. 12.66 on the Eh–pH diagram will have a negative slope of −0.1773 v/pH and will meet with the line given by Eq. 12.65.

What remains to be accomplished includes establishing the phase boundaries between $<Fe>$ and $<Fe_3O_4>$, $<Fe_3O_4>/<Fe_2O_3>$, and $\{Fe^{2+}\}/<Fe_3O_4>$:

$$3<Fe_2O_3> + 2\{H^+\} + 2e \rightarrow 2<Fe_3O_4> + \{H_2O\}$$

$$E = 0.221 - 0.059\,pH \quad \textbf{(Eq. 12.67)}$$

$$<Fe_3O_4> + 8\{H^+\} + 8e \rightarrow 3<Fe> + 4\{H_2O\}$$

$$E = -0.085 - 0.059\,pH \quad \textbf{(Eq. 12.68)}$$

$$<Fe_3O_4> + 8\{H^+\} + 2e \rightarrow 3\{Fe^{2+}\} + 4\{H_2O\}$$

$$E = 0.98 - 0.2364\,pH - 0.0886 \log \{Fe^{2+}\} \quad \textbf{(Eq. 12.69)}$$

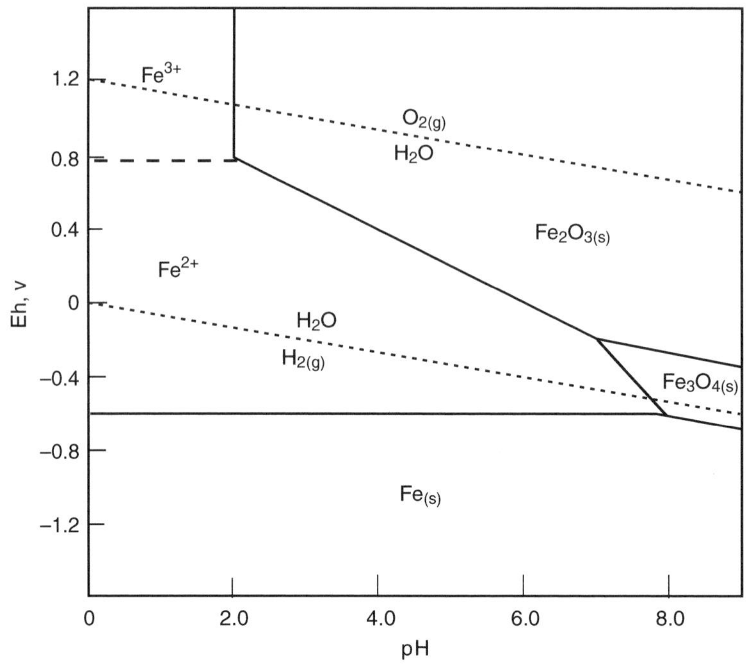

Note: The activity of dissolved ion is assumed to be 10^{-6} and partial pressure of gas is 1 atm.

FIGURE 12.3 Eh–pH diagram for the system of iron-water at 25°C

It should be noted that the two lines described by Eqs. 12.67 and 12.68 are parallel to each other and also to the lines given by Eqs. 12.58 and 12.60. The phase boundary line represented by Eq. 12.69 on the Eh–pH diagram will have a negative slope of −0.2364 v/pH and will meet with the line given by Eq. 12.66.

We now have completed the Eh–pH diagram for the iron-water system, and this diagram is shown in Figure 12.3. It should be noted that the acidity and alkalinity of the solution could easily be adjusted by using an acid (such as sulfuric, nitric, or hydrochloric) or an alkali (such as sodium hydroxide, sodium carbonate, or ammonia). To adjust the oxidation/reduction potential of the solution, various oxidants—such as oxygen, sodium chlorate, manganese dioxide, potassium permanganate, sodium hypochlorite, ferric compounds, and nitric acid—can be used, as well as reductants such as sulfur dioxide, hydrogen gas, carbon monoxide, carbohydrates, and ferrous salts.

When complexing agents such as ammonia, cyanide, or even chloride are added into the solution, Eh–pH diagrams will become very complicated. Researchers interested in this area of information should consult other references (Vu and Han 1977; Bhuntumkomol, Han, and Lawson 1980; Meng and Han 1996).

ELECTROCHEMISTRY

Electrode Processes*

Leaching of metals involves two half-electrochemical reactions: oxidation and reduction. An oxidation reaction, often referred to as an anodic reaction that gives up electrons, and a reduction reaction, often

*This section draws from Bockris and Reddy (1970); Newman (1973); and Levich (1962).

referred to as a cathodic reaction that consumes electrons, may be described as follows for metallic element M:

Oxidation (anodic reaction: electrons are generated)

$$\langle M\rangle\ (\{M^{2+}\} + 2e \qquad \textbf{(Eq. 12.70)}$$

Reduction (cathodic reaction: electrons are consumed)

$$2\{H^{+}\} + 2e \rightarrow (H_2)$$

$$2\{H^{+}\} + \tfrac{1}{2}(O_2) + 2e \rightarrow \{H_2O\}$$

Overall reaction

$$\langle M\rangle + 2\{H^{+}\}\ (\ \{M^{2+}\} + (H_2) \qquad \textbf{(Eq. 12.71)}$$

Let us assume that zinc metal is subject to dissolution in acidic medium. It should be noted that, as shown in Eq. 12.71, zinc could easily be leached in an acid reaction without additional oxidant such as oxygen because zinc is located in the electromotive series (Table 12.11) below the reaction given by Eq. 12.71.

Let us examine what may happen with the dissolution of zinc in an acidic medium. The anodic reaction for the dissolution of zinc would be, following Eq. 12.70:

$$\langle Zn\rangle \rightarrow \{Zn^{+2}\} + 2e \qquad \textbf{(Eq. 12.72)}$$

and the cathodic reaction could be written in the form of either Eq. 12.71 or Eq. 12.72:

$$2\{H^{+}\} + 2e \rightarrow (H_2)$$

$$2\{H^{+}\} + \tfrac{1}{2}(O_2) + 2e \rightarrow \{H_2O\}$$

When Eq. 12.52 is the cathodic reaction for the dissolution of zinc, the overall leaching process may consist of

- Hydrogen ion adsorption onto the zinc substrate
- Zinc dissolution with transfer of two electrons
- Consumption of two transferred electrons by two hydrogen ions
- Formation of hydrogen gas

This process is depicted in Figure 12.4. On the other hand, if Eq. 12.59 is the cathodic reaction, the process becomes more complicated, as depicted in Figure 12.5.

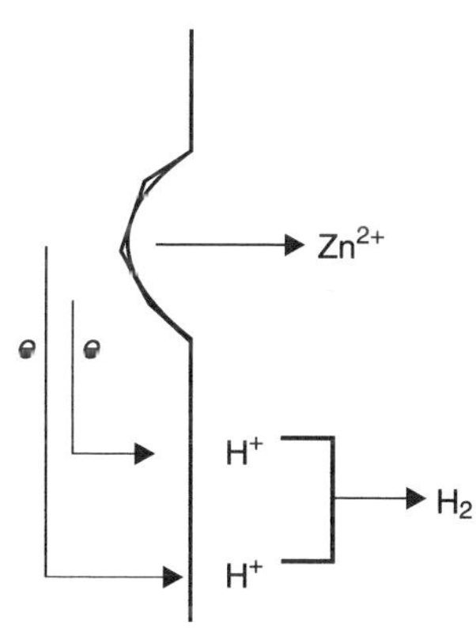

FIGURE 12.4 Dissolution of zinc accompanying hydrogen discharge

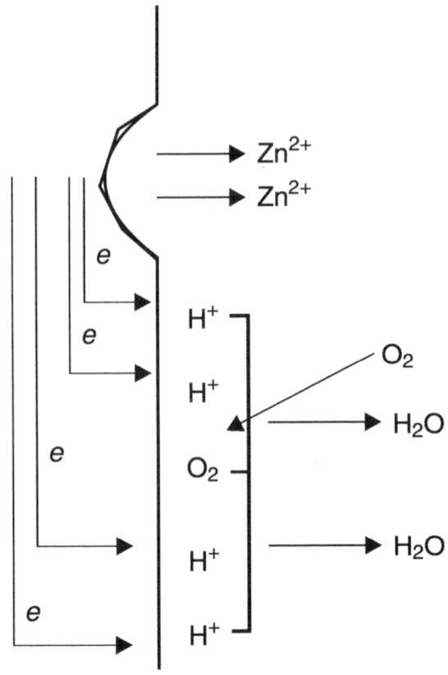

FIGURE 12.5 Dissolution of zinc with oxygen as an oxidant

The overall reactions of these two reaction processes are

$$\langle Zn \rangle + 2\{H^+\} \rightarrow \{Zn^{2+}\} + (H_2) \qquad \Delta G_R^o = -35.184 \text{ kcal/mol} \qquad \textbf{(Eq. 12.73)}$$

$$\langle Zn \rangle + 2\{H^+\} + \tfrac{1}{2}(O_2) \rightarrow \{Zn^{2+}\} + \{H_2O\} \qquad \Delta G_R^o = -91.874 \text{ kcal/mol} \qquad \textbf{(Eq. 12.74)}$$

Although the thermodynamic calculations indicate that the reaction with oxygen as an oxidant is more favorable, the steps involved in the reaction are more complicated; hence, slower kinetics are expected with this reaction.

A similar approach could be used for copper dissolution in an acidic medium. However, there are some differences between the two systems:

$$\langle Cu \rangle + 2\{H^+\} \rightarrow \{Cu^{2+}\} + (H_2) \qquad \Delta G_R^o = 15.530 \text{ kcal/mol} \qquad \textbf{(Eq. 12.75)}$$

$$\langle Cu \rangle + 2\{H^+\} + \tfrac{1}{2}(O_2) \rightarrow \{Cu^{2+}\} + \{H_2O\} \qquad \Delta G_R^o = -41.16 \text{ kcal/mol} \qquad \textbf{(Eq. 12.76)}$$

Unlike the case of zinc dissolution, thermodynamic considerations do not favor the dissolution of copper in the absence of oxygen. An equilibrium calculation for Eq. 12.75 indicates that the maximum concentration of cupric ion at pH 1 would be about 4.07×10^{-10} mol/L. Compared to this numerical value, Eq. 12.76 indicates that the equilibrium activity of cupric ion at pH 1 and in the presence of oxygen would be 1.54×10^{28}.

Polarization Curves*

As the preceding discussion shows, leaching of metals is represented by an electrochemical process. The electrochemical process can easily be visualized and understood when cathodic and anodic reactions are investigated independently.

Concentration Overpotential. Let us consider a cathodic reaction of zinc being deposited on a zinc cathode. This process can be represented as follows:

$$\{Zn^{2+}\} + 2e \rightarrow \langle Zn \rangle \qquad \textbf{(Eq. 12.77)}$$

*This section draws from Bockris and Reddy (1970); Newman (1973); and Levich (1962).

The flux of the zinc ion deposition throughout the diffusion boundary layer can be presented by Fick's first law:

$$j = -D\ (dc/dx) = D\ \frac{C_b - C_s}{\delta} \qquad \textbf{(Eq. 12.78)}$$

where

j = the flux, in mol/cm^2·s
D = the diffusivity of Zn^{2+}, in cm^2/s
c = molar concentration
x = x-coordinate
C_b = the bulk concentration
C_s = the surface concentration
δ = the diffusion boundary layer thickness

The deposition current density is calculated as

$$i_d = z_i\ j\ F\ \frac{C_b - C_s}{\delta} \qquad \textbf{(Eq. 12.79)}$$

where

i_d = deposition current density, in A/cm^2
z_i = the valence of the zinc ion (+2)
F = the Faraday constant = 96,500 coulomb/eq

By rearranging this equation, we get

$$1 - \frac{C_s}{C_b} = \frac{i_d\ \delta}{z_i DFC_b} \quad \text{or} \quad \frac{C_s}{C_b} = 1 - \frac{i_d\ \delta}{z_i DFC_b} \qquad \textbf{(Eq. 12.80)}$$

When diffusion of zinc ion through the mass transfer boundary layer is limiting, i_d becomes i_l, the limiting current density, and hence,

$$\frac{C_s}{C_b} = 1 - \frac{i_d}{i_l} \qquad \textbf{(Eq. 12.81)}$$

If there is no current flowing, and if the deposition reaction is reversible, that reaction may be written as

$$E_r = E^o + \frac{RT}{nF}\ ln\ a_{Zn^{2+},b} \qquad \textbf{(Eq. 12.82)}$$

where

E_r = the reversible potential
n = the number of electrons involved in the deposition reaction (n is positive when electrons appear on the right-hand side of the reaction and negative when they appear on the left-hand side)
$a_{Zn^{2+}}$ = the activity of the zinc ion in the bulk solution

Note that $a_{Zn^{2+},b}$ is equivalent to the C_b term shown in Figure 12.6.

On the other hand, when there is a current i flowing, Eq. 12.82 will become:

$$E_{obs} = E^o + \frac{RT}{nF}\ ln\ a_{Zn^{2+},s} \qquad \textbf{(Eq. 12.83)}$$

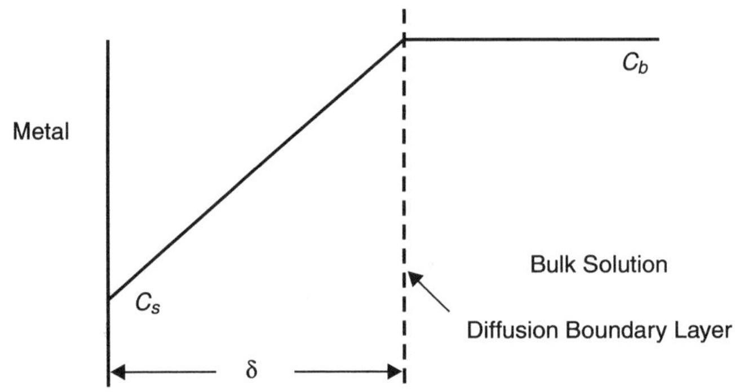

FIGURE 12.6 Metal ion deposition onto a cathodic metal electrode

Here, E_{obs} is the observed potential, which could be different from E_r of Eq. 12.82, and the activity of zinc ion at surface, $a_{Zn^{2+},s}$, is equivalent to the C_s term shown in Figure 12.6. The difference between E_{obs} and E_r may be expressed as

$$\Delta E = E_{obs} - E_r = \eta_c \qquad \textbf{(Eq. 12.84)}$$

where η_c is known as the concentration overpotential and has the following form:

$$\eta_c = \frac{RT}{nF} \ln \frac{a_s}{a_b} \qquad \textbf{(Eq. 12.85)}$$

Because $a_s = \gamma_s C_s$ and $a_b = \gamma_b C_b$, where γ_s and γ_b are activity coefficients for zinc ion at the surface and bulk, respectively, Eq. 12.85 can be written as

$$\eta_c = \frac{RT}{nF} \ln \frac{\gamma_s C_s}{\gamma_b C_b} \qquad \textbf{(Eq. 12.86)}$$

By assuming $\gamma_s = \gamma_b$ and substituting Eq. 12.83 into Eq. 12.86, we get

$$\eta_c = \frac{RT}{nF} \ln \left(\frac{i_l - i}{i_l}\right) = \frac{2.303RT}{2F} \log \left(\frac{i_l - i}{i_l}\right) \qquad \textbf{(Eq. 12.87)}$$

When potential is plotted as a function of log i, a curve such as that shown in Figure 12.7 will be obtained based on Eq. 12.87.

Note that because $i_l = nFDC_b/\delta$, the limiting current density will be influenced by the bulk concentration and by the mass transfer boundary layer thickness, the latter of which is in turn affected by the stirring speed of a rotating disk (in cases where a rotating disk is used as the cathode).

Activation Overpotential. When a metal ion is subject to deposition on a cathode, it is believed that the metal ion will be faced with a potential barrier, E_a, as indicated in Figure 12.8. E_a is an activation energy barrier because of the chemical reaction contribution alone. However, if there is an additional barrier caused by an added potential, the overall activation energy, E_a', will be

$$E_a' = E_a + f(VF)$$

where

V = the voltage
F = the Faraday constant

The term $f(VF)$ is believed to have the following expression:

$$f(VF) = \beta' VF$$

where β' is the symmetry factor, which is frequently assumed to be 1/2. Hence, $E_a' = E_a + \beta' VF$ for the cathodic reaction, and $E_a' = E_a - (1 - \beta')VF$ for the anodic reaction representing metal dissolution.

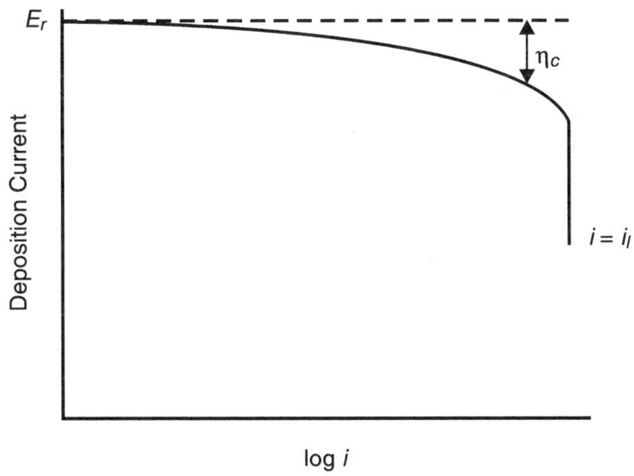

FIGURE 12.7 Deposition current versus logarithm of the current

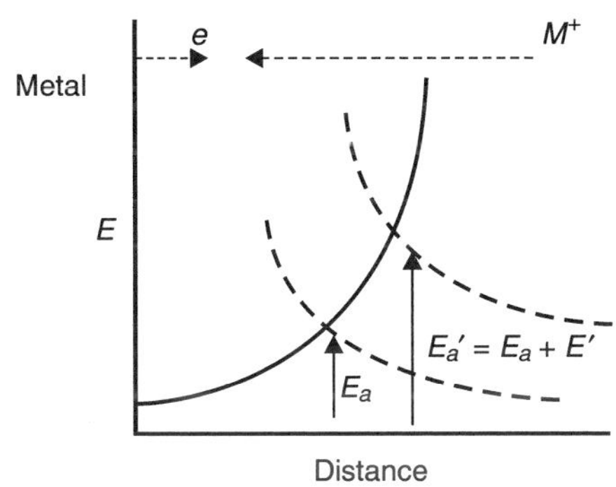

FIGURE 12.8 Potential barrier for metal deposition

Therefore, the cathodic rate equation will become

$$\text{rate (cathode)} = r_c = k_c C_+ \exp(-\beta' VF/RT) \qquad \textbf{(Eq. 12.88)}$$

where

$k_c = A \exp\{-E_a/RT\}$, which is the rate constant in the absence of the applied potential V

C_+ = the concentration of the metal ion in the bulk solution

Similarly, the anodic reaction rate can be written as

$$\text{rate (anode)} = r_a = k_a\, \theta \exp[(1 - \beta')VF/RT] \qquad \textbf{(Eq. 12.89)}$$

where

$k_a = A \exp\{E_a/RT\}$, which is again the rate constant in the absence of the applied potential V

θ = the fractional site occupied by the adsorbed species

Let us define the activation overpotential as $\eta_a = v - V_{rev}$, where $V_{rev} = E_r$, the reversible electrical potential. Because $i_c = nFr_c$, where i_c is the cathodic current density, we have

$$\begin{aligned} i_c &= nFC_+\theta k_c \exp(-\beta FV/RT) \\ &= nFC_+\theta k_c \exp(-\beta FV_{rev}) \exp(-\beta F\eta_a/RT) \end{aligned} \qquad \textbf{(Eq. 12.90)}$$

Similarly for the anodic current density, i_a becomes

$$i_a = nFk_a \exp[(1-\beta)FV/RT\}$$
$$= nFk_a \exp[(1-\beta)FV_{rev}/RT] \exp[(1-\beta)F\eta_a/RT] \quad \textbf{(Eq. 12.91)}$$

At the rest potential, when $\eta_a = 0$, we have $i_a = i_c$; therefore,

$$nFk_a \exp[(1-\beta)FV_{rev}/RT] = nFC_+ k_c \exp(-\beta FV_{rev}/RT) = i_o \text{ (exchange current density)} \quad \textbf{(Eq. 12.92)}$$

Finally, the net current density becomes

$$i = i_a - i_c = i_o \{\exp[(1-\beta)F\eta_a/RT] - \exp(-\beta F\eta_a/RT)\} \quad \textbf{(Eq. 12.93)}$$

Equation 12.93 is often known as the Butler–Volmer equation. Note that when η_a is relatively large—say, greater than 0.052 v—the second term within the braces in the equation above becomes far less than the first term and can be ignored. Therefore, in that case, the Butler–Volmer equation becomes

$$i = i_o \exp[(1-\beta)F\eta_a/RT] \quad \textbf{(Eq. 12.94)}$$

If we take the logarithm of Eq. 12.94 and rearrange, we may obtain either of the following expressions:

$$\eta_a = \frac{RT}{(1-\beta)F} \ln\left(\frac{i}{i_o}\right) \quad \textbf{(Eq. 12.95)}$$

$$\eta_a = B \log\left(\frac{i}{i_o}\right) \quad \textbf{(Eq. 12.96)}$$

where $B = [2.303RT/(1-\beta)F]$, which is equal to 0.12 v if $\beta = 0.5$ and the temperature is 25°C. A number of examples of activation overpotential for selected systems are shown in Table 12.13.

Equation 12.96 can be rearranged to obtain the following expression:

$$\eta_a = a + b \ln i \quad \textbf{(Eq. 12.97)}$$

This equation indicates that when the overpotential is plotted as a function of $\ln i$, a straight line is obtained; this form of the equation is often referred to as the Tafel equation.

TABLE 12.13 Activation overpotential for selected systems

Metal	Temperature, °C	Solution	B, v	i_o, A/cm^2	η_a, v
Hydrogen on Metals					
Pt	20	1N HCl	0.03	10^{-3}	0.00
W	20	5N HCl	0.11	10^{-5}	0.22
Ni	20	0.1N HCl	0.1	5×10^{-7}	0.31
Fe	25	4% NaCl	0.10	10^{-7}	0.40
Cu	20	0.1N HCl	0.12	2×10^{-7}	0.44
Hg	20	0.1N HCl	0.12	7×10^{-13}	1.10
Pb	20	0.1N HCl	0.12	2×10^{-13}	1.16
Oxygen on Metals					
Pt	20	0.1N H_2SO_4	0.10	9×10^{-12}	0.81
Au	20	0.1N NaOH	0.05	5×10^{-13}	0.47
Metal on Metal					
Zn	25	1*M* $ZnSO_4$	0.12	2×10^{-5}	0.20
Cu	25	1*M* $CuSO_4$	0.12	2×10^{-5}	0.20
Fe	25	1*M* $FeSO_4$	0.12	10^{-8}	0.60
Ni	25	1*M* $NiSO_4$	0.12	2×10^{-9}	0.68

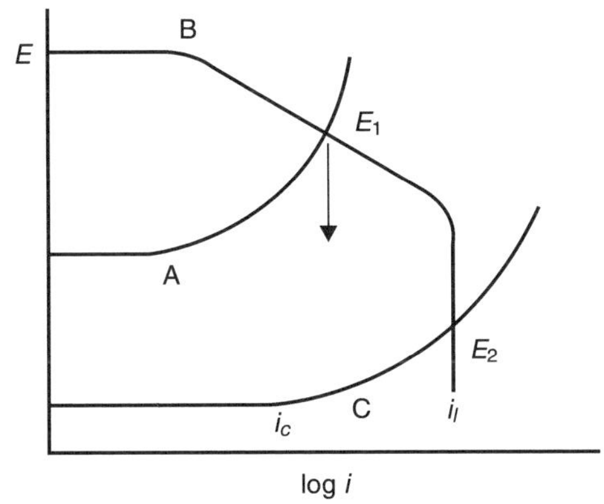

FIGURE 12.9 Potential-current relationship for mixed electrode systems

Mixed Potential and Leaching Current*

Let us suppose an iron bar is immersed in an air-free sulfuric acid solution and is subject to dissolution. The relevant anodic and cathodic reactions describing this dissolution process are

Reaction A, anodic: <Fe> $\rightarrow$ $\{Fe^{2+}\}$ + 2e

Reaction B, cathodic: 2 $\{H^{+}\}$ + 2e $\rightarrow$ (H_2)

These anodic and cathodic reactions are shown schematically in Figure 12.9.

The potential of cathodic reaction B can be theoretically calculated when there is no current flowing by use of the Nernst equation, such as Eq. 12.82 (see Figure 12.9). The potential, then, drops as the current increases, in accordance with the Tafel equation (Eq. 12.97). The relationship between the potential and current for anodic reaction A is shown in Figure 12.9; the potential in this case increases with current. Note that the potential where these two curves meet, E_1, is called the mixed potential, representing the potential at which the dissolution of iron is taking place. The current at that intersection, i_c, is referred to as the corrosion current or leaching current. Note also that the cathodic curve eventually reaches the limiting current, as shown in Eq. 12.87.

If zinc metal instead of iron is subjected to dissolution, a different anodic reaction–reaction C in Figure 12.9–will take place. Note that the zinc dissolution will intersect the cathodic curve (curve B) at the limiting current, meaning that the dissolution reaction of zinc will be controlled by mass transfer of hydrogen ion across the diffusion boundary layer.

Solution IR Drop

In addition to the concentration overpotential and activation overpotential, there can be a significant potential drop resulting from resistance through the solution medium. This IR drop can be estimated if the solution conductivity, κ (in units of $ohm^{-1}cm^{-1}$)–often referred to as the solution specific conductance–is known. This solution conductivity can be estimated if the individual ionic equivalent conductivity, λ (in units of $ohm^{-1}cm^{2}eq^{-1}$), is known. The ionic equivalent conductivity and the solution conductivity are related through the following equation:

$$\lambda = \frac{1{,}000\ \kappa}{c_{eq}} \qquad \textbf{(Eq. 12.98)}$$

where $c_{eq}/1{,}000$ has units of equivalents per cubic centimeter.

*This section draws from Kudryk and Kellogg (1954); Cathro and Koch (1964); Guan and Han (1994); and Sun, Guan, and Han (1996).

When the solution conductivity is known, the resistance *R* of the solution can be calculated–given the distance between two electrodes, *d*, and the cross-sectional area of the electrode, *A*–from the following equation:

$$R = d/(\kappa A) \quad \textbf{(Eq. 12.99)}$$

It should be noted that 0.1 mol/dm^3 of $Fe_2(SO_4)_3$ solution is equivalent to 0.6/1,000 eq/cm^3.

Example:

Calculate the IR drop between two electrodes that are separated by 10 cm and have a cross-sectional area of 1 cm^2. The solution contains 10^{-4} mol/dm^3 of KCl (λ = 147 ohm^{-1}cm^{2}eq^{-1}), and the current is flowing at 10^{-8} A between these two electrodes.

Solution:

$$\kappa = 147 \times 10^{-4} \times 10^{-3} = 147 \times 10^{-7}\ \text{ohm}^{-1}\ \text{cm}^{-1}$$

$$R = 10\ \text{cm}/\ (1\ \text{cm}^2 \times 1.47 \times 10^{-5}\ \text{ohm}^{-1}\ \text{cm}^{-1}) = 6.8 \times 10^5\ \text{ohm}$$

$$\text{IR} = 10^{-8} \times 6.8 \times 10^5 = 0.0068\ \text{v}$$

REACTION KINETICS

As discussed earlier, the thermodynamics of any given reaction do not tell how fast the reaction would occur. Therefore, there is a need to understand the kinetics of the leaching process. Unfortunately, the dissolution kinetics cannot be predicted from basic principles. The leaching kinetics are traditionally calculated based on experimental results.

Let us consider a general reaction consisting of reactants A and B and products R and S. The overall stoichiometric equation is given by the following equation:

$$a\text{A} + b\text{B} = r\text{R} + s\text{S} \quad \textbf{(Eq. 12.100)}$$

where *a*, *b*, *r*, and *s* are stoichiometric coefficients of species of A, B, R, and S, respectively.

The usual convention of the rate expression for reactants A and B and products R and S can be given by

$$r_R = -\frac{1}{a}\frac{dC_A}{dt} = -\frac{1}{b}\frac{dC_B}{dt} \quad \textbf{(Eq. 12.101)}$$

$$r_P = -\frac{1}{r}\frac{dC_R}{dt} = \frac{1}{s}\frac{dC_S}{dt} \quad \textbf{(Eq. 12.102)}$$

where

r_R and r_P = the rate of reaction for reactants and products, respectively

C_A, C_B, C_R, and C_S = concentrations of A, B, R, and S, respectively, in mass/volume (mole/liter)

Equation 12.101 represents that the rate of disappearance of A is equivalent to that of B with an adjustment of the stoichiometric coefficients involved. Similarly, the rate of appearance of the product follows the same format as given in Eq. 12.102. Additionally, at steady state, when the rate of disappearance of the reactant is exactly balanced by the rate appearance of the product,

$$r_R = r_P = -\frac{1}{a}\frac{dC_A}{dt} = \frac{1}{r}\frac{dC_R}{dt} \quad \textbf{(Eq. 12.103)}$$

For example, the rate of dissolution of a metal oxide, MO, with an acid can be given by the following expression:

$$\text{<MO>} + 2\ \{\text{H}^+\} \rightarrow \{\text{M}^{+2}\} + \{\text{H}_2\text{O}\} \quad \textbf{(Eq. 12.104)}$$

The reaction rates for this leaching system can be given by

$$r_R = -\frac{dC_{MO}}{dt} = -\frac{1}{2}\frac{dC_{H^+}}{dt} \quad \textbf{(Eq. 12.105)}$$

or

$$r_P = \frac{dC_{M^{2+}}}{dt} = \frac{dC_{H_2O}}{dt} \quad \textbf{(Eq. 12.106)}$$

It should be noted that the units of the rate of reaction may be moles per liter-second. Also to be noted are that the signs of the stoichiometric coefficients of the reactants are negative and that those of the products are positive, as indicated by Eqs. 12.104 and 12.105.

Leaching Data Analysis*

One of the most important objectives in studying leaching kinetics is the establishment of the rate expression that can be used in design, optimization, and control of metallurgical operations. The parameters that need to be established include the numerical value of the rate constant and the order of reaction with respect to reactants and products whose concentrations are subjected to change during the course of the leaching reaction. For example, if a chemical reaction involves A and B as reactants and C and D as products, the stoichiometric reaction can be written as follows:

$$a\text{A} + b\text{B} \underset{k_2}{\overset{k_1}{\rightleftarrows}} c\text{C} + d\text{D} \quad \textbf{(Eq. 12.107)}$$

where

a, b, c, and d = stoichiometric coefficients of species A, B, C, and D, respectively
k_1, k_2 = reaction coefficients in the forward and reverse directions, respectively

The rate expression of this stoichiometric reaction can be written in a more general way:

$$-\frac{1}{a}\frac{dC_A}{dt} = -\frac{1}{b}\frac{dC_B}{dt} = \frac{1}{c}\frac{dC_C}{dt} = \frac{1}{d}\frac{dC_D}{dt} = k_1 C_A^n C_B^m - k_2 C_C^p C_D^q \quad \textbf{(Eq. 12.108)}$$

where

C_A, C_B, C_C, and C_D = concentrations of species A, B, C, and D, respectively
m, n, p, and q = orders of reaction

However, if the reaction given in Eq. 12.107 is irreversible, as in most leaching systems, Eq. 12.108 is reduced to the following form:

$$-\frac{1}{a}\frac{dC_A}{dt} = k_1 C_A^n C_B^m$$

or

$$-\frac{dC_A}{dt} = k_1{'} C_A^n C_B^m \quad \textbf{(Eq. 12.109)}$$

where $k_1{'} = k_1 \times a$.

For this system, the rate constant, $k_1{'}$, and the orders of reaction, n and m, should be determined with the aid of leaching experimental data. The rate expression given in Eq. 12.109 can be further reduced if the reaction is carried out in such a way that the concentration of A is kept constant. For such situations, the rate expression is reduced to

$$-\frac{dC_A}{dt} = k_1{''} C_B^m \quad \textbf{(Eq. 12.110)}$$

where $k_1{''} = k_1{'} \times C_A^n$. It should be noted that the rate constant and the order of reaction are constant as long as the temperature of the system is maintained constant.

*This section draws from Levenspiel (1972) and Smith (1970).

Let us consider the dissolution of zinc in acidic medium as given earlier in Eq. 12.73:

$$\langle Zn \rangle + 2\,\{H^+\} \rightarrow \{Zn^{2+}\} + (H_2)$$

We can see that the rate of disappearance of hydrogen ion is directly related to the rate of appearance of zinc ion; that is,

$$\frac{dC_{Zn^{2+}}}{dt} = -\frac{1}{2}\frac{dC_{H^+}}{dt} = k'C^m_{Zn}C^n_H = kC^n_H \qquad \textbf{(Eq. 12.111)}$$

A very important assumption is made in formulating Eq. 12.111—the concentration of zinc metal is assumed to be constant. It should be noted that surface area is usually used instead of mass concentration as the concentration of the solid because the contribution of the solid in the rate of dissolution is via surface area. The dissolution rate of 1 g of zinc rod would be quite different from that of 1 g of zinc powder even though the total mass of these two systems is identical.

Let us examine what kinds of integral rate expressions would be expected when the order of reaction, n, in Eq. 12.111 is known. For purposes of discussing the integral rate expressions, Eq. 12.111 is generalized to be

$$-\frac{dC_A}{dt} = kC^n_A \qquad \textbf{(Eq. 12.112)}$$

The order of reaction, n, can be any real number (0, 1, 2, 1.3, etc.). In this discussion, however, we will look at the cases where $n = 0$, 1, and 2 as typical examples that could be encountered in practice.

When $n = 0$, the reaction is referred to as zero order with respect to the concentration of A.

$$-\frac{dC_A}{dt} = kC^0_A \qquad \textbf{(Eq. 12.113)}$$

(where the concentration is expressed in moles per liter-second). Integrating Eq. 12.113 yields

$$\int_{C^o_A}^{C_A} dC_A = C_A - C^o_A = k\int_0^t dt = -kt \qquad \textbf{(Eq. 12.114)}$$

or

$$X_A = \frac{k}{C^o_A}\,t \qquad \textbf{(Eq. 12.115)}$$

In Eqs. 12.114 and 12.115, C^o_A represents the concentration of A at $t = 0$, and X_A represents the fractional conversion, i.e., $X_A = [(C^o_A - C_A)/C^o_A]$.

If the plot of X_A versus t gives a straight line as shown in Figure 12.10, the zero-order assumption is consistent with experimental observations. Therefore, the k value can be obtained from the slope of the plot.

When $n = 1$, the order of reaction is first with respect to the concentration of A:

$$-\frac{dC_A}{dt} = kC_A \qquad \textbf{(Eq. 12.116)}$$

Here, the rate constant, k, has units of per second when the rate is expressed in terms of moles per liter-second. Integrating Eq. 12.116 yields

$$\int_{C^o_A}^{C_A} \frac{dC_A}{C_A} = -k\int_0^t dt = -kt \qquad \textbf{(Eq. 12.117)}$$

$$ln\,\frac{C_A}{C^o_A} = -kt \qquad \textbf{(Eq. 12.118)}$$

$$ln\,(1 - X_A) = -kt \quad \text{or} \quad X_A = 1 - e^{-kt} \qquad \textbf{(Eq. 12.119)}$$

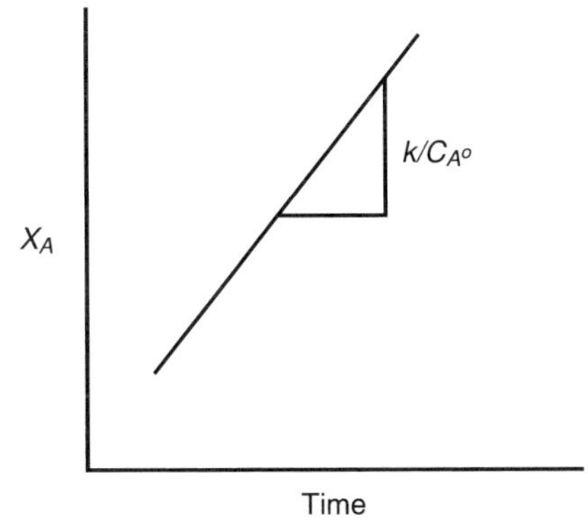

FIGURE 12.10 Plot of X_A versus *t* for zero-order reaction

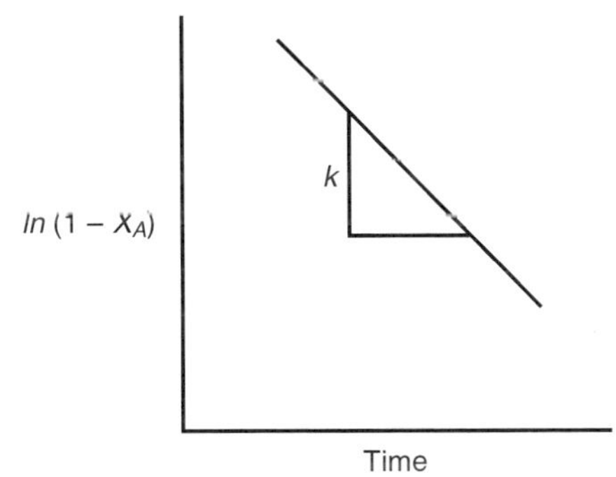

FIGURE 12.11 Plot of *ln* (1 – X_A) versus *t* for first-order reaction

The rate constant, k, can be found by plotting $ln(1 - X_A)$ versus t as shown in Figure 12.11, provided the first-order assumption is correct. It should be noted that the slope is independent of the initial concentration of A, which is a characteristic of first-order reactions.

When $n = 2$, the reaction is referred to as second order with respect to the concentration of A in the solution:

$$-\frac{dC_A}{dt} = kC_A^2 \quad \textbf{(Eq. 12.120)}$$

The units of the rate constant, k, will be in liters per mole-second when the rate is expressed in moles per liter-second. Upon integration of Eq. 12.120, Eqs. 12.122 and 12.123 are obtained:

$$\int_{C_A^o}^{C_A} \frac{dC_A}{C_A^2} = -k\int_0^t dt = -kt \quad \textbf{(Eq. 12.121)}$$

$$\frac{1}{C_A} - \frac{1}{C_A^o} = -kt \quad \textbf{(Eq. 12.122)}$$

$$\frac{X_A}{1-X_A} = C_A^o kt \quad \textbf{(Eq. 12.123)}$$

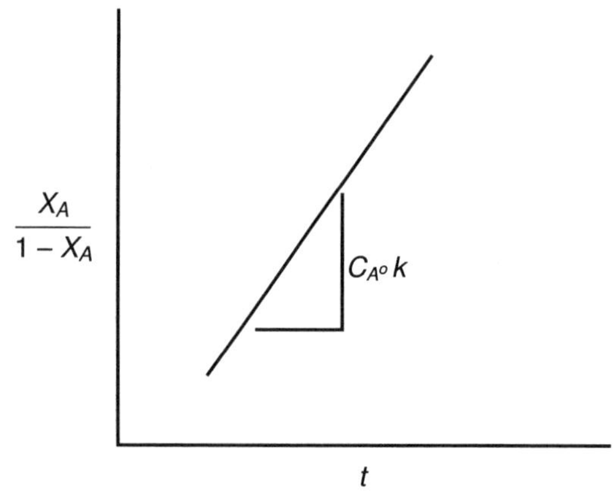

FIGURE 12.12 Plot of *ln* $[X_A/(1 - X_A)]$ versus *t* for second-order reaction

If the second-order assumption is valid, we obtain a straight line from a plot of $X_A/(1 - X_A)$ versus t, and the rate constant will be determined from the slope of the plot as shown in Figure 12.12.

Temperature Effect on the Reaction Rate

Reaction rate increases markedly with increasing temperature. It has been found empirically that temperature affects the rate constant in the manner shown in the following equation:

$$k = k^o e^{-E/RT} \quad \textbf{(Eq. 12.124)}$$

Equation 12.124 is often known as the Arrhenius law, where E is the activation energy and k^o is a constant known as the frequency factor, frequently assumed to be independent of temperature. The activation energy, E, is an important parameter that provides information on the effect of temperature on the rate of dissolution.

The way to determine the activation energy for a dissolution system is first to run a number of experiments, at least at three or four different temperatures, with all other variables being kept constant. The next step is to calculate the rate constant for each temperature as discussed in the previous section. This gives us four rate constant values for four different temperatures. We then rearrange Eq. 12.124 by taking the logarithm of both sides of the equation:

$$ln\, k = ln\, k^o - \frac{E}{R}\frac{1}{T} \quad \textbf{(Eq. 12.125)}$$

or

$$\log k = \log k^o - \frac{E}{2.303R}\frac{1}{T} \quad \textbf{(Eq. 12.126)}$$

As indicated in Figure 12.13, a plot of $ln\, k$ versus $1/T$ yields a straight line from which the activation energy, E, can be calculated.

Mass Transfer*

The equation of continuity for component i in metallurgical systems can be written as

$$\frac{\partial C_i}{\partial t} + \nabla \cdot N_i = R_i \quad \textbf{(Eq. 12.127)}$$

*This section draws from Geiger and Poirier (1973); Sherwood, Pigford, and Wilke (1975); and Bird, Stewart, and Lightfoot (1962).

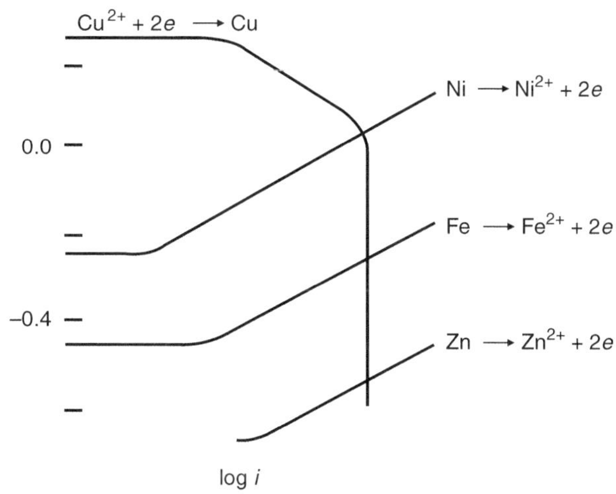

FIGURE 12.13 Three anodic systems of nickel, iron, and zinc crossing the cathodic curve of copper at various mixed potentials

where

C_i = concentration of i
N_i = flux of i
R_i = reaction term, which is zero for a fluid medium if the chemical reaction takes place at the solid–fluid interface

For hydrometallurgical systems, N_i frequently consists of a molecular diffusion term, electromigration term, and convective diffusion term, as indicated in the following expression:

$$N_i = -D_i \nabla C_i - z_i \mu_i F C_i \nabla \phi + C_i V \quad \textbf{(Eq. 12.128)}$$

where

D_i = concentration of i
∇C_i = concentration gradient
z_i = valence of the ion in question
μ_i = ionic mobility
F = the Faraday constant
C_i = concentration
$\nabla \phi$ = electrical potential gradient
V = net velocity of the fluid of the system

It should be noted that if N_i consists of the molecular diffusion term only, Fick's first law results in

$$N_i = -D_i \nabla C_i \quad \textbf{(Eq. 12.129)}$$

Therefore, Eq. 12.127 without R_i becomes

$$\frac{\partial C_i}{\partial t} + \nabla \cdot (-D_i \nabla C_i) = 0$$

and

$$\frac{\partial C_i}{\partial t} + D_i \nabla^2 C_i = 0 \qquad \text{(Fick's second law)} \quad \textbf{(Eq. 12.130)}$$

However, if the flux includes the convection term, C_iV, then

$$N_i = D_i\nabla C_i + C_iV$$

Therefore,

$$\frac{\partial C_i}{\partial t} + \nabla \cdot \nabla C_i = D_i\nabla^2 C_i \quad \text{(Eq. 12.131)}$$

Let us introduce dimensionless parameters as follows:

$$C_i^* = \frac{C_i}{C_i^o}$$

$$x^* = x/L$$

$$t^* = t/(L/V)$$

$$\nabla^* = L\nabla$$

$$\nabla^{*2} = L^2\nabla^2$$

where

C_i^o = the initial concentration of i

L = the characteristic length of the system

Therefore, Eq. 12.131 becomes:

$$\frac{DC_i^*}{Dt^*} = \frac{D_i}{LV}\nabla^{*2}C_i^* \quad \text{(Eq. 12.132)}$$

where D/Dt^* is the substantial derivative.

In Eq. 12.132, the parameter LV/D_i is known as the Peclet number and can be separated into two other parameters:

$$\frac{LV}{D_i} = \left(\frac{LV\rho}{\mu}\right)\left(\frac{\mu}{\rho D_i}\right)$$

where

μ = the viscosity of the fluid

ρ = the density of the fluid

The term $LV\rho/\mu$ is known as the Reynolds number, and $\mu/\rho D_i$ is the Schmidt number. It is interesting to note that the Peclet number has a practical meaning; that is,

$$\frac{LV}{D_i} = \frac{\nabla C_iV}{D_i(\nabla C_i/L)} = \frac{\text{convective diffusion}}{\text{molecular diffusion}} \quad \text{(Eq. 12.133)}$$

Therefore, the Peclet number is regarded as a measure of the role of convective diffusion against molecular diffusion. For most hydrometallurgical systems, for example, the Schmidt number is on the order of 1,000 because the diffusivity of ions and kinematic viscosity of water are, respectively, on the order of 10^{-5} cm^2/s and 10^{-2} cm^2/s. Therefore, if the Reynolds number is greater than 10^{-3}, the Peclet number is greater than 1, and consequently, convective diffusion is more dominating than molecular diffusion in such systems.

Mass Transfer Coefficients for Convective Diffusion

For systems with large Peclet numbers, it is frequently assumed that there is a diffusion boundary layer at some distance from the solid surface (see Figure 12.14 later in this chapter). For such systems, it is quite common to write the mass flux from the bulk solution to the solid surface as follows:

$$N_i = k_m\,(C_b - C_s) \quad \text{(Eq. 12.134)}$$

where

N_i = mass flux of species i

k_m = mass transfer coefficient, in cm/s

C_b = concentration of species i in the bulk solution, in mol/cm^3

C_s = concentration of species i at the solid surface, in mol/cm^3

Because the units of measure of k_m are the same as those of (D/δ), where δ is the diffusion boundary layer thickness, k_m is often substituted by this ratio. Therefore,

$$N_i = \frac{D}{\delta}(C_b - C_s) \quad \textbf{(Eq. 12.135)}$$

It should be noted that the diffusion boundary layer thickness is often estimated by the relationship $k_m = D/\delta$, provided k_m is known. It should also be noted that because k_m is a function of geometry and is strongly affected by the hydrodynamics of the system, the notion that the diffusion boundary layer thickness is fixed at about 10^{-3} cm is incorrect. To elaborate on this point, let us next consider two of the most common geometries as examples.

Mass Transfer from or to a Flat Plate. The mass transfer coefficient for a flat plate where fluid is flowing over the plate at a velocity V_o has been well documented. The mass transfer coefficient for such a system can be estimated from first principles and has the following form:

$$k_m = 0.664\, D^{2/3}\, \nu^{-1/6}\, L^{-1/2}\, V_o^{\,1/2} \quad \textbf{(Eq. 12.136)}$$

where

D = the diffusivity of the diffusing species

ν = the kinematic viscosity of the fluid

L = the length of the plate

Note that Eq. 12.136 is applicable as long as the Reynolds number, defined by V_oL/ν, is less than 10^6. Let us equate Eq. 12.136 to D/δ as indicated earlier and evaluate the value δ, the diffusion boundary layer thickness, for the value of V_o over the range of 1 to 10,000 cm/s. In this calculation, D is assumed to be 10^{-5} cm^2/s; L is 1 cm; and ν is 0.01 cm^2/s. The diffusion boundary layer thickness defined, as such, for this hypothetical—but practical—system is estimated to be anywhere between 10^{-2} and 10^{-4} cm.

Rotating Disk. Although it is not a practical geometry, because the mathematical representation of the system is exact and follows very closely to the experimental data, a rotating disk is frequently used to determine the mass flux and the mass transfer coefficient. The mass transfer coefficient for this system is as follows:

$$k_m = 0.62\, D^{2/3}\, \nu^{-1/6}\, \omega^{1/2} \quad \textbf{(Eq. 12.137)}$$

This relationship is valid as long as the Reynolds number, $r^2\omega/\nu$ is less than 10^5, where r and ω are, respectively, the radius and the angular velocity of the disk. A similar analysis can be made as in the case of the flat plate for the diffusion boundary layer thickness. This layer is found by calculation to be in the range of 1.6×10^{-2} to 5.1×10^{-4} cm over the angular velocity range of 1 to 10^3 rad/s if the radius is 1 cm.

Particulate System*

It has been demonstrated in the literature that the mass transfer coefficient for particulate systems can be given by the following equation:

$$k_m = \frac{2D}{d} + 0.6\, V_t^{\,1/2}\, d^{-1/2}\, \nu^{-1/6}\, D^{2/3} \quad \textbf{(Eq. 12.138)}$$

where

d = the diameter of the particle

V_t = the slip velocity, which is often assumed to be the terminal velocity of the particle in question

*This section draws from Vu and Han (1979, 1981).

The terminal velocity of a particle can be calculated using the following equation depending on the Reynolds number of the system, which is defined by $dV_t\rho/\mu$, where ρ is the density of the fluid:

$$V_t = \frac{2r^2(\rho_s - \rho)g}{9\mu} \quad \textbf{(Eq. 12.139)}$$

where ρ_s is the density of the particle. The preceding equation is often referred to as the Stokes' equation and is valid as long as the Reynolds number is less than 1. However, when the Reynolds number is between 1 and 700, the following equations are used (Han 1984):

$$V_t = \frac{\mu}{d\rho}10^A \quad \textbf{(Eq. 12.140)}$$

$$A = 5.0\,(0.66 + 0.4 \log K)^{1/2} - 5.55 \quad \textbf{(Eq. 12.141)}$$

$$K = \frac{4gd^3\rho(\rho_s - \rho)}{3\mu^2} \quad \textbf{(Eq. 12.142)}$$

where g is the gravitational coefficient.

It has been shown that the mass transfer coefficient for particles is reasonably insensitive to the size of particles (Vu and Han 1981).

Example:

A cementation reaction, $Zn + Cu^{2+} \rightarrow Cu + Zn^{2+}$, is taking place at the surface of a zinc plate of 10 cm × 10 cm area. Feed flowing parallel to the plate at a velocity of 1 m/s contains copper at 1 mol/dm^3. Suppose we want to estimate the rate of deposition assuming that the mass transfer of Cu^{2+} to the zinc plate is limiting. The diffusivity of Cu^{2+} is 7.2×10^{-6} cm^2/s, and the kinematic viscosity of water is 0.01 cm^2/s.

$$-\frac{1}{S}\frac{dN_{Cu^{2+}}}{dt} = k_m\,(Cu_b^{2+} - Cu_s^{2+}) = k_m\,Cu_b^{2+} \quad \textbf{(Eq. 12.143)}$$

where

S = the surface area of the plate

$N_{Cu^{2+}}$ = the number of moles of Cu^{2+} ion

Cu_b^{2+} = the concentration of Cu^{2+} in the bulk

Cu_s^{2+} = the concentration of Cu^{2+} at the interface

From Eq. 12.136,

$$\begin{aligned} k_m &= 0.664\,(7.2 \times 10^{-6})^{2/3}\,(0.01)^{-1/6}\,(10)^{-1/2}\,(100)^{1/2} \\ &= 0.664 \times 3.7 \times 10^{-4} \times 2.15 \times 0.316 \times 10 \\ &= 1.7 \times 10^{-3} \text{ cm/s} \end{aligned}$$

Therefore,

$$-\frac{1}{S}\frac{dN_{Cu^{2+}}}{dt} = 1.7 \times 10^{-3} \times 1{,}000 = 1.7 \text{ mol/cm}^2 \cdot \text{s}$$

and

$$\text{Re} = \frac{100 \times 10}{0.01} = 10^5$$

Example:

Consider the situation from the previous example, except that instead of a zinc plate, zinc particles 100 μm in diameter are suspended in a 1-mol/dm^3 Cu^{2+} solution. Suppose we want to estimate the rate of deposition of Cu^{2+}. (Note that the density of Zn is 7.14 g/cm^3.)

From Eqs. 12.142 and 12.141,

$$K - \frac{4 \times 981 \times (0.01)^3 \times 1 \times (7.14 - 1)}{3 \times 10^{-4}} = 80.3$$

$$A = 5(0.66 + 0.4 \log 80.3)^{1/2} - 5.55 = 0.412$$

Therefore,

$$V_t = 10^{0.412} = 2.58 \text{ cm/s}$$

$$\text{Re} = (10^{-2} \times 2.58/10^{-2}) = 2.58$$

As a result,

$$\begin{aligned} k_m &= \frac{2 \times 7.2 \times 10^{-6}}{0.01} + 0.6 \times 2.58^{1/2} \times 0.01^{-1/2} \times 0.01^{-1/6} \times (7.2 \times 10^{-6})^{2/3} \\ &= 1.44 \times 10^{-3} + 7.75 \times 10^{-3} \\ &= 9.19 \times 10^{-3} \text{ cm/s} \end{aligned}$$

Finally,

$$k_m = -\frac{1}{S}\frac{dN_{Cu^{2+}}}{dt} = 9.19 \times 10^{-3} \times 1{,}000 = 9.19 \text{ mol/cm}^2 \cdot \text{s}$$

Effect of Temperature on Mass Transfer Coefficient

It is frequently asked what effect the temperature has on the mass transfer coefficient. This is rather a difficult question to answer because this effect is very much a function of the system. In other words, the effect of temperature on the mass transfer coefficient varies from system to system. For instance, consider the temperature effect on the mass transfer coefficient for a rotating disk. As the mass transfer coefficient for a rotating disk is given by Eq. 12.137, combining this equation and the Arrhenius equation (Eq. 12.124) results in

$$k_m = 0.62\, \omega^{1/2} D_o e^{-2E_0/3RT} \nu_o e^{-E_V/6RT} \quad \textbf{(Eq. 12.144)}$$

$$k_m = k_o e^{-\frac{4E_{D_o} + E_v}{6}/RT} = k_o e^{-E_{app}/RT} \quad \textbf{(Eq. 12.145)}$$

where

ω = the angular velocity
D_o = the standard diffusivity
^{E}D = 3,000 cal/mol
ν_o = the standard kinematic viscosity
E_v = 3,600 cal/mol
k_o = the standard mass transfer coefficient
E_{app} = $(4E_o = E_v)/6$

Therefore, E_{app} is calculated to be about 2,600 cal/mol (Levich 1962; Rubicumintara and Han 1990a,b). This means that if mass transfer is limiting for a rotating system, the overall activation is expected to be in the neighborhood of 2,600 cal/mol.

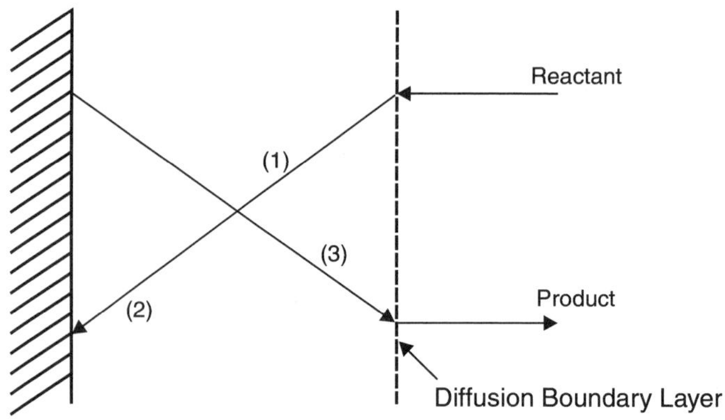

FIGURE 12.14 Leaching of metal oxide in acid

Limiting Reaction Step*

Hydrometallurgical processes involve a series of rate processes. The sequence of reaction depends on the type of reaction under consideration. For example, the leaching reaction of metal oxides in acidic solution can be viewed as a sequential reaction that consists of the mass transport of H^+ through the bulk solution to the solid–liquid interface followed by heterogeneous reaction (i.e., the reaction involving two phases, in this case solid and liquid). The product, soluble metal ion(s), then, will diffuse out to the bulk solution. This process is shown schematically in Figure 12.14. The diffusion of the reactant, H^+, is denoted by path 1; the chemical reaction between the metal oxide and H^+ is denoted by path 2; and path 3 represents the mass transfer of the product metal ion, which is diffusing out into the bulk solution. In most practical cases, one of these steps is slower than the others. If the time of the slowest step is much greater than those required for the other steps, the slowest step alone practically determines the overall reaction. This step, then, is referred to as the rate-limiting step or the rate-determining step.

As a way of demonstrating this point, let us assume that step 1 takes 1 s, step 2 takes 1,000 s, and step 3 takes 2 s. Step 2 is by far the slowest step and, therefore, is the rate-limiting step. The overall reaction would therefore take at least 1,000 s, as the other steps are relatively fast.

Frequently, one of the most important objectives of studying metallurgical kinetics is to identify the slowest step in the sequential reaction. The question often asked is whether the heterogeneous chemical reaction or the mass transfer of reactants and products is the rate-limiting step. If both of these steps are equally important, terms representing both steps should appear in the final rate expression. This section examines a simple system that comprises both mass transfer and heterogeneous reaction.

Suppose a solid particle, B, reacts with a dissolved species, A, and the reaction is irreversible:

$$\langle B \rangle + \{A\} \rightarrow \text{products}$$

Furthermore, the rate of the heterogeneous reaction is assumed to be first order with respect to C_{As}, the concentration of A at the surface (i.e., at the solid–liquid interface):

$$-\frac{dC_A}{dt} = kC_{As} \quad \textbf{(Eq. 12.146)}$$

*This section draws from Vu and Han (1981); Meng and Han (1993); and Meng, Sun, and Han (1995).

The mass transfer rate of the reactant A diffusing in through the diffusion boundary layer is

$$-\frac{dC_A}{dt} = k_m\frac{A}{V}(C_{Ab} - C_{As}) \quad \textbf{(Eq. 12.147)}$$

$$-\frac{dC_A}{dt} = k'_m(C_{Ab} - C_{As}) \quad \textbf{(Eq. 12.148)}$$

where

C_{Ab} = the concentration of A in the bulk solution
k'_m = $k_m A/V$
A = the surface area of the solid
V = the volume of the solution

At steady state, Eq. 12.146 should be equal to Eq. 12.148; therefore,

$$kC_{As} = k'_m C_{Ab} - k'_m C_{As}$$

which yields the following equation:

$$C_{As} = \frac{k'_m}{k + k'_m}C_{Ab} \quad \textbf{(Eq. 12.149)}$$

Substituting Eq. 12.149 into Eq. 12.146 yields

$$-\frac{dC_A}{dt} = \frac{kk'_m}{k + k'_m}C_{Ab} \quad \textbf{(Eq. 12.150)}$$

In practice, we prefer the rate expression in the following form using the observed rate and the observed rate constant, k_{obs}:

$$\text{rate}_{obs} = k_{obs}\, C_{Ab} \quad \textbf{(Eq. 12.151)}$$

From Eqs. 12.150 and 12.151:

$$k_{obs} = \frac{kk'_m}{k + k'_m} \quad \text{or} \quad \frac{1}{k_{obs}} = \frac{1}{k'_m} + \frac{1}{k} \quad \textbf{(Eq. 12.152)}$$

As Eq. 12.152 shows, the mass transfer rate constant and the heterogeneous reaction rate constant are analogous to electrical resistance, and the addition procedure is the same way.

As discussed earlier, the rate constants are a function of temperature; i.e.,

$$k = k_o e^{-\frac{E_r}{RT}} \quad \text{for heterogeneous reaction} \quad \textbf{(Eq. 12.153)}$$

$$k_m = k_{mo} e^{-\frac{E_m}{RT}} \quad \text{for mass transfer} \quad \textbf{(Eq. 12.154)}$$

Therefore,

$$k_{obs} = k^o_{obs}\, e^{-\frac{E_{obs}}{RT}} \quad \text{for the overall rate} \quad \textbf{(Eq. 12.155)}$$

Substituting Eqs. 12.153, 12.154, and 12.155 into Eq. 12.152 yields

$$k^o_{obs}\, e^{-\frac{E_{obs}}{RT}} = \frac{k_o k^o_m\, e^{-\frac{(E_r + E_m)}{RT}}}{k_o e^{-\frac{E_r}{RT}} + k^o_m\, e^{-\frac{E_m}{RT}}} \quad \textbf{(Eq. 12.156)}$$

Figure 12.15 shows a plot of log k_{obs} versus $1/T$. As the figure shows, when the reaction is mixed controlled at intermediate temperatures T′ and T″, both mass transfer and heterogeneous reaction are important; the activation energy is of intermediate value. At lower temperatures, namely in region III,

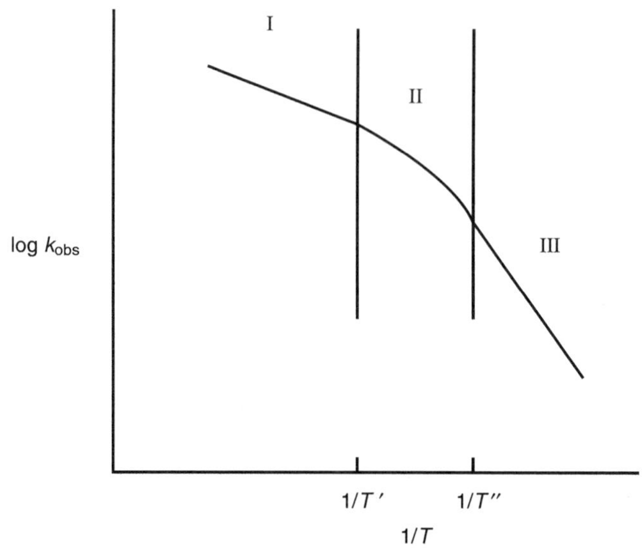

FIGURE 12.15 Plot of log k_{obs} versus 1/T, giving the activation energy

the slope of log k_{obs} versus $1/T$ becomes steeper, giving a high activation energy (for example, greater than 10 kcal/mol). On the other hand, when the temperature is increased in order to push the reaction into region I, the corresponding activation energy becomes low (2–4 kcal/mol), which is characteristic of diffusion processes. Therefore, such plots prove to be very useful because they shed some light on whether the overall reaction in question is chemically or mass transfer controlled.

SHRINKING CORE MODELS*

In many practical situations, leaching of solids produces a layer that is insoluble but permeable to ion diffusion. For example, the leaching of galena in a ferric chloride solution produces sulfur through which ferric ion can diffuse, enabling the leaching reaction to proceed further. This situation is represented pictorially in Figure 12.16.

Consider an example where reactant solid B reacts with aqueous reactant A, producing solid product S and dissolved ion R:

$$A_{(l)} + b\,B_{(s)} \rightarrow r\,R_{(l)} + s\,S_{(s)} \qquad \textbf{(Eq. 12.157)}$$

where b, r, and s are coefficients of the reaction.

It should be noted that the equation is always written so that the stoichiometric coefficient of reactant A is always 1. The reaction is assumed to be irreversible. It is assumed that the particle is spherical; that the radius, R, is constant throughout the reaction; and that the reaction interface at r_c shrinks uniformly.

Three resistances are identifiable in such a reacting system: (1) diffusion through the mass transfer boundary layer (film diffusion); (2) chemical reaction at the reactant–product interface, $r = r_c$; and (3) diffusion through the porous product layer (product layer diffusion).

Whichever of these steps is slowest is the limiting step for the overall reaction; therefore, identifying this step is of utmost importance. In this discussion, these three limiting steps will initially be discussed independently. Then the mixed controlled situation will be discussed when all three steps contribute to the overall reaction.

*This section draws from Levenspiel (1972) and Sohn and Wadsworth (1979).

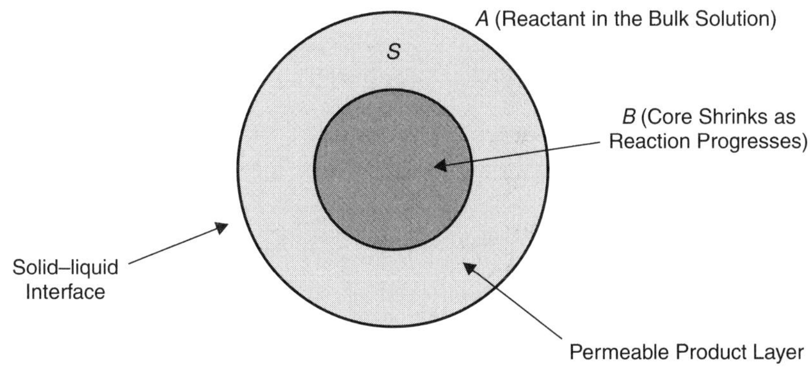

FIGURE 12.16 Reactant *B* reacting with reactant *A*, producing insoluble product layer *S*

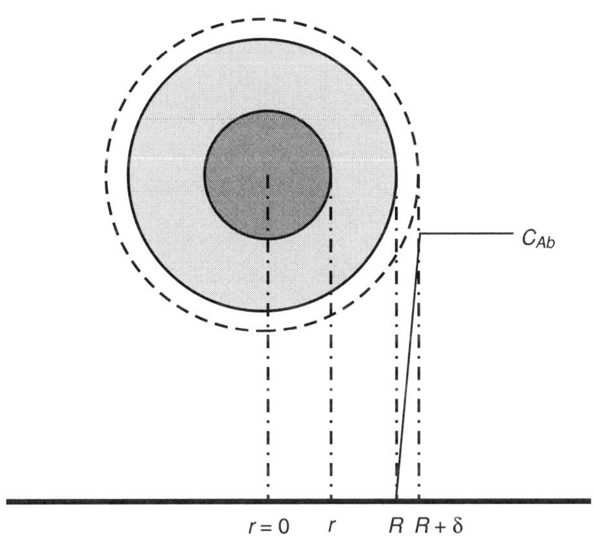

FIGURE 12.17 Shrinking core model when film diffusion is limiting

Film Diffusion as the Limiting Step

When film diffusion of reactant A is the limiting step, the concentration of A will be uniform up to $r = R + \delta$ and practically zero at $r = R$. The concentration profile is given in Figure 12.17. Here, δ is the diffusion boundary layer thickness. Therefore, the reaction rate of B can be described as follows:

$$-\frac{1}{S}\frac{1}{b}\frac{dN_B}{dt} = -\frac{1}{4\pi R^2}\frac{1}{b}\frac{dN_B}{dt} = -\frac{1}{S}\frac{dN_A}{dt} \qquad \textbf{(Eq. 12.158)}$$

where

S = the surface area of the particles = $4\pi R^2$
b = the stoichiometric coefficient for reactant B (see Eq. 12.157)
N_A = the number of moles of A
N_B = the number of moles of B

From the stoichiometric relationship, the following equation is valid:

$$-\frac{1}{S}\frac{dN_B}{dt} = -\frac{b}{4\pi R^2}\frac{dN_A}{dt}$$

Because

$$-\frac{1}{S}\frac{dN_B}{dt} = k_m(C_{Ab} - C_{As})$$

where

k_m = the mass transfer coefficient
C_{Ab} = the bulk concentration of A
C_{As} = the surface concentration of A

we have

$$-\frac{1}{S}\frac{dN_B}{dt} = bk_m(C_{Ab} - C_{As}) \qquad \textbf{(Eq. 12.159)}$$

For film diffusion controlling, $C_{As} \approx 0$; therefore,

$$-\frac{1}{S}\frac{dN_B}{dt} = bk_m C_{Ab} = \text{constant at steady state}$$

It should be noted that, at any given time, the following relationships hold:

$$N_{B_t} = \rho_B \frac{4}{3}\pi r_c^3 \qquad \textbf{(Eq. 12.160)}$$

$$-dN_B = -4\pi\rho_B r_c^2\, dr_c \qquad \textbf{(Eq. 12.161)}$$

where

ρ_B = molar density, mol/vol
r_c = the radius of the core

Therefore, Eq. 12.159 becomes

$$-\frac{1}{S}\frac{dN_B}{dt} = -\frac{\rho_B r_c^2}{R^2}\frac{dr_c}{dt} = bk_m C_{Ab} \qquad \textbf{(Eq. 12.162)}$$

By integrating Eq. 12.162, we obtain

$$-\frac{\rho_B}{R^2}\int_R^t r_c^2\, dr_c = bk_m C_{Ab}\int_0^t dt$$

Therefore,

$$t = \frac{\rho_B R}{3bk_m C_{Ab}}\left[1 - \left(\frac{r_c}{R}\right)^3\right] \qquad \textbf{(Eq. 12.163)}$$

Equation 12.163 gives the time required for a reaction to proceed from particle radius R to r_c. If t_{comp} is defined as $t = t_{comp}$ when $r_c = 0$,

$$t_{comp} = \frac{R\rho_B}{3bk_m C_{Ab}} \qquad \textbf{(Eq. 12.164)}$$

Therefore,

$$\frac{t}{t_{comp}} = 1 - \left(\frac{r_c}{R}\right)^3 \qquad \textbf{(Eq. 12.165)}$$

Also note that

$$X_B = \text{fractional conversion}$$
$$= \frac{N_B^o - N_B}{N_B^o} = \frac{V_B^o - V_B}{V_B^o}$$
$$= \frac{R^3 - r_c^3}{R^3} = 1 - \left(\frac{r_c}{R}\right)^3 \quad \textbf{(Eq. 12.166)}$$

where

N_B^o = the number of moles of B at t = 0
N_B = the number of moles of B at t = t
V_B^o = the volume of B at t = 0
V_B the volume of B at t = t

Substituting Eq. 12.166 into Eq. 12.165 yields

$$X_B = \frac{t}{t_{\text{comp}}} \quad \textbf{(Eq. 12.167)}$$

Example:

Consider a metal sulfide MS (molecular weight = 100; diameter = 1 mm; density = 5 g/cm^3) that is subjected to leaching.

$$<MS> + 2\{H^+\} + \tfrac{1}{2}\{O_2\} \; (\; \{M^{2+}\} + <S> + \{H_2O\} \quad \textbf{(Eq. 12.168)}$$

If the diffusion of O_2 through the diffusion boundary layer is limiting, how long would it take to complete the reaction? Assume k_m = 0.1 cm/s; $\{O_2\} = 2.7 \times 10^{-4}$ mol/dm^3. The solution is as follows:

$$\rho_B = \frac{5 \text{ g/cm}^3}{100 \text{ g/mol}} = 0.05 \text{ mol/cm}^3$$

$$C_{Ab} = 2.7 \times 10^{-4} \times 10^{-3} \text{ mol/cm}^3 = 2.7 \times 10^{-7} \text{ and } b = 2$$

Therefore, from Eq. 12.164,

$$t_{\text{comp}} = \frac{0.05 \times 0.05}{3 \times 2 \times 10^{-1} \times 2.7 \times 10^{-7}} = 1.54 \times 10^4 = 4 \text{ h, } 16 \text{ min, } 40 \text{ s}$$

Product Layer Diffusion as the Limiting Step

When the reactant diffusion through the product layer is limiting, the concentration of reactant A is uniform up to $r = R$ and approaches zero at the unreacted core surface, $r = r_c$ (see Figure 12.18). Therefore, at steady state,

$$-\frac{dN_A}{dt} = -4\pi r^2 J_{A_r} = -4\pi r_c^{\,2} J_{A_r} = \text{constant} \quad \textbf{(Eq. 12.169)}$$

where J_A is the flux of A, or

$$JA = -D_o \frac{dC_A}{dr} \quad \textbf{(Eq. 12.170)}$$

where

D_e = effective diffusivity
C_A = concentration of reactant A

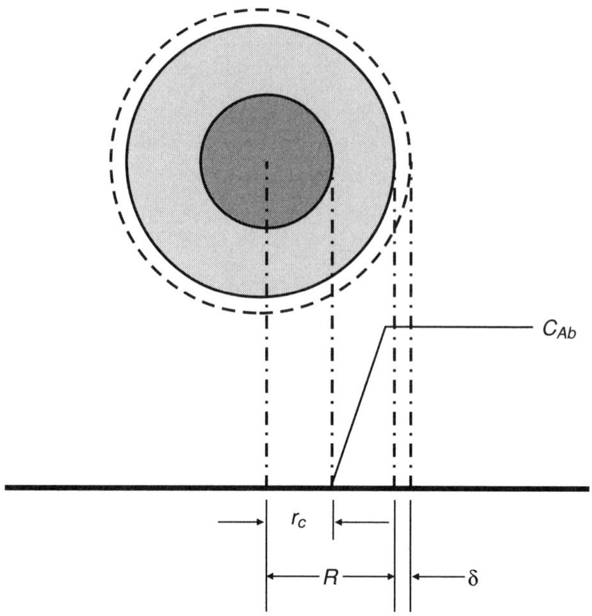

FIGURE 12.18 Shrinking core model when product layer diffusion is limiting

Note that the effective diffusivity, D_e, is used in the flux expression. The effective diffusivity for a species diffusing through porous media is often related to the porosity, θ, and tortuosity, τ, of the porous structure in the following manner:

$$D_e = D\ (\theta/\tau)$$

where the porosity is always less than 1 and the tortuosity is greater than 1.

Substituting Eq. 12.170 into Eq. 12.169 and integrating yields

$$-\frac{dN_A}{dt}\int_R^{r_c}\frac{dr}{r^2} = 4\pi D_e\int_{C_{Ab}}^{C_{Ac}=D} dC_A$$

Therefore,

$$\frac{dN_A}{dt}\left(\frac{1}{r_c}-\frac{1}{R}\right) = -4\pi D_e C_{Ab} \qquad \textbf{(Eq. 12.171)}$$

Note that $b\ dN_A = dN_B = 4\pi\rho_B r_c^{\ 2}\ dr_c$. Therefore, Eq. 12.171 can be rewritten:

$$\frac{4\pi\rho_B r_c^2}{b}\frac{dr_c}{dt}\left(\frac{1}{r_c}-\frac{1}{R}\right) = -4\pi D_e C_{Ab} \qquad \textbf{(Eq. 12.172)}$$

Upon integration,

$$-\rho_B\int_R^{r_c}\left(\frac{1}{r_c}-\frac{1}{R}\right)r_c^2\,dr_c = bD_eC_{Ab}\int_0^t dt \qquad \textbf{(Eq. 12.173)}$$

Therefore,

$$t = \frac{\rho_B R^2}{6bD_eC_{Ab}}\left[1-3\left(\frac{r_c}{R}\right)^2+2\left(\frac{r_c}{R}\right)^3\right] \qquad \textbf{(Eq. 12.174)}$$

Because $t = t_{comp}$ when $r_c = 0$,

$$t_{comp} = \frac{\rho_B R^2}{6bD_e C_{Ab}} \quad \text{(Eq. 12.175)}$$

or

$$\frac{t}{t_{comp}} = 1 - 3\left(\frac{r_c}{R}\right)^2 + 2\left(\frac{r_c}{R}\right)^3 \quad \text{(Eq. 12.176)}$$

Therefore,

$$\frac{t}{t_{comp}} = 1 - 3(1 - X_B)^{2/3} + 2(1 - X_B) \quad \text{(Eq. 12.177)}$$

Example:

Consider the circumstances given in the previous example. If the rate-limiting step is diffusion of a reactant through the product layer, what should t_{comp} be? (Assume the porosity of the product layer is 0.5 and the tortuosity is 10.) To solve this problem, first determine D_e:

$$D_e = 2.5 \times 10^{-5} \times (0.5/10) \times 10^{-6}\ \text{cm}^2/\text{s}$$

Therefore,

$$t_{comp} \frac{0.05 \times (0.05)^2}{6 \times 2 \times 1.25 \times 10^{-6} \times 2.7 \times 10^{-4} \times 10^{-3}} = 3.09 \times 10^7 = 357\ \text{days, 4 h, 13 min, 20 s}$$

Chemical Reaction as the Limiting Step

When heterogeneous chemical reaction is the limiting step (see Figure 12.19), the concentration of A at the unreacted core surface is the same as that of the bulk solution: C_{Ab}. Therefore, at steady state,

$$-\frac{1}{4\pi r_c^2}\frac{dN_B}{dt} = -\frac{b}{4\pi r_c^2}\frac{dN_A}{dt} = bk_r C_{Ab} \quad \text{(Eq. 12.178)}$$

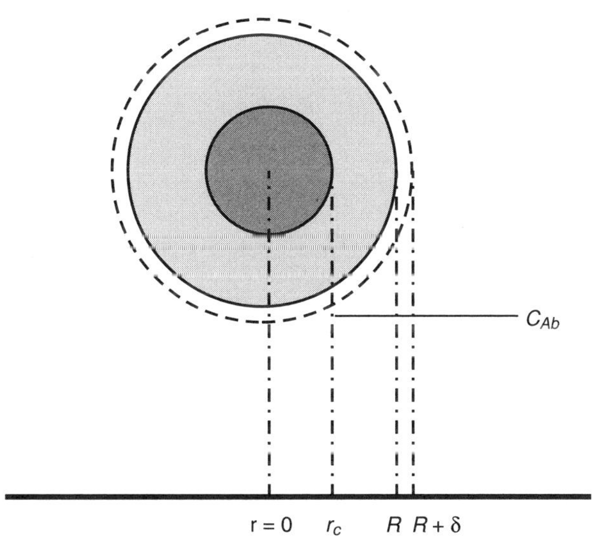

FIGURE 12.19 Shrinking core model when chemical reaction is limiting

In the preceding equation, the heterogeneous reaction is assumed to be first order and irreversible, and k_r is the first-order rate constant. Substituting $dN_B = 4\pi\rho_B r_c^2\, dr_c$ yields

$$-\rho_B dr_c = bk_r C_{Ab} dt \qquad \textbf{(Eq. 12.179)}$$

Upon integration we have

$$-\rho_B \int_R^{r_c} dr_c = bk_r C_{Ab} \int_0^t dt$$

Therefore,

$$t = \frac{\rho_B R}{bk_r C_{Ab}}\left(1 - \frac{r_c}{R}\right) \qquad \textbf{(Eq. 12.180)}$$

Because $t = t_{comp}$ when $r_c = 0$,

$$t_{comp} = \frac{\rho_B R}{bk_r C_{Ab}} \qquad \textbf{(Eq. 12.181)}$$

and

$$\frac{t}{t_{comp}} = 1 - \frac{r_c}{R} = 1 - (1 - X_B)^{1/3} \qquad \textbf{(Eq. 12.182)}$$

It should be noted that t_{comp} for three different mechanisms can be summarized as follows:

$$t_{comp} = \frac{\rho_B R}{3bk_m C_{Ab}} \propto \frac{R}{k_m} \quad \text{for film diffusion}$$

$$t_{comp} = \frac{\rho_B R^2}{6bD_e C_{Ab}} \propto \frac{R^2}{D_e} \quad \text{for product layer diffusion}$$

$$t_{comp} = \frac{\rho_B R}{bk_r C_{Ab}} \propto \frac{R}{k_r} \quad \text{for heterogeneous reaction}$$

Effect of Particle Size

It is interesting to note that t_{comp} is directly proportional to the size of particles for film diffusion and heterogeneous reaction. It should be noted, however, that the mass transfer coefficient (k_m) is, in general, a function of particle size. For the most common size range of particles, k_m is found to be independent of size (Vu and Han 1981). The effect of size on t_{comp} is more pronounced when diffusion through a product layer is limiting.

Effect of Temperature

The temperature effect will appear mainly through k_m, D_e, or k_r. It is generally agreed that the effect of temperature on k_m or D_e is moderate, whereas the effect on k_r is significant. In general, the activation energy for k_m or D_e is of the order of 2–4 kcal/mol; for k_r the value is greater.

Generally speaking, predicting the value of t_{comp} for film diffusion and product layer limiting cases is possible within the range of experimental error. However, t_{comp} for situations where the chemical reaction is limiting is much greater than in either of the other cases. Hence, the effect of temperature is commonly used to determine whether chemical reaction is the limiting step.

On the other hand, the effect of hydrodynamics, such as stirring of the impeller, is a good indicator of whether or not the film diffusion is the rate-limiting step. The reason is that k_m, unlike D_e or k_r, is very much affected by the hydrodynamics of the system.

If all three different mechanisms are in effect for a system, the system should be handled accordingly. This situation is presented schematically in Figure 12.20. Note that the concentrations of the

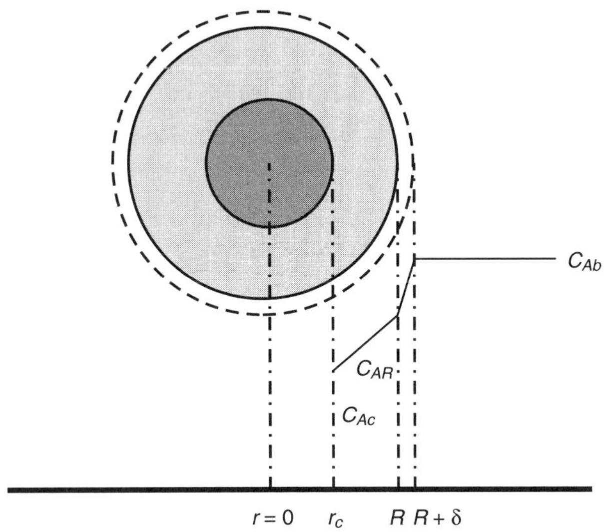

FIGURE 12.20 Shrinking core model when all three mechanisms play an important role

reactant A at various points are finite and nonzero. Therefore, the following rate expressions for each stage can be formulated:

$$\text{rate (film)} = r\pi R^2 b k_m (C_{Ab} - C_{AR}) \quad \text{(Eq. 12.183)}$$

$$\text{rate (pore)} = \frac{4\pi D_e b}{\frac{1}{r_c} - \frac{1}{R}}(C_{AR} - C_{Ac}) \quad \text{(Eq. 12.184)}$$

$$\text{rate (chem)} = 4\pi r_c^2 \, b k_r C_{Ac} \quad \text{(Eq. 12.185)}$$

By rearranging Eqs. 12.183, 12.184, and 12.185, we obtain Eqs. 12.186, 12.187, and 12.188:

$$C_{Ab} - C_{AR} = \text{rate} \frac{1}{4\pi R^2 b k_m} \quad \text{(Eq. 12.186)}$$

$$C_{AR} - C_{Ac} = \text{rate} \frac{1}{4\pi D_e b}\left(\frac{1}{r_c} - \frac{1}{R}\right) \quad \text{(Eq. 12.187)}$$

$$C_{Ac} = \text{rate} \frac{1}{4\pi r_c^2 \, b k_r} \quad \text{(Eq. 12.188)}$$

Adding the preceding three equations yields

$$C_{Ab} = \frac{\text{rate}}{4\pi R^2 b}\left[\frac{1}{k_m} + \frac{(R - r_c)R}{r_c D_e} + \frac{R^2}{r_c^2 \, k_r}\right] \quad \text{(Eq. 12.189)}$$

Because rate = $-dN_B/dt$, we have

$$-\frac{1}{4\pi R^2}\frac{dN_B}{dt} = \frac{bC_{Ab}}{\left[\frac{1}{k_m} + \frac{(R-r_c)R}{r_c D_e} + \frac{R^2}{r_c^2 k_r}\right]}$$

$$-\frac{4\pi\rho_B r_c^2}{4\pi R^2}\frac{dr_c}{dt} = \frac{\frac{bC_{Ab} r_c^2}{R^2}}{\left[\frac{1}{k_m} + \frac{(R-r_c)R}{r_c D_e} + \frac{R^2}{r_c^2 k_r}\right]}$$

Therefore,

$$-\frac{dr_c}{dt} = \frac{\frac{bC_{Ab}}{\rho_B}}{\left[\frac{r_c^2}{R^2 k_m} + \frac{(R-r_c)r_c}{RD_e} + \frac{1}{k_r}\right]} \quad \textbf{(Eq. 12.190)}$$

REACTOR DESIGN*

One of the main objectives of studying metallurgical kinetics is to develop competence in designing metallurgical reactors. Once the information on reaction kinetics is known, we should be able to predict the overall conversion of the reaction in a given reactor, provided the characteristics of the reactor are well understood. The efficiency of the overall conversion will vary if the reactor type to be used for the reaction changes. In this section, we will look at various types of reactors and their effects on a given reaction.

Ideal Reactor

Broadly speaking, there are two types of ideal reactors: the ideal stirred tank reactor and the ideal plug flow reactor. The ideal stirred tank reactor may be operated as a steady-state flow type (continuously stirred flow reactor, CSFR), as a batch type, or as a non-steady-state type (see Figure 12.21). The typical characteristics of the ideal stirred tank reactor are that mixing is complete and, therefore, the properties of the fluid in the system are uniform in all parts of the vessel. The characteristics of the ideal plug flow reactor, on the other hand, are that the feed enters the end of the reactor uniformly and that the product stream leaves at the other end with an identical residence time.

Batch Reactor

Let us make a material balance for component A for the reaction A → product. In ideal reactors, because the composition is uniform throughout the reactor, we could make the material balance over the whole reactor:

$$\underbrace{\text{input}}_{0} - \underbrace{\text{output}}_{0} - \begin{matrix}\text{rate of disappearance}\\ \text{because of reaction}\end{matrix} = \begin{matrix}\text{rate of}\\ \text{accumulation}\end{matrix}$$

*This section draws from Levenspiel (1972) and Smith (1970).

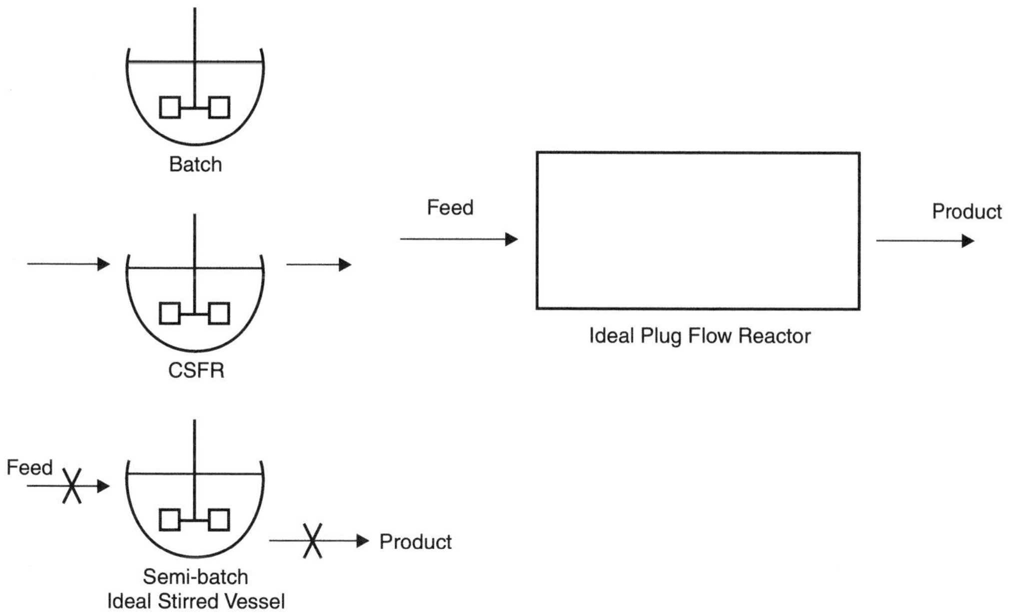

FIGURE 12.21 Ideal reactors

Therefore,

[rate of disappearance because of reaction] = – [rate of accumulation of A]

$$(-r_A)V = -\frac{dN_A}{dt} = N_A^o \frac{dx_A}{dt}$$

where

r_A = the rate of disappearance of A
V = the volume of the reactor
N_A = the number of moles of A
N_A^o = the number of moles of A at t = 0
x_A = the fractional conversion of A

Upon integration,

$$t = N_A^o \int_0^{x_A} \frac{dx_A}{(-r_A)V} \qquad \textbf{(Eq. 12.191)}$$

If v is constant–that is, if the volume of the reacting system does not change during the reaction process–we have

$$t = \frac{N_A^o}{V} \int_0^{x_A} \frac{dx_A}{-r_A} = C_A^o \int_0^{x_A} \frac{dx_A}{-r_A} \qquad \textbf{(Eq. 12.192)}$$

Example 1:

Let us examine the reactor performance for a first-order reaction:

$$-r_A = kC_A = kC_A^o(1 - x_A)$$

From Eq. 12.192,

$$t = C_A^o \int_0^{x_A} \frac{dx_A}{kC_A^o(1 - x_A)} = -\frac{1}{k} \ln(1 - x_A) \qquad \textbf{(Eq. 12.193)}$$

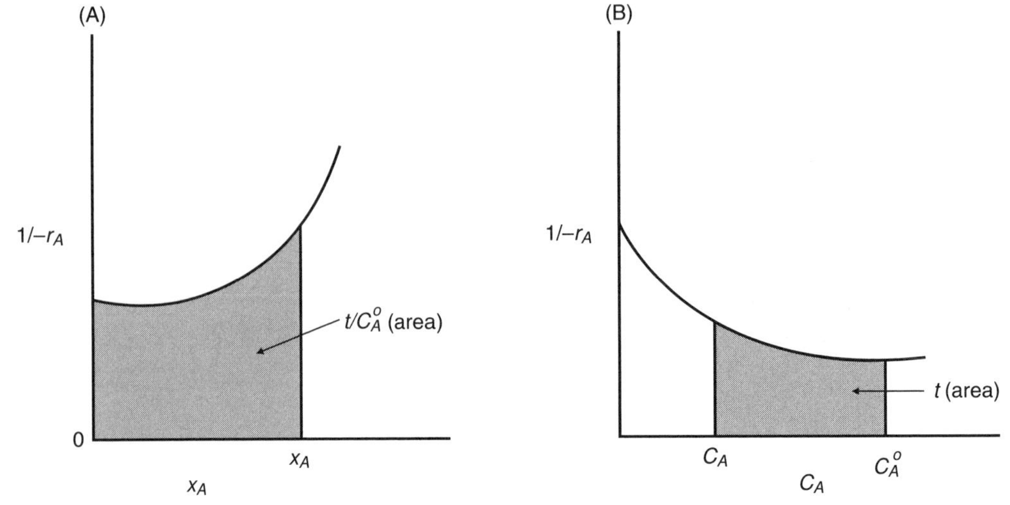

FIGURE 12.22 Graphical representation of the design of a batch reactor

It should be noted that the time required for a given conversion x_A is independent of C_A^o. Consequently, it is independent of the volume of the reactor if the reaction is first order. This is the unique property of a first-order reaction.

Example 2:

Let us examine the reactor performance for a second-order reaction:

$$-r_A = kC_A^2 = k(C_A^o)^2(1 - x_A)^2$$

From Eq. 12.192,

$$t = C_A^o \int_0^{x_A} \frac{dx_A}{k(C_A^o)^2(1 - x_A)^2} = \frac{1}{kC_A^o} \frac{x_A}{(1 - x_A)} \quad \textbf{(Eq. 12.194)}$$

Example 3:

In general, for nth order, the following analysis can be made:

$$t = C_A^o \int_0^{x_A} \frac{dx_A}{k(C_A^o)^n(1 - x_A)^n} = \frac{1}{k(C_A^o)^{n-1}} \frac{1 - (1 - x_A)^{n-1}}{(1 - x_A)^{n-1}} \quad \textbf{(Eq. 12.195)}$$

It can be shown from Eq. 12.192 that the area given by the plot of $-1/r_A$ versus x_A represents t/C_A^o, as shown in Figure 12.22A. Similarly, the area under the curve of $-1/r_A$ versus C_A between C_A^o is equivalent to t (see Figure 12.22B).

Example 4:

Suppose we wish to find the time required for a batch reaction of A → P to be processed to achieve x_A = 0.8. The reaction is first order; the initial concentration of A (i.e., C_A^o) is 1 mol/L, and the first-order rate constant, k, is 0.1 min^{-1}. The solution to this problem, from Eq. 12.193, is

$$t = -\frac{1}{0.1} = ln\ (1 - 0.8) = 16.09 \text{ min}$$

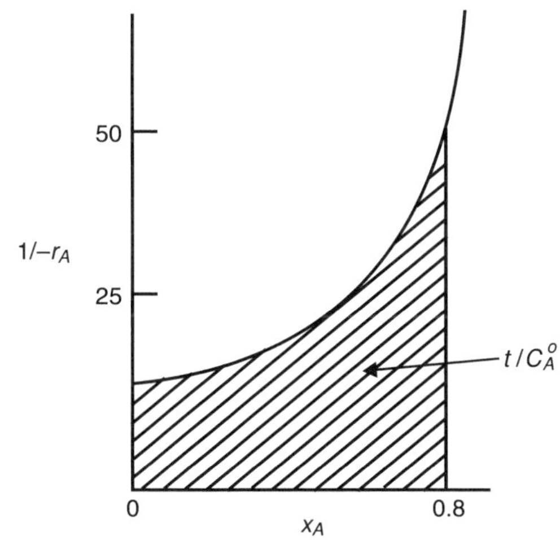

FIGURE 12.23 Graphical analysis of the size determination of a first-order reaction

This problem can also be solved graphically. First, a table of $-1/r_A$ versus x_A is established. The area under the curve bound by $x_A = 0$ and $x_A = 0.8$ is then calculated (see Figure 12.23).

x_A	$-r_A$	$-1/r_A$
0	0.1	10.0
0.2	0.08	12.5
0.4	0.06	16.67
0.6	0.04	25.0
0.8	0.02	50.0

From the area under the curve of this figure, t = 16.1 min. It should also be noted that the area under the curve could also be obtained by using a numerical technique such as Simpson's rule.

Plug Flow Reactor

In a plug flow reactor, the composition of the fluid varies along the flow path. Therefore, the material balance should be made for a differential element as given in Figure 12.24. In this figure, C_A, F_A, x_A, and Q refer, respectively, to concentration (in moles per liter), mass flow rate (in moles per second), conversion (in liters per second), and volume flow rate (in liters per second). The subscripts o and f refer to inlet and exit conditions, respectively.

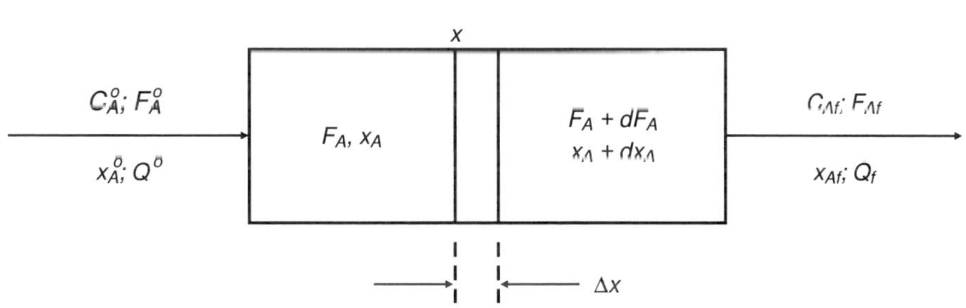

FIGURE 12.24 Plug flow reactor showing change in concentration along the axial direction

The material balance on A over the differential element bound by x and $x + \Delta x$ is as follows:

$$\text{input} - \text{output} - \begin{matrix}\text{rate of disappearance}\\ \text{because of reaction}\end{matrix} = \underbrace{\begin{matrix}\text{rate of}\\ \text{accumulation}\end{matrix}}_{\text{0 for steady state}}$$

Therefore,

$$\text{input} - \text{output} = \text{rate of disappearance}$$

$$F_A - (F_A + dF_A) = (-r_A)dv$$

Hence,

$$F_A^o dx_A = (-r_A)dv$$

Upon integration,

$$\int_o^V \frac{dv}{F_A^o} = \int_0^{x_A} \frac{dx_A}{-r_A}$$

therefore,

$$\frac{V}{F_A^o} = \int_0^{x_A} \frac{dx_A}{-r_A}$$

or

$$\frac{V}{Q_o} = C_A^o \int_0^{x_A} \frac{dx_A}{-r_A}$$

Note that $F_A{}^o = C_A{}^o Q_o$. Also, if we define τ as $\tau \equiv v/Q_o$, we have

$$\tau = C_A^o \int_0^{x_A} \frac{dx_A}{-r_A} \quad \textbf{(Eq. 12.196)}$$

It is interesting to note that Eq. 12.196 is very similar to Eq. 12.192 for batch reactors.

Continuously Stirred Flow Reactor

In the CSFR, materials inside the reactor are well mixed and, hence, uniform throughout. Furthermore, the conditions at the exit port can be assumed to be the same as those inside the reactor. Figure 12.25 shows the inlet and outlet conditions of a CSFR.

$$\text{input} - \text{output} - \text{rate of disappearance} = \underbrace{\begin{matrix}\text{rate of}\\ \text{accumulation}\end{matrix}}_{\text{0 at steady state}}$$

$$C_A^o Q_o - C_{Af} Q_f = (-r_A)V$$

Because we can assume that $Q_o = Q_f$, we have

$$\frac{V}{Q} = \tau = \frac{C_A^o x_A}{-r_A} \quad \textbf{(Eq. 12.197)}$$

or, when $x_A^o \neq 0$,

$$\tau = \frac{C_A^o (x_{Af} - x_A^o)}{-r_A} \quad \textbf{(Eq. 12.198)}$$

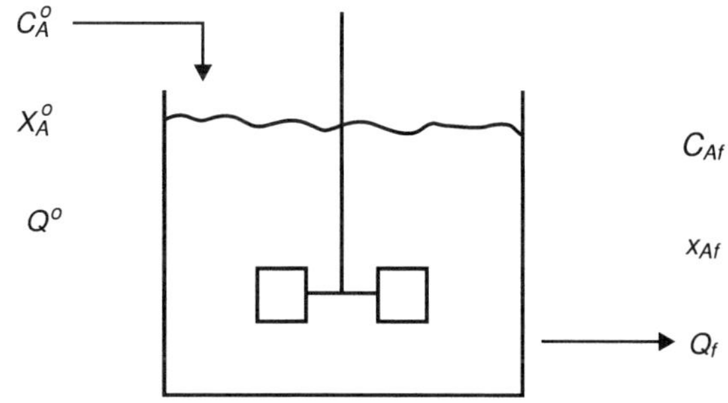

FIGURE 12.25 CSFR showing conditions of inlet and outlet

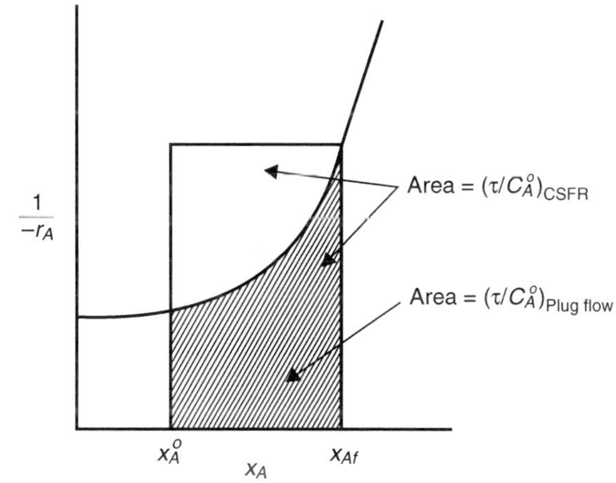

FIGURE 12.26 Graphical presentation of the difference in residence time between CSFR and plug flow reactor

Figure 12.26 shows a rate curve in which the rate of A decreases as the reactant A decreases. For such reactions, we can see that a CSFR always requires a larger volume than does a plug flow reactor.

Example 1:

Suppose we wish to find the residence time for $x_A = 0.8$. The reaction, A → P, is first order in a CSFR and plug flow reactor. We also have $C_A^o = 1$ mol/L and $k = 0.1$/min.

Solution:

From Eq. 12.196 for plug flow, we have

$$\tau = C_A^o \int_0^{x_A} \frac{dx_A}{kC_A^o(1 - x_A)}$$

$$= \frac{1}{0.1}\, ln\,(1 - 0.8)$$

$$= 16.09 \text{ min}$$

From Eq. 12.198 for a CSFR, we have

$$\tau = \frac{C_A^o(0.8-0)}{0.1\ C_A^o(1-x_A)} = 40 \text{ min}$$

Example 2:

A homogeneous reaction A → 3R has a reported rate of $-r_A = 0.01\ C_A^{1/2}$ mol/(L–s) at 215°C and 5 atm. Suppose we want to find the residence time needed for 80% conversion of a feed to a plug flow reactor. The initial concentration of A is 0.0625 mol/L. In this case, from Eq. 12.192, we have

$$\tau = C_A^o \int_0^{x_A} \frac{dx_A}{-r_A} = \frac{C_A^o}{k(C_A^o)^{1/2}} = \int_0^{x_A} \frac{dx_A}{(1-x_A)^{1/2}}$$

$$= \frac{0.0625^{1/2}}{10^{-2}} \int_0^{0.8} \frac{dx_A}{(1-x_A)^{1/2}} = 25 \times (-2) \times [(1-x_A)^{1/2}]^{0.8}$$

$$= 27.6 \text{ min}$$

Example 3:

Liquid flowing at 1 L/min contains many different sulfides and gangue minerals. Measurements were made to find that the feed consisting of 0.1 mol/L of A (copper sulfide powder) and 0.01 mol/L of B (sulfuric acid) flows into a 1-L CSFR. The materials react in a complex manner for which the stoichiometry is unknown. The outlet stream from the reactor contains 0.02 mol/L of A, 0.03 mol/L of B, and 0.04 mol/L of C (zinc ion), which was absent in the feed. Suppose we want to find the rate of reaction of A, B, and C for the conditions within the reactor.

Solution:

From Eq. 12.193, we have

$$\tau = \frac{C_A^o x_{Af}}{-r_A} = \frac{C_A^o - C_{Af}}{-r_A}$$

Therefore,

$$-r_A = \frac{C_A^o - C_{Af}}{\tau} = \frac{0.1-0.02}{1} = 0.08 \text{ mol/(2 min)}$$

$$-r_B = \frac{C_B^o - C_{Bf}}{\tau} = \frac{0.01-0.03}{1} = 0.02 \text{ mol/(2 min)}$$

$$-r_C = \frac{C_C^o - C_{Cf}}{\tau} = \frac{0-0.04}{1} = 0.04 \text{ mol/(2 min)}$$

Multiple-Reactor Systems

Suppose we have *n* plug flow reactors connected in series and the reaction A → P occurring in these reactors:

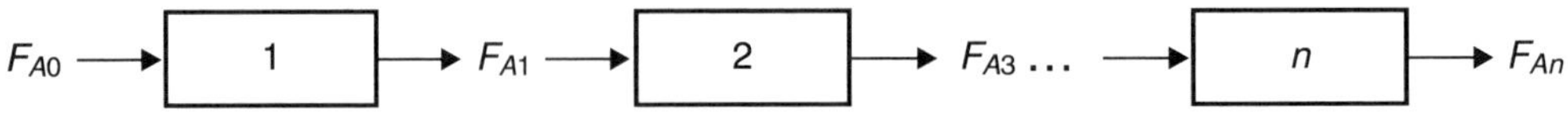

Therefore, the residence times can be calculated using Eq. 12.192:

$$\tau_1 = C_A^o \int_0^{x_{A1}} \frac{dx_A}{-r_A}; \quad \tau_2 = C_A^o \int_{x_{A1}}^{x_{A2}} \frac{dx_A}{-r_A} \ \ldots$$

The total residence time is

$$\tau = \tau_1 + \tau_2 + \tau_3 + \ldots + \tau_n$$

Therefore,

$$\tau = C_A^o \int_0^{x_{A1}} \frac{dx_A}{-r_A} + \int_{x_{A1}}^{x_{A2}} \frac{dx_A}{-r_A} + \ldots + \int_{x_{An-1}}^{x_{An}} \frac{dx_A}{-r_A}$$

$$= C_A^o \int_0^{x_{An}} \frac{dx_A}{-r_A} \qquad \textbf{(Eq. 12.199)}$$

The resulting equation, Eq. 12.199, shows that the overall conversion of n plug flow reactors is identical to one plug flow reactor having a volume of $v = V_1 + V_2 + \ldots + V_n$.

Let us examine n identical CSFRs in series so that the overall residence time is

$$\tau = \tau_1 + \tau_2 + \tau_3 + \ldots + \tau_n = \sum_1^n \tau_i = \sum \frac{C_A^o(x_i - x_{i-1})}{-r_A} = \sum \frac{C_A^o \Delta x_A}{-r_A}$$

By taking $\Delta x \rightarrow 0$, we get

$$\tau = \lim_{\Delta x} \rightarrow 0 \sum \frac{C_A^o \Delta x_A}{-r_A} = C_A^o \int \frac{dx_A}{-r_A} \qquad \textbf{(Eq. 12.200)}$$

Note that Eq. 12.200 is identical to Eq. 12.192. Graphically, this result can be demonstrated as shown in Figure 12.27.

Another useful relationship for n equal-size CSFRs in series can be shown. The residence time, τ_i, for the ith CSFR can be written as

$$\tau_i = \frac{C_{A_{i-1}} - C_{A_i}}{-r_A}$$

If first-order and irreversible reactions are assumed, we have $-r_A = kC_A$. Therefore,

$$\tau_i = \frac{C_{A_{i-1}} - C_{A_i}}{C_{A_i}}$$

and

$$\frac{C_{A_i}}{C_{A_{i-1}}} = \frac{1}{1 + k\tau} \qquad \textbf{(Eq. 12.201)}$$

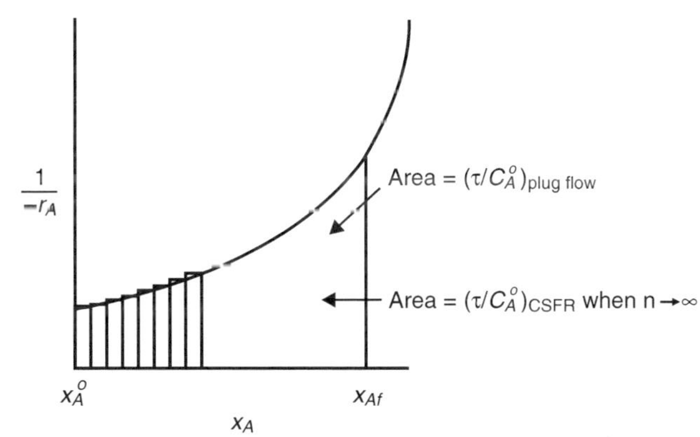

FIGURE 12.27 Graph showing that the size of a CSFR is the same as that of a plug flow reactor when an infinite number of CSFRs are connected in series

The overall fractional conversion, x_{AT}, is

$$x_{AT} = 1 - \frac{C_{A_n}}{C_A^o}$$

Therefore,

$$\frac{C_{A_n}}{C_A^o} = 1 - x_{AT} = \frac{C_{A_1}}{C_A^o} \frac{C_{A_2}}{C_{A_1}} \frac{C_{A_3}}{C_{A_2}} \cdots \frac{C_{A_n}}{C_{A_{n-1}}}$$

$$= \left(\frac{1}{1+k\tau}\right)\left(\frac{1}{1+k\tau}\right)\left(\frac{1}{1+k\tau}\right) \cdots \left(\frac{1}{1+k\tau}\right)$$

$$= \frac{1}{(1+k\tau)^n}$$

Finally,

$$x_{AT} = 1 - \frac{1}{(1+k\tau)^n} \quad \textbf{(Eq. 12.202)}$$

Example 1:

Consider a three-part example. First, calculate the fractional conversion of A (i.e., x_A) for five 1-L plug flow reactors in series given these parameters: $r_A = kC_A$; $k = 0.1/\text{min}$; $C_A^o = 1$ mol/L; and $\tau = 16.1$ min:

$$\tau = C_A^o \int_0^{x_A} \frac{dx_A}{-r_A}$$

$$16.1 = \frac{1}{0.1} \, ln\,(1 - x_A)$$

$$x_A = 0.8$$

Next, calculate x_A for one 5-L CSFR:

$$\tau = 16.1 = \frac{x_A}{0.1(1 - x_A)}$$

$$x_A = 0.617$$

Finally, calculate x_A for five 1-L CSFRs:

$$x_{A_n} = 1 - \frac{1}{(1+k\tau)^n}$$

$$x_A = 0.752$$

Example 2:

Consider a second-order reaction ($-r_A = kC_A{}^2$) that is occurring in a CSFR (CSFR 1) yielding 90% conversion. If this same reaction is to take place in two additional identical CSFRs (CSFR 2 and CSFR 3) having the same size as CSFR 1, what should the final fractional conversion be in the two-CSFR case? (See Figure 12.28.)

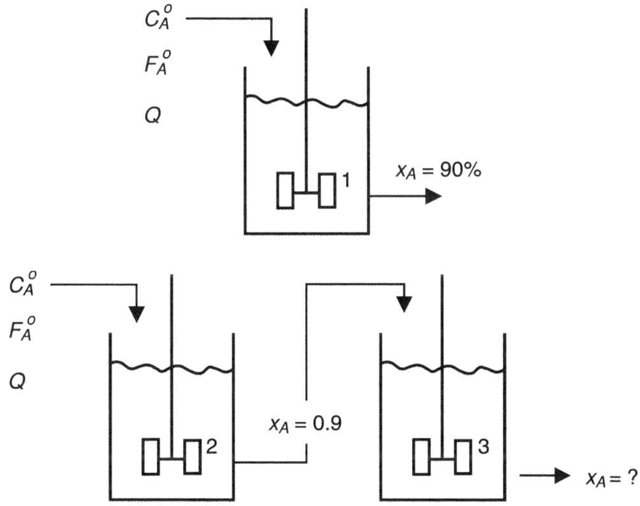

FIGURE 12.28 Diagram for the example involving CSFRs

First, note that $V_1 = V_2 = V_3$. The retention times for the CSFRs are then calculated as follows:

$$\tau_1 = \frac{C_A^o x_A}{k(C_A^o)^2(1-x_A)^2} = \frac{0.9}{k(C_A^o)^2(0.1)^2} = 90\frac{1}{kC_A^o}$$

$$\tau_2 = \tau_2 = \tau_3 = \frac{90}{kC_A^o}$$

$$\tau_3 = \frac{x_{Af} - 0.9}{kC_A^o(1-x_{Af})^2}$$

From the latter two of these equations, we get

$$\frac{90}{kC_A^o} = \frac{x_{Af} - 0.9}{kC_A^o(1-x_{Af})^2}$$

Therefore,

$$x_{Af} - 2.011x_{Af} + 1.01 = 0$$

$$x_{Af} = 0.974$$

Graphical Analysis

The efficiency of a reactor arrangement is easily represented by graphical illustration. Consider an example. Suppose that we would like to achieve a final conversion, x_{Af}, by arranging two CSFRs in series. An infinite number of two different-size reactor combinations could give the same final conversion (see Figure 12.29). It should be noted, however, that the arrangement that gives the maximum area of polygon ABCD is the most efficient one.

Similar analysis can be made for any number and type of reactors connected in series. For example, in Figure 12.30, two CSFRs and one plug flow reactor are in series, with the plug flow reactor between the two CSFRs. The final conversion for this arrangement can be estimated graphically as shown.

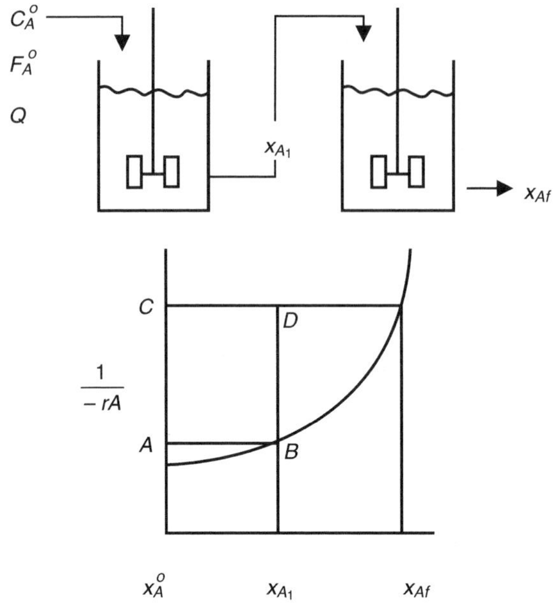

FIGURE 12.29 Two-reactor arrangement that gives the final conversion x_{Af}

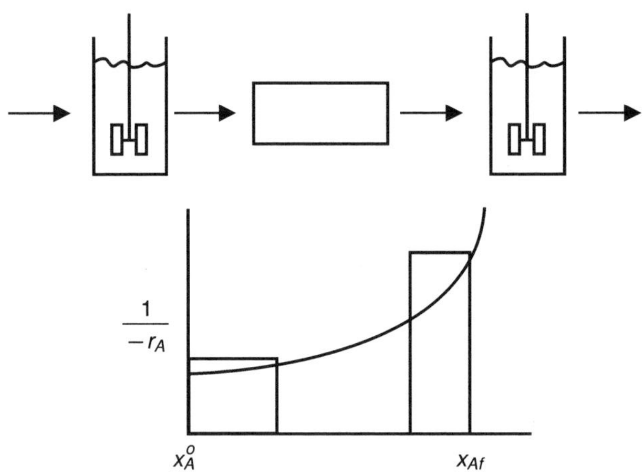

FIGURE 12.30 An arrangement of two CSFRs and a plug flow reactor

Nonideal Reactors

In real systems, reactors frequently behave nonideally. There may be dead space resulting from nonuniform mixing of fluids in the system. It is convenient to define the reduced time, θ:

$$\theta \equiv \frac{t}{\tau}$$

where

t = time

$\tau = V/Q$ = volume/flow rate

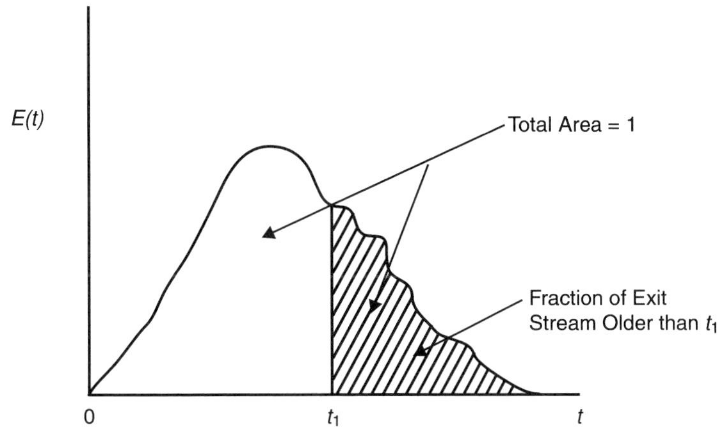

FIGURE 12.31 The exit age distribution

Exit Age Distribution Function: *E*-curve

The elements of fluid take different paths in a vessel from the inlet to exit port. Some elements take longer or shorter time than others. It is not practical to follow each element's exact route taken inside the reactor. It is more practical to tag the elements at the inlet port at any given time and inspect them at the exit port. Therefore, we define the exit age distribution function $E(t)$ such that $E(t)dt$ represents the fraction of material in the exit stream with age between t and $t + dt$.

Figure 12.31 shows the exit age distribution curve for a fluid flowing through a vessel. It should be noted that

$$\int_0^\infty E\,dt = 1 \tag{Eq. 12.203}$$

The fraction younger than t_1 is given by the following integration:

$$\int_0^{t_1} E\,dt = 1 \tag{Eq. 12.204}$$

whereas the fraction older than t_1, shown as the shaded area in Figure 12.31, is

$$\int_{t_1}^\infty E\,dt = 1 - \int_0^{t_1} E\,dt \tag{Eq. 12.205}$$

Experimentally, there are two tracer input methods to identify the exit age distribution: step input and pulse input. The response functions are, respectively, the *F*-curve and the *C*-curve.

F-Curve

To determine an *F*-curve via the step input tracer method, a tracer is introduced in the inlet port at an initial concentration of C_o, and this concentration is kept constant at this level during the tracer investigation.

As shown in Figure 12.32, the exit distribution of a tracer step input is known as the *F*-curve, which rises from 0 to 1. It should also be noted that the residence time, τ (which equals V/Q, where V is volume and Q is flow rate), occurs at a time less than when $F = C/C_o = 1$.

C-Curve

The tracer exit distribution curve for a pulse tracer input is referred to as the *C*-curve (see Figure 12.33). For practical purposes, the *C*-curve is the exit age distribution, *E*. To relate *E* with *F*, however, let us

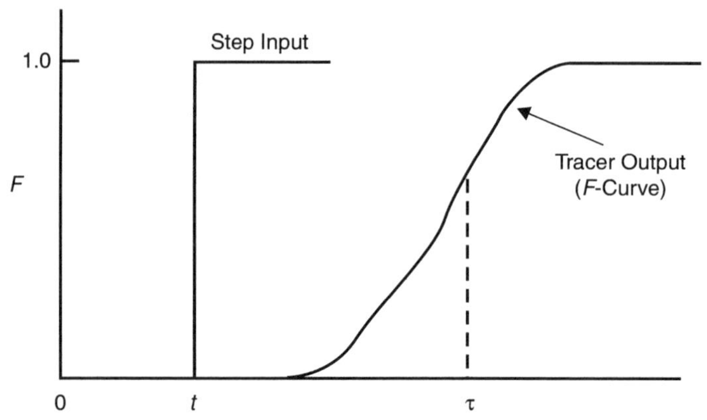

FIGURE 12.32 Step input and *F*-curve

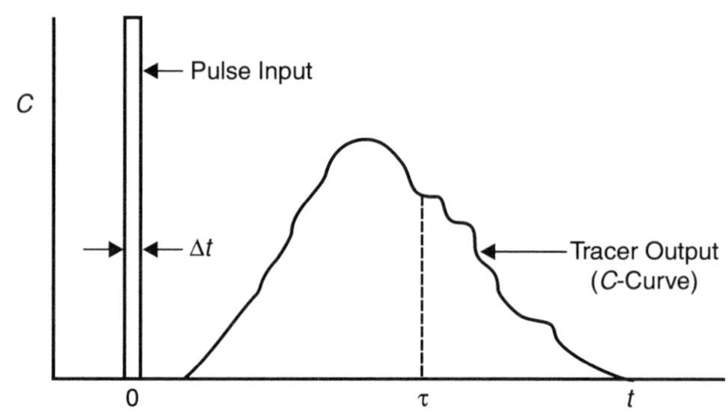

FIGURE 12.33 Pulse input and *C*-curve

examine the following situation. Suppose that we have a well-stirred vessel to which pure water is flowing in and out at a constant rate. Imagine that at $t = 0$, we have switched to a purple fluid and recorded the concentration buildup of the purple fluid corresponding to the *F*-curve. Therefore, the following mass balance holds.

At any time t,

fraction of purple fluid in the exit stream	=	fraction of exit stream younger than age t
F	=	$\int_0^t E\,dt$

Therefore,

$$\frac{dF}{dt} = E \quad \textbf{(Eq. 12.206)}$$

It should also be noted that the mean residence time, τ, can be expressed in terms of the E function as

$$\tau = \int_0^\infty t\,E\,dt \quad \textbf{(Eq. 12.207)}$$

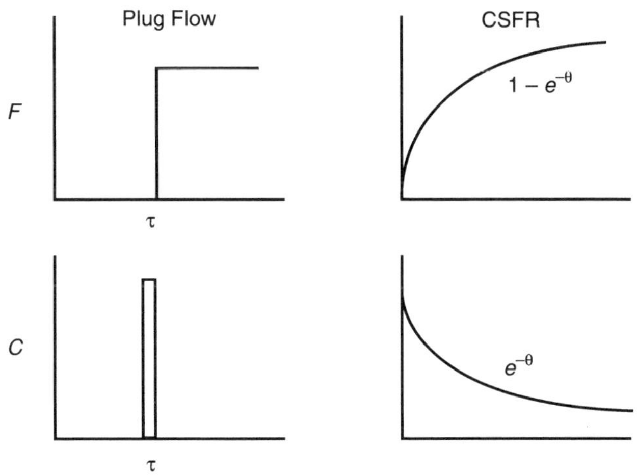

FIGURE 12.34 ***F*- and *C*-curves for plug flow reactor and CSFR**

Figure 12.34 shows the exit age distribution for two ideal reactors. To illustrate the tracer response for a CSFR, let us make a material balance for the step tracer input:

$$\text{input} \quad - \quad \text{output} \quad = \quad \text{rate of accumulation}$$

$$C_oQ \quad - \quad CQ \quad = \quad \frac{dc}{dt}V$$

$$\int_0^C \frac{dc}{C_o - C} = \frac{Q}{V}\int_0^t dt$$

The following substitutions may be made:

$$\frac{Q}{v} = \tau$$

$$\frac{t}{\tau} = \theta$$

Therefore,

$$1 - \frac{C}{C_o} = e^{-\theta} \text{ or } \frac{C}{C_o} = \text{F} = 1 - e^{-\theta} \qquad \textbf{(Eq. 12.208)}$$

$$C(\theta) = \frac{dF}{d\theta} = e^{-\theta} \qquad \textbf{(Eq. 12.209)}$$

$$C(t) = \frac{1}{\tau}e^{-t/\tau} \qquad \textbf{(Eq. 12.210)}$$

Example 1:

A real reactor is being operated to extract metal values by leaching a slurry containing valuable minerals. It is desired to find the mean residence time by introducing magnetite particles as a pulse input. The magnetic particles are nonreactive in the vessel and can be collected in the exit stream by applying a magnetic field. The magnetic particles are collected in the exit stream, and the results are given in Table 12.14. Estimate the mean residence time of the reactor, and establish the *E*-curve.

TABLE 12.14 Mass of magnetite as a function of time for example problem

Time, min	Amount of Magnetite, g/L of fluid
0	0
5	3
10	5
15	5
20	4
25	2
30	1
35	0

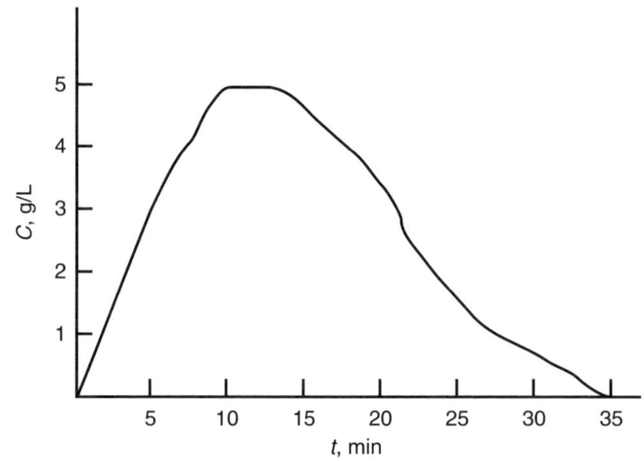

FIGURE 12.35 The *C*-curve from Table 12.14

The data from Table 12.14 are plotted in Figure 12.35. The area under this concentration-time curve is

$$M = \Sigma = (3 + 5 + 5 + 4 + 2 + 1) \times 5 = 100 \text{ g-min/L}$$

The mean residence is calculated as

$$\tau = \frac{\Sigma C \cdot t}{\Sigma C} = \frac{15 + 50 + 75 + 80 + 50 + 30}{20} = 15 \text{ min}$$

Note also that $E = C/M$, so

t	0	5	10	15	20	25	30
$E = \frac{C}{M}$	0	0.03	0.05	0.05	0.04	0.02	0.01

Therefore,

$$\int_0^\infty E \, dt = 5 \times (0.03 + 0.05 + 0.05 + 0.04 + 0.02 + 0.01) = 1.0$$

Estimating the conversion of a reaction in a nonideal reactor is possible if the tracer information is given and the reaction is linear. In this case, the average concentration of a reactant, A, may be expressed as

$$<C_A> = \int_0^\infty C_A \, E \, dt \qquad \textbf{(Eq. 12.211)}$$

The following substitutions may be made:

$$-r_A = kC_A$$
$$C_A = C_A^o \, e^{-kt}$$

Finally, we have

$$<C_A> = C_A^o \int_0^\infty e^{-kt} \, E \, dt \qquad \textbf{(Eq. 12.212)}$$

Example 2:

A first-order irreversible reaction has the form $-r_A = kC_A$, where $k = 0.307/\text{min}$. Find the fractional conversion for a plug flow and for a real reactor having the same tracer information as given in Table 12.14.

For the plug flow case, the expression to use is

$$\tau = C_A^o \int_0^{x_A} \frac{dx_A}{-r_A} = \frac{1}{k} \ln \frac{C_A^o}{C_A}$$

Therefore,

$$\frac{C_A}{C_A^o} = e^{-kt} = e^{-0.307 \times 15} = 0.01$$

$$x_A = 0.99$$

For a real reactor, we have

$$<C_A> = C_A^o \int_0^\infty e^{-kt} \, E \, dt$$

Therefore, for the real reactor, we have $x_A = 1 - 0.0469 = 0.953$. Table 12.15 shows a rate analysis for the real reactor.

Design of Reactors for Mixture of Particles

Plug Flow Reactor. In real systems where slurries are subjected to processing in a reactor just as in the case of hydrometallurgical operations, the reaction analysis becomes complicated particularly because of nonuniform particle size. There will be a size distribution over the wide range of sizes. In such cases, the information on the size distribution is of utmost importance. Suppose we have the size distribution so that the overall feed rate is equal to $M = M(R_i)$. The reaction given in Eq. 12.213 represents the fluid/solid reaction,

$$b <B> + \{A\} \rightarrow c <C> + d \{D\} \qquad \textbf{(Eq. 12.213)}$$

TABLE 12.15 Rate analysis for the real reactor

t, min	0	5	10	15	20	25	30
E	0	0.03	0.05	0.05	0.04	0.02	0.01
e^{-kt}	0	0.2154	0.0464	0.01	0.0021	0.0005	0.0001
$e^{-kt}E\delta\tau$	0	0.0323	0.0116	0.0025	0.0004	0.0001	0

The material balance of B can be formulated for a plug flow reactor as below:

$$\begin{pmatrix}\text{mean value for}\\ 1-x_B\end{pmatrix} = \sum_{\text{all sizes}} \begin{pmatrix}\text{fraction } B\\ \text{unconverted in size } R_i\end{pmatrix} \begin{pmatrix}\text{fraction of}\\ \text{feed size of } R_i\end{pmatrix}$$

$$1-x_B = \sum[1-x_B(R_i)]\frac{M(R_i)}{M} \quad \textbf{(Eq. 12.214)}$$

Example:

A metal sulfide is being leached in an acid solution. The leaching of the sulfide consisting of three size fractions is believed to be by chemical reaction. Calculate the overall conversion. Assume a plug flow reactor.

Size, µm	%wt	t_{comp}, min
50	30	5
100	40	10
200	30	20

The mean residence time is 8 min.

Solution:

It should be noted that in a plug flow reactor, the size fraction of 50 µm will react completely in 8 min.

$$1-\langle x_b\rangle = (1-x_B(50\ \mu\text{m}))\frac{M(50)}{M} + (1-x_b(100\ \mu\text{m}))\frac{M(100)}{M} + (1-x_B(200\ \mu\text{m}))\frac{M(200)}{M}$$

Note:

$$1-x_B(R_i) = \left(1-\frac{t}{t_{comp}}\right)^3$$

Therefore:

$$1-\langle x_B\rangle = \left(1-\frac{8}{10}\right)^3 0.4 + \left(1-\frac{8}{20}\right)^3 0.3 + 0 = 0.068$$

$$\langle x_B\rangle = 0.932$$

CSFR. When the reactor behaves as a CSFR, the material balance will yield:

$$1-\langle x_B\rangle = \int_0^\infty (1-x_B)\ E\,dt \quad \textbf{(Eq. 12.215)}$$

In Eq. 12.215, the upper limit of the time should be t_{comp} for all practical applications. To solve Eq. 12.215, the information on E should be available. If the reactor operates as an ideal CSFR, it was shown earlier that

$$E(t) = \frac{1}{\tau}e^{-t/\tau}$$

Therefore, Eq. 12.215 becomes

$$1-\langle x_B\rangle = \int_0^{t_{comp}} (1-x_B)\ \frac{e^{-t/\tau}}{\tau}dt$$

Equations 12.216, 12.217, and 12.218 are obtained, respectively, for film diffusion, chemical reaction, and product layer diffusion limiting cases.

Film diffusion.

$$1-\langle x_B\rangle = \int_0^{t_{comp}} \left(1-\frac{t}{t_{comp}}\right) \frac{e^{-t/\tau}}{\tau}dt$$

Therefore,

$$\langle x_B \rangle = \frac{\tau}{t_{comp}} (1 - e^{-t_{comp}/\tau})$$

or in equivalent expanded form for large τ/t_{comp}

$$1 - \langle x_B \rangle = \frac{1}{2}\frac{t_{comp}}{\tau} - \frac{1}{3!}\left(\frac{t_{comp}}{\tau}\right)^2 + \frac{1}{4!}\left(\frac{t_{comp}}{\tau}\right)^3 \ldots \qquad \textbf{(Eq. 12.216)}$$

Chemical reaction.

$$1 - \langle x_B \rangle = \int_0^{t_{comp}} \left(1 - \frac{t}{t_{comp}}\right)^3 \frac{e^{-t/\tau}}{\tau} dt$$

Therefore,

$$\langle x_B \rangle = \left[3\frac{\tau}{t_{comp}} - 6\left(\frac{\tau}{t_{comp}}\right)^2 + 6\left(\frac{\tau}{t_{comp}}\right)^3\right](1 - e^{-t_{comp}/\tau})$$

or in equivalent expanded form for large τ/t_{comp}

$$1 - \langle x_B \rangle = \frac{1}{2}\frac{t_{comp}}{\tau} - \frac{1}{20}\left(\frac{t_{comp}}{\tau}\right)^2 + \frac{1}{120}\left(\frac{t_{comp}}{\tau}\right)^3 \ldots \qquad \textbf{(Eq. 12.217)}$$

Product layer diffusion.

$$1 - \langle x_B \rangle = \frac{1}{5}\frac{t_{comp}}{\tau} - \frac{19}{420}\left(\frac{t_{comp}}{\tau}\right)^2 + \frac{41}{4{,}620}\left(\frac{t_{comp}}{\tau}\right)^3 \ldots \qquad \textbf{(Eq. 12.218)}$$

RECOVERY OF METAL IONS FROM LEACH LIQUOR

When metals are extracted from an ore by dissolution, usually more than one kind of metal ions is dissolved in the solution. Therefore, these metals are individually recovered from the solution. There are many ways of achieving this goal. Some of these processes are

1. Solvent extraction
2. Ion exchange
3. Electrowinning
4. Cementation
5. Chemical precipitation
6. Solvent extraction

Solvent extraction is one of the most common methods of recovering metal ions from leach liquor. In this process, organic chemicals are introduced in the solution. The organic chemicals used should have a chemical affinity for the metal ion to be separated. The way to take these organic chemicals out of the solution is to add water to an immiscible oil such as kerosene, which will then absorb this organic chemical. This is shown in Figure 12.36.

In Figure 12.36, the organic chemical (known as solvent) is dissolved in the organic phase, and metal ions are attracted to the solvent. When the organic phase is completely loaded with the metal ion, then the organic phase is separated from the water phase. The metals within the organic phase could be subjected to stripping, for example, by contact with high acid concentration. Hydrogen ion will then replace metal ion attached to the organic moiety. Eq. 12.219 shows an example of chelation of uranyl ion, UO_2^{2+} with monoalkyl phosphoric acid.

$$2[\mathrm{RO-\overset{\displaystyle O}{\overset{|}{\underset{\underset{\displaystyle OH}{|}}{P}}}-OH}] + \mathrm{UO_2^{2+}} \rightarrow \mathrm{RO-\overset{\displaystyle O}{\overset{|}{\underset{\underset{\displaystyle OH}{|}}{P}}}-O-UO_2-O-\overset{\displaystyle O}{\overset{|}{\underset{\underset{\displaystyle OH}{|}}{P}}}-OR} + 2\mathrm{H^+} \qquad \textbf{(Eq. 12.219)}$$

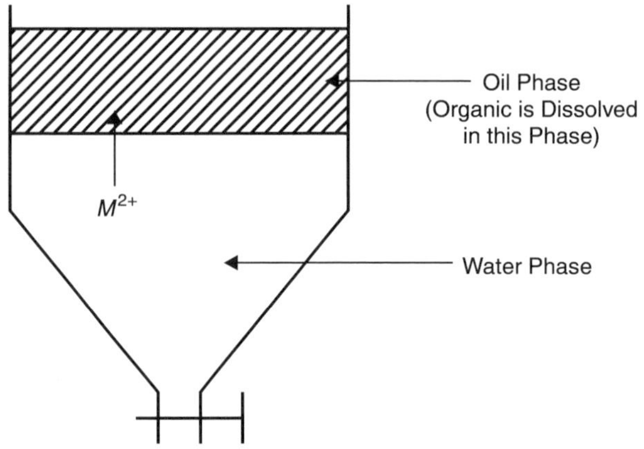

FIGURE 12.36 Schematic showing solvent extraction process

Here, R represents an alkyl chain, C_nH_{n+1}. Because this reaction is too long, the same equation is frequently shortened as

$$2RH + UO_2^{2+} \rightarrow R_2UO_2 + 2\,H^+ \qquad \textbf{(Eq. 12.220)}$$

It should be noted that the reverse of the reaction shown in Eq. 12.219 will represent the stripping step.

The selection of the right organic solvent is the key for the effective separation of the desired metal element from leach liquor. The criteria used to select the most desirable solvent are based on: (1) selectivity, (2) high extraction capacity, (3) ease of stripping, (4) ease of water separation, (5) safety in handling (nontoxic and nonflammable), and (6) cost.

Selectivity of a desired metal ion is frequently the key to the success of the solvent extraction process. The selectivity is often described using the term distribution ratio. The distribution ratio, D, is defined by Eq. 221.

$$D = \frac{(w - w_1)/V_o}{w_1/V_a} \qquad \textbf{(Eq. 12.221)}$$

where

w = original weight of the solute in the aqueous phase
w_1 = final weight of the solute in the aqueous phase

Rearranging Eq. 12.221 will yield:

$$\frac{w_1}{w} = \frac{V_a}{V_oD + V_a} = \frac{1}{1 + D(V_o/V_a)} \qquad \textbf{(Eq. 12.222)}$$

Because percent extraction can be described by

$$\%\ \text{extraction} = \frac{w - w_1}{w} \times 100 = \frac{D}{D + V_a/V_o} \times 100 \qquad \textbf{(Eq. 12.223)}$$

When solvent extraction is taking place in series, the following analysis is valid.

In the first tank, the final weight of the solute, w_1 would have the following expression according to Eq. 12.222:

$$w_1 = w\left(\frac{1}{1 + D(V_o/V_a)}\right) \qquad \textbf{(Eq. 12.224)}$$

The final weight in the second tank could be expressed similarly,

$$w_2 = w_1\left(\frac{1}{1 + D(V_o/V_a)}\right) = w\left(\frac{1}{1 + D(V_o/V_a)}\right)^2 \quad \text{(Eq. 12.225)}$$

Similarly, for the nth tank, the final weight of the solute, w_n could be evaluated as

$$w_n = w\left(\frac{1}{1 + D(V_o/V_a)}\right)^n \quad \text{(Eq. 12.226)}$$

It should be noted that the ultimate recovery of metal ions in the nth tank can be calculated by Eq. 12.227.

$$\%\ \text{extraction} = \left(1 - \frac{w_n}{w}\right) \times 100 \quad \text{(Eq. 12.227)}$$

Let us take an example where D for a metal ion is 10, V_o is 10 L, and V_a is also 10 L. How many tanks would be required to extract at least 99% of this metal ion? For 99% recovery, $w_n/w = 0.01 = (1/1 + 10)^n$. Therefore, n is calculated to be 1.92. However, n cannot be fractional. Consequently, we need two tanks, and the overall recovery could be recalculated using Eq. 12.227 to arrive at 99.2%.

McCabe–Thiele Diagram. Solvent extraction operations are often run continuously in a countercurrent mode as shown in Figure 12.37.

Equation 12.228 shows the mass balance for a metal ion being extracted from the aqueous phase to the organic phase through this process:

$$A x_o + O y_{n+1} = A x_n + O y_1 \quad \text{(Eq. 12.228)}$$

where A and O represent, respectively, flow rates of aqueous and organic phases; x_o and x_n represent the fractional composition of the ion in the aqueous phase entering and exiting the countercurrent extraction operation; and y_{n+1} and y_1 represent the fractional composition of the ion in the organic phase entering and exiting the countercurrent extraction process.

Equation 12.228 can be rearranged to yield Eq. 12.229:

$$y_1 = \frac{A}{O}(x_o - x_n) + y_{n \neq 1} \quad \text{(Eq. 12.229)}$$

Figure 12.38 shows the distribution isotherm line for an ion between the aqueous phase and organic phase together with the operating line given by Eq. 12.229.

The y-axis of Figure 12.38 represents the concentration of the ion in the organic phase and the x-axis represents the concentration of the same ion in the aqueous phase. The entering aqueous solution contains the ion to be extracted at x_o denoted by the port (a) in the graph. As this solution enters the first tank, tank 1, the ion will be subjected to transfer into the oil phase, and the final concentration of this ion in the tank will approach x_1 given a sufficient time, which is denoted by the port (b) in the diagram. This represents the concentration of the ion in the new feed solution entering the tank 2. This process will continue until a satisfactory composition of this metal has been reached in the nth tank. Such a plot is often called a McCabe–Thiele diagram.

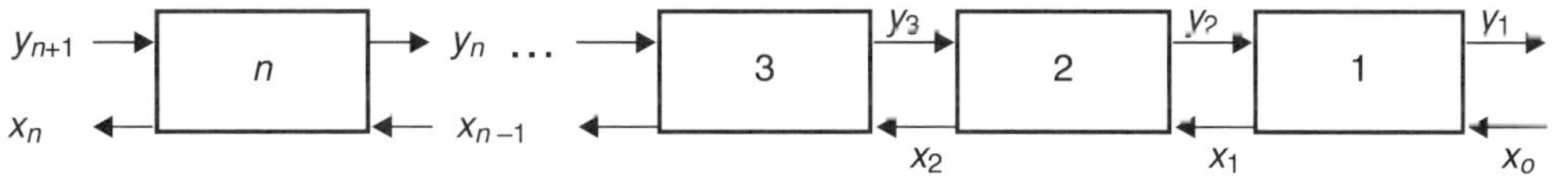

FIGURE 12.37 Schematic of countercurrent extraction process

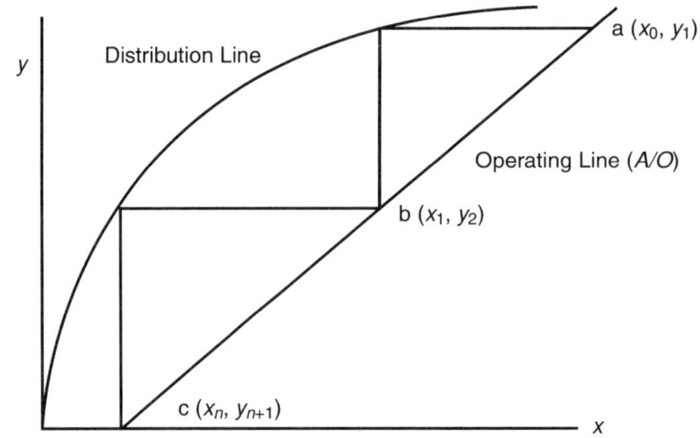

FIGURE 12.38 McCabe–Thiele diagram showing relationship between distribution isotherm and operating line

Ion Exchange

Removal and recovery of ions from solution can also be achieved using ion exchange technology. The recovery mechanism is very similar to that of solvent extraction; in fact, solvent extraction is frequently referred to as "liquid ion exchange." The major difference is that in ion exchange, solid substrates are involved. The substrates may be inorganic or organic in nature. Ion exchangers with inorganic substrate include zeolites, clays, and inorganic phosphates. Ion exchangers utilizing organic substances are activated carbon and ion exchange resins.

Ion exchange resins are elastic, three-dimensional hydrocarbon networks containing attached ionizable groups. The network is usually formed by co-polymerizing styrene and divinyl benzene. Frameworks with this composition have been shown to provide maximum resistance to oxidation, abrasion and breakage, and are insoluble in most common solvents (Anon. 1958).

The nature of the attached groups determines the exchange characteristics of the resin. Strong and weak acid resins (cation exchangers) and strong and weak base exchangers (anion exchangers) are available. The active group on strong acid resins is sulfonate; the active group on strong base resins is quaternary amine. Weak base resins have primary, secondary, and tertiary amine groups attached. The active group on weak acid resins can be carboxylate.

Ion exchange is a process in which ions contained in solution diffuse into a resin bead (0.5–1.0 mm in size) and exchange on a stoichiometric basis for counter ions within the resin bead. This phenomenon is illustrated in Figure 12.39.

Exchange Reactions

Typical ion exchange reactions are

$$2\,\underline{\mathrm{Na}X} + \mathrm{Ca}^{2+}{}_{(aq)} \leftrightarrow \underline{\mathrm{Ca}X_2} + 2\,\mathrm{Na}^{+}{}_{(aq)} \qquad \textbf{(Eq. 12.230)}$$

$$4\,\underline{X\mathrm{Cl}} + \mathrm{UO_2(SO_4)_3^{4-}}{}_{(aq)} \leftrightarrow \underline{\mathrm{UO_2(SO_4)_3}X_4} + 4\,\mathrm{Cl}^{-}_{(aq)} \qquad \textbf{(Eq. 12.231)}$$

where X represents the fixed site on the ion exchanger, solid phases are underlined, and (aq) indicates ions in aqueous solution.

The forward reaction describes adsorption in these systems. The back reaction occurs in desorption and elution.

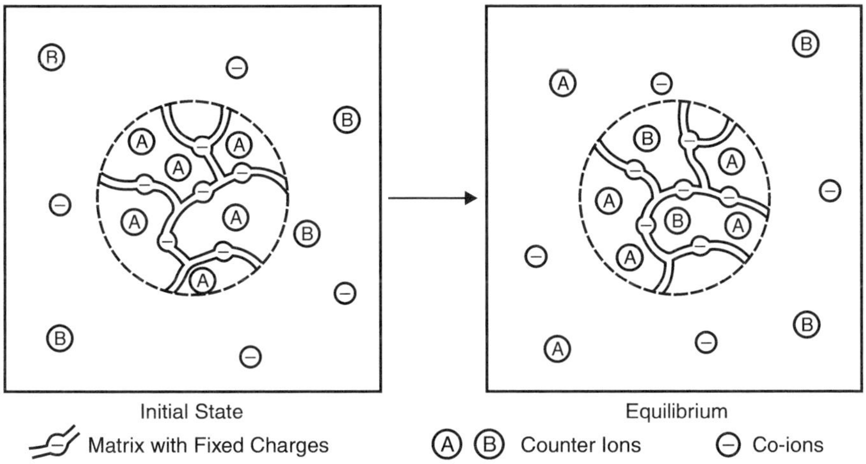

FIGURE 12.39 Schematic of ion exchange. Cation exchange resin containing counter ion A is placed in a solution containing counter ion B. Counter ions are redistributed by diffusion until equilibrium is attained.

Selectivity

In general, ion exchangers prefer (Helfferich 1962):

- The counter ion of higher valence
- The counter ion with the smaller solvated volume
- The counter ion with the greater polarizability
- The counter ion that interacts more strongly with the fixed ionic group

As shown in Figure 12.40, selectivity in ion exchange systems is increased with dilution of solution with ion exchangers of high internal molality. This has been explained on the basis of the Donnan potential that will be present between the interior solution of the bead and the external aqueous solution (Helfferich 1962). This phenomenon may also be explained on the basis of the differences in activity coefficient of species in the interior of the bead and the external solution.

The preference of an ion over another can be described in a number of ways; for example, by separation factor or selectivity coefficient. Separation factor, α_B^A, is defined as

$$\alpha_B^A = \frac{\bar{m}_A\, m_B}{\bar{m}_B\, m_A} = \frac{\bar{C}_A\, C_B}{\bar{C}_B\, C_A} = \frac{\bar{x}_A\, x_B}{\bar{x}_B\, x_A} \qquad \textbf{(Eq. 12.232)}$$

where

$\bar{m}_A$ = molality of A in ion exchanger

m_A = molality of A in aqueous solution

$\bar{C}_A$ = concentration of A in ion exchanger

C_A = concentration of A in aqueous solution

$\bar{x}_A$ = equivalent ionic fraction of A in ion exchanger

x_A = equivalent ionic fraction of A in aqueous solution

These phenomena are shown graphically in Figure 12.41. The isotherm is shown as a heavy line. For any ionic composition, the separation factor equals the ratio of the two rectangular areas I and II

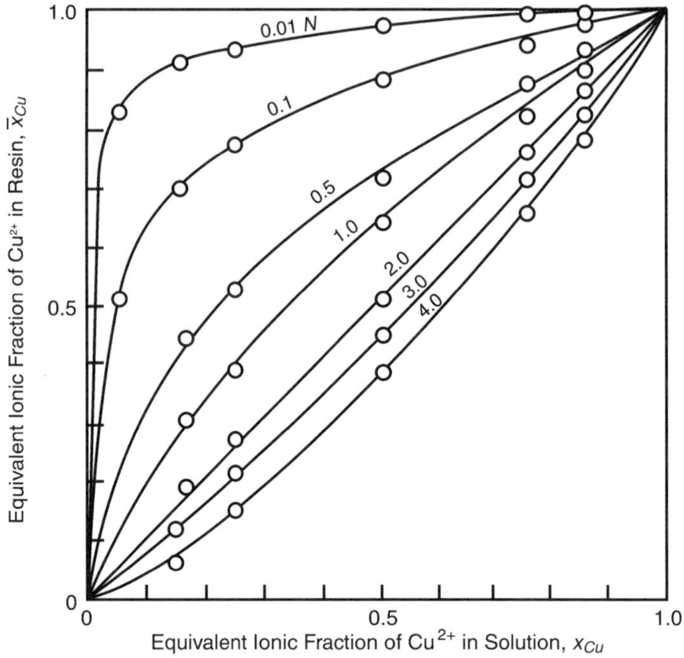

Source: Helfferich 1962.

FIGURE 12.40 Selectivity of a cation exchanger for Cu^{2+}/Na^{+}

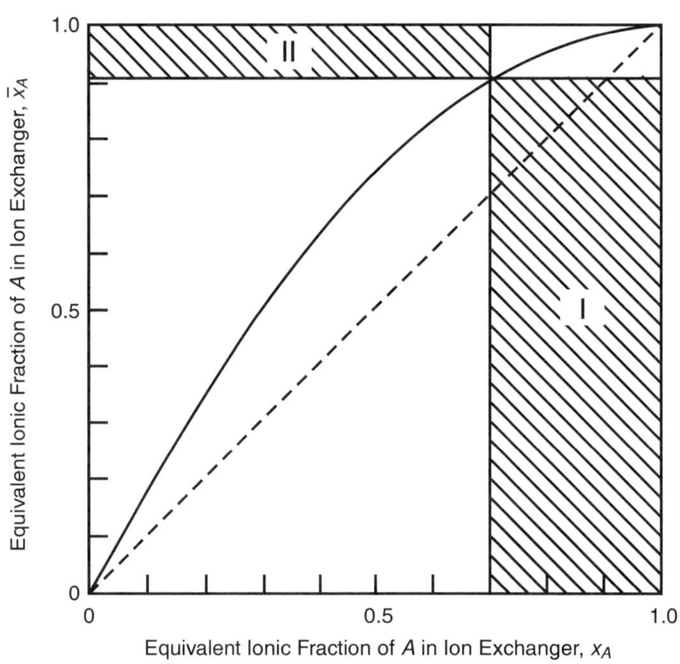

Source: Helfferich 1962.

FIGURE 12.41 Schematic of ion exchange isotherm and separation factor

touching one another in the corresponding point on the isotherm. The broken line is the isotherm of a fictitious ion exchanger that has no preference for either counter ion (Helfferich 1962).

Equivalent ionic fraction, x_A, of ion A in a solution containing species A and B is defined by (Helfferich 1962):

$$x_A = \frac{z_A m_A}{z_A m_A + z_B m_B} \quad \textbf{(Eq. 12.233)}$$

where

z = valence
m = molality

Capacity

The exchange capacity of ion exchange resins is in the range of several milliequivalents per gram (e.g., about 5 meq/g in the case of sulfonated polystyrenes).

In practical operations, theoretical capacity is rarely achieved. Depending on operating conditions with column operations, breakthrough of valuable species may occur before a maximum is reached in ion exchange. This "breakthrough" capacity is the capacity of importance in industrial practice.

Electrowinning

Metal ions can be recovered from leach liquors by applying an electromotive force to the system. Positively charged metal ions will migrate toward a negatively charged pole. It should be noted that by adjusting the potential, selective deposition of metal ions is possible. For example, in a solution containing zinc ion and copper ion, if the electrical potential is gradually increased, the copper ion will be deposited first at a lower electrical potential because copper is a more noble metal than zinc.

The efficiency of the deposition of any metal can be evaluated if we know the amount of electricity drawn for the deposition of the metal and the total current consumed. The current efficiency, η, is defined by the following equation.

$$\eta = \frac{I_m}{I_T} = \frac{\text{current used for metal deposition}}{\text{total current}} \quad \textbf{(Eq. 12.234)}$$

In electrowinning of zinc, the cathodic and anodic reactions are
Cathodic reaction:

$$Zn^{2+} + 2e \rightarrow Zn$$

Therefore,

$$\begin{aligned} E &= E^o + (0.059/2) \log \{Zn^{2+}\} \\ &= -0.763 + (0.059/2) \log \{Zn^{2+}\} \end{aligned}$$

If the activity of the zinc ion is known, the required potential, E, can be calculated.
Anodic reaction:

$$H_2O \rightarrow 2H^+ + \tfrac{1}{2} O_2 + 2e$$

Therefore,

$$E = E^o + (0.059/2) \log \{O_2\}^{1/2} - 0.059 \text{ pH}$$

If the activity of hydrogen ion and the partial pressure of oxygen are known, the required potential, E, can be calculated. Therefore, the total voltage required to deposit zinc would be $E_{anodic} - E_{cathodic}$. However, other factors come into play, such as IR drop in the solution and other losses of potential (including anodic overpotential). Let us examine a real situation for recovering copper by electrowinning. Leach solutions of copper usually contain 30 to 60 kg/m^3 of copper dissolved in the leach liquor. Further, the solution may also contain other metal ions such as Mn^{2+}, Ni^{2+}, Zn^{2+}, Co^{2+}, and Fe^{2+}.

TABLE 12.16 Total voltage required for the deposition of various metals

Reaction couple	E^o, v
Cu^{2+}/Cu	0.89
Mn^{2+}/Mn	2.35
Ni^{2+}/Ni	1.45
$Zn^{2+}/Zn1$	1.99
Fe^{2+}/Fe	1.67

Desired cathodic reaction	E^o, v
$Cu^{2+} + 2e \rightarrow Cu$	0.337
Anodic reaction	
$H_2O \rightarrow 1/2\, O_2 + 2H^+ + 2e$	1.23
Other cathodic reactions	
$Mn^{2+} + 2e \rightarrow Mn$	−1.18
$Ni^{2+} + 2e \rightarrow Ni$	−0.25
$Zn^{2+} + 2e \rightarrow Zn$	−0.763
$Fe^{2+} + 2e \rightarrow Fe$	−0.44

Therefore, the total voltage for the reactions of five metals can be calculated by:

$$\text{Total voltage} = \text{anodic} - \text{cathodic}$$

The results for the above five metals are given in Table 12.16.

As can be seen in Table 12.16, as long as the electrical potential is maintained around 0.89 v, the amounts of deposition of all other metals will be minimal. It should be noted, however, that the electrical potentials listed here are the minimum potentials required based on the thermodynamic calculations. There are other potentials required for the overall deposition process to occur. These additional potentials will include overpotential, and IR drop mainly through the solution, among others. Therefore, the overall potential required would be typically greater than 2 v:

	v
Decomposition potential	0.89
Anodic overpotential	0.6
IR drop	0.5
Other loss	0.1
	2.1

The IR drop can be easily calculated if the conductivity of the solution is known. This was discussed earlier.

The amount of metal deposition and the current used for the deposition are related through Faraday's law:

$$\frac{It}{z_i F} \times \text{atomic weight} = \text{grams of metal deposition} \quad \textbf{(Eq. 12.235)}$$

where

I = current in amperes used for the deposition

t = time duration in seconds

z_i = valance of the metal ion

F = Faraday constant (96,487 coulomb/eq)

For example, let us say that 16 g of zinc have been deposited during an hour of deposition. The current density was observed to be 15 amp/ft^2, and the cathodic area was 1.2 ft^2. We would like to calculate the current efficiency of the deposition process. The total current used can be calculated to be $15 \times 1.2 = 18$ amps. From Eq. 12.231, the theoretical amount of current used can be calculated:

$$I = \frac{16 \times 2 \times 96{,}500}{3{,}600 \times 65.4} = 13.12 \text{ amps}$$

Therefore, the current efficiency,

$$\%\, \eta = \frac{13.12}{18} \times 100 = 73\%$$

Cementation

Cementation is one of the old technologies used in the recovery of metal ions from solution. When relatively noble metals such as copper ion or gold cyanide are present in solution, the elemental state of a less noble metal or active metal, such as zinc or iron, is added into this system to remove the noble metal. For example, copper ion in heap leaching solution is subjected to cementation by iron scrap or zinc powder. The chemistry of this process can be examined as follows:

$$\{Cu^{2+}\} + \langle Fe\rangle = \langle Cu\rangle + \{Fe^{2+}\} \qquad \textbf{(Eq. 12.236)}$$

for copper deposition on an iron substrate, or

$$\{Cu^{2+}\} + \langle Zn\rangle = \langle Cu\rangle + \{Zn^{2+}\} \qquad \textbf{(Eq. 12.237)}$$

for copper depostion on a zinc substrate.

The equilbrium constants for these reactions at 25°C are 1.90×10^{26} and 1.57×10^{37}, respectively, for Eqs. 12.232 and 12.233. These reactions have been found to be mass transfer controlled and both reactions are equally effective. As can be noted here, more noble metal ions in solution are easily deposited on less noble, or more reactive metal substrate in the cementation reaction. Therefore, theoretically all the metals below Cu/Cu^{2+} in Table 12.11 should be potential host metals to deposit Cu^{2+} from the solution, but Zn/Zn^{2+} and Fe/Fe^{2+} are the ones frequently used (not Pb/Pb^{2+} nor Ni/Ni^{2+}). This is because when the former two metals are used, mass transfer is the limiting step, which means that fast deposition is possible. When the latter metals are used, chemical reaction is the limiting step. Therefore, the overall reaction is slower than the case of mass transfer limiting. This phenomenon can be explained using Figure 12.13.

In Figure 12.13, a cathodic curve for copper is plotted for potential versus log current density. In addition, three anodic curves for nickel, iron, and zinc are plotted for potential versus log current density. As expected, copper is the most noble metal of the four metals listed here. Therefore, copper metal ions in solution can be deposited on any of the other three metals. However, the anodic curve of nickel crosses at the Tafel region of the copper cathodic curve indicating that the overall cementation reaction is chemically controlled. On the other hand, the cathodic curves for iron and zinc both cross at the limiting current of the copper cathodic curve. As a result, both metals will give nearly the same rate of cementation dictated by the copper limiting current.

Industrially dissolved copper is usually recovered from solution by adding scrap iron from a relatively dilute solution. The Merill-Crowe process is a well-known technology that is widely used to recover gold and silver from cyanide leach liquor by adding zinc powder as seen in the following equations:

$$\{Au(CN)_2^-\} \rightarrow \{Au^+\} + 2\{CN^-\}$$

$$2\,\{Au^+\} + \langle Zn\rangle \rightarrow 2\,\langle Au\rangle + \{Zn^{2+}\}$$

$$\{Zn^{2+}\} + 4\,\{CN^-\} \rightarrow \{Zn(CN)_4^{2-}\}$$

The overall reaction leads to

$$2\ \{Au(CN)_2^-\} + \langle Zn\rangle \rightarrow 2\ \langle Au\rangle + \{Zn(CN)_4^{2-}\} \quad \textbf{(Eq. 12.238)}$$

Although cementation is an old technology, it is still widely used in the mineral industry and it is an effective and relatively inexpensive way of recoverying valuable metals from leach liquor.

BIBLIOGRAPHY

Anonymous. 1958. *Dowex: Ion Exchange*. Chicago: The Lakeside Press, R.R. Donnelley & Sons Company.

Bhuntumkomol, K., K.N. Han, and F. Lawson. 1980. Sec. C. *Trans. IMM*, 89:C7.

Bird, R.B., W.E. Stewart, and E.N. Lightfoot. 1962. *Transport Phenomena*. New York: John Wiley & Sons.

Bockris, J. O'M., and A.K.N. Reddy. 1970. *Modern Electrochemistry*. Vol. 1 & 2. New York: Plenum Press.

Butler, J.N. 1964. *Ionic Equilibrium*. Reading, Mass.: Addison-Wesley.

Cathro, K.J., and D.F.A. Koch. 1964. The Dissolution of Gold in Cyanide Solution. *Australasian Inst. Min. Metall. Proc.*, Vol. 210, 111–126.

Criss, C.M., and J.W. Cobble. 1964a. The Thermodynamic Properties of High Temperature Aqueous Solutions IV. *J. Am. Chem. Soc.*, 86:5385–5390.

——. 1964b. The Thermodynamic Properties of High Temperature Aqueous Solutions VI. *J. Am. Chem. Soc.*, 86:6394–6401.

Curthoys, G., and J.G. Mathieson. 1970. Partial Molal Volume of Ions. *Trans. Far. Soc.*, 66:43–50.

Derry, R. 1972. Pressure Hydrometallurgy—A Review. *Miner. Sci. Eng.*, 4:3–24.

FACT Web Programs. 1997. <www.crct.polymtl.ca/FACT/web/factweb.htm>.

Garrels, R.M., and C.L. Christ. 1965. *Solutions, Minerals, and Equiligria*. New York: Harper and Row.

Geiger, G.H., and D.R. Poirier. 1973. *Transport Phenomena in Metallurgy*. Boston, Mass.: Addison-Wesley.

Gokcen, N.A. 1979. Determination and Estimation of Ionic Activities of Metal Salts in Water. In *Bureau of Mines*, Report of Investigation (RI 8372).

——. 1982. *Hydrometallurgy–Research, Development and Plant Practice*. Edited by K. Osseo-Asare and J.D. Miller. New York: AIME.

Guan, Y., and K.N. Han. 1994. An Electrochemical Study on the Dissolution of Gold and Copper from Gold-Copper Alloys. *Met. Trans. B.*, 25B:817–827.

Han, K.N. 1984. A Simple and Accurate Method of Determining the Free Settling Velocity. *J. Korean Inst. Mineral and Mining Eng.*, 21:237–240.

Han, K.N., and C. Vu. 1981. A Kinetic Model for Leaching of Cobalt Metal Powder in NH_3-H_2O System. *Hydromet.*, 6:227–238.

Helfferich, F. 1962. *Ion Exchange*. New York: McGraw-Hill.

Dow Chemical Company. 1985. *JANAF Thermochemical Tables*. Midland, Mich.: Thermal Research Laboratory, Dow Chemical.

Kelley, K.K. 1960. Contributions to the Data on Theoretical Metallurgy. In *XIII, High-temperature Heat-content, Heat-capacity, and Entropy Data for the Elements and Inorganic Compounds*. Bulletin 584. U.S. Bureau of Mines, 232.

Kestu, D.R., and R.M. Pytkowicz. 1970. Effect of Temperature and Pressure on Sulfate in Association in Sea Water. *Geochim Et Cosmo. Acta.* 34:1039–1051.

Kielland, J. 1937. Individual Activity Coefficients of Ions in Aqueous Solutions. *J. Am. Chem. Soc.*, 59:1675–1678.

Kubaschewski, O., and E.L. Evans. 1979. *Metallurgical Thermochemistry*. 5th ed. London: Pergamon Press.

Kudryk, V., and H.H. Kellogg. 1954. Mechanism and Rate Controlling Factors in the Dissolution of Gold in Cyanide Solutions. *J. Metals*, 6:541–548.

Kusik, C.L., and N.P. Meissner. 1975. Calculating Activity Coefficients in Hydrometallurgy. *Int. J. of Miner. Process.*, 2:105–115.

Kwok, O.J., and R.G. Robins. 1972. In *International Symposium on Hydrometallurgy*. Edited by D.J.I. Evans and R.S. Shoemaker. New York: AIME.

Latimer, W.M. 1952. *Oxidation Potentials*. New York: Prentice-Hall.

Levenspiel, O. 1972. *Chemical Reaction Engineering*. New York: John Wiley & Sons.

Levich, V.G. 1962. *Physicochemical Hydrodynamics*. New York: Prentice-Hall.

Lowson, R.T. 1971. *Potential-pH diagrams above 298.16 K. Part I, Theoretical background.* Atomic Energy of Canada. AAEC/E 219.

Macdonald, D.D. 1972. *The Thermodynamics of Metal-water Systems at Elevated Temperatures, Parts 1, 2, 3, and 4*. AECL-report series (Atomic Energy of Canada Limited).

Martell, E.A., and R.M. Smith. 1974. *Critical Stability Constants*. Vols. 1–5. New York: Plenum Press.

Meissner, N.P., C.L. Kusik, and J.W. Tester. 1972. Activity Coefficients of Strong Electrolytes in Aqueous Solutions. *AIChE J.*, 18 (3):661–662.

Meng, X., and K.N. Han. 1993. A Kinetic Model for Dissolution of Metals—The Role of Heterogeneous Reaction and Mass Transfer. *Miner. Metall. Proc.* 10:128–134.

——. 1996. The Principles and Applications of Ammonia Leaching of Metals—A Review. *Min. Proc. & Ext. Met. Review*, 16:23–61.

Meng, X., X. Sun, and K.N. Han. 1995. A Dissolution Kinetic Model for Metals in Solutions. *Miner. Metall. Proc.* 12:97–102.

Narita, E., K.N. Han, and F. Lawson. 1982. A Colorimetric Determination of Dissolved Oxygen by an Improved Indigo Carmine Technique. *J. Chem. Soc. Japan*, 3:530–533.

——. 1983. Solubility of Oxygen in Aqueous Electrolyte Solutions. *Hydrometallurgy*, 10:21–37.

Newman, J.S. 1973. *Electrochemical Systems*. New York: Prentice-Hall.

Perry, J.H., ed. 1969. *Chemical Engineers' Handbook*. New York: McGraw-Hill.

Pourbaix, M. 1966. *Atlas of Electrochemical Equilbria in Aqueous Solutions*. New York: Pergamon Press.

Rubicumintara, T., and K.N. Han. 1990a. Metal Ionic Diffusivity: Measurement and Application. *Min. Proc. & Ext. Met. Rev.*, 7:23–47.

——. 1990b. The Effect of Concentration and Temperature on Diffusivity of Metal Compounds. *Met. Trans. B*, 21B:429–438.

Sherwood, T.K., R.L. Pigford, and C.R. Wilke. 1975. *Mass Transfer*. New York: McGraw-Hall.

Smith, J.M. 1970. *Chemical Engineering Kinetcis*, New York: McGraw-Hill.

Sohn, H.Y., and M.E. Wadsworth. 1979. *Rate Processes of Extractive Metallurgy*. New York: Plenum Press.

Sun, X., Y. Guan, and K.N. Han. 1996. Electrochemical Behavior of the Dissolution of Gold-Silver Alloys in Cyanide Solutions. *Met. Trans. B*, 27B:355–361.

Vu, C., and K.N. Han. 1977. Leaching Behavior of Cobalt in Ammonia Solutions. *Trans. IMM. Sec. C.* 86:C119–125.

——. 1979. Effect of System Geometry on the Leaching Behavior of Cobalt Metal. Mass Transfer Controlling Case. *Met. Trans. B.*, 10B, 57.

Weast, R.C., M.J. Astle, and W.H. Beyer. 1985. *Handbook of Chemistry and Physics*. 66th ed. Boca Raton, Fla.: CRC Press.

Xu, X., X. Meng, and K.N. Han. 1996. The Adsorption Behavior of Gold from Ammoniacal Solutions on Activated Carbon. *Min. Metall. Proc.*, 13:141–146.

Zena, R., and E. Yeager. 1967. Ultrasonic Vibration Potentials and Their Use in the Determination of Ionic Partial Molal Volumes. *J. Phys. Chem.*, 71:521–536.

CHAPTER 13

Mineral Processing Wastes and Their Remediation

Ross W. Smith and Stoyan N. Groudev

Various solid and liquid wastes can be discharged from mineral processing concentrators. The handling, treatment, remediation, and disposal of such wastes are important and sometimes monumental tasks for today's mineral industries. In many cases, current technology does not allow for an economical recovery of value from these materials. However, if such substances are toxic, they may nonetheless need to be removed from aqueous or solid wastes that are to be discharged from the plant. Given the current likelihood that extensive soil/tailings remediation will be needed when a minerals operation is decommissioned, designing the plant so that costs are minimized when the operation is terminated is an important concern.

LIQUID WASTES

Liquid effluents are primarily aqueous solutions, although in some cases organic fluids can be discharged. In addition, the use of heavy mechanical equipment in and around a plant can lead to spillage of petroleum products, which can contaminate soils.

In flotation plants, a number of reagents are added at various stages of the processes to carry out the flotation separation. Among these reagents are collectors, nonionic extenders, frothers, organic and inorganic activators and depressants, dispersing agents, and flocculating agents. The reagents can potentially remain in waste solutions and ultimately be discharged from the plants; they can also remain associated with solid wastes. In either case, the waste solutions can enter ground waters and react with various substances in soils, contaminating the soils with the waste metals, oils, or other substances. In addition, some of these substances can dissolve from the minerals being processed and can exit the mill in aqueous effluent streams. Examples of reagents and other chemicals that are potentially discharged from mineral processing plants are noted in Table 13.1.

Typical quantities of reagents used in flotation plants are summarized in Table 13.2 for collectors, Table 13.3 for modifiers, and Table 13.4 for frothers and hydrocarbon oils. The actual fate of reagents and contaminants both within and from flotation circuits is imperfectly known. There have, however, been a few measurements of concentration of some of the substances at various points in flotation circuits. Davis, Hyatt, and Cox (1976), citing data from a number of sources, reported that for sulfide flotation plants, tailings concentrations of sodium ethyl xanthate (a typical thiol collector) can range from a trace (<0.1 mg/dm^3) to 1.7 mg/dm^3. They also cited an occurrence of about 0.1 mg/dm^3 of dithiophosphates in tailings water from dithiophosphate flotation of sphalerite and <0.1 mg/dm^3 fatty acid in tailings water from scheelite flotation. Woodcock and Jones (1970) and Jones and Woodcock (1984) studied the concentrations of xanthates, thiophosphates, and thiocarbamates in operating base metal concentrators. Xanthate and thiophosphates tend to adsorb almost completely onto minerals, whereas Z-200 often remains in relatively high concentrations in water streams. Thus, there is greater

TABLE 13.1 Chemicals potentially discharged from flotation plants

Collectors: Sulfide Mineral Flotation	Collectors: Oxide, Silicate, Salt-type Mineral Flotation	Frothers	Modifiers (Activators and Depressants)	Flocculants and Dispersants	Chemicals Derived from the Ore Itself
Mercaptans	Monoamines (primary, secondary, tertiary)	Pine oil	Copper sulfate	Various starches	Copper ions
Thiourea	Quaternary ammonium compounds	Methylisobutylcarbinol (MIBC)	Chromates	Dextrin	Lead ions
Thiocarboxylates	Diamines	Various alkyl alcohols	Permanganates	Polyacrylamide flocculants	Chromates
Thiocarbamates	Amphoteric collectors (such as alkyl amino propionic acids)	Cresylic acid	Sodium sulfide	Polyethylene oxides	Arsenic compounds
Thiocarbonates	Alkyl phosphoric acids	Polypropylene glycols	Ferrocyanides	Sodium aluminate	Antimony compounds
Nonionic oils	Alkyl sulfates	Alkoxy-substituted paraffins	Sodium silicates	Aluminum sulfate	Nickel ions
Thiophosphate	Alkyl sulfonates (petroleum sulfonates)	Other frothers	Zinc sulfate	Ferric chloride	Selenium compounds
Mercaptobenzothiazole	Fatty acids and related compounds		Lime	Clays	Fluorides
Thiocarbanilide	Stearic acid		Aluminum sulfate	Condensed phosphates	Ferric ions
Dixanthogens	Oleic acid		Aluminum chloride	Soluble silicates	Ferrous ions
Other thio collectors	Linoleic acid		Soda ash	Polyimines	Phosphates
	Linolenic acid		Sodium sulfite	Polysaccharides	Cobalt ions
	Palmitic acid		Sodium carbonate	Aluminum polymers	Zinc ions
	Rosin acids		Lead acetate	Ferric sulfate	Cadmium ions
	Arsonic acids		Lead nitrate	Guar gum	Other ions
	Nonionic oils		Citric acid	Others, organic and inorganic	
	Lauric acid and myristic acid		Tannic acid		
	Other thio collectors		Ferricyanides		
			Quebracho		
			Hydrosulfites		
			Sodium cyanide		
			Fluorides		
			Lignin sulfonates		
			Calcium sulfite		
			Ammonium hydroxide		
			Various acids		
			Various multivalent metal ions		
			Others		

TABLE 13.2 Addition ranges for collectors

Collector	Flotation Type	Addition Range, kg/t*
Xanthates	Sulfide flotation	0.006–0.71
Xanthates	Nonsulfide flotation	0.14–2.36
Dithiophosphates	Sulfide or native metal flotation	0.006–0.20
Thionocarbamates	Sulfide flotation	0.01–0.12
Mercaptobenzothiazole	Sulfide, oxidized ore flotation	0.03–0.30
Thiocarbanilide	Sulfide flotation	0.03–0.09
Xanthogen formates	Sulfide flotation	0.01–0.15
Fatty acids (includes tall oil)	Nonsulfide flotation	0.12–1.81
Amines	Various flotation operations	0.06–1.77
Sulfonates	Various flotation operations	0.12–0.89

Source: Davis, Hyatt, and Cox 1976; Arbiter et al. 1985; and Jarrett and Kirby 1978.

*Multiply values by 2 to obtain equivalents in U.S. customary units (pounds per short ton).

TABLE 13.3 Addition ranges for modifiers (depressants and activators)

Additive	Usual Function	Usual Addition Range, kg/t*
Cyanide	Depressant	0.006–0.21
Sodium dichromate	Depressant	0.04–0.50
Sodium ferrocyanide	Depressant	0.16–1.77
Fluorosilicic acid	Depressant	0.12–0.59
Calgon (condensed phosphate)	Depressant	0.07
Citric acid	Depressant	0.30
Nokes reagent (phosphorus or arsenic pentasulfide reacted with lime or soda ash)	Depressant	2.5–5.9
Sodium hypochloride	Depressant	0.89–8.3
Sodium silicate	Depressant	0.30–2.36
Sodium sulfide	Depressant	0.41–23.6
Sodium sulfite	Depressant	0.35–0.65
Zinc sulfate	Depressant	0.05–1.00
Guar products	Depressant	0.05–0.18
Lignin sulfonates	Depressant	0.0006–5.9
Quelbracho	Depressant	0.12–0.18
Starch and dextrin	Depressant	0.18–3.54
Hydrofluoric acid	Depressant or activator	0.50
Calcium carbonate	Activator	1.2
Copper sulfate	Activator	0.12–0.71
Lead acetate	Activator	0.89–4.1

Source: Davis, Hyatt, and Cox 1976; Arbiter et al. 1985; and Jarrett and Kirby 1978.

*Multiply values by 2 to obtain equivalents in U.S. customary units (pounds per short ton).

potential for thiocarbamates to be discharged from flotation circuits. Ranges of measured quantities of various substances discharged from flotation plants are listed in Table 13.5; various other reported physical and chemical characteristics of such waters are listed in Table 13.6.

The U.S. Environmental Protection Agency (USEPA) publishes drinking water regulations containing tables that give data on maximum contaminant level goals (MCLGs) and maximum contaminant levels (MCLs) allowable for a number of the substances cited in Tables 13.5 and 13.6 (USEPA 1994).

TABLE 13.4 Addition ranges for frothers and hydrocarbon oils

Substance	Addition Range, kg/t*
MIBC	Trace–0.30
Pine oil	0.006–0.18
Cresylic acid	0.006–0.21
Polyglycols	Trace–0.15
Other alcohols	0.002–0.06
Triethoxybutane	0.012–0.08
Fuel oil	0.06–1.77
Kerosene	0.02–0.53

Source: Davis, Hyatt, and Cox 1976; Arbiter et al. 1985; and Jarrett and Kirby 1978.

*Multiply values by 2 to obtain equivalents in U.S. customary units (pounds per short ton).

These standards are frequently revised; the ones listed in this chapter were published in July 1994. Some states have their own water standards, which may be more stringent than those of the USEPA.

Toxicity of Mill Reagents and Discharges

Data on the toxicity of substances potentially discharged from a mineral processing or hydrometallurgy operation are extensive. For example, Hawley (1977) conducted an exhaustive study of the toxic nature of substances discharged from Ontario mines and mills. The study concluded that mine-mill reagents vary greatly in toxicity; some are highly toxic and others are relatively nontoxic. Also, some reagents are unstable and readily break down to relatively nonharmful substances in mill tailings, whereas others are quite persistent. Work performed by the Ontario Ministry of the Environment (Hawley 1977) studied the effect of such reagents on *Daphnia magna,* a small crustacean; *Ontropis atherinoides,* an emerald shiner; and *Pimephales promelas,* a flathead minnow. Among the flotation collectors in general, thiol, sulfonate, and amine collectors tend to exhibit moderate to high toxicity toward these species. To a large degree, frothers vary in toxicity, with polypropylene glycols being rather nontoxic, shorter chain alcohols slightly more toxic, and cresylic acid yet more toxic. Long-chain polymers are usually less toxic. Modifiers vary greatly in toxicity, with some substances (such as cyanide and some of the heavy metal ions) being highly toxic.

There is little doubt about the potential toxicity of cyanide that reaches water systems outside the plant or mill site; there have been numerous reports of massive fish kills caused by cyanide. Although laboratory test results cannot be directly related to field conditions, it is of some interest to note that some reports (Doudoroff 1976; Cardwell et al. 1976) indicate that, in general, concentrations greater than 0.10 mg/L of HCN can be expected to kill sensitive fish species in either freshwater or marine environments.

Recycling and Treatment of Mill Water

Successful recycling of mill or tailings water depends on successful treatment to remove inorganic and organic contaminants to a sufficient degree. Many schemes are used or have been devised to allow for partial or total recycle of mill or tailings water. Unusual approaches include the use of mineral slimes as adsorbate in the purification of industrial and municipal wastewaters (Broman 1975); a floto-flocculation method (Khavskij et al. 1975; Khavskij and Tokarev 1981); spherical agglomeration of flocculated slimes (Neczaj-Hruzewisz et al. 1981); and various bioremediation schemes (Smith 1989; Smith, Dubel, and Misra 1991; Smith and Misra 1993).

Tailing Ponds. A common type of treatment used for wastewaters today in the mineral-milling industry is the use of tailing ponds (Williams 1975; Brawner 1979; Klohn and Dingeman 1979;

TABLE 13.5 Some reported concentrations and ranges of ions in flotation waters, in milligrams per liter

Ion	Iron Ore Flotation	Copper Sulfide Flotation	Lead-Zinc Flotation	Other Sulfide Flotation	Nonsulfide Flotation
Al	0.009–5.0	<0.5	—	6.2–7.8	210–552
Ag	—	<0.1	—	<0.02	0.04
As	—	<0.02–0.07	—	0.02–3.50	<0.01–0.15
B	—	—	—	<0.01	<0.01–0.65
Be	—	—	—	<0.002	36
Ca	55–250	—	—	<0.6	43–350
Cd	—	0.05–3.0	1.2–16.4	<0.01–0.74	<0.002–0.01
Co	—	1.68	—	—	—
Cr	—	—	9.8–40	0.03–0.04	0.02–0.35
Fe	<0.02–10.0	550–18,800	2,900–35,000	<0.5–2,800	0.06–500
Hg	—	0.0006–0.006	—	0.0008–27.5	—
K	—	—	—	—	77
Pb	0.045–5.0	<0.01–21	76–560	<0.02–9.8	0.02–0.1
Mg	—	—	—	1.93	320
Mn	0.007–330	31	295–572	0.12–56.5	0.19–49
Mo	—	29.3	—	<0.05–21	<0.2–0.5
Na	—	—	—	—	270
Ni	0.01–0.20	2.8	—	0.05–2.4	0.15–1.19
Sb	—	<0.5	—	<0.2–64	–
Se	—	<0.003	—	0.144–0.155	0.06–0.13
SiO_2	—	46.8	—	—	—
Te	—	—	—	<0.08–0.3	<0.2
Ti	—	—	—	—	<0.5–2.08
Tl	—	—	—	—	<0.05
V	—	—	—	<0.5	<0.2–2.0
Zn	0.006–10	4.8–310	160–3,000	0.02–76.9	<0.02–19
Rare earths	—	—	—	—	4.9
Chloride	0.35–180	—	—	1.5	57–170
Fluoride	—	—		4.8–11.7	1.3–365
Nitrate	—	—	—	—	1.25
Phosphate	—	20.8	—	—	0.8
Sulfate	5–475	—	—	—	9–10,600
Cyanide	0.008–0.02	<0.01–0.17	—	<0.01–0.45	<0.01
Sulfide	—	—	—	<0.5	<0.5
NH_3	—	—	—	—	1.4

Source: Davis, Hyatt, and Cox 1976; Jarrett and Kirby 1978a; Woodcock and Jones 1970; Carta, Ghiani, and Del Fa 1977; Ilie and Tutsek 1977; and Hawley 1977.

TABLE 13.6 Some reported properties of wastewaters from flotation mills

Property	Iron Ore Flotation	Copper Sulfide Flotation	Lead, Zinc Sulfide Flotation	Other Sulfide Flotation	Nonsulfide Flotation
Conductivity, microohms	130–375	—	—	—	650–17,000
Total dissolved solids, mg/L	0.3–1,090	395–4,300	—	68–2,600	192–18,400
Total suspended solids, mg/L	0.4–1,900	114,000–465,000	20,500–269,000	2–550,000	4–360,000
Chemical oxygen demand, mg/L	0.2–36	—	—	15.9–238	<1.6–39.7
Total organic carbon, mg/L	—	—	—	7.8–290	9–3,100
Oil and grease, mg/L	0.03–90	<0.05–10	—	2.0–11.4	<1–3.4
pH	5–10.5	8.1–10.1	7.9–11	6.5–11	5–11

Source: Woodcock and Jones 1970; Carta et al. 1977; Ilie and Tutsek 1977; and Hawley 1977.

Vick 1983). The primary function of a tailing pond is to remove suspended solids. For such purposes the pond must be properly designed to provide sufficient surface area, adequate retention time, and quiescent conditions. Oxidation and destruction of some noxious substances can also be provided for in a pond that is properly designed. Retention time in the pond may vary greatly depending on climatic conditions, and the size of solids present, among other factors. Water may leave the pond through a number of means, including simple overflow, seepage through or under the dam, evaporation, and pumping and recycling to the mill. Zero discharge from the pond can, in principle, be achieved either by chemical treatment and recycling of all water or, if the environment is sufficiently dry, by evaporation from the pond. Unless the bottom of the pond is sealed by natural or artificial means, a primary problem with tailing ponds is the possibility of seepage into aquifers used by cities or agriculture. However, the low cost–and the fact that tailing ponds are often the only means for suspended solids removal–make such ponds indispensable in many mineral industry operations. Settling ponds may sometimes be used rather than tailing ponds if solids concentrations are low and a major quantity of water is to be recycled.

Thickening, Hydrocycloning, Centrifugation, and Filtration. Thickeners are often used to remove all or most of the solids from an effluent stream. Although the thickened waste stream must ultimately be placed in tailing ponds, the use of thickeners often has advantages over the use of ponds without prethickening. For instance, less land space is required if a thickener is used. In addition, ponds that use prethickening can be placed at or near the mill; thus, clarified water can easily be sent back into the mill circuit, thereby reducing problems with rainfall.

Hydrocyclones can be used in a similar manner to thickeners and can, in principle, allow for an even greater savings in floor and land space. However, removing very fine particles with hydrocyclones is very difficult; hence, the use of these devices for treatment of slimy waste streams is limited.

Either centrifugation or filtration can be used for solids removal from waste streams. Two main problems with either technique are the cost and the difficulty of handling the large tonnages discharged. However, these techniques are suited to handling waters that contain only small amounts of solids.

Flocculation. The addition of flocculating agents can greatly increase the efficiency of thickeners, tailing ponds, or other treatment processes. Classes of flocculating agents that can be used include ionic compounds, polyelectrolytes, and various starches. Ionic compounds that have been used include lime, magnesium carbonate, ferric salts, and aluminum salts. These compounds function by reducing the electrostatic charge on particles, thereby reducing the repulsion between similarly charged particles and resulting in flocculation of the suspension. Studies indicate that when Al(III) is present in polynuclear form, flocculation is particularly strong (Dempsey, Ganko, and O'Melia 1984). The polymeric substances (polyelectrolytes and starches) function by physically entwining particles and by forming bridges between particles.

A significant problem with flocculating agents is their cost. Lower cost ionic flocculants are normally used in the concentration range of 10 to 100 mg/L. More expensive flocculants are usually used in the range of 2 to 20 mg/L.

Base or Acid Addition. A simple and common treatment practiced in the mineral industries is neutralization of acidic or basic solutions. Such neutralization, particularly for acidic effluents, is often practiced not only to bring the wastewater to a pH near neutrality but also to precipitate heavy metals from solution. Lime, limestone, dolomite, sodium hydroxide, soda ash, and ammonium hydroxide are basic substances used for neutralization of acidic waters; lime is the substance most often used. Basic effluent solutions are less common than acidic solutions. These solutions can be neutralized by the addition of an acid, such as sulfuric acid. Under optimum conditions, neutralization of an acidic stream can reduce Cu(II) concentrations to about 0.2 mg/L at pH 7 and Zn(II), Cd(II), and Ni(II) concentrations to about 1 mg/L. A problem with such treatment is that the heavy metal hydroxide is often difficult to filter.

Metals in Aqueous Streams

In both hydrometallurgy and mineral processing, the potential for release of toxic metallic cations into the off-site water regime presents a source of concern in terms of impacts on the surrounding environment. Today, several technologies can be considered for metals separation from aqueous streams. Patterson (1986) subdivided the technologies into three categories:

1. Conventional treatment technologies (in order of approximate frequency of application in metals pollution control):
   - Precipitation (including primary coprecipitation)
   - Oxidation/precipitation
   - Reduction/precipitation
   - Concentration/precipitation
   - Coprecipitation
2. Recognized recovery technologies (unranked):
   - Evaporative recovery
   - Ion exchange
   - Membrane separation
   - Reductive electrolysis
3. Emerging recovery technologies:
   - Differential precipitation
   - Various froth and dissolved air flotation schemes
   - Selective adsorption
   - Bioadsorption and other bioprocesses

Some of the technologies are discussed in the following sections.

Precipitation of Metals. Precipitation of metals from aqueous solution can often be accomplished by simple pH adjustment. As a first approximation, the amount of metal ions removable from solution can be estimated by considering the stability constant ($*k_{so}$) values for metal hydroxides with respect to the following general reaction:

$$M(OH)_n + nH^+ \leftrightarrow M^{n+} + nH_2O$$

where M is the metal in question.

TABLE 13.7 Values of log($*k_{so}$) for selected metal hydroxides

Metal	log($*k_{so}$)	Metal	log($*k_{so}$)
Al(III)	9.2	Mg(II)	18
Ba(II)	24	Mn(II)	15.4
Be(II)	6.9	Ni(II)	13.3
Bi(III)	4.4	Pb(II)	13.0
Ca(II)	23.0	Sn(II)	2.0
Cd(II)	6.5	Sr(II)	24
Ce(III)	20.1	Th(IV)	11.4
Co(III)	12.8	Ti(IV)	–0.5
Cr(III)	12.7	Tl(III)	3.35
Cu(II)	8.85	U(IV)	5
Fe(II)	13.05	U(VI)	6
Fe(III)	2.5	V(III)	7.6
Hf(IV)	0.3	Zn(II)	12.1
La(III)	19.7	Zr(IV)	–0.4

Source: Kragten 1978.

Table 13.7 lists $*k_{so}$ values for a number of metal hydroxides. Such a table can be misleading, however, for a number of reasons: possible hydrolysis of cations; the existence of less soluble polynuclear species; an inherent solubility of an uncharged species; and the possible presence of differing solid oxides and hydroxides of the metal in equilibrium with the solution. A more complete picture of the solubility of metals can be obtained from log concentration diagrams calculated from thermodynamic data on the stability of the various metal species (e.g., see Baes and Mesmer [1976]). Note, however, that some metal ions can be present in solution in supersaturation concentrations (Smith 1971).

In the case of some substances, such as calcium and magnesium salts, the use of $*k_{so}$ values and log concentration diagrams is of little benefit because the formation of metal carbonates exercises considerable control over aqueous calcium and magnesium solutions. The presence of carbonate species results from equilibration of solutions with the atmosphere. Table 13.8 lists solubility products (k_{sp}) for selected metal carbonates. In this table the k_{sp} values are the stability constants for the following general reaction:

$$M_mX_n(s) \leftrightarrow mM^{n+} + nX^{m-}$$

$$k_{sp} = [M^{n+}]^m\,[X^{m-}]^n$$

where

M = metal

X = carbonate

The solubility of metal oxides, hydroxides, and carbonates at any particular pH value in the absence of other ions and at constant temperature will be controlled by the following equilibria, assuming a divalent cation (adapted from Stumm and Morgan [1970]):

$H_2O \leftrightarrow H^+ + OH^-$

$CO_2(g) + H_2O \leftrightarrow H_2CO_3(aq)$

$CO_2(g) + H_2O \leftrightarrow HCO_3^- + H^+$

$H_2CO_3(aq) \leftrightarrow HCO_3^- + H^+$

$HCO_3^- \leftrightarrow CO_3^{2-} + H^+$

TABLE 13.8 Values of $\log k_{sp}$ for metal carbonates at 25°C

Carbonate	$\log k_{sp}$
$AgCO_3$	–11.07
$BaCO_3$	–8.58
$CaCO_3$ (aragonite)	–8.22
$CaCO_3$ (calcite)	–8.34
$CaMg(CO_3)_2$ (dolomite)	–16.7
$CdCO_3$	–11.21
$FeCO_3$ (siderite)	–10.4
Li_2CO_3	–3.0
$MgCO_3$ (magnesite)	–4.9
$MnCO_3$	–10.63
$NiCO_3$	–6.84
$PbCO_3$	–12.83
$SrCO_3$	–9.03

Source: Aquatic Chemistry, Stumm, W., Morgan, J.J.; Copyright © 1970 John Wiley & Sons. This material is used by permission of John Wiley & Sons, Inc.

$$MCO_3(s) \leftrightarrow M^{2+} + CO_3^{2-}\ (k_{sp})$$
$$MCO_3(s) + H^+ \leftrightarrow M^{2+} + HCO_3^-$$
$$MCO_3(s) + H_2CO_3(aq) \leftrightarrow M^{2+} + 2HCO_3^-$$
$$MCO_3(s) + H_2O + CO_2(g) \leftrightarrow M^{2+} + 2HCO_3^-$$
$$MCO_3(s) + 2H^+ \leftrightarrow M^{2+} + CO_2(g) + H_2O$$
$$MCO_3(s) + 2H^+ \leftrightarrow M^{2+} + H_2CO_3(aq)$$

Sulfide precipitation of heavy metals can be more effective than using a base, although the precipitate obtained may have low chemical purity and may not be amenable to physical separation from the aqueous waste stream. Either sodium sulfide or hydrogen sulfide may be used in the process. Sulfide precipitation can be applied only when the pH is sufficiently high to result in at least partial formation of sulfide ion, because the pK_a values for hydrogen sulfide at 25°C lie at pH 6.97 and pH 13.8 (Perrin 1982).

Table 13.9 lists solubility products of metal sulfides. Work at the University of Nevada, Reno, by J.L. Hendrix and J. Nelson (personal communication) has indicated that some of the problems inherent in sulfide precipitation can be overcome if the sulfide can be generated in place through a precursor such as thiourea or thioacetamide. In this method the precipitating agent (sulfide) is produced within the solution at a rate comparable to the rate of crystal growth. The method should allow for production of purer precipitates through careful control of conditions of pH, concentration, and time; in addition, it should allow the particle size of the precipitate to be controlled.

Coprecipitation. Some inorganic aqueous substances, such as molybdenum in molybdate form and vanadium in several anionic forms, cannot be effectively removed from solution by direct precipitation using hydroxide or sulfide ions. However, these substances can be removed from solution by coprecipitation with ferric ion (Jarrett and Kirby 1978b). In the case of molybdate ion, the molybdate is incorporated into ferric hydroxide precipitates at acid pH values. Such precipitates can be removed by filtration or flotation, and effluents containing as little as 0.2 mg/L molybdenum can be obtained. In the case of vanadium, a ferric metavanadate can be coprecipitated with ferric hydroxide.

Recovery of Metal Waste by Hydrometallurgy. Although the first major unit operation associated with hydrometallurgy is extraction of metals by leaching, many of the subsequent unit operations are used to concentrate and remove metals from aqueous streams. These unit operations are candidates

TABLE 13.9 Values of k_{sp} for metal sulfides at 25°C

Sulfide	$\log k_{sp}$
Ag_2S	–50.83
Bi_2S_3	–98.74
CdS	–28.85
CoS	–17.5
CuS	–35.89
Cu_2S	–45.05 (50°C)
FeS	–18.8
HgS	–52.3
MnS	–13.34
NiS	–17.8
PbS	–26.1
SnS	–27.49
ZnS	–21.35

Source: Chang 1985 and Hogfeldt 1982.

for use as purification operations of waste streams. Concentration techniques can include liquid membrane technology; ion exchange, including liquid ion exchange; or organic solvent extraction.

Adsorption. Many substances–including activated carbon, alumina, silica gel, and biomaterials–can be used as adsorbents for various metal ions. Of these, activated carbon is perhaps the best. In addition, very inexpensive and readily available substances, such as mineral slimes (Broman 1975), serpentine, and fly ash (Smith and Hwang 1978), have also been suggested as adsorbents for aqueous metal ions. Such substances could therefore–in principle at least–be used to remove metal ions from effluent streams. The key to the successful application of any adsorbent for metal ions will be a demonstration of the adsorbent's ability to selectively adsorb specific metals from complex wastewater matrices, followed by elution or other recovery of a metal-rich concentrate.

Various adsorbents can also be used for removing organic substances, such as collectors and frothers (both of which are organic flotation reagents), from mill effluents. If the mill is operating under optimum conditions, collectors should primarily exit the flotation mill adsorbed onto floated minerals. Frothers should primarily remain in the aqueous phase, although some should vaporize and exit into the atmosphere. At any rate, both collectors and frothers can adsorb from aqueous solution onto various adsorbents. Read and Manser (1975), for example, noted that both activated carbon and an anionic resin can be used to remove xanthate from a mill stream.

Bioremediation of Liquid Wastes. It has long been known that microorganisms can concentrate heavy metal ions from aqueous solution to a remarkable extent (Smith 1989). Sakaguchi and Nakajima (1991) investigated the ability of 135 different species of microorganisms (42 bacteria, 26 yeasts, 34 fungi, and 33 actinomycetes) to accumulate U, Co, Mn, Ni, Cu, Zn, Cd, Hg, Th, and Pb. Particularly good at accumulating UO_4^{2+} from solution were *Pseudomonas stutzeri, Neurospora sitophilia, Streptomyces albus,* and *Streptomyces viridochromogenes.*

Of primary importance in the commercialization of metal cleanup and recovery via microorganisms as sorbents are the handling and harvesting of the microorganisms after metal accumulation. An interesting process involves the use of BIOFIX® beads, wherein organic materials such as microorganisms or sphagnum moss are incorporated in a porous polysulfone matrix (Bennett and Jeffers 1990; Jeffers, Ferguson, and Bennett 1991). Inexpensive materials are used in the makeup of the beads, and the beads can be used either in column contactors or in very simple low-maintenance systems. The beads have been shown to be particularly useful in removing very low levels of heavy metal ions from aqueous solution and to function very well in the presence of large concentrations of Ca(II) and Mg(II) ions. Because the metal ions are readily eluted from the beads, the beads can easily be recycled.

In another approach, froth flotation has been used experimentally to harvest microorganisms after heavy metal accumulation (Smith, Yang, and Wharton 1991; Sadowski, Golab, and Smith 1991). Such a system may prove attractive in heavy metal recovery schemes where the metals are present in quite dilute concentrations.

Other schemes of promise incorporate immobilized living microorganisms on various fibers (Clyde and Whipple 1983; Clyde 1986) and in a calcium alginate, sodium alginate, or another similar matrix (Brierly and Brierly 1993). The alginate itself has metal-binding properties (Apel and Torma 1993). These procedures can run in a continuous manner as long as a certain amount of nutrients is continuously added to the system to maintain cell growth.

Biomass obtained from higher plants–such as *Eichhornia crassipes* (water hyacinth), *Typha latifolia* (common cattail), and *Potomogenton luscens*–has an extremely large biosorption capacity for heavy metals; for example more than 100 mg Hg/g biomass (Schneider et al. 1995; Robichaud et al. 1995). Further, some plant biomass has demonstrated the ability to go through more than 75 loading and elution cycles without a loss of loading capacity.

Yet another interesting process is to use organic mats on pond surfaces to remove either metal ions (such as Pb(II) ions) or anions (such as Se(VI) ions) from water (Bender et al. 1989; Bender et al. 1991). The scheme devised to remove selenium (in selenate form) from water uses a mat composed of blue-green algae, *Anabaena*, as the top layer; organic material, such as grass cuttings, as the middle layer; and selenate-reducing bacteria as the bottom layer. The bacteria use the algae as their food source and reduce the selenate to elemental selenium. The mats, containing the orange-brown selenium, have considerable structural integrity after selenium reduction and can readily be harvested from the pond surface. Whether or not a similar scheme could be employed for remediation of arsenate-containing waters is an interesting source of speculation.

The contamination of aquatic ecosystems by mercury wastes is a serious environmental problem. Mercury-reducing bacteria can be used to remediate such contamination (Ogunseitan and Olson 1991). Also, Apel and Turick (1991) have investigated the effectiveness of three chromate-reducing bacterial strains in the reduction of Cr(VI) to the less toxic Cr(III). All three strains appeared to be good reducers of Cr(VI) and have potential in the remediation of such wastes.

Considering organic flotation wastes, Carta, Ghiani, and Rossi (1980) studied the effect of different bacterial strains and consortia–along with collector, oxygen, and carbon dioxide concentrations and various physicochemical factors–on the biodegradation of three flotation collectors. The collectors studied were sodium hexadecylsulfate (SHS), sodium oleate (SOL), and dodecylammonium acetate (DAC). The researchers found that sodium dodecyl sulfate (SDS) and SOL were readily broken down by microorganisms. DAC was less readily biodegraded. In another study, Solozhenkin and Lyubavina (1980) investigated the biodegradation of three thiol collectors–potassium 2,2',6,6'-tetramethyl-1-iminoxyl-4 xanthate (KTIX), potassium butyl xanthate (KBX), and sodium diethyldithiocarbamate (NaEC)–by the bacterium *Pseudomonas fluorenscens*. In the presence of bacteria, KTIX was completely destroyed within 45 minutes. In the absence of bacteria, only 45% of the compound was destroyed in this time. Solozhenkin and Lyubavina also cited a study that determined that a residual xanthate concentration of 0.12 mg/L in the wastewater from a lead concentrator was completely destroyed in 5 minutes after treatment with a bacterial suspension. These studies indicate that flotation collectors can be effectively destroyed by bacterial degradation and that the procedure may be a viable treatment method for such wastes.

Other organic chemicals potentially discharged from flotation plants include various frothers and modifiers. The frothers, in particular, pose problems because they often pass through flotation plants into effluent water streams (much more so than collectors, which adsorb onto minerals); further, frothers are often quite toxic. Although biodegradation studies have been carried out on substances similar in structure to frothers, no studies have been performed to date specifically on flotation frothers.

Certain microorganisms or enzymes derived from microorganisms are known, under some conditions, to be able to break down cyanides; thus, such organisms can be used in the bioremediation of cyanide wastes discharged from precious metal hydrometallurgy plants (Noel, Fuerstenau, and Hendrix 1991; Chapatwala et al. 1995) Whitlock (1987) described the success of a biological cyanide degradation wastewater treatment plant used by a mining company in South Dakota; at this plant, 90%–98% of the metal-complexed cyanides are removed. Further, it may even be possible that fatty acids can be biologically produced from cyanide wastes (Finnegan 1992).

Sulfate-reducing Microbial Processes. In the early 1990s, a process was developed for simultaneously removing both heavy metal ions and sulfate from a dilute aqueous solution by using anaerobic bacteria to convert sulfate to sulfide (Barnes et al. 1992; Scheeren et al. 1992). Thus, in this process, the metals are precipitated as very insoluble metal sulfides. The overall reaction for a metal, M, in such a system (in the presence of suitable anaerobic bacteria) is

$$MSO_4 + \text{C substrate} + \text{starter bacterial cells} \rightarrow MS + CO_2 + \text{additional growth bacterial cells}$$

The carbon source can be a substance such as ethanol. Suitable sulfate-reducing organisms include *Desulfovibrio vulgaris, Desulfomonas pigra, Desulfobulbus propionicus, Desulfococcus multivorans,* and *Desulfobacter postgatei.* Proper operation of the process requires control of several parameters: pH (near neutrality), temperature (35°C), redox potential (Eh), carbon source, nutrients, residence time, and buildup of contaminants in the system. In a unit tested in the United Kingdom, a sludge-blanket reactor was used in which a sludge blanket containing the biomass was raked or suspended by liquid recycle of the high-density sludge (Barnes et al. 1992). Operation of the test unit determined that a wide range of heavy metals and sulfate can be removed from solution by the sulfate-reducing bacteria process. Energy consumption of the process is quite low.

Floto-flocculation

Khavskij et al. (1975) and Khavskij and Tokarev (1981) described a process for clarifying mineral industry waste solutions by using high-molecular-weight water-soluble surface-active substances. In this process, known as floto-flocculation, the surface-active polymers are introduced into the fluid to be clarified. The dispersed particles are flocculated and, at the same time, rendered partly hydrophobic by interaction with the polymers. This step is followed by flotation and subsequent separation of the recovered particles in the form of an unstable froth. The polymers can be amphoteric, cationic, or anionic in nature; they must contain sufficient hydrophobic groups in addition to adsorptive groups to render the overall particle-polymer flocs somewhat hydrophobic. According to Khavskij et al. (1975), water-soluble polymers having molecular weights in excess of $3 \times 10^4 - 5 \times 10^4$ and at a concentration of 0.01% can lower the surface tension of water at least 3 dyne/cm below that of pure water; such polymers should be effective reagents in the floto-flocculation process. Floto-flocculation has been used on a commercial scale, and final effluent solids contents as low as 0.001 to 0.03 g/L have been claimed.

Oxidation

A number of organic contaminants can be degraded by oxidation through exposure to aeration. In addition, cyanide-containing solutions can be rendered less harmful by aeration. In the process, free cyanide is oxidized to cyanate and, ultimately, to carbon dioxide and nitrogen.

Water Recycling

Although complete or near-complete recycling of mineral processing waters often poses many problems, such recycling is now often accomplished in operating concentrators (Ilie and Tutsek 1977; Joe 1984; Read and Manser 1975). The concentrations of both inorganic and organic substances in the recycled water must be reduced to a level such that the plant operations are not affected. In general, to achieve water of this quality, several of the treatment operations noted earlier—such as flocculation, settling, and removal of solids and removal of inorganic and organic contaminants by one or more methods (precipitation, adsorption, etc.)—must be utilized.

CONTAMINATED SOILS

Soils can become contaminated by discharges from mineral processing, hydrometallurgy, and chemical plant operations. A listing of possible soil contaminants is essentially the same as for contaminated water and includes heavy metals and toxic oils.

Characteristics of Soils

Soil is a complex environment composed of three main phases—solid, liquid, and gas—which are variously arranged to produce the different soil types (Finkl 1982; Lozet and Mathieu 1991). Soil is formed by the physical, chemical, and biological weathering of rocks to small particles; the mineral component, along with the organic material, forms the solid phase. The chemical composition of the soil reflects that of the rock from which the soil was derived. Often, the dominant component is silica, which is present in sand, silt, and clays; however, some soils, such as peats, have very high organic levels but virtually no inorganic mineral matter. The organic matter in soils consists of more or less unaltered fragments of plants, animals, microorganisms, and their metabolites, as well as humus, which is a product of organic decomposition. Spaces between the solid particles are filled with water and gases.

The soil water is a weak solution of salts; it is the solvent system from which the plants and microorganisms take up mineral nutrients. The amount of soil water is inversely related to the amount of soil atmosphere. The soil atmosphere fills those pores not occupied by water and is usually saturated with water vapor. It usually has 10 to 100 times more carbon dioxide (CO_2) than the air has, as well as somewhat less oxygen (O_2). Volatile organic substances, such as methane, hydrogen sulfide, ammonia, and hydrogen, may also be present in greater concentrations than in air.

Soil is composed of distinct layers called soil horizons, and the sequence of horizons from the surface down is known as the soil profile. The upper horizon, the A horizon, is rich in organics and can be subdivided into distinct layers that represent progressive stages of humification. These layers are designated (from the surface downward) as litter (A-0), humus (A-1), and leached zone (A-2). The next horizon, the B horizon, is composed of mineral soil in which organic compounds have been converted by microorganisms/decomposers into inorganic compounds by the process of mineralization; the compounds are thoroughly mixed with finely divided parent rock material. The bottom-most horizon, the C horizon, consists of more or less unmodified but finely divided parent material.

Soil profiles and relative thicknesses of the horizons vary greatly depending on climatic and topographic regions. In fact, in some cases there may be more variation between the horizons in a single soil than there is between corresponding horizons in different soils.

The soil biocenoses consists of three different size groups:

1. The microbiota, which include algae, cyanobacteria, bacteria, fungi, and protozoa. The heterotrophic bacteria are the most numerous microorganisms, and their numbers depend on the amount of organic material present.
2. The mesobiota, which include the nematodes, the small oligochaete worms, the smaller insect larvae, and microarthropodes (such as mites)
3. The macrobiota, which include the roots of plants, the larger insects, earthworms, and other larger organisms, including the burrowing vertebrates such as moles, voles, and gophers. The macrobiota are important for mixing the soil and giving it a “spongy” consistency.

In Situ Remediation Technologies for Soils Contaminated with Heavy Metals

Normal unpolluted soils contain a large variety of heavy metals, though in low concentrations. Plants and soil microorganisms use some of these metals as micronutrients. However, almost all countries, both industrialized and nonindustrialized, have serious contamination of soils by heavy metals. This contamination originates from a number of different sources, including mining, industrial, agricultural, and military operations. Some sources of the pollution, such as the mining and processing of heavy metal ores, date back to antiquity. However, increased industrialization and the demands of modern military forces have greatly accelerated heavy metal accumulation in the environment (Forstner 1995).

The metals can exist in soils in a number of different forms depending on the source of the metals, the anions and other cations present, the organic matter present (both living and nonliving), the pH and Eh of the soil, and the possible speciations of a particular metal ion. The metals present have the potential of being mobilized and subsequently transported in ground water and surface water, thereby contaminating these waters. Microorganisms can immobilize, mobilize, or transform metals by several means: extracellular precipitation reactions; oxidation and reduction reactions; methylation and demethylation; extracellular binding; and complexation and intracellular accumulation (Hughes and Poole 1989; Brierly 1990; Tuovinen, Kelly, and Groudev 1991; Groudev 1995a; Gaylarde and Videla 1995; Groudev 1995b). Additionally, various plants concentrate heavy metals in their biomass; if these metals reach sufficient concentration, the biomass can be toxic to animal life, thus rendering some soils unsuitable for agricultural use. Of course, on the other hand, specific plants can potentially be used to concentrate the metals in this way and thereby decontaminate specific soils.

As a result of these various reactions and conditions, the heavy metals can be present in soil in a number of different forms: as free ions (mainly cations) in the pore solutions; as inorganic or organometallic soluble complexes; as ions or molecules adsorbed on the soil particles; or as different solid-phase metal-containing compounds such as oxides, hydroxides, sulfides, and carbonates. Note that only some water-soluble forms (bioavailable forms) are toxic for living organisms in the soil above certain concentrations. However, the metals can be turned from inert to toxic forms as a result of chemical or biological leaching.

Various schemes have been proposed for the decontamination or stabilization of soils contaminated with heavy metals. Most of the known technologies for remediation of such soils have been applied only under laboratory or pilot-scale conditions. These technologies are of three general types: off-site, on-site, and in situ (Groudev 1995c; Groudev 1995b).

The off-site technologies involve removing the contaminated soil and transporting it to another place, where it is treated by a suitable method using equipment especially intended for that purpose. The cleaned soil is then returned to its original site. Most current industrial treatment of soils involves off-site treatment, where the soil, either cleaned or not, is disposed of off-site in such a manner that the metal contaminants cannot enter surface water or ground water. The problems with this approach have been destruction of the soils, difficulty in finding a suitable storage site, and high transportation costs.

The on-site technologies also involve removing the contaminated soil; however, treatment is carried out by means of some mobile or stationary unit located near the site. After treatment, the cleaned soil is returned to its original location.

The in situ technologies do not involve removing the contaminated soil. These technologies are of two different types: delivery and recovery. In other words, they are based on processes that facilitate the transport of the metals either into or out of the subsurface. In principle, the in situ technologies are likely to be less efficient in metal removal; however, they are economically attractive in that they may, in some cases, avoid causing serious damage to the treated soil, especially with respect to the soil's physical-mechanical and chemical properties and microflora. In some cases, the in situ treatment involves detoxifying the contaminants rather than removing them. This detoxification is achieved either (1) by immobilization of the metals in the form of different insoluble substances at the location (in which case the soil may not be damaged) or (2) by encapsulation of the contaminants into stable substances. If only the removed contaminants are immobilized, the soils will not be damaged; however, if the soils plus the contaminants are encapsulated, the soils will be obviously destroyed. This section addresses in situ technologies in detail.

The mechanism of the in situ treatment may be physical, chemical, biological, or some combination thereof (Groudev 1995b; Sims 1990; Gee and Wing 1991; Means and Hinchee 1993; Hinchee, Means, and Burris 1995; Groudeva, Groudev, and Ivanova 1995). Liquids are the principal phase in most in situ operations. In the process, an effective collection system is required to prevent metal migration and pollution of ground water and surface water. The various in situ treatment technologies can be placed into six main groups:

1. Soil flushing by means of suitable leach solutions
2. Immobilization of metals inside the soil by converting them into their least soluble or toxic forms or by encapsulating them in solids of high structural integrity
3. Separation of metals from soil particles by physical or chemical means, such as permeable barriers or electrokinetics
4. Soil capping to prevent the infiltration of oxygen or water that causes solubilization and migration of metals
5. Vitrification, in which the combination of soil plus contaminants is transformed into a glassy matrix
6. The use of plants or microorganisms to accumulate the metals in their biomass. The existence of such plants and microorganisms is demonstrated by the developed resistance of plant communities found on old mine sites and dumps to the toxicity of metal ions including Cu(II), Zn(II), and Cd(II).

Several of these in situ technologies will be considered in more detail in the following paragraphs.

Soil Flushing of Heavy Metals. Soil flushing is the removal of metal ions from soils by washing with a suitable solvent, such as water or other aqueous or nonaqueous solutions. The most suitable solvent is pure water or dilute sulfuric acid. Fresh water is applied in the cases when the net neutralization potential of the soil is positive and where the heavy metals are mainly present in the soil as (1) free ions in the pore solution or as ions adsorbed on the soil particles, (2) certain precipitated hydroxides in the soil, or (3) easily soluble sulfates or other compounds. In these cases, the pH of the leach solutions is either neutral or mildly alkaline. The removal of metals in these cases is thus connected to the metals' solubility in water. However, the solubility of most heavy metals at neutral or slightly alkaline pH values is quite limited. Apart from pH, the metal solubility is affected by factors such as the redox potential, temperature, the presence of other metallic and nonmetallic ions, and the presence of different organic compounds in solution. Because the removal of metal ions or metals otherwise available for removal takes place when the concentrations are higher in the pore solutions than in the leach solution, the efficiency of removal strongly depends on the differences between these concentrations. A high concentration ratio can be maintained by continually adding the flushing fresh water. Under such treatment, the most efficient system will be one in which the metals are removed from the effluent water and then the cleaned water is returned for flushing.

Microbial Solubilization of Heavy Metals. In addition to the chemical solubilization of the metals and the desorption of the adsorbed ions and molecules from the soil particles to the flushing water or water solution, some biological processes affecting metal removal also take place. Some heterotrophic microorganisms—bacteria (species of the genus *Arthrobacter, Bacillus, Pseudomonas, Aerobacter, Clostridium,* etc.), streptomycetes, and fungi (species of the genus *Aspergillus, Fusarium,* etc.)—are able to reduce the ferric ion to the ferrous state and thus solubilize iron. Some bacteria (*Arthrobacter, Bacillus, Aeromonas, Clostridium*) are able to solubilize manganese by reducing Mn(IV) to Mn(II).

Some microorganisms are able to solubilize metals from different minerals (mainly from oxides and hydroxides but also from carbonates and silicates) by means of secreted metabolites, primarily organic acids. These organic acids, as well as some other microbial metabolites, facilitate the solubilization of metals and prevent their precipitation by forming soluble organometal complexes (Noble et al. 1991; Rogers and Wolfram 1991).

The solubilization of metals from sulfide minerals by various chemolithotrophic bacteria (*Thiobacillus* and *Leptospirillum* species) is not significant in these metal solubilization systems because the bacteria are acidophilic and unable to grow in neutral or alkaline environments.

At the same time, some microorganisms cause dissolved metals to precipitate and will impede the removal of these metals from soil. Heterotrophic bacteria (*Sphaerotilus, Leptothrix*) oxidize ferrous ions to the ferric state by the peroxide oxidative mechanism. The ferric ions then react with water and are precipitated as ferric hydroxide. In a similar manner, some bacteria precipitate manganese by oxidizing Mn(II) to Mn(IV). The greatest problems, however, involve the activity of anaerobic sulfate-reducing bacteria (*Desulfovibrio, Desulfobacterium, Desulfobacter,* etc.), which precipitate the heavy metals as metal sulfides through the production of hydrogen sulfide (H_2S).

The mobilization of metal ions and subsequent flushing of soils that have a high sulfur content and a negative net neutralization potential is carried out by means of acidic aqueous solutions. The acid generation in the soil is the result of oxidation of sulfide minerals in the soil. The oxidation is caused by several electrochemical, chemical, and biological reactions occurring in the presence of oxygen, water, and acidophilic chemolithotrophic bacteria possessing iron- and sulfur-oxidizing abilities (*Thiobacillus ferrooxidans, Leptospirillum ferrooxidans*). The sulfide minerals are oxidized to metal sulfates and are thus generally solubilized. (An important exception is $PbSO_4$, which possesses only very limited solubility. In this system, Pb(II) can be mobilized in the form of organometallic complexes.)

The soil flushing can accomplish a permanent removal of contaminants and, in conjunction with suitable bioremediation, may be a cost-effective treatment of metal-contaminated soils.

Biological and Chemical Immobilization. Various techniques can be used to immobilize metals within the soils by converting the metals into their least soluble forms (Groudev 1995b). The techniques involve precipitation of the metals. In some cases the precipitation is a pure chemical process and is the result of changes in some environmental factors, mainly the soil pH. The required increase in pH is accomplished by adding an alkalizing agent, such as lime or limestone, and the metals are precipitated as their corresponding hydroxides. A problem with such treatment is that many metal hydroxides possess relatively large solubility products, and the metal ions can relatively easily be released back into the environment, especially if acid generation develops in the soil. Thus, repeated treatment of the soil at frequent intervals may be necessary, with accompanying increases in costs.

The precipitation of some metals can be facilitated by prior oxidation or reduction. Examples are the oxidation of Fe(II) to Fe(III), Mn(II) to Mn(IV), and As(III) to As(V), as well as the reduction of Cr(VI) to Cr(III) and U(VI) to U(IV). Some of these reactions can proceed as chemical processes, but in soil systems their rates can be accelerated by various microorganisms. The most efficient precipitation of metals is achieved by microbial sulfate reduction. The process is carried out by sulfate-reducing bacteria (*Desulfovibrio, Desulfobacterium, Desulfobacter,* etc.), which under anaerobic conditions use various organic compounds as sources of energy for their growth. The electrons removed from these organic compounds are transferred to the sulfate ions, which are reduced to free hydrogen sulfide. Alkalinity is produced during this process:

$$2CH_2O + SO_4^{2-} \leftrightarrow H_2S + 2HCO_3^-$$

The hydrogen sulfide generated in situ dissolves in the water, and the resulting sulfide ion reacts with heavy metals:

$$M^{2+} + S^{2-} \leftrightarrow MS$$

where M represents a divalent metal ion, such as Cu^{2+} or Zn^{2+}. The overall reaction can be represented as follows:

$$\text{metal sulfate} + \text{carbon substrate} \leftrightarrow \text{metal sulfide} + CO_2 + H_2O + \text{bacterial biomass}$$

Proposed Biotechnical Detoxification Scheme. This section describes a proposed biotechnical method for in situ detoxification of soils contaminated with heavy metals; it draws heavily from

Groudev (1995b). The detoxification is achieved by the joint action of microflora and of bacterial and chemical reagents added to the soil. The procedure consists of two consecutive phases:

1. Mobilization of the metals initially located in the top soil horizons (A and B) by leaching of the metal-bearing minerals (which may be oxides, hydroxides, sulfides, silicates, and sulfates). This mobilization is achieved by inoculating the soil with different bacteria or fungi that are able to solubilize oxide and silicate minerals; such solubilization occurs either through the production of complexing organic acids (such as citric acid, ketoglutaric acid, or oxalic acid; see Noble et al. [1991] and Rogers and Wolfram [1991]) or with chemolithotrophic bacteria that are able to solubilize sulfide minerals (e.g., *Thiobacillus ferrooxidans*). Various other environmental factors must also be controlled (optimum humidity, pH range 4.0–4.5, enhanced natural aeration, presence of nutrients). The soil is washed periodically with slightly acidified water (sulfate present) of pH 4.0–4.5. At the end of this phase, most of the bioleachable metals are mobilized and transferred to the deeply located soil horizons.
2. Precipitation of the metals in the deepest horizon (C) in the form of metal sulfides through the use of anaerobic sulfate-reducing bacteria growing in the lowest horizons. Such bacteria will likely have to be introduced into this location. The precipitated metals are further immobilized by sorption of clay minerals present in the lowest horizons.

The time required for such treatment will likely vary with the character and degree of contamination of the soil.

Agglomeration and Encapsulation of Solid Wastes

Agglomeration and encapsulation of reactive, sulfide-containing flotation tailings and other acid-water-generating materials promises safe disposal of these materials. The agglomerated and encapsulated wastes can be disposed of by backfilling in the mine.

Various attempts have been made to fix hazardous wastes in glassy matrices by using a high-temperature electric furnace to produce the glassy matrix (Misra, Kumar, and Neve 1993). However, such processes are energy-intensive and, hence, of limited practical use for fixing large quantities of reactive wastes.

Low-temperature agglomeration and encapsulation procedures show more promise. In one scheme that has been suggested (Misra, Kumar, and Neve 1993; Kumar, Neve, and Misra 1994), the finely divided reactive wastes are agglomerated in conventional balling drums. A binding agent, such as type II portland cement, is added during pelletizing to give the pellets sufficient strength. Subsequent to the addition of the portland cement, a sodium silicate sol is added in the final agglomeration stage. The sodium silicate is added to make the pellet "glassy" and to prevent further oxidation and weathering. Tests completed to date (i.e., as of 1998) indicate that strong and stable agglomerates are readily formed with a number of different mining industry wastes. Leaching experiments show that the dissolution of heavy metals from the reactive materials can be prevented and that the silicate encapsulation keeps pyrite particles isolated within the pellet structure. In addition, the agglomerates can incorporate a bactericide to inhibit bacterial catalyzed conversion of sulfides to sulfate (Misra, Kumar, and Neve 1993).

In Situ Remediation Technologies for Petroleum-contaminated Soils

Petroleum and its products are mixtures of hydrocarbons and are known to include more than 100 different chemical compounds. The following are some of the most common and toxic compounds (Preslo et al. 1989):

benzene	benz(a)anthracene
ethylbenzene	benzo(a)pyrene
n-heptane	naphthalene
pentane	phenanthrene

n-hexane	acenaphthylene
o-xylene	acenapaphthene
toluene	fluorene
phenol	benzofluoranthenes

As with remediation of heavy metal-contaminated soils, the remediation of petroleum-contaminated soils can be on-site, off-site, or in situ. The off- and on-site technologies described earlier can also generally be applied to petroleum-contaminated soils. Such technologies include land treatment (allowing natural degradation to take place after removal of the soil from its original location), thermal treatment (heating and volitalizing petroleum components), incorporation in asphalt, solidification/stabilization, and extraction of contaminant oils through chemical or physical means (Preslo et al. 1989). However, this section deals primarily with in situ remediation.

The in situ technologies include volatilization, biodegradation, soil flushing with chemicals (particularly surfactants), vitrification, passive remediation (allowing natural degradation to take place while monitoring the process), and isolation and containment. Several of these processes are described in the sections that follow.

Volatilization. Because many of the components of petroleum and its products are relatively easily vaporized, volatilization can sometimes be used in clean-up efforts (Preslo et al. 1989; Malot 1989; Ying et al. 1989). In particular, vacuum extraction technology can be successfully used in many different types of soils. When low-vacuum, fan-type systems are employed, the procedure is sometimes referred to as "soil venting," although the same term may also be applied to systems in which no vacuum is applied and the soil is merely dug up and spread out or plowed for the "venting." In the soil-venting process, vacuum extraction wells and vacuum-monitoring wells are placed at designated locations about the contaminated sites and a vacuum is applied to the extraction wells. In a pilot test in Florida, a vacuum process operating for a total of 150 days was able to reduce benzene levels of about 30 ppm to less than 1 ppm (Malot 1989).

In Situ Vitrification. In situ vitrification is a thermal treatment process that converts contaminated soil into a stable glassy product (Preslo et al. 1989; Timmerman, Buelt, and FitzPatrick 1989). In the process an array of electrodes is inserted into the ground to the required depth. A conductive mixture of flaked graphite and glass frit is placed around the electrodes to act as a conducting starter path, and an electrical potential is applied to the electrodes. The electrical current in the starter path heats the starter path and the surrounding soil to temperatures on the order of 1,500°–2,000°C. The starter path is consumed by oxidation, and the current is transferred to the increasing amount of molten soil, which is conductive. As the vitrification increases, the vitrified soil incorporates nonvolatile contaminants and destroys the organics by pyrolysis. The pyrolyzed materials float to the surface, where they are consumed by combustion. A hood is required over the processing area to accumulate the off gases for further treatment (Timmerman, Buelt, and FitzPatrick 1989).

This method may be the only viable treatment for relatively low quantities of very highly contaminated wastes. However, it is costly and destroys the soil being treated.

Soil Flushing/Washing of Hydrocarbons. Although the terms "soil flushing" and "soil washing" are sometimes used interchangeably, in general, soil flushing refers to in situ treatment and soil washing refers to ex situ treatment. Both terms refer to the mobilization of hydrocarbons by introducting aqueous surfactant solutions that solubilize the hydrocarbons in surfactant micelles. The soil-flushing/washing process will not work on every type of substance. For example, the contaminants must be hydrophobic, which includes most petroleum hydrocarbons (Wilson and Clarke 1994). Further, the effectiveness of soil flushing/washing of hydrocarbon-contaminated soils depends on how strongly the hydrocarbons are sorbed onto soil components (Nash and Traver 1989). Also, the surfactant solutions must be able to be delivered to the contaminants; this task may be difficult if the soil permeability is low. Finally, all of the surfactant solutions must be able to be contained and recovered (Wilson and Clarke 1994). The process is not yet fully developed, and its future is uncertain. An interesting variation is to use a combination of surfactant flushing and bioremediation (Page et al. 1997; Brown, Guha, and Jaffe 1997).

Bioremediation. Both in situ and ex situ bioremediation of soils contaminated by petroleum and petroleum products have now been well studied; they have often proven to be effective (Groudeva et al. 1995b; Dineen et al. 1989; Ying et al. 1989). The design of successful in situ remediation depends on the following subsurface parameters (Dineen et al. 1989):

1. Soil microbiology: Petroleum-degrading microorganisms must be present throughout the zone where petroleum hydrocarbons are to be cleaned.
2. Soil chemistry: Concentration of soil nutrients must be adequate to maintain microbial growth, and no toxic levels of salts and toxic metals can be present.
3. Soil physics: Soil air permeability must be adequate to allow movement of added oxygen and nitrogen to the contaminated soil, as well as movement of carbon dioxide away from the soil.
4. Soil morphology: Soil stratification throughout the affected zone should be well characterized in order to design an effective delivery system (for oxygen and nitrogen).
5. Hydrogeology: The depth to ground water, the ground water flow direction and gradient, the presence or absence of floating products, and the petroleum hydrocarbon concentrations in ground water should be understood before bioremediation is implemented, so as to avoid recontamination of the cleaned soil from the ground water.

The crucial steps in the breakdown of petroleum hydrocarbons are the oxidation of straight- or branched-chain alkanes and the breakage of aromatic rings by oxygenase enzymes. Only a few microorganisms possess an enzyme system capable of producing an oxygenase enzyme (Dineen et al. 1989). Thus, successfully implementing a hydrocarbon bioremediation scheme requires determining whether microorganisms are present in the soil at the hydrocarbon locations and whether any of these microorganisms are capable of degrading the petroleum hydrocarbons. Alternately, it may be possible to introduce natural or engineered organisms to the contaminated plot, along with the proper nutrients, of course (Groudeva, Groudev, and Ivanova, 1995; Groudeva et al. 1995). It may be possible to enhance the hydrocarbon-degrading ability of microorganisms, preferably taken from the petroleum-contaminated area, by culturing the organisms initially in an environment where the carbon source is some nontoxic material (such as glucose) and then, slowly as function of time, replacing the glucose with the contaminating hydrocarbon. It may ultimately be possible to develop strains of microorganisms that can use the contaminant as their sole carbon source.

Groudeva et al. (1995) describe a pilot plant study carried out on petroleum-contaminated soil in and around the Tulenovo oil field in Bulgaria. Table 13.10 shows the makeup of an adapted microbial culture introduced into the oil-contaminated soil, as well as laboratory degradation rates. The consortium was selected from a large number of different organisms and consortia on the basis of its superior ability to degrade the Tulenovo oil. The results of the study are summarized in Table 13.11, and the microorganisms in selected plots are shown in Table 13.12. These data show that petroleum oxidizing microorganisms can be effective in cleaning a petroleum-contaminated soil.

SOLIDS DISPOSAL AND LONG-TERM MANAGEMENT OF TAILINGS IMPOUNDMENTS

Suspended solids are perhaps the most critical pollutants in the effluent from concentrators. However, provided there is adequate space available to construct a large enough pond, proper handling of the solid (as well as liquid) component of a waste stream can be achieved if the tailing pond is designed with proper retention time and stability in mind.

Thus, in general, solid tailings from a mineral processing concentrator will be handled in tailings impoundments. However, other schemes, such as discharge of tailings directly into a lake or marine environment or disposal in abandoned operating mines, are sometimes used.

TABLE 13.10 Composition of the microbial consortium inoculated into the oil-contaminated soil of the Tulenovo deposit in Bulgaria

Microorganism	Relative Portion in the Inoculum, %	Maximum Rate of Degradation of Oil from Tulenovo under Laboratory Conditions, mg/(L-h)
Bacillus species 1	15	36.9
Bacillus species 2	15	33.0
Mixed culture of bacteria and yeasts	20	37.4
Sporosarcima species	15	20.3
Pseudomonas species	15	24.2
Mixed bacterial culture	20	42.4

Source: Groudeva, Groudev, and Ivanova 1995 and Groudeva et al. 1995.

TABLE 13.11 Characteristics of different test sections subjected to microbial treatment of petroleum-contaminated soil: Tulenovo, Bulgaria

	Experimental Sections		Control Sections		
Characteristic	A (without Zeolite Addition)	B (with Zeolite Addition)	A	B	C
Duration of treatment, months	5	5	10	10	10
Initial oil content, g/kg dry soil	99	95	88	87	91
Final oil content, g/kg dry soil	33	30	82	64	48
Oil degradation, %	67	68	6	26	47
Maximum oil degradation rate, g oil/(kg dry soil·month)	27	28	1.7	8	15
Ammonium phosphate consumption, kg/(ton dry soil·month)	0.6	0.45	—	—	0.6

Source: Groudeva, Groudev, and Ivanova 1995 and Groudeva et al. 1995.

TABLE 13.12 Microbial counts, in cells/g dry soil, during the most active phase of treatment of a petroleum-contaminated soil

	Experimental Sections		Control Sections		
Microorganism*	A (without Zeolite Addition)	B (with Zeolite Addition)	A	B	C
Oil oxidizers	3×10^6	3×10^6	3×10^3	9×10^4	9×10^5
Aerobic heterotrophic bacteria	1×10^8	3×10^8	1×10^5	3×10^5	1×10^8
Oligocarbophiles	9×10^6	8×10^6	1×10^5	5×10^5	5×10^6
Spore-forming bacteria	3×10^6	5×10^6	5×10^4	8×10^4	3×10^6
Nitrogen-fixing bacteria	3×10^5	5×10^5	8×10^3	3×10^4	9×10^4
Anaerobic heterotrophic bacteria	3×10^4	4×10^4	3×10^3	1×10^4	4×10^4
Molds	1×10^4	1×10^4	5×10^2	1×10^3	9×10^3

Source: Groudeva, Groudev, and Ivanova 1995 and Groudeva et al. 1995.

Surface tailings impoundments are of two general types: water-retention-type dams and raised embankments. Water-retention-type dams are similar to conventional water storage dams in that they are constructed to their full height before tailings are discharged into the impoundment. Thus, construction is similar to construction of other earth dams. Earth fill of the dam is usually composed of natural soil borrow materials. This type of dam is usually used where water storage requirements are high, such as when the dam must be placed in a high-runoff location or when substantial recycle of the water to the mill is not possible.

Raised embankment structures are more commonly used. These structures begin with a starter dike, normally constructed of soil borrow of sufficient size to impound the first several years' worth of mill tailings with allowances for runoff storage. Periodically, the embankment is raised to keep pace with the rising level of tailings in the impoundment. The raises can be composed of a number of materials, such as soil borrow, various waste materials, or cycloned sand tailings.

Three general methods of embankment raising are employed: upstream, downstream, and centerline. In the upstream method, the tailings are discharged from the starter dike, which gives rise to a beach extending upstream from the dike. This beach acts as a foundation for the next dike, which is built on the beach. Subsequent raises are handled in the same manner. In this scheme exceptional care must be employed to ensure that the beach forms a competent support for the subsequent dikes; thus, sand should account for at least half of the tailings.

In the downstream method, subsequent raises are constructed by placing fill on the downstream slope of the previous raise. Often an impervious zone is placed on the upstream side of the raises, and an internal drain may be placed in the embankment. More storage of water is possible with this method of construction.

The centerline method is basically a compromise between the other two methods. The upstream part of the raiser rests on the tailing beach; the downstream portion resembles downstream construction. Internal drainage zones can be incorporated into the overall structure. This method cannot be used to store as large of quantities of water as with the downstream-type structure.

Overall, the downstream-type construction is the most expensive because a greater fill volume is required; the upstream type is the least expensive. However, the downstream-type construction has better water storage ability, better seismic resistance, and more flexibility in raising rates. The centerline construction is generally intermediate between the other two in terms of these characteristics.

During milling operations, aquifer contaminants are derived mainly from drainage of tailings placed in the impoundments and subsequent seepage into ground waters. After closure of the milling operations, contaminants can also be derived from precipitation that leaches the in-place tailings.

Existing impoundments from closed operations may have no seepage controls built into them. Evaluation of such seepage requires the use of one or several methods to assess the saturated and unsaturated hydraulic properties of mill tailings (Larson and Stephens 1985; Lewis and Stephens 1985). Also, mobilization of contaminants (such as heavy metal ions) depends on the pH both of the water in the impoundments and of rainwater falling onto or running into the impoundments. Such items must be evaluated to assess long-term treatment of seepage from the impoundments.

At new or existing operations, seepage controls can be built into new impoundments or sometimes placed in existing impoundments. Several types of systems are possible, including (1) systems where effluent is partially retained although some seepage loss is expected and (2) systems where seepage is completely restricted by structural barriers. These latter systems use impoundment liners to achieve zero or near-zero discharge of contaminants.

Partial containment seepage controls use barriers such as cutoff trenches, slurry walls (containing bentonite slurry plus, sometimes, portland cement), and grout curtains. Using these barriers requires that the embankment incorporate an internal impervious fill zone to which the barrier can connect.

These barriers usually cannot be used with upstream-type embankments. They function by restricting the lateral migration of seepage; they do not prevent downward migration and thus are most effective when the impoundment is underlain by an impervious stratum.

Seepage return systems can also be employed. These systems attempt to collect seepage and return it rather than restrict its flow. They utilize either (1) collector ditches plus sumps or (2) wells surrounding or downstream from the impoundment.

Containment liners are expensive and are used only where stringent ground water protection is indicated. Liners completely line both the bottom and sides of an impoundment; thus, their effectiveness is independent of subsurface strata. Several different types have been used, including clay liners, tailing slimes, and various types of synthetic liners. In all cases, the effluent streams should not chemically react with the liner; the liners should exhibit sufficient chemical stability with time; and the liner should not be too easily broken, cracked, or torn. Tailing slime liners should be relatively inexpensive and not greatly subject to cracking; they must be very carefully placed. Clay liners can be effective but are subject to cracking and to increased seepage with increased hydrostatic head on the liner. They also may be subject to chemical alteration at certain pH values. Synthetic liners are effective, although expensive. Some are subject to cracking or tearing, and some tend to deteriorate with time or when in contact with various hydrocarbons.

BIBLIOGRAPHY

Apel, M.L., and A.E. Torma. 1993. Diffusion Kinetics of Metal Recovery Biosorption Processes. In *Biohydrometallurgy Technologies*, Vol. II. Edited by A.E. Torma, M.L. Apel, and C.L. Brierley. Warrendale, Pa.: The Minerals, Metals, and Materials Society.

Apel, W.A., and C.E. Turick. 1991. Bioremediation of Hexavalent Chromium by Bacterial Reduction. In *Mineral Bioprocessing*. Edited by R.W. Smith and M. Misra. Warrendale, Pa.: The Minerals, Metals, and Materials Society.

Arbiter, N., H. Cooper, M.C. Fuerstenau, C.C. Harris, M.C. Kuhn, J.D. Miller, and R.F. Yap. 1985. Section 5: Flotation. In *SME Mineral Processing Handbook*. Edited by N.L. Weiss. New York: AIME.

Baes, C.F., Jr., and R.E. Mesmer. 1976. *The Hydrolysis of Cations*. New York: John Wiley & Sons.

Barnes, L.J., F.J. Janssen, P.J.H. Scheeren, J.H. Versteegh, and R.O. Koch. 1992. Simultaneous Microbial Removal of Sulphate and Heavy Metals from Waste Water. *Trans. Instn. Min. Metall.,* 101:183–189.

Bender, J.A., E.R. Archibold, V. Ibeanusi, and J.P. Gould. 1989. Lead Removal from Contaminated Water by a Mixed Microbial Ecosystem. *Water Sci. Tech.,* 21:1661–1664.

Bender, J.A., J.P. Gould, Y. Vatcharapijarn, and G. Saha. 1991. Uptake, Transformation and Fixation of Se(VI) by a Mixed Selenium Tolerant Ecosystem. *Water, Air Soil Pollut.,* 59:359–367.

Bennett, P.G., and C.R. Jeffers. 1990. Removal of Metal Contaminants from a Waste Stream Using Bio-Fix Beads Containing Sphagnum Moss. In *Mining and Mineral Processing Wastes*. Edited by F.M. Doyle. Littleton, Colo.: SME.

Brawner, C.O. 1979. Concepts and Experience for Subsurface Storage of Tailings. In *Tailings Disposal Today*, Vol. 2. Edited by G.A. Argall Jr. San Francisco, Calif.: Miller Freeman.

Brierley, C.L. 1990. Bioremediation of Metal-contaminated Surface and Ground Water. *Geomicrobiology Journal,* 8:201–223.

Brierley, C.L., and J.A. Brierley. 1993. Immobilization of Biomass for Industrial Application of Biosorption. In *Biohydrometallurgy Technologies*, Vol. II. Edited by A.E. Torma, M.L. Apel, and and C.L. Brierley. Warrendale, Pa.: The Minerals, Metals, and Materials Society.

Broman, G. 1975. Purification of Industrial and Municipal Waste Water by Means of Mineral Slimes. In *XI International Mineral Processing Congress*. Cagliari, Sardegna, Italy: Instituto de Arti Mineraria. 1371–1396.

Brown, D.G., S. Guha, and P.R. Jaffe. 1997. Modeling Biodegradation of Phenanthrene in the Presence of Non-ionic Surfactant. In *In Situ and On-Site Bioremediation*, Vol. 2. Edited by B.C. Alleman and A. Leeson. Columbus, Ohio: Battelle Press.

Cardwell, R.D., D.G. Foreman, T.R. Payne, and D.J. Wilbur. 1976. *Acute Toxicity of Selected Toxicants to Six Species of Fish.* Report EPA-600/3-76-008. Cambridge, Mass.: U.S. Environmental Protection Agency.

Carta, M., M. Ghiani, and C. Del Fa. 1977. Problems of Purification and Recycling of Residual Waters in the Flotation of Complex Minerals with Sulphides, Barite and Fluoride. In *Meeting 8–Paper 2, XII International Mineral Processing Congress.* Sao Paulo, Brazil: Ministéria das Minas e Energia.

Carta, M., M. Ghiani, and G. Rossi. 1980. Biochemical Beneficiation of Mining Industry Wastes. In *Biochemistry of Ancient and Modern Environments* (Proceedings of the 4th International Symposium on Environmental Geochemistry). Edited by P.A. Trudinger, M.R. Walter, and B.J. Ralph. New York: Springer-Verlag.

Chang, J.C. 1985. Solubility Product Constants. In *CRC Handbook of Chemistry and Physics.* 66th ed. Boca Raton, Fla.: Chemical Rubber Company.

Chapatwala, K.D., G.R.V. Babu, E.R. Armstead, and J.H. Wolfram. 1995. Degradation of the Metals Tailing Pond Waters Containing Cyanides by Alginate Immobilized Cells of *Pseudomonas putida*. In *Minerals Bioprocessing II.* Edited by D. Holmes and R.W. Smith. Warrendale, Pa.: The Minerals, Metals, and Materials Society.

Clyde, R.A. 1986. *Apparatus for Harvesting Cells*. United States Patent 4,600,694. Washington, D.C.: U.S. Department of Commerce, Patent and Trademark Office.

Clyde, R.A., and A. Whipple. 1983. *Method for Treating Waste Fluid with Bacteria*. United States Patent 4,530,763. Washington, D.C.: U.S. Department of Commerce, Patent and Trademark Office.

Davis, F.T., D.E. Hyatt, and C.H. Cox. 1976. Environmental Problems of Flotation Reagents in Mineral Processing Plants Tailings Water. In *Flotation: A.M. Gaudin Memorial*, Vol. 2. Edited by M.C. Fuerstenau. New York: AIME.

Dempsey, B.A., R.M. Ganko, and C.R. O'Melia. 1984. The Coagulation of Humic Substances by Means of Aluminum Salts. *Jour. AWWA,* 76(4):141–150.

Dineen, D., J.P. Slater, P. Hicks, J. Holland, and L.D. Clendening. 1989. In Situ Biological Remediation of Petroleum Hydrocarbons in Unsaturated Soil. In *Petroleum Contaminated Soils*, Vol. III. Edited by P. Kostecki and P. Calabrese. Chelsea, Mich.: Lewis Publishers.

Doudoroff, P. 1976. *Toxicity to Fish of Cyanides and Related Compounds: A Review.* Report EPA-600/3-76-038. Cambridge, Mass.: U.S. Environmental Protection Agency.

Finkl, C.W., Jr. 1982. *Soil Classification.* Stroudsburg, Pa.: Hutchinson Ross.

Finnegan, I. 1992. Microbial Bio-remediation of One Carbon Wastes Excluding Photosynthesis. In *Wastecom '92.* Johannesburg, South Africa: Institute of Waste Management, Rand Afrikaans University.

Forstner, U., 1995. Land Contamination by Metals: Global Scope and Magnitude of the Problem. In *Heavy Metals, Problems and Solutions.* Edited by W. Salomons, U. Forstner, and P. Mader. New York: Springer-Verlag.

Gaylarde, C.O., and H.A. Videla. 1995. *Bioextraction and Biodeterioration of Metals.* Cambridge, United Kingdom: Cambridge University Press.

Gee, G.W., and N.R. Wing. 1991. *In Situ Remediation: Scientific Basis for Current and Future Technologies.* Columbus, Ohio: Battelle Press.

Groudev, S.N. 1995a. Microorganisms and Acid Mine Drainage. In *TEMPUS JEP N3567.* Sofia, Bulgaria: University of Mining and Geology.

——. 1995b. Microbial Treatment of Metal-contaminated Liquid Effluents. In *Proceedings of the Symposium on Environmental Engineering.* Sofia, Bulgaria: Bulgarian Academy of Sciences.

——. 1995c. Microbial Detoxification of Heavy Metals in Soils. In *TEMPUS JEP N3567.* Sofia, Bulgaria: University of Mining and Geology.

Groudeva, V.I., S.N. Groudev, and I.S. Ivanova. 1995. Microbial Remediation of Oil-polluted Soils in the Tulenovo Oil Deposit, Bulgaria. In *Biohydrometallurgical Processing.* Edited by C.A. Jerez, T. Vargas, H. Toledo, and J.V. Wiertz. Santiago, Chile: University of Chile.

Groudeva, V.I., S.N. Groudev, G. Uzunov, and I.A. Ivanova. 1995. Microbial Removal of Oil from Polluted Soils in a Pilot Scale Operation in the Tulenovo Deposit, Bulgaria. In *Minerals Bioprocessing (II)*. Edited by D.S. Holmes and R.W. Smith. Warrendale, Pa.: The Minerals, Metals, and Materials Society.

Hawley, J.R. 1977. *The Use, Characteristics and Toxicity of Mine-mill Reagents*. Ontario, Canada: Province of Ontario, Ministry of the Environment.

Hendrix, J.L., and J.H. Nelson. Personal communication.

Hinchee, R.E., J.I. Means, and D.R. Burris. 1995. *Bioremediation of Inorganics*. Columbus, Ohio: Battelle Press.

Hogfeldt, E. 1982. *Stability Constants of Metal Ion Complexes, Part A: Inorganic Ligands*. IUPAC Chem. Data Series, No. 21. New York: Pergamon Press.

Hughes, M.N., and R.K. Poole. 1989. *Metals and Microorganisms*. London: Chapman and Hall.

Ilie, P., and P. Tutsek. 1977. Flotation of Lead-zinc Ores with Total Water Recycling. In *Meeting 8–Paper 1, XII International Mineral Processing Congress*. Sao Paulo, Brazil: Ministério das Minas e Energia.

Jarrett, B.M., and R.G. Kirby. 1978a. *Development Document for Effluent Limitations Guidelines and New Source Performance Standards for the Ore Mining and Dressing Point Source Category*, Vol. I. Report 440/1 78/061-d. Cambridge, Mass.: U.S. Environmental Protection Agency.

——. 1978b. *Development Document for Effluent Limitations Guidelines and New Source Performance Standards for the Ore Mining and Dressing Point Source Category*, Vol. II. Report 440/1-78/061-e. Cambridge, Mass.: U.S. Environmental Protection Agency.

Jeffers, T.H., C.R. Ferguson, and P.G. Bennett. 1991. Biosorption of Metal Contaminants from Acidic Mine Waters. In *Mineral Bioprocessing*. Edited by R.W. Smith and M. Misra. Warrendale, Pa.: The Minerals, Metals, and Materials Society.

Joe, E.G. 1984. Water and Solution Recycling Practice in the Canadian Mineral Industry. In *Mineral Processing and Extractive Metallurgy: Proceedings of Conference Held in Kunning, Peoples Republic of China*. London: Institution of Mining and Metallurgy.

Jones, M.H., and J.T. Woodcock. 1984. Applications of Pulp Chemistry to Regulation of Chemical Environment in Sulphide Mineral Flotation. In *Principles of Mineral Flotation: The Wark Symposium*. Edited by M.H. Jones and J.T Woodcock. Melbourne, Australia: Australasian Institute of Mining and Metallurgy.

Khavskij, N.N., and V.D. Tokarev. 1981. Clarification of Sewage and Effluents from Mineral Processing and Metallurgical Plants by Floto-flocculation. In *Mineral Processing, XIII International Mineral Processing Congress*, Vol. IIB. Edited by J. Laskowski. Warsaw, Poland: Polish Scientific.

Khavskij, N.N., V.D. Tokarev, B.V. Nevski, A.M. Kotov, M.P. Vilianski, and I.A. Yakulovitch. 1975. Clarification of Water and Aqueous Solutions by the Floto-flocculation Method. In *XI International Mineral Processing Congress*. Cagliari, Sardegna, Italy: Instituto de Arti Mineraria.

Klohn, E.J., and D. Dingeman. 1979. Tailings Disposal for Reserve Mining Company. In *Tailings Disposal Today*, Vol. 2. Edited by G.A. Argal Jr. San Francisco, Calif.: Miller Freeman.

Kragten, J., 1978. *Atlas of Metal-ligand Equilibria in Aqueous Solution*. New York: John Wiley & Sons.

Kumar, S., C. Neve, and M. Misra. 1994. Safe Disposal of Reactive Flotation Tailings and Retorted Fines of Eastern Oil Shales by an Agglomeration Process. *Fuel*, 73(9):1472–1475.

Larson, M.B., and D.B. Stephens. 1985. A Comparison of Methods to Characterize Unsaturated Hydraulic Properties of Mill Tailings. In *Low Level Waste and Hazardous Waste*, Vol. 2. Fort Collins, Colo.: Colorado State University.

Lewis, G.J., and D.B. Stephens. 1985. Analysis of Infiltration Through Mill Tailings Using a Bromide Tracer. In *Low Level Waste and Hazardous Waste*, Vol. 2. Fort Collins, Colo.: Colorado State University.

Lozet, J., and C. Mathieu. 1991. *Dictionary of Soil Science*. 2nd ed. Rotterdam, The Netherlands: A.A. Balkema.

Malot, J.J. 1989. Cleanup of a Gasoline Contaminated Site Using Vacuum Extraction Technology. In *Petroleum Contaminated Soils*, Vol. II. Edited by P. Calabrese and P. Kostecki. Chelsea, Mich.: Lewis Publishers.

Means, J.I., and R.E. Hinchee. 1993. *Emerging Technology for Bioremediation of Metals.* Columbus, Ohio: Battelle Press.

Misra, M., S. Kumar, and C. Neve. 1993. Mitigation of Acid Mine Drainage by Agglomeration. In *Extraction and Processing Division Congress, 1993*. Edited by J. Hager. Warrendale, Pa.: The Minerals, Metals, and Materials Society.

Nash, J.H., and R.P. Traver. 1989. Field Studies of In Situ Soil Washing. In *Petroleum Contaminated Soils*, Vol. I. Edited by P. Kostecki and P. Calabrese. Chelsea, Mich.: Lewis Publishers.

Neczaj-Hruzewisz, J., R. Sprycha, J. Janusz, and A. Monies. 1981. Treatment of Flotation Tailings by the Spherical Agglomeration of Flocculated Slimes. In *Mineral Processing, XIII International Mineral Processing Congress*, Vol. IIB. Edited by J. Laskowski. Warsaw, Poland: Polish Scientific.

Noble, E.G., E.G. Baglin, D.L. Lampshire, and J.A. Eisele. 1991. Bioleaching of Manganese from Ores Using Heterotrophic Microorganisms. In *Mineral Bioprocessing.* Edited by R.W. Smith and M. Misra. Warrendale, Pa.: The Minerals, Metals, and Materials Society.

Noel, D.M., M.C. Fuerstenau, and J.M. Hendrix. 1991. Degradation of Cyanide Utilizing Facultative Anaerobic Bacteria. In *Mineral Bioprocessing.* Edited by R.W. Smith and M. Misra. Warrendale, Pa.: The Minerals, Metals, and Materials Society.

Odum, E.P. 1971. *Fundamentals of Ecology.* Philadelphia, Pa.: W.B. Saunders Co.

Ogunseitan, O.A., and B.H. Olson. 1991. Potential for Genetic Enhancement of Bacterial Detoxification of Mercury Wastes. In *Mineral Bioprocessing.* Edited by R.W. Smith and M. Misra. Warrendale, Pa.: The Minerals, Metals, and Materials Society.

Page, C.A., J.S. Bonner, T.J. McDonald, and R.L. Autenrieth. 1997. Enhanced Solubility of Petroleum Hydrocarbons Using Various Biosurfactants. In *In Situ and On-Site Bioremediation*, Vol. 2. Edited by B.C. Alleman and A. Leeson. Columbus, Ohio: Battelle Press.

Patterson, J.W. 1986. Metals Separation and Recovery. In *Proceedings, International Symposium on Metal Speciation, Separation, and Recovery.* Edited by J.W. Patterson and R. Passino. Chelsea, Mich.: Lewis Publishers.

Perrin, D.D. 1982. *Ionization Constants of Inorganic Acids and Bases in Aqueous Solution.* 2nd ed. IUPAC Chem. Data Series, No. 29. New York: Pergamon Press.

Preslo, L., M. Miller, W. Suyama, M. McLearn, P. Kostecki, and E. Fleischer. 1989. Available Remedial Technologies for Petroleum Contaminated Soils. In *Petroleum Contaminated Soils*, Vol. I. Edited by P. Kostecki and P. Calabrese. Chelsea, Mich.: Lewis Publishers.

Read, A.D., and R.M. Manser. 1975. Residual Flotation Reagents: Problems in Effluent Disposal and Water Recycle. In *XI International Mineral Processing Congress.* Cagliari, Sardegna, Italy: Instituto de Arti Mineraria.

Robichaud, K., M. Misra, R.W. Smith, and I.A.H. Schneider. 1995. *Adsorption of Heavy Metals by Aquatic Plant Roots.* In *Water Rock Interaction.* Edited by O. Chudaev and Y.A. Kharaka. Rotterdam, The Netherlands: A.A. Balkema.

Rogers, R.D., and J.H. Wolfram. 1991. Biological Separation of Phosphate from Ore. In *Mineral Bioprocessing.* Edited by R.W. Smith and M. Misra. Warrendale, Pa.: The Minerals, Metals, and Materials Society.

Sadowski, Z., Z. Golab, and R.W. Smith. 1991. Flotation of *Streptomyces pilosus* After Lead Accumulation. *Biotech. Bioeng.*, 37:955–959.

Sakaguchi, T., and A. Nakajima. 1991. Accumulation of Heavy Metals Such as Uranium and Thorium by Microorganisms. In *Mineral Bioprocessing.* Edited by R.W. Smith and M. Misra. Warrendale, Pa.: The Minerals, Metals, and Materials Society.

Scheeren, P.J.H., R.O. Koch, C.J.N. Buisman, L.J. Barnes, and J.H. Versteegh. 1992. New Biological Treatment Plant for Heavy Metal-contaminated Groundwater. *Trans. Instn. Min. Metall.*, 101:190–199.

Schneider, I.A.H., J. Rubio, M. Misra, and R.W. Smith. 1995. Stems and Roots of *Eichlornia* as Biosorbents for Heavy Metal Ions. *Minerals Engineering,* 8:979–988.

Sims, R.C. 1990. Soil Remediation Techniques at Uncontrolled Hazardous Waste Sites. *Jour. Air and Waste Management Association,* 40:704–737.

Smith, R.W. 1971. Relations Among Equilibrium and Nonequilibrium Aqueous Species of Aluminum Hydroxide Complexes. In *Nonequilibrium Systems in Natural Water Chemistry.* ACS Advances in Chemistry Series, No. 106. Washington, D.C.: American Chemical Society.

——. 1989. Flotation of Algae, Bacteria and Other Microorganisms. *Mineral Process. Extract. Metall. Rev.,* 4:277–299.

Smith, R.W., J. Dubel, and M. Misra. 1991. Mineral Bioprocessing and the Future. *Minerals Eng.,* 4:1127–1141.

Smith, R.W., and M.Y. Hwang. 1978. Phosphate Adsorption of Magnesium Silicates. *J. Water Pol. Control Fed.,* 50:2189–2197.

Smith, R.W., and M. Misra. 1993. Recent Developments in the Bioprocessing of Minerals. *Mineral Process. Extract. Metall. Rev.,* 12:37–60.

Smith, R.W., Z. Yang, and R.A. Wharton Jr. 1991. Flotation of *Chlorella vulgaris* with Anionic, Cationic and Amphoteric Collectors. *Minerals Metall. Proc.,* 7:22–26.

Solozhenkin, P.M., and L. Lyubavina. 1980. The Bacterial Leaching of Antimony and Bismuth Bearing Ores and the Utilization of Sewage Waters. In *Biochemistry of Ancient and Modern Environments* (Proceedings of the 4th International Symposium on Environmental Geochemistry). Edited by P.A. Trudinger, M.R. Walter, and B.J. Ralph. New York: Springer-Verlag.

Stumm, W., and J.J. Morgan. 1970. *Aquatic Chemistry.* New York: Wiley Interscience.

Timmerman, C.L., J.L. Buelt, and V.E. FitzPatrick. 1989. In Situ Vitrification of Soils Contaminated with Hazardous Wastes. In *Petroleum Contaminated Soils*, Vol. I. Edited by P. Kostecki and P. Calabrese. Chelsea, Mich.: Lewis Publishers.

Tuovinen, O.H., B.C. Kelly, and S.N. Groudev. 1991. Mixed Cultures in Biological Leaching Processes and Mineral Biotechnology. In *Mixed Cultures in Biotechnology.* Edited by G. Zeikus, and E.A. Johnson. New York: McGraw-Hill.

USEPA (U.S. Environmental Protection Agency). 1994. *Code of Federal Regulations, Protection of the Environment,* 40:749–753. Washington, D.C.: Office of the Federal Register, National Archives and Records Administration.

Vick, S.G. 1983. *Planning, Design and Analysis of Tailings Dams.* New York: John Wiley & Sons.

Whitlock, J.L. 1987. Performance of the Homestake Mining Company Biological Cyanide Degradation Wastewater Treatment Plant, August 1984–August 1986. In *Preprint 87-36 of AIME 116th Annual Meeting.* Littleton, Colo.: SME.

Williams, R.E. 1975. *Waste Production and Disposal in Mining, Milling and Metallurgical Industries.* San Francisco, Calif.: Miller Freeman.

Wilson, D.J., and A.N. Clarke. 1994. Soil Surfactant Flushing/Washing. In *Hazardous Waste Site Soil Remediation.* Edited by D.J. Wilson and A.N. Clarke. New York: Marcel Dekker.

Woodcock, J.T., and M.H. Jones. 1970. Chemical Environment in Australian Lead-Zinc Flotation Plant Pulps. *Proc. Austr. Inst. Min. Met.,* 235:46–76.

Ying, A., J. Duffy, G. Shepherd, and D. Wright. 1989. Bioremediation of Heavy Petroleum Oil in Soil at a Railroad Maintenance Yard. In *Petroleum Contaminated Soils*, Vol. III. Edited by P. Kostecki and P. Calabrese. Chelsea, Mich.: Lewis Publishers.

CHAPTER 14

Economics of the Minerals Industry

Matthew J. Hrebar and Donald W. Gentry

The economics of the minerals industry is characterized by a unique supply-demand relationship for many commodities. The industry also has a number of distinctive features when compared to other industries. These fundamental relationships and features are essential to understanding the economics of the industry and in analyzing investment opportunities within that industry.

SUPPLY-DEMAND RELATIONSHIPS

In broad terms, many metals are sold in either a competitive market or a producer market. A competitive market is one in which (1) producers do not control prices (i.e., the producers are price takers), (2) prices change frequently and at times significantly, and (3) short-term prices are generally influenced by the equilibrium of short-term supply and demand. Examples of such markets are the London Metal Exchange (LME) copper market and the U.S. Ferrous Scrap market. In a producer market, (1) prices are quoted by individual producers and an individual company is often the price leader or setter, and (2) prices change less frequently than in a competitive market. When demand is low, companies often will discount the quoted price; when demand is high, the product is rationed but sold at the producer price. Examples of producer markets are the aluminum and steel markets of past decades.

In the following discussion, competitive metal markets are utilized to illustrate mineral commodity supply-demand relationships.

Supply

The supply made available to the market varies with the planning horizon. In the short term (e.g., 0–2 years), the industry's mining, milling, and smelting capacity cannot increase beyond the capacity of existing operations. In the intermediate term (e.g., 3–10 years), the capacity of the industry can be increased by investment and development of known deposits. In the very long term (e.g., greater than 10 years), the capacity of the industry may be expanded by developing deposits that are discovered as a result of success in future exploration programs. The supply of a metal that an industry can provide to the market is a function of the time available to respond to a perceived or actual increase in demand (Tilton 1981).

The supply curve shown in Figure 14.1 is a short-term curve; it indicates the quantities per period that the total industry is willing to provide to the market at given price levels per unit. The industry will continue to supply product as long as the price is adequate to cover the additional or marginal costs of producing the additional product. Note that the supply curve has a flat slope at low prices, which means that the elasticity (i.e., percent change in quantity per 1% change in price) is high. In other words, at low prices, slight price changes cause a significant change in the quantity supplied. At higher price levels, when the industry is operating near capacity, the slope of the supply curve is quite steep, or elasticity is low. A significant increase in price causes a very small increase in the quantity of the metal supplied. The low price elasticity is a result of the inability of the individual mines and mills to provide

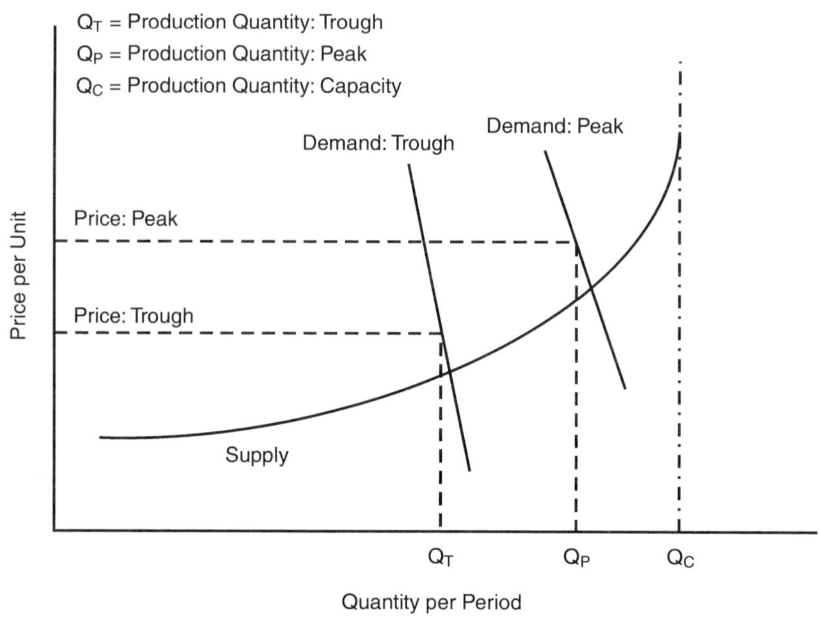

FIGURE 14.1 Short-term supply-demand relationship: Price versus quantity

additional output at any price. The supply curve becomes asymptotic at the total capacity of the industry.

Demand

The demand curves shown in Figure 14.1 have very steep slopes and indicate the quantities that the consuming industries are willing to purchase at various price levels. The demand for metals is a derived demand because metals are usually an intermediate product in a final consumer product and constitute a small percentage of the final cost of the consumer product. For example, the amount of copper utilized in automobiles is quite small, and an increase in copper price would have little impact on the demand for automobiles. Because metals are an intermediate product, the demand for metals is often quite inelastic (i.e., has a low elasticity of demand) over the short term because it is difficult for the producer of the consumer product to substitute for a high-priced metal. Substitution would require changing the production method or retooling the process. Such revisions are usually possible only in the long term, although they can eventually affect pricing through a reduction in demand.

The other major characteristic of the demand for metals relates to changes in demand as a result of changes in the general level of business activity. Approximately two-thirds of the metals industry output enters the automobile, transportation, construction, and consumer durable sectors of the economy. Because consumption in these sectors is usually postponable to a time when business activity is strong and funds are available to make the associated large expenditures, activity in these sectors is highly volatile and tied closely to the general level of economic activity. Consequently, the short-run elasticity of demand with respect to national income (i.e., the percent increase in metal demand per 1% increase in national income) is quite high (Tilton 1981; Kaufmann 1984). In other words, slight changes in the national level of economic activity cause major changes in demand for metals. As the economic activity level decreases, the demand for final products (e.g., automobiles) drops, and the demand for metals drops as reflected by a shift in the demand curve.

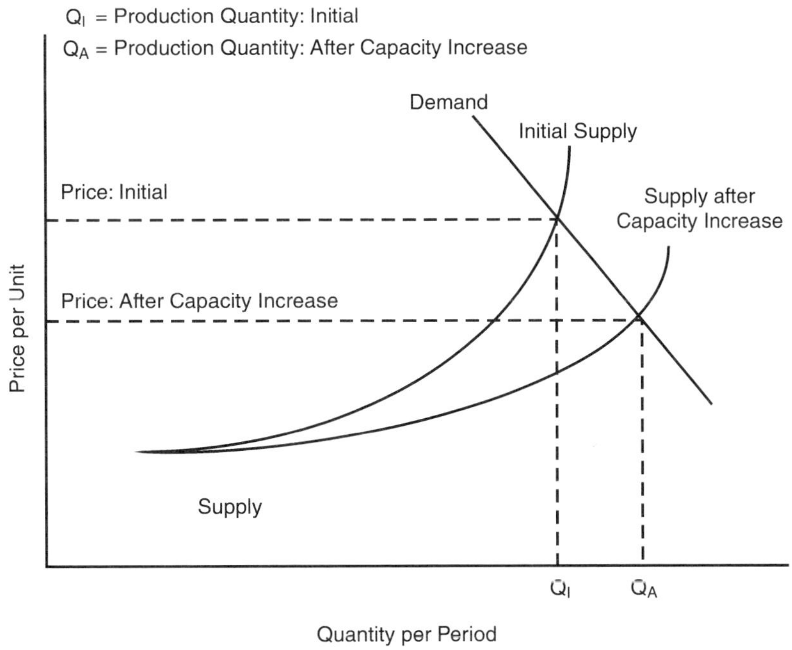

FIGURE 14.2 Industry supply-demand relationship: Price versus quantity

The net result of a combination of the characteristics of metal supply and demand is a cyclic short-term price as indicated in Figure 14.1. When the business cycle is near its peak and demand for the metal is high, prices are high and the industry is operating near capacity. When the business cycle is near its trough, capacity is low and prices are low. Consequently, there is a doubly negative impact on the minerals industry in depressed economic times, because revenue is lowered not only by a decrease in quantity sold but also by a decreased price level per unit.

Long-term Pricing

If demand for a metal remains high and prices remain high, there may be an incentive for a producer to increase capacity by developing new deposits. An increase in capacity shifts the supply curve right, as shown in Figure 14.2. If the price after addition of the new capacity is greater than the price required to provide an adequate return on investment in the new capacity, the investment should be made. The price level required to justify the investment is referred to as the incentive price. The long-run price should, in theory, remain near the incentive price (Radetzki 1983).

This simplified theoretical treatment ignores the fact that the incentive price is a perceived price and that numerous events can alter that price level. For example, on the supply side, a number of events could occur that would result in a price other than that perceived. A number of producers may decide to develop new mines and add capacity. This move would lower the supply curve and reduce the equilibrium price. Depending on the magnitude of the capacity increase, the equilibrium price may be reduced below the incentive price. On the demand side, technologic change may cause a reduction in demand for the metal, resulting in a shift of the demand curve to the left, with a concomitant reduction in equilibrium price. Many other factors can affect both supply and demand relationships. These simple examples are meant to illustrate the uncertainties about long-term commodity price that confront a producer about to embark on the development of a mineral venture.

DISTINCTIVE FEATURES OF THE MINERALS INDUSTRY

The mining industry operates within a rather unique economic environment. The features discussed in the following paragraphs are characteristic of the environment, and although some of these features may be held in common with other industries, the combination of the factors leads to a unique business environment. The features are presented from the supply side and the demand side.

Supply

Features on the supply side include the following:

- Capital intensity
- Unique cost structure
- Long preproduction periods
- Unique deposits
- Aging technology
- Depletable assets
- International competition
- Recycling

Capital intensity relates to the high investment required per employee and the large absolute dollar investments required for the large, low-grade deposits currently being operated. It is not uncommon to have multibillion dollar investments in a mine/mill/port facility. Even small, high-grade precious metals operations employing a small number of employees may require a multimillion dollar investment (Gentry and O'Neil 1984). These capital costs can be further increased by the need to provide infrastructure as deposits are identified in remote locations, such as the Arctic, Indonesia, or Papua New Guinea. Mining and processing methods, production capacity, and other parameters also influence project investment.

This capital intensity leads to a unique cost structure. The average cost of production per unit is often higher than the marginal cost per unit. Average cost includes a high fixed-cost component for capital recovery. Consequently, in periods of low demand and price, the mining operation may be covering marginal cost but actually losing money if average cost per unit is considered. The operations often fail to "recover" capital in periods of low demand (Kaufmann 1984).

Once the existence of an orebody is established, a long preproduction period is required to perform all the activities required to bring the operation into full production. This period may range from 3 to 12 years, depending on the mining and processing methods, size and location of the deposit, complexity of the operating and environmental permitting procedures, and other factors. Expenditures of capital are required throughout the period, and, generally speaking, the longer the preproduction period, the greater the returns required to offset the lost investment opportunities during the preproduction period. The capital intensity of the industry results in large sums of investment capital. The long preproduction period allows ample time for the economic environment to change, which, in turn, may cause a significant difference in actual versus anticipated financial results.

A mineral deposit tends to be unique because of its geometry, geology, mineralogy, size, location, and other factors. Consequently, each deposit must be specifically tested, and mining and processing methods must be devised to meet the unique set of parameters associated with that deposit. As a result, preproduction periods must be lengthened to include these activities, and funding the longer periods contributes to capital intensity must be provided. Operating problems or complete technologic and economic failures can result from attempting to use "off-the-shelf" technologies on a unique mineral deposit.

The mining industry has witnessed a slow evolution of mining and processing technology; few changes, if any, can be viewed as revolutionary. Many observers (e.g., Kaufman 1984) suggest that the industry is utilizing aging technology and requires an influx of revolutionary new technologies to significantly reduce costs and revitalize the industry.

Mining and processing operations are unique in that the major asset, the orebody, is consumed or depleted in the process of mining. The concept of a depletable or nonrenewable resource has numerous implications. First, the investment plus an adequate return on that investment must be recovered within the finite life of the orebody. Second, a company must conduct ongoing exploration to replace the orebody currently being mined and to sustain its own existence. As a result of depleting the asset in the process of mining, many countries provide for special tax treatment in the form of a depletion or similarly named allowance to recognize this unique feature. In addition, the developer of an orebody is faced with choosing the rate of extraction that will maximize the value of the asset within market constraints. Finally, the fact that a mining operation has a finite life may cause government agencies to require special concessions from the developer because of the finite nature of the benefits from mine development. Government-provided infrastructure will no longer be required when the deposit is inevitably exhausted. Thus, the depletable nature of mineral deposits introduces a special set of constraints that must be taken into account in mine development.

International competition is a major feature in many sectors of the minerals industry. Although this feature is not particularly unique to mining and processing activities, its impact is quite significant. Some countries have a competitive advantage over other countries as a result of having higher grade deposits and lower labor costs while utilizing basically the same level of technology. In addition, where these deposits are under government control, the overall objectives may be different from those found in the private, free-enterprise sector; the basic supply-demand concepts described previously may not be the governing doctrine. In the case where employment and foreign exchange are the major objectives, long periods of oversupply can cause long periods of depressed prices. In addition, unfavorable exchange rates can result in a further advantage to a foreign mineral exporter. These factors, which are by and large beyond the control of a domestic mineral producer, can result in considerable economic hardship.

The final feature on the supply side deals with recycling, which results from the indestructibility of many metals. Recycling results in a secondary market and a reduction in the amount of primary ore that must be mined to provide required supply. It offers considerable economic advantages in terms of energy and other cost savings, and it contributes significant percentages to the domestic consumption of such metals as aluminum, iron and steel, copper, and lead. Any new mining venture should consider recycling projections when contemplating prospective future supply.

Demand

Distinctive features on the demand side include

- Derived demand
- Undifferentiated nature of metals
- Slow growth

In the prior discussion of supply-demand relationships, the concepts of derived demand, inelastic demand over the short term, and high income elasticity were discussed. In addition, metal inventories play an important role in demand. As a result of the business cycle heading into a trough, demand drops and consumers of metals begin working off excess inventories and cancel or fail to place new orders for metals. The miners' demand drops significantly as a result of this inventory effect. When economic activity increases in a recovery, the actual demand for metals increases because of the increased demand for product, as well as the goods producer's desire to build inventory. This inventory effect further accentuates the cyclic nature of demand and related commodity price fluctuations.

Most metals tend to be an undifferentiated product; that is, there is little difference between the metals produced in one location versus those produced at another location. Consequently, metals sell primarily on the basis of price.

Finally, on the demand side, most metals are characterized by relatively slow growth in demand. Many markets are mature and have passed through periods of rapid percentage increases in their growth history. Consequently, growth rates are often limited and largely related to population growth.

Summary

As a result of the numerous features relating to supply and demand, the minerals industry is often classified as a high-risk industry. Inherent to the business are long lead times and significant capital investment, which must be made well in advance of the returns generated from production. The returns are often influenced by volatile markets that cause cyclic prices and thus cyclic returns. In addition, geologic uncertainty about the actual asset and technologic uncertainty derived from the unique nature of each mineral deposit may contribute to a variation in returns from the investment. Uncertainties in the political arena–ranging from changing tax structures to expropriation in a foreign investment–contribute to additional uncertainty about returns.

The entire process is overlain by the exploration uncertainties associated with finding new deposits. With these potential investment pitfalls, created by the features and characteristics of the industry, the economic analysis of a new venture must be a thorough and painstaking procedure. All the above parameters and other features must be addressed, and the evaluator must be satisfied that satisfactory economic results can be achieved. Mineral project evaluation provides a definite challenge to those involved in the process.

MINERAL PROJECT EVALUATION*

Historically in the minerals industry, there has been little interchange among individuals in the geological, mining, metallurgical, and financial disciplines during the evaluation stage of new mineral ventures. Characteristically, each discipline has concentrated on its own unique set of problems and has ignored most of the problems faced by the others. Unfortunately, this segregated approach to mine evaluation has led to some poor investment decisions.

There is no doubt that the evaluation of new mining projects in today's investment environment is much more complex than it was a decade ago. There are myriad variables that are either directly or indirectly associated with the mine evaluation process. Consequently, mine evaluation has become truly interdisciplinary in nature. Rarely is an individual knowledgeable in all the areas involved in the evaluation process. Consequently, most organizations prefer to establish multidisciplinary groups that perform the evaluation function for new investment opportunities. These evaluation groups typically consist of individuals with expertise in each of the major areas associated with the evaluation process (geology, mining, processing, economics, environment, regulations, etc.). They represent the preferred approach to the problem.

The Iterative Process

The term "mine evaluation" deals with assessing the relative economic viability of a single mining project or investment opportunity. In this regard, estimates of costs, benefits, expected returns, and associated risks are made for each project or investment alternative available to the firm. Appropriate decision criteria are calculated for each project; these projects are then ranked according to the investment criteria and incorporated into the corporate capital-budgeting process.

The process for evaluating mine investment opportunities is usually iterative in nature. The general process may be represented as shown in Figure 14.3. The quantity of ore reserves is an important variable in determining optimum mine size. Mine size, in turn, affects production costs (capital and operating) because economies of scale are enjoyed with larger production rates. Finally, the level of production costs determines what material can be mined at a profit (i.e., the cutoff grade) and therefore determines the

*This section draws heavily from sections in Gentry and O'Neil (1984).

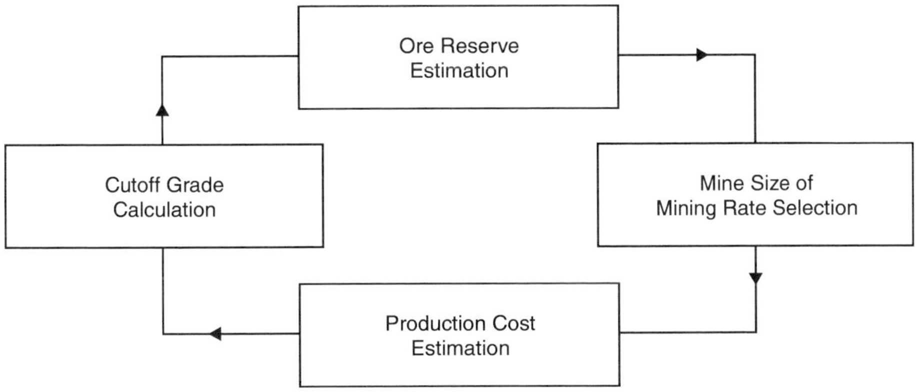

FIGURE 14.3 General prices for evaluating mine investment opportunities

quantity of ore reserves. The unique relationship between cutoff grade and ore reserves is discussed in more detail later in this chapter.

The important point here is to recognize that each time a variable changes, the analyst must assess the impact of this change on the other project variables and on the financial results. This iterative process must be repeated until the most economical design is achieved for the project being analyzed. This process is indeed time-consuming, but it represents the essence of engineering in the course of evaluation.

Factors for Consideration

Nothing improves the output of an engineering/economic evaluation of a mining investment opportunity more than good input data. Unfortunately, analysts preparing feasibility studies for mining properties or projects never have all the data they would like. In addition to inadequacy or unavailability of some needed data, care must be taken not to overlook any variable that may influence project viability. In this regard, analysts often find it helpful to compile a list of factors that should be considered when preparing feasibility studies on mining properties. Outline 14.1 is an outline of some—but certainly not all—of the pertinent factors that evaluators must consider, study, and analyze when assessing mining properties. (Outlines are grouped at the end of this chapter.) Obviously, the significance of each factor will be a function of the specific property being investigated and the mineral commodity (metallic, nonmetallic, fuel) involved. For example, Outline 14.2 illustrates the salient factors requiring consideration for feasibility studies in coal; it shows clearly that the same variables are not of equal importance for all commodities. Nonetheless, all these factors should be assessed to some degree during preparation of at least one of the feasibility studies conducted throughout the evaluation period.

A quick review of Outlines 14.1 and 14.2 suggests there are some fundamental areas of concern that are applicable to all mining property evaluations:

- Estimating the magnitude and quality of the ore reserves
- Estimating and projecting sales revenues
- Estimating technological advancements
- Estimating project capital and operating costs
- Estimating the overall operating environment relative to environmental and other regulatory requirements

Types of Costs

The costs associated with mining activities are a source of considerable debate and even more misunderstanding. There are many categories of costs; consequently, it is essential that the analyst define the cost terms being used in analyses. Although various categories of costs may have precise meanings to accountants, these categories often do not lend themselves to making efficient cash-flow-based decisions in the investment process. In addition, accounting interpretations can vary to a significant degree from one mining company to another. As a consequence, when a copper company is reported to have production costs of, for example, $0.65 per pound, very little useful information is communicated unless those "production costs" are further defined. Thus, specific accompanying definitions of cost information and categories are necessary if useful information is to be conveyed to those analysts involved in the evaluation of mining properties.

Outline 14.3 provides an excellent classification of total costs of production. The major headings of the outline may be summarized as follows:

- Operating costs
  - Direct costs
  - Indirect costs
  - Contingencies
  - Distribution costs
- General expenses
  - Marketing expenses
  - Administrative expenses

Although industrial minerals are often exceptions, distribution costs for most minerals are not sufficiently large to justify a separate cost categorization; hence, they would be combined with the other operating cost categories.

Operating costs are considered to represent all expenses at the plant site, whereas general expenses represent off-site management or corporate-level expenditures. General expenses may be related directly to plant-site activity, or they may be indirect headquarters' items that are allocated across all production divisions.

Direct Versus Indirect Costs. Direct costs, or variable costs, are items such as labor, materials, and supplies that are consumed directly in the production process and are used roughly in direct proportion to the level of production. Indirect costs, on the other hand, represent fixed costs and constitute expenditures that largely are independent of the level of production—at least over certain ranges.

In the limit, there are, obviously, few truly fixed costs. If the mining operation is terminated, for example, most fixed costs are eliminated; and in cases where production is severely curtailed or greatly expanded, some indirect costs (e.g., insurance) would change. Nonetheless, the concept of fixed versus variable costs is valid in a general sense and is useful in understanding some of the characteristics of the mining industry.

The mining industry is characterized by a high degree of capital intensity. In the category of assets per unit of sales, mining ranks near the top of all industrial sectors. In addition to depreciation, other items of indirect (fixed) costs, such as taxes, also are higher than average for mining. As a result, the relatively high level of fixed costs in mining usually means that the break-even production level for mining operations is closer to capacity than for firms with lower fixed costs. This is a major contributing factor in why mine operators attempt to run mines at capacity, often employing three-shift, 7-days-per-week work schedules.

Capital Costs. In addition to operating costs, the mine investment decision clearly must also consider capital costs (also known as first cost or capital investment). Capital costs are expenditures made to acquire or develop capital assets, the benefits from which will be derived over several years in the future. The largest share of capital costs is incurred at project start-up, but some capital expenditures are made yearly throughout the life of the mine.

Capital costs fall into one of three classes, depending on the treatment of the cost for income tax purposes:

1. Depreciable investment: This form of investment applies to a capital asset and is allocated over the useful life of the asset according to some formula acceptable to the tax authorities. All types of mining machinery and equipment fall into this category.
2. Expensible or amortizable investment: Expenditures in this class can, at the taxpayer's option, either be charged off against revenue immediately or be capitalized and amortized over some reasonable time period. Mine development is a good example here; the amortization option can be exercised by charging off such development at the same rate at which the ore, which is thus exposed, is mined.
3. Nondeductible investment: Included here are capital expenditures that cannot be deducted for tax purposes. Examples are successful exploration and property acquisition–which become the basis for the depletion allowance–and working capital, which is recovered at the end of the mine's life.

Obviously, the attractiveness of a mining investment is affected by the amount of capital investment involved. It may not be quite so obvious, however, that the *types* of capital expenditures involved can also be very important in evaluating a prospective new project. This significance is primarily the result of different tax treatments accorded different types of capital expenditures.

Other Cost Concepts. Other cost concepts frequently arise in investment analysis, and some of these are described here for reference:

1. Cash versus noncash costs: Cash costs are those that represent actual monetary outlays. Noncash costs do not directly represent such outlays; instead, they are permissible deductions from revenue, the sole impact of which is to reduce the income tax liability. Depreciation and depletion are two important examples of noncash costs.
2. Sunk costs: A sunk cost is simply an expenditure that has already been made. Sunk costs are irrelevant to a capital investment decision, which must weigh only future benefits against future costs. Although there may be a strong personal commitment to some previous capital investment, those funds have been irrevocably spent; therefore, that prior decision should have a bearing in subsequent investment decisions only to the extent that some tax monies can be recovered from these transactions in the future.
3. Marginal costs and benefits: Only those costs and benefits to be experienced by the firm as a result of the contemplated investment are relevant to an investment decision. These marginal cash flows do not include, for example, allocated corporate overhead, which would be incurred regardless of whether or not the new project were accepted.
4. Cost of capital: Capital costs (which were discussed previously) and the cost of capital are two entirely different concepts. Basically, the term "cost of capital" is used to refer to the cost of raising funds for capital investment. The cost of capital is expressed as a percent and is usually determined by combining the costs of specific sources of capital (debt and equity) into a single value based on the firm's relative use of the various sources.
5. Opportunity cost: This cost refers to the yield or rate of return foregone on the most profitable investment opportunity rejected by a firm. These costs are generally experienced when capital-rationing constraints are imposed in the capital-budgeting process. When budget ceilings are imposed, projects that are otherwise profitable may be rejected. The resulting cost to the firm associated with rejecting these projects is the opportunity foregone on the most profitable investment alternative that remains unfunded.

Cash Flow Analyses

Cash flow analyses and accounting concepts depict investments differently. The primary difference between these approaches is the timing of costs. Because there are generally major differences between accounting profits and actual net cash benefits derived from an investment, investors are increasingly using cash flow as the primary measure of benefits produced from a capital investment. This approach is predicated on the belief that the proper method for evaluating a capital investment is to compare the present outlay with the anticipated positive net cash flows that will accrue from the project in the future. In making this comparison, it is essential that the timing of the various cash flows be recognized through the use of an appropriate interest rate.

As the preceding paragraph suggests, cash flow analyses relate the expenditures associated with investments to the subsequent revenues or benefits generated from these investments. Cash flows are normally calculated on an annual basis for evaluation purposes and are determined by subtracting annual outflows from annual inflows resulting from the investment. Therefore, a cash flow analysis may be made for any investment that has income and expenses associated with it.

Annual cash flows resulting from an investment may be either positive or negative. Net cash flows for a new mining property will be negative during the preproduction years as a result of large capital expenditures. After production begins, the cash flow usually will be positive as an inflow of cash results from the investment in the project.

Net cash flow is basically a combination of two components: (1) the return *on* the investment and (2) the recoupment *of* the investment. In the mineral industry, net cash flow is defined as net income after taxes, plus depreciation and depletion, minus capital expenditures and working capital. The net income after taxes represents the return on the investment; depreciation and depletion represent the recoupment of the investment.

The fact that depreciation and depletion are added back in the cash flow calculation often causes confusion. In a cash flow analysis, each investment receives credit for income taxes saved. Because depreciation and depletion allowances reduce the amount of taxable income (and therefore reduce the amount of taxes paid), they have the effect of saving the organization money. Therefore, they are a credit to the cash flow calculation and are added to net income after taxes. It is important to realize, however, that depreciation and depletion are noncash items and do not actually *flow* anywhere.

Table 14.1 illustrates the components and basic calculation procedure for determining annual cash flows for a mining property. Table 14.2 lists some of the more important factors relating to preproduction, production, and postproduction mining activities that need to be considered in the course of preparing cash flow analyses. The appropriate use and manipulation of these input variables represent an extremely important facet of the cash flow.

In spite of the foregoing explanation, there is often a great deal of confusion surrounding the idea that cash flow is more important than profit. It is important to remember that profit is an accounting concept, subject to an extensive set of fairly rigid rules established by the accounting profession. In the final analysis, however, an investor is simply concerned with how much cash surplus a project will generate in relation to how much cash outlay the project required. Unlike the accountant, the investor is not particularly interested in the method for determining the level of net cash flow from a project. His or her major concern is to estimate whether or not the "cash in" will exceed the "cash out" by a sufficient amount.

This is not to say that profit is irrelevant. Profit is often the largest component of cash flow, but depreciation, amortization, and depletion also often account for a large share of a project's cash flow. As mentioned previously, these three items are noncash expenses because they do not represent cash transactions during the current tax year. The financial impact of these noncash expenses is simply to reduce the amount of income taxes that would otherwise be paid.

The income statement outlined in Table 14.1 is designed to promote rapid calculation of annual cash flows. This type of calculation is illustrated in the example shown in Table 14.3. The income statement

TABLE 14.1 Components of an annual cash flow calculation

Calculation Operator	Component
—	Revenue
Less	Royalties
Equals	Gross income from mining
Less	Operating costs
Equals	Net operating income
Less	Depreciation and amortization allowance
Equals	Net income after depreciation and amortization
Less	Depletion allowance
Equals	Net taxable income
Less	State income tax
Equals	Net federal taxable income
Less	Federal income tax
Equals	Net profit after taxes
Plus	Depreciation and amortization allowances
Plus	Depletion allowance
Equals	Operating cash flow
Less	Capital expenditures
Less	Working capital
Equals	Net annual cash flow

TABLE 14.2 Factors for consideration in cash flow analysis of a mining property

Preproduction Period	Production Period	Postproduction
Exploration expenses	Price	Salvage value
Water rights	Processing costs	Mine closure
Mine and plant capital requirements	Recovery	Contractual and reclamation expenditures
Sunk costs	Postconcentrate cost	
Working capital	Reserves and percent removable	
Land and mineral rights	Grade	
Environmental costs	Investment tax credit	
Development costs	State taxes	
Financial structure	Depletion rate	
Administration	Depreciation	
	Capital investment—replacement and expansions	
	Royalty	
	Mining cost	
	Development cost	
	Exploration cost	
	General and admission	
	Insurance	
	Production rate—tons per year	
	Percent production not sent to processing plant	
	Operating days per year	

Source: Laing 1977.

TABLE 14.3 Example of calculating annual cash flows

	Item	Amount, thousand $		Cash Items, thousand $
	Gross sales (net smelter return)	100		100
Less:	Royalties	2		2
	Gross income from mining	98		98
Less:	Mining cost	24		24
	Beneficiation costs	20		20
	General expense	16		16
	Net operating income	38		38
Less:	Fixed charges, general and administrative	8		8
	Depreciation, amortization	10		
	Depletion	4		
	Pretax net income	16		
	Income tax at 50%	8		8
	Net profit	8		8
Add back:	Depreciation, amortization	10		
	Depletion	4		
	Net operating cash flow	22	=	22

format is needed to calculate the income tax liability, which is often one of the largest expenses of the venture. In this regard, it might be worthwhile to reiterate the fact that in a cash flow analysis, each investment receives a credit for income taxes saved. Therefore, for profitable organizations, it is advantageous to maximize pretax deductions and thereby reduce the amount of taxable income and, consequently, income taxes paid. To take advantage of these tax savings as soon as possible, the firm would opt to expense all possible expenditures in the year incurred, as opposed to capitalizing them followed by subsequent write-offs over the amortization period. Although the total amount of the pretax deduction would be the same in either case, by expensing as soon as possible the firm will realize an earlier return of the resulting tax savings. This early return of tax savings enables the firm to utilize these dollars sooner than would otherwise be possible.

Time Value of Money

The old adage "Time is money" is an accurate statement because of the existence of interest rates. If interest did not exist, the analysis of investment opportunities–particularly in mining–would be greatly simplified. In the absence of interest, investors would be indifferent as to when cash outlays were made or cash inflows were received. It would, in fact, be irrelevant whether the outlays preceded or followed inflow, as long as both amounts were known with certainty.

However, it does make a considerable difference whether, for example, a project anticipates receiving $1 million now or 5 years from now. The reason is that money does indeed have a value that is a function of time. Interest is how this time value is measured.

Interest is generally defined as money paid for the use of borrowed money. In other words, interest is the rental charge for using an asset over some specific time period. The rate of interest is the ratio of the interest chargeable at the end of a specific period of time to the money owed, or borrowed, at the beginning of that period.

The history, philosophy, and theoretical foundations of interest are covered exhaustively in a large number of other books and references; this material need not be repeated here. It is sufficient at this point to recognize simply that money has earning power; that is, the timing of when payments are made and earnings are received in a capital project is very important.

TABLE 14.4 Future dollar values of a $1 investment over various time periods at stipulated annual compound interest rates

Annual Compound Interest Rate, %	Time of Investment, years				
	1	5	10	20	30
5	1.05	1.28	1.63	2.65	4.32
10	1.10	1.61	2.59	6.73	17.45
15	1.15	2.01	4.05	16.37	66.21
20	1.20	2.49	6.19	38.34	237.38

TABLE 14.5 Present dollar value of a promised $1 future payment at various time periods and stipulated annual compound discount rates

Compound Annual Discount Rate, %	Time of Future Payment, years				
	1	5	10	20	30
5	0.95	0.78	0.61	0.38	0.23
10	0.91	0.62	0.39	0.15	0.06
15	0.87	0.50	0.25	0.06	0.02
20	0.83	0.40	0.16	0.03	0.004

The importance of timing of project payments and earnings is perhaps best illustrated by the statement "A dollar today is worth more than a dollar tomorrow." This statement stems from the fact that because interest exists, someone possessing a dollar today could invest that sum in an activity that would earn interest at some rate and thus would yield more than a dollar at some future time period. This is the well-known problem in engineering economics of finding the future sum of money that would result from the investment of a present sum of money over some time period if interest were earned at some specified rate.

Table 14.4 shows the future values that would result from the investment of $1 over specified time periods for various annual compound interest rates. This tabulation clearly shows what everyone already knows—that higher interest rates over longer investment periods maximize future values of an investment.

Of course, the reverse rationale suggests that the promise of having a dollar tomorrow is worth less than having it today. In other words, what is the amount of money an investor would be willing to take today (present value) in lieu of a sum of money promised at some future date? Again, given that interest exists, the investor would be willing to take fewer present-value dollars today—because he or she could in turn invest these dollars at some interest rate over the intervening period and generate an equivalent amount of money to that promised at the future date. This process of finding the present value of a sum to be received in the future is known as discounting.

To illustrate the concept of discounting, Table 14.5 shows the present values that would result from the promise of $1 at various times in the future under specified compound annual interest or discount rates. This tabulation shows the extreme sensitivity of future project earnings or annual cash flows to the discounting process—particularly when cash flows occur far into the future and discount rates are high. For example, if an analyst estimated that the preproduction period for a new mining venture would require 10 years before any positive cash flows were generated and that the discount rate was 15%, the future benefits derived from the investment would start by generating only $0.25 for each dollar of positive cash flow and would decline thereafter. This phenomenon is common in mining investment opportunities and clearly illustrates one of the unique characteristics or risks associated with new mine projects: long preproduction periods.

Obviously, then, the concept of the time value of money as it relates to future values and present values is important in assessing the economic viability of mining investment opportunities. Indeed, the need to incorporate time-value-of-money calculations is fundamental to the income approach to determining mine value or worth. Essentially, the evaluation of a mineral project is predicated on estimating the future net annual cash flows resulting from an investment in the project and then discounting this earnings stream back to the present time by using an appropriate interest rate. This process is sometimes referred to as the capitalized income approach to mining investment decision making.

Decision-making Criteria

Given that the feasibility studies of a new mining investment opportunity produce estimates of relative benefits, costs, and annual cash flows for the project, it then becomes necessary to convert these estimates into measures of relative desirability or attractiveness to the organization contemplating the investment decision. These decision-making criteria are intended to assist the firm in making investment decisions in concert with the primary objective of the organization: to maximize the value of the firm to its owners (i.e., to maximize stockholders' wealth). This value or wealth is represented by the market price of the firm's common stock; consequently, the primary objective of the firm can be restated as one of maximizing the value (price) of the firm's common stock in the marketplace over the long term. Thus, the value of the firm is ultimately related to the firm's investment decisions of the past and present, which are in turn dependent on the criteria the firm employs in the decision-making process.

A major aspect of the investment decision—capital budgeting—deals with evaluating the attractiveness of various investment proposals under consideration, as well as the problem of selecting among alternative projects for optimum allocation of capital. Any evaluation criterion should give company management a means of distinguishing between acceptable and unacceptable projects in a consistent manner. In other words, the criterion should help answer the question "Is project A and/or project B good enough to justify capital investment by the company?" To provide this necessary information for investment decision making, any satisfactory evaluation criterion must respect two basic principles (Quirin 1967):

1. Larger benefits are preferable to smaller benefits.
2. Early benefits are preferable to later benefits.

In addition, it is desirable for the evaluation criterion to provide a ranking of the proposals under consideration in the order of their desirability. Consequently, the problem is one of asking not only "Are projects A and B acceptable to the firm?" but also "Is project A better or worse than project B?" If the financial objective of the firm is to maximize stockholder wealth, as previously stated, the ranking of capital investment opportunities (capital-budgeting decisions) should be based on the following basic principles (Stevens 1979):

1. Every increment of capital expenditure must justify itself.
2. An acceptable investment proposal today is better than the speculation that a better proposal will become available in the future.

The criteria and techniques typically utilized to determine project viability or desirability to investing organizations are, then, the topics of interest here. The project evaluation criteria presented in this section are not intended to represent an exhaustive list available to the analyst, nor are they discussed and illustrated in detail. Instead, they represent the major evaluation criteria utilized for evaluating investment proposals within the minerals industry. It should be recognized, however, that many variations of these basic techniques exist within the industry. Typically these variations have evolved as the result of companies trying to assess, prioritize, and quantify what they perceive to be the most critical parameters affecting an investment decision. Those interested in a more exhaustive treatment of this subject are referred to Gentry and O'Neil (1984).

Accounting Rate of Return. One of the more common calculations of the accounting rate of return is often referred to as the average rate of return. The average rate of return is determined by dividing average annual profits after taxes by the average investment in the project (average book value after deducting depreciation). The principal disadvantages of this method are that (1) it is based on accounting profits rather than cash flows and (2) it does not take into account the timing of these profits. These are very serious disadvantages in that they violate the basic concepts and requirements set forth earlier in this section.

Payback (Payout) Period. One of the most common evaluation criteria used by mining companies is the payback or payout period. Although it was once used as a primary investment criterion, the payback period today is generally used in conjunction with other, more informative methods. The payback period is simply the number of years required for the cash income from a project to return the initial cash investment in the project.

The investment decision criterion for this technique suggests that if the calculated payback period for an investment proposal is less than some maximum value acceptable to the company, the proposal is accepted; if not, the proposal is rejected. In other words, an investment proposal having a payback period of 3 years is acceptable to a company having a hurdle value of 5 years and is preferable to a second project having a payback period of 4 years.

When analyzing the effectiveness of using the payback period for investment decision making, some significant disadvantages of the criterion become apparent. Briefly, these drawbacks are

1. The payback period method fails to consider project cash flows occurring after the payback period.
2. The payback period method does not consider the magnitude or timing of cash flows during the payback period.
3. Establishing the appropriate hurdle rate, or maximum acceptable value for the payback period, is a subjective determination.

An objective appraisal of the payback period indicates that it can offer some useful information to a decision maker considering investment proposals. However, the technique has too many drawbacks to be used in isolation. It should *not* be used as the sole quantitative tool for making investment decisions; instead, it should play a supplementary role to other, more sophisticated methods. Many firms use the payback period criterion as a hurdle that investment proposals must clear before progressing to more rigorous and sophisticated forms of analyses. The payback period should appropriately be regarded as a *constraint* on the acceptability of an investment proposal, not as a criterion to be optimized.

Net Present Value. The present value (PV), or present worth, method of measuring investment proposal desirability is a widely used technique. The term "present value" simply represents an amount of money at the present time ($t = 0$) that is equivalent to some sequence of future cash flows discounted at a specified interest rate. In other words, this technique recognizes the time value of money and provides for the calculation of a present-time amount that is equivalent in value to a series of future cash flows.

Present value calculations are most frequently performed to determine the present worth of income-producing property, such as an existing mining operation. Thus, if the future annual cash flows can be estimated by selecting an appropriate interest rate, the present value of the property can be calculated. This value should provide a reasonable estimate of the price at which the property could be bought or sold.

In the more general case of investment proposal evaluation, the analyst is interested in determining the difference between cash outflows and cash inflows associated with the proposal on a present value basis. This calculation procedure is called the net present value (NPV) method and simply determines the difference between the sum of the present value of all cash inflows and the sum of the present value of all cash outflows. NPV can be expressed as follows:

$$\text{NPV} = \sum \text{present value of cash benefits} - \sum \text{present value of cash costs}$$

If the NPV of the proposal is a positive value (NPV > 0), the project should be accepted. A positive NPV indicates that the investment proposal will provide for (1) the recovery of invested capital, (2) a return on the unrecovered capital each year throughout the project life at the stipulated interest rate utilized in the calculation, and (3) some surplus amount as well. In other words, the project promises to yield a return in excess of that rate used in the calculation procedure. If the rate used in the calculation is the rate of return investors expect the firm to earn on investments, proposals having a positive NPV should increase the wealth of the firm. Similarly, proposals yielding a negative NPV at the required discount rate should be rejected.

The present value technique has a number of characteristics that make the present value suitable as an accept/reject criterion for proposal evaluation. First, the method takes into account the time value of money by utilizing a specified interest rate in the calculation. Second, it provides a single number, or cash equivalent, that can be used as an index for comparison at a specific point in time ($t = 0$). Third, the present value amount is always a unique quantity for a given interest rate.

Benefit/Cost Ratio. The benefit/cost ratio (B/C ratio), often referred to as the profitability index (PI), is generally defined as the ratio of the sum of the present value of future benefits to the sum of the present value of present and future investment outlays and other costs (Quirin 1967). This ratio is expressed as follows:

$$\text{B/C ratio (or PI)} = \frac{\Sigma\text{PV of net cash inflows}}{\Sigma\text{PV of net cash outflows}}$$

For this calculation to be performed, an interest rate must be specified before present value determinations are made. If the calculation results in a PI > 1.0, the investment proposal should be accepted; if not, the proposal should be rejected. This is the same as saying the project should be accepted if NPV > 0. Indeed, the only difference between the NPV calculation and the PI calculation is that the NPV is the difference between the present value of inflows and outflows, whereas the PI is the ratio between the two.

For any given project, the NPV and PI methods will provide the same accept/reject decision, assuming the calculations are performed at the same interest rate. However, if a choice must be made between two investment proposals, these methods may provide inconsistent project rankings.

A rigorous comparison between the NPV and the PI suggests that the NPV method is preferable for determining the absolute expected economic contribution of a project. However, in many cases (particularly capital-rationing situations), analysts are interested in the *relative* profitabilities of projects; in these circumstances, the project-ranking capability of the B/C ratio is more appropriate.

Internal Rate of Return. When evaluators in the minerals industry speak of a rate of return on an investment proposal, they are almost always referring to the so-called discounted cash flow return on investment (DCF-ROI) or the discounted cash flow rate of return (DCF-ROR). These terms are special versions of the more generic term "internal rate of return" (IRR), or "marginal efficiency of capital." This criterion is employed more often in the minerals industry for investment proposal evaluation than perhaps any other criterion.

The internal rate of return is defined as that interest rate that equates the sum of the present value of cash inflows with the sum of the present value of cash outflows for a project. In other words, the IRR is that rate that satisfies each of the following expressions (all of which are equivalent to each other):

$$\Sigma\text{PV cash inflows} - \Sigma\text{PV cash outflows} = 0$$

$$\text{NPV} = 0$$

$$\text{PI} = 1.0$$

$$\Sigma\text{PV cash inflows} = \Sigma\text{PV cash outflows}$$

In general, the calculation procedure involves a trial-and-error solution unless the annual cash flows subsequent to the investment take the form of an annuity.

The acceptance or rejection of a project based on the IRR criterion is made by comparing the calculated IRR with the required rate of return, or cutoff rate, established by the firm. If the IRR exceeds the required rate, the project should be accepted; if not, the project should be rejected. If the required rate of return is the return investors expect the organization to earn on new projects, accepting a project with an IRR greater than the required rate should result in an increase in the price of common stock (i.e., an increase in shareholders' wealth) in the marketplace.

There are several reasons for the widespread popularity of the IRR as an evaluation criterion. Perhaps the primary advantage offered by the technique is that it provides a single figure that can be used as a measure of project value. Further, this figure is expressed as a percentage value. Most managers and engineers prefer to think of economic decisions in terms of percentages, as compared with absolute values provided by present, future, and annual value calculations.

Another advantage offered by the IRR method is related to the calculation procedure itself. As its name suggests, the IRR is determined internally for each project and is a function of the magnitude and timing of that project's cash flows. Some evaluators find this superior to selecting a rate before calculating the criterion, such as in the profitability index and the present, future, and annual value determinations. In other words, the IRR eliminates the need to have an external interest rate supplied for calculation purposes.

In spite of the popularity of this evaluation criterion throughout the minerals industry, the IRR is not without some significant problems in terms of providing appropriate information for investment decision making. For instance, even though the IRR provides for the determination of an internal percentage rate, that rate still must be compared with the hurdle, cutoff, or required rate of return established by the firm before the accept/reject decision can be made. Presumably, this stipulated required rate of return is related to the firm's cost of capital or required cutoff rate and carries with it the implicit borrowing and reinvestment assumptions of any discounting process.

Perhaps the most serious problem associated with the IRR lies in what engineers and managers perceive it to mean. In this regard, it is important to note that "rate of return" is defined as the percentage or rate of interest earned on the *unrecovered* portion of the investment, such that the payment schedule makes the unrecovered investment equal to zero at the end of the investment's life. This is significant because it recognizes that the initial investment declines annually and not all of the investment is working on an annual basis. Thus, the "rate of return" is on the unrecovered investment, and the investment must be recovered at the end of the project life. Other troublesome issues associated with the IRR technique are (1) the question of whether reinvestment of the cash inflow is inherently implied in the technique; (2) the fact that, in some cases, there can be more than one solution to the equation that defines the IRR ("the multiple roots question"); and (3) the fact that the IRR and NPV techniques can provide inconsistent project rankings for mutually exclusive investment opportunities.

Wealth Growth Rate. To overcome some of the disadvantages associated with some of the evaluation criteria previously discussed particularly the internal rate of return the wealth growth rate (WGR) was developed. Berry (1972) defines the wealth growth rate as that interest rate that equates the *future* value of the capital investment with the *future* value of the cash flows resulting from the project. The time horizon for both future values is the termination date of the project. The positive net annual cash flows subsequent to the investment are assumed to be reinvested at the firm's reinvestment rate to the termination date of the project. If investment occurs over several years (i.e., preproduction development), these negative cash flows are discounted to time 0 (initial investment) through the same reinvestment rate. This approach recognizes that the discounting process assumes that the borrowing and reinvesting rates are the same. Thus, the WGR is the compound rate at which the cumulative discounted capital investment must grow in order to equal the future wealth generated by the project. The reinvestment rate is specified by the firm (external to the calculation), and, therefore, the WGR determination is a rather simple process.

The accept/reject decision is determined by comparing the calculated wealth growth rate with the firm's required target rate. This target or threshold rate may be stipulated by the firm or may be the reinvestment rate used in the calculation procedure. If the WGR is equal to or exceeds the required or target rate, the project should be accepted; if not, the project should be rejected.

There are several unique properties of the WGR that give this criterion some distinct advantages over some other criteria. First, the WGR uses annual cash flows as opposed to profits, and it recognizes the time value of money. The technique enables the firm to specify the actual or anticipated reinvestment rate that the firm can reasonably expect during a project's life. Thus, when used to rank projects, the WGR provides a uniform and consistent reinvestment rate for *all* projects, rather than a different rate (the IRR) for each project. Another advantage is that the WGR provides a unique solution that quantifies the rate of wealth growth; furthermore, the WGR expresses this solution in terms of an annual rate that can be directly compared with the firm's reinvestment rate. In other words, this criterion determines the average rate of growth of the firm's accumulated wealth resulting from a capital project.

An inherent assumption associated with the WGR calculation procedure is the reinvestment rate for use at project termination when project alternatives with different lives are being compared. The WGR assumes reinvestment rates for projects under consideration up to the termination date of the longest lived project. As previously stated, the rate assumed during the life of each project is the firm's reinvestment rate, which is externally supplied. However, this technique then uses the calculated WGR to compound the cumulative value at the end of an individual project's life to the termination date of the longest lived project. This assumption is defended on the grounds that near the conclusion of the project, management, because of its experiences and learning with the project in question, should be able to search for and implement a replacement project with an equivalent or superior WGR. Some take exception to this assumption and view it as a disadvantage of the WGR.

Growth Rate of Return. The growth rate of return (GRR) is identical to the wealth growth rate except that a common, arbitrary terminal date (i.e., time horizon t) is used for calculating the GRRs of multiple projects. As developed by Capen, Clapps, and Phelps (1976), the GRR is calculated by first compounding all the positive cash flows forward to some time horizon t years in the future. Any cash flows occurring after time t are discounted back to time t. The rate at which these cash flows are compounded or discounted is the reinvestment rate or opportunity rate of the firm and is externally supplied for calculation purposes. This total amount of money determined for time horizon t then represents the expected revenue from the project plus the earnings or interest generated by the reinvestment rate (reinvestment in future projects). The negative cash flows resulting from the investment decision are discounted to a present value ($t = 0$) in order to obtain an equivalent investment at this point in time. The discount rate is again the same as the reinvestment (opportunity) rate supplied by the firm. By definition the GRR is that interest rate at which the investment would have to grow in order to equal the total amount of money accumulated by the project at time t.

The virtues of the GRR are similar to those of the WGR in that it utilizes the firm's actual reinvestment rate in the calculation procedure and results in a percentage return that represents a measure of investment efficiency. The significant difference, however, between the GRR and the WGR is in the time horizon or common terminal date for calculation of the criterion. The GRR uses a common terminal date for all projects, so that the same reinvestment rate assumption for cash flows is made for all projects. This approach eliminates the potential problem with the WGR, for which the longest lived project is taken as the base for comparative purposes and projects with shorter lives are assumed to have two reinvestment rates (the stipulated rate to project termination and the WGR from project termination to the end of the longest lived project).

Some people prefer not to use the GRR because the calculated rate for a project depends on the time horizon t chosen. This observation is absolutely correct. However, when this technique is used by a firm to evaluate investment proposals, all projects are compared under the same values of t and of r, the reinvestment rate. Under these conditions the GRR provides the same accept/reject decisions as present value determinations and the same project rankings as the profitability index. Also, the relative rankings of projects do not change as the time horizon changes, nor does the accept/reject decision change.

TABLE 14.6 Comparison of financial measurement techniques

	Technique							
Characteristic	**Accounting Rate of Return**	**Payback Period**	**Discounted Payback Period**	**NPV**	**PI**	**IRR**	**WGR**	**GRR**
Uses profit or cash flow?	Profit	Either	Either	Cash flow	Cash flow	Cash flow	Cash flow	Cash flow
Recognizes time value of money?	No	No	Yes	Yes	Yes	Yes	Yes	Yes
Requires reinvestment rate in calculation?	No	No	No	Yes	Yes	No	Yes	Yes
Assumes a sinking fund?	No	No	No	No	No	No	No	No
Are results in the form of a rate of return?	Yes	No	No	No	No	Yes	Yes	Yes
Can yield multiple solutions?	No	No	No	No	No	Yes	No	No
Compares different investment requirements?	Yes	Maybe	Maybe	No	Yes	Yes	Yes	Yes
Accounts for benefits after payback period?	Yes	No	No	Yes	Yes	Yes	Yes	Yes
Appraises market value of project?	No	No	No	Yes	No	No	Yes	Yes
May have varying rankings with different reinvestment rates?	No	No	Yes	Yes	Yes	No	Yes	Yes
Explicitly recognizes life of the project?	No	No	No	No	No	No	Yes	Yes

Source: Berry 1972.

Comparison of Evaluation Criteria. Table 14.6 represents a comparison of the major criteria discussed in this chapter. Although general in nature, this comparison illustrates some of the major differences that exist among criteria and leads the analyst to some tentative conclusions about the appropriateness of specific criteria in mining project evaluation work.

Summary. An extremely important point to remember is that project evaluation criteria do not, by themselves, provide investment *decisions*. They provide only *guidelines* for making decisions. Ultimately, managers must make the actual investment decision after considering all the engineering/economic analyses, the large amount of relevant qualitative information that affects any major decision, and the unique risk and uncertainty possessed by each investment alternative. Investment decision making is a complex process in which quantitative economic studies are of considerable assistance. However, in a world where future values of critical variables are subject to large estimating errors, there is no substitute for sound managerial judgment.

Project Example

The case study used here to illustrate the concepts of economic analysis involves a hypothetical underground lead/zinc/silver vein property located in Colorado; it is adapted from Hrebar and Nilsen (1985).* This case study assumes that a company is evaluating the viability of acquiring a lease on the

*Note that because this example was adapted from another source, units of measure are expressed in terms of the U.S. customary system rather than the International System (SI). Pertinent unit conversions are as follows: 1 metric ton = 1.1 short tons (i.e., 1 t = 1.1 st); 1 meter = 3.28 feet (i.e., 1 m = 3.28 ft); 1 kilogram = 2.2 pounds (i.e., 1 kg = 2.2 lb).

property; conducting an exploration drilling program; performing a feasibility assessment; and then, if all phases are successful, proceeding into mine development. Based on an assessment of the existing geologic information, there is potential for a wide (15 ft), steeply dipping vein with a strike length of approximately 1,500 ft and a potential depth of at least 1,700 ft. Such a deposit is assumed to contain a recoverable reserve of 3,750,000 st and to provide a mill head grade of 10% lead, 5% zinc, and 15 oz/st silver after mine dilution. The property probably would be mined via mechanized cut-and-fill methods at a rate of 750 st of ore per day, 250 days per year (i.e., 187,500 st/year). The ore would be processed in a two-circuit conventional flotation mill at a rate of 600 st of ore per day, 312.5 days per year (i.e., 187,500 st/year), producing both a lead concentrate and a zinc concentrate that would be shipped to their respective custom smelters.

The property consists of 23 unpatented claims held by an individual who requires a $550,000 property payment at the end of the first exploration year and wishes to retain a 5% net smelter return royalty. The company maintains the right to decide whether or not to proceed with the exploration program. The company is assumed to be profitable at the federal level and, therefore, expenses all appropriate expenditures at the federal level whenever possible. The company has no other Colorado operations, so any expensed items in the preproduction period are carried forward to the production period at the state level.

This project example illustrates the procedures employed in deriving the cash flow information necessary for making an investment decision. Determination of project cash flows requires preliminary calculation of the major components, namely, capital and operating costs, revenues, and taxes. In addition, the role of major processing-related variables is considered.

Capital and Operating Costs. In this case study, various cost-estimating techniques will be utilized to establish capital and operating costs. Capital costs are shown in Table 14.7 and are arranged in a fashion that allows the analyst to distinguish among the various categories of costs for subsequent tax calculation purposes.

When capital costs are being calculated, it is imperative that the initial capital include equipment, equipment installations, and engineering and contingency allowances. In the example, the average total engineering plus contingency allowance is 29.8% for mine and mill buildings and 27.4% for mine and mill equipment. These allowances are included in the mine and mill buildings and equipment line items shown in Table 14.7. In addition, the estimate must allow for adequate replacement capital to sustain the operation over the projected life of the project. Replacement equipment for the mine and mill is estimated at 25% of initial capital costs in project years 10, 13, and 16 (as reflected in the capital expenditure line item in Table 14.13, which appears later in this chapter). Working capital, required for the initial parts and supplies inventory, as well as the operating funds to build product inventories and bridge the span until smelter returns are actually received, is estimated at 4 months' worth of operating costs, totaling $4,431,000.

In addition to the estimate of the magnitude of these capital expenditures, estimating the timing of these expenditures over the project's preproduction period is extremely important. These estimates are shown in Table 14.8 under the capital expenditure section. Note that the table includes an estimate of exploration costs and that mine and mill buildings and equipment are combined into single line items. In addition, an estimate of property taxes during preproduction is included. The working capital expenditure of $4,431,000 is divided between the last preproduction year and the first production year. Of the total amount reflecting spares and supplies, $975,000 is assigned to the last preproduction year, with the remaining $3,456,000 expended in the first production year (as shown later in Table 14.13).

Operating costs have been calculated by using various cost-estimating techniques and are shown in Table 14.9.

Revenues. Estimation of revenues requires consideration of metal prices, mill recovery, concentrate grade, smelter terms, and other selling costs such as freight and insurance (Lewis and Streets 1978).

TABLE 14.7 Project capital costs

Category	Cost, thousand $
Mine Costs	
Development	
Project overhead	8,560
Shafts and mine openings	21,000
Total	29,560
Real property	
Hoist room	330
Personal property	
Hoist	3,200
Headframe	574
Compressor	665
Underground equipment	3,960
Underground maintenance equipment	780
Total	9,179
Mill Costs	
Real property	
Clearing and excavation	755
Foundations	1,510
Concentrator building	2,030
Thickener and filter	573
Concentrate storage	305
Tailings impoundment	850
General plant services	1,180
Access road	1,615
Total	8,818
Personal property	
Crushing	2,345
Grinding	2,360
Flotation	875
Electrical	1,785
Water supply	850
Total	8,215
Working capital*	4,431
Total	60,533

*[4 months/(12 months per year)] × (187.5 st of ore per year) × $70.90/st (total operating cost per short ton).

Metal price forecasts can be determined through one of four approaches: naive methods, econometric modeling, rational pricing, and supply/demand schedules (Gentry and O'Neil 1984). For the purposes of the present example, consider a naive no-change model, which assumes that today's spot prices are a reasonable estimate of future prices. The prices assumed are shown in Table 14.10.

Mill recovery and concentrate grade are often estimated on the basis of analogy, empirical data, bench-scale metallurgical testing, or pilot-plant testing, depending on the stage in the exploration project. For the example, the metallurgical balance shown in Table 14.11 has been used.

TABLE 14.8 Preproduction project cash flow

	Cash Amount, thousand $					
Category	**Year 1**	**Year 2**	**Year 3**	**Year 4**	**Year 5**	**Total**
Capital Expenditure						
Property payment	550	0	0	0	0	550
Exploration and feasibility study	1,300	1,400	0	0	0	2,700
Preproduction development	0	0	7,390	7,390	14,780	29,560
Mine and mill buildings	0	0	2,969	2,969	5,938	11,876
Mine and mill equipment	0	0	5,539	5,539	11,078	22,156
Property tax	0	0	0	0	613	613
Working capital	0	0	0	0	975	975
Total	1,850	1,400	15,898	15,898	33,384	68,430
Tax savings						
Exploration and feasibility study	346	400	57	57	57	916
Preproduction development	0	0	1,966	2,121	4,241	8,328
Property tax	0	0	0	0	275	275
Total cash generated	346	400	2,023	2,178	4,573	9,519
Net cash flow	–1,504	–1,000	–13,876	–13,720	–28,809	–58,910

TABLE 14.9 Project operating costs

Category	**Unit Cost, $/st ore**	**Annual Cost, thousand $**
Mining and development		8,850*
Labor	39.10	
Supplies	8.10	
Total	47.20	
Processing		2,194†
Labor	6.95	
Supplies	4.75	
Total	11.70	
General		2,250‡
General and administration	7.50	
Plant and general and administration supplies	2.70	
Electric power	1.80	
Total	12.00	
Total operating costs	70.90	13,294

*750 st/day × 250 day/year × $47.20/st.
†600 st/day × 312.5 day/year × $11.70/st.
‡750 st/day × 250 day/year × $12.00/st.

The smelter terms, including smelting and refining charges, are the final key element in revenue determination. The typical smelter contract includes provisions for payments, treatment charges, refining charges, deductions (unit or percentage), escalation, penalties, credits, and participation–all of which influence net smelter returns. In addition, contracts include administrative provisions, such as payment terms, length of contract, currency consideration, umpire-assaying provisions, and other terms (Mineral Economics Group 1981). Smelter terms tend to be dynamic, as shown by Lewis and Streets (1978).

TABLE 14.10 Net smelter return and revenue calculations

	Assumptions		
	Lead	**Zinc**	**Silver**
Prices	$0.46/lb	$0.79/lb	$4.73/lb
Ore grades	10%	5%	15 oz/st
Lead concentrate grades*	53%	1.13%	74.4 oz/st
Zinc concentrate grades*	3.79%	53%	7.58 oz/st
	Net Smelter Returns per Ton: Lead Concentrate		
		$/st (dry)	
Payments			
Lead	2,000 lb/st × (0.53 – 0.01) × 0.95 × $(0.46 – 0.02)/lb	434.72	
Silver	(74.4 – 0.5) oz/st × 0.95 × $(4.73 – 0.25)/oz	314.52	
Zinc	No payments	0	
Total			749.24
Deductions			
Smelter chargers		190.00	
Freight		35.35	
Total			225.35
Net smelter return—lead			523.88
	Net Smelter Returns per Ton: Zinc Concentrate		
		$/st (dry)	
Payments			
Lead	2,000 lb/st × (0.0379 – 0.015) × 0.65 × $(0.46 – 0.05)/lb	12.21	
Silver	(7.58 – 1.0) oz/st × 0.70 × $(4.73 – 0.30)/oz	20.40	
Zinc	2,000 lb/st × (0.53 – 0.08) × 1.00 × $(0.79 – 0.12)/lb	603.00	
			635.61
Deductions			
Smelter charges		160.00	
Freight		25.00	
Price adjustment	(79 – 40)¢ × $3.50/¢†	136.50	
Total			321.50
Net smelter return—zinc			314.11
	Revenue Calculations		
		thousand $/year	
Lead concentrate	187,500 st/year × 0.1774 × 0.99 × $523.88/st ÷ 1,000	17,251	
Zinc concentrate	187,500 st/year × 0.0792 × 0.99 × $314.11/st ÷ 1,000	4,618	
Total revenue		21,869	

*Transit loss is 1% for both the lead and zinc concentrates.
†For every $0.01 (i.e., 1¢) that the zinc price is over $0.40, the price deduction increases by $3.50 per ton of concentrate.

TABLE 14.11 Assumed metallurgical balance for 100 st of hypothetical ore

	% Weight	Assays			Distribution, %		
	Recovery	% Pb	% Zn	Ag, oz/st	Pb	Zn	Ag
Lead concentrate	17.74	53.00	1.13	74.40	94	4	88
Zinc concentrate	7.92	3.79	53.00	7.58	3	84	4
Tails	74.34	0.40	0.81	1.61	3	12	8
Heads	100.00	10.00	0.00	1.50	100	100	100

The actual terms negotiated with a custom smelter depend on the quality of the concentrate, the supply-and-demand situation for the particular concentrate, and the amenability of the concentrate to being blended with the current smelter feed. All of these factors, in addition to freight differentials, dictate that the mine operator directly investigate the smelting alternatives to maximize net smelter returns. For initial evaluations, Minerals Economics Group (1981) and Lewis and Streets (1978) provide typical smelter terms. Annually updated smelter terms can be found in the smelter section of *Mining Cost Service,* published by Western Mine Engineering, Inc. (Spokane, Wash.). The smelter schedule assumed in the case study is shown in Table 14.12, with net smelter returns and revenues calculated in Table 14.10.

TABLE 14.12 Smelter schedules

	Lead Smelter Schedule	Zinc Smelter Schedule
Payments		
Lead	Deduct 1.0 unit and pay for 95% of the lead content at the "MW US Producers" quotation for common domestic lead, as published in *Metals Week* for the second calendar month following delivery, less $0.02/lb.	Deduct 1.5 units from lead assay, and pay for 65% of the remainder at the "MW US Producers" quotation for common domestic lead as published in *Metals Week* for the calendar month following delivery, less $0.05/lb payable lead. No payment for less than 3% lead.
Silver	Deduct 0.50 troy ounce per short ton (dry) from the silver content, and pay for 95% of the remaining silver content at the "Handy and Harmon NY" quotation for refined silver as published in *Metals Week* for the second calendar month following delivery, less $0.25 per troy ounce.	Deduct 1.0 troy ounce, and pay for 70% of remainder at "H&H" quotation of silver in *Metals Week* averaged for the calendar month following delivery, less $0.30 per troy ounce.
Zinc	No payment.	Deduct 8 units, and pay for 100% of the zinc content at the delivery price as published in *Metals Week,* averaged for the calendar month following delivery, less $0.12/lb.
Deductions		
Smelter charge	$190.00/st (dry)	$160.00/st (dry)
Freight	$35.35/st (dry)	$25.00/st (dry)
Price adjustment	—	Increase smelter charge by $3.50/st for each $0.01 by which the zinc quotation exceeds $0.40/lb. Fractions in proportion.

Tax Considerations and Cash Flows. As previously noted, taxes, in addition to capital and operating costs, are a deduction from revenue in the calculation of net annual cash flow. Federal, state, and local taxes must be considered. It is important to note that U.S. tax regulations change rather frequently; these changes require continual monitoring in order for a firm to be aware of its current tax status. Similarly, state and local tax codes are frequently changed, although their impact is seldom as great as those on the federal level. Publications of the National Mining Association (Washington, D.C.), Commerce Clearing House (Chicago, Ill.), Internal Revenue Service (Washington, D.C.), and numerous accounting houses should be consulted to ensure that the latest tax regulations and procedures are integrated into the analysis.

Details of the calculations for depreciation, exploration and development deductions, state and local taxes, investment tax credits, minimum tax, and federal tax are beyond the scope of this chapter. The reader is referred to Gentry and O'Neil (1984) and Hrebar and Nilsen (1985) for the details of this lengthy and involved process. Results of the process for the present example are shown for the preproduction period in Table 14.8 and for the production period in Table 14.13.

Calculations of the various investment criteria are shown in Table 14.14. As the table shows, the internal rate of return for the project is only 4.9%, with a payback from first production of 18.2 years. At this point in the analysis, additional cases would be investigated to determine if other throughput capacities might enhance project economics. In addition, a number of other price/cost escalation scenarios also would be investigated. Consideration of these scenarios would provide decision makers with information necessary to make a final determination on this project.

Impact of Major Variables. At this point in the analysis, considerable attention should be given to the variables influencing profitability—particularly those variables controllable by management (e.g., throughput capacity). Sensitivity analysis is often conducted to demonstrate the effect of a change in major variables on project economics. This type of analysis involves varying one parameter through a range of values while holding all other parameters constant; the resultant returns are then calculated for each of the parameters in sequence. This approach indicates which of the variables has the greatest potential to change the forecasted economic results. If controllable, these "sensitive" variables can be studied further in an effort to devise the means to change the parameter and improve the economics.

Table 14.15 shows the results of a sensitivity analysis performed by using appropriate spreadsheets on the case study data. This sort of information is often presented in graphic form, as shown in Figure 14.4. The price graph, for example, is useful in the determination of minimum price levels required to achieve a minimum rate of return.

The information developed in a sensitivity analysis is often further processed to rank the parameters in terms of absolute change in the investment criterion for a fixed percent change in the parameter. Table 14.16 shows the results of ranking various parameters this way for a 10% change in each. As the table shows, the project economics are most sensitive to changes in parameters that affect revenue, such as price, grade, and mill recovery. The project is moderately sensitive to changes in operating costs and capital cost and slightly sensitive to changes in ore reserves and production rates. Note that, for changes that increase the IRR, the effect of mill recovery is less than that of ore grade, because a mill recovery of 100% is the highest value possible.

A review of Tables 14.15 and 14.16 shows that price has a far greater effect than mill recovery in this situation. The reason for this difference is that an increase in price causes a greater percentage increase in the net smelter returns. Whereas increases in grade and recovery cause a proportional increase in concentrate produced, a price increase is magnified by the smelter contract because many of the deductions (e.g., smelter charge, freight, etc.) are constant. Table 14.17 shows a comparison of net smelter returns and revenues. Note that a 10% price increase results in a 15% revenue increase, a 10% grade increase results in a 17% revenue increase, and a 10% recovery increase results in a 14% revenue increase.

The interrelationship of head grade and recovery can, in some cases, be significant in assessing project economics. A hypothetical recovery-versus-grade curve for lead and zinc is shown in Figure 14.5.

TABLE 14.13 Project production cash flows: December 1997

Cash Flow Item	Project Year/Production Year									
	6/1	7/2	8/3	9/4	10/5	11/6	12/7	13/8	14/9	15/10
Revenue	21,869	21,869	21,869	21,869	21,869	21,869	21,869	21,869	21,869	21,869
Royalty at 5%	1,093	1,093	1,093	1,093	1,093	1,093	1,093	1,093	1,093	1,093
Net mine revenue	20,776	20,776	20,776	20,776	20,776	20,776	20,776	20,776	20,776	20,776
Operating costs	13,294	13,294	13,294	13,294	13,294	13,294	13,294	13,294	13,294	13,294
Property tax	1,077	1,026	975	925	974	913	853	892	822	751
Severance tax	93	93	93	93	93	93	93	93	93	93
Net after costs	6,313	6,364	6,414	6,465	6,416	6,476	6,537	6,498	6,568	6,639
Exploration and development deductions	1,858	1,774	1,330	887	0	0	0	0	0	0
Depreciation mine and mill	3,317	5,730	4,179	3,071	3,074	3,639	3,251	2,775	2,155	1,767
Net after depreciation	1,138	–1,140	905	2,506	3,341	2,837	3,285	3,722	4,413	4,871
Depletion	0	0	0	354	1,628	1,382	1,600	1,813	2,150	2,373
Loss forward	0	0	–1,140	–235	0	0	0	0	0	0
Net after depletion	1,138	–1,140	–235	1,917	1,714	1,455	1,685	1,909	2,263	2,498
Colorado state tax at 5%	57	0.0	0	96	86	73	84	95	113	125
Federal taxable income	1,081	0	0	1,821	1,628	1,382	1,601	1,813	2,150	2,373
Federal income tax at 35%	378	0	0	637	570	484	560	635	752	831
Alternative minimum taxable income	1,313	–142	372	341	352	398	703	1,074	1,584	3,236
Minimum tax at 20%	263	0	74	68	70	80	141	215	317	647
Federal income tax	378	0	74	637	570	484	560	635	752	831
Net profit	702	0	–74	1,184	1,058	898	1,040	1,179	1,397	1,542
Exploration and development deductions	1,858	1,774	1,330	887	0	0	0	0	0	0
Depreciation mine and mill	3,318	5,730	4,179	3,072	3,074	3,639	3,252	2,776	2,155	1,768
Depletion	0	0	0	354	1,628	1,382	1,600	1,813	2,150	2,373
Loss forward	0	0	1,140	235	0	0	0	0	0	0
Operating cash flow	5,878	7,504	6,575	5,732	5,760	5,920	5,892	5,767	5,702	5,683
Capital expenditures	0	0	0	0	5,538	0	0	5,538	0	0
Working capital	3,456	0	0	0	0	0	0	0	0	0
Net cash flow	2,422	7,504	6,575	5,732	222	5,920	5,892	229	5,702	5,683

Note: All cash flow amounts are in thousand dollars.

(Table continues on next page)

TABLE 14.13 Project production cash flows: December 1997 (continued)

Cash Flow Item	Project Year/Production Year									
	16/11	17/12	18/13	19/14	20/15	21/16	22/17	23/18	24/19	25/20
Revenue	21,869	21,869	21,869	21,869	21,869	21,869	21,869	21,869	21,869	21,869
Royalty at 5%	1,093	1,093	1,093	1,093	1,093	1,093	1,093	1,093	1,093	1,093
Net mine revenue	20,776	20,776	20,776	20,776	20,776	20,776	20,776	20,776	20,776	20,776
Operating costs	13,294	13,294	13,294	13,294	13,294	13,294	13,294	13,294	13,294	13,294
Property tax	820	779	739	698	658	627	596	566	545	525
Severance tax	93	93	93	93	93	93	93	93	93	93
Net after costs	6,569	6,610	6,651	6,691	6,732	6,762	6,793	6,824	6,844	6,865
Exploration and development deductions	0	0	0	0	0	0	0	0	0	0
Depreciation mine and mill	2,282	2,402	1,767	1,490	1,046	799	799	551	304	304
Net after depreciation	4,287	4,208	4,883	5,200	5,686	5,963	5,994	6,272	6,540	6,561
Depletion	2,088	2,050	2,379	2,533	2,770	2,905	2,920	3,056	3,186	3,196
Loss forward	0	0	0	0	0	0	0	0	0	0
Net after depletion	2,199	2,158	2,504	2,667	2,916	3,058	3,074	3,217	3,354	3,364
Colorado state tax at 5%	110	108	125	133	146	153	154	161	168	168
Federal taxable income	2,089	2,050	2,379	2,534	2,770	2,905	2,920	3,056	3,186	3,196
Federal income tax at 35%	731	717	833	887	970	1,017	1,022	1,070	1,115	1,119
Alternative minimum taxable income	3,833	4,561	4,700	4,831	5,172	5,437	5,467	5,733	5,988	6,009
Minimum tax at 20%	767	912	940	966	1,034	1,087	1,093	1,147	1,198	1,202
Federal income tax	767	912	940	966	1,034	1,087	1,093	1,147	1,198	1,202
Net profit	1,322	1,138	1,439	1,568	1,736	1,818	1,827	1,909	1,988	1,995
Exploration and development deductions	0	0	0	0	0	0	0	0	0	0
Depreciation mine and mill	2,282	2,402	1,768	1,491	1,046	799	799	552	305	305
Depletion	2,088	2,050	2,379	2,533	2,770	2,905	2,920	3,056	3,186	3,196
Loss forward	0	0	0	0	0	0	0	0	0	0
Operating cash flow	5,693	5,590	5,585	5,592	5,552	5,522	5,546	5,516	5,479	5,495
Capital expenditures	5,538	0	0	0	0	0	0	0	0	0
Working capital	0	0	0	0	0	0	0	0	0	+4,431
Net cash flow	155	5,590	5,585	5,592	5,552	5,522	5,546	5,516	5,479	9,926

Note: All cash flow amounts are in thousand dollars.

TABLE 14.14 Investment criteria calculations

Project Year	Net Cash Flow, thousand $	Cumulative Cash Flow, thousand $	Present Value of Cash Flow at 5% Interest Rate, thousand $	Present Value of Cash Flow at 3% Interest Rate, thousand $
1	–1,504	–1,504	–1,433	–1,460
2	–1,000	–2,505	–907	–943
3	–13,876	–16,380	–11,986	–12,698
4	–13,720	–30,100	–11,288	–12,190
5	–28,809	–58,910	–22,573	–24,851
6	2,422	–56,488	1,807	2,028
7	7,504	–48,984	5,333	6,101
8	6,575	–42,409	4,450	5,190
9	5,732	–36,676	3,695	4,393
10	222	–36,454	136	165
11	5,920	–30,534	3,461	4,276
12	5,892	–24,642	3,281	4,133
13	229	–24,413	122	156
14	5,702	–18,711	2,880	3,770
15	5,683	–13,028	2,734	3,648
16	155	–12,873	70	97
17	5,990	–6,883	2,613	3,624
18	5,585	–1,298	2,321	3,281
19	5,592	4,294	2,213	3,189
20	5,552	9,846	2,092	3,074
21	5,522	15,368	1,982	2,968
22	5,546	20,914	1,896	2,894
23	5,516	26,430	1,796	2,795
24	5,479	31,909	1,699	2,695
25	9,926	41,835	2,931	4,741
		Net present value =	–675	11,077

Notes: Internal rate of return = 3 + [11,077/ (11,077 + 675)] × (5 – 3) = 4.89%.
Payback period = 18 + [1,298/ (1,298 + 4,294)] × 1 = 18.23 years.

This type of data is often developed in bench- or pilot-scale testing and is useful in the evaluation stages when capacity/grade/life determination studies are undertaken. As the data relationship in the figure shows, the operation receives a double benefit from increased head grades–production is increased in proportion to grade and is also increased as a result of higher recoveries. The effect on project economics varies depending on the slopes of the curves.

Concentrate grade can have a significant effect on project economics. Higher concentrate grades result in lower transportation costs and smelting costs. With the same payment schedule, the resulting higher revenues lead to higher returns. To illustrate, concentrate grades in the example were increased from 53% to 60% for lead and from 53% to 57% for zinc. Percent distributions were held constant as in Table 14.11, resulting in the new metallurgical balance shown in Table 14.18. The weight recovery decreased significantly, while net smelter returns increased at a greater rate. The net result is a significant increase in project revenue and project returns. Revised project economics show that return on the project would increase from 4.9% to 7.8% as a result of the increase in concentration.

TABLE 14.15 Sensitivity analysis results: Values of internal rate of return for various percent changes in given parameters

Parameter	Percent Change in Parameter						
	–30%	–20%	–10%	0%	+10%	+20%	+30%
Commodity price	–8.7	–6.7	–1.7	4.9	8.9	12.5	15.7
Ore grade	8.9	7.2	0	4.9	9.4	13.0	17.0
Mill recovery	–4.9	–2.8	1.5	4.9	7.6	9.9	12.3
Capital investment	9.0	7.3	6.1	4.9	3.8	2.8	2.0
Operating costs	9.9	8.3	6.5	4.9	2.8	0.5	–0.3
Production rate	2.9	3.5	4.3	4.9	5.0	5.0	5.9
Ore reserves	3.0	4.0	4.7	4.9	5.7	6.2	6.6
State tax rates	4.9	4.9	4.9	4.9	4.8	4.8	4.8

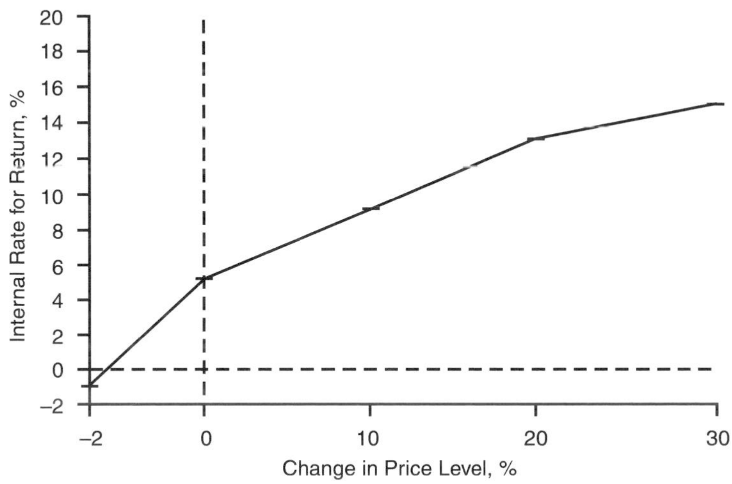

FIGURE 14.4 Internal rate of return versus change in price level

TABLE 14.16 Ranks of various parameters in terms of their effect on IRR for a 10% change in the parameter

Rank	Parameter	Increasing IRR		Decreasing IRR	
		Change in Parameter	Change in IRR, number of percentage points	Change in Parameter	Change in IRR, number of percentage points
1	Commodity price	+10%	81.9	–10%	103.4
2	Ore grade	+10%	92.9	–10%	99.9
3	Mill recovery	+10%	55.7	–10%	68.4
4	Capital investment	–10%	24.3	+10%	22.6
5	Operating costs	–10%	33.7	+10%	43.4
6	Production rate	+10%	2.4	–10%	12.9
7	Ore reserves	+10%	16.5	–10%	4.1
8	State tax rate	–10%	0.4	+10%	0.9

TABLE 14.17 Net smelter returns and revenue when price, grade, and recovery are each separately increased by 10%

	Net Smelter Return for Lead Concentrate, $/st (dry)		Net Smelter Return for Zinc Concentrate, $/st (dry)		Revenue		
Variable Factor	Base Value	After Increase in Variable Factor	Base Value	After Increase in Variable Factor	Base Value, thousand $	After 10% Increase in Variable Factor, thousand $	% Increase After Increase in Variable Factor
Price	524	603	314	361	21,869	25,151	15
Grade	524	600	314	389	21,869	25,480	17
Recovery	524	649	314	317	21,869	25,056	14

Note: Base weight recovery values are 17.74% for lead, 7.92% for zinc.

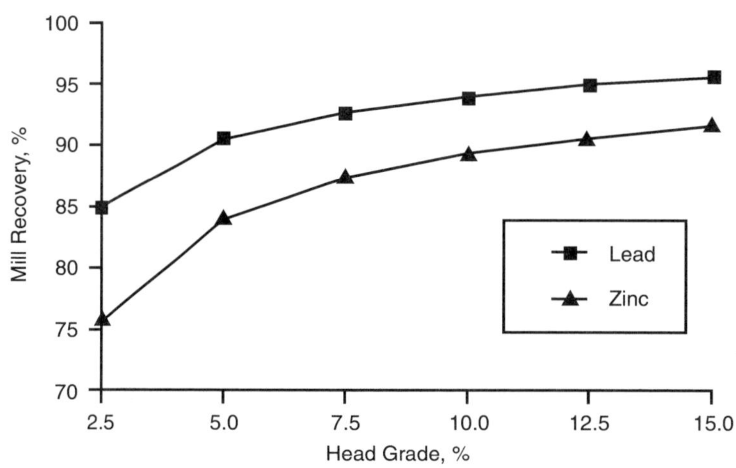

FIGURE 14.5 Hypothetical curves of recovery versus head grade

TABLE 14.18 Revised metallurgical balance for 100 st of hypothetical ore

	% Weight Recovery	Assays			Distribution, %		
		% Pb	% Zn	Ag, oz/st	Pb	Zn	Ag
Lead concentrate	15.67	60.00	1.28	84.30	94	4	88
Zinc concentrate	7.37	4.07	57.00	8.14	3	84	4
Tails	76.96	0.39	0.78	1.56	3	12	8
Heads	100.00	10.00	5.00	15.00	100	100	100

Summary. The preceding analyses generally dealt with increases or decreases in a single variable while all others were held constant. In reality, there are trade-offs resulting from the many interrelationships. For example, higher concentrate grades may be achieved only by sacrificing recovery or by incurring higher capital and/or operating costs. Higher recoveries might be achieved with higher capital and/or operating costs.

Role of Mineral Processing Alternatives

The project example in the previous section illustrated the procedure of developing cash flows for mining projects, as well as the value of some of the investment criteria associated with the example. Allowing key project parameters to vary within selected ranges illustrated the impact of those parameters on project viability. Project sensitivity to these variable changes is extremely important in the investment decision-making process.

Mining investment opportunities appear to be particularly sensitive to certain variables or parameters. Among these are variables associated with mineral processing alternatives. For instance, the independent and interdependent relationships among ore grade, mineral recovery, ore reserve base, concentrate grade, processing capital and operating costs, and project attractiveness are good examples. These trade-offs associated with mineral processing alternatives are briefly illustrated in the following sections.

Mutually Exclusive Processing Alternatives. One of the classic examples of mutually exclusive alternatives facing some mineral producers deals with the decision of whether to ship ore directly to a smelting facility or whether to first send the ore to a concentrator and then ship concentrate to the smelter. These alternatives are nicely illustrated by a simple example. Before proceeding with the example, however, consider a few comments on smelter schedules. A smelter schedule, as illustrated in the example given earlier in this chapter, represents an agreement between a custom smelter and a minerals producer that articulates the conditions under which further processing of minerals occurs. Whereas long-term (10- to 20-year) contracts were common during the 1960s, contracts today have much shorter durations, generally less than 2 to 3 years. Economic uncertainty is the principal reason contract lengths have been reduced.

Actual smelter contracts are often quite lengthy and consist of several parts. Some of the most important clauses they include are

1. Weighing, sampling, and moisture determination: describes procedures to be used, including resolution of disputes.
2. Assaying: describes procedures to be used; specifies the splitting limits, or maximum permissible difference between buyer's and seller's assays before requiring an umpire assayer to help resolve the difference.
3. Loading and unloading of concentrates: specifies which party pays for the many costs arising from concentrate shipment. Penalties for shipping in nonstandard vessels or cars can be substantial.
4. Title and risk of damage or loss: specifies responsibilities of the parties and procedures to be followed if cargo is lost or damaged in transit.
5. Force majeure: specifies events beyond the control of either the seller or the buyer for which the failure of either party to meet the provisions of the contract is excusable.
6. Settlement of disputes: describes procedures by which the parties agree to resolve any controversy or claim. Arbitration is often specified.
7. Environmental matters: specifies conditions under which seller may be required to assume responsibility for disposal of sulfuric acid derived from the seller's concentrates. Seller may bear some of the risk for environmentally mandated expenditures in the future.

The following example, adapted from Bull (1979), illustrates the two mutually exclusive mineral processing alternatives available to some mineral producers.* Consider the case where a mine operator contemplates directly shipping its lead-silver ore to a custom smelter. Assume the ore assays 10% lead

*Again, because this example is adapted from another source, the units of measure are given in the U.S. customary system.

and 5 oz silver per short ton. Mining costs are \$14/st and shipping costs to the smelter are \$30/st. The smelter treatment charge is assumed to be \$55/st.

The costs associated with this alternative are as follows. If we consider 100 st of ore,

mining cost = 100 st × \$14/st	=	\$1,400
shipping cost = 100 st × \$30/st	=	\$3,000
smelter cost = 100 st × \$55/st	=	\$5,500
total costs		\$9,900

If we assume that the ore contains no elements above the levels at which penalties are invoked, the payments derived from smelting the ore can be calculated under the following assumptions:

- The smelter deducts two units from the lead assay and pays for 95% of the remainder at the market price for lead less \$0.065/lb (market price = \$0.46/lb).
- The smelter deducts 1.0 oz/st from the silver assay and pays for 95% of the remainder at the market price for silver less \$0.10/oz (market price = \$6.30/oz).

Hence, the payments can be calculated as follows:

payment for lead = 100 × (10 − 2) × 20 × 0.95 × (\$0.46 − \$0.065)	=	\$6,004
payment for silver = 100 × (5 − 1) × 0.95 × (\$6.30 − \$0.10)	=	\$2,356
total payments		\$8,360

The difference between costs and payments is a *loss* of \$1,540, or \$15.40/st of ore. Because of this unpleasant prospect, let us now suppose that the mine operator considers the concentration of his ore followed by the shipping of concentrate to the smelter. Assume that the ore is shipped at a cost of \$2/st to a mill where, at a cost of \$20/st milled, a concentrate is recovered containing 85% of the lead and 80% of the silver, assaying 72% lead and 33.9 oz silver per short ton. This concentrate would weigh 11.81 st dry and, at 12% moisture, 13.42 st wet. If transport from mill to smelter is \$21/st (wet), the costs are as follows:

mining cost (as before) = 100 st × \$14/st	=	\$1,400
transport to mill = 100 st × \$2/st	=	\$200
milling cost = 100 st × \$20/st	=	\$2,000
shipping to smelter = 13.42 st × \$21/st	=	\$282
smelting charge = 11.81 st × \$55/st	=	\$650
total cost		\$4,532

Assuming the concentrate contains no elements above the levels at which they are penalized, the resulting smelter payments may be calculated as follows:

payment for lead = 11.81 × (72 − 2) × 20 × 0.95 × (\$0.46 − \$0.065)	=	\$6,204
payment for silver = 11.81 × (33.9 − 1) × 0.95 × (\$6.30 − \$0.10)	=	\$2,289
total payments		\$8,493

The difference between costs and payments now results in a profit of \$3,961, or \$39.61/st of ore. It is interesting to note that this significant difference results primarily from changes in

1. Total transportation costs (\$3,000 versus \$482)
2. Smelter charges (\$5,500 versus \$650)
3. Additional milling cost (\$0 versus \$2,000)

Although overly simplified, this example illustrates one of the typical mutually exclusive processing alternatives that face many mine operators. Another example would be the decision about whether to produce additional concentrate products or to install special circuits for the recovery of certain minerals from the ore, either for sale or for the removal of contaminants from saleable concentrates. In any event, these are important mineral processing decisions that can have a significant impact on the investment decision-making process for mineral properties.

Special Role of Cutoff Grade and Ore Reserves. The techniques most often utilized to assess project sensitivity to variable changes implicitly assume that the random variables are independent; that is, that the value of any one parameter is not affected by the value of any other parameter. In actuality, however, this is an oversimplification because some of these variables are related. Perhaps the best illustration of variable interdependence is that between ore grade and mill recovery. Virtually every mineral deposit exhibits some unique but clearly characteristic relationship between these two variables. Invariably, the economic viability of ore deposits—particularly metallic deposits—is significantly dependent on each of these variables. As such, the trade-offs between the grade of ore delivered to the mill and the resulting mineral recovery also are extremely important, as demonstrated previously in this chapter. This relationship needs to be defined carefully within acceptable limits if the evaluation process is to accurately incorporate the potential ramifications of these trade-offs into the analyses.

Another excellent example of interdependence among project variables for most mining ventures is the unique relationship between ore grade and ore reserve tonnage. As discussed earlier in this chapter, ore grade (as described by mining cutoff grade) and ore reserve tonnage constitute two of the major components in the iterative process of evaluating mine investment opportunities. Because of this special and important relationship with respect to both the investment decision process and mineral conservation, determination of the appropriate mining cutoff grade has been the subject of considerable discussion and debate in the minerals industry for many years.

Conceptually, the problem is one of balancing the mining cutoff grade, the ore reserves available for mining, the mining rate (mine size), and the associated capital and operating costs such that the total value derived from mining the deposit is maximized. Consequently, the cutoff grade has normally been employed in mining to represent the criterion by which ore and waste are distinguished in an ore deposit. In other words, a unit of material possessing mineralization estimated at or above the cutoff grade is considered to be ore, whereas that material possessing mineralization below the cutoff grade is considered to be waste and is not sent to the treatment plant for further processing. Because of this unique feature of mining, mine profitability is directly affected by the choice of cutoff grade. In theory, mining cutoff grade is a dynamic variable, changing in response to variations in product price and production costs. In practice, however, few mining operations have the ability to alter mine plans and production areas with the flexibility necessary to rapidly adjust to changing market conditions. Therefore, determination of the cutoff grade is a very important concept because it effectively determines short- to intermediate-term mining decisions and significantly influences the economics of the operation.

One of the more interesting and intriguing approaches to cutoff grade determination is described by Lane (1963). This technique provides for a procedure to determine the economically optimum cutoff grade based on maximizing the present value of future cash flows from the project. In choosing the optimum cutoff grade for a mining property, Lane recognizes that most mining operations involve three basic production processes: mining, concentrating, and refining. The mathematical procedure developed assumes (1) that each of the three stages has its own associated costs and a limiting capacity, (2) that the operation as a whole incurs continuing fixed costs, and (3) that prices and costs remain stable. Although this latter assumption is a severe limitation in theory, the deficiency is no more constraining than with other procedures currently in use. Certainly, more research in developing a general theory for determining cutoff grades in a fluctuating market environment is needed.

The model developed for calculating cutoff grade is based on the following variables:

M = Maximum throughput of material (ore plus waste) for the mine per period within the limits of the deposit

C = Maximum ore throughput per period for the concentrator

R = Maximum output of final product per period for the refinery, assuming concentrate at a fixed grade

m = Unit costs of material mined, either ore or waste

c = Unit costs of ore processed

r = Unit costs of product refined

f = Fixed costs per period (maintenance of billings, rents, administration, etc.)

s = Selling price per unit of product

y = Recovery as an overall value for the concentrator and the refinery (final material product as a proportion of the mineral content of the original ore feed)

On the basis of these variables—along with other developed expressions for the quantity of material mined, the time factor, and the maximum present value of future cash flows from the operation—three cutoff grade formulas are derived by assuming that each of the three stages (mining, concentrating, and refining) alone limits the total capacity of the operation. Interestingly, these derived grades, called the limiting economic grades, depend directly on product price and costs but only indirectly on the actual grade distribution of the deposit.

However, none of the three limiting economic cutoff grades derived is necessarily the optimum cutoff grade to utilize in the operation. The reason is that the capacity of the operation is not necessarily limited by any one stage; instead, the capacity is sometimes limited simultaneously by two and, exceptionally, by all three stages. As such, balancing cutoff grades must be defined. These are cutoff grades that cause each pair of stages to be in balance at their maximum capacities. In essence, these balancing cutoff grades are completely independent of economic factors and are determined by the grade distribution within the deposit. Because they depend on the grade distribution of the material ahead, they are dynamic and can vary widely and rapidly as mining progresses through an irregular orebody. The degree of variation, of course, is determined largely by the mine plan and by the sequence in which different positions of the orebody are developed. Thus, given the mining sequence, an optimum cutoff grade can be determined.

It can be shown that the optimum cutoff grade is always one of these six cutoff grades; that is, either a limiting economic grade or a balancing grade. Lane (1963) offers a graphical procedure for identifying which of the various cutoff grades is the optimum in terms of maximizing present value of cash flows.

Lane draws an important conclusion from this approach to cutoff grade determination: Decisions about cutoff grades cannot be derived by the application of some simple cost formula that equates marginal revenue to marginal cost. In fact, the optimum cutoff grade is influenced by the economics of present value, the capacities of the various stages in the mining operation, and the grade distribution of the deposit. These three influences often interact in ways that cause the optimum cutoff grade to change, sometimes widely, during the life of the mining operation.

Decisions about mine cutoff grades can significantly affect the overall economics of mining operations. As such, the cutoff grade should be carefully determined in a way that accounts for the economics of the concentrating and the smelting portions of the mining activity, in addition to the economics of the mine itself. Furthermore, the balances among these activities, along with the grade distribution within the deposit, are important variables within Lane's process for determining the optimum cutoff grade. Certainly, the simple relationships between cutoff grade and ore reserve estimation, as once used in the mining industry, appear to be inadequate for the complexity of the investment decision-making process being used today. Further refinements in this particular area of variable interdependence give mine analysts the ability to determine dynamic cutoff grades that reflect changing market prices, operating costs, and other key variables.

The Impact of Inflation on Mining Investments

Inflationary pressures, characterized by a continuing upward spiral of prices, are obvious to nearly all consumers. In a capital-intensive industry like mining, where construction periods for new projects are exceptionally long, inflationary increases can be breathtaking. Indeed, under certain conditions, inflation may become the most important factor in a mining investment; it can rarely be safely ignored in capital investment analyses.

Handling Inflation. Although most analysts recognize the need to integrate inflation into investment analysis studies to avoid potentially serious errors, few organizations have developed consistent approaches to handling this problem. The reasons for this inconsistency are not entirely clear, but certainly a contributing factor is the inherent difficulty in accurately predicting inflation rates.

Gentry and O'Neil (1984) point out that several options are available to the analyst for integrating inflation into investment analyses. One such option, which results in consistently correct answers, is to observe the following fundamental rule: *Convert all net annual cash flows into constant dollars before applying any investment criterion.*

According to Gentry and O'Neil, using current or inflated dollars to calculate, for example, a payout period yields a meaningless result. If the currency value changes from year to year, this calculation approach is no better than stating one year's cash flow in German marks, the next in British pounds, and so forth. Annual cash flows have relative meaning only if they are expressed in units of constant value. Otherwise, an elastic yardstick is being used to measure investment value. As a consequence, all cash flows should first be converted to the same currency, usually constant present-day dollars, before investment criteria are applied.

When faced with the uncertainties of estimating future inflation rates, many firms adopt the position that, over the long run, the rate of increase of production costs will be matched by the rate of increase of product sales prices. Assuming that this approach compensates for inflation, these firms then often use some market-derived cost of capital for the discount rate, or required project rate of return. This approach to the problem leads to at least two significant errors:

1. If the prices for all goods and services were rising at the same rate, and if there were no income taxes, the quantitative impact of inflation on investment decisions would be small. However, escalation and income taxes do exist and are not likely to disappear. Therefore, mainly because of the depreciation deduction, ignoring inflation (i.e., assuming cost and price rises are perfectly offsetting) results in *overvaluing* an investment project on an after-tax basis. In other words, when costs and price are assumed to rise together uniformly at the general rate of inflation, an evaluation that compensates for inflation will always yield a lower after-tax interest rate of return than an evaluation that ignores inflation.
2. A market-determined cost of capital includes a component that is based on investors' perceptions of future inflation. If this rate is applied to project cash flows that are not adjusted for inflation, the project will be seriously *undervalued*.

As a result of these potentially significant errors, this commonly used procedure for handling inflation is usually unacceptable.

In essence, inflation must be included in any analysis where major capital requests are involved. Perhaps the best method is to separate cost and revenue items into categories that track well with published cost index series. Individual escalation rates can then be applied to each of these categories for an appropriate time period, followed by some uniform inflation rate thereafter.

Constant- Versus Current-dollar Analyses. Even though it is necessary to integrate inflation into financial analyses, a continuing need exists for constant-dollar studies. Because engineers and scientists often direct and coordinate mining project analyses and because many of these individuals do not possess financial analysis skills, requiring these individuals to adjust their analyses for inflation could result in two potential problems:

1. "Not having expertise in analyzing economic trends, these analysts might generate results of insufficient credibility to the decision-makers and, furthermore, the process might divert attention away from the crucial technical analysis required" (Gentry and O'Neil 1984, p. 317).
2. "Technical analysts need to constantly examine their projections from the standpoint of common sense. At the evaluation stage, it is important to be able to distinguish between future cost changes created by technical factors and those created by inflation assumptions. Inflation-adjusted future cost projections tend to obscure the underlying technical relationships" (Gentry and O'Neil 1984, p. 317).

As a consequence, the need to consider inflation in investment studies does not normally eliminate the need to conduct constant-dollar analyses. The latter permit greater technical insight into the project and allow management to better analyze the source of risk to the project. In the final presentation, both types of analyses are important.

BIBLIOGRAPHY

Berry, C.W. 1972. *A Wealth Growth Rate Measurement for Capital Investment Planning*. Ph.D. diss. Pennsylvania State University, University Park.

Bull, R. 1979. Custom Milling and Smelting: Their Influence on Small Mining Operations. In *1979 Mining Yearbook*. Denver, Colo.: Colorado Mining Association.

Capen, E.C., R.V. Clapp, and W.W. Phelps. 1976. Growth Rate: A Rate-of-Return Measure of Investment Efficiency. *Journal of Petroleum Technology,* May:531–534.

Gentry, D.W., and M.J. Hrebar. 1978. Economic Principles for Property Valuation of Industrial Minerals. Short course at SME-AIME Fall Meeting. New York: AIME.

——. 1980. Planning and Economic Aspects: Surface Coal Mining. Short course notes (August). Golden, Colo.: Colorado School of Mines.

Gentry, D.W., and T.J. O'Neil. 1984. *Mine Investment Analysis.* New York: AIME.

Hrebar, M.J., and M.J. Nilsen. 1985. *Economic Analysis.* Unpublished manuscript.

Jelen, F.C. 1970. *Cost and Optimization Engineering.* New York: McGraw-Hill.

Kaufmann, T.D. 1984. Metals and Their Ores. Course notes. Golden, Colo.: Colorado School of Mines, Minerals Economics Department.

Laing, G.J. 1977. Effects of State Taxation on the Mining Industry in the Rocky Mountain States. *Colorado School of Mines Quarterly,* 72(1):1–126.

Lane, K.F. 1963. Choosing the Optimum Cut-off Grade. *Colorado School of Mines Quarterly,* 59(4): Part B.

Lewis, P.M., and G.C. Streets. 1978. An Analysis of Base Metal Smelter Terms. Paper presented at the 11th Commonwealth Mining and Metallurgical Congress, Hong Kong.

Mineral Economics Group. 1981. The Smelter Contract. *Mine Development Monthly,* April:1–5.

Quirin, D.G. 1967. *The Capital Expenditure Decision.* Homewood, Ill.: Richard D. Irwin.

Radetzki, M. 1983. Long Run Price Prospects for Aluminum and Copper. *Natural Resources Forum,* 7(1):22,35.

Stevens, G.T., Jr. 1979. *Economic Financial Analysis of Capital Investments.* New York: John Wiley & Sons.

Tilton, J.E. 1981. The Causes of Market Instability: An Overview. *Materials and Society,* 5(3):247–255.

OUTLINE 14.1 Salient factors requiring consideration in a mining project feasibility study
Source: Gentry and Hrebar 1978

I. Information on Deposit
   A. Geology
      1. Mineralization: type, grade, uniformity
      2. Geologic structure
      3. Rock types: physical properties
      4. Extent of leached or oxidized zones
      5. Possible genesis
   B. Geometry
      1. Size, shape, and attitude
      2. Continuity
      3. Depth
   C. Geography
      1. Location: proximity to population centers, supply depots, services
      2. Topography
      3. Access
      4. Climatic conditions
      5. Surface conditions: vegetation, stream diversion
      6. Political boundaries
   D. Exploration
      1. Historical: mining district, property
      2. Current program
      3. Reserves
         a. Tonnage-grade curve for deposit, distribution classification; computation of complete mineral inventory (geological resources and mining reserves) segregated by orebody, ore type, elevation, and grade resources categories
         b. Derivation of dilution and mining recovery estimates for mining reserves
      4. Sampling: types, procedures, spacing
      5. Assaying: procedures, check assaying
      6. Proposed program

II. Information on General Project Economics
   A. Markets
      1. Marketable form of product: concentrates, direct shipping ore, specifications, regulations, restrictions
      2. Market location and alternatives: likely purchasers, direct purchase versus toll treatment
      3. Expected price levels and trends: supply-demand, competitive cost levels, new source of product substitutions, tariffs
      4. Sales characteristics: further treatment; sales terms; letters of intent; contract duration; provisions for amendments and cost escalations; procedures/requirements for sampling, assaying, and umpiring
   B. Transportation
      1. Property access
      2. Product transportation: methods, distance, costs
   C. Utilities
      1. Electric power: availability, location, ownership right-of-way, costs
      2. Natural gas: availability, location, costs
      3. Alternatives: on-site generation

- D. Land, Water, and Mineral Rights
  1. Ownership: surface, mineral, water, acquisition or securement by option or otherwise, costs
  2. Acreage requirements: concentrator site, waste dump location, tailing pond location, shops, offices, change houses, laboratories, sundry buildings, etc.
- E. Water
  1. Potable and process: sources, quantity, quality, availability, costs
  2. Mine water: quantity, quality, depth and service, drainage method, treatment
- F. Labor
  1. Availability and type: skilled/unskilled in mining
  2. Rates and trends
  3. Degree of organization: structure and strength
  4. Local/district labor history
  5. Housing and transport of employees
- G. Government Considerations
  1. Taxation: federal, state, local
     - a. Organization of the enterprise
     - b. Tax authorities and regimes
     - c. Special concessions, negotiating procedures, duration
     - d. Division of distributable profits
  2. Reclamation and operating requirements and trends: pollution, construction, operating and related permits, reporting requirements
  3. Zoning
  4. Proposed and pending mining legislation
  5. Legal issues: employment laws, licenses and permits, currency exchange, expatriation of profits, agreements among partners, type of operating entity for tax and other purposes
- H. Financing
  1. Alternatives: sources, magnitudes, issues of ownership
  2. Obligations: repayment of debt, interest, project guarantees
  3. Type of operating entity: organizational structure
  4. Division of profits: legal considerations

III. Mining Method Selection

- A. Physical Controls
  1. Strength: ore, waste, relative
  2. Uniformity: mineralization, blending requirements
  3. Continuity: mineralization
  4. Geology: structure
  5. Surface disturbance: subsidence
  6. Geometry
- B. Selectivity
  1. Dilution, ore recovery estimates
  2. Waste mining and disposal
- C. Preproduction Requirements
  1. Preproduction development or mining requirements: quantity, methods, time required
  2. Layout and plans: schedule
  3. Capital requirements
- D. Production Requirements
  1. Relative production (rate, procedure)
  2. Continuing development: methods, quantity, time requirements
  3. Labor and equipment requirements
  4. Capital requirements versus availability

IV. Processing Methods
   A. Mineralogy
      1. Properties of ore: metallurgical, chemical, physical
      2. Ore hardness
   B. Alternative Processes
      1. Type and stages of extraction process
      2. Degree of processing: nature and quality of products
      3. Establish flowsheet: calculation of quantities flowing, specification of recovery and product grade
      4. Production schedule
   C. Production Quality Versus Specifications of Product
   D. Recoveries and Product Quality
      1. Estimate effects of variations in ore type or head grade (trade-offs, e.g., recovery versus grade)
   E. Plant Layout
      1. Capital requirements
      2. Space requirements
      3. Proximity to deposit

V. Capital and Operating Cost Estimates
   A. Capital Costs
      1. Exploration
      2. Preproduction development (may also be considered operating costs)
         a. Site preparation
         b. Development of deposit for extraction
      3. Working capital
         a. Spares and supplies (inventory)
         b. Initial operations
         c. Financing costs (when appropriate)
      4. Mining
         a. Site preparation
         b. Mine buildings
         c. Mine equipment: freight, taxes and erection costs, replacement schedule
         d. Engineering and contingency fees
      5. Mill
         a. Site preparation
         b. Mill buildings
         c. Mill equipment: freight, taxes and erection costs, replacement schedules
         d. Tailings pond
         e. Engineering and contingency fees
   B. Operating Costs
      1. Mining
         a. Labor: pay rates plus fringes
         b. Maintenance and supplies: quantities, unit costs
         c. Development
      2. Milling
         a. Labor: pay rates plus fringes
         b. Maintenance and supplies: quantities, unit costs
      3. Administrative and supervisory
         a. Overhead charges
         b. Irrecoverable social costs

OUTLINE 14.2 Salient factors requiring consideration in coal property feasibility studies
Source: Gentry and Hrebar 1980

I. Information on Deposit
   A. Geology
      1. Overburden
         a. Stratigraphy
         b. Geologic structure
         c. Physical properties (highwall and spoil characteristics, degree of consolidation)
         d. Thickness and variability
         e. Overall depth
         f. Topsoil parameters
      2. Coal
         a. Quality (rank and analysis)
         b. Thickness and variability
         c. Variability of chemical characteristics
         d. Structure (particularly at contacts)
         e. Physical characteristics
   B. Hydrology (Overburden and Coal)
      1. Permeability
      2. Porosity
      3. Transmissivity
      4. Extent of aquifer(s)
   C. Geometry
      1. Coal
         a. Size
         b. Shape
         c. Attitude
         d. Continuity
   D. Geography
      1. Location (proximity to distribution centers)
      2. Topography
      3. Altitude
      4. Climate
      5. Surface conditions (vegetation, stream diversion)
      6. Drainage patterns
      7. Political boundaries
   E. Exploration
      1. Historical (area, property)
      2. Current program
      3. Sampling (types, procedures)

II. General Project
   A. Market
      1. Customers
      2. Product specifications (tonnage, quality)
      3. Locations
      4. Contract agreements
      5. Spot sale considerations
      6. Preparation requirements

- B. Transportation
  1. Property access
  2. Coal transportation (methods, distance, cost)
- C. Utilities
  1. Availability
  2. Location
  3. Right-of-way
  4. Costs
- D. Land and Mineral Rights
  1. Ownership (surface, mineral, acquisition)
  2. Average requirements (on- and off-site)
  3. Location of oil and gas wells, cemeteries, etc.
- E. Water
  1. Potable and preparation
  2. Sources
  3. Quantity
  4. Quality
  5. Costs
- F. Labor
  1. Availability and type (skilled and unskilled)
  2. Rate and trends
  3. Degree of organization
  4. Labor history
- G. Governmental Considerations
  1. Taxation (local, state, federal)
  2. Royalties
  3. Reclamation and operating requirements
  4. Zoning
  5. Proposed and pending mining legislation

III. Development and Extraction

- A. Compilation of Geologic and Geographic Data
  1. Surface and coal contours
  2. Isopach development (thickness of coal and overburden, stripping ratio, quality, and costs)
- B. Mine Size Determination
  1. Market
  2. Optimum economic
- C. Reserves
  1. Method(s) of determination
  2. Economic stripping ratio
  3. Mining and barrier losses
  4. Burned, oxidized areas
- D. Mining Method Selection
  1. Topography
  2. Refer to previous geologic/geographic factors
  3. Production requirements
  4. Environmental considerations
- E. Pit Layout
  1. Extent of available area
  2. Pit dimensions and geometry

        3. Pit orientation
        4. Haulage, power, and drainage systems
    F. Equipment Selection
        1. Sizing, production estimates
        2. Capital and operating cost estimates
        3. Repeated for each unit operation
    G. Project Cost Estimation (Capital and Operating)
        1. Mine
        2. Mine support equipment
        3. Office, shop, and other facilities
        4. Auxiliary facilities
        5. Human resources requirements
    H. Development Schedule
        1. Additional exploration
        2. Engineering and feasibility study
        3. Permitting
        4. Environmental approved
        5. Equipment purchase and delivery
        6. Site preparation and construction
        7. Start-up
        8. Production

IV. Economic Analysis

V. Sensitivity Analysis
    A. Sections III and IV repeated for various alternatives

OUTLINE 14.3 Production cost components
Source: Adapted from Jelen 1970

- I. Operating Costs or Manufacturing Costs
  - A. Direct Production Costs
    - 1. Materials
      - a. Raw materials
      - b. Processing materials
      - c. By-product and scrap credit
      - d. Utilities
      - e. Maintenance materials
      - f. Operating supplies
      - g. Royalties and rentals
    - 2. Labor
      - a. Direct operating labor
      - b. Operating supervision
      - c. Direct maintenance labor
      - d. Maintenance supervision
      - e. Payroll burden on all labor charges
        - Federal Insurance Compensation Act tax
        - Workers' compensation coverage
        - Contributions to pensions, life insurance, etc.
        - Vacations, holidays, sick leave, overtime premium
        - Company contribution of profit sharing
  - B. Indirect Production Costs
    - 1. Plant overhead or burden
      - a. Administration
      - b. Indirect labor
        - Laboratory
        - Technical service and engineering
        - Shops and repair facilities
        - Shipping department
      - c. Purchasing, receiving, and warehouse
      - d. Personnel and industrial relations
      - e. Inspection, safety, and fire protection
      - f. Automotive and rail switching
      - g. Accounting, clerical, and stenographic
      - h. Plant custodial and plant protective
      - i. Plant hospital and dispensary
      - j. Cafeteria and clubrooms
      - k. Recreational activities
      - l. Local contributions and memberships
      - m. Taxes on property and operating licenses
      - n. Insurance: property, liability
      - o. Nuisance elimination: waste disposal
  - C. Contingencies
  - D. Distribution Costs
    - 1. Containers and packages
    - 2. Freight
    - 3. Operation of terminals and warehouses

a. Wages and salaries plus payroll burden
b. Operating materials and utilities
c. Rental or depreciation

II. General Expenses
- A. Marketing or Sales Expenses
  1. Direct
     a. Sales personnel salaries and commissions
     b. Advertising and promotional literature
     c. Technical sales service
     d. Samples and displays
  2. Indirect
     a. Sales supervision
     b. Travel and entertainment
     c. Market research and sales analysis
     d. District office expenses
- B. Administrative Expenses
  1. Salaries and expenses of officers and staff
  2. General accounting, clerical, and auditing
  3. Central engineering and technical
  4. Legal and patent
     a. Within company
     b. Outside company
     c. Payment and collection of royalties
  5. Research and development
     a. Own operations
     b. Sponsored, consultant, and contract work
  6. Contributions and dues to associations
  7. Public relations
  8. Financial
     a. Debt management
     b. Maintenance of working capital
     c. Credit functions
  9. Communications and traffic management
  10. Central purchasing activities
  11. Taxes and insurance

Index

Note: *f* indicates figure; *t* indicates table.